Elements
of
Molecular
Neurobiology

Third Edition

For Rosemary

Always in my heart

Elements
of
Molecular
Neurobiology

Third Edition

C. U. M. SMITH

Department of Vision Sciences
Aston University
Birmingham, UK

JOHN WILEY & SONS, LTD

Other Wiley Editorial Offices

John Wiley & Sons Inc., 111 River Street, Hoboken, NJ 07030, USA

Jossey-Bass, 989 Market Street, San Francisco, CA 94103-1741, USA

Wiley-VCH Verlag GmbH, Boschstr. 12, D-69469 Weinheim, Germany

John Wiley & Sons Australia Ltd, 33 Park Road, Milton, Queensland 4064, Australia

John Wiley & Sons (Asia) Pte Ltd, 2 Clementi Loop #02-01, Jin Xing Distripark, Singapore 129809

John Wiley & Sons Canada Ltd, 22 Worcester Road, Etobicoke, Ontario, Canada M9W 1L1

British Library Cataloguing in Publication Data

A catalogue record for this book is available from the British Library

ISBN 0 470 84353 5 (case)
ISBN 0 471 56038 3 (paper)

Typeset in 10/11½ pt Times from the author's disks by Dobbie Typesetting Ltd, Tavistock, Devon
Printed and bound in Great Britain by TJ International, Padstow, Cornwall
This book is printed on acid-free paper responsibly manufactured from sustainable forestry in which
at least two trees are planted for each one used for paper production.

CONTENTS

PREFACE

Another six years have passed since I wrote the preface to the second edition and the subject matter of molecular neurobiology has continued its explosive development. President Clinton did well to designate the 1990s 'the decade of the brain'. Once again I have found it necessary to rewrite large sections of the text to incorporate new developments and to design over fifty new and revised illustrations. In particular, the publication in 2001 of the first draft of the human genome and the genomes of a number of other organisms merited the insertion of a new chapter (chapter 6). The great advances in unravelling the structures (at the atomic level) of some of the voltage-gated channels has also meant that chapter 11 has been completely redesigned. Otherwise the overall organisation of the book remains unchanged. I have taken the opportunity to reproduce the intricately beautiful representations of some of the great molecules which lie at the root of molecular neurobiology. These are collected in a colour section and my thanks are due to the scientists who gave permission. Nowhere, it seems to me, is the truth of Schelling's dictum that 'architecture is frozen music' more apparent than in these magnificent structures.

Prefaces although placed at the beginning are generally (as is this) the last item to be written. They provide an opportunity for a concluding overview. Having just read and corrected page proofs an author has, transiently, the whole book in his head. I have been impressed once again by the sheer complexity in depth of animal and human brains. We no longer have the telephone exchange image of the early twentieth century, but much more a picture of an ever-changing quilt of chemical activity, bound together via synapses and gap junctions and second and third messengers leading to subtle modifications of a host of channels, growth factors and neurochemistry. There is ample scope for the multitudinous states of consciousness we all live through. Through it all runs the thread of evolution and the work of the genes. More than ever we recognise that we are bound into a seamless web of living matter. Solutions found to biological problems half a billion years ago in sea squirt, worm and fly are still at work in us today. This is truly remarkable: a confirmation of Charles Darwin's insight and a revolution in our understanding of our place in Nature.

The huge value of the comparative approach is confirmed by the finding that when the genomes of *Drosophila* and *Homo sapiens* are compared, 177 of the 289 known human disease genes are also found in the fly. The medical significance of molecular neurobiology is stressed throughout the following pages. Recent advances in our knowledge of channel proteins gives insight into the causes of a number of troubling conditions and neural stem cell research gives hope to those suffering from damaged nervous systems and even to those facing the neurodegenerations of old age. Knowledge, as ever, gives power. Our increasing ability to control and manipulate can, nevertheless, be used for ill as well as good. At the outset of the twenty-first century we are just beginning to develop techniques for subtly altering the functioning of the brain. In experimental animals it has become possible to switch genes controlling the activity of specific groups of nerve cells on and off. We begin to see how, in the years to come, we may gain presently unthinkable ability to control the operation of the brain. The ethical issues involved are already beginning to trouble forward thinkers. Neurobiology, especially molecular neurobiology, is becoming too important just to be left to the experts.

Even more than in previous editions this one can only be an introduction. It is impossible to place within the confines of a manageable book all the details of the burgeoning subject. I have been only too well aware of how much I have left out and of how many alternative assessments have had to be passed over. I have accordingly developed the bibliographies by including not only printed sources but also relevant web sites. I hope that students will be sufficiently intrigued with what they find in the following pages to follow up their interest through these references.

Finally, as in previous editions, I have many debts to acknowledge. Once more I have to thank the many scientists who have given their permission to reproduce their illustrations. Once again I have to thank my publishers and their illustrators and proof-readers for turning a complex typescript into a presentable text. But, finally, I have to say once more that the final responsibility for the accuracy or otherwise of the following pages remains with its author.

CUMS
July 2002

PREFACE TO THE FIRST EDITION

This book is intentionally entitled 'elements'. It is intended as an introductory account of what is now a vast and rapidly expanding subject. Indeed so rapid is the advance that any writer finds difficulty in steering between the Scylla of up-to-dateness (with its danger of rebuttal) and the Charybdis of received understanding (with its danger of obsolescence). I hardly expect to have safely navigated between these twin sirens at first attempt. But I hope to have avoided shipwreck to the extent that further attempts can be made in subsequent editions. To this end I would welcome critical (I hope constructively critical) comments so that the text can be updated and improved in the years ahead.

The elements upon which I have based my account have been relevant parts of molecular biology, biophysics and neurobiology. Several themes have wound their way through the book as if they were leitmotivs. Any biologist must see his subject from an evolutionary perspective and this theme is never far from the surface. Any biophysicist must recognise that the operation of nervous systems depends on the flows of ions across membranes; this theme, also, recurs throughout the text. Any molecular biologist must approach the subject in terms of the structure and function of great and complex molecules. From the beginning to the end of the book the operation of these intricately beautiful structures is a central concern. They are shown to underly not only action potentials and synaptic transmission but also, multiplied up through the architecture of the brain, to determine such holistic phenomena as memory and psychopathology.

Because of the interdisciplinary nature of the subject I have tried to make the book accessible to as broad a readership as possible. It is for this reason that I have started with an introductory account of animal brains, in particular mammalian brains, and it is for this reason that I have included an extensive glossary and a list of the acronyms with which the subject abounds. After the introductory chapter I have attempted to start at the beginning, at the molecular level, and work upwards through considerations of membrane, ion fluxes, sensory transduction, nerve impulses and synaptic biochemistry to end with such higher level phenomena as neuroembryology, memory and neuropathology. I have hoped to show that the molecular approach is beginning to provide a coherent theory of the brain's structure and functioning. At the same time I have hoped to emphasise that the complexity of the 'two handfuls of porridge' within our skulls precludes any crass and over-hasty reductionism. Molecular approaches to the brain are, nevertheless, beginning to give us considerable power: in order to use it well our decisions must be informed with an understanding of the underlying science.

Leonardo da Vinci annotated one of his anatomical drawings thus: 'O Writer, with what words will you describe with like perfection the entire configuration as the design here makes . . . and the longer you write, minutely, the more you will confuse the mind of the auditor . . .' (trans. Keele). Accordingly I make no apology for supplementing my text with numerous illustrations. This, moreover, is the place to repay a debt of gratitude to the illustrator at my publishers who was able to transform my pencil sketches into finished and stylistically consistent figures. I hope that these, as Leonardo insisted, go some way to clarifying the written descriptions. Equally I owe an immense debt of gratitude to the many scientists who kindly allowed me to reprint their half-tones and line drawings. These latter debts are acknowledged in the figure legends.

Last, but far from least, I would like to acknowledge the anonymous reviewers who read the first drafts of many of my chapters. I have benefited greatly from their comments though hardly dare to hope that all my errors have thereby been eliminated. This is also the place to thank the editorial staff at John Wiley who provided indispensable help in integrating a complicated typescript. I cannot finish, however, without acknowledging the generations of students who have listened to my lectures (without too much complaint) and who by their conscious and unconscious reactions have taught me what little I know of developing a subject in a consistent and coherent fashion. Nor can I finish without acknowledging the help of my wife who, as with previous books, has put up with absences of mind and company and remained the most loyal of critics.

CUMS
February 1989

PREFACE TO THE SECOND EDITION

In the six years since I wrote my preface to the first edition the subject matter of molecular neurobiology has undergone explosive development. In attempting to incorporate the most important of these new understandings I have found myself rewriting large sections of the text and designing well over a hundred new illustrations. In particular the exciting progress in developmental neurobiology seemed to merit an entirely new chapter. Nevertheless, in spite of the huge accession of knowledge since the late 1980s I have (with the exception of this new chapter) kept the overall structure of the book unchanged. I have started with the molecular biology of nucleic acids, proteins and membranes and proceeded to those all-important elements, the multitudinous channels and receptors, with which neuronal (and neuroglial) membranes are studded. Here the developments since the 1980s have been astonishing. Somewhere approaching a hundred of these great molecular complexes have been isolated and analysed, often in great detail. This fascinating topic leads naturally to a consideration of membrane biophysics and this, in turn, to an account of the molecular biology of sensory cells and the biophysics of nerve impulse propagation. An outline of the transmission of the impulse along a nerve fibre leads naturally to a group of chapters on the synapse, that most central of the brain's organelles. A final group of chapters then deals with the development of the brain, its genetic control, and the closely associated topic of memory. The book ends, as before, with a consideration of what can go wrong. Increasingly, today, neuropathologies are being traced to the molecular

level. The hope strengthens that with ever greater understanding of molecular neurobiology effective therapies can be developed to ameliorate and/or prevent these devastating conditions.

My approach to the subject matter of the book remains the same as in the first edition. Molecular neurobiology is not written in tablets of stone, a fossilised unchanging body of facts. It is a living, developing subject. I have, accordingly, sought to show something of the excitement of the chase, of how neurobiologists have isolated and analysed the crucial molecular elements of the brain and how they have used a wide spectrum of techniques to investigate their function. Throughout the book, too, I have retained the emphasis on the evolutionary dimension. Indeed this dimension has become yet more prominent in the years since the first edition was printed and now forms a major and recurring theme. I have also retained the stress on the molecular causes of many neuropathologies, not only in Chapter 20, but throughout the book. Finally, I have sought to integrate our understanding of molecular neurobiology so that the book does not present a mere sequence of disparate chapters and sections but strives to provide a coherent theory of the brain in health and disease. I have also introduced a number of boxes to deal with topics branching out from the main narrative or with areas of historical and philosophical interest. The bibliography has been expanded and updated and if the book does nothing else I hope it can provide an entry to the vast journal literature.

As in the first edition I have innumerable debts to acknowledge. Once again I have to thank the many scientists who have given me permission to reproduce or adapt their illustrations and, of course, more generally, for the uncountable hours in the laboratory from which the data and interpretations described in the following pages have emerged. I have also to renew thanks to my publishers and their illustrators who have once again transformed a complicated and many-sided typescript into a unified text. Thanks are also due to the anonymous referees who read and commented on an early version of the revision. I have gained much from their advice and have wherever possible incorporated their suggested improvements. Much help has also been provided by colleagues at Aston, both academic staff and students. Professor Richard Leuchtag at Texas Southern University has very kindly combed the first edition for mistakes, typographical and other, and I have greatly profited by his comments, especially on the biophysical areas. But, as is customarily said, the final responsibility for the accuracy or otherwise of the text must ultimately remain with its author. I cannot finish without referring once again to my wife to whom this second edition is dedicated.

CUMS
January 1996

1

INTRODUCTORY ORIENTATION

Origins of molecular neurobiology – outline of nervous systems – significance of invertebrates – developmental introduction to vertebrate nervous systems – cellular structure of brains – neurons – glia – nature and organisation of synapses – organisation of neurons in the mammalian brain – complexity of the cortex – modular structure – columns – integrality

The nervous system and, in particular, the brain is commonly regarded as the most complex and highly organised form of matter known to man. Indeed it has sometimes been said that if the brain were simple enough for us to understand, we ourselves would be too simple to understand it! This, of course, is a play on the word 'simple' and, moreover, seems in the long perspective of scientific history unnecessarily pessimistic.

Our task in this text is, anyway, far less ambitious. We do not hope to achieve any total 'understanding' of the brain in the following pages. All we shall attempt is an exposition of the elements of one very powerful approach to its structure and functioning – **the molecular approach**. It is always important to bear in mind that this is but one of several approaches. A full understanding (if and when that comes) will emerge from a synthesis of insights gained from many different disciplines and from different techniques applied at different 'levels' of the brain's structure and functioning (see Figure A, Appendix 1). In this respect the brain is very like a ravelled knot. Indeed Arthur Schopenhauer, in the nineteenth century, famously alluded to the mind/brain problem as '**the world knot**'.

Molecular neurobiology is a young subject. But, like all science, its roots can be traced far back into the past. It has emerged from the confluence of a number of more classical specialisms: **neurophysiology**, **neurochemistry**, **neuroanatomy**. While neurophysiology and neuroanatomy may be traced back into the mists of antiquity, neurochemistry originated comparatively recently. Thudichum is generally regarded as having founded the subject in 1884 with the publication of his book *The Chemical Constitution of the Brain*. This comparatively recent origin has, of course, to do with the great difficulty of studying the chemistry of living processes, especially those occurring in the brain. Biochemistry itself, although originating in the nineteenth century, only began to gather momentum in the middle decades of the twentieth.

Perhaps the decisive moment came almost exactly midway through the twentieth century when, in 1953, James Watson and Francis Crick published their celebrated solution to the structure of DNA. From this date may be traced a vast and still explosively developing science – molecular biology – which has informed the work of all biologists, not least those who have been concerned with the biology of the nervous system.

Molecular biology itself originated by the coming together of two very different strands of scientific endeavour. It combined the work of **biophysicists** interested in the molecular structure of biological materials, especially the structure of proteins and nucleic acids, with the work of **geneticists**, especially microbial geneticists, concerned with understanding the nature of heredity and the genetic process. Although molecular biology has undergone a huge development and diversification in the decades since 1953 these concerns still remain at its core. The conjunction of these two apparently dissimilar interests has led in the 1980s and 1990s to a new high-tech industry – **biotechnology**. Biology is no longer a descriptive subject: the understandings flowing from molecular biology are beginning to allow us to manipulate living material in powerful and fascinating ways. The first company to be founded explicitly to exploit this manipulative ability (Genentech) was valued at over $200 million by the New York Stock Exchange in 1981; in 1987 the world-wide sales of genetically engineered chemicals were upwards of $700 million and, although few gene companies have yet to show a profit (except those manufacturing scientific instruments), a hundred-billion-dollars-a-year industry is confidently predicted for the twenty-first century.

This new-found ability to manipulate has very recently begun to be applied to the nervous system. It is this development which lies at the root of the subject to be outlined in this book – **molecular neurobiology**. It is beginning to be possible to manipulate basic features of the nervous system both to aid understanding and, as knowledge is often power, to bring about desirable change. The brain is man's most precious possession and to a large extent makes him what he is and can become. The birth of molecular neurobiology thus brings prospects of enormous practical importance – for good or ill. We have every reason to study it carefully.

1.1 OUTLINE OF NERVOUS SYSTEMS

There are many excellent accounts of the nervous system. Some recommended texts are indicated in the Bibliography. This introductory section is merely designed to present some of the salient points in a convenient form.

It is possible to argue that the nervous system developed to serve the senses. Heterotrophic forms such as animals necessarily have to seek out their nutriment. The information gathered by the sensory cells has to be collated and appropriate responses computed. Hence the nervous system. It also follows that, to an extent, the nature of the nervous system which an animal possesses reflects its life-style. Active animals develop large and elaborate nervous systems; quiescent forms make do with minimal nervous tissue. In general animals cannot afford to carry more nervous system than they actually need.

A glance at any zoology text is enough to remind us of the huge variety of animals with which we share the globe. It follows that there is a huge number of different nervous system designs. Many of these designs provide opportunities to investigate neurobiological problems which are difficult to solve in mammalian systems. An awareness of the wealth of different systems presented by the animal kingdom is a valuable asset for any neurobiologist and, in particular, as we shall see, for any molecular neurobiologist.

One general design feature is found in all nervous systems above the level represented by the Porifera (sponges) and Cnidaria (jelly fish, sea anemones, hydroids). This is the separation of the nervous system into a **central** 'computing' region and a **peripheral** set of nerve fibres carrying information to and from the centre. In the chordates the 'central region', or **central nervous system (CNS)**, consists of the brain and spinal cord (Figure 1.1), and the **peripheral nervous system (PNS)** consists of the cranial and spinal nerves.

Other animals show other designs. Often we can dimly discern evolutionary reasons for these differences. One major difference which is worth mentioning at this stage is that which obtains between the chordates and the great assemblage of heterogeneous forms grouped for convenience under the title 'invertebrates' or 'animals without backbones'. The CNS of chordates (this phylum includes all the vertebrates) always develops in the dorsal position whilst that of the invertebrates develops in the ventral position (Figure 1.2). It is believed that this striking difference is due to the fact that chordates originated in the warm upper layers of palaeozoic seas whilst invertebrates originated as forms crawling over the bottoms of equally or yet more ancient seas and lagoons. The major sensory input for the chordates would have thus come from above, that for the invertebrates from below. Hence the

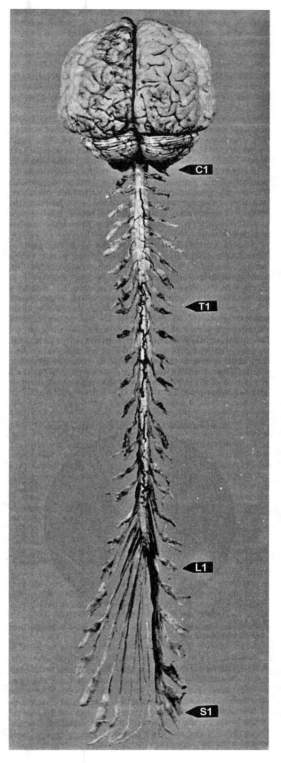

different positioning of their central nervous systems. We shall see, in later chapters, that evolutionary considerations also play a significant role in molecular neurobiology, indeed they form one of the major themes of this book. Here, as elsewhere, they help us answer the question of why things are as they are.

Whilst the nervous systems of all animal phyla are of great interest, neurobiologists have tended to concentrate their attention on a few phyla in particular. The phylum **Nematoda** (roundworms) provides forms with extremely simple nervous systems and quick generation times. The worm *Caenorhabditis elegans* has provided a nervous system simple enough (just 302 neurons subdivided into 118 classes, some 5600 synapses and about 2000 neuromuscular junctions) to have its genetics, development and anatomy mapped in its entirety. This very simple nervous system nevertheless supports a wide variety of behaviours. Neurobiologists using genetics, laser ablation and chemical analysis are well on the way towards running these behavioural patterns into the neural 'wiring diagram'. The phylum **Annelida** (segmented worms) contains forms such as the leech *Hirudo* whose ganglionated CNS has also provided a simple system for intensive investigation. The phylum **Mollusca** has also been much studied. The squid *Loligo* has provided invaluable experimental preparations. More recently, the sea-hare *Aplysia* has been the focus of a great deal of interest at the molecular level. The phylum **Arthropoda** provides many insect and crustacean preparations, including perhaps the simplest system of all, the 28-neuron crustacean **stomatogastric ganglion** which controls

Figure 1.1 Human brain and spinal cord showing roots of the spinal nerves. The central nervous system is viewed from behind. The posterior view of the brain shows the two large cerebral hemispheres resting on top of the two cerebellar hemispheres. Pairs of spinal nerves emerge between the vertebrae of each segment of the cord (8 cervical, 12 thoracic, 5 lumbar and 5 sacral). The spinal cord ends between the twelfth thoracic and the second lumbar segment and continues as the cauda equina. In the figure the latter has been fanned out on the left and left undisturbed on the right. C1=first cervical vertebra; T1=first thoracic vertebra; L1=first lumbar vertebra; S1=first sacral vertebra. From Warwick and Williams (1973), *Gray's Anatomy*, reproduced by permission of Churchill Livingstone, Edinburgh.

A

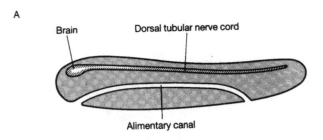

B

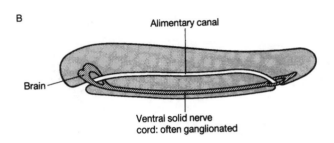

Figure 1.2 (A) Schematic sagittal section through idealised chordate to show position of the CNS. (B) Schematic diagram to show the position of the CNS in a typical non-chordate. It should be borne in mind that whereas chordates form a homogeneous group, sharing a common design principle, non-chordates are many and various. The schematic diagram in (B) fits the worms and the Arthropoda but is quite inappropriate for radial symmetric groups such as the Cnidaria and Echinodermata and can only with difficulty accommodate the Mollusca.

the rhythmical action of the gastric mill, whilst *Limulus*, the 'king' or horsehoe 'crab' (in fact an arachnid), has been much studied by visual physiologists. In recent years the fruit fly *Drosophila*, long a favourite with geneticists, has become central to those interested in the genetics and embryology of the nervous system. The '**mushroom bodies**' or corpora pedunculata, in its nervous system, deeply involved in olfactory learning and memory, consist of only 2500 neurons. Finally, of course, we come to the phylum **Chordata** – the phylum to which we, along with all the other vertebrates, belong. Here many species have provided important opportunities for neurobiological research. The simplest of all, the larva of the urochordate *Ciona intestinalis*, consists of only 2600 cells and its nervous system, which controls typical sinuous swimming movements, is made up of fewer than 100 cells. Three vertebrates deserve special mention: *Xenopus laevis*, the South African clawed frog; *Danio rerio*, the zebra fish; and *Mus musculus*, the mouse. Each of these species has proved valuable for the investigation of particular neurobiological problems.

Although disinterested curiosity has always motivated scientists, and animal nervous systems are worth investigating in their own right 'because

they're there', the major thrust of neurobiological endeavour (and its funding agencies) has always been to illuminate the workings of the human brain. Invertebrates, as indicated above, frequently provide particularly convenient preparations for investigating problems which are difficult to tackle in mammalian and *a fortiori* human brains, but at the end of the day it is an understanding of the human nervous system which is sought.

Further information about invertebrate nervous systems can be obtained from the books listed in the Bibliography. Here we shall confine ourselves to a very brief résumé of the mammalian and, especially, the human CNS.

1.2 VERTEBRATE NERVOUS SYSTEMS

One of the best ways of getting a grip on the structure of the vertebrate nervous system is to follow its development. There has been an enormous increase in our understanding of this process in the last decade or so. This new understanding often goes under the provocative title 'evo-devo'. This draws attention to the fact that investigations of early developmental processes often throw light on early phases of animal evolution. An

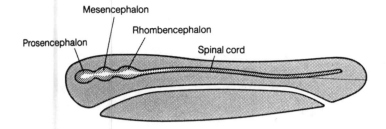

Prosencephalon
Mesencephalon
Rhombencephalon
Spinal cord

Figure 1.3 Embryology of the vertebrate brain: idealised sagittal section of three-vesicle stage.

analogy might be drawn with fundamental physics. Research in high energy physics, at CERN and elsewhere, assists astrophysicists in their researches on the beginnings of the universe and vice versa. We shall look more deeply at the developmental genetics of the vertebrate nervous system in Chapter 18. Here a quick sketch will suffice.

The vertebrate CNS originates as a longitudinal strip of **neurectoderm** (=**neural plate**) which appears on the dorsal surface of the very early embryo. Does embryology recapitulate phylogeny here as Ernst Haeckel long ago suspected? This strip of neurectoderm soon sinks beneath the surface of the embryo, first forming a gutter and then rolling up to form a neural tube. At the anterior end of this tube three swellings (or vesicles) appear (Figure 1.3). These constitute the embryonic fore-, mid- and hindbrains (**prosencephalon, mesencephalon** and **rhombencephalon**). Again, does embryology recapitulate phylogeny? All bilaterally symmetrical animals move with one end of their bodies entering new environments first. It follows that sense organs to pick up information from and about the environment tend to be concentrated on that anterior end. It also follows that specialisation of these sense organs to pick up the principal types of information is likely to occur. Thus animals tend to develop detectors for chemical substances (**chemoreceptors**), electromagnetic radiation (**photoreceptors**) and mechanical disturbance (**mechanoreceptors**). It turns out that the three primary vesicles are initially concerned with the analysis of these three primary senses: olfaction, vision (although the eye itself originates from the posterior part of the forebrain) and vibration, respectively.

As embryological development continues, the early three-vesicle brain subdivides to form a five-vesicle structure. This happens by the hindbrain (the **rhombencephalon**) subdividing into a posterior **myelencephalon** and a more anterior **metencephalon** and the forebrain (the **prosencephalon**) also subdividing into an anterior **telencephalon** and a more posterior **thalamencephalon** (or **diencephalon**). The midbrain remains undivided. The cavity within the metencephalon now expands somewhat to form the **fourth ventricle** joined by a narrow canal, the **iter**, to the **third ventricle** within the thalamencephalon, which in turn communicates with two **lateral ventricles** within the **cerebral hemispheres** which develop from the telencephalon.

Further development of the brain does not involve any further major subdivision. The fundamental architecture of the brain remains essentially as shown in Figure 1.4. Great developments, however, occur principally in the roof of this five-vesicle structure. From the roof of the metencephalon grows the **cerebellum**. This structure, as it is involved in the orchestration of the muscles to produce smooth behavioural movements, is always large in active animals. In primates, such as ourselves, it is thus extremely well developed. Survival of thirty million years or so of arboreal life demanded an extreme of neuromuscular coordination. In ourselves it is the second largest part of the brain. Associated with the cerebellum, in the floor of the metencephalon, is another large structure, the **pons**. The pons acts as a sort of junction box where fibres to and from the cerebellum can interact with fibres running to and from other parts of the CNS.

The roof of the midbrain forms the **tectum** in the lower vertebrates. It is to this region, as indicated above, that the visual information is directed. This information is so important that, in the fish and amphibia, it attracts fibres carrying information

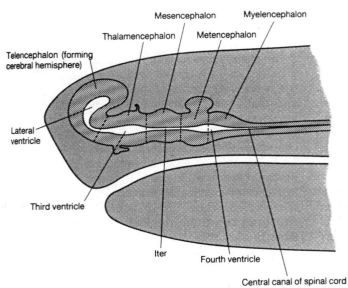

Figure 1.4 Embryology of the vertebrate brain: idealised sagittal section of five-vesicle stage. The figure shows the telencephalon growing backwards over the surface of the thalamencephalon. This only occurs in animals which develop large cerebral hemispheres, such as mammals. From the roof of the thalamencephalon grows the pineal gland while from its floor develops the neural part of the pituitary. The cerebellum grows from the roof of the metencephalon while the floor of this region expands to form the pons. The whole structure contains a cavity continuous with the central canal of the spinal cord and filled with cerebral spinal fluid (CSF).

from the other senses so that the tectum becomes the major brain area for association and cross-correlation of sensory information. The tectum in these animals is perhaps the most important part of the brain. In the mammals, however, this importance is lost. Visual information, as we shall see, is mostly directed to the cerebral cortex. The roof of the midbrain in mammals is thus quite poorly developed. Four smallish swellings can be detected there – two inferior and two superior **colliculi**. The inferior colliculi are part of the auditory pathway from the cochlea whilst the superior colliculi still play a small, though important, role in the analysis of visual information.

It is the forebrain, however, which has undergone the most dramatic development in the mammals and especially in the primates. A number of important nerve centres are located in the thalamencephalon (the **lateral geniculate**, **medial geniculate** and **thalamic nuclei**) which act as 'way stations' for fibres running from the senses towards the cerebrum. From the roof of this region grows the **pineal** organ (in the mammals an important endocrine gland of which we shall have more to say later), and from the floor (the hypothalamus) grows the neural part of the **pituitary**.

But by far the greatest development occurs in the telencephalon. This grows enormously and

becomes reflected back over the thalamencephalon which it ultimately covers and encloses (Greek *thalamos*=inner room) (Figure 1.5). It divides into two great 'hemispheres', the **cerebral hemispheres**, each of which contains a ventricle – the lateral ventricle. In the mammals information from all the senses is brought to the cerebrum and it is here that it is collated and analysed.

In *Homo sapiens* the cerebrum has become gigantic and overgrows and obscures the other (more ancient) regions of the brain. The anatomy is also made more difficult to understand by man's assumption of an upright stance. This causes the brain to bend through nearly a right angle – a characteristic called **cerebral flexure** (Figure 1.6).

One other feature of the general anatomy of the human brain should be mentioned. This is the existence of a series of structures which lie between the cerebrum and the thalamencephalon. These structures constitute the **limbic system** (Figure 1.7) – so called from the Latin *limbus* meaning 'edge' or 'border', as in the Dantean limbo which was conceived as a region between earth and hell. The limbic system is not only situated between the cerebrum and the thalamencephalon but is also believed to be involved in emotions and emotional responses. Some have therefore seen this region as a relic from our infra-human evolutionary past.

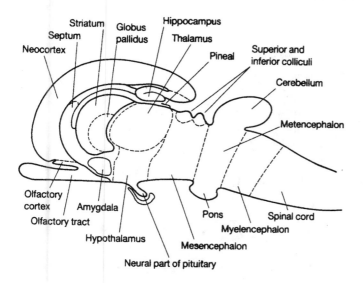

Figure 1.5 Ground plan of the mammalian brain. The schematic figure shows the basic architecture of the mammalian brain. Notice how the neocortex has grown back over the thalamencephalon. In humans this enlargement reaches a climax so that the neocortex grows back as far as the cerebellum and hides the more ancient parts of the brain. After Nauta and Feirtag (1986), *Fundamentals of Neuroanatomy*, New York: Freeman.

It must be emphasised, once again, that all that has been attempted in the preceding paragraphs is a very brief outline of the brain's overall anatomy. It is important, however, that in their study of minute particulars molecular neurobiologists do not lose sight of the fact that the brain is a great, complex and intricate system. Further details of the anatomy may be found in the books listed in the Bibliography.

1.3 CELLS OF THE NERVOUS SYSTEM

The nervous system is built of two major types of cell: **neurons** and **neuroglia** (=**glia**). Both play essential roles in the life of the system. It is only the neurons, however, that are able to transmit messages from one part of the CNS to another or out of the system altogether to the muscles and glands, and vice versa from the sense organs into the CNS. Let us consider each type of cell in turn.

1.3.1 Neurons

Neurons constitute some of the most interesting and intensively studied of all the cells in the body. One of their most distinctive features (which they share with cardiac muscle cells and auditory hair cells) is their permanence. With the exception of olfactory neurons mammalian neurons do not

divide and proliferate after an initial burst during embryological life (see Chapter 19). Instead, in many cases, they grow enormously in size. Indeed, the ratio of cytoplasm to DNA increases by a factor of 10^5 in some neurons during development (see also Appendix 3). Nor do they easily die except in old age and neurodegenerative conditions (see Chapter 21). Programmed death (apoptosis) does, however, play an important role during the development of the nervous system. That 'many are called and few are chosen' seems to be a common feature of neurobiology. It is easy to speculate that the longevity and stability of the neurons which survive to maturity has evolved because of the need to maintain signalling pathways through the brain. Perpetual scrambling of connections by the birth and death of cellular units would most likely be inconsistent with efficient information processing and memory.

Histologists have described many different types of neuron: pyramidal cells, stellate cells, Purkinje cells, Martinotti cells, mitral cells, granule cells etc. Szentágothai recognises over fifty major types and there are many subtypes. All, however, share a common basic design. All possess a metabolic centre (**cell body/cyton/perikaryon**) from which one or more processes spring. The number of processes provides a useful classification. Thus we can distinguish between **monopolar**, **bipolar** and **multipolar** neurons (Figure 1.8).

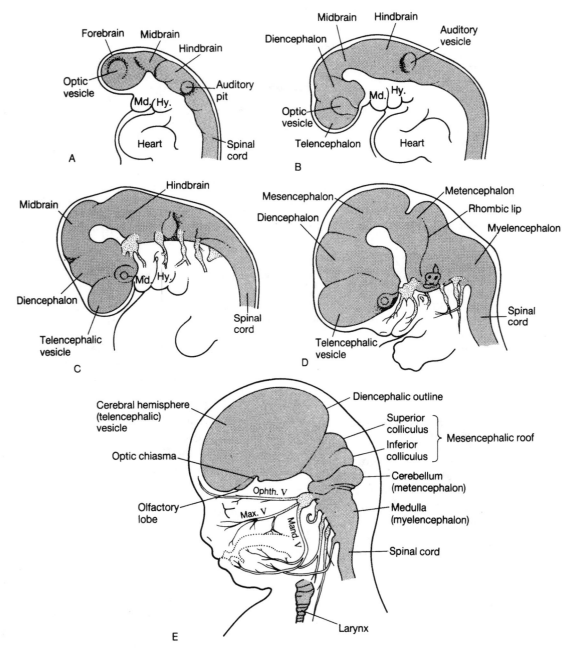

Figure 1.6 Development of the human brain showing flexure. Hy. = hypothalamus; Mand. V = mandibular branch of Vth cranial nerve; Max. V = maxillary branch of Vth cranial nerve; Md. = midbrain; Ophth. V = ophthalmic branch of Vth cranial nerve. From Patten and Carlson (1974), *Foundations of Embryology*, New York: McGraw Hill; with permission.

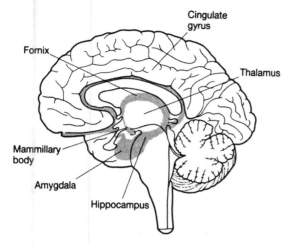

Figure 1.7 Parasagittal section through the human brain to show some elements of the limbic system. From *Biological Psychology*, 2nd edn, by James W. Kalat © 1984, 1982, by Wadsworth, Inc., Belmont, CA. Reprinted by permission of the publisher.

Another useful classification of neurons is into **principal** or **projection** neurons and **local circuit** or **interneurons**. Principal neurons transmit messages out of the local region where their cell bodies are located, whilst local circuit neurons interact closely with their near neighbours.

As much of the remainder of this book is concerned with the molecular biology of neurons it is important to give an introductory outline of the major features of a typical neuron. The **multipolar neuron** is by far the commonest type of neuron in animal nervous systems. Let us therefore look at it in a little detail.

Figure 1.9 shows that two types of process emerge from the perikaryon: the short, branching **dendrites** and the long, unbranched (except at its terminal) **axon**. Both the foregoing statements (as most statements in biology) have many exceptions. Monopolar neurons have unbranched dendrites, and in many cases the axons of multipolar neurons branch.

Neurons are **physiologically 'polarised'**. Messages flow down the dendrites to the perikaryon and away from the perikaryon along the axon. Furthermore in the multipolar neuron only the axon transmits the message by means of action potentials (impulses). The dendrites, as we shall see, do not *in general* develop action potentials.

The perikaryon itself shows all the ultrastructural features of intense biochemical activity.

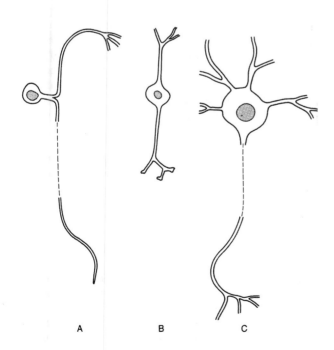

Figure 1.8 Classification of neurons. A simple way of classifying neurons is by noting the number of processes springing from the perikaryon. The figure shows (A) unipolar neuron (e.g. mammalian somaesthetic sensory neuron); (B) bipolar neuron (e.g. retinal bipolar neuron); (C) multipolar neuron (e.g. mammalian motor neuron).

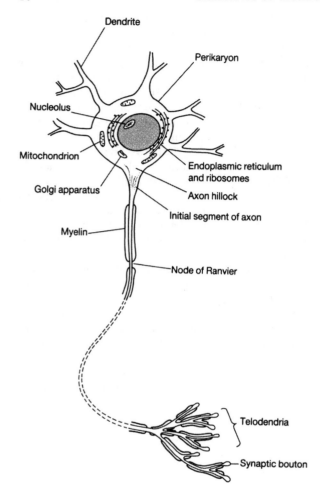

Figure 1.9 Multipolar neuron.

that neurons are tiny computers in their own right or, at the least, much more like multi-functional silicon chips than simple yes/no gates.

In many neurons, especially large multipolar neurons, the axon (=nerve fibre) emerges from a conical part of the perikaryon termed the **axon hillock**. It then generally runs without varying in diameter to its final destination. In many cases, as shown in Figure 1.9, it is encased in a **myelin sheath**. The myelin, as we shall see, is formed by neuroglial cells and plays a vital role in determining the rate of impulse conduction. The junctions between the neuroglial cells constitute the **nodes of Ranvier**. It is only at these junctions that the axonal membrane is exposed to the intercellular medium. At its termination the axon branches into a more or less large number of **telodendria**. The endings of these telodendria make synaptic 'contact' with other neurons or, if the axon is a motor fibre leading out of the CNS, with muscle fibres.

The axon again must not be mistaken for a passive conducting 'wire'. It is true, as we shall see in detail later, that an impulse once initiated at the **initial segment** (see Figure 1.9) runs without decrement to the telodendrial terminations, yet the axon itself has an intricate ultrastructure. It has been shown to possess a complex and dynamic cytoskeleton in which are embedded mitochondria, vesicles of transmitter substances en route to the synaptic termini, and numerous other biochemical entities. All these elements are moving more or less slowly (**axoplasmic flow**) in both directions, either towards the telodendria or vice versa from the telodendria back to the perikaryon. Again we shall have much more to say about the ultrastructure of axons and axoplasmic flow later in the book (Chapter 15).

Following the axon out to its termination we ultimately arrive at the **synaptic 'bouton'**, **'knob'** or **'end foot'** (Figure 1.10). In some cases this termination is far more elaborate than a simple swelling and may form a complicated claw or other intricate structure. Within the termination the electron microscopist can usually detect **mitochondria** and **synaptic vesicles**; other organelles are, however, scarce. We shall return to the structure of synapses later in this chapter and, in much more detail, in Chapters 15, 16 and 17.

Finally, in this introductory section on neurons, let us turn our attention to those other processes

There is a large nucleus, well-developed nucleolus (sometimes more than one), rich rough endoplasmic reticulum, prominent Golgi apparatus (again sometimes more than one) and abundant mitochondria, whilst lysosomes, peroxisomes and multivesicular bodies are frequently visible. In addition to this wealth of organelles, neurons exhibit a well-developed cytoskeleton. Neurotubules, neurofilaments and intermediate filaments are all present and, as we shall see, play vital roles in the life of the neuron. The neuron should never be mistaken for the simple on/off relays of which computers are made. Indeed it has been pointed out

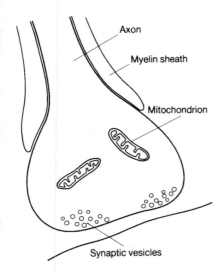

Figure 1.10 Synaptic bouton.

which emerge from the perikaryon – the dendrites. In many multipolar neurons these have a much greater diameter than the axon. We shall see the reason for this in Chapter 12 where we discuss electrotonic conduction. Again, as Figure 1.9 shows, dendrites unlike axons are extensively branched. Indeed the dendrites of the large Purkinje cells of the cerebellum resemble nothing so much as the branches of an espaliered fruit tree. In addition to arboraceous branching, dendrites often develop tiny protuberances commonly known as spines. These, as again we shall see, are the sites of synaptic 'contact'. Lastly, it is worth emphasising once again that dendrites are in no way passive or inert. Like axons they possess a complex ultrastructure formed, in this case, principally of neurotubules.

1.3.2 Glia

Glial cells outnumber neurons ten to one in many parts of the CNS. They were first identified by Virchow in 1856, who considered them to form a structural 'glue' (from which the name derives) holding together the other elements of the nervous system. We now know they have many other important roles (Figure 1.14). Moreover, unlike neurons they have not lost the ability to multiply after birth. This means that they are able to invade

damaged regions and clear away necrotic material and in so doing they leave a glial scar.

On the other hand, glia resemble neurons in showing a large number of different structural forms. It is usual to recognise three major types: **astroglia**, **oligodendroglia** and **microglia**. Each type has an important role in the life of the nervous system. Let us review each in turn.

Astroglial cells (=astrocytes), as the name implies, possess a number of radiating (star-like) processes from a large central cell body (c. 20 μm in diameter) which contains the nucleus (Figure 1.11). There is evidence that astrocytes are profusely interconnected by 'gap junctions' (see Section 7.9) which allow the interchange of molecules and ions. It is frequently the case that some astroglial processes end on the endothelial walls of cerebral blood vessels whilst others are closely adposed to neurons. In other cases (or sometimes the same case) the feet of astroglial cells abut the ependymal cells lining a cerebral ventricle or, alternatively, the cells of the innermost of the brain's meningeal membranes – the pia mater. It has been suggested, in consequence, that astroglial cells are involved in the movement of materials between **cerebrospinal fluid (CSF)**, **blood** and **neuron** – perhaps with some metabolic elaboration en route. However, although the close metabolic symbiosis between astrocytes and neurons is undisputed it is now thought unlikely that they actively transfer metabolites from blood to neuron.

Another important feature of astrocytes is the strong development of filaments (=**glial filaments**) in the cytoplasm. Generally speaking these filaments are more strongly developed in the astrocytes located in the white matter than in those located in grey matter. These two types of astrocyte are consequently called **fibrous** and **protoplasmic** astrocytes, respectively (Figure 1.12). The filaments are believed to confer a certain tensile strength and as astrocytes are often firmly bound to each other and to neurons by way of **tight junctions** (see Chapter 7) they may be regarded as giving structural support to nervous tissue.

Astrocytes invade injured regions of the CNS (reactive gliosis) and are consequently responsible for the formation of glial scars (as mentioned above). There is evidence (as we shall see in Box 19.2) that astrocytes, at some stages in their life, are able to manufacture and secrete some

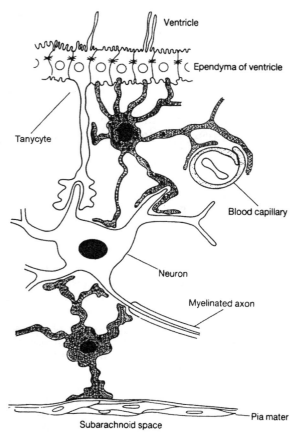

Figure 1.11 Astroglial cells. The schematic diagram shows two astrocytes (stippled). The upper astrocyte stretches from the ependymal epithelium lining the cavity of the ventricle to the perikaryon and dendrites of a neuron. It also invests a blood capillary. The lower astrocyte reaches from the flattened epithelium of the pia mater (which abuts the subarachnoid space) to the neuron. Note that this is a schematic diagram: it is unlikely that a neuron will have astrocytic connections with both the ventricle and the subarachnoid space. After Warwick and Williams (1973), *Gray's Anatomy*, Edinburgh: Churchill Livingstone.

neurotransmitters. In the developing brain these act as growth factors. It is not impossible that in the damaged brain they may play a somewhat analogous role as repair factors. There is, furthermore, evidence that astrocytes in the subventricular area retain proliferative potential into the adult brain where they play a part as **neural stem cells** (NSCs) (see Section 19.2).

Finally, it should be noted that recent work suggests that one of the most important functions of astrocytes is the formation and maintenance of synapses in the CNS. This work shows that astrocytes secrete both cholesterol and a large lipoprotein, apolipoprotein E (ApoE), into the intercellular space. The latter assists neurons in taking up the secreted cholesterol which, in turn, is essential for large-scale synapse formation. We shall meet ApoE again in Chapter 21 when discussing Alzheimer disease. We shall see that one of its isoforms, ApoE-ε4, is defective in its function of assisting neurons to retrieve cholesterol from the extracellular space. It is this isoform which is one of the best predictors of susceptibility to this much feared neurodegeneration.

Oligodendroglial cells constitute a second class of glial cells found in the CNS (Figure 1.13). As the name indicates (*oligos* = few) these cells have fewer processes radiating from the cell body than do astrocytes and the cell body is itself much smaller (c. 5 mm in diameter). Oligodendroglia also differ from astrocytes in having few if any microfilaments but large numbers of microtubules in their cytoplasm. These cells are found in both the grey matter and the white matter. In the white matter, as we shall see, they have the very important role of investing axons in their myelin sheaths; in the grey matter they may be involved in close metabolic interactions with neuronal perikarya.

It is appropriate at this point to indicate that glial cells, known as **Schwann cells**, although not classified as oligodendroglial cells, carry out the business of enveloping **peripheral** axons in their myelin. Peripheral and central myelin is not laid down in precisely the same way, as we shall see, but the end result is much the same.

Microglial cells constitute the third major class of glia to be found in the adult nervous system. They differ from the preceding two classes of glia in originating not in the neural plate (neurectoderm) but in the bone marrow. Their cell bodies are smaller than the other types of glia – seldom exceeding 3 μm in diameter. They make up for their lack of size by their large numbers. They probably have numerous functions. It has been suggested that they are of importance in maintaining the **ionic environment** surrounding neurons – of the greatest significance to the biophysics of the action potential. It is probable, also, that they are

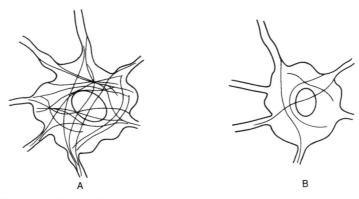

Figure 1.12 Fibrous and protoplasmic astrocytes. (A) Fibrous astrocyte. (B) Protoplasmic astrocyte.

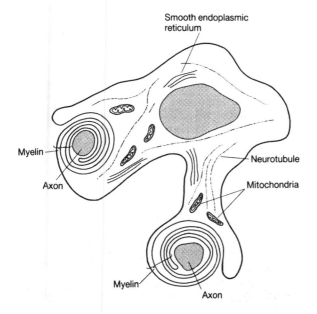

Figure 1.13 Oligodendroglial cells. The cell is shown with two processes each of which has wrapped a central axon in its myelin sheath. This process is described in Chapter 7 (see Figure 7.15).

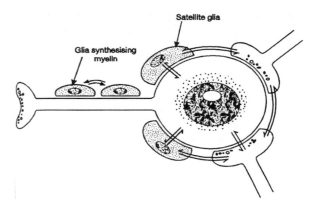

Figure 1.14 Interrelationships between glia and neurons. The arrows show the interactions between glia and glia and neurons and glia. After Vermadakis (1988), *Annual Review of Neurobiology*, **30**, 149–224.

involved in the **uptake and disposal** of unwanted end-products of synaptic activity. But perhaps their most important role in the CNS lies in their ability to proliferate, enlarge into macrophages, and invade any site of injury to **phagocytose** necrotic tissue.

Before completing this introductory section it is worth noting that in the embryonic nervous system glial cells play many other important roles (Figure 1.14). For instance, where cortices are destined to develop **radial glia** appear (Figure 1.15). These cells develop long processes, sometimes extending across the whole width of the brain, from the cerebral ventricle to the pial surface, and guide the migration of neurons during embryonic development. Radial glia, for the most part, disappear or are transformed into astroglia in adult brains. However, they remain virtually unchanged in two regions – the retina, where they are known as **Müller cells**, and the cerebellum, where they are called **Bergmann glia**.

In recent years it has become clear that glia also play a vital role in forming the boundaries around, and thus defining, many structures in the

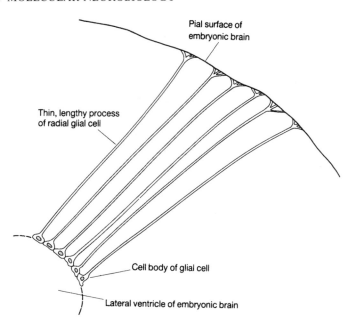

Figure 1.15 Radial glia in the early development of the telencephalon. The figure shows that these glial cells develop extraordinarily lengthy processes which extend from the cell body (next to the ventricle) right across the width of the developing brain to the pial surface.

developing CNS: for instance, they are believed to be instrumental in forming boundaries between segments of the CNS; borders of nuclei such as the lateral geniculate body; peripheries of smaller structures such as olfactory glomeruli and mouse whisker 'barrels'; and, smaller still, sealing off synaptic boutons. This boundary-forming activity of glial cells tends to disappear in the adult CNS but can be made to reappear during recovery from injury. We shall return to these topics in Chapter 18, where we consider brain development.

1.4 ORGANISATION OF SYNAPSES

The structure and function of synapses forms one of the most important areas of research in molecular neurobiology. We shall discuss the molecular detail in Chapters 15, 16 and 17. In this section we shall merely look in an introductory way at their organisation in the brain.

Figure 1.16 shows the structure of a typical synapse in the CNS. The termination of the axon swells to form a 'bouton', as we noted in Section 1.3.1. The bouton contains a number of small (20–40 nm) vesicles which are believed by most workers (there are some exceptions) to contain the molecules of a transmitter substance. The **presynaptic membrane** is separated from the

postsynaptic (= subsynaptic) membrane by a gap of some 30–40 nm. Characteristically the postsynaptic membrane appears denser and thicker in the electron microscope than the presynaptic membrane. The presence of synaptic vesicles and this **postsynaptic thickening** enables the physiological polarity of the synapse to be determined; i.e. transmission always occurs across the synaptic gap in one direction – **from** presynaptic **to** postsynaptic membrane.

Just as there are many different types of neuron and many different types of glia, so there are many different types of synapse. Indeed it would be somewhat strange if there were not for, at a conservative estimate, there are some 10^{14} synapses in the human brain. The structural and biochemical diversity is gigantic. Some of the different structures and arrangements are shown in Figure 1.17. The structures range from simple **electrical synapses** (='**gap junctions**') (see Chapter 7), through classical synapses, to synapses made **en passant** (sometimes called '**varicosities**'), to **reciprocal** synapses and complicated groups of synapses.

One simplifying feature of synaptic appositions was first proposed by the pharmacologist Henry Dale in the 1930s. '**Dale's principle**' states that any given neuron synthesises only one type of transmitter molecule – hence all the terminations

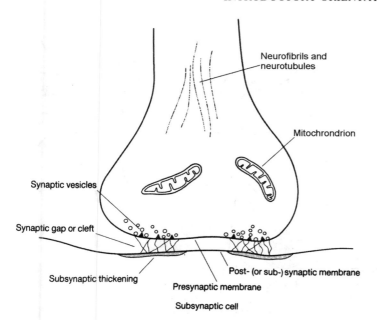

Neurofibrils and neurotubules

Mitochrondrion

Synaptic vesicles

Synaptic gap or cleft

Subsynaptic thickening

Post- (or sub-) synaptic membrane

Presynaptic membrane

Subsynaptic cell

Figure 1.16 'Classical' synapse. Description in text. The figure shows that there is often evidence of some material filling the synaptic cleft. The figure also shows ridges projecting upwards from the presynaptic membrane. These ridges form part of the presynaptic grid (see Chapter 15).

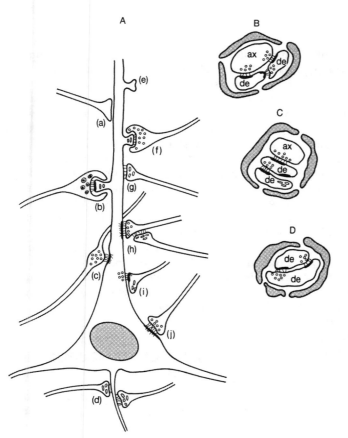

Figure 1.17 Varieties of synapse. (A) (a) Electrically conducting synapse; (b) spine synapse containing dense-core vesicles; (c) 'en passant' synapse or synaptic varicosity; (d) inhibitory synapse (note ellipsoidal vesicles) on initial segment of axon; (e) dendritic spine; (f) spine synapse; (g) inhibitory synapse; (h) axo-axonic synapse; (i) reciprocal synapse; (j) excitatory synapse. (B) Transverse sections through three neuronal processes: one axon (ax) and two dendrites (de) showing complex organisation. The stippled profiles around the group represent glial cells. (C) Transverse section through three neuronal processes: one axon (ax) and two dendrites (de). The two dendrites form a reciprocal pair. They are arranged in a negative feedback loop so that excitation of the lower switches off the upper. (D) Reciprocal synapse made between two dendrites (de). In this case there is positive feedback. Excitation of the lower dendrite re-excites the upper.

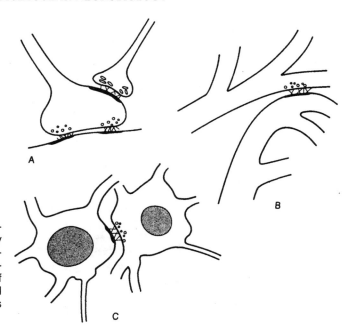

Figure 1.18 'Non-classical' synapses. (A) Axo-axonic synapse. The termination of one axon may control the activity another terminal. (B) Dendro-dendritic synapse. Synaptic appositions are sometimes found between the dendritic processes of neighbouring neurons. (C) Perikaryo-perikaryal synapse. Very occasionally synaptic junctions are made between adjacent perikarya.

of that neuron contain only that one type of transmitter. Although many exceptions to Dale's principle are nowadays known (see Chapter 16), it remains a good first approximation. It is also to some extent possible to relate the transmitter molecules present in a synaptic terminal to the form of the presynaptic vesicles. Thus small spherical translucent vesicles are believed to contain excitatory transmitters such as acetylcholine or glutamate, whilst small translucent ellipsoidal vesicles are thought to contain inhibitory transmitters such as glycine or γ-aminobutyric acid (GABA). Larger, dense-cored, vesicles contain catecholamine transmitters, whilst large translucent vesicles probably contain peptide transmitters.

Classical neurophysiologists understood the connectivity of the nervous system to be one way only – from axon to dendrite or perikaryon. It remains true that most synapses are **axo-dendritic** or **axo-perikaryal** (=**axo-somatic**), but in recent years other arrangements have been discovered. **Axo-axonic** synapses are quite common. This arrangement allows one neuron to control the synaptic activity of another. More recently it has been shown that dendrites also make synapses. **Dendro-dendritic** synapses have been demonstrated

in the olfactory bulb, in the retina, in the superior colliculi and elsewhere. Finally it appears that synapses are sometimes made between perikarya. It seems, therefore, that all the possible permutations between neuronal processes are made somewhere or other in the brain. Some of these 'non-classical' arrangements are shown in Figure 1.18.

It is clear from the foregoing paragraphs that the synaptic organisation of the brain is exceedingly complex and as yet far from completely understood. The dendritic and perikaryal surfaces of many neurons are densely covered with synaptic endings of various sorts. It has been computed that the large Purkinje cells of the cerebellum are exposed to over 100 000 synaptic appositions. The dense investment by synaptic endings of various different sizes of the perikaryon of a spinal motor neuron is shown in Figure 1.19.

1.5 ORGANISATION OF NEURONS IN THE BRAIN

To the naked eye a section of the mammalian brain seems to reveal two types of substance: **grey matter** and **white matter**. White matter is composed of

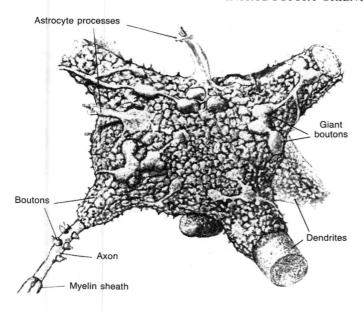

Astrocyte processes

Giant
boutons

Boutons

Dendrites

Axon

Myelin sheath

Figure 1.19 Synaptic contacts on the perikaryon of a spinal motor neuron. This reconstruction from serial electron micrographs shows how densely covered the perikaryon of a motor neuron is with large and small synaptic endings. From Poritsky (1969), *Journal of Comparative Neurology*, **135**, 423–452; with permission.

huge numbers of nerve fibres. In bulk they appear white because the myelin sheaths with which the majority are enveloped reflect and glisten in the light. Grey matter, on the other hand, consists of the dendrites and perikarya of the neurons plus numerous glial cells. These are not surrounded by myelin and hence in bulk appear greyish.

In the early embryo the grey matter is situated in the centre of the CNS immediately surrounding the central fluid-filled cavity (central canal in spinal cord, ventricle in brain). It retains this primitive position throughout life in the spinal cord, but in the brain many of the neurons migrate during embryological development along the processes of radial glia to form surface cortices or 'rinds' (see Figure 1.15 and Chapter 18). This occurs especially in the cerebrum and the cerebellum and gives rise to the cerebral and cerebellar cortices. Other groups of perikarya, however, remain deep within the brain, forming islands of grey matter amongst the fibre tracts: these constitute nuclei and ganglia. An outline of this organisation is shown in Figure 1.20.

Grey matter, especially that of the cerebral cortex, has an extremely complex and little-understood organisation. Silver staining by the Golgi–Cox technique shows an elaborate inter-connexity (Figures 1.21, 1.22). It is known, more-over, that this staining technique impregnates only

about 1% (at random) of the neurons present. The true interconnexity is thus almost unimaginably intricate. Electron micrographs of the cortex reveal densely packed masses of cells and cell processes with apparently rather little intercellular space (Figure 1.21B). The 'wiring' of the cortex remains one of the most difficult research frontiers in neuroscience. The pattern of synaptic connection and interaction is of almost inconceivable complexity. Indeed the cortex has been compared with a hologram, implying that information is not held in discrete localities but 'smeared' throughout.

Against this idea of a 'randomised' cortex there has always been a strong tradition which envisages the cortex as consisting of a number of functionally and structurally distinct units or **modules**. This tradition reached a *reductio ad absurdum* in the early nineteenth century in the phrenological crazes started by Gall and Spurzheim. Although phreno-logy quickly fell into scientific disrepute, the idea that the cortex could be subdivided into discrete organs was never entirely lost and reappeared at the end of the nineteenth and beginning of the twentieth century in the functional topography of Ferrier and in the cortical architectonics of Brodmann and the Vogts. This initiative also fell into disrepute due to its seeming over-elaboration. Von Economo's atlas of the cerebral cortex, for

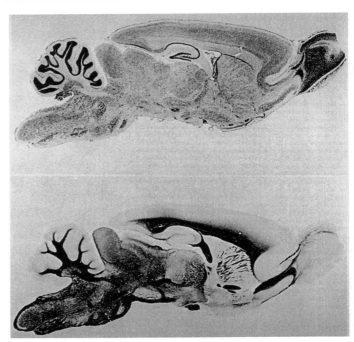

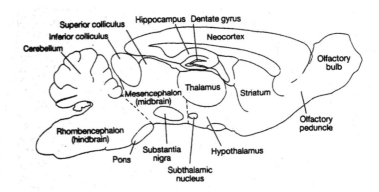

Figure 1.20 Parasagittal section of mammalian (rat) brain to show arrangement of grey and white matter. The top section has been subjected to a staining technique (Nissl stain) which stains the perikarya of neurons. Each dot represents a cell body. The Loyez technique has been used to stain the middle section. This technique stains myelin but does not affect perikarya. The middle section thus shows the white matter fibre pathways. The bottom section maps the anatomical structures delineated by the two stains. From 'The organization of the brain', by W.J.H. Nauta and M. Feirtag. Copyright © 1979 by Scientific American Inc. All rights reserved.

instance, delineated over 200 histological areas. In the mid-twentieth century this detailed architectonics was replaced by a more functional modularisation. Neurophysiologists interested in sensory cortices (Mountcastle – somaesthetic cortex; Hubel and Wiesel – visual cortex) showed that the neocortex consists of **functional columns** or **slabs**.

The most obvious feature of the neocortex when viewed under the optical microscope is, however, its layered stratification. This layering is shown in Figure 1.22. Traditionally **six laminae** have been

distinguished. Layer four is conventionally further subdivided into three sublayers: a, b and c. The stratification of the neocortex is more obvious in some regions (e.g. visual cortex) than others (e.g. association cortex). Cortical columns lie **orthogonal** to this stratification. The first histological hints of this vertical organisation were provided by Lorente de No in his classical research during the 1930s. Nowadays cortical columns are believed to have a diameter of about 300 mm and to contain some 7500 to 8000 neurons (subdivided into about 80

A

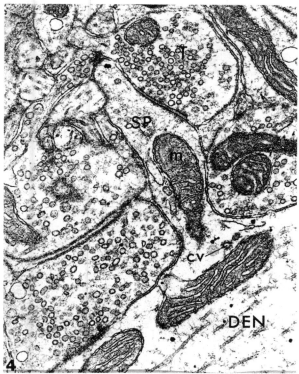

B

Figure 1.21 Structure of grey matter. (A) Silver-stained section of cerebral cortex (×300). A pyramidal neuron can be seen on the right-hand side of the picture and three large, vertically-running, dendrites to its left. The surfaces of the dendrites are covered in spines. (B) Electron micrograph of the oculo-motor nucleus of the cat (×52 000). In the lower right-hand corner the pale expanse is a dendrite (DEN) from which springs a spine (SP). Synaptic boutons filled with vesicles surround the spine and the dendrite. The bouton labelled T makes a particularly well-imaged synaptic contact with the spine. m=mitochondrion; cv=cytoplasmic vesicle. From Pappas and Waxman (1972), in *Structure and Function of Synapses*, ed. by G.D. Pappas and D.P. Purpura, Amsterdam: North Holland; with permission.

minicolumns which are believed to be the recurring unit). The neocortex appears to be a mosaic of such columns which, moreover, vary very little in diameter throughout the mammals, from mouse to man.

Figure 1.22B shows that the output from the cerebral cortex is carried by axons leaving the bases of the pyramidal cells. These axons may course through the white matter and re-enter the cortex at some more-or-less distant location, or may cross

Figure 1.22 Stratification of the cerebral cortex. (A) Cortical neurons stained by the Golgi–Cox silver technique (×100). The figure shows that incoming axons terminate in complex ramifications in layers IVa and IVc. These ramifications are some 350–450 μm in diameter. One complete ramification is shown in the centre of the figure flanked by two half ramifications. From Rakic (1979), in *The Neurosciences: Fourth Study Program*, ed. by F.O. Schmitt and F.G. Worden, Cambridge, MA: MIT Press, pp.109–127; with permission. (B) Output from the cortex. The figure shows that the axons (ax) from the small pyramidal cells (Py) in layers II and III mostly pass out of the cortex to run in the subcortical white matter to re-enter the cortex at some other place. The axons from pyramidal cells in layers V and VI, however, run to subcortical nuclei or out of the cerebrum altogether to the brain stem or spinal cord.

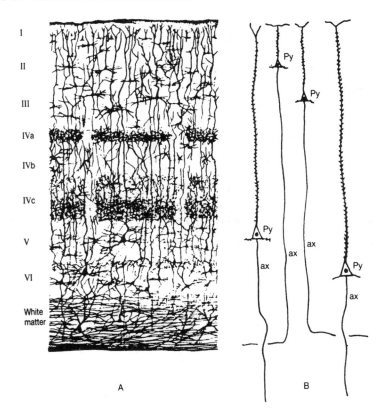

from one hemisphere to the other via the corpus callosum and re-enter the cortex on the opposite side. Other axons run out of the cortex altogether and terminate in some distant part of the brain or spinal cord. It is interesting to note, however, that the latter are in the minority. Of axons leaving or entering the cortex by far the greater number go to or come from other parts of the cortex. Indeed Braitenberg estimates that cortico-cortical fibres outnumber non-cortico-cortical fibres by a factor approaching 10 000 : 1. Each part of the cortex is thus influenced by every other part – each module, it has been argued, contains, like the fragment of a holograph, a fuzzy representation of the whole.

Figure 1.23 shows the neuronal structure of a cortical module as envisaged by Szentágothai. This is not the place to enter into a detailed description of this neuronal meshwork. Interested readers should examine Szentágothai's account. Figure 1.23 is included merely to give some 'feel' for the

complexity which undoubtedly exists at the histological level.

Figure 1.23 emphasises that the intricate juxtaposition of neurons and neuroglia in the cortex provides innumerable possibilities for the synaptic contacts discussed in the previous section and for the neurochemistry to be outlined in the subsequent chapters of this book. Neurons cannot be regarded as discrete, 'introspective', units such as the transistors and resistors of a circuit board but as interacting together in rich and diverse ways. The long-axon 'principal' neurons of classical neurophysiology are not typical of the intricate webs of dendrites, short axons and perikarya, neuroglia of various sorts, dendritic spines and tortuous intercellular spaces, which are characteristic of grey matter. Here the full complexity of sub-millivolt cable conduction (Chapter 12), of subtle shifts of base-level resting potentials and postsynaptic sensitivities (Chapter 17), of heterogeneous membrane

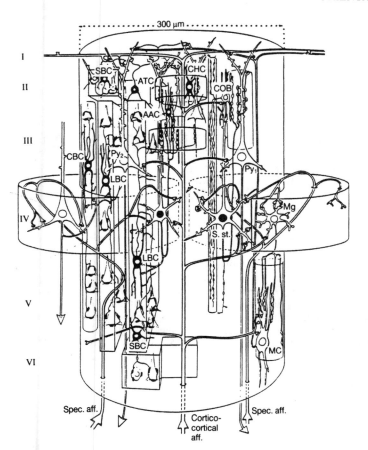

Figure 1.23 Neocortical module. The diagram represents a cortico-cortical column 300 μm in diameter. The six horizontal layers of the cortex are numbered to the left of the figure. The two flat cylinders in lamina IV correspond to the termination territory of a specific afferent. From Szentágothai (1979), in *The Neurosciences: Fourth Study Program*, ed. by F.O. Schmitt and F.G. Worden, Cambridge, MA: MIT Press, pp. 399–415; with permission.

patches (Chapter 7), of molecular transfer between cells via gap junctions (Chapter 7), of changes in ambient ion concentration, of complicated sculpturing of electric fields by the three-dimensional geometries of dendritic arbors and spine morphologies and so forth can occur. The state of matter in the cerebral cortex is of mind-boggling complexity.

Perhaps, in spite of the analyses of the structural biologists – the anatomists, histologists, cytologists and molecular biologists – the cortex can best be regarded, to quote Szentágothai again, 'as something of a continuous medium'. As in all areas of scientific endeavour so with the cerebral cortex: analysis comes first. The reconstruction of the

whole from its constituent fragments follows later – in this case very much later, some time in the yet-unforeseeable future. The observer surveying the cortex naturally wishes to see edges, modules, demarcations, levels – this is the only hope of progress. In reality, however, there is an immensely complex, extended, pattern of material activity, a flow of activity comparable, as Freeman puts it, to the 'continuum of a chemical reaction'. Moreover the cortex, as we have already emphasised, is linked together so that each part of the immense sheet is affected by what is happening in every other part. It is this complex interconnexity which makes the brain unique among living tissues.

2

THE CONFORMATION OF INFORMATIONAL MACROMOLECULES

Definition of informational macromolecules. **Proteins**: levels of structure. Primary structure – amino acids and their properties – polypeptides – neuroactive peptides – glycosylation. Secondary structure – β-pleated sheets, hairpins and barrels – collagen three-stranded ropes – AchE – the α-helix – amphipathic helices. Tertiary structure – modularisation, domain structures and motifs – allosteric flexures. Quaternary structures – haemoglobin – fragility – necessity for synthesis to be chaperoned – heat shock proteins (hsps). **Nucleic acids**: DNA – units of structure – base pairing. RNA – types of RNA – mRNA – tRNA – rRNA – synthesis and assembly. **Conclusion**: function from structure – complementary surfaces – molecular symbioses

In Chapter 1 we reminded ourselves of the great structural complexity of the brain, from its 'naked-eye' anatomy to its cellular and subcellular detail. There was no need to labour the point that this intricacy of structure has to do with its central function: information processing – the computing of life-preserving responses to the challenges of the biotic and abiotic environment.

In this chapter we start from the other end of the scale of neurobiological magnitudes, the molecular end. But once again the same feature stands out. Biological molecules often have extremely complicated structures and these structures are basic to their biological function. This, indeed, is the fundamental insight of molecular biology.

Two types of biological macromolecule are central to molecular biology: the **proteins** and the **nucleic acids**. These molecules are sometimes called **informational macromolecules**. Other molecular species also play a crucial role – e.g. the carbohydrates and the lipids – but they are of lesser importance. We shall discuss the role of the latter in the formation of biomembranes in Chapter 7. In this chapter and the next we shall concentrate on the proteins and the nucleic acids. These are the molecules which bring about the information processing upon which the cell, just as much as the brain, depends. Their structures, unlike those of other macromolecules, encode information, principally genetic information refined by two or three billion years of evolutionary trial and error.

2.1 PROTEINS

Proteins, like all informational macromolecules, are **polymers** built of a large number of monomeric units. In the case of proteins these monomeric units are **amino acids**. Molecular biologists recognise twenty different types. These are shown in Table 2.1. Except for **proline** they all share a common structure, differing only in their side chains (symbolised by the letter R). This structure is shown below:

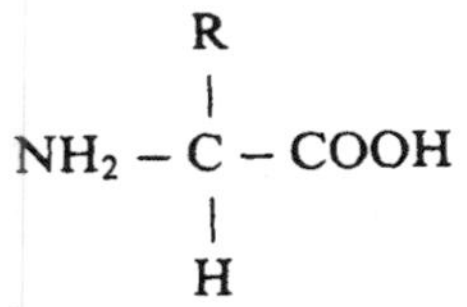

$$\begin{array}{c} R \\ | \\ NH_2 - C - COOH \\ | \\ H \end{array}$$

The side chains of the amino acids are, however, of great importance. Table 2.1 shows some of their more important properties. The side chain of proline bends round, as shown, and unites with the amino group by removing one of its hydrogens. This, as we shall see, has significant consequences for some of the conformations into which amino acid chains are twisted. Another even more significant feature so far as the conformation of proteins is concerned is the **hydrophilicity** or **hydrophobicity** of the side chains. We shall see that the solubility or insolubility of different parts of a protein in water is of great importance in molecular neurobiology. The sheer bulkiness of the side chains is also significant. **Glycine**, with just a single hydrogen, has the smallest side chain of all, and **alanine** with a methyl group is not much bigger; in contrast **tyrosine**, **phenylalanine** and in particular **tryptophan** have bulky aromatic side chains. Finally two amino acids, **methionine** and **cysteine**, contain sulphur atoms in their side chains. Both play important roles in molecular biology. The sulphydryl (SH) group of cysteine is particularly important. The three-dimensional form of many proteins is stabilised by the formation of disulphide linkages between the SH groups of neighbouring cysteines. Table 2.1 also shows that each amino acid has been assigned a three-letter abbreviation and a one-letter symbol.

The conformation of protein molecules can be extremely complex. For convenience it is usual to treat it as if it had four different levels: **primary**, **secondary**, **tertiary** and **quaternary**. Only the first of these levels can be determined by conventional biochemical techniques (chromatography, electrophoresis, etc.); in recent years, as we shall see, the techniques of genetic engineering have been increasingly used to determine this level of structure. Higher levels are investigated principally by the techniques of **X-ray diffraction** and **nuclear magnetic resonance (NMR)** although electron microscopy also plays a part. The determination of protein structure is laborious and time-consuming although as the years have passed automation has relieved much of the tedium and

greatly speeded the process. More than 100 000 primary structures and more than 6000 secondary and higher structures are now known. These structures are collected in various data banks which can be accessed through the internet. Amino acid sequences (primary structures) are listed at **www-nbrf.georgetown.edu/pir/** (Georgetown University, Washington DC, USA) and **www.expasy.org/prosite/** (Geneva, Switzerland and Hinxton, UK). Higher structures are archived at the Protein Data Bank, **www.rcsb.org** (Rutgers University, New Jersey, USA).

2.1.1 Primary Structure

It is well known that amino acids are able to link together by the elimination of the elements of water between the carboxylic acid (—COOH) group and the amino (—NH_2) group. In this way it is possible to form long chains of amino acids (often called **residues** when incorporated in a chain) linked by **peptide bonds**. Peptide bonds are covalent forces caused by the sharing of valency electrons and are hence very strong. An energy input of about **70 kcal/mol** is required to break them. Hence the primary structure of a protein or polypeptide is tough and difficult to disrupt.

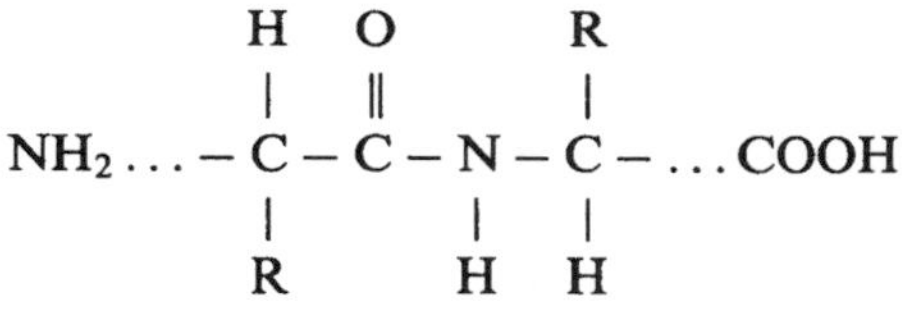

$$\begin{array}{c} \quad\quad H \quad\ O \quad\quad\ R \\ \quad\quad | \quad\quad \| \quad\quad\ | \\ NH_2 \ldots - C - C - N - C - \ldots COOH \\ \quad\quad | \quad\quad\quad | \quad\ | \\ \quad\quad R \quad\quad\quad H \quad H \end{array}$$

It is worth noting at this stage that resonance occurs between the double covalent bond of the carbonyl (CO) group and the single covalent bond of the imino (NH) group. In other words valency electrons are shared between the O, C and N atoms. This seeming detail ensures that all the atoms in the amide group (shown below) are co-planar. This, in turn, restricts the number of conformations into which the amino acid chain can be twisted.

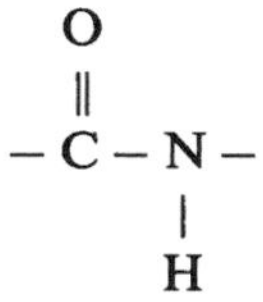

$$\begin{array}{c} O \\ \| \\ - C - N - \\ | \\ H \end{array}$$

Planar amide group

The sequence in which the amino acids are linked together is called the **primary sequence** or **primary**

Table 2.1 Amino acids

Name	Abbreviation	Letter	Formula	Comment
1. Polar (hydrophilic) side chains				
Glycine	Gly	G	$$\begin{array}{c} NH_2 \\ | \\ H-C-H \\ | \\ COOH \end{array}$$	Simplest and smallest side chain
Aspartic acid	Asp	D	$$\begin{array}{c} NH_2 \quad\quad O \\ | \quad\quad\quad\; \| \\ H-C-CH_2-C \\ | \quad\quad\quad\; | \\ COOH \quad\; O^- \end{array}$$	Acidic, negatively charged side chain
Asparagine	Asn	N	$$\begin{array}{c} NH_2 \quad\quad O \\ | \quad\quad\quad\; \| \\ H-C-CH_2-C \\ | \quad\quad\quad\; | \\ COOH \quad NH_2 \end{array}$$	
Glutamic acid	Glu	E	$$\begin{array}{c} NH_2 \quad\quad\quad\quad\; O \\ | \quad\quad\quad\quad\quad\quad \| \\ H-C-CH_2-CH_2-C \\ | \quad\quad\quad\quad\quad\quad | \\ COOH \quad\quad\quad\; O^- \end{array}$$	Acidic, negatively charged side chain
Glutamine	Gln	Q	$$\begin{array}{c} NH_2 \quad\quad\quad\quad\; O \\ | \quad\quad\quad\quad\quad\quad \| \\ H-C-CH_2-CH_2-C \\ | \quad\quad\quad\quad\quad\quad | \\ COOH \quad\quad\quad NH_2 \end{array}$$	
Cysteine	Cys	C	$$\begin{array}{c} NH_2 \\ | \\ H-C-CH_2-SH \\ | \\ COOH \end{array}$$	SH (sulphydryl) group frequently involved in disulphide (S–S) linkages
Serine	Ser	S	$$\begin{array}{c} NH_2 \\ | \\ H-C-CH_2-OH \\ | \\ COOH \end{array}$$	Hydroxyl group often phosphorylated by protein kinases
Threonine	Thr	T	$$\begin{array}{c} NH_2 \; CH_3 \\ | \quad\; | \\ H-C-CH-OH \\ | \\ COOH \end{array}$$	Hydroxyl group often phosphorylated by protein kinases

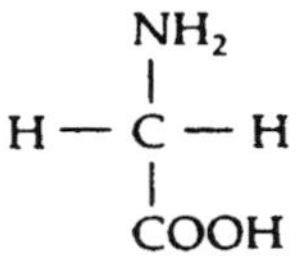

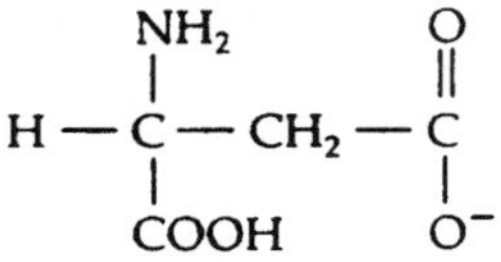

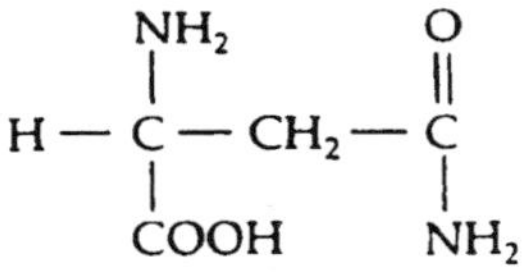

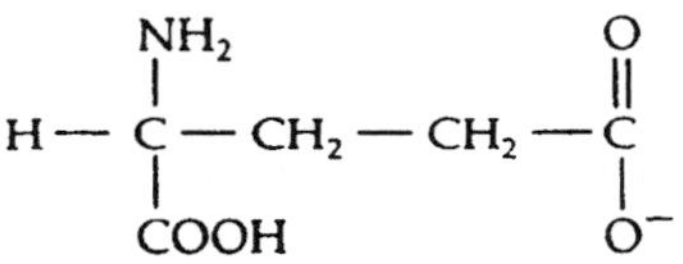

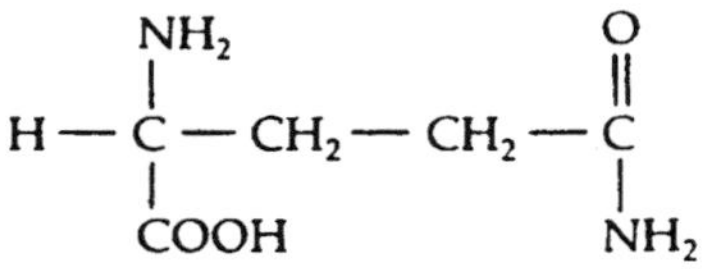

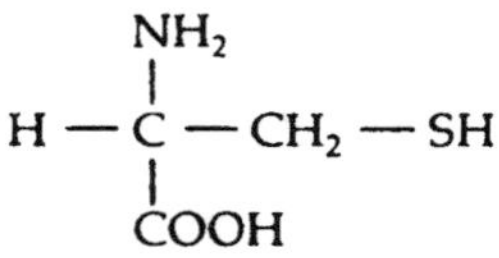

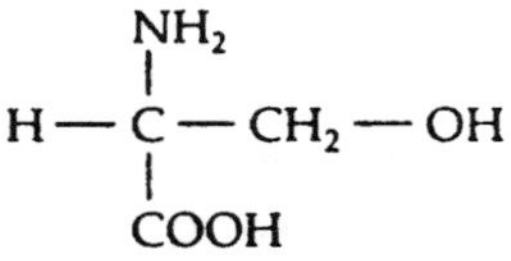

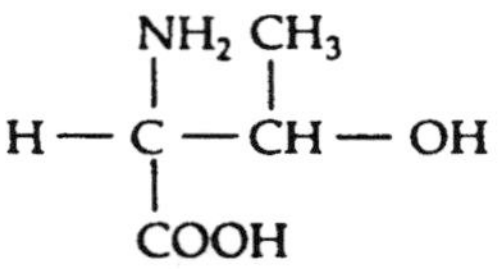

Continued

Table 2.1 *Continued*

Name	Abbreviation	Letter	Formula	Comment
Tyrosine	Tyr	Y	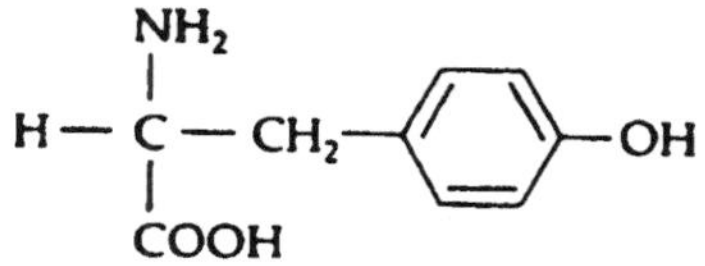	Bulky side chain, hydroxyl group
Lysine	Lys	K	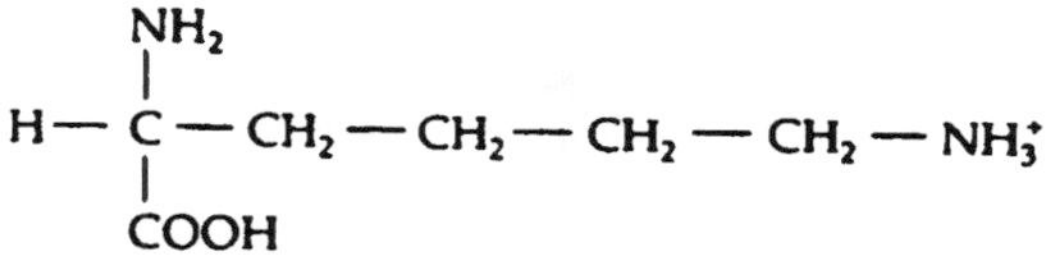	Basic, positively charged side chain
Arginine	Arg	R	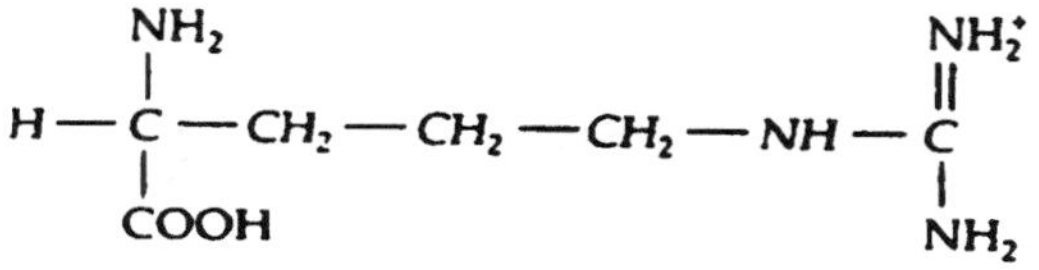	Basic, positively charged side chain
Histidine	His	H	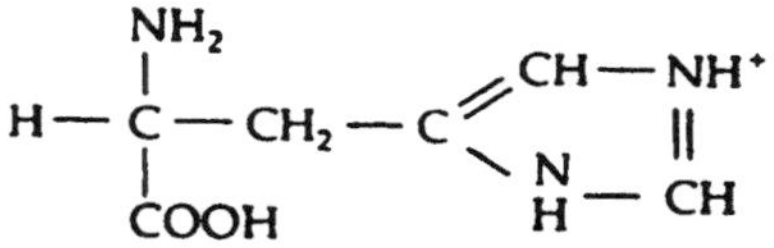	Basic, bulky, positively charged side chain

2. Non-polar (hydrophobic) side chains

Name	Abbreviation	Letter	Formula	Comment
Alanine	Ala	A	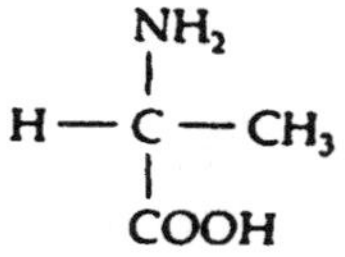	Small side chain
Valine	Val	V	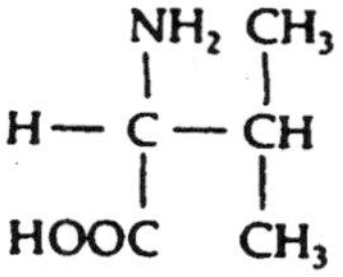	
Leucine	Leu	L	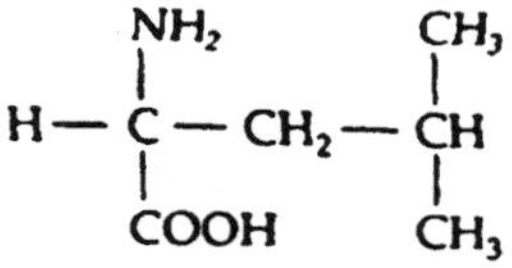	

Continued

Table 2.1 *Continued*

Name	Abbreviation	Letter	Formula	Comment
Isoleucine	Ile	I		
Proline	Pro	P		Note lack of amino group; in fact an imino acid
Phenylalanine	Phe	F		Bulky side chain
Methionine	Met	M		Side chain contains a sulphur atom
Tryptophan	Trp	W		Bulky side chain

structure. The amino acids in a primary sequence are numbered from the **N-terminal** end of the chain. The distinction between **polypeptides** and **proteins** is rather arbitrary. Traditionally sequences of more than about fifty residues were regarded as proteins; sequences of less than that number of residues were termed polypeptides. That tradition has broken down. Nowadays there seems to be little or no distinction between what is termed a polypeptide and what is termed a protein. Many proteins have considerably more than 100 amino acids in their primary structure. As there are twenty different amino acids it follows that primary sequences could, theoretically, be almost infinitely various. The number of different sequences possible for a primary structure of 100 amino acids is 20^{100}, i.e.

1 followed by about 130 zeros! In other words, there are more possible primary sequences than there are atoms in the universe. Needless to say only a very small subset of this immense number of sequences is synthesised by living cells. The primary sequences of some neuroactive peptides are shown in Table 2.2.

Glycoproteins

Before completing this section it is important to describe the structure of **glycoproteins**. We shall see in subsequent chapters that many of the most important neurobiological proteins are **glycosylated**. This means that attached to the polypeptide chain are oligosaccharide side chains. Such proteins

Table 2.2   Primary structure of some neuroactive peptides

Leu-enkephalin	Try-Gly-Gly-Phe-Leu
Met-enkephalin	Try-Gly-Gly-Phe-Met
β-Endorphin	Try-Gly-Gly-Phe-Met-Thr-Ser-Glu-Lys-Ser-Gln-Thr-Pro-Leu-Val-Thr-Leu-Phe-Lys-Asn-Ala-Ile-Lys-Ile-Lys-Asn-Ala-Tyr-Lys-Lys-Gly-Glu
Cholecystokinin 8 (CCK8)	Asp-Tyr-Met-Gly-Trp-Met-Asp-Phe
Cholecystokinin 4 (CCK4)	Trp-Met-Asp-Phe
Neurotensin (NT)	Glu-Leu-Tyr-Glu-Asn-Lys-Pro-Arg-Arg-Pro-Tyr-Ile-Leu
Angiotensin (AT)	Asp-Arg-Val-Tyr-Ile-His-Pro-Phe
Substance P (SP; NK1)	Arg-Pro-Lys-Pro-Glu-Gln-Phe-Phe-Gly-Leu-Met
Neurokinin A (NKA; NK2)	His-Lys-Thr-Asp-Ser-Phe-Val-Gly-Leu-Met
Neurokinin B(NKB; NK3)	Asp-Met-His-Asp-Phe-Phe-Val-Gly-Leu-Met
Bradykinin (BK)	Arg-Pro-Pro-Gly-Phe-Ser-Pro-Phe-Arg
Vasopressin (VP)	Cys-Tyr-Phe-Gln-Asn-Cys-Pro-Arg-Gly
Oxytocin (OT)	Cys-Tyr-Ile-Gln-Asn-Cys-Pro-Leu-Gly
Bombesin (BB)	Glu-Gln-Arg-Leu-Gly-Asn-Glu-Trp-Ala-Val-Gly-His-Leu-Met
Somatostatin 14 (ST)	Ala-Gly-Cys-Lys-Asn-Phe-Phe-Trp-Lys-Thr-Phe-Thr-Ser-Cys
Neuropeptide Y (NPY)	Tyr-Pro-Ser-Lys-Pro-Asp-Asn-Pro-Gly-Glu-Asp-Ala-Pro-Ala-Glu-Asp-Met-Ala-Arg-Tyr-Tyr-Ser-Ala-Leu-Arg-His-Tyr-Ile-Asn-Leu-Ile-Thr-Arg-Gln-Arg-Tyr

The table shows that several neuroactive peptides share common amino acid sequences. We shall see in Chapters 3 and 4 that this is no coincidence. Families of neuroactive peptides are often derived from a single mother polypeptide precursor. The sequence for neuropeptide Y (NPY) is for the human version. Although the 36-residue sequence is well preserved through evolutionary time it nevertheless varies somewhat in different organisms.

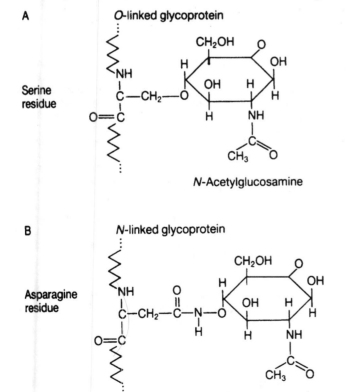

A   *O*-linked glycoprotein

Serine residue

CH₂OH

N-Acetylglucosamine

B   *N*-linked glycoprotein

Asparagine residue

CH₂OH

N-Acetylglucosamine

Figure 2.1   Glycoproteins and glycosidic links. In (A) *N*-acetylglucosamine is linked to a serine residue by an *O*-glycosidic bond; in (B) *N*-acetylcglucosamine is linked to an asparagine residue by an *N*-glycosidic bond.

are defined as glycoproteins. The most common saccharide units in glycoproteins are **galactose (Gal)**, **mannose (Man)** and **fucose (Fuc)** (=**6-deoxy-galactose**). In addition the amino sugars *N*-acetylgalactosamine (GalNAc) and *N*-acetylglucos-amine (GlcNAc) are very common, as is **sialic acid** (e.g. *N*-acetylneuraminic acid (NANA)) (for further discussion see Chapter 7).

Only certain amino acids form points of attachment for oligosaccharide chains. *O*-glycosidic links are formed with the hydroxyl terminals of side chains of **serine**, **threonine**, **hydroxylysine (Hyl)** and **hydroxyproline (Hyp)**, whilst *N*-glycosidic links are made to the terminal amino group of **asparagine**'s side chain. These linkages and structures are shown in Figure 2.1. The cell biology of glycosylation will be described in Chapter 15.

2.1.2 Secondary Structure

Secondary structure is somewhat difficult to define. In effect it consists of the structure conferred by **hydrogen bonding** between contiguous parts of a primary structure. H-bonds can occur between the hydrogen atoms of imino groups and the oxygen atoms of carbonyl groups. This is because electro-negative atoms such as N attract hydrogen's lone electron, leading to a fractional negative charge ($\delta-$) on N and a fractional positive charge ($\delta+$) on H. H is thus open to electrostatic attraction from another electronegative atom such as O:

$$C=O \;\|\|\|\; H-N$$
$$\delta+ \;\; \delta- \quad\; \delta+ \;\; \delta-$$

The electrostatic force between the partial charges is very weak. It is computed to be about **1% of the covalent-bond force**, i.e. **1–5 kcal/mol**.

Carbonyl and imino groups are, as we saw in Section 2.1.1, repeated features of a protein's primary structure. Consequently it is possible for very large numbers of H-bonds to form if amino acid chains are correctly aligned. It follows that, although each H-bond may be easy to break, the force exerted by large numbers may be quite strong. But, as indicated, this does depend on correct alignments of the bonded chains. Here we meet for the first time one of the themes which run throughout molecular biology: the **'stickiness' of complementary surfaces**.

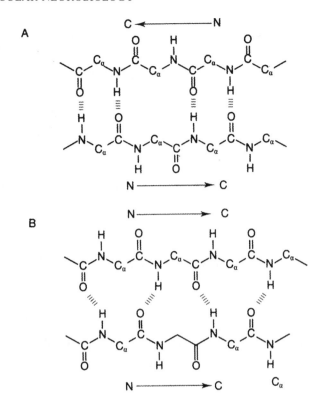

Figure 2.2 Antiparallel and parallel β-pleated sheets. (A) Antiparallel pleated sheet. The figure shows that if the two polypeptide chains run in opposite directions they can be aligned so that their hydrogen-bonding potentialities are easily satisfied. The side chains attached to the α-carbon atoms are not shown: they project in the third dimension – above and below the plane of the paper. (B) Parallel pleated sheet. The figure shows that if two polypeptide chains both run in the same direction they can nevertheless be aligned so that hydrogen bonds are formed between their imino hydrogens and carbonyl oxygens. The alignment between carbonyl oxygen and imino hydrogen is, however, not so precise as it is in the antiparallel sheet. In consequence the parallel sheet tends to develop a right-hand twist.

There are three important secondary structures: β-pleated sheets, 'collagen'-like triple helices and α-helices. Let us briefly consider each in turn.

β-Pleated Sheets

β-pleated sheets come in two varieties: **parallel** and **antiparallel**. The antiparallel type is the most

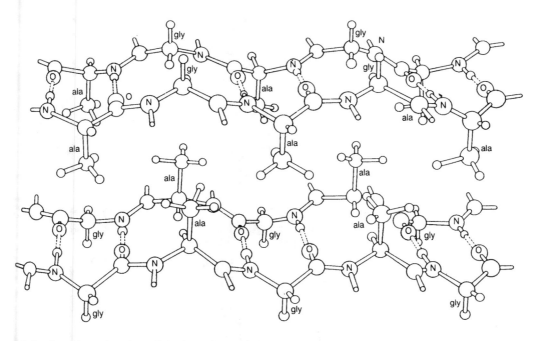

Figure 2.3 Stacking of antiparallel pleated sheets in silk fibroin. The figure gives a three-dimensional view of a small portion of the antiparallel pleated β-sheet structure of silk fibroin. The hydrogen bonding between imino nitrogen and carbonyl oxygen (in contrast to Figure 2.2) is shown in the third dimension – above and below the plane of the paper. The small side chains of alanine and glycine nestle neatly between the sheets.

stable. Here amino acid chains run in opposite directions. In Figure 2.2A the chain from N to C runs right to left above and from left to right below. When this alignment occurs the imino hydrogens and the carbonyl oxygens are all optimally positioned for H-bond formation.

Figure 2.2B shows that when the amino acid chains run in the same direction the hydrogen bonding potentialities are not quite so easily satisfied. Nonetheless alignment is possible and the result is the 'parallel' β-pleated sheet. Because of the slight strain in the hydrogen bond alignments the sheet tends to twist in a right-handed sense. This twist can be seen in the β/β tertiary structure shown in Figure 2.8.

The best-known example of an extensive antiparallel β-pleated sheet is found in **silk fibroin** (Figure 2.3). Silk is built up from layer upon layer of pleated sheet. It is clear that if the pleated sheets are to fit snugly together the amino acid side chains cannot be too bulky. It is thus no surprise to find

that silk fibroin consists almost entirely of Gly and Ala residues.

β-Pleated sheets are by no means restricted to fibrous proteins such as silk. We shall see that they often form major structural domains within large globular proteins. Here the β-strands are frequently involved in tight turns known as **β-hairpins**. This allows a strand to turn back on itself, forming either parallel or antiparallel hydrogen bonds with its initial length. Another common form which β-sheets take in globular proteins is that of a hollow cylinder, or **β-barrel** (often known as a '**TIM barrel**'). Here eight or so linked β-strands curve round so that the eighth strand hydrogen bonds with the first to form a hollow cylinder. If these barrels are organised such that only hydrophobic amino acids (such as alanine) project outwards they will be soluble in the lipid cores of biomembranes (Chapter 7). It is thus not surprising to find β-barrels, often associated with α-helices, as significant structural motifs in membrane proteins.

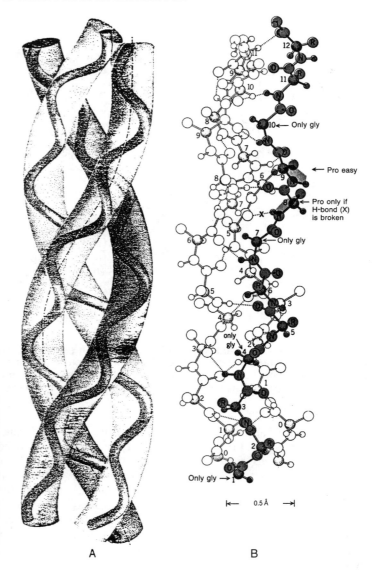

Figure 2.4 Collagen. (A) This shows the three-stranded 'rope' of a portion of a tropocollagen molecule. Each strand is itself a twisted amino acid chain. (B) The detailed molecular structure of the three-stranded rope shown in (A). The numbers refer to α-carbon atoms; two of every three imino hydrogens and carbonyl oxygens are hydrogen-bonded to another chain (dotted lines). This gives a structure with a high tensile strength. After Dickersen and Geis (1969), *The Structure and Action of Proteins*, Menlo Park, CA: Benjamin/Cummings.

A   B

Collagen and Collagen-like Structures

A second well-understood example of secondary structure is provided by **collagen**. Collagens are found throughout the Metazoa and in the mammals they constitute some 25% of all the body's proteins. There are several different types of collagen. As this molecular structure plays rather little part in neurobiology we shall consider only its general features. Unlike silk fibroin the amino acid chain is not extended in the β-conformation. Instead each

amino acid is rotated through 120 degrees with respect to its predecessor in the chain. This produces a twisted thread. Next, three of these twisted threads are wound together and held by intra-chain hydrogen bonds (Figure 2.4). For this complex structure to be possible only certain amino acids can be incorporated into the chains. Glycine, because it is small, forms every third residue and there are large quantities of hydroxyproline and hydroxylysine. Hydroxyproline is believed to be involved in the formation of the intra-chain

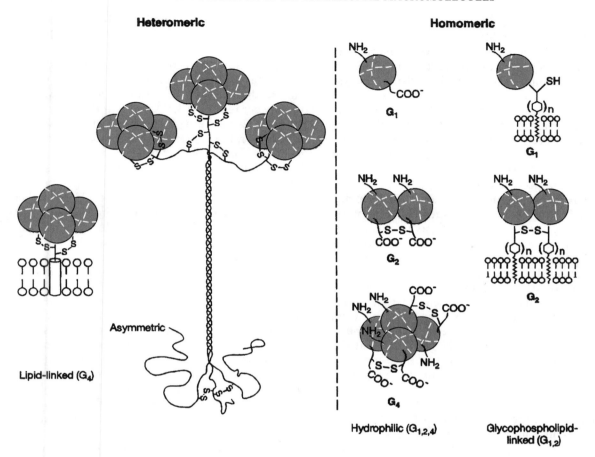

Figure 2.5 Molecular structures of acetylcholinesterase. The homomeric forms of AChE are shown on the right of the figure. On the far right are monomers and dimers linked into the bilayer structure of a cell membrane by glycophospholipid tails. To the left of these are the hydrophilic members of the group, either monomers or disulphide-linked dimers or tetramers. G1–G4 indicate the number of catalytic subunits. To the left of the dashed line are the heteromeric forms, either linked through disulphide bonds to lipid or to a lengthy collagen-like tail. After Taylor and Radic. Reproduced with permission from the *Annual Review of Pharmacology and Toxicology*, **34**, 281–320, © 1994, by Annual Reviews Inc. www.AnnualReviews.org.

H-bonds, whilst the hydroxylysine is frequently glycosylated by the addition of two carbohydrate residues – galactose and glucose. Collagen can thus be classified at a 'glycoprotein'.

Although, as we have already noted, collagens play little part in the structure and functioning of the brain, another molecule which shares the collagen structure does: this is the synaptically important enzyme – **acetylcholinesterase (AChE)**. In recent years many AChEs have been cloned and their structures solved by the molecular biological techniques described in Chapter 5.

We shall consider the important role of this enzyme in Chapter 16 and again in Chapter 17. Here we shall restrict ourselves to noting that AChE is found in two major configurations: **homomeric** and **heteromeric**. The homomeric class consists of globular units sometimes linked together to form dimers and tetramers. Some of these, as Figure 2.5 shows, are hydrophilic and others are amphipathic, having a hydrophilic globular region and glycophospholipid (hydrophobic) tail which inserts into cell membranes (see Chapter 7). The heteromeric class consists of four globular subunits linked through disulphide bonds either to a short lipophilic subunit or a lengthy collagen-like tail.

The globular subunits are known as the **catalytic units**. It is this part of the molecule which possesses the acetylcholinesterase activity. X-ray analysis shows them all to possess a narrow (20 Å) 'gorge' into which the substrate, acetylcholine, fits. The amino acid side chains constituting the walls of this gorge have been determined (at least 14 aromatic residues are present) and biochemists are well on the way to understanding how they facilitate the hydrolysis of acetylcholine.

The reason for considering the AChEs in this section is, however, the collagen-like tail of the heteromeric form. This consists, as in collagen, of three polypeptide chains, each with a 'three-fold screw' axis, twisted around each other to form a three-stranded 'rope'. Further confirmation of the collagen-like nature of the tail comes from the finding that, like collagen itself, there is much hydroxyproline and hydroxylysine and that the whole structure is sensitive to enzymic digestion by collagenase. Finally, it should be noted that this 'asymmetrical' form of AChE is a huge molecule, the molecular weight being computed at roughly 1000 kDa. It is believed that the collagen-like tail is inserted into the basement membrane which is particularly strongly developed in the folds of the neuromuscular junction (see Figure 15.17). In this way the catalytic units are held in place, optimally positioned to carry out their enzymic activity.

The α-Helix

Last, but very far from least, of these 'secondary structures' is the **α-helix**. This is an important conformation not only in many fibrous proteins but also (like the β-pleated sheet) in regions of numerous globular proteins. It differs radically from the previous structures in that the H-bonds are made between the imino hydrogens and the carbonyl oxygens of the same chain. All possible H-bonds are made so that the structure is energetically very stable and, most importantly, the side chains of the amino acids project outwards away from the longitudinal axis of the helix. Thus, with the sole exception of proline, the structure can incorporate every one of the twenty different amino acids. Proline, it will be remembered, does not possess an imino hydrogen atom and hence is unable to form the requisite H-bond with the carbonyl oxygen in the spiral above or below its position.

Figure 2.6 shows a right-handed α-helix. This is more stable than the left-handed form and is almost always found in nature. It also shows that the rise per residue is 1.5 Å and that there are about 3.6 residues per turn. It turns out that an isolated α-helix is rather unstable in an aqueous environment. In consequence α-helices are usually stabilised by the interaction of their side chains with neighbouring groups. These neighbours may be other regions of the same molecule or other molecules altogether. It is sometimes found that every seventh residue in an α-helix is non-polar. Because there are about 3.6 residues per turn this so-called **heptad** repeat results in a non-polar stripe forming at a slight inclination to the long axis of the helix. It follows that if another α-helix with a non-polar heptad repeat is laid alongside the first, twisting gently round it in a left-handed sense, then its non-polar residues can fit against the first's non-polar stripe. In this way **two-stranded** ropes, and **coiled coils**, can form (Figure 2.7). These coils are quite commonly found in fibrous proteins and particularly important cases for the neurobiologist are the **intermediate filament (IF)** proteins of neurons and, especially, axons. We shall consider these more fully in Chapter 15.

α-Helical domains are also often found in transmembrane proteins. In these cases an α-helical run of some twenty hydrophobic residues is believed to span the lipid core of the membrane. These domains are usually termed hydropathic. Much use is made of these hydropathic sequences in predicting the disposition of a polypeptide chain in a membrane. We shall see in Chapters 8 to 11 that numerous neurobiologically important proteins are predicted to zigzag across the membrane. These predictions are made on the strength of hydropathic analysis of the amino acid sequence in the polypeptide chain. It should, however, be borne in mind, when examining these important membrane-bound structures, that our knowledge of their disposition in the membrane is based mainly on this hydropathy analysis. Nevertheless, this analysis is now so consistent with other conceptions of the disposition of neurobiologically significant membrane proteins that it is unlikely to be very far amiss. Furthermore, a number of

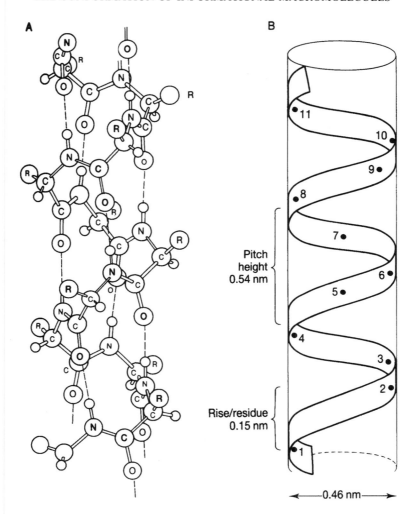

Figure 2.6 α-Helix. (A) Atomic structure of the right-handed α-helix. Note that the hydrogen bonds are aligned with the longitudinal axis of the molecule and that the amino acid side chains (R) project outwards. (B) Schematic diagram of a right-handed α-helix. The diameter is 0.46 nm, the rise per residue is 0.15 nm and the pitch height is 0.54 nm (i.e. 3.6 × 0.15 nm). The α-carbon atoms are numbered.

membrane-embedded proteins have, as we shall see in later chapters, been successfully solved by X-ray diffraction. These solutions show that the indirect reasoning from biochemistry is remarkably accurate (see Plates 1 and 2).

In other cases it is found that an α-helix does not occur in isolation. It is quite common to find a number of helices stacked so that their long axes are at a small angle with each other. Each helix may have both hydrophobic and hydrophilic side chains. It can then be arranged that each helix projects its hydrophobic side chains into the lipid environment of the membrane's interior and maintains its hydrophilic side chains pointing inwards towards its neighbours. Helices which distribute their hydrophobic and hydrophilic residues in this way are called **amphipathic**. In consequence, it is possible to see how hydrophilic pores may be developed across the lipid bilayer of a biomembrane. We shall return to this concept when we consider the structure of membranes and channel proteins in later chapters. We shall also

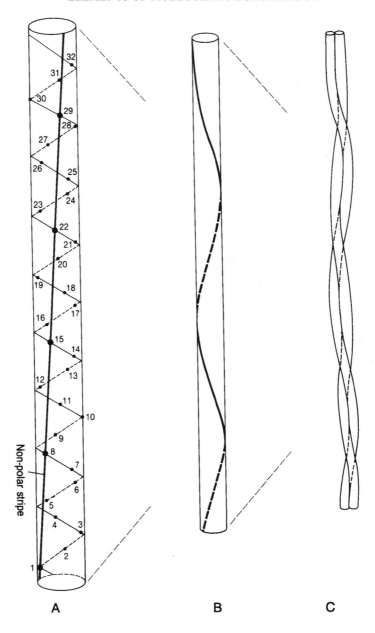

A   B   C

Figure 2.7   Non-polar 'stripes' and two-stranded α-helical 'ropes'. (A) Schematic diagram to show a 32-residue length of α-helix. If every seventh residue possesses a non-polar side chain a non-polar 'stripe' develops along the length of the helix. (B) A lengthy segment of α-helix showing the way in which a non-polar stripe twists around the molecule. (C) Schematic to show how two α-helices each possessing a non-polar stripe are able to twist around each other, being held together by hydrophobic forces in the ambient aqueous environment.

Table 2.3 Energies of chemical bonds

Single covalent bonds	$\Delta E = -50$ to -110 kcal/mol
Double covalent bonds	$\Delta E = -120$ to -170 kcal/mol
Triple covalent bonds	$\Delta E = -195$ kcal/mol
Ionic interactions ('salt bridges')	$\Delta E \approx -4$ kcal/mol
Hydrogen bonds	$\Delta E = -1$ to -5 kcal/mol
Van der Waals interactions	$\Delta E \approx -1$ kcal/mol
Hydrophobic interactions	$\Delta E \approx -1$ kcal/mol

It is largely the last four forces that hold the higher structures of biological macromolecules together.

find that seven transmembrane helices are a leitmotiv of G-coupled receptor molecules in neuronal membranes.

2.1.3 Tertiary Structure

Tertiary structure is the name given to the three-dimensional conformation of the globular proteins. It is often extremely intricate and always extremely fragile and flexible. As enzymatic and other biological activities depend on this three-dimensional conformation it is also extremely important.

In general the tertiary structure of a protein is believed to be determined by its primary structure (but see Section 2.1.5 below). The most important forces involved are **hydrophobic** and **hydrophilic** forces. We have already emphasised the significance of hydrophobic and hydrophilic amino acid side chains above. In globular proteins situated in an aqueous environment (i.e. most) hydrophobic residues will tend to end up in the interior, whilst hydrophilic residues will cluster on its surface where they can enter the surrounding water structure. Some have therefore likened a globular protein to a tiny oily droplet covered by hydrophilic hairs. In addition non-specific **van der Waals** forces between complementary 'docking' surfaces also play an important role in maintaining the three-dimensional structure. Finally, **disulphide linkages** between neighbouring cysteine residues and **ionic (salt) linkages** between neighbouring polar side chains may be formed and thus help to stabilise the conformation (Table 2.3). Even so the structure is often, as indicated above, on the edge of unravelling into instability. The energy state associated with the conformation of a typical small globular protein such as egg-white lysozyme is only

40 kJ/mol below that of a random organisation; in other words only that quantity of energy is required to **denature** the molecule.

A large protein molecule frequently consists of several different regions: these are known as **domains**. The domains normally consist of structurally well-formed regions and normally show evolutionary conservation. Well-known domains consist, for instance, of runs of β-pleated sheets, lengths of α-helix, TIM barrels, etc. But there are many others. Indeed some workers recognise nearly 2000 distinct domains in human proteins. Excellent collections of these structures may be found in the two books by Lesk listed in the Bibliography. As he remarks, quoting Schelling, 'architecture is frozen music', and nowhere is this more true than in the intricate thematic structures of the great biological molecules. We shall return to domains in Chapters 4 and 6 when we consider the evolution of proteins and the human genome. They are connected to each other by lengths of nondescript polypeptide chain. Some instances of common domain structures are shown in Figure 2.8.

It begins to seem probable that many of the large globular proteins synthesised by cells have evolved by the union of genes which program the synthesis of individual domains. We shall return to this concept in later chapters where we discuss molecular evolution.

In adddition to recurring domain structures many globular proteins exhibit recognisable 'motifs'. Motifs consist of similar groupings of amino acid residues. These often occur at binding and/or active sites. For instance, the amino acid sequence of nucleotide binding sites is generally G*G**G (where G symbolises glycine and * another amino acid), serine proteinase active sites are

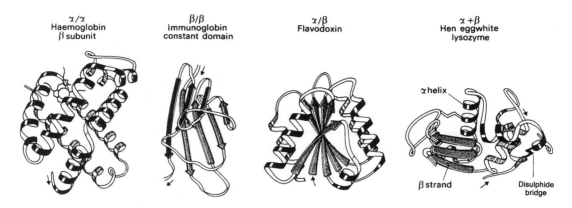

Figure 2.8 Domain structure of globular proteins. Globular proteins can be classified into four major classes depending on the organisation of the internal α-helical and β-sheet domains. In the figure α-helices are represented by coiled ribbons, β-pleated sheets by flat arrows which indicate the direction in which the polypeptide chain runs. In the α/α organisation the protein consists almost entirely of α-helices; the β/β structure, in contrast, consists almost entirely of β-pleated strands (in the figure, as the arrows indicate, antiparallel) and no α-helix; the α/β structure is built from alternating regions of α-helix and β-sheet (regions of α-helix commonly surround the β-sheet); finally in the α + β structure the α-helices and β-sheets tend to develop in different parts of the molecule. From Rees and Sternberg (1984), *From Cells to Atoms*, Oxford: Blackwell Scientific Publications, with permission.

characterised by GDSGG, the selectivity gate of K$^+$ channels presents GYG, and the special transcription factors which we shall discuss in the next chapter have a number of very special conformations which allow them to recognise regulatory sequences of DNA.

Before leaving the topic of tertiary structure we must emphasise once more that the integrity of this fragile three-dimensional conformation is essential for the biological activity of the protein. The active site of an enzyme is normally some specific region of the protein's surface which is rather precisely 'tailored' to fit its substrate. This accurate stereochemical fit depends on the maintenance of tertiary structure. Any denaturation (= degradation) destroys biological activity. But this seeming disadvantage is made use of in living cells for the purposes of control. Because the tertiary structure is so fragile small molecules (allosteric effectors) can be used to change it and thus alter the catalytic or other properties of the active site. In a sense, perhaps, one can detect at this level a primordial instance of sensitivity, of response to environment. The activity of an enzyme's active site changes when ambient conditions affect some other part of the molecule. This very important process is shown in Figure 2.9.

This sensitivity of tertiary conformation to 'environmental' influences is, as we shall see, crucial to numerous neurobiological actions. Indeed conformational changes are responsible for a whole raft of fundamental neurobiological phenomena. We shall see, as we proceed through the following chapters, that phosphorylation alters the conformation of proteins, especially membrane channel proteins, thus altering their functional state; that transmitter/modulator molecules act by changing the conformation of their protein receptor molecules; that photons affect the conformation of opsins; that cyclic adenosine monophosphate (cAMP) and other 'second messengers' affect the three-dimensional (3D) form of membrane channels; that conformational changes in G-proteins signal switches between 'on' and 'off' states; and that voltage changes across neuronal membranes alter the conformation of embedded membrane ion channels. Controlled alteration of the 3D conformation of proteins is a basic and continuing theme in molecular neurobiology. Some of the most fascinating X-ray analyses that have been published

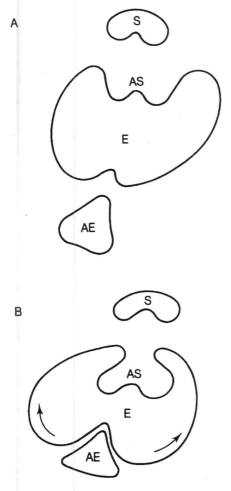

Figure 2.9 Allosteric control of tertiary structure. Conceptual diagram to show the effect of an allosteric effector on the activity of an enzyme. S=substrate; AS=active site; E=enzyme; AE=allosteric effector. When the allosteric effector binds to the enzyme a change is induced in the latter's three-dimensional structure so that the active site is no longer accessible to the substrate molecule. In other cases the opposite happens: the binding of the allosteric effector opens the active site to the substrate.

have to do with exactly how, at the atomic level, these conformational changes come about.

2.1.4 Quaternary Structure

Many proteins consist of more than one subunit. The organisation of the subunits to form one coherent molecule constitutes the molecule's quaternary structure. Unlike the domain structure discussed above, the subunits of a quaternary structure are not joined by a continuous polypeptide chain. Such proteins are termed multimeric. In most cases the subunits are held (or pushed!) together merely by hydrophobic forces. In many significant neurobiological cases, however, the multimeric structure is embedded in a lipid membrane. This is the case with the membrane-embedded receptors, pumps and channels which we shall discuss in Chapters 8, 9, 10 and 11. In these cases the intricate structure is held together by the complementarity of internal surfaces and by hydrophobic residues holding the molecule in its lipid environment away from aqueous internal and external solutions. In both cases, however, whether free in the aqueous cytosol or confined to a lipid membrane, the quaternary structure is very fragile.

The best-known quaternary structure is still that of **haemoglobin** – the earliest to be solved. It consists of four subunits, two identical α and two identical β chains. Because it is so well known it has come to form a model for quaternary structures. It is shown in Figure 2.10.

Haemoglobin's subunits each have a molecular weight of about 16 000 Da and consist of 141 amino acids (α-chains) or 146 amino acids (β-chains). The molecular weight of the entire molecule is about 68 000 Da. Many multimeric globular proteins are far bigger. The nicotinic acetylcholine receptor (nAChR), which we shall discuss in Chapter 10, consists of five large subunits, whilst ferritin is built of 20 identical subunits each consisting of 200 amino acids and weighs in at a total molecular weight of 480 000 Da. We shall look in some detail at the tertiary and quaternary structures of some proteins of neurobiological importance in later chapters. Once again, when we come to consider the activity of these huge structures, we shall find that their extreme fragility allows conformational changes in response to their immediate environments and that this lies at the root of much neurophysiology. With the first X-ray determination of a channel protein, the KcsA K$^+$ channel (Chapter 11; Plate 3), we are at last getting a fundamental insight into neurobiological structure–function relations at the atomic level.

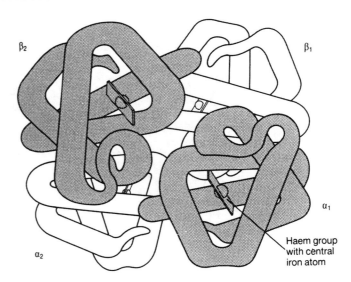

Figure 2.10 Quaternary structure of haemoglobin. The α_1 and β_1 chains face towards the viewer in this representation of oxyhaemoglobin. The identical α_2 and β_2 chains are partially hidden. The figure indicates that the major part of the polypeptide chain in each of the four subunits is in the form of an α-helix. The dimensions of the molecule are 6.4 × 5.5 × 5.0 nm. After I. Geis, in Dickersen and Geis (1969), *The Structure and Action of Proteins*, Menlo Park, CA: Benjamin/Cummings.

The principle of building larger and larger structures by making use of the complementary surfaces of smaller units can of course be extended from proteins to other molecules. The structures of viruses and such ubiquitous organelles as ribosomes consist of assemblages of globular proteins and nucleic acids. Eukaryotic chromosomes are built on similar principles and the structure of cell membranes (as we shall see in Chapter 7) can be understood in much the same way.

2.1.5 Molecular Chaperones

It was suggested in Section 2.1.3 that tertiary and higher structures are the automatic outcome of primary structure. This concept derives from the early work of Anfinsen on the structure of ribonuclease. Ribonuclease is a 128-residue chain held together by four intra-chain disulphide linkages. If the enzyme is denatured by exposure to 8 M urea, not only is its 3D conformation and enzymatic activity destroyed but also its four internal disulphide linkages. If, now, the denatured ribonuclease is placed back into its normal physiological environment it is possible to rejoin the SH groups of the denatured molecule and regain nearly 100% of the molecule's enzymic activity. Now, it is easy to calculate that it is possible to rejoin the eight SH groups in 105 different ways. It follows that in physiological conditions the primary sequence automatically winds up into its proper 3D form so that the 'correct' SH groups abut each other and, on oxidation, the correct disulphide linkages are formed. The enzyme regaining its correct 3D conformation regains its catalytic power. Similar experiments and similar conclusions have been drawn from other and more complex structures: tobacco mosaic virus, bacterial ribosomes, etc.

The concept of self-assembly has stood the test of time until fairly recently. There were, however, always some difficulties with it: surely the complementary surfaces of some tertiary structures and of multimeric quaternary structures in particular might align incorrectly during self-assembly, entering local energy minima before the global energy minimum of the entire structure could be reached? Moreover, many large protein structures, especially multimeric structures, undergo steric changes as they pass through membrane channels or participate in biochemical activities such as protein biosynthesis. Once again it is difficult to see how their 'complementary' or 'interactive' surfaces can be kept from forming 'incorrect' liaisons. Such liaisons become more likely the higher the temperature and the nearer a protein approaches the denatured state.

It is now believed that these improper liaisons are prevented by the aptly named molecular **chaperones**. The molecular chaperones form a group of unrelated proteins which bind non-sterically to exposed interactive surfaces during self-assembly or denaturation–renaturation, thus preventing the occurrence of incorrect unions. The binding is reversed when conditions are once again favourable to correct liaisons between complementary surfaces. The group of molecular chaperones includes the **chaperonins**, an interrelated family of largely prokaryotic proteins and the **heat shock proteins (hsp)** or stress proteins. In the latter case, stress of any kind, for instance increased temperature, amplifies synthesis of the proteins which, acting as chaperones, protect the interactive surfaces of partially denatured tertiary and quaternary structures until the stress is removed. Numerous hsps are known, generally classified into two large groups, hsp 70 and hsp 90, and all can now be regarded as being members of the class of molecular chaperones. It should be emphasised, finally, that chaperones themselves contain no information about the final conformation of their protein charges. They merely act 'catalytically', allowing the proteins with which they interact to find their correct folding pattern. This is emphasised by the observation that a given chaperone can assist a number of quite different proteins to 'discover' their correct configuration.

2.2 NUCLEIC ACIDS

There are two major classes of nucleic acid: **deoxyribonucleic acid**, **DNA** (largely but not exclusively confined to the nucleus in eukaryotic cells), and **ribonucleic acid**, **RNA** (largely but not exclusively confined to the cytoplasm). Both, like the proteins, are polymers. The monomers of which they are constituted are called nucleotides. Since at least 1944, when Avery, MacLeod and McCarty showed that DNA was responsible for pneumococcal transformation, and certainly from 1953, when Watson and Crick published their solution to the structure of DNA, it has been clear that nucleic acids store and transmit the cell's genetic information.

2.2.1 DNA

The nucleotides of which nucleic acids are built consist of three parts: a **phosphate group**, a **sugar** and a **nitrogenous base**. In the case of DNA the sugar is deoxyribose (hence the nucleotide is a

deoxyribonucleotide) and four different bases are involved, two purines and two pyrimidines. The purine bases are **adenine (A)** and **guanine (G)** and the pyrimidines are **cytosine (C)** and **thymine (T)**. The structure of these molecules is shown in Figure 2.11. It should be noted that the purine bases are considerably bigger than the pyrimidines, that both purines and pyrimidines show considerable **conjugation** (i.e. alternation of single and double covalent bonds), which means (as with the amide group in amino acids (Section 2.1.1) that they are all **planar** and **hydrophobic**.

The deoxyribonucleotides are strung together to form long **polynucleotide** chains. The connection between one nucleotide and the next is made through the phosphate group of one bonding (by elimination of the elements of water) with the 3' carbon of the deoxyribose sugar of another. The phosphate group thus connects the 3' carbon of one deoxyribose with the 5' carbon of the next deoxyribose. This bonding, a **phosphodiester** bonding, is shown in Figure 2.12.

It is clear that this bonding can be continued (analogously to peptide linkages) to build up long chains of nucleotides. Figure 2.13 shows that the resulting polynucleotide consists of a **–phosphate– sugar–phosphate–sugar–** · · · 'backbone' to which are attached the nitrogenous bases. It is important to note that the backbone (again analogous to the polypeptides) has a **polarity**. Conventionally one proceeds along the chain from 5' carbon to 3' carbon. The initial nucleotide is thus located at the 5' end of the sequence and the terminal nucleotide at the 3' end.

The great contribution of the X-ray diffraction analyses of the early 1950s (Franklin, Wilkins, Watson, Crick) was to establish the conformation of these polynucleotide strands in nucleic acids, especially DNA.

The Watson–Crick double helix (Figure 2.14) is now almost a cliché. Two polynucleotide strands wind around each other, one proceeding in the 5' to 3' direction, the other in the 3' to 5' direction. In other words (analogous to the amino acid structure of silk fibroin), the two strands are 'antiparallel'. The bases project inwards toward each other, pyrimidines always partnering purines: thus ensuring that the two phosphate–sugar backbones are always at a constant distance (1.085 nm) from each other. The structure is a brilliant solution to

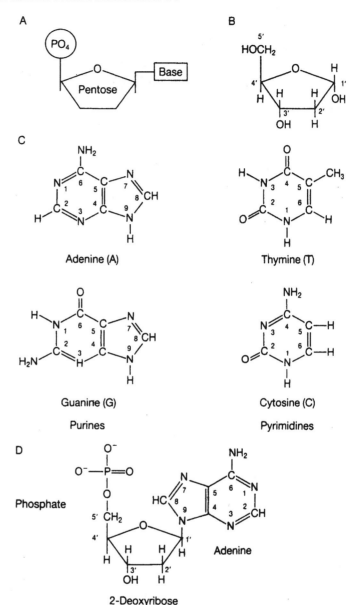

Figure 2.11 Structure of deoxyribonucleotides. The figure shows how deoxyribonucleotides are built from a phosphate
group, a sugar (deoxyribose) and one of
four nitrogenous bases. The latter come in
two sizes: the purines and the smaller
pyrimidines. The purines are attached to
the pentose sugar through nitrogen atom 9
whilst the pyrimidines are attached through
nitrogen atom 1. (A) Schematic diagram of a
nucleotide. (B) The pentose sugar, 2-deoxy-
ribose. (C) The four different bases. (D) A
typical deoxyribonucleotide: deoxyadeno-
sine monophosphate (dAMP).

the X-ray diffraction data and the requirement of
chemical stability (the hydrophobic bases project
inwards and the hydrophilic sugar–phosphate
backbone outwards toward the aqueous environment); and last, but very far from least (as Watson
and Crick note in their famous understatement 'it
has not escaped our notice'), it brilliantly satisfies
the requirements of a genetic molecule.

Hereditary information is stored in the sequence
of bases. Figure 2.15 shows that **adenine (A) always
partners thymine (T)** and **guanine (G) always
partners cytosine (C)**. These pairings (the so-called
Watson–Crick pairings) are established by the
selective stickiness of hydrogen bonds – two such
bonds, as the figure shows, being formed between
A and T and three between G and C.

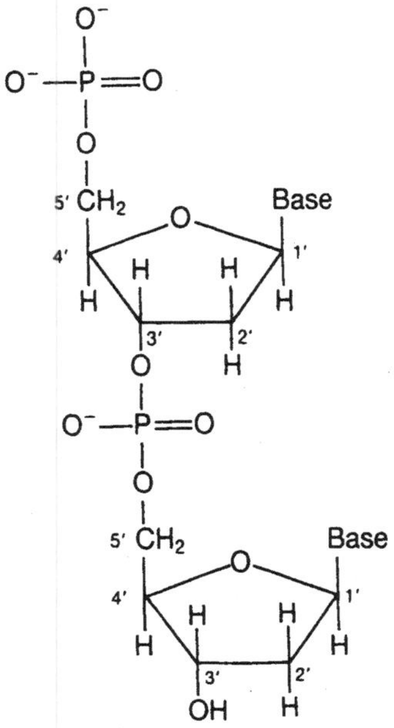

Figure 2.12 Phosphodiester bond between two deoxyribonucleotides. Note how the phosphate group connects the 3′ carbon of one pentose sugar to the 5′ carbon of the next.

The processes of information transfer in the living cell in which DNA plays so central a role will be discussed in Chapter 3.

2.2.2 RNA

Although the DNA double helix can exist in three different forms (A, B and Z) it mostly exists in the classic Watson–Crick A-form in the living cell. The forms which RNA takes in the cell are, however, far more varied. Let us first look at its 'primary' structure. Like DNA it is composed of nucleotides strung together by phosphodiester linkages. However, the nucleotides differ from those constituting DNA in two respects. First the sugar is not deoxyribose but **ribose**. As Figure 2.16 shows, it carries a bulky hydroxyl group at the 2′ position instead of deoxyribose's single hydrogen atom. Again, this is more than a detail. Because of this bulkiness at the 2′ position RNA is unable to stack its bases perpendicular to the long axis of a polynucleotide double helix as can DNA. If a double helix is formed the bases are stacked awkwardly at 20 degrees to the long axis, as they

are in the B-configuration of DNA. Second, one of the pyrimidine bases differs from those found in DNA. Instead of thymine (T), **uracil (U)** (having the same Watson–Crick pairing properties) is found. This, too, is not an accidental detail. We shall see in the next chapter that T, although it requires more energy to synthesise, is necessary if repair enzymes are to keep DNA's message undegraded. Because of its far more transient existence this is not important in RNA. The latter can thus make do with the less energy-demanding U.

Three major types of RNA are found in the cell: **messenger RNA (mRNA), transfer RNA (tRNA)** and **ribosomal RNA (rRNA)**. A fourth type, **heteronuclear RNA (hnRNA) (primary transcript RNA)**, is found in the nuclei of eukaryotic cells. This latter type, as we shall see in the next chapter, is in fact a precursor of mRNA. We shall see that all these varieties of RNA are involved in the information transfer process (central to molecular biology) whereby the message held in the structure of DNA is expressed in the structure (and hence activity) of proteins. Here we shall look briefly at their conformations

mRNAs hardly have a conformation at all. They are but single-stranded transcripts of the appropriate base sequences of DNA. They range in length from 75 to over 3000 ribonucleotides.

tRNAs, on the other hand, do have a complex conformation. As we shall see in the next chapter, each amino acid has a specific tRNA assigned to it. Thus we refer to $tRNA_{phe}$, $tRNA_{leu}$, etc. It follows that there are at least twenty different types of tRNA molecule. However, they all have much in common. To begin with, their molecular weights are all about 25 000 Da, which means that they all consist of about 75 nucleotides. Second, they are all believed to have a somewhat similar tertiary structure – rather like a clover leaf. Third, at the end of the middle clover leaf there is always a group of 'free' (i.e. non-hydrogen-bonded) bases. These form the so-called **anticodon** which, as we shall see in the next chapter, recognises the genetic message carried by mRNA from the nuclear DNA. Finally all tRNAs have a group of three ribonucleotides (C, C and A) at the opposite end of the molecule from the anticodon. The appropriate amino acid is attached to the final ribonucleotide of this triplet. This attachment is made via the 3′ carbon of the adenosine's ribose

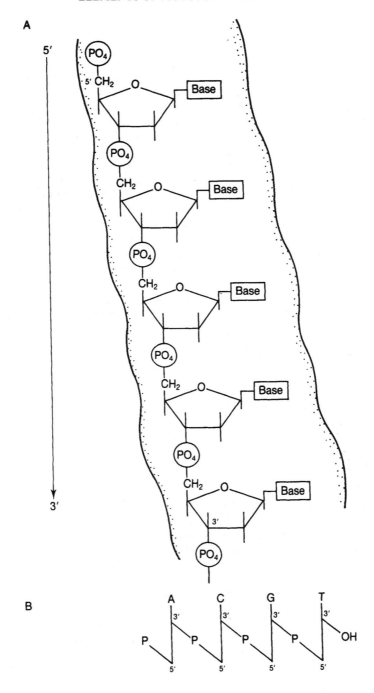

Figure 2.13 (A) Portion of a polynucleotide chain. Note the direction of the chain: from 5′ to 3′. (B) Short-hand form for writing the formula of a polynucleotide strand. P represents the phosphate group; a vertical line represents the pentose sugar; an appropriate letter represents the base.

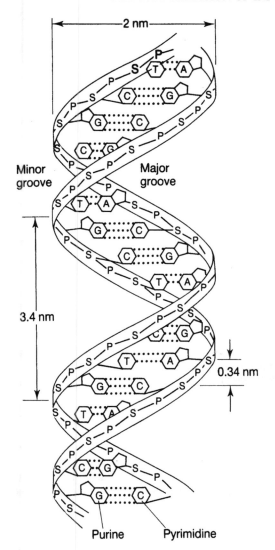

Figure 2.14 DNA double helix. The figure shows the dimensions of the double helix and the pairing properties of the nucleotide bases.

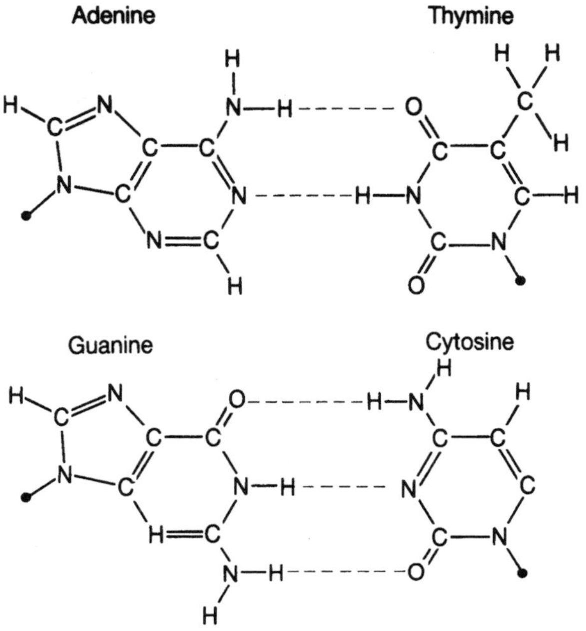

Figure 2.15 Watson–Crick pairings of nucleotide bases. The figure shows that when properly aligned two hydrogen bonds can be formed between A and T and three between C and G. The 'cocktail sticks' attached to the nitrogen atoms of the four bases indicate where the bond uniting them to the deoxyribose and thus the rest of the polynucleotide strand is formed.

moiety. An example of this complicated conformation is shown in Figure 2.17.

The complexity of the tertiary structure of tRNAs reflects the difficulty of the job they have to do in the cell. They act as go-betweens betwixt the proteins and the nucleic acids. Each type of tRNA requires a complex and highly individual conformation so that it can be 'recognised' by the active site of an enzyme – an **aminoacyl-tRNA synthetase enzyme**. Yet each tRNA has also to recognise an appropriate triplet of nucleotide bases in the mRNA transcript. Proteins and nucleic acids are two very different types of molecule. Although both make use of complementary surfaces for recognition, the sorts of complementary surface (as must by now be very clear) are radically different. tRNAs must be able to recognise, or be recognised by, both. The means by which this is achieved is considered in the next chapter (Section 3.3.2).

rRNAs are transcribed from nuclear DNA as very large molecules (about 13 000 ribonucleotides in length) known as **45S rRNA**. S is an abbreviation for the **svedberg**, a sedimentation coefficient. The larger the value of this coefficient the greater the molecular weight. 45S rRNA is, however, soon

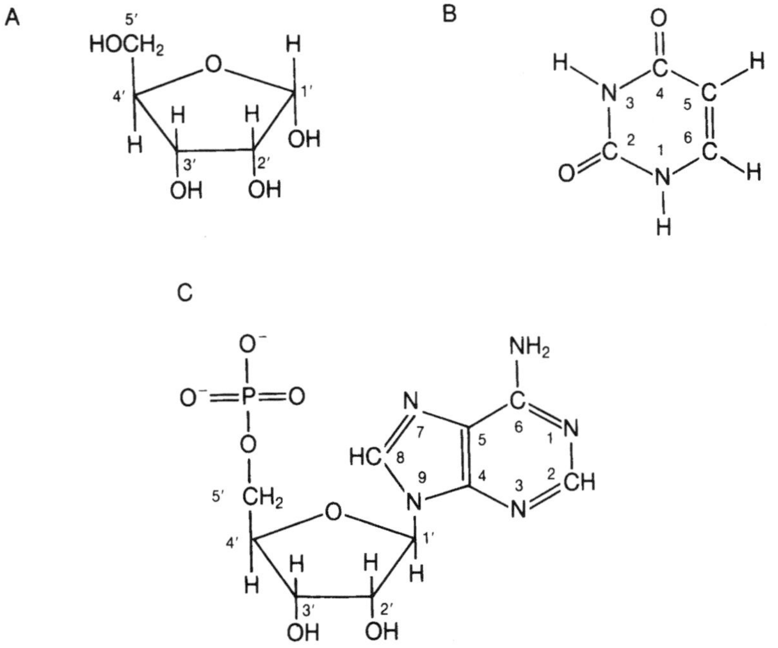

A

B

C

Figure 2.16 Structure of ribonucleotides. The figure shows how ribonucleotides are built from a pentose sugar (ribose), a nitrogenous base and a phosphate group. (A) The pentose sugar (ribose). (B) Uracil (U) replaces DNA's thymine. (C) A typical ribonucleotide, adenine monophosphate (AMP).

cleaved into three smaller fragments – 28S (c. 5000 ribonucleotides), 18S (c. 2000 ribonucleotides) and 5.8S (c. 160 ribonucleotides). In addition to 45S rRNA and its cleavage products an entirely separate **5S rRNA** is also synthesised. All of these various types of rRNA are assembled with a large array of different proteins to form the two subunits of the **ribosome**. This intricate assembly occurs in the **nucleolus** and the finished products (the two ribosomal subunits) are passed separately out of the nucleus into the cytoplasm (Figure 2.18). It is notable, as mentioned in Chapter 1, that neurons are distinguished by their large nuclei and prominent nucleoli. It is clear that neurons are very active in the synthesis of ribosomes and thus in the whole business of protein biosynthesis.

2.3 Conclusion

In this chapter we have looked, all too briefly, at the conformation of the most important of the

molecules which we shall meet as we proceed with our subject. The biological activity of these molecules emerges from their three-dimensional conformation, just as the biological activity of the brain emerges from its three-dimensional anatomy. It is also worth noting once again the deep significance of complementary surfaces and selective 'stickiness'. The three-dimensional structure responsible for these complementary surfaces and selective stickinesses depends on the presence of multitudinous in themselves negligible forces – the H-bonds, hydrophobic forces, van der Waals attractions, etc. Finally, the great difference in the conformation of the two types of macromolecule we have been considering in this chapter should be borne in mind. Although it is now known that some RNAs have enzymatic activity and that prion proteins can replicate (see Chapter 21), these are at best inefficient and primitive processes. The smooth and rapid biology of all contemporary cells depends on the ability to translate between

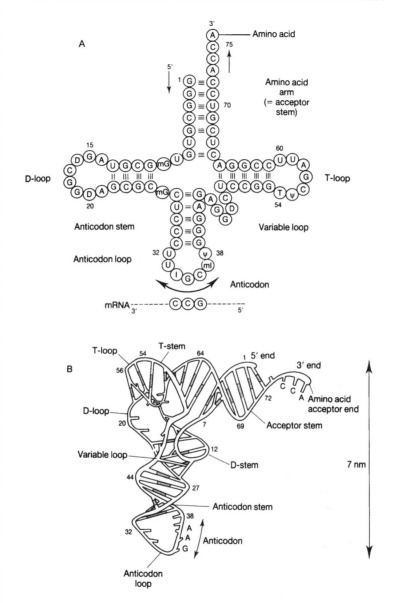

Figure 2.17 The conformation of yeast phenylalanine tRNA (tRNA$_{phe}$). (A) The conventional 'clover-leaf' representation. The majority of the nucleotides are the classical Watson–Crick U, G, C or A but there are a few non-classical types present. These are often found in the loops, where they may play a role in protein synthesis. The non-classical bases are ψ=pseudouridine; D=dihydrouridine; I=inosine; T=thymine; m=methyl group. Inosine (I) in the anticodon recognises C (it will also recognise A and U). It will also be noted that in positions 3 and 13 G/U pairs are formed. From Rees and Sternberg (1984), *From Cells to Atoms*, Oxford: Blackwell Scientific Publications, with permission. (B) The three-dimensional form as deduced by X-ray diffraction. The compact and intricate L-shaped structure is stabilised not only by Watson–Crick hydrogen bonding between bases but also by numerous hydrophobic forces. From Rees and Sternberg (1984), *From Cells to Atoms*, Oxford: Blackwell Scientific Publications, with permission.

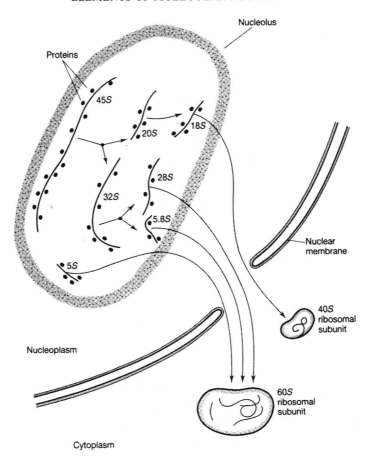

Figure 2.18 Assembly of ribosomes from rRNA and proteins in the nucleolus. Explanation in text. The figure shows that the 45S rRNA is progressively split into three fragments: 18S, 28S and 5.8S rRNA. A separate 5S rRNA is transcribed independently from the DNA. The figure shows that all of these rRNA moieties are from the beginning associated with proteins. The two ribosomal subunits are assembled within the nucleolus and passed (probably through a pore in the nuclear membrane) into the cytoplasm. From *Molecular Cell Biology* by James E. Darnell *et al.*, Copyright © 1986 Scientific American Books, Inc. Reprinted with permission.

nucleic acids and proteins. Over evolutionary time each type of informational macromolecule has come to depend on the other; they are caught up in a profound symbiotic relationship. On their own

they would be comparatively ineffective. Nucleic acids make very poor enzymes (if they possess any catalytic power at all); proteins (except for prions) cannot replicate.

3

INFORMATION PROCESSING IN CELLS

The central dogma – symbiosis of proteins and nucleic acids – the genetic code. **Replication** – always in 5′ to 3′ direction – replication fork and Okazaki fragments. **Transcription**: polymerases – regulatory sites (promoters/enhancers/silencers) – palindromic sequences – transcription factors (general and specific) – palindromic sequences and transcription sites – CRE, CREBs and CREMs. Post-transcriptional processing: introns and exons – spliceosomes, excision, splicing – pathologies due to defective splicing: thalassaemias – jimpy, Pelizaeus–Merzbacher disease – mRNA editing. **Translation**: aminoacyl synthetases – eukaryotic ribosomes – initiation, elongation, termination – fate of synthesised protein. **Control of gene expression**: cloning – genomic control – trinucleotide repeat pathologies; transcriptional control: DNA chips – the *lac* operon – immediate early genes (IEGs) – *c-jun*, *c-fos* – the AP1 site; post-transcriptional control: substance P and substance K – calcitonin and CGRP; translational control: synthesis of globin; post-translational control: insulin – enkephalins – endorphins. **Conclusion**: multiplicity of control points – the significance of the control of gene expression in multicellular organisms

In Chapter 2 we briefly reviewed the nature of the cell's 'informational macromolecules' and also emphasised that cells, like brains, are deeply involved in information processing. In contrast to the brain, however, most (though by no means all) of the information available to the cell is hereditary information. It has been accumulated over two or three billion years of trial and error interaction with the environment. By far the greatest amount is stored in the base sequences of DNA.

We also emphasised in Chapter 2 that the two informational macromolecules – the nucleic acids and the proteins – were very different from each other and yet interacted symbiotically in the life of the cell. Each type of molecule is especially good at one of life's necessities – the nucleic acids at preserving and transmitting genetic information (in particular information which determines the structure of proteins), the proteins at catalysing

metabolic reactions (including those which lead to the replication of DNA) – but both are required if the cell is to survive.

The central area of molecular biology is thus the study of the interaction of nucleic acids and proteins. Francis Crick summarised the essence of this area in what he memorably called the **central dogma of molecular biology**: 'DNA makes RNA and RNA makes protein.' We now know that the 'dogma' has exceptions. As we shall see, we now know that in some cases information flows from RNA to DNA. A complete reversal of the dogma has, however, never been observed; indeed as we shall see there appear to be good molecular reasons for thinking that a flow of information all the way from protein to nucleic acid is impossible. Thus Crick's 'dogma' remains essentially uncontroverted: the major flow of information progresses from nucleic acids to proteins and never vice versa.

It might be added in parenthesis here that Crick's 'dogma' merely states in molecular terms what orthodox biology has been saying since the publication of Charles Darwin's *Origin of Species* in 1859. Information flows from the 'germ plasm', or 'genotype', to the 'somatoplasm', or 'phenotype', and never in the contrary direction. Any reversal of the direction of this information flow would amount to an instance of Lamarckism – the inheritance of acquired characteristics. In spite of much effort and ingenuity Lamarckian inheritance has never yet been observed.

3.1 THE GENETIC CODE

As we saw in Chapter 2, there are up to twenty different amino acids in a protein but only four different nucleotides in DNA. How can four different nucleotides specify twenty different amino acids? This was the essence of the 'coding problem' of the late 1950s and early 1960s.

It is clear that four different nucleotides, taken on their own, could only specify four different amino acids; it is also clear that taken in groups of two there would be $4^2 = 16$ different pairs of nucleotides, still not sufficient to specify 20 different amino acids; if, however, they were grouped in threes then there would be $4^3 = 64$ different possible triplets, more than enough to code for the twenty amino acids. Clever experiments by Crick, Brenner and others in the early

1960s established that the DNA genetic code was indeed a triplet code. Each amino acid was specified by one or usually more than one nucleotide triplet. These triplets became known as **codons**. Figure 3.1 shows the genetic code as it is nowadays understood.

There are several features of the code which should be noted. First it is read continuously from the 3' to the 5' end of the polynucleotide strand. That is say the mRNA strand is synthesised in the 5' to 3' direction. There are, moreover, no commas, colons or semi-colons separating one codon from the next. There are, however, start and stop signals ('capital letters' and 'full stops'). Figure 3.1 shows that **AUG**, in virtue of the fact that it codes for Met, often (but not always) acts as a **start signal** (this is sometimes also the case with **GUG**), whilst **UAA**, **UAG** and **UGA** act as full stops and terminate the message. Apart from the 'stop' codons all the others specify an amino acid. There is thus considerable redundancy or **degeneracy** – amino acids such as Arg, Leu and Ser being assigned no less than six codons, whilst Thr, Pro, Ala, Gly and Val are assigned four codons each. In this connection it is worth noting that the first two nucleotides in the triplet are the most important – suggesting, perhaps, that the triplet code has evolved from a primordial doublet code. Finally, the chemical nature of the amino acid is to some extent reflected in its codon. Every codon with a **U** in the second position specifies a **hydrophobic**

First position (5' end)	Second position of codon				Third position (3' end)
	U(T)	C	A	G	
U(T)	UUU $\}$ Phe UUC UUA $\}$ Leu UUG	UCU $\}$ UCC $\}$ Ser UCA UCG $\}$	UAU $\}$ Tyr UAC UAA $\}$ Term UAG $\}$	UGU $\}$ Cys UGC UGA Term UGG Trp	U(T) C A G
C	CUU $\}$ CUC $\}$ Leu CUA CUG $\}$	CCU $\}$ CCC $\}$ Pro CCA CCG $\}$	CAU $\}$ His CAC CAA $\}$ Gln CAG $\}$	CGU $\}$ CGC $\}$ Arg CGA CGG $\}$	U(T) C A G
A	AUU $\}$ AUC $\}$ Ile AUA • AUG Met	ACU $\}$ ACC $\}$ Thr ACA ACG $\}$	AAU $\}$ Asn AAC AAA $\}$ Lys AAG $\}$	AGU $\}$ Ser AGC AGA $\}$ Arg AGG $\}$	U(T) C A G
G	GUU $\}$ GUC $\}$ Val GUA • GUG $\}$	GCU $\}$ GCC $\}$ Ala GCA GCG $\}$	GAU $\}$ Asp GAC GAA $\}$ Glu GAG $\}$	GGU $\}$ GGC $\}$ Gly GGA GGG $\}$	U(T) C A G

Figure 3.1 The genetic code. Bases are given as ribonucleotides. For deoxyribonucleotides substitute T for U.

* is initiation

amino acid and every (meaningful) codon with an **A** in the middle specifies a **hydrophilic** residue.

3.2 REPLICATION

DNA has, of course, two roles: replication and transcription. In parentheses here, and looking towards the final chapter of this book, we can note that the prions which are believed to cause the neurodegenerations of Creutzfeldt–Jakob disease (CJD), kuru, scrapie, bovine spongiform encephalopathy (BSE or 'mad cow disease'), etc. seem only to have one of these roles: replication. We shall consider this in more detail in Chapter 21. In this section we shall briefly consider DNA replication. No more than an outline is warranted as, with the exception of olfactory neurosensory cells and small numbers of stem cells, neurons do not replicate in the adult brain.

Replication depends on the action of DNA-polymerase. In fact, three DNA-polymerases have been identified in both prokaryocytes and eukaryocytes. In prokaryocytes these are classified as polymerase 1, 2 and 3; in the eukaryocytes they are referred to as α, β and γ. Eukaryocyte polymerase α is analogous to bacterial polymerase 3 in being the most important in the synthesis of new polynucleotide strands. In all cases the polymerase acts by adding fresh nucleoside triphosphates to the 3′ end of the growing polynucleotide chain. Synthesis is thus always in the 5′ to 3′ direction (Figure 3.2).

Now it will, of course, be objected that because of the antiparallel nature of the DNA double helix one of the unwinding strands will progress in the 3′ to 5′ direction whilst the other proceeds in the opposite sense, from 5′ to 3′. How (it will be asked) can a DNA polymerase use the latter strand as a template? The answer to this conundrum has to do with short lengths of DNA named after their discoverer, **Okazaki fragments**. These fragments are synthesised in the conventional sense from 5′ to 3′ and later joined up on the 5′ to 3′ parental strand in the conventional direction.

We shall leave consideration of DNA replication at this point. Readers interested in pursuing the subject further can refer to the Bibliography and will find Watson *et al.* particularly informative. The subject of DNA replication will be raised again when we consider molecular evolution and its mutational basis in Chapter 4, especially Sections

4.1.1 and 4.1.2. Let us next turn our attention to the complexities of transcription and translation.

3.3 'DNA MAKES RNA AND RNA MAKES PROTEIN'

Let us take Francis Crick's 'central dogma' a step at a time. The process whereby 'DNA makes RNA' is called transcription; the process by which 'RNA makes protein' is called translation. The idea behind these two terms is that whilst DNA and RNA are variants of the same nucleic acid 'language', RNA and protein are two quite different languages.

3.3.1 Transcription

We saw in Section 3.1 above that the genetic code is an uninterrupted sequence of nucleotides read continuously from one end to the other. It will probably already have occurred to you that as DNA consists of two polynucleotide chains both cannot carry the message. After all one is in effect the 'mirror image' of the other. You are right. Only one strand, the **(−) strand**, is read; the other, the **(+) strand**, is left alone. The (+) strand only comes into its own at replication. Then, of course, it is required as a template for the formation of a daughter double helix. It is, however, conventional to refer to this, the (+) strand, as the coding or 'sense' strand because, as Figure 3.3 shows, the message transcribed on to the mRNA is identical to this strand saving only that U substitutes for T.

Transcription in Prokaryocytes

The process of joining ribonucleotides together using the (−) strand as a template is catalysed by a **DNA-dependent RNA polymerase** enzyme. Prokaryocytes possess just one form of this enzyme. In *Escherichia coli* it is a huge multimeric protein consisting of five subunits – two α, one β, one β′ and a σ (i.e. αββ′σ) – and has a molecular weight of about 500 kDa. The σ subunit is essential for the recognition of the promoter region of the DNA double helix. This is a region of the DNA double helix immediately 'upstream' (in the 5′ direction) of the gene to be transcribed (Figure 3.4).

After the σ subunit has done its job of recognising the promoter region it falls away and

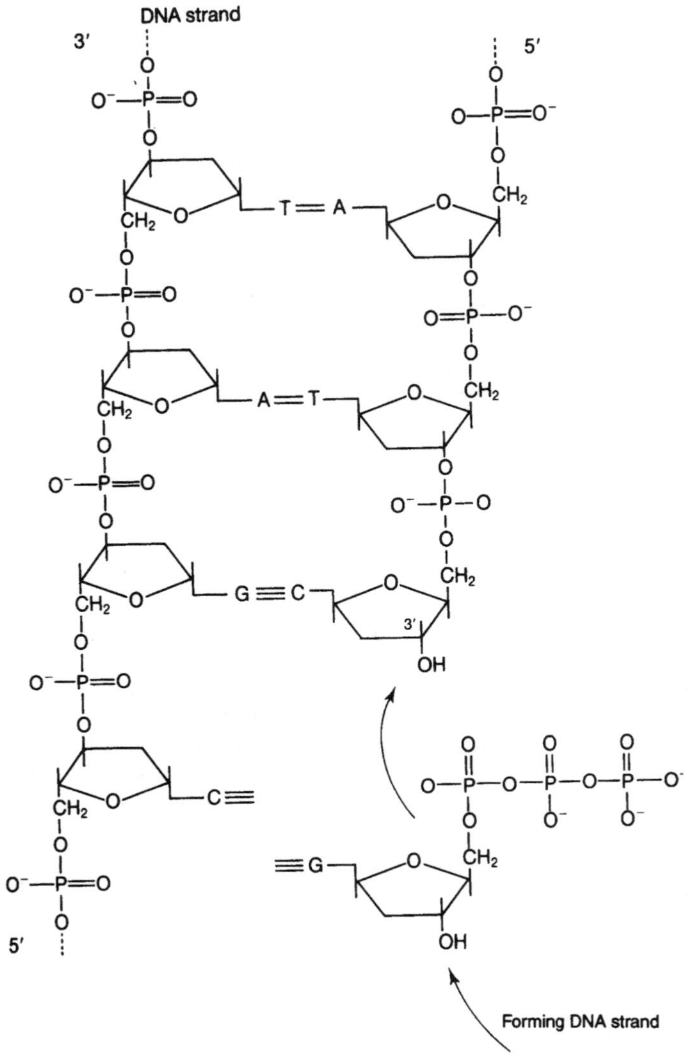

Figure 3.2 Growth of a polynucleotide strand in the 5′ to 3′ direction. Synthesis of a daughter polydeoxyribonucleotide strand alongside the (−) DNA strand. Deoxyguanosine triphosphate (dGTP) is shown approaching at the bottom of the figure. Cytosine on the DNA template ensures that it is acceptable. The DNA polymerase enzyme is then able to connect it by way of a phosphodiester link to the nucleotide immediately above. The energy for this synthesis comes from breaking the energy-rich phosphate bond in the triphosphate. The two terminal phosphate groups of the triphosphate and the terminal hydrogen of the hydroxyl group form pyrophosphate, which is released.

the remaining polymerase is able to unwind the double helix and to synthesise on the template provided by the (−) strand, a complementary strand of RNA. In order to carry out this elongation phase of the transcription process the polymerase, of course, requires raw materials and these take the form of ribonucleoside triphosphates (rNTPs). The (−) strand is transcribed from the 3′ to the 5′ end. In other words, the fresh ribonucleotides are added to the 3′ end of the growing mRNA. This means that the strand is synthesised in the 5′ to 3′ direction and, as we shall shortly see, is read in this direction at the ribosome.

The process of elongation continues until the polymerase enzyme reaches a termination sequence on the DNA. Whilst the RNA chain is growing, earlier (upstream) parts of it become displaced from the (−) strand. This occurs because the RNA–DNA duplex is thermodynamically less stable (see remark on difference between ribose and deoxyribose in Chapter 2) than the DNA–DNA duplex. When a termination sequence is reached, the RNA becomes completely detached from the DNA template and the polymerase enzyme also falls free, able to bind another σ factor and initiate another synthesis.

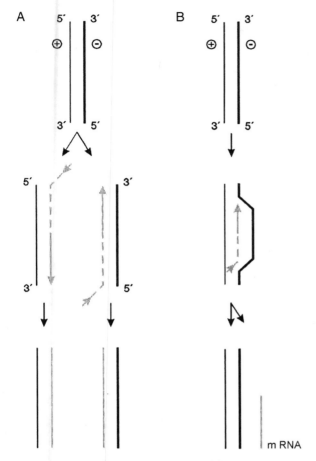

Figure 3.3 Replication and transcription: the (+) and (−) strands of DNA. Column A shows DNA replication when both (+) and (−) strands of DNA act as templates for the synthesis of two daughter double helices. The dashed lines represent deoxyribonucleotides entering the synthetic process. Column B shows transcription where only the (−) DNA strand is used. The dashed lines in this case represent ribonucleotides lining up to form mRNA alongside the (−) strand.

The process of transcription is now over. The genetic message encoded in the base sequence of the DNA (+) strand has now been transferred to a sequence of RNA bases. In prokaryocytes this message passes directly to the ribosomes, where it is 'translated' into protein structure.

Transcription in Eukaryocytes

In eukaryotic cells transcription is far more complex. There are four different kinds of DNA-dependent RNA polymerases. Three are located in the nucleus and one in the mitochondria. Of the three located in the nucleus, **polymerase 1** is found in the nucleolus and is responsible for transcribing the 45S rRNA of ribosomes (see Chapter 2). **Polymerase 2**, the eukaryotic analogue of the single prokaryotic enzyme, is located in the nucleoplasm and transcribes hnRNA (i.e. primary transcript mRNA). **Polymerase 3**, also found in the nucleoplasm, transcribes 5S rRNA and the tRNAs. In what follows we shall deal only with **polymerase 2 (pol 2)**.

Although, as we have just noted, pol 2 is responsible for synthesising primary transcript mRNA, using the (−) strand of DNA as a template, the process cannot start until a hugely complex initiating mechanism comes together. It is obviously vitally important that all eukaryotic cells, especially nerve cells, only 'switch on' the genes they need. This initiating mechanism ensures that only these appropriate genes are switched on. It is to this hugely complex process that we turn next.

Full accounts of this intricate and precise mechanism can be found in texts of molecular biology and genetics (see Bibliography). Nevertheless it is important for the neurobiologist to have an outline understanding as the control of transcription (gene expression) is crucial to the differentiation of cells and hence brain development (see Chapters 18 and 19), the consolidation of memory (Chapter 20) and also, in many cases, to the synaptic control of neurons in the mature brain.

In essence, control of the transcriptional process is brought about by DNA regions, which as in prokaryocytes, are 'upstream' (i.e. in the 5' direction) of the gene. These regions, known as regulatory regions, are placed both **proximally** and **distally** to the gene. The proximal regulatory region is known (again as in prokaryocytes) as the **promoter**. The distal regulatory regions are of two types known as **enhancers** and **silencers** (Figure 3.5).

The promoter is itself a complex region. It consists of a **'core' promoter**, generally about 30 bp 'upstream' of the gene, and some **promoter-proximal** regions about 100 bp further upstream. The core promoter contains the **TATA box**. The name derives from the finding that this region

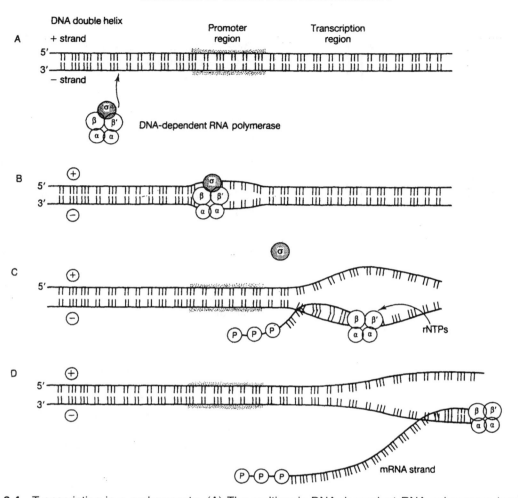

Figure 3.4 Transcription in a prokaryocyte. (A) The multimeric DNA-dependent RNA polymerase 'explores' the DNA double helix until it happens upon the promoter region (about 40 bp in length) of a gene. (B) The σ subunit recognises this region and the polymerase enzyme binds to the DNA. (C) The σ subunit is released and the two strands of the double helix unwind. The remainder of the polymerase enzyme moves along the (−) strand using it as a template for synthesising a complementary RNA strand from rNTPs. The DNA double helix reforms behind. (D) The synthesis of mRNA continues until a terminator sequence on the (−) strand is reached. The polymerase enzyme then detaches. In prokaryocytes translation of the mRNA message into protein structure then occurs directly without any further modification of the mRNA strand.

almost always contains the quartet of bases, T-A-T-A.

The distal regulatory regions are specific to particular genes and particular tissues. They are located much further away, usually over a hundred and sometimes several thousand base pairs upstream. An enhancer greatly increases the activity of the promoter in initiating transcription of its gene. Silencers, on the contrary, reduce transcription.

Both the proximal and distal regulatory regions are recognised by proteins that bind to the particular sequences they expose.

A complex of **general transcription factors**, **TFIID** and **TFIIB**, bind to the TATA promoter site. Another three general transcription factors,

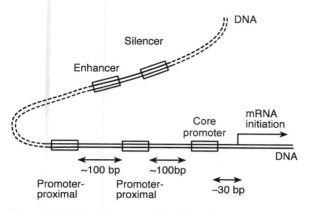

Figure 3.5 Regulatory regions in eukaryotic DNA.

TFIIE, **TFIIF** and **TFIIH**, bind to the polymerase enzyme, **pol 2**, enabling it to bind to TFIIB on the TATA promoter site. TFIIH is then able to use phosphate groups from ATP to phosphorylate pol 2 and thus release its transcriptional function. Primary transcript RNA is accordingly synthesised. This rather complicated sequence of events is shown in Figure 3.6.

The **rate** of transcription is controlled by **special transcription factors** which bind to the proximal and distal regulatory regions of the DNA.

These factors are proteins of several different conformations which have been given memorable names. Four major families have been distinguished: **zinc fingers**, **leucine zippers**, **helix–loop–helix (HLH)** and **helix–turn–helix (HTH)** (Figure 3.7). In general they are designed so that one surface recognises a specific nucleotide sequence in the promoter region and other surfaces present sites for binding to other proteins essential for initiating transcription.

The HTH family of transcription factors includes the homeodomain proteins which, as we shall see in Chapter 18, play crucial roles in cell differentiation during embryology. In the HTH motif α-helix 3 lies across and recognises a specific major groove of the promoter region of DNA (Figure 3.8); helices 2 and 1 contact specific proteins of the transcription complex. Their precise *modus operandi* remains somewhat obscure. But, as we shall see in later chapters, their role is crucial. Control of transcription factors gives control of gene expression.

It is usually the case that transcription factors recognise short lengths of **palindromic DNA**; that is,

like the words 'radar' and 'civic', it makes the same sense read forwards or backwards. Two of the best known of these palindromic sequences are the so-called **AP-1 site**, 5′TGACTCA3′, and the **cAMP responsive element** or **CRE site**, 5′TGACGTCA3′. We shall meet both of these transcription sites later in this book. It is likely that the significance of palindromic sequences is that they allow the development of short hairpin loops in the DNA strand. These loops probably provide the recognition sites for some of the transcription factors.

But how, it will be asked, can special transcription factors attached to enhancer and/or silencer sites several thousand base pairs upstream affect the core promoter and its complex of general transcription factors? The best answer, at present, relies on the 'bendiness' of the DNA double helix. It is believed that the helix loops round so that the special transcription factors come into contact with the promoter region (Figure 3.9A).

Some of the most important of these special transcription factors in neurobiology are the proteins which bind to the **CRE** sequence mentioned above. These proteins, for obvious reasons, are called CRE binding proteins, or **CREBs**. They comprise a group of at least ten proteins. Not all members of the family are, in fact, activators. Other members, known as **CRE modulators (CREMs)**, block CRE-dependent transcription. The crucial event in the activation of a CREB is the phosphorylation of a serine residue. This allows another protein, **CREB binding protein (CBP)**, to attach to CREB and the complex is then able to act on the general transcription factors over the TATA box and initiate transcription (Figure 3.9B). We shall see the significance of this intricate biochemistry when we come to discuss the molecular biology of memory in Chapter 20.

Finally, it should be noted that enhancers (and silencers) usually act in a tissue-specific manner. Special transcription factors which bind to specific enhancers are found only in certain cells; vice versa, factors which bind to specific silencers are similarly restricted to certain cell types. This huge intricacy (especially when compared with the relative simplicity of the control mechanisms in prokaryocytes) provides a combinatorial control so that genes can be switched on and off with great delicacy. This, of course, is indispensable if the huge variety of differently structured, and differently functioning, cells of a

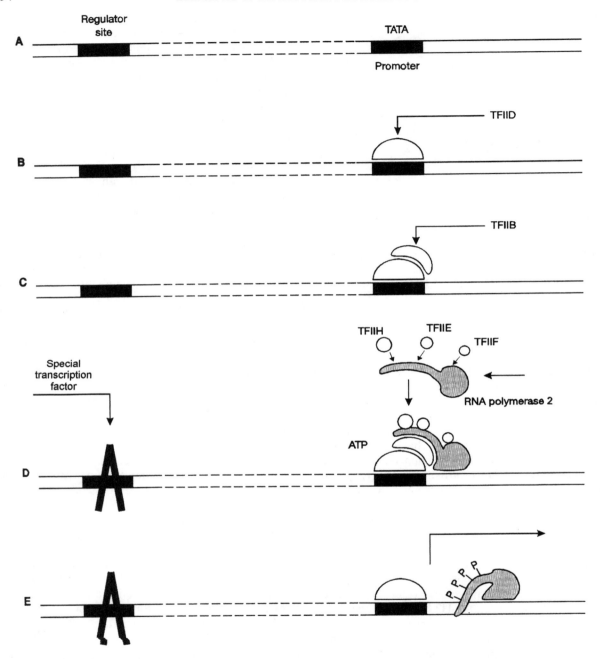

Figure 3.6 Initiation of transcription at the TATA box. (A) A stretch of DNA showing promoter and regulator sites. The regulator may be tens of thousands of base pairs distant from the promoter. (B) The first general transcription factor (TFIID) arrives and attaches to the TATA box. (C) The second general transcription factor (TFIIB) attaches. (D) RNA polymerase 2 complexes with TFIIF, TFIIE and TFIIH and then attaches to TFIIB and TFIID. The initiation complex is now in place. TFIIH uses phosphate from ATP to phosphorylate (and thus activate) pol 2. A 'special' transcription factor may attach itself to the regulator site and influence the rate of transcription. (E) Pol 2 commences transcription. Modified from Alberts *et al.*, 1994, *Molecular Biology of the Cell* (3rd edn), New York: Garland.

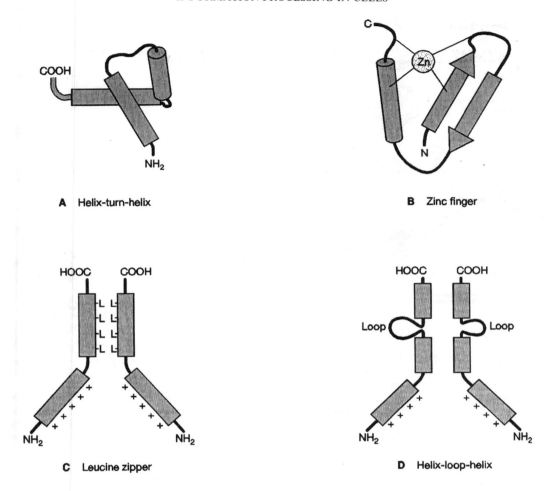

A Helix-turn-helix

B Zinc finger

C Leucine zipper

D Helix-loop-helix

Figure 3.7 Major structural features of 'special' transcription factors (TFs). (A) Helix–turn–helix. This family includes the homeodomain proteins which play important roles in cell differentiation (Chapter 18). (B) Zinc finger. Characterised by a zinc-coordinated binding motif. They, too, develop a helical segment which interacts with specific base sequences in the DNA major groove. (C) Leucine zipper. Like most of the special transcription factors this family interacts with DNA as dimers. In this case helices of the two units are held together by hydrophobic forces between amino acids such as (usually) leucine (L), hence the name. The positively charged domain straddles the DNA promoter site. Important instances of these TFs are the *fos* and *jun* immediate early genes (IEGs) (see Figure 3.18). (D) Helix–loop–helix. The HLH structure consists of just two helical regions (cf. the three of the HTH structure). Dimerisation allows two HLH structures to straddle the appropriate DNA transcription site. In this sense they are somewhat similar to the leucine zippers. From Shepherd (1994), *Neurobiology* (3rd edn), Oxford: OUP, with permission.

multicellular organism is to develop and survive. It has also been made use of, as we shall see in Section 5.18 and in later parts of this book, to switch genes on and off at different stages in development and in different parts of the nervous system.

The result of all this intricate biochemistry is the primary transcript. This accumulates in the nucleoplasm. But before 'translation' into protein occurs, a large amount of post-transcriptional processing occurs.

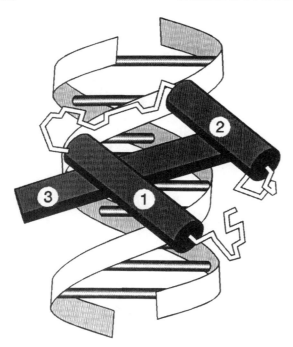

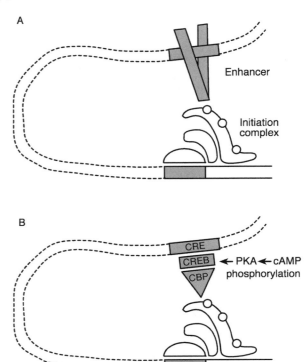

Figure 3.8 Interaction of an HTH homeodomain transcription factor with DNA. Helix 3 makes contact with a specific base sequence in the major groove of DNA. Amino acid side chains from helices 1 and 2 interact (in a lesser way) with nucleotide bases in the lesser groove. After Wolberger *et al.*, *Cell*, **67**, 517–528, © 1991, Cell Press.

Figure 3.9 Control of transcription by special transcription factors. A: general; B: control from the CRE site. CBP can only bind to CREB and initiate transcription when it is phosphorylated via PKA. Explanation in the text.

3.3.2 Post-transcriptional Processing

Post-transcriptional processing of all three types of RNA – mRNA, tRNA and rRNA – occurs in eukaryotic cells. We shall restrict ourselves to the post-transcriptional processing of mRNA.

Differential Splicing

Unlike the situation in prokaryotic cells, eukaryotic DNA contains large stretches where the base sequences code nonsense. These nonsensical stretches are called **introns** (an abbreviation for 'intervening sequences'). They separate the meaningful sections of DNA, or **exons**, from each other.

Figure 3.10 *(Opposite)* Post-transcriptional processing of eukaryotic mRNA. (A) The DNA double helix is schematised as two parallel lines with the exons coloured grey and the introns coloured black. DNA-dependent RNA polymerase 2 is represented as a stippled circle. (B) RNA polymerase attaches to the promoter region and opens the DNA double helix. Transcription commences. The newly synthesised mRNA strand is capped by guanosine triphosphate and methylated. (C) The process of transcription continues. Both exons and introns are transcribed. (D) On reaching a termination signal on the DNA strand the RNA polymerase detaches. A lengthy 'tail' of adenosines (150–200) is attached to the 3′ end of the newly transcribed hnRNA. Spliceosomes now cut out the introns and splice together the cut ends of the exons to form the finished mRNA (shown at E). In some cases this may be altered by a process known as 'editing' (see text). The remains of the introns form small loops known (from their shape) as 'lariats'.

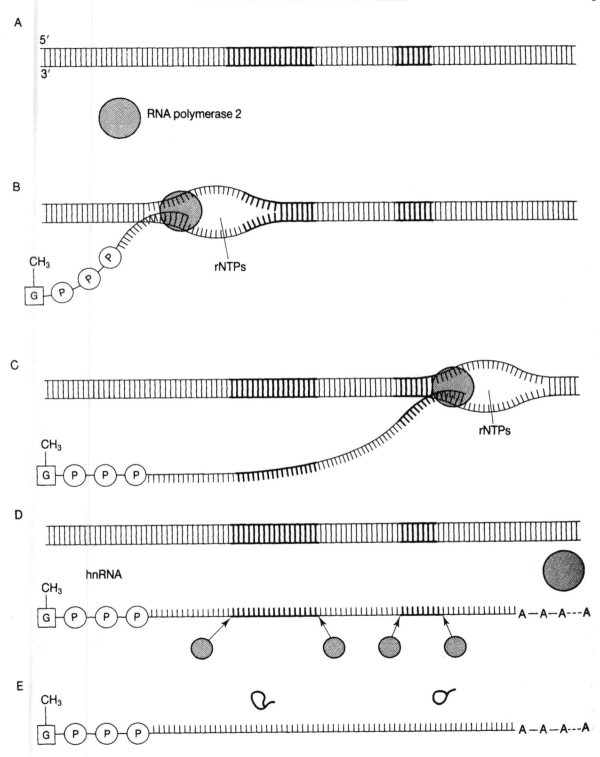

Practically all eukaryotic genes have at least one intron and some, for instance the collagen gene, have as many as 50 whilst the 250 kb *tintin* has 234.

What is the significance of this eukaryotic characteristic? Is it that prokaryocytes have been around longer than eukaryocytes and, moreover, usually have much more rapid generation times? Has this allowed prokaryocytic DNA time to rid itself of useless segments? Or do the seemingly senseless sections of eukaryotic DNA have an important function? It seems more likely that the latter proposition is correct. For, as we shall see, they allow important post-transcriptional control of gene expression. Different combinations of exon transcripts allow one gene to program the production of several different proteins. These different proteins may be characteristic of different tissues and/or of different stages in the development of one tissue. Perhaps, also, the more complicated structure of the eukaryotic gene has something to do with the more complicated cytogenetics of the eukaryotes (mitosis, meiosis, crossing over, etc.) and the evolutionary possibilities opened up by recombination during sexual reproduction. Only further investigation will give us a complete answer.

Figure 3.10 shows that the mRNA transcribed from eukaryotic DNA first of all forms the heteronuclear RNA (hnRNA) or primary transcript we mentioned above. Both the exon and the intron regions of the DNA are transcribed. The mRNA chain is modified by the addition of a **cap** to the 5′ end (usually a guanine (G) nucleotide) and a lengthy sequence of about 150 to 200 adenosine nucleotides to the 3′ end. Enzyme complexes known as **spliceosomes** cut out the intron regions from the primary transcript. This is done with great accuracy. The enzymes recognise specific '**consensus sequences**' of nucleotide bases at the 5′ and 3′ ends of the introns. They cut the introns at these sites and splice together the cut ends of the exons to give the mature mRNA. This is now ready for translation. The excised introns curl up to form '**lariats**' and play no further part in the process.

It is, of course, very important to get this post-transcriptional splicing right. It has been shown that several human hereditary blood diseases known as **thalassaemias** are due to the incorrect intron excision and resplicing of β-globin primary transcript. More relevant to molecular neuro-

biology has been the demonstration that the *jimpy* (*jp*) mutation which affects central **myelination** in the mouse is also due to defective post-transcriptional processing. It is found that the genetic defect causes the splicing process to omit a 74-base sequence from the mRNA coding for a proteolipid component of central (though not peripheral) myelin. The resulting proteolipid protein, which normally consists of 277 amino acids, consequently lacks residues 208–232. Histology shows that the white matter in the CNS is severely affected and very little myelin develops. Behavioural symptoms first appear as body tremor on the eleventh postnatal day. The tremor increases, leading ultimately to general convulsions and death in week five. This mutation will be considered again in Section 7.7, where the structure of myelin is discussed.

The gene for central myelin is carried on the X chromosome in both mouse and man. The jimpy mutation is regarded as analogous to **Pelizaeus–Merzbacher disease** – a sex-linked recessive leucodystrophy (i.e. white matter deficiency) – which affects humans. The symptoms first appear shortly after birth – nystagmus and inability to control head alignment – and these are followed by seizures and ataxia, usually leading to death before the age of three.

mRNA Editing

In recent years it has been found that the mature mRNA produced by the excision of introns and resplicing of exons described above is subjected to yet further modification. This so-called **editing** was first discovered in trypanosomes. As these organisms are some of the most ancient of all existing eukaryotes it has been suggested that this process is a relic from a primordial 'RNA world' before the evolution of DNA genetics. Be that as it may, it has been found that RNA editing still plays a bit-part in controlling the information flow in higher organisms.

In the trypanosomes the editing is carried out by a complex biochemical mechanism involving short stretches of RNA known as 'guides'. In a mammalian case where the mature transcript from the apolipoprotein B gene is edited, a simpler mechanism obtains: an enzyme deaminates a C to a U, thus creating a 'stop' codon in the middle of the mRNA strand. More interesting,

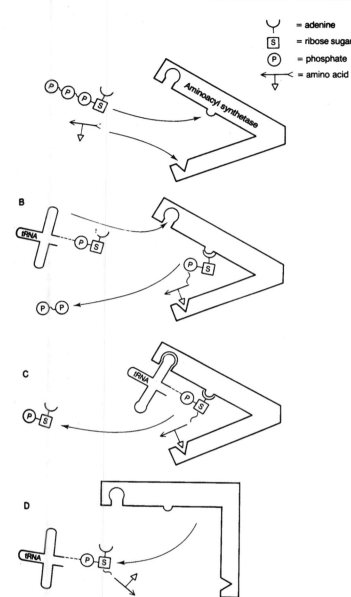

= adenine
S = ribose sugar
P = phosphate
= amino acid

Figure 3.11 Recognition and coupling of tRNA and appropriate amino acid by an aminoacyl tRNA synthetase. (A) The aminoacyl tRNA synthetase has 'active' sites. One of these sites fits a specific amino acid (symbolised by an arrow with a triangle representing its side chain). Another site fits adenine (represented by a semicircular cup). The third accepts a specific tRNA (symbolised by the conventional cross-shaped clover leaf). (B) The adenine of ATP has occupied the adenine site and the amino acid has taken up position in its site. The amino acid is activated by an aminoacyl-AMP by displacing pyrophosphate (P–P) from the ATP. (C) An appropriate tRNA arrives and finds its complementary site in the aminoacyl tRNA synthetase. The adenosine which is always present at the end of the amino acid acceptor stem of tRNA displaces AMP's adenine and the 3' OH of its ribose moiety accepts the energy-rich bond to the amino acid. AMP is released. (D) The aminoacyl–tRNA complex is released from the aminoacyl tRNA synthetase enzyme. The figure is highly schematic. Until recently very little was known of the molecular structure of aminoacyl tRNA synthetases.

however, from our point of view is the finding that editing alters the mRNA code for some **glutamate** receptors (Section 10.4), dramatically changing their Ca^{2+} conductivity. It has been shown that this conductivity depends on the presence of a positively charged arginine residue in the lining of the channel. This residue is not specified in the DNA but is incorporated during post-transcrip-tional processing by RNA editing. Whether this editing is carried out by complex mechanisms analogous to those found in trypanosomes or by some variant of the simpler operation at work in apolipoprotein B is not at present known. It is likely that RNA editing, whatever the mechanism, will be found at work creating diversity in other neurobiologically important structures.

3.3.3 Translation

The business of translating the mRNA message into protein structure is just as complicated as the mechanisms of transcription described above. Once again, it involves a large and heterogeneous group of cooperating factors.

The most significant members of this group are **ribosomes**, **tRNAs**, **aminoacyl tRNA synthetases**, **mRNA**, **amino acids** and numerous protein **initiation** and **elongation** factors.

Again the process is best known in prokaryotic systems such as *E. coli* and again it is believed that the eukaryotic mechanisms, although more complex, are basically similar.

Translation begins by the attachment of the appropriate amino acid to its designated tRNA. This vital step is catalysed by an **aminoacyl tRNA synthetase enzyme** and energy is provided by the hydrolysis of ATP.

$$\text{amino acid} + \text{tRNA} + \text{ATP} =$$
$$\text{aminoacyl tRNA} + \text{AMP} + \text{PP}_i$$

We noted in Chapter 2 that just as there is a specific tRNA molecule for each of the twenty different amino acids so there is a specific aminoacyl synthetase enzyme for each amino acid. Where an amino acid is coded by more than one DNA codon (as is usually the case (see Figure 3.1)) then there will be more than one synthetase enzyme assigned. These enzymes play a central role in the whole complicated business. It is upon their ability to recognise both the specific amino acid and the specific tRNA that the entire operation of translation depends. Indeed, so important is their place in protein biosynthesis that they have been said to constitute a 'second genetic code'. Until quite recently very little was known of their structure. In 1989, however, X-ray analysis succeeded in solving the structure of glutaminyl-tRNA-synthetase at 2.8 Å resolution, and since then the three-dimensional structures of most of the others have been elucidated. We now know that they can be divided into two major groups which differ in the structure of the domain containing the active site. Each groups consists of ten members.

Figure 3.11 shows, schematically, the recognition of tRNA and its designated amino acid by an

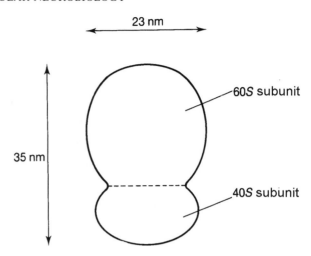

Figure 3.12 Eukaryotic ribosome. The figure shows the 'cottage-loaf' structure of a eukaryotic ribosome. The 60S and the 40S subunits together constitute an 80S (about 4.3 MDa) organelle with a maximum diameter of about 35 nm.

aminoacyl tRNA synthetase enzyme and the formation of an aminoacyl–tRNA complex.

First, an ATP molecule finds its way to the adenine site on the synthetase molecule and an amino acid with a side chain which fits the amino acid site occupies the amino acid site.

Next, the synthetase molecule catalyses the formation of aminoacyl-AMP, using the energy of one of ATP's energy-rich bonds, and releasing pyrophosphate. The amino acid is said to be 'activated'.

Lastly, the aminoacyl-AMP reacts with tRNA to form an aminoacyl–tRNA complex, releasing AMP in the process, and itself being released from the synthetase enzyme into the cytosol.

It is clear that the distinctive conformation of tRNA, which we described in Chapter 2, is of great importance. It is this conformation which the tRNA site of the synthetase enzyme is designed to recognise. It is this which ensures that the correct tRNA becomes bonded to the correct amino acid.

The synthetase enzyme has one final remarkable property. It is able to 'proof-read'. It recognises any incorrect pairing of tRNA and amino acid and decomposes it back into free tRNA and amino acid. This striking feature highlights once again the crucial importance of uniting an amino acid with its

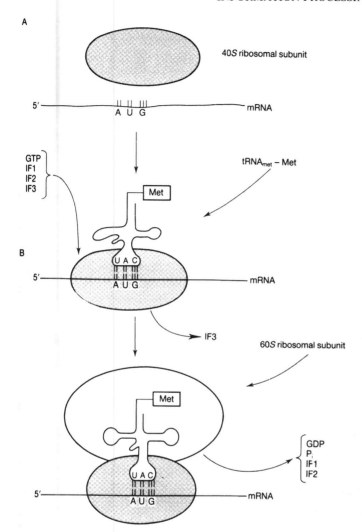

A

40S ribosomal subunit

5'——— A U G ——— mRNA

GTP
IF1
IF2
IF3

tRNA_{met} – Met

Met

B

U A C
5'——— A U G ——— mRNA

IF3

60S ribosomal subunit

Met

GDP
P_i
IF1
IF2

U A C
5'——— A U G ——— mRNA

Figure 3.13 Initiation of translation at the ribosome. (A) The 40S ribosome and a length of mRNA approach each other. (B) In the presence of a group of initiation factors (IF1, IF2 and IF3) and GTP the 40S ribosome binds to the start signal (AUG) towards the 5' end of mRNA. A tRNA_{met}–Met complex binds through its anticodon to AUG. IF3 is released. (c) The complex is now joined by the 60S subunit, GTP is hydrolysed to GDP and P_i, IF1 and IF2 are released. The stage is now prepared for chain elongation.

correct tRNA. Alanine is coupled to tRNA_{ala}, serine to tRNA_{ser}, etc.; any mispairings are eliminated.

The amino acid–tRNA complex (or charged tRNA) now encounters a ribosome. This occurs by random 'thermal' motion in the cytosol.

Eukaryotic ribosomes resemble minute cottage loaves (Figure 3.12; for images of accurate molecular models see web site in Bibliography). They have a sedimentary coefficient of 80S and easily dissociate into a 60S and a 40S subunit. The larger subunit consists of 28S rRNA, 5.8S rRNA and 5S rRNA and about 45 different proteins whilst the smaller subunit is built of 18S rRNA and about 33 different proteins.

In the cytosol the two subunits of the ribosome exist independently of each other. The coming together of the two subunits and the initiation of translation is a very intricate affair. It requires the interaction of a large number of factors.

First the smaller ribosomal (40S) subunit comes into contact with the 5' end of an mRNA strand. An initiation factor (IF3) is also involved at this stage. Next the mRNA–40S complex picks up a **tRNA–met** complex. This is due to the presence of the initiation signal, **AUG**, close to the 5' end of mRNA, and is assisted by another initiation factor (IF2). The anticodon of tRNA_{met} (UAC) recognises AUG by Watson–Crick base pairing. The 40S

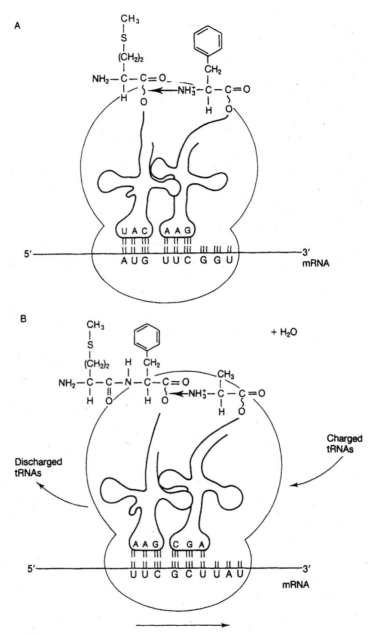

Figure 3.14 Peptide elongation. (A) tRNA$_{met}$–Met occupies the 'P' site in the ribosome. The mRNA codon beneath the 'A' site is UUC. This is complementary to AAG. The anticodon of tRNA$_{phe}$ is AAG. Hence this can occupy the 'P' site. When it does so the NH$_3^+$ group of its Phe residue is brought into the near neighbourhood of the energy-rich bond attaching met to tRNA$_{met}$. (B) With the help of elongation factors and ribosomal enzymes a peptide bond is formed between phenylalanine and methionine. A complex series of biochemical events now ensures that the tRNA$_{met}$ is released from the 'P' site, that the ribosome moves three bases in the 3' direction along the mRNA and that the tRNA$_{phe}$ comes to occupy the 'P' site. Another tRNA–amino acid complex (in the diagram tRNA$_{ala}$–Ala) can now occupy the 'A' site and the cycle begins again. Charged tRNAs enter from the right of the figure and discharged tRNAs leave from the left.

BOX 3.1 Antisense and triplex oligonucleotides

An understanding of the ways in which genes express themselves in animal cells opens ways of control for the biologist and in the future the physician. In theory, any of the complex of steps between the DNA genotype and the protein phenotype could be (and indeed has been) targeted. The best-known and most used technique is, however, to target either mRNA's translation or DNA's transcription step. Let us look briefly at each in turn.

The easiest way to block the translation of mRNA is to introduce carefully designed antisense oligoribonucleotides. These short sequences of ribonucleotides, synthesised in the laboratory by standard techniques, are designed to hybridise with the relevant mRNA and thus prevent translation at the ribosome. To be effective, to be sure of recognising the targeted mRNA, the strands should be at least 15 nucleotides in length. This leads to some difficulties in inserting the strands into a cell and there is always the problem of preventing the insert being digested by cellular enzymes before it can hybridise with its target. Biochemists have developed various techniques to overcome these problems but the method of choice nowadays is to substitute one of the oxygens in the phosphate groups with a sulphur atom. These phosphorothioate oligonucleotides – a jaw-breaking name, best abbreviated as S-oligos – are both water soluble and resistant to enzymic digestion. Once inside the cell the S-oligos not only bind to the correct mRNA but also stimulate a ribonuclease enzyme to attack the mRNA. They thus have a double chance of destroying the message before it is translated into protein. Finally, another technique for blocking the mRNA message must be mentioned. This consists of introducing plasmid containing the targeted sequence in reverse order and under the control of a powerful promoter. This will program the synthesis of antisense strands by the cell's own synthetic machinery and once again inactivate the targeted native mRNA by hybridisation.

The other approach for preventing the expression of deleterious genes is to block transcription. This can be done by synthesising an oligonucleotide complementary to the transcription site in DNA. In appropriate conditions this oligonucleotide strand will recognise the transcription site and form a triplex structure by slotting into the major groove of the DNA double helix. It has been found, furthermore, that this triplex formation occurs most readily when one strand of the double helix consists (in the transcription region) of purine bases: adenine and guanine. If the oligonucleotide strand consists of CCT triplets, for instance, it will recognise a transcription site with a GGA sequence. It has also been found that oligos containing these bases will line up parallel to the purine sequences in the DNA strand. Oligos rich in pyrimidines, however, line up in an antiparallel fashion. Taking all these features into account it is beginning to be possible to design oligonucleotide sequences which home in and block the transcription of specified genes.

Antisense and triplex technology are at present in their infancy. It is clear, however, that they show great promise for future research and therapy.

---

initiation complex so formed now binds to the 60S ribosomal subunit with the help of yet another initiation factor (IF1), the energy being provided by GTP. This series of events is schematised in Figure 3.13.

The stage is now set for the translation of the mRNA message into the amino acid sequence of a protein or polypeptide.

The 60S ribosomal subunit contains two sites which can accept charged tRNA molecules. The first site (see Figure 3.13) is called the **'P' site** (polypeptide or protein site), the second the **'A' site** (amino acid site). So far in our account, Met–tRNA$_{met}$ occupies the 'P' site. Which tRNA can occupy the 'A' site depends on the codon immediately to the right of AUG in the mRNA strand. Suppose (as in Figure 3.14) it is UUC. AAG is the Watson–Crick complement of UUC. The anticodon domain of Phe–tRNA$_{phe}$ happens to be AAG. Hence this aminoacyl–tRNA complex

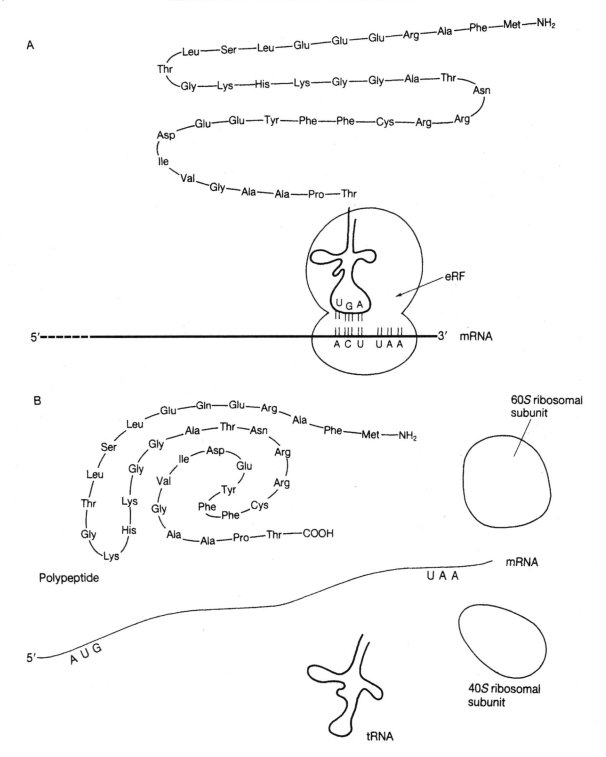

will be able to slot into this site. With the help of a number of other factors including various 'elongation factors' (EFs), a peptidyl transferase present in the 60S subunit and a further input of energy from GTP, the NH_2 group of Phe forms a peptide link with the COOH group of Met. When this occurs Met is released from $tRNA_{met}$ and the latter in turn is released from the ribosome. Phe–$tRNA_{phe}$ now shifts into the 'P' site in the 60S subunit as the entire ribosome moves three nucleotides (i.e. one codon) in the 3' direction along the mRNA strand. This brings another codon into the 'A' site, specifying another tRNA, and the whole process repeats itself.

Figure 3.15 shows that a repetition of the sequence of events outlined in the previous paragraph will result in a steadily growing peptide chain. The growth will continue until one of the 'stop' codons (UAA, UGA or UAG) comes under the 'A' site. When this happens no tRNA molecule can occupy the 'A' site and instead a release factor (eRF) alters the activity of the peptidyl transferase so that instead of catalysing the formation of peptide links it hydrolyses the final amino acid from its tRNA. The polypeptide chain is thus freed from the ribosome. The ribosome meanwhile detaches from the mRNA and dissociates into its 60S and 40S subunits ready for another bout of translation.

The freed polypeptide chain twists itself into its more or less complex tertiary structure as its amino acid residues interact with each other and with the environment in which it finds itself. The genetic code only **specifies primary structure**. In general the higher structures form automatically although, as we noted in Section 2.1.5, molecular **chaperones** are often required to navigate local energy minima.

Finally, it is worth noting that an mRNA strand normally supports several ribosomes. These are all occupied in translating the genetic message and move down the mRNA strand in line one behind the other. The complex is known as a **polyribosome** or **polysome**.

We shall return to consider what happens to the protein manufactured by this complex machinery in Chapter 15. We shall see that in eukaryotic cells such as neurons the growing polypeptide chain may suffer one of several fates. If the ribosome (as is the case with the majority) is attached to the endoplasmic reticulum (ER) then the polypeptide is either passed directly into the cisterna of the **endoplasmic reticulum** and from there via the **Golgi apparatus** into the axon, where it may find itself caught up and carried along in the axoplasmic flow. Alternatively, because of its hydrophobic characteristics, it may be caught in the membrane of the ER and carried as a membrane-bound protein to its final destination in the axon, dendrite or soma. Finally a small number of ribosomes remain 'free' in the cytosol, unattached to ER membranes, and these will deliver their polypeptide directly into the perikaryon. These polypeptides, when they reach their mature conformation, are involved in the so-called 'housekeeping' activities of the perikaryon.

3.4 CONTROL OF THE EXPRESSION OF GENETIC INFORMATION

Some fifty years ago John Gurdon (now Sir John Gurdon) showed that in some amphibia (e.g. the clawed frog, *Xenopus*) it is possible to remove the nucleus from a differentiated cell (e.g. a gut epithelial cell) and introduce it into an enucleated oocyte and induce that oocyte to develop into an adult. In spite of a large number of spontaneous abortions perfectly normal frogs often result.

As we begin the twenty-first century this technique has been vastly developed and has now been successfully applied to many infra-human animals, including several types of farm animal (famously 'Dolly the sheep') and there is even the prospect (highly controversial) of applying it to humans. The techniques and ethical implications, etc. of **cloning** are set out in the Roslin Institute's web site listed in the Bibliography.

Figure 3.15 *(opposite)* Termination of translation. (A) The ribosome has moved to the 3' end of mRNA where it encounters a 'stop' signal (UAA, UAG or UGA). The release factor (eRF) ensures that the polypeptide chain falls free from its tRNA and that the ribosome dissociates into its two subunits and detaches from mRNA. (B) mRNA, tRNA, the two subunits of the ribosome and the polypeptide are now disassociated. The polypeptide begins to wind itself into a three-dimensional conformation and is often subjected to post-translational processing. In particular, the N-terminal Met residue is normally excised.

Cloning carries the implication that the DNA of a specialised cell (epithelial, mammary, etc.) still carries all the information necessary to program the features of all the many different types of cell (muscle, neuron, fibroblast, hepatocyte, etc.) which make up an adult's body. Yet, of course, a gut epithelial or mammary cell is nothing like a neuron or a muscle fibre. It follows that much of the information in the DNA of an adult differentiated cell must be dormant, switched off, repressed, or in some other way unexpressed.

This conclusion has been confirmed by molecular biology. Cloned DNA is used to recognise specific mRNA transcripts in a cell's cytosol. It is not difficult to show that different cells generate very different mRNA transcripts although all contain much the same nuclear DNA. Which genes are expressed, and when, is crucial in the life of a multicellular organism.

The control of gene expression has been much studied in prokaryotic systems, especially (once again!) E. coli. The Nobel-prize-winning work of Jacob and Monod in the early 1960s has been followed by a great deal of brilliant molecular biology, the upshot of which has been to show how intricate molecular feedback loops control the expression of the prokaryotic genome. Unfortunately it does not seem that the molecular mechanisms at work in prokaryotic cells can be generalised in any very straightforward manner to eukaryotic cells. It seems that at this level of complexity E. coli loses its pre-eminence as a model. Although Monod, in a vivid phrase, once suggested that the elephant is merely E. coli writ large, it seems that he was wrong!

In fact the control of gene expression in eukaryotes remains, at the time of writing, a field of intense interest and research. The eukaryotic cell is far more complex and far larger than the typical prokaryote. By definition it contains a distinctive nucleus. This means that DNA and the processes of transcription are segregated by a membrane (the nuclear membrane) and by an appreciable distance (compared with the sizes of molecules) from the ribosomes and the processes of translation. Furthermore the DNA is coiled in an intricate fashion and complexed with **histones** to form chromosomes, unlike the comparatively naked DNA of prokaryotes. It seems that the control of genetic information is correspondingly complex.

Instead of the comparatively simple direct control of transcription found in prokaryotes, the eukaryotes have developed a host of different mechanisms operating at different stages in the flow of information from DNA to protein.

Figure 3.16 shows the major points at which control of the expression of genetic information in eukaryotes is believed to be exerted. The figure shows that there seem to be at least five major levels: **genomic**, **transcriptional**, **post-transcriptional**, **translational** and **post-translational**. We shall look briefly at each of these in turn.

3.4.1 Genomic Control

One way in which the quantity of a particular gene product may be varied is by loss or amplification of the amount of DNA present in the genome. Genes seldom seem to be lost but there are well-known instances where they are multiplied many times over. The best-known example of this type of amplification is found in *Xenopus*, where the genes which code for 18S and 28S rRNA (already present in some five hundred copies) are multiplied some 4000-fold during oogenesis so that the mature oocyte ultimately comes to possess about two million copies. This is apparently necessary to ensure a sufficiency of ribosomes to sustain the intense protein synthesis characteristic of early embryogenesis. However, although some other instances are known, gene amplification does not at present seem to be a major mechanism of gene control in adult eukaryotes.

But if genes are seldom lost altogether and infrequently multiplied, multiplication of codons is more common. For example, we shall see later, in Chapter 21, that multiplication of a nucleotide triplet, CGG, on the long arm of the X chromosome has been found to be responsible for the mental retardation of fragile X syndrome, and multiplication of a CAG triplet on chromosome 4 is responsible for Huntington's disease. At least 13 other neurological syndromes are now known to be associated with **trinucleotide multiplications**. Why neurons should be especially sensitive to such DNA pathologies is not known. It may have something to do with the fact that they do not replicate in postnatal life.

Finally, it must be noted that the above remarks apply only to ontogeny. When we come, in

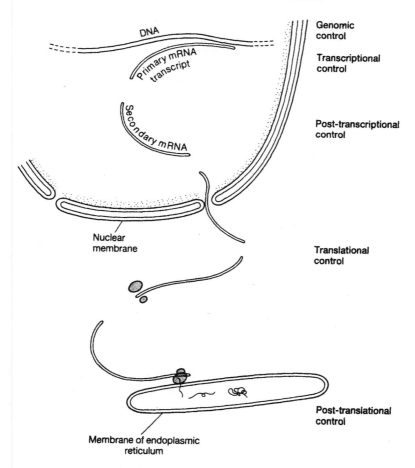

Figure 3.16 Multi-level control of the expression of genetic information in a eukaryotic cell. The figure distinguishes five levels at which the information held by nuclear DNA may be controlled. Genomic control involves loss or amplification of DNA; transcriptional control consists in the 'switching' on and off of structural genes; post-transcriptional control involves the manipulation of the primary mRNA transcript by excising introns and resplicing exons; RNA editing (where it exists) will occur in the cytosol; translational control influences the highly complex sequence of events by which the message in the mRNA determines the primary structure of the protein; and, finally, post-translational control allows the alteration of the protein or polypeptide by proteolytic enzymes. This last process commonly occurs within the cisternae of the endoplasmic reticulum and/or Golgi body. It should be borne in mind that these are just some (by no means all) of the control mechanisms available to the eukaryotic cell. Further explanation in text. Partly after Becker (1986), *The World of the Cell*, Menlo Park, CA: Benjamin/Cummings.

Chapter 4, to consider phylogeny we shall find that gene duplication and 'exon shuffling' play major roles.

3.4.2 Transcriptional Control

Transcriptional control is (as we shall see below) by far the most important (and well-understood) control mechanism in both prokaryocytes and eukaryocytes. Indeed, it is basic to the processes of embryogenesis (see Chapter 18). It is not difficult to show (as mentioned above) that different cells of a single organism contain very different mRNA transcripts. The brain, furthermore, is the most histologically diverse of all the body's tissues. It consists of a great variety of different cells. It is not surprising, therefore, to find that at least **125 000 different mRNA** transcripts are expressed at

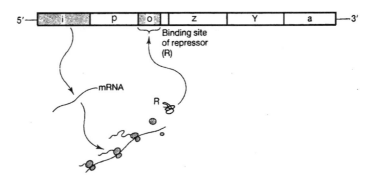

Figure 3.17 Control of transcription at the *lac* operon. The *lac* operon consists of the operator gene (o) and three structural genes (z, y and a). The structural genes code for β-galactosidase, lactose permease and transacetylase, respectively – all enzymes involved in the metabolism of lactose by the bacterial cell. The figure shows the operator gene consists of a short sequence of nucleotides (21) towards one end of the promoter sequence. 'Upstream' from the promoter (p) is another gene, the regulator gene (i). The regulator gene programs the synthesis of a repressor protein (R). This binds to the operator sequence. When the repressor is attached to the operator DNA-dependent RNA polymerase cannot gain access to the promoter and start transcribing the structural genes. In the presence of lactose, however, the three-dimensional form of the repressor is altered so that it can no longer attach itself to the operator. In this circumstance the structural genes can be transcribed.

different times and in different cells in the brain: three to five times greater than in any other tissue.

The new technique of constructing and using **DNA chips** (for description see Chapter 5) is revolutionising the study of gene expression. These chips can detect in one operation which of several thousand mRNA molecules are being expressed in a given tissue, at a given developmental stage, in response to changes in ambient conditions or during episodes of disease.

Because of the great importance of transcriptional control it will be useful, first of all, to outline the well-known mechanisms at work in prokaryocytes. The widely used distinction between **regulator**, **operator** and **structural gene** was first developed in these systems. An understanding of transcriptional control in prokaryocytes is, moreover, of importance if some of the techniques used in genetic engineering (Chapter 5) are to be grasped.

Figure 3.17 shows the best-known of all prokaryotic control systems – the *lac* **operon** in *E. coli*. The bacterial DNA contains four relevant regions: a regulator gene, a promoter sequence, an operator sequence (which partially overlaps the promoter) and a set of structural genes (z, y and a). The regulator gene programs the synthesis of a tetrameric repressor protein which normally attaches itself to the operator sequence thus

blocking the insertion of RNA polymerase. In consequence the structural genes cannot be transcribed – in other words they are switched off, or repressed. In the presence of an inducer molecule (in this case lactose) which binds to the repressor protein, the latter undergoes an allosteric change in its conformation which makes it unable to stick to its site on the operator. In consequence the RNA polymerase can insert itself and transcribe the structural genes. These genes program the synthesis of a group of enzymes involved in the entry of lactose into the cell and its subsequent metabolism: β-galactosidase, galactoside-permease and thiogalactoside acetyl-transferase. We shall see in Chapter 5 that if a foreign gene is spliced into the operon it too will by controlled by the operator. Hence it can be switched on (induced) by adding lactose to the medium.

Because of the much greater molecular complexity of the eukaryotic chromosome it has proved much more difficult to determine the molecular mechanisms which control transcription in this system. It seems likely that the first event is the unravelling of the eukaryotic chromosome from its complex union with histones. This process is known as **decondensation**. There is evidence that histone **acetylation** may trigger this process. Decondensation allows selective lengths of DNA

BOX 3.2   Oncogenes, proto-oncogenes and IEGs

Cell proliferation is held in delicate balance by excitatory and inhibitory factors. This balance can be upset by mutation of genes which either induce cell growth and multiplication or which hold such proliferation in check and perhaps induce cell death (apoptosis). In this box we shall only consider the genes responsible for growth and cell division. They are called **proto-oncogenes**. When a proto-oncogene mutates it may result in an **oncogene** which programs unregulated cell division, i.e. cell division without regard to the needs of the rest of the organism. Oncogenes thus lead to cancerous growths. The study of proto-oncogenes and oncogenes has been of great importance in cancer research.

It has been known for some ninety years that some viruses can cause cancer. It was shown by Rous in 1911 that a virus could cause connective tissue tumours in chickens. This is the **Rous sarcoma virus**. Later work has shown it to be an RNA retrovirus. A retrovirus injects its RNA into an appropriate cell which then, by means of the virus's reverse transcriptase, copies itself first into the host's mRNA and from there into the host's DNA and inserts this copy into the host genome. In due course the host cell transcribes the inserted DNA into mRNA and the cell's own molecular machinery ensures that this is translated back into virus which ultimately erupts from the cell. The retrovirus thus very efficiently parasitises the eukaryocyte's genetic machinery. But how does this cause a tumour? So far we have only outlined a retrovirus life-cycle and means of multiplication, nothing about cancerous proliferation of the host cell.

The answer to this question is that some retroviruses (the Rous sarcoma virus amongst them) have picked up a proto-oncogene from the eukaryotic host cell during the countless life-cycles of evolutionary time. The proto-oncogene has subsequently mutated or its expression otherwise altered so that when reincorporated into the eukaryotic genome and transcribed it acts as a cancer-forming oncogene. Biologists prefix the viral oncogene with 'v' and its eukaryotic homologue with 'c'.

How is it that mutation of a proto-oncogene can have such catastrophic effects on the host cell and its neighbours? The answer to this question lies in the processes which proto-oncogenes control. These include cell membrane receptors for growth factors, cell signalling systems (such as G-protein systems), protein kinases and nuclear transcription factors. All of these processes assist in ensuring that the cell's growth, differentiation and division are appropriate to its place in the multicellular body. We shall return to this topic in Chapter 18. In this chapter we see how *v-fos* and *v-jun*, viruses which cause feline osteosarcomas and avian sarcomas (respectively), have their cellular counterparts in the proto-oncogenes, *c-fos* and *c-jun*. These proto-oncogenes are sensitive to a large number of extracellular influences (see Chapter 17) and because they are rapidly reacting are called immediate-early genes (IEGs). Their transcripts encode transcription factors which act on DNA's AP-1 sites. It has been found that both *c-fos* and *c-jun* are members of small families of genes with similar AP-1 activity: *c-fos, fra-1, fos-B*; *c-jun, jun-B, jun-D*. The AP-1 sites control the expression of secondary (sometimes called 'late onset') genes. It is clear that mutated forms of these IEGs (i.e. oncogenes) might well cause uncontrollable syntheses of the proteins which the secondary genes encode.

to become accessible for transcription by DNA-dependent RNA polymerase.

We noted in Section 3.3.1 that transcription of these decondensed regions is a far more intricate process than in prokaryotic cells. This allows many subtle control mechanisms to come into play. Not only this, but it has also been shown that in many neurobiologically important cases (for instance, many of the channel protein genes) there is more than one promoter region. These different promoter segments are differentially accessible in different neural tissues. Hence the primary mRNA transcript synthesised in one brain region may differ from the primary transcript synthesised from the same gene in

A

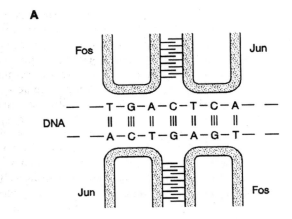

B

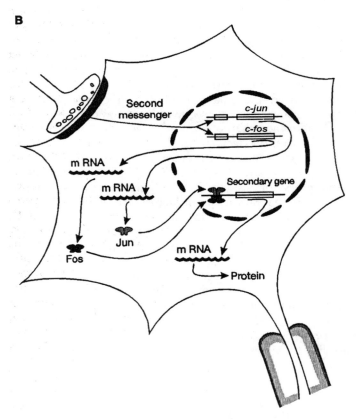

Figure 3.18 Immediate early genes. (A) Heterodimer of the Fos and Jun proteins on the AP-1 site. The figure shows how the two proteins 'zip' together by a number of leucine side chains represented by horizontal lines between the proteins. (B) The figure shows how an extracellular stimulus interacts with a receptor so that a second messenger is synthesised and released into the cytosol. This penetrates the nuclear membrane (through the pores) and activates the IEGs *c-jun* and *c-fos*. The mRNA from these two IEGs diffuses back out into the cytosol, where it is translated to form the dimeric proteins Jun and Fos. These, in turn, pass back into the nucleoplasm and attach themselves as a heterodimeric complex to the consensus site (TGACTCA) of a 'late onset' or 'secondary' gene.

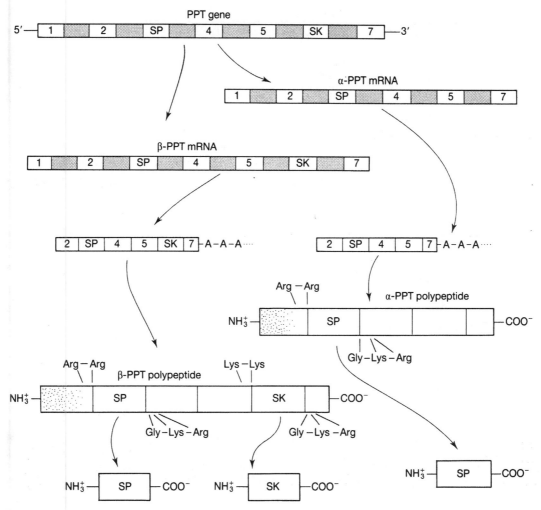

Figure 3.19 Post-transcriptional processing of PPT gene. The preprotachykinin (PPT) gene consists of seven exons and six introns (cross-hatched). Two primary mRNAs are transcribed: α-PPT mRNA, which lacks the substance K (SK) exon, and β-PPT, which is a complete transcript. After polyadenylation and excision of the introns the two PPT mRNAs are ready for translation. Two PPT polypeptides are formed: α-PPT (112 amino acids in length) and β-PPT (130 amino acids). The first twenty or so amino acids in both polypeptides form a 'signal sequence'. This sequence, as we shall see in Chapter 15, is required to attach the ribosome to the ER. It is composed of mainly hydrophobic residues so that it can pass through the membrane into the cisterna of the ER. It is then excised. In post-translational processing the substance P (SP) peptide (11 residues) and the substance K peptide (10 residues) are excised from the PPT polypeptides. The cutting points are marked by basic residues – Arg, Lys. The end result of all this processing are two copies of the SP peptide and one copy of the SK peptide. Partly after Karpati (1984), *Trends in Neurosciences*, **7**, 57–59, with permission from Elsevier Science; and Nawa, Kotani and Nakashani (1984), *Nature*, **312**, 729–734, with permission.

another. It has become customary to refer to these more complex genetic regions as '**transcription units**'.

In recent years new molecular biological techniques have been applied to the analysis of transcription factors which (as we saw in Section 3.3.1) play a crucial role in initiating transcription. Valuable insights into the control of transcription in eukaryotic cells have come from the study of the

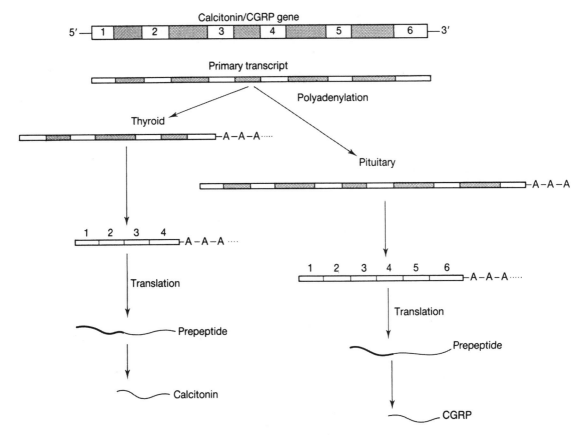

Figure 3.20 Differential post-transcriptional processing to yield calcitonin and CGRP. The calcitonin/CGRP gene has six exons separated from each other by introns. The primary mRNA transcript is cut and polyadenylated at different places in the thyroid and the pituitary. After excision of the introns, translation yields two different prepeptides. These, in turn, are further processed to give calcitonin and CGRP. See text for further details.

so-called **immediate early genes (IEGs)** or **primary response genes** first analysed in cancer cells. For our purposes the most significant of these genes are *c-fos* and *c-jun*, both of which are members of small families of IEGs (see Box 3.2). They are switched on very rapidly (within two or three minutes, hence the name) by 'second messengers' (see Chapter 8) which are themselves initiated by extracellular stimuli of many kinds. In the context of neurobiology these extracellular stimuli may include neurotrophic factors, neurotransmitters and neuromodulators, Ca^{2+} influx, etc. We shall return to this in Chapter 17.

Once switched on, the mRNA transcripts from the IEGs accumulate in the cytoplasm and are translated into dimeric proteins, **Fos** and **Jun**. These proteins (sometimes regarded as 'third messengers') turn out to be transcription factors of the leucine zipper type (see Section 3.3.1). The mRNAs decay away in one or two hours so that the presence of the Fos and Jun proteins is only transient. But whilst they are present they are able to diffuse back into the nucleus where they complex together to form a heterodimer which recognises the palindrome (5'TGACTCA3') present in the regulatory regions of several structural genes (Figure 3.18A). This motif, or **consensus site**, is, as we noted in Section 3.3.1, often referred to as the **AP-1 site**. In the CNS the best known of the so-called 'late onset' or 'secondary genes' controlled

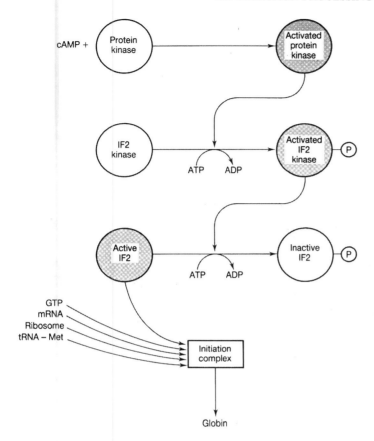

Figure 3.21 Regulation of translation in erythrocytes by phosphorylation of a protein kinase. Cyclic AMP (cAMP) activates a protein kinase. This, in turn, catalyses the phosphorylation of a kinase for initiation factor 2 (IF2). Lastly, this kinase catalyses a reaction leading to the phosphorylation-dependent inactivation of IF2. IF2 (see Figure 3.13) is an essential component of the initiation complex required to initiate protein synthesis. It is interesting to note that haemin, an essential constituent of haemoglobin, is required if the first step in the cascade is to proceed.

from an AP-1 site are those which program the synthesis of neuropeptides. The binding of the Fos–Jun heterodimer to the AP-1 consensus site controls the expression of these genes. This whole rather intricate control system is schematised in Figure 3.18B.

Other work on the control of transcription in eukaryotic genes indicates that regulator genes often omit the protein stage in the regulatory circuit and exert their effects on 'downstream' genes directly through their transcript mRNA. There is undoubtedly a hugely complex cascade controlling the expression of eukaryotic genes. We shall return to this topic in Chapter 18 where we consider the early development of the brain.

3.4.3 Post-transcriptional Control

We described post-transcriptional processing and the possibilities it affords for controlling the genetic message in Section 3.3.2. It will be recalled that the introns in the primary transcript are cut out and the remaining exons rejoined. Clearly this provides considerable scope for altering the mRNA which finally arrives at the ribosomes for translation. Let us look at some neurobiologically relevant instances.

First, let us consider the post-transcriptional processing which occurs during the synthesis of an important neuropeptide: **substance P**. We shall see later that substance P is a significant neurotransmitter in the spinal cord and brain. It is believed to be involved (along with the enkephalins) in nerve pathways mediating pain. Figure 3.19 shows that substance P is derived from a **preprotachykinin (PPT)** gene. The figure also shows that two distinct mRNAs are derived from this gene: α-PPT mRNA and β-PPT mRNA. These two types of mRNA are derived from differential splicing of the primary transcript. After translation the α-PPT polypeptide is cut open to release the 11-amino-acid peptide,

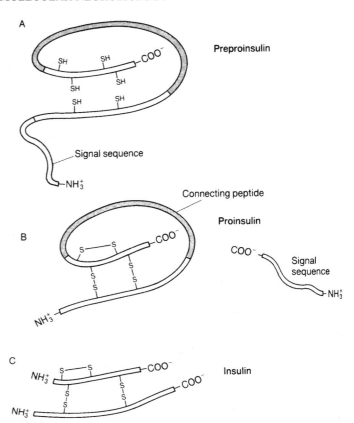

Figure 3.22 Post-translational modification of insulin. (A) The ribosome initially translates the insulin mRNA to form 'preproinsulin'. The 24 amino acids at the N-terminal end of this molecule constitute a hydrophobic 'signal sequence' which enables the ribosome to attach to an ER membrane and the protein to be inserted into the ER lumen. (B) Once inside the ER, the signal sequence is excised leaving the 84-residue 'proinsulin'. This orientates itself so that the correct disulphide linkages are formed. (C) The 33-residue connecting or C peptide is excised leaving the familiar disulphide linked α- and β-chain structure of insulin. The α-chain consists of 21 amino acid residues; the β-chain of 30 residues.

substance P, whilst the β-PPT is cleaved to give both substance P and another neuroactive peptide, **substance K**.

Another well-known instance of post-transcriptional control in molecular neurobiology is that which enables the same primary transcript to program the synthesis of both **calcitonin** and **CGRP (calcitonin-gene-related peptide)**. Here the situation is a little different. Figure 3.20 shows that the primary mRNA transcript is cut and polyadenylated at one site in thyroid cells and at another site in pituitary and some nerve cells. In both cases the resulting mRNA strand has its introns cut out and the remaining exons spliced together to form the final strand from which translation takes place. Whilst calcitonin is a well-known calcium-controlling hormone, the function of CGRP is, at the time of writing, obscure: it is an addition to the steadily growing list of neuropeptides found in the brain. It is also important to note, as Figure 3.20 shows, that in both cases the

mRNA is once again translated as a prepeptide which undergoes post-translational processing before yielding the biologically active end product.

3.4.4 Translational Control

We saw in Section 3.3.3 above that translation of mRNA into polypeptide is an enormously complicated affair involving not only tRNA, mRNA and ribosome but also numerous initiation, elongation and termination factors. There is evidently great scope for control at this level. In some cases (e.g. sea-urchin oocyte) it can be shown that although all the necessary mRNA is present in the cytoplasm very little, if any, translation, occurs until the egg is fertilised. Presumably some triggering factor is released by sperm entry. In other cases the **stability** of the mRNA strand may be affected so that it persists for a longer or shorter time in the cytoplasm and hence programs the synthesis of more or less polypeptide. In yet other cases it has

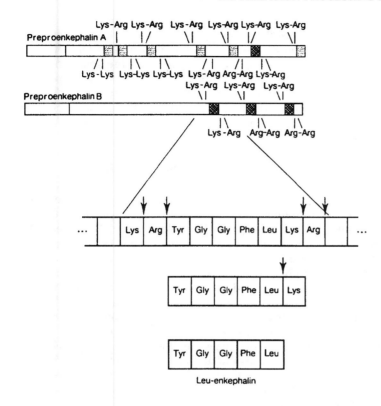

Figure 3.23 Post-translational production of met- and leu-enkephalin. Proenkephalin A contains six copies of met-enkephalin (stippled) and one copy of leu-enkephalin (cross-hatched). Proenkephalin B contains three copies of leu-enkephalin. The excision points are signalled by basic amino acids (Lys and/or Arg). The bottom part of the diagram shows that the enkephalins are cut out of the propeptide in two steps. The arrows indicate that the enzymes first attack to the right of each signal residue yielding a six-residue peptide. Then a second enzyme cuts to the left of the C-terminal residue leaving the pentapeptide leu-enkephalin.

been shown that control of the synthesis of a protein is effected by control of the activity of an **initiation factor (IF)**. In erythrocytes, for instance, it appears that the synthesis of globin is controlled through the phosphorylation (and hence inactivation) of one of the initiation factors. This phosphorylation in turn depends on the activation (by phosphorylation) of an **initiation factor kinase** which then catalyses the phosphorylation of the IF. Because second messenger systems in neurons often control protein kinases this erythrocyte mechanism may be of general relevance. It is schematised in Figure 3.21.

3.4.5 Post-translational Control

Many proteins are released from the ribosome as precursors requiring further biochemical change before they assume their mature and biologically active form. One well-known example is provided by insulin. This is synthesised as **preproinsulin** – a single continuous amino acid chain of 108 residues.

The 24 N-terminal residues constitute a 'signal sequence'. They are necessary for the attachment of a ribosome to the ER membrane and for the initial insertion of the protein into the lumen of the ER (see Chapter 15). Once inside the ER the signal sequence is cut away leaving the 84-residue **proinsulin**. After folding into a specific shape, proinsulin is stabilised by the formation of disulphide linkages. A protease then removes a large run of the amino acid chain (33 residues), leaving the mature **insulin** as two separate chains joined together by disulphide linkages (Figure 3.22).

An interesting neurobiological instance of post-translational control is provided by the so-called 'natural opioids': the **enkephalins** and **endorphins**. There are two enkephalins – **met-enkephalin** and **leu-enkephalin**. Both consist of five amino acids. They have been dubbed natural opioids because they seem able to inhibit the synaptic transmission of impulses in the brain's pain pathways. Their action will be discussed in more detail in Chapter 16.

leu-enkephalin: Tyr-Gly-Gly-Phe-Leu

met-enkephalin : Tyr-Gly-Gly-Phe-Met

It is relevant to note in this section, however, that the two enkephalins arise from two much larger precursor peptides – **preproenkephalin A** and **preproenkephalin B**. It is found that preproenkephalin A contains six copies of met-enkephalin and one of leu-enkephalin whilst preproenkephalin B contains three copies of leu-enkephalin and none of met-enkephalin. In addition, it has been shown that the enkephalin sequences hidden in the precursors are marked at each end by signal amino acids. These may be either two lysine residues or two arginine residues or a lysine and an arginine residue. These signals mark where post-translational enzymes can cut the precursor. The first enzyme always cuts to the right of the signal residue (i.e. towards the C-terminal). The second cuts the extra residue off the C-terminal end to give the final enkephalin structure. These steps are shown in Figure 3.23.

The other natural opioids – the endorphins – are produced by a similarly complicated post-translational processing of a large (265-residue) precursor known as **pro-opiomelanocortin (POMC)**. It is interesting to note that this large precursor contains within it both **α-endorphin** and **β-endorphin** and also several pituitary hormones, e.g. **ACTH**, **α-MSH** and **β-MSH**, etc. It is clear that the informational machinery within cells economises wherever possible. A large and sometimes diverse assemblage of protein and polypeptide end products may be derived form a single primary transcript. These insights into the relatedness of neuropeptides and peptide hormones also have evolutionary implications. Molecular evolution forms the subject matter of the next chapter and we shall consider these peptides again there.

3.5 CONCLUSION

Thus, in conclusion, we can see that there are many different ways in which the expression of the genetic information held in the DNA can be controlled. The nervous system is, as we have already noted, the most heterogeneous of the body's tissues. Its cells assume a great variety of shapes and sizes. It is not difficult to see that the ability to control gene expression is of great importance. This ability has become more significant in recent years as the surprisingly small number of genes in the genome has become apparent. Without this genetic flexibility whereby one gene can lie at the root of a large number of different protein and polypeptide products, it would be difficult to account for the extreme heterogeneity of the brain. We shall return to this topic in later chapters when, for instance, we discuss the development of the brain and the formation of nerve pathways.

4

MOLECULAR EVOLUTION

Evolutionary depth of the molecular realm: new molecular classifications – new insights into neuropathologies – new possibilities for therapy – new recognitions of the relatedness of molecular processes. **Point mutations**: mutability of DNA – synonymous and non-synonymous mutations – conservative and radical substitutions – chemical mutagens – transitions – transversions – frameshift – triplet expansion diseases (TREDs). **Proof-reading and repair**: DNA polymerases – exonuclease and endonuclease activity – nick translation. **Chromosome mutations** – gene duplication – intragenic duplication – exon shuffling and transposons. **Protein evolution** – orthologous and paralogous – cytochrome c and differential rates of evolution – globins and gene duplications – pseudogenes – exon shuffling – nAChRs – relatedness of receptor proteins. **Families and superfamilies of proteins** – mosaic proteins – complexity of the brain at the molecular level. **Evolution by post-translational mechanisms**: relationships of polyproteins – POMC, proenkephalin A and B – PCKK and PPT. **Conclusion**: the multitudinous ways in which the structure and thus functioning of protein molecules are controlled leads ultimately to variations in the behaviour of the organism which is presented for selection by the Darwinian forces

On first consideration it seems almost impossible to believe that the remarkable molecules and mechanisms described in the last two chapters could conceivably have originated by, as Jacques Monod put it, 'chance and necessity': blind variation and selective retention. We feel an awe similar to that felt by the natural theologians of the nineteenth century who considered that the superb design, the exact fitness for purpose, of living organisms could not but imply a designer, a creator. Yet since 1859, the publication date of the *Origin of Species*, all orthodox biologists have worked on the assumption that the living world did in fact come to be by the Darwinian mechanism of random variation and environmental (both abiotic and biotic) selection.

One of the most valuable contributions of molecular biology has been to support the neo-Darwinian synthesis. It has enabled us for the first time to quantify evolutionary change. In addition, the evolutionary approach is beginning to illuminate relationships between molecules which otherwise seem very dissimilar. And not only individual molecules: the ubiquity of gene duplication (see Section 4.1.3) and 'exon shuffling' leads to the development and persistence of functional modules consisting of many gene products, holding together and interacting with each other, over great periods of evolutionary time. Garcio-Bellido has coined the term 'syntagmata' for these interacting multiplexes. Investigation of the molecular biology of *Caenorhabditis elegans*, whose divergence from phylogenies leading to the mammals is lost in the mists of the pre-Cambrian, over half a billion years ago, has, for example, revealed a number of these syntagmata. It appears, for instance, that several mutations within a family of genes, the **degenerins** (see Section 13.3.2), located on one of the chromosomes of *C. elegans*, lead to the onset of neuromuscular disorders which, at the cellular level, appear similar to some human neurodegenerations, including Huntington's disease and amyotrophic lateral sclerosis (ALS). We shall meet other instances of these thought-provoking persistencies as we proceed through this book and especially when we discuss the homeobox genes in Chapter 18.

The recognition that biological macromolecules and syntagmata have massive evolutionary histories is beginning to provide us with classificatory schemes, just as the evolutionary insights of the late nineteenth century led to improved (so-called 'natural') classifications of the animal and plant kingdoms. Furthermore, just as nineteenth-century evolutionary thought changed our perception of the living world and suggested new questions to ask, so the viewing of biomolecules and interacting groups of biomolecules, in an evolutionary context, suggests a variety of new possibilities and questions. Are there molecular fossils within us dating back to the beginning of life? Are there vestigial molecules which have no function in modern cells? Are evolutionary 'advances' at first 'reactionary' as Romer suggested? Does the distinction between 'analogy' and 'homology' apply at the molecular level in the same way as it does at the organismic level? Can we, by making reasonable assumptions about the rate of molecular evolution, propose a date for the common ancestor of two dissimilar but evolutionarily related molecules? But, perhaps of more importance than anything else, this dawning recognition of the evolutionary relatedness at the molecular level immeasurably deepens our understanding of biology.

The study of molecular evolution has become possible because of our rapidly increasing understanding of the primary and higher structures of proteins, of the structure of DNA and its organisation to form the genetic material, and finally of the interrelationship between DNA and proteins. Because these insights have been revolutionised in recent years, our understanding of molecular evolution has accelerated at a phenomenal pace. Our knowledge of gene structure has, for instance, leapt ahead due to new methods of DNA sequencing. This, in turn, due to the discovery of 'reverse transcriptases', has led to a great increase in our knowledge of protein primary structure. We noted in Chapter 2 that protein tertiary structure very often consists of 'domains' or 'motifs' which reappear in different combinations in different proteins. We saw in Chapter 3 that eukaryotic DNA consists of meaningful 'exons' separated by apparently unmeaningful 'introns'. In this chapter we shall see that there is good evidence that some genes (transposons) can be moved from one site to another in the eukaryotic genome. In Chapter 6 we shall see how the elucidation of increasing numbers of animal genomes is providing deep insights into the relationships between genes in different organisms. All of these developments are beginning to coalesce to greatly increase our understanding of molecular evolution.

It may perhaps be asked: what has all this molecular biology to do with neurobiology? A few years ago that might have been a somewhat difficult question to answer. Not now. Profound interconnections between molecular and neurobiology are becoming more and more apparent. It is now not only possible to understand the relatedness of some of the most important of the proteins of which the nervous system consists or which it uses but also, as we shall see in subsequent chapters, to use some of the central techniques in genetic engineering to investigate and understand the structures and functions of crucial neurobiological molecules: membrane receptors, ion channels, neurotransmitters and modulators, cytoskeletal proteins, etc. Indeed, it is becoming possible to use molecular biological techniques to investigate such seemingly higher level phenomena as the organisation of the retina and the processes of memory acquisition, retention and access (Chapter 20). The molecular approach shows, above all, how apparently unrelated aspects of the brain's biology have, in fact, a unitary foundation.

In this chapter we shall proceed as follows. First we shall look at the molecular basis of mutation and the vital processes of gene duplication and exon shuffling. Then we shall briefly consider the phenomenon of gene transposition. Finally we shall consider the evolution of some important neurobiological polypeptides and proteins.

4.1 MUTATION

Mutations may be divided into two large categories: **point mutations** and **chromosomal mutations**. Point mutations are changes in single base pairs; chromosomal mutations are changes in large stretches of DNA including exon shuffling and the deletion and/or duplication of entire genes. Let us examine point mutations first.

4.1.1 Point Mutations

Point mutations may be caused by a large number of agents: some **chemical** (e.g. nitrous acid, 5-bromouracil), some **physical** (UV irradiation, X-rays, radioactive emissions). Furthermore, the DNA molecule itself has an inherent tendency to mutate. This is yet another feature which makes it a good genetic molecule. Without mutation living forms could not be selected to fit their environments: evolution could not have occurred. Indeed DNA's mutability is perhaps too great for its own good. It has to be held in check (as we shall see) by repair mechanisms.

Let us look first at DNA's spontaneous mutability. Consider the cytosine/guanine pair:

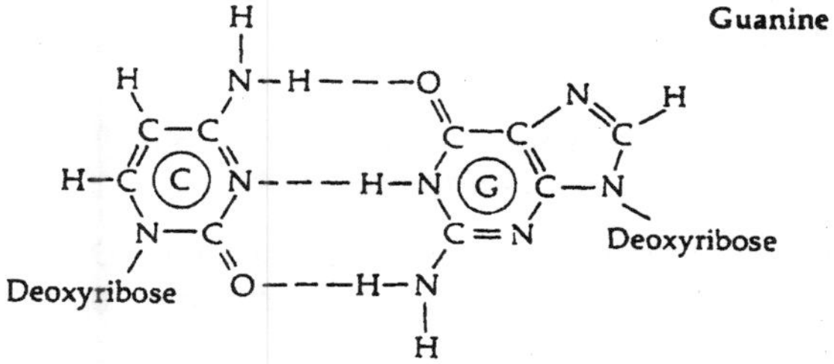

In about one case in every 10^4 or 10^5 cytosine rearranges into another tautomeric form:

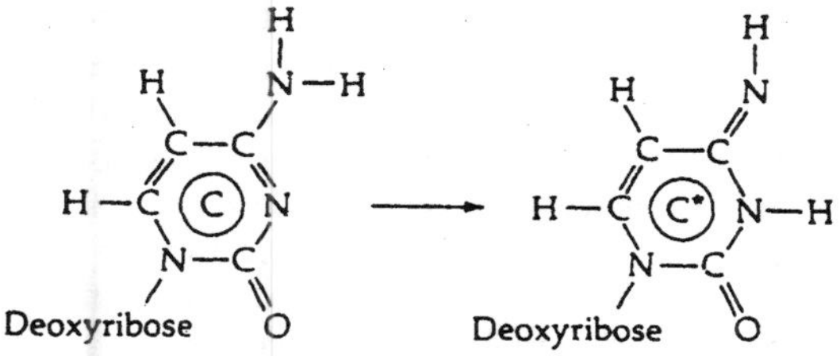

This clearly can no longer pair with guanine. It can, however, pair with adenine:

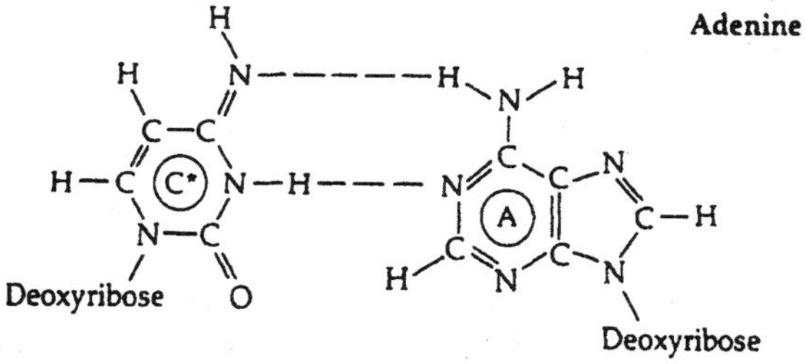

Hence, at replication, A is specified instead of G. At a subsequent replication cytosine is likely to return to its more stable tautomer, but A will remain to pick up T. Hence an **A/T pair** is substituted for a **G/C pair**:

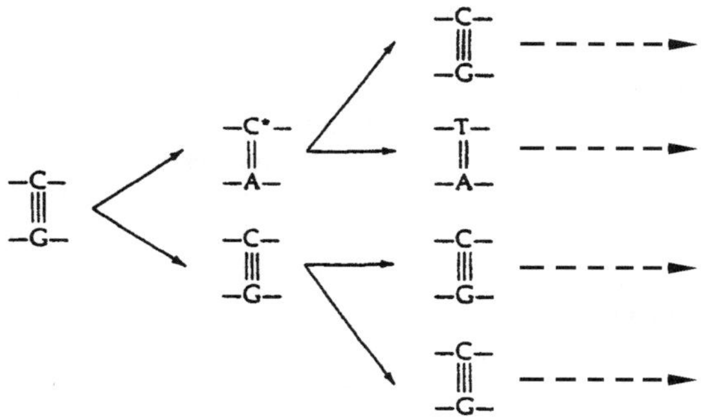

Reference back to Figure 3.1 will show that such a substitution, especially if it occurs in the first or second position of a triplet, could have a profound effect on the amino acid specified. In other cases, especially if the substitution occurs in the third position of codon, there be no effect at all. For instance reference to Figure 3.1 shows that glycine is coded by GGT, GGC, GGA and GGG. Any substitution in the first two positions leads to a change in the amino acid specified. Vice versa, any change in the third position has no effect: glycine remains the amino acid specified. The first two positions are called **non-synonymous** sites and the third a **synonymous** site. Substitutions in non-synonymous sites are said to be **non-synonymous mutations** and those at synonymous sites are called **synonymous mutations**.

Let us consider **non-synonymous substitutions**. In some cases the amino acid newcomer specified by the substitution will not vary greatly in its physico-chemical characteristics from the residue it replaced (see Table 2.1). If a hydrophobic residue is replaced by another hydrophobic residue, an aromatic side chain by a similar bulky side chain, a basic group by another basic group, an acidic by another acidic, or a hydrophilic side chain by another hydrophilic side chain, the three-dimensional conformation of the protein may not be too greatly upset. These acceptable newcomers are referred to as **conservative** substitutions. **Radical** substitutions where, for instance, a hydrophobic residue is replaced by a hydrophilic residue, or a basic group by an acidic group, are, however, much more difficult to accept. Most difficult of all is substitution of **cysteine** by another amino acid lacking, as all the rest do, the sulphydryl group which is so important in forming the disulphide bonds that stabilise many tertiary structures.

We noted in Chapter 2 that all the classical Watson–Crick bases are conjugated structures and hence liable to tautomeric change. Hence the DNA molecule has built-in mutability. Nevertheless, it has been calculated that in mammals only one error ultimately appears in every 10^9 base pairs. DNA's mutability is held in check by **proof-reading** and **repair** mechanisms. Before we look at these let us briefly consider the effect of chemical mutagens.

As an example let us take **nitrous acid (HNO_2)**. The reactions of HNO_2 with adenine and cytosine are shown in Figure 4.1. In the first case a C/G pair is substituted for an A/T pair and in the second case an A/T pair is substituted for a C/G pair.

It is worth noting that in all the cases considered so far a **pyrimidine** has been replaced by another **pyrimidine** or a **purine** by a **purine**. This type of point mutation is known as a **transition**. The exchange of a **purine** for a **pyrimidine** or vice versa is much rarer but may occur very occasionally by mispairing during replication. This second (much rarer) type of point mutation is called a **transversion**.

A final type of point mutation, known as a '**frameshift mutation**', occurs when one or more base pairs (not three or a multiple of three) is inserted into or deleted from the DNA sequence. This is termed a **gap event**. When a gap event occurs all the codons to the 5′ side of the insertion or deletion are changed. The so-called 'reading frame' thus reads out an altered set of codons (remember there are no punctuation marks (Section 3.1)) and

is, indeed, likely to find a new stop codon (UAA, UAG, UGA) before too long. Gap events are caused by polcyclic molecules (present in many foodstuffs) which bind to, or intercalate between, adjacent nucleotide bases when the polynucleotide strands separate during DNA replication. Particularly troublesome gap events are responsible for **triplet repeat expansion diseases (TREDs)** such as Huntington's disease and fragile X syndrome. We shall return to these diseases in Chapter 21.

Before leaving the topic of point mutations it will be useful to introduce the convenient short-hand molecular biologists have developed. A point mutation in DNA is referred to by **position** on the nucleotide strand and the **substitution** which has occurred, for instance $T_{1387} \Rightarrow A$. This is sometimes written, more simply, as T1387A. Similarly, an alteration in a polypeptide sequence is again given by its position and the substitution, for instance $Gln_{213} \Rightarrow Arg$ or, and more commonly, the more compact Q213R.

Finally, it is worth noting that if a point mutation occurs at a splice junction between an exon and an intron or on a **regulator** rather than a structural gene the consequences can be far more dramatic. It has been suggested that such mutations might well account for 'sudden' evolutionary changes.

4.1.2 Proof-reading and Repair Mechanisms

It was emphasised above that although DNA is very mutable, replication is nevertheless normally carried out with extremely high fidelity. This is because the DNA polymerase enzymes responsible for laying down a daughter polynucleotide strand alongside the parent polynucleotide template eliminate errors by 'proof-reading' the new strand.

In Section 3.2 we noted that there were three DNA polymerases in both prokaryotic and eukaryotic cells. We noted also that they all act in the 5′ to 3′ direction. It turns out all three polymerases also have an **exonuclease** activity. Polymerases 1 and 3 show this activity in both the 3′ to 5′ and the 5′ to 3′ directions; polymerase 2 is only able to act in the 3′ to 5′ direction. In bacterial systems it has been shown that polymerase 3 cannot join a new deoxyribonucleotide to the 3′ end of the preceding deoxyribonucleotide if the latter is not **securely**

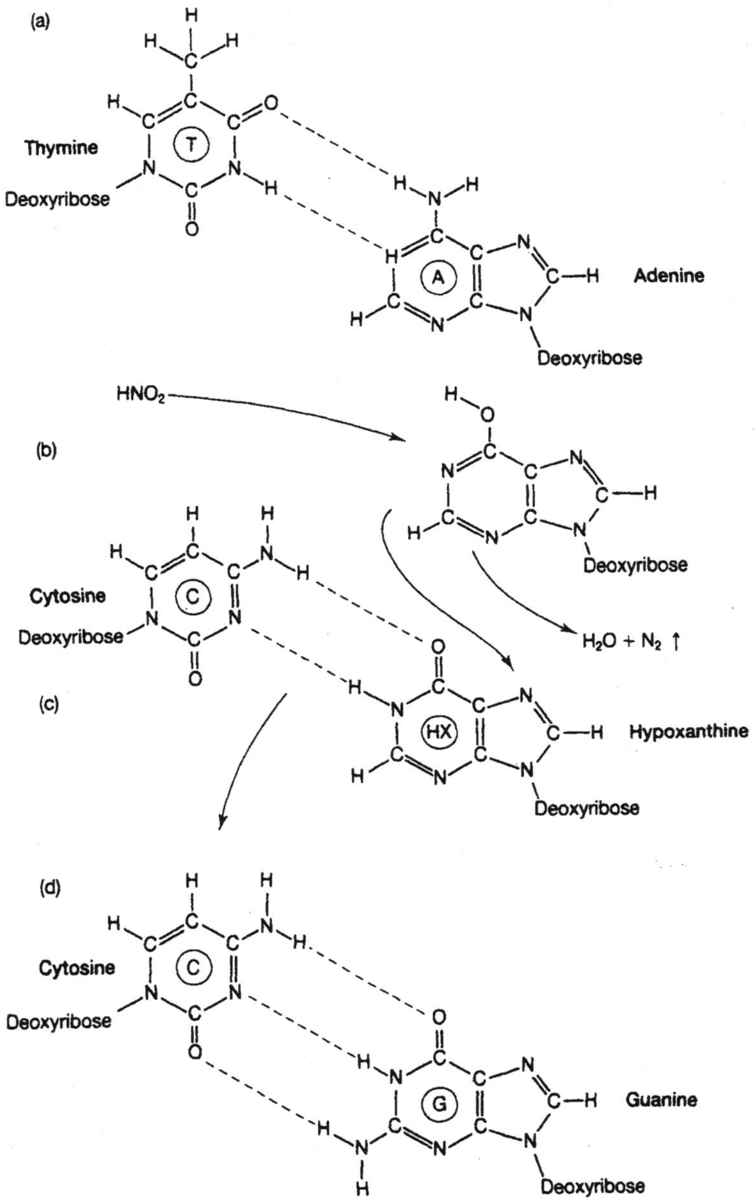

Figure 4.1 The mutagenic effect of nitrous acid (HNO$_2$). (A) Substitution of a C–G pair for a T–A pair. In (a) thymine and adenine are shown paired as in the DNA double helix; in (b) HNO$_2$ is shown to deaminate adenine, which after rearrangement of the hydroxyl group forms (c) hypoxanthine (HX), which partners cytosine; (d) at the next DNA replication cytosine acts as a template for guanine.

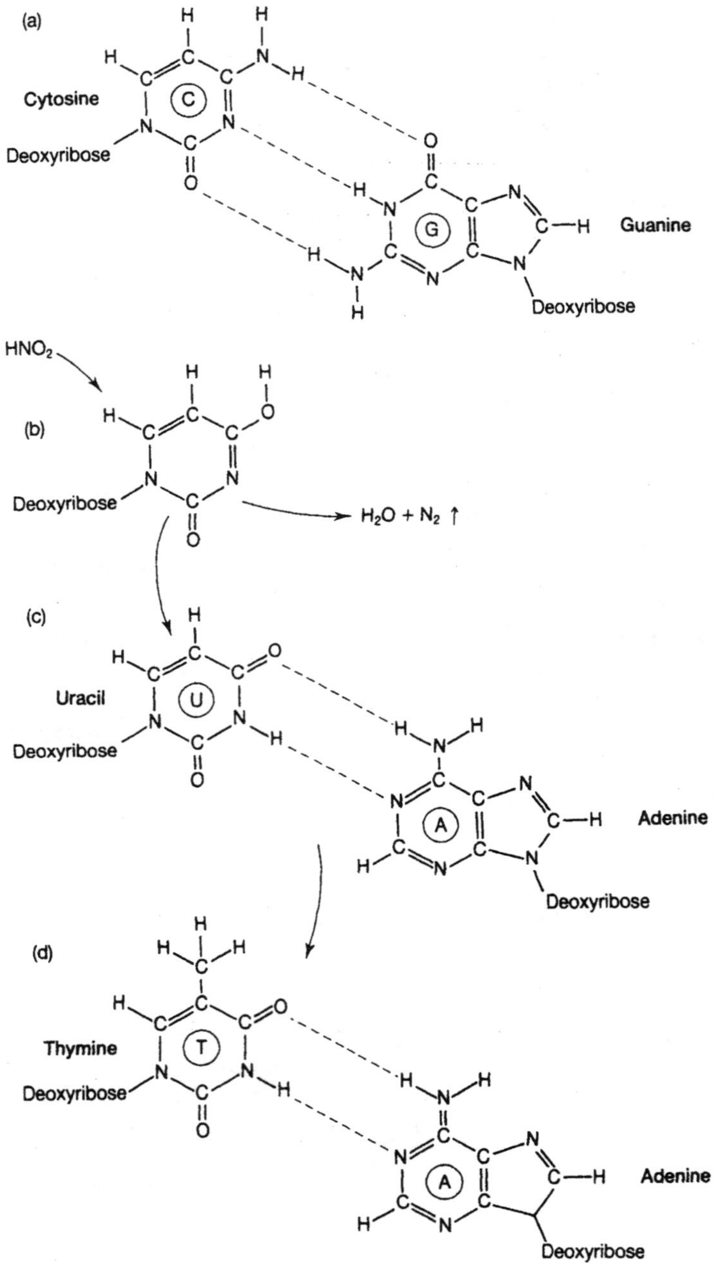

Figure 4.1   (B) Substitution of a T–A pair for a C–G pair. In (a) cytosine and guanine are shown paired as in the DNA double helix; in (b) HNO_2 deaminates cytosine, which after rearrangement of the hydroxyl group (c) forms uracil (U), which partners adenine; (d) at the next DNA replication adenine acts as a template for thymine.

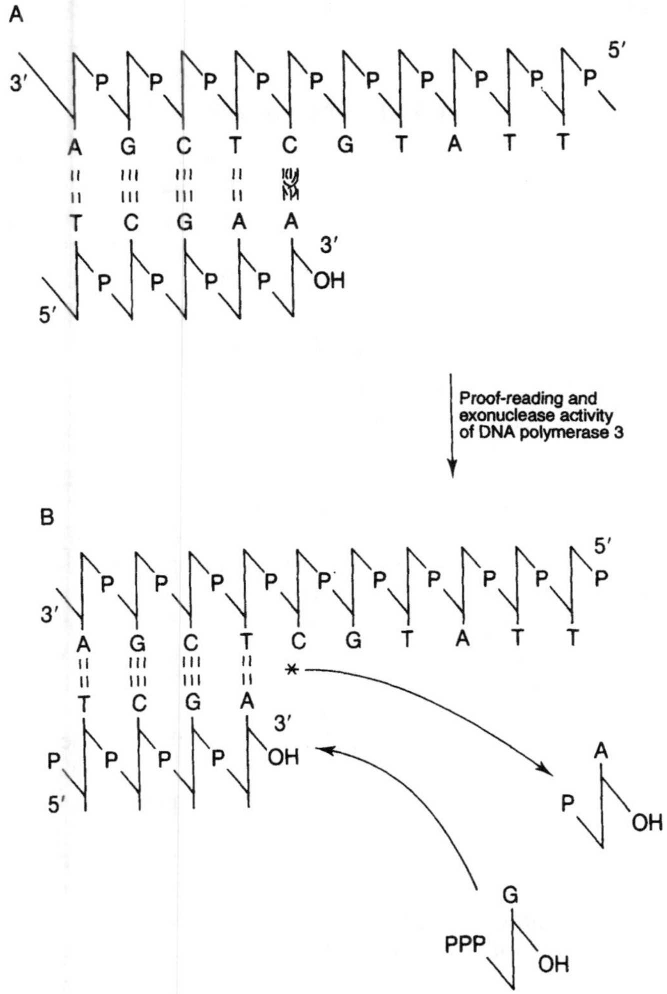

Figure 4.2 Proof-reading activity of DNA polymerase 3. (A) An incorrect adenosine has been added to the 3′ end of the growing chain. (B) DNA polymerase 3 detects the faulty base pairing, excises the incorrect residue and replaces it with a correct deoxyribonucleotide.

base-paired to its parent polynucleotide template. Its 3′ to 5′ exonuclease activity is switched on when it discovers an incorrectly matched base pair. The offending nucleotide is clipped off. Thus polymerase 3 acts as a **self-correcting enzyme** and '**proof-reads**' out errors as it goes along. A similar process is believed to occur with polymerase in eukaryocytes. The prokaryotic process is shown diagrammatically in Figure 4.2.

In parentheses here it worth noting that the dual polymerase/exonuclease activity of bacterial DNA polymerase 1 has been made use of in the technique of '**nick translation**'. A deoxyribonuclease (DNase 1) enzyme derived from the pancreas is used to break open (i.e. 'nick') a DNA double helix at random points. This leaves free 3′-hydroxyl and 5′-phosphate groups. DNA Pol 1 is simultaneously used to progressively remove nucleotides from the free 5′ end and to add fresh nucleotides to the 3′ end of the nicked chain. In other words the polynucleotide strand is 'chewed' back in the 5′ to 3′ direction and fresh nucleotides added in the same direction. If one or more of the nucleoside triphosphates being added is radiolabelled or biotinylated the DNA can be effectively tagged.

It is also worth noting at this stage that the necessity for high-fidelity DNA replication explains the existence of **thymine** in DNA but **uracil** in RNA. Thymine, as we noted in Chapter 2, possesses a methyl group at a position where uracil only has a hydrogen atom. In consequence it requires appreciably more energy to synthesise than uracil. It turns out. however, that cytosine is not only open to nitrous acid deamination to uracil (Figure 4.1) but that this may happen spontaneously by hydrolytic reaction with ambient water molecules. Such deaminated bases (Figure 4.1 shows that cytosine is not the only possibility) are recognised by specific enzymes – **DNA glycosidases** – which remove the base from the nucleotide by cleaving the glycosidic bond which links it to the deoxyribose. This leaves a hole in the base sequence usually known as an **AP (apurinic or apyrimidinic) site**. Another enzyme, a repair or **AP endonuclease**, detects the defective nucleotide and removes it entirely from the poly-nucleotide strand leaving a '**nick**' in the phosphodiester backbone. DNA polymerase 1 now comes into play. It detects the nick and its endonuclease activity works back in the 5′ to 3′ direction removing a short run of nucleotides and replacing them with fresh nucleotides. Finally the phosphodiester bond is sealed again by DNA ligase.

This complicated set of events is shown in Figure 4.3. We shall refer to some of these enzymes and processes again in Chapter 5 when we look at some of the techniques and enzymes involved in genetic engineering. But, returning to the topic of this section, it is clear that if uracil were present in DNA in the first place it would be impossible for a DNA glycosidase to distinguish between what should be present and what should not; between correct uracils and those due to the deamination of cytosine.

Nonetheless, in spite of all this ingenious molecular machinery to ensure that the genetic message is not degraded, in spite of a fidelity of **one part in a billion**, point mutations, especially transitions, are bound to creep in during the countless generations of geological time and in the vast number of DNA replications occurring in biological populations. These mutations ensure that all possible amino acid sequences are tried out over the vast stretches of evolutionary time. They are responsible for the slow, gradual and finely tuned changes of neo-Darwinian evolution.

4.1.3 Chromosomal Mutations

Chromosomal mutations take a number of different forms. The most important are gene duplications and deletions, exon and transposon shuffling. They all play important roles in molecular evolution. Let us consider each in turn.

Gene Duplications

These occur by incorrect crossing over at meiosis. Several instances of this have been thoroughly studied in *Drosophila*. If two non-sister chromatids line up somewhat imprecisely at the beginning of meiotic division, chiasmata may occur in such a way that one of the daughter nuclei resulting from the division contains two copies of a gene and the other daughter nucleus does not contain the gene at all (Figure 4.4). The latter will probably not survive, but the former, as we shall see, has great evolutionary possibilities ahead of it.

There are other possibilities. The cross-over may occur **within** a gene. This is especially possible in the multi-intron genes of eukaryotes. If this happens then one of the 'daughter' genes may contain two initially identical exons whilst the other will by missing that exon altogether. Such **intragenic** duplication may be important in the evolution of **multimeric quarternary structures**. Yet another possibility is the duplication by the same mechanisms of misaligment during meiosis of **multigene segments** of the entire chromosome.

Once a duplication event has occurred, whether intragenic, genic or multigenic, the likelihood of further such events is increased. When homologous chromosomes line up at meiotic division incorrect matches may well be made between the duplicated regions. Hence the chromosomes become mis-aligned and the possibility of further duplications and deletions enhanced.

We shall return to consider the consequences for molecular evolution of gene duplication in Section 4.2. We shall see that one of the duplicated genes is released from life and death selection pressures. Whilst its sister gene continues programming its essential product, the other is able to 'drift' towards 'fresh woods and pastures new'.

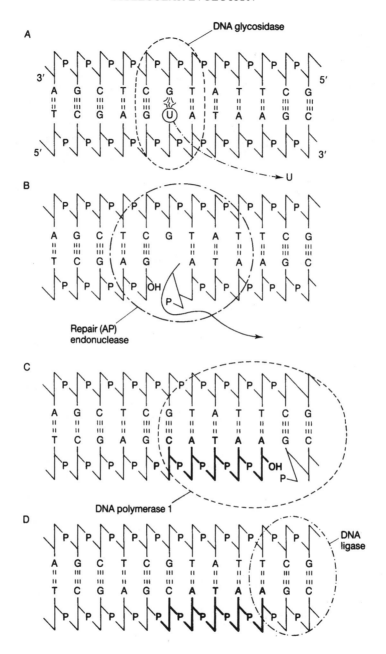

Figure 4.3 Repair of deaminative point mutations. (A) Portion of a DNA double helix in which cytosine has been deaminated to uracil (see Figure 4.1). DNA glycosidase recognises the deaminated base and removes it by cleaving its glycosidic linkage to deoxyribose. (B) A repair or AP endonuclease recognises the AP (i.e. apurinic or apyrimidinic) site and cuts the phosphodiester backbone. (C) DNA polymerase 1 now cuts back a few nucleotides from the 'nick' in the phosphodiester backbone and fills in the gap so created with fresh nucleotides. This important activity is known as 'nick translation'. (D) Finally DNA ligase seals the 'nick' by reforming the phosphodiester linkage.

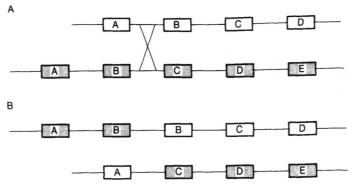

Figure 4.4 Imprecise alignment of chromatids leading to gene duplication at cross-over. (A) Chromatids misalign themselves at the commencement of meiotic division. (B) Recombination leads to gene duplication in one chromatid and gene deletion in the other. This figure could equally well represent intragene duplication/deletion of exons where the boxes represent exons and the connecting lines introns.

Exon Shuffling

We saw above that intragenic duplication may lead to duplication and deletion of exons. This phenomenon plays a significant role in protein evolution. In many cases exons program the synthesis of specific 'domains' or 'motifs' within a protein. Proteins, in this way, can be seen as 'modular' structures. It is clear that by combining and recombining exons a huge variety of different modular proteins may be synthesised. We shall meet examples of this in Section 4.2.2. It has been proposed that all the proteins in the human body are built of combinations selected from a few thousand 'domains'. The shuffling of exons thus allows new protein structures to appear and be tested by the do-or-die examination of natural selection.

Transposons

Evidence for the existence of '**jumping genes**' was first published by Barbara McClintock in the 1940s although she had to wait until the 1980s to receive recognition for her work by the award of a Nobel prize. This long wait was partly because the molecular methods required to establish her interpretation were not available until fairly recently.

There is, however, overwhelming evidence today that the genomes of microbial, plant and animal organisms are not as unchanging as classical geneticists thought. Several different types of moveable elements or **transposons** are nowadays known to exist in eukaryotic chromosomes. They range in length from a few hundred to several tens of thousands of base pairs. At each end of a transposon are near-identical nucleotide sequences

(20–40 base pairs), running in the opposite sense to each other, which are involved in the processes of excision and reinsertion (Figure 4.5). Transposons also carry a gene which programs the synthesis of a transposase enzyme which recognises these inverted repeats and is involved in the excision and reintegration of a transposon into a new site. Transposons are triggered every so often (by an as yet unknown mechanism) to 'jump' from one location in the genome to another. Finally, it is found that the mechanism by which some transposons move from one location to another involves replication.

Transposons are remarkably common. It has been estimated that up to half the mutations observed in *Drosophila* are due to the insertion of a transposon near the mutant gene. This is because when a transposon enters or exits from a chromosome it causes and leaves behind short nucleotide duplications. These are likely to affect the expression of the gene. Transposons are also deeply implicated in the **exon shuffling** mentioned above. Two transposons located at each end of a segment of DNA containing an exon may be excised together and carry the exon to a completely different stretch of DNA, perhaps inserting the

| Inverted | Transposase | Inverted |
| sequence | | sequence |

Figure 4.5 Transposon. The main structural features of a typical transposon are shown. At each end of the transposon is a nearly identical sequence of bases which are orientated in the opposite sense to each other. Further explanation in text.

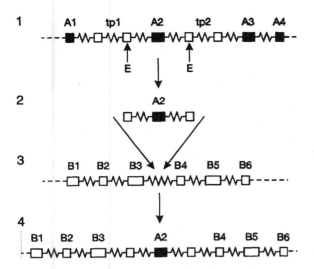

Figure 4.6 Exon shuffling. (1) Two transposons (tp1 and tp2) insert themselves into gene A, between exons A1 and A2 and between exons A2 and A3. If the end sequences of both transposons are similar the transposase may excise (E) a segment containing an exon, A2, instead of the two ends of the same transposon. (2) Excised fragment containing exon flanked by two tp end sequences. (3) As the introns in most mammalian genes are very extensive it is probable that the fragment will insert itself into another gene. (4) Insertion of the transposon fragment into gene B. As a consequence gene B now has a new exon. Exon A2 has been shuffled from gene A to gene B. Modified from Alberts *et al.* (1994), *Molecular Biology of the Cell* (3rd edn), New York: Garland.

exon into a completely different gene (Figure 4.6). The ubiquity of transposons ensures that exon shuffling plays a significant role in protein evolution. Finally, the insertion of a transposon will sometimes bring with it a regulator site and these sites sometimes act as enhancers for genes many hundred of nucleotides distant (see Section 3.3.1).

4.2 PROTEIN EVOLUTION

We are now in a position to consider the evolution of proteins. There are several different aspects to this study. First, we can look at **orthologous** proteins. These are proteins appearing in different species but which can be traced back to a common ancestor. In other words we can examine the amino acid sequences of orthologous proteins in a variety of organisms and by noting degrees of similarity

deduce a phylogenetic relationship. Second, we can consider **paralogous** proteins. These are proteins which have different functions in a given organism but have marked sequence and/or structural similarities. This suggests that we can trace their origin to a duplication event and the two genes now occupy different loci in the genome of the same organism. Of course the classification into orthologous and paralogous is not exclusive. If the paralogous duplication occurred before the split into two or more species occurred, proteins are both paralogous and orthologous. Finally, we can observe that proteins (or polypeptides) having very different primary sequences and different functions may be derived from the same mRNA transcript by differential post-transcriptional or post-translational processing.

We touched on the last of these aspects in Chapter 3, where we considered the generation of the enkephalin and endorphin neurotransmitters and some of the pituitary hormones from single 'mother' primary transcripts. We shall return to this topic at the end of this chapter. To begin with, however, let us consider the first two aspects of protein evolution listed above. We shall examine many neurobiological instances of molecular evolution in succeeding chapters (see, for instance, Box 10.1 and Section 11.8) but it is well to set the scene with some well-worked-out 'classical' examples from the general field of molecular biology.

4.2.1 Evolutionary Development of Protein Molecules and Phylogenetic Relationships

We shall see in the following chapters that the evolutionary dimension of molecular neurobiology is becoming more and more evident. It is becoming apparent that nearly all the great protein molecules upon which so much of the functioning of the nervous system depends have ancient and interesting evolutionary lineages. In this section, however, we introduce the topic by reviewing two classical examples from molecular biology: cytochrome *c* and haemoglobin. The ubiquitous respiratory electron-carrier **cytochrome *c***, whose primary structure has been determined in more than 80 different species from *Neurospora* to man, is often taken as an example of the use of sequence data to suggest

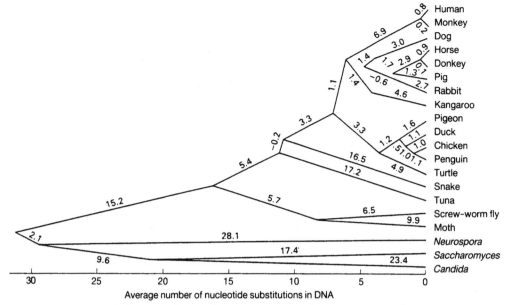

Figure 4.7 Phylogenetic tree for cytochrome *c*. The respiratory haem-containing coenzyme cytochrome *c* is found throughout the living world from prokaryotes to man. Although the number of amino acids in its primary structure varies considerably (e.g. 134 in *Paracoccus*, 82 in *Pseudomonas*, 122 in *Homo*) its tertiary structure remains very much the same. By comparing the amino acid differences in cytochromes *c* derived from a wide variety of organisms, relating them to the underlying nucleotide changes in the codons, and incorporating data on the observed rates at which different nucleotides change, it is possible to construct an evolutionary tree. This tree is shown in the figure. The numbers on its branches indicate the number of nucleotide substitutions required to join one branch point to another. As it has been estimated that in cytochrome *c* a 1% change in amino acid sequence takes about 20 million years, it is possible to gain a rough idea of the evolutionary time elapsing between the branch points. From *Molecular Cell Biology*, by C. Avers. Copyright © 1986 by Addison-Wesley Publishing Company; with permission.

phylogenetic relationships. Another favourite example is provided by the **globins**: the α and β chains of haemoglobin and the single globin chain of myoglobin. These two molecules have been studied in depth over many years and provide a valuable and well-understood introduction to the subject of molecular evolution.

An initial question which requires an answer is whether change in nucleotide base sequence is related to number of generations or simply to elapsed time. Clearly this is crucial to any phylogenetic interpretation. It turns out that, as would be expected, base sequence changes are related more closely to number of replications, that is to generation time, rather than to historical time.

Furthermore, different amino acids, different polypeptides and different proteins evolve at different rates. It turns out that Asn, Ser and Ala are the most mutable amino acids and Trp, Cys, Tyr and Phe the least. We saw the reason for this difference when discussing conservative and radical substitutions in Section 4.1.1. So far as peptides and proteins are concerned there is once again a wide range of evolutionary velocity. **Fibrino-peptides**, for instance, evolve comparatively rapidly, **histones** comparatively slowly. This, of course, has to do with the importance of the exact primary sequence for biological function. If it is very significant, as it is in the histones, little change is possible and vice versa. Taking all these factors into account, it is possible to calculate how far back in time a common ancestor must have existed. An example of the molecular phylogeny of an ortho-logous protein is provided by cytochrome *c*. A cladogram is shown in Figure 4.7.

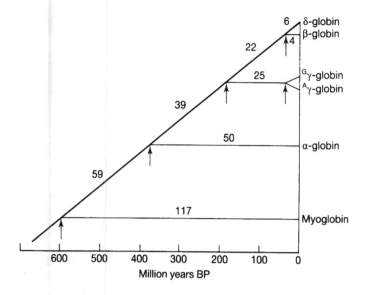

Figure 4.8 Phylogeny of human globins. Duplications are represented by arrows. The figures indicate the number of base substitutions required to transform one globin into another. Further explanation in text.

Another very important 'model' for molecular evolution is provided by the **globins**: haemoglobin and myoglobin. Probably more is known about the molecular biology of these molecules than any other protein. The globin gene may be traced back 600 million years to its origin in the earliest invertebrates. Unlike the cytochrome *c* gene its evolution is marked by a number of **duplications**. In this, as we shall see, it resembles a number of neurobiological proteins, in particular, the **visual pigments** (see Section 8.13). A phylogeny of the globins is shown in Figure 4.8.

Figure 4.8 shows that the earliest globin is believed to have been a single unit, such as myoglobin is now. The earliest duplication, occurring some 600 million years (Ma) ago, freed one gene from severe selection pressure. This is an example of the great importance of gene duplication in molecular evolution. So long as one gene continues to program the synthesis of a viable protein the other can 'experiment' creatively. Many instances of this trick have occurred in protein evolution. In the case of the globin gene another duplication occurred, as Figure 4.8 shows, some 400 Ma ago, producing the precursors of the α- and **β-globin** genes. More recently, the β-globin gene has duplicated several times (as indicated), producing genes coding for ε-globin (**embryonic**), γ-globin (**fetal**) and most recently δ-globin. The

γ-globin gene, furthermore. has duplicated (perhaps about 25 Ma ago) into $^{G}\gamma$ and $^{A}\gamma$ genes differing by only one nucleotide.

It is clear from the foregoing that the genetic representation of the globins is rather complicated. A whole **cluster** of genes – α β, ε, δ, $^{G}\gamma$ and $^{A}\gamma$ – is involved. In addition there are a number of **pseudogenes**, i.e. inactive duplicates of functional genes. Pseudogenes are symbolised by the prefix ψ, e.g. ψβ. The β-like cluster is carried on human chromosome 11. It is spread over a length of about 60 000 base pairs (60 kbp). About 95% of this length consists, however, of non-coding DNA. The gene programming for α-globin along with some pseudogenes is carried on chromosome 16.

Can we relate the globin gene structure to the exon/intron organisation of the genome which we discussed in Chapter 3? Interestingly it does seem that the exon/intron organisation of all the globin genes is very similar (Figure 4.9). It is particularly interesting to note that even though they are on different chromosomes, the organisation of the α- and β-globin genes is almost identical (although the magnitude of the introns in the β-globin cluster is much greater than that of the α-globin gene). The 'splice' junctions (where the exon is joined to the intron) are, moreover, virtually identical throughout. This observation adds weight to the argument that the presence of introns in eukaryotic

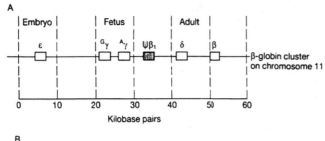

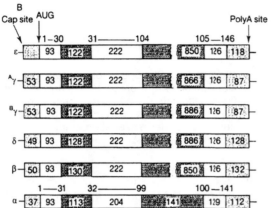

Figure 4.9 Exon/intron organisation of the human globin genes. (A) The β-globin cluster on chromosome 11. Functional genes are white; the pseudogene is stippled. (B) Exon/intron organisation of human globin genes. The transcribed sequences are compared beginning with the 5′ cap site. The numbers within the boxes indicate the number of nucleotides present in each region whilst the numbers above the boxes indicate the amino acid residues in the resulting polypeptide. The stippled boxes at the beginnings and ends of the sequences show regions which are transcribed but not translated. The second introns in the β-cluster globins are broken to align them with the much shorter intron in the α-globin gene. Exons white; introns stippled. From *Molecular Cell Biology* by James E. Darnell *et al.*, Copyright © 1986 Scientific American Books, Inc. Reprinted with permission.

genomes has something to do with crossing over, gene duplication, gene shuffling and molecular evolution.

It is likely that the globin model applies to the receptor proteins which play so important a role in animal nervous systems. One of the best characterised of these protein complexes is the **nicotinic acetylcholine receptor (nAChR)**. We have already seen that it consists of five subunits (Chapter 2). At the vertebrate neuromuscular junction the nAChR has been found to consist of two identical α-subunits and single copies of β-, γ- and ε-subunits (see Chapter 10). Although the precise nature of the subunits varies, the pentameric structure is believed to occur throughout the vertebrates and in those invertebrates which have been examined.

It is important to note, however, that, like the globins, the amino acid sequences of nAChR subunits, and hence the pharmacological properties of the receptor, differ from species to species and indeed vary from one part (e.g. central nervous system) of an organism to another (e.g. peripheral nervous system). It seems likely that the original

gene coded for an ancestral α-subunit and that, as with the globins, gene duplication brought about the subsequent appearance of β-, γ-, δ- and ε-subunits. Early in the evolutionary development of vertebrates, the gene clusters coding for peripheral nAChRs began to diverge from those coding for central nAChRs. Indeed it is now known that brain nAChR pentamers consist of several types of α- and non-α-subunit (see Section 10.1.5).

This developing insight into the evolution of nicotinic acetylcholine receptors is strengthened by the finding that nAChRs isolated from insect nervous systems resemble in size and complexity vertebrate brain nAChRs. Breer has isolated a pentameric nAChR from the cockroach *Periplaneta americana* which consists of just one type of subunit. He suggests that this homo-oligomer of units resembling the vertebrate α-subunit may resemble the ancestral nAChR from which all the others have evolved. Finally, turning from phylogeny to ontogeny, it is interesting to note that nAChRs resemble globins in yet another respect: the subunits of fetal nAChRs differ from those of the mature functioning pentamer.

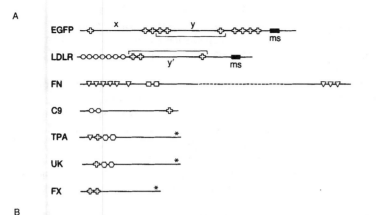

Figure 4.10 Mosaic proteins. (A) Seven disparate vertebrate proteins possessing homologous 'modules'. (B) The nature of the modules represented by symbols in (A). The nature of the disulphide arrangement in the C9-type unit is unknown. EGFP= epidermal growth factor; LDLR= low-density lipoprotein receptor; FN=fibronectin; C9=complement component 9; TPA=tissue plasminogen activator; UK=urokinase; FX= blood clotting factor X; *=active site of serine proteases; x, y and y'= homologous sequences; black rectangles labelled 'ms' represent membrane-spanning segments of the proteins. From Doolittle (1985), *Trends in Biochemical Sciences*, **10**, 233–237; © 1985 with permission from Elsevier Science.

We shall return to the nicotinic acetylcholine receptor in Chapter 10 where we examine it in detail. In addition, we shall see that many of the other receptors with which neural membranes are studded have similarly complex evolutionary biologies. The globins thus form a useful model with which to understand the molecular evolution of neural receptor molecules.

4.2.2 Evolutionary Relationships of Different Proteins

In the case of the globin and nAChR genes duplication has resulted in the production of separate molecules which have then undergone independent evolution. This is not always the case. In many instances duplication simply results in a doubling of the original protein's size. In other cases crossing over and hence duplication may occur within introns and thus, as we noted above, result in exon duplication and deletion. In yet other cases transposons may move exons around the genome. These latter instances of 'exon shuffling' produce what are known as **'mosaic proteins'**. Such proteins show evidence of the combination and recombination of exons coding for valuable 'domains'. Doolittle gives an example of an otherwise apparently unrelated group of vertebrate proteins which appear to consist of just such combinations of well-tried domains. These are shown in Figure 4.10. We saw in Chapter 2 that up to 2000 different domains have been recognised in proteins. It follows that there are nearly endless combinations and permutations and some large proteins may contain dozens of structural domains.

But, more interesting to the neurobiologist, is the emerging understanding that groups of neurobiologically important proteins are evolutionarily related to each other. They are, to use the terminology introduced in the last section, **orthologues**. We shall note several of these interesting relationships in Chapters 7–11. We shall see in

Chapter 7 that the **gap junction** protein shows considerable resemblance to the **nicotinic acetylcholine receptor** whilst the latter is related to the **GABA$_A$** and **glycine** receptors. The latter three receptors are discussed in Chapter 10. In Chapter 8 we shall note the similarities between the **β-adrenergic receptor**, the **muscarinic acetylcholine receptor**, the **substance K receptor** and the photopigment protein, **opsin**. In this case the evolutionary relationship is made more obvious by all four proteins acting through a similar 'collision-coupling' mechanism. In Chapter 9 we shall see that the vital membrane-embedded pump proteins, the **Na$^+$+K$^+$ ATPase** and the **Mg^{2+}-dependent Ca^{2+}-ATPase** show striking homologies and, finally, in Chapter 11 we shall look at the molecular similarities of the **sodium**, **potassium** and **calcium voltage-gated channels**. These are far from all the evolutionary relationships which molecular neurobiology is revealing. It seems that the evolutionary process here, as elsewhere, has modified a number of basic structures to serve somewhat different functions. It seems that many of the molecules at the basis of neurobiology fall into a small number of 'superfamilies'.

A **superfamily** of proteins is conventionally defined as a group whose primary sequences resemble each other with **greater than chance probability** and whose tertiary structure is obviously similar. The members of a superfamily are often themselves families of proteins. It has turned out to be an error to suppose that there is, for instance, *a* muscarinic acetylcholine receptor or *a* Na$^+$+K$^+$ ATPase. Instead these names denote **families** of proteins having a similar (not identical) function and resembling each other in **50%** or more of their amino acid residues. At the level of these great protein molecules we leave behind the simplicities of small molecule biochemistry where, for example, the names glucose or phenylalanine always denote the same chemical structure. The further research proceeds the more heterogeneous do neurobiological proteins turn out to be. Different cells express subtly different subtypes of a protein family. The remorseless slow drip of evolutionary change (Section 4.1) ensures that proteins and polypeptides gradually change over 'geological' time. The brain retains its complexity and functional differentiation all the way down to the molecular level.

4.2.3 Evolution by Differential Post-transcriptional and Post-translational Processing: the Opioids and Other Neuroactive Peptides

So far in this chapter we have seen how protein evolution can occur through mutation of the DNA genetic blueprint. However, as we saw in Chapter 3, the genetic instructions are open to radical modification in the processes of transcription and translation. This provides another point where evolutionary forces can operate. A neurobiological example is provided by the neuroactive peptides.

We have already met some of the neuroactive peptides when we discussed the post-translational processing of the natural opioids in Section 3.4.5. We also noted the similarity of their amino acid sequences in Table 2.2. They are found throughout the living world, from *Tetrahymena* to man. What are their evolutionary relationships?

The opioids are a group of peptide neurotransmitters and hormones which are involved in responses to stress. In Section 3.4.5 we concentrated on the two enkephalins: **met-** and **leu-enkephalin**. There are, however, a number of other related peptides: **substance P**, **α, β- and γ-melanocyte-stimulating hormone (MSH)**, **α- and β-endorphin**, **dynorphin**, **adrenocorticotrophic hormone (ACTH)** and **cortico-releasing factor (CRF)**. In addition there is a long list of other peptides which show opioid activity. These turn out to be amino acid extensions of the carboxy-terminal of met- or leu-enkephalin. Many members of this great collection of neuroactive peptides can be shown to be derived from just three different precursors: **pro-opiomelanocortin (POMC)** (265 amino acid residues); **preproenkephalin A** (263 amino acid residues); **preproenkephalin B** (256 amino acid residues). These three precursor proteins are schematised in Figure 4.11.

It is clear that the three precursors are all nearly the same length. They are examples of a class of proteins sometimes called **polyfunctional proteins** or **polyproteins** because they contain two or more copies of a bioactive protein or polypeptide, or perhaps more than one type of bioactive protein or polypeptide.

Figure 4.11 shows that the neuroactive peptides are mostly found towards the carboxy-terminal end of the polyprotein precursor. Furthermore, it has

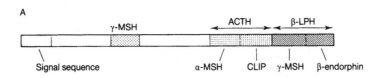

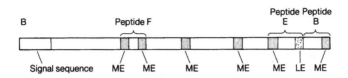

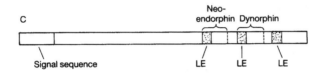

Figure 4.11 Precursor proteins of the opioid peptides. (A) **Pro-opiomelanocortin** (**POMC**). This precursor protein contains amino acid sequences for γ-MSH, ACTH and β-LPH. It is processed differently in different lobes of the pituitary. In the intermediate lobe ACTH is cleaved into α-MSH and CLIP (corticotrophin-like intermediate lobe protein), whilst β-LPH (β-lipotrophic hormone) is divided into another molecule of γ-MSH and β-endorphin. A different post-translational processing occurs in the anterior lobe. The cleavage sites are marked by pairs of basic amino acids. (B) **Proenkephalin A**. This precursor contains six copies of met-enkephalin (ME) and one of leu-enkephalin (LE). Cleavage sites are again marked by basic amino acids. The initial processing releases peptide F, peptide E and peptide B. Subsequently the enkephalins are cut free from these larger peptides. (C) **Proenkephalin B**. This slightly smaller precursor contains three copies of leu-enkephalin (LE). Cleavage sites are signalled by basic amino acids. Again initial post-translational processing releases two larger peptides – neo-endorphin and dynorphin – and leu-enkephalin pentapeptides are cut from these. All three precursor proteins have N-terminal signal sequences which allow secretion from the ribosome through the ER membrane into the ER cisterna. After Douglass, Civelli and Herbert (1984), *Annual Review of Biochemistry*, **53**, 665–715 (with permission, © 1984 by Annual Reviews. www.Annual Reviews.org); and Lynch and Snyder (1986), *Advances in Biochemistry*, **55**, 773–799.

been shown that the neuroactive 'domains' of the precursors are marked by pairs of basic amino acid residues (see Section 3.4.5) that form potential cleavage sites for trypsin-like enzymes. The legend to Figure 4.11 describes the post-translational processing which extracts various neuroactive peptides from these precursor polyproteins.

It is found that the same precursor protein occurs in several different tissues (for instance anterior and intermediate lobes of the pituitary,

hypothalamus, placenta, intestine) but undergoes different post-translational processing. This is evidently a neat way of achieving adaptive variety from a single transcript. It also emphasises the fact that genes are expressed differently in different tissues. It emphasises the fact that it is misleading to state, as is so often stated, that there are genes for this and genes for that: that, in other words, genes stand in one-to-one relationship with phenotypic characters. However, the interesting question

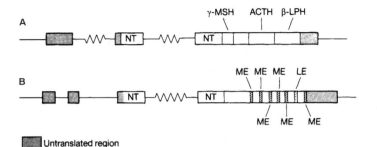

Untranslated region

Figure 4.12 Exon/intron structure of (A) human POMC and (B) preproenkephalin genes. The exons are represented by boxes, the introns by lines. NT=N-terminal end; other abbreviations as in Figure 4.11. After Douglass, Civelli and Herbert (1984), *Annual Review of Biochemistry*, **53**, 665–715; with permission, © 1984 by Annual Reviews. www.Annual Reviews.org.

in this chapter is whether the three precursors are evolutionarily related.

It seems likely that the differences in the three precursor proteins – **POMC, preproenkephalin A** and **preproenkephalin B** – arise from differences in the splicing of the mRNA chains after the excision of introns. The base-sequence similarity of the two preproenkephalins shows that they are closely related. That POMC and the preproenkephalins are related is suggested by the similarity of their exon/intron structure (Figure 4.12). The prepro-enkephalin gene, however, is located on human chromosome 12 whilst the POMC gene is on chromosome 2. Is this (like the similar situation obtaining for the α- and β-globins) a relic of some past episode involving transposon shuffling?

Comparing different species, it is found that the DNA sequences for the human, rat and mouse POMC and preproenkephalin genes are highly conserved. Taking an evolutionarily greater jump to the preproenkephalin gene of the clawed frog, *Xenopus*, sharing a common ancestor with man some 350 Ma ago, we find that whilst the exon/intron structure remains very similar and the nucleotide sequences of the exons are conserved, the nucleotide homology in the intron sequences is sometimes quite dissimilar, ranging from 36 to 91%. It is interesting to note, furthermore, that although met-enkephalin is represented there is no leu-enkephalin sequence at the carboxy-terminal end. Perhaps **met-enkephalin** is the primordial unit and **leu-enkephalin** has appeared more recently.

The opioids are far from being the only neuroactive peptides which have evolved through differential post-transcriptional and post-translational processing. The **procholecystokinin (PCKK)** polyprotein is also subject to tissue-specific processing (Figure 4.13). In the intestine a large 33-

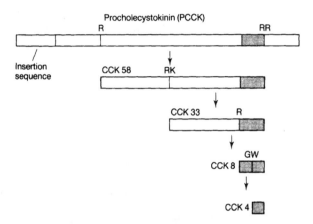

Figure 4.13 Processing the PCCK polyprotein. The 115-residue PCCK precursor is cut first to a 58-residue unit and then to the 33-residue CCK which is found in the intestine. In the brain CCK is further cut first to CCK 8 and finally, in the nerves innervating the pancreas, to CCK 4. R, K, G and W symbolise the amino acids arginine, lysine, glycine and tryptophan which mark cleavage sites. After Lynch and Snyder (1986), *Advances in Biochemistry*, **55**, 773–799.

residue polypeptide (CCK 33) is excised from the 115-unit precursor, in the brain an eight-residue unit (CCK 8) is cut out, whilst in the nerves innervating the pancreas a yet smaller fragment is used (CCK 4).

Another interesting case of differential processing is provided by the **preprotachykinin (PPT)** gene. This gene consists of seven exons of which six code for preprotachykinin. We saw in Chapter 3 (Figure 3.19) that differential resplicing of the primary transcript leads to two mRNAs: α-PPT mRNA and β-PPT mRNA. Translation then occurs in the usual way but the resulting polypeptides are then subject to further processing. The α-PPT polypeptide is processed to give substance P (SP) whilst the β-PPT is differentially processed to give both SP and substance K (SK).

It can be seen in these peptide examples that the evolutionary process is as parsimonious at the molecular level as it is at the organismic level. Just as at the organismic level gill arches become jaws and, later, auditory ossicles, so at the molecular level the same precursor protein is modified to carry out different functions. The evolutionary process works always by modifying existing structures, never by creating entirely new ones. Slight variations on what exists are seized upon by natural selection and over the generations diverge ever further from their ancestral form.

4.3 CONCLUSION

This chapter and the two which preceded it have outlined the mechanisms of, and evidence for, evolution at the molecular level. We shall meet many more examples and suggestive hints of this important theme as we proceed through the pages of this book. But we can already see that molecular biology provides a host of means by which molecules can change their structure and consequently alter their biological function. Beyond the molecular level other structures and functional complexes supervene before the integrated organism presents itself to the processes of Darwinian selection. The nervous system is arguably the most intricate of these higher-level complexes. Variations in the molecular basis of the nervous system can work through the levels of structure to induce momentous consequences. Some of these consequences, and we shall look at some in Chapter 21, are totally disabling. Infra-human organisms could not survive. Other consequences are more subtle. Nonetheless even slight variations in behaviour can have profound selective outcomes. The nervous system, as much as, if not more than, any other part of the organism, is under Darwinian control.

5

MANIPULATING BIOMOLECULES

Knowledge is power – power brings responsibility – the development of a new technology. **Restriction endonucleases** break DNA at specific points – the fragments are separated on agarose gels, allowing the construction of constriction maps and the synthesis of recombinant DNA. Some basic **cloning techniques** – plasmids – phage – cosmids – BACs – YACs. The creation of genomic and/or cDNA **gene libraries** and isolation of sought-after genes. Positional cloning. An alternative technique: amplification and extraction of DNA sequences by the **polymerase chain reaction (PCR)**. Techniques for determining the base sequence of isolated polynucleotide strands. **Expression vectors** for isolated eukaryotic DNA – prokaryocytes and their defects – *Xenopus* oocyte – use of the oocyte to express neuronal channel proteins. **Site-directed mutagenesis** allows controlled alteration of the DNA code – gene targeting and knockout genetics allows insertion and/or deletion of genes from mammalian chromosomes. Transfection and the immortalisation of cells. The engineering of transgenic mammals – dangers of drawing conclusions across species: the Lesch–Nyan syndrome. **Targeted gene expression** – the GAL4-UAS$_G$ system. **Hybridisation histochemistry**. **Gene chips** – a means of detecting the expression profiles of cellfuls of genes. **Conclusion**: the beckoning prospect of gene therapy and its ethical dimension

Francis Bacon said, long ago, that knowledge is power. The recent vast increase in our understanding of molecular biology is beginning to give us the power to manipulate living processes. As with every field of scientific-technological endeavour this is a two-way process. As we begin to be able to engineer organisms and biochemicals we achieve new insights into their structure and activity. These new insights, in their turn, feed back into the design of yet more powerful manipulative techniques. We are at the beginning of a rising spiral of biotechnological expertise.

The nervous system as much as any other part of biology is open to these new approaches. Already fundamental new understandings have been reached or are on the horizon. In this chapter a brief account of the more important and relevant of these new techniques will be presented.

Chapters 2, 3 and 4 have already laid the groundwork for the subject matter of this chapter. The manipulations of the modern molecular engineer depend essentially on an understanding of the structure and activity of the informational macromolecules we have been considering. They have to do essentially with **molecular recognition**, with the **transfer of sequence information**, with the **multiplication of specific molecular structures** by the cellular mechanisms we have discussed and, most important of all, with the **expression of information** held in nucleotide sequences where the scientist rather than the cell wants to find it. The molecular biologist is beginning to be able to reach into the information processing machinery which organisms have evolved over thousands of millions of years and tweak it towards his or her own ulterior purpose.

The key operations of genetic engineering, so far as molecular neurobiology is concerned, are the isolation of the stretches of nucleic acid (i.e. genes) that code for proteins and polypeptides of neurobiological interest and the elucidation of their base sequences. A knowledge of the relevant base sequences enables the primary structure of the resultant protein or polypeptide to be deduced. In parentheses, here, it is worth recalling the many control mechanisms between gene and finished protein which we discussed in Chapter 3: this makes prediction from nucleotide sequence to amino acid sequence an inexact science. Nevertheless, the powerful computer-based bioinformatic techniques now becoming available, together with our knowledge of an increasing number of genomes (see Chapter 6), enables reasonably reliable predictions from nucleotide sequence to protein to be made. Knowledge of primary sequence can, in turn, be used to provide at least some hints as to the conformation and function of the molecule. If the molecule is embedded in a membrane (as are so many of the most important neurobiological proteins) a knowledge of the distribution of hydrophobic and hydrophilic residues allows its disposition in the lipid bilayer to be predicted.

The methods used in the manipulation of biomolecules combine genetics and biochemistry in complex and intricate ways. The first step is often to break up DNA molecules into more manageable fragments. These fragments may then be 'cloned' to create a 'gene library'. The library has then to be screened and the gene of interest 'fished' out. It may then be sequenced and/or set to work in an 'expression vector' to produce the protein for which it codes. Alternatively a reverse transcriptase may be used to create a cDNA library from neuronal mRNA. This library again must be screened to find the gene of interest. There are numerous side-lines to this powerful and rapidly developing methodology. **DNA chips** can be used to detect the mRNAs being expressed in a given tissue, in particular circumstances and/or at a particular stage in development. The **polymerase chain reaction (PCR)** can be used to amplify minute amounts of DNA to quantities sufficient for sequence determination. In what follows we shall merely look at the essentials.

5.1 RESTRICTION ENDONUCLEASES

DNA molecules, especially eukaryotic DNA molecules, although comparatively simple in structure (as we have seen), are horrendously lengthy: the haploid human genome consists of 3.2 thousand million base pairs. For many years after the Watson–Crick breakthrough it seemed almost an impossible dream to hope to home in on an interesting gene and establish its base sequence. All this changed with the discovery in the late 1960s of **restriction endonucleases**, which cut the DNA strand at specific points into smaller, manageable, fragments. These enzymes are produced by bacteria as a defence against bacteriophages. Nowadays well over a hundred different restriction endonucleases are known and commercially available. Let us look at a few examples:

EcoR1: this is one of the best-known and most used restriction endonucleases. It is found in *Escherichia coli* RY13, hence the name. This endonuclease recognises a portion of the double helix which reads:

polynucleotide-G-A-A-T-T-C-polynucleotide
polynucleotide-C-T-T-A-A-G-polynucleotide

It then makes a very precise cut in the helix between the G and the A in each strand to give:

polynucleotide-G
polynucleotide-C-T-T-A-A

and

A-A-T-T-C-polynucleotide
G-polynucleotide

EcoR1 will cut DNA wherever this particular base sequence turns up. It can be calculated that if the four nucleotide bases appear in random order in the double helix then this particular sextuplet will occur once every 4096 base pairs (i.e. 4^6). EcoR1 is thus likely (left long enough) to cut the hundreds of millions or thousands of millions of base pairs in a eukaryotic DNA strand into much more manageable 4000–5000 bp segments.

There are a couple of other points to note about the action of EcoR1. Firstly the sequence which it recognises is **palindromic**. We have already noted the significance of such sequences in Section 3.3.1 when considering the sites which transcription factors recognise. It turns out that this is a general feature applying to nearly all restriction endonucleases. The sites on the DNA double helix which they recognise are invariably of this nature. The second important feature to note is that the polynucleotide fragments resulting from the enzyme's action have '**sticky ends**'. The A-A-T-T- end projecting from the DNA fragment will very readily join with its Watson–Crick partner.

Far from all restriction endonucleases produce fragments with 'sticky ends'. An endonuclease derived from another bacterium – *Haemophilus parainfluenzae* – known as **Hpa1** recognises the following DNA sequence:

$$\downarrow$$
polynucleotide-G-T-T-A-A-C-polynucleotide
polynucleotide-C-A-A-T-T-G-polynucleotide
$$\uparrow$$

and cuts it in the position shown by the arrows into

polynucleotide-G-T-T
polynucleotide-C-A-A

and

A-A-C-polynucleotide
T-T-G-polynucleotide

We shall see later that these so-called 'blunt-ended' restriction fragments have their own uses.

Finally many restriction endonucleases recognise not a sequence of six but a sequence of four nucleotides. An example of this is an endonuclease derived from *Haemophilus haemolyticus* – **Hha1**. This recognises the sequence:

$$\downarrow$$
-G-C-G-C-
-C-G-C-G-
$$\uparrow$$

Elementary statistics tell us that groups of four nucleotides are much more likely to turn up by chance than groups of six. They should indeed appear every 4^4, i.e. 256, nucleotides. Hence the application of these restriction enzymes results in much shorter fragments than those obtained with endonucleases recognising sextuplets.

The length of time the restriction endonuclease is allowed to act on the DNA double helix materially affects the number of fragments obtained. Only by allowing an endonuclease such as Hha1 to act on a DNA strand for a very long time will *all* the

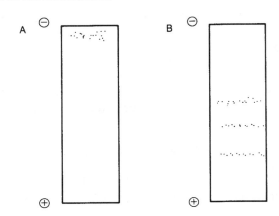

Figure 5.1 Separation of restriction fragments by electrophoresis. A suspension of restriction fragments is placed at the top of an agarose or polyacrylamide gel column (A). Application of a voltage causes the fragments to move through the gel towards the anode. The smaller fragments move faster than the larger (B). After staining for visualisation they can be eluted and analysed.

appropriate -G-C-G-C- sites be found. Only then will fragments of about 250 bp result.

5.2 SEPARATION OF RESTRICTION FRAGMENTS

Having broken the DNA strands into smaller, more manageable fragments it is next necessary to separate the mixture. This can be done by a number of methods, of which electrophoresis on agarose gels is the favourite (Figure 5.1). The restriction fragments travel unharmed (still double-helical) through the agarose gel at different rates according to their size. They can be visualised on the gel by staining and can then be eluted and sequenced. Techniques have now been developed which allow the routine determination of sequences of up to and beyond 5000 bp in length (see Section 5.12 below).

5.3 RESTRICTION MAPS

By subjecting DNA to different restriction endo-nucleases, or the same endonuclease for different durations, assortments of fragments of different lengths are obtained. By carefully examining the fragments it is possible to find overlapping sequences and hence determine the order in which

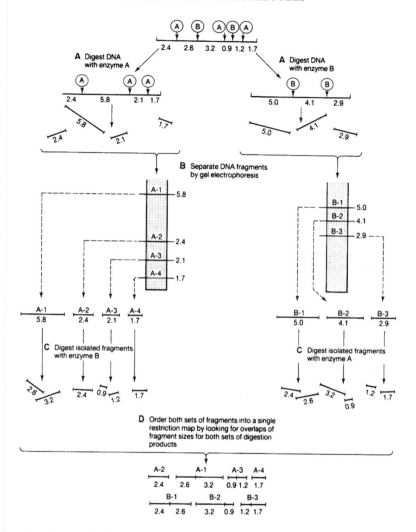

Figure 5.2 Construction of a restriction map. The DNA molecule in the figure contains 12 000 nucleotide base pairs (i.e. (2.4+2.6+3.2+0.9+1.2+1.7) kbp). It contains three recognition sites for endonuclease A and two for endonuclease B. The numbers between the recognition sites indicate kilobase pairs but these are, of course, initially unknown. (A) Digestion with endonuclease A yields four restriction fragments and digestion with endonuclease B yields three. (B) The restriction fragments are separated by agarose or polyacrylamide gel electrophoresis. The fragments are conventionally numbered from largest to smallest. (C) Each fragment is eluted from the gel and digested with the other restriction enzyme. In some cases, e.g. A1 and A3 and all three B fragments, this results in further cleavage as they contain restriction sites for the other endonuclease. (D) The two sets of fragments can now be examined for overlapping sequences and the best fit determined. The result is a restriction map which shows the sites at which the two endonucleases attack. From Becker (1986), *The World of the Cell*, Menlo Park, CA: Benjamin/Cummings; with permission.

the fragments were present in the original DNA (Figure 5.2). As the fragments themselves can be sequenced (as mentioned above) one can begin to build up the nucleotide sequence of the entire DNA strand. Alternatively (and more usually) one is able to determine on which restriction fragment a gene of interest is located. We shall outline the techniques by which this can be achieved later. The

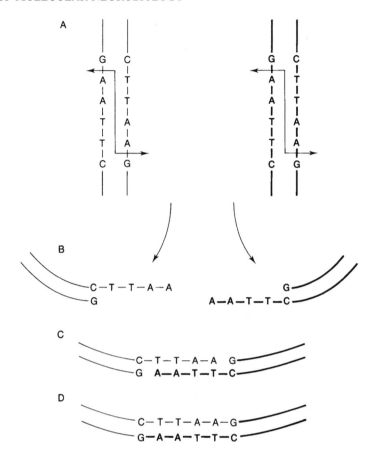

Figure 5.3 Production of recombinant DNA molecules. (A) A restriction enzyme (in the example EcoR1) cleaves DNA from two different sources. The cleavage is shown by the arrow. (B) A mixture of fragments with complementary 'sticky ends' is generated. (C) Mixing the two sets of cleaved fragments allows recombinant DNA strands to form by complementary base pairing. (D) The phosphodiester link between G and C is formed by incubation with DNA ligase. The recombinant DNA fragments will be separated from the homologous fragments which will also have been generated by subsequent cloning procedures.

upshot of this work is a 'restriction map' of the DNA, showing where different endonucleases attack and where genes of interest are located.

5.4 RECOMBINATION

The construction of restriction maps is not, of course, the only consequence of the discovery of restriction endonucleases. Of even greater importance is the opportunity to use these enzymes to combine DNA fragments from different sources. In order to do this it is of course necessary that the various DNA fragments have matching 'sticky ends'. This will be the case if the DNA obtained from different sources has been digested with the same restriction endonuclease (Figure 5.3). If so, the sticky ends will find each other by the usual processes of complementary base pairing. Addition

of the enzyme **DNA ligase** (see Section 4.1.2) will seal the union by covalent bonding.

An alternative approach is to use 'blunt-ended' restriction fragments. On their own these, of course, have no complementary base-pairing properties. However, an enzyme – **terminal transferase** – is known which is able to add nucleotides to the 3′ end of such fragments. Thus if a sequence A-A-A-A-A- - - is added to one group of fragments and a sequence of -T-T-T-T-T- - - to the other then, once again, the helices will find each other and stick by complementary base pairing. As before DNA ligase is used to seal the union. For obvious reasons this technique is called '**homopolymer tailing**' (Figure 5.4).

We are now in a position to consider the ways in which genes and their products may be amplified. There are two important techniques: **cloning** and the **polymerase chain reaction (PCR)**.

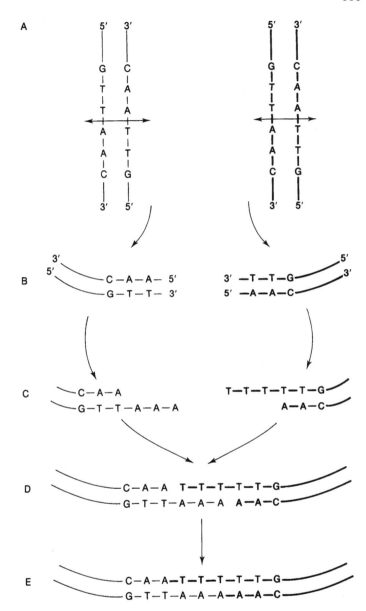

Figure 5.4 Recombination by homopolymer tailing. (A) A restriction enzyme such as Hpa1 cleaves the DNA from two different sources in the position indicated by the arrow. (B) Blunt-ended fragments are generated. (C) Incubation with terminal transferase and either dATP or dTTP adds A or T to the 3′ terminal end of the fragments. (D) The two sets of fragments are mixed and recombination occurs by complementary base pairing. (E) Addition of DNA ligase generates the missing phosphodiester bond between T and A and A and A.

5.5 CLONING

In essence 'cloning' requires that the gene of interest is spliced into a **replication** or **cloning vector**. Although the technique is much more complex and time-consuming than PCR it is still valuable and much used, especially when it is important to investigate the gene product. There are a number of different cloning vectors: **plasmid**, **phage**, **cosmid** or

bacterial and **yeast artificial chromosomes** (**BACs** and **YACs**). Let us consider each in turn.

5.5.1 Plasmids

Plasmids are tiny circlets of DNA, seldom more than 2000 bp in length, present in many bacteria and yeasts (Figure 5.5). They replicate independently of the bacterium's major DNA strand,

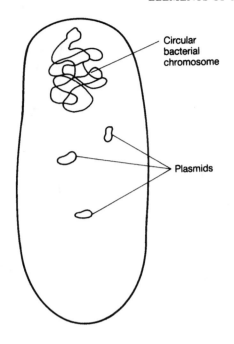

forming anything up to 200 copies. The reason for this is that they often carry the genes which confer antibiotic resistance on the bacterium. They provide excellent cloning vectors for the genetic engineer as not only do they replicate within the bacterium but they are also replicated each time the bacterium itself replicates.

Because plasmid DNA is so much smaller than the chromosomal DNA it is easily separated and purified. In the presence of Ca^{2+} plasmids are readily taken up by plasmid-free bacteria and replicated. It is clear that such an organelle provides enormous opportunity for gene cloning. All the cloner has to do is rupture a bacterium and obtain a plasmid circlet, break it open with a restriction endonuclease leaving sticky ends, provide a length of the DNA he or she wishes to clone prepared with complementary sticky ends, add DNA ligase, and reintroduce to a population of bacteria in the presence of Ca^{2+} (Figure 5.6).

5.5.2 Phage

Bacteriophages (=bacterial viruses) or phage provide an alternative to plasmid vectors. They have,

Figure 5.5   Plasmids within a bacterium.

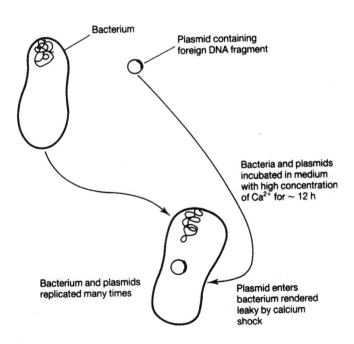

Figure 5.6   Cloning DNA using a plasmid vector.

as we shall see, some advantages. The phage injects its DNA into its bacterial host much as if it were a tiny syringe. Some of this DNA is copied by the host cell's mRNA and translated into protein in the usual way, whilst other copies of the phage DNA are incorporated into the host's DNA by its own DNA polymerase enzymes. In the case of the phage most used for cloning (**λ-phage**) the injected DNA first circularises and is then inserted into the host's chromosome (Figure 5.7).

The great advantage of phage as a vector for cloning is that it is able to carry far longer stretches of DNA than a bacterial plasmid. Plasmids, being very small, generally become unstable if DNA lengths of more than a thousand or so base pairs are spliced in. The DNA of phage vectors, such as λ-phage, being much lengthier (48 513 bp), can carry fragments up to 15 000 bp (15 kbp) with ease. Still larger fragments (35–45 kbp) can be cloned using a specially modified λ-phage known as a **cosmid**.

5.5.3 Cosmids

The preparation of a cosmid is a somewhat complex affair. In essence what is done is to make use of the discovery that λ-phage first makes its 'head' capsid and then has to find a way to package its 48.5 kbp DNA within it. A length of λ-phage DNA consists of a 35 or 45 kbp coding sequence spliced between two short stretches of 'sticky' single-stranded polynucleotide. Because these two ends are complementary they join so that the DNA forms a circle when injected into a bacterium. The join is referred to as the *cos* site. During the lytic stage of the phage's life-cycle hundreds of copies of λ-DNA are synthesised and their *cos* sites join together end to end to form a long chain or **concatemer**. In the assembly of the next generation of λ-phage a group of enzymes, the λ-packaging enzymes, recognise the *cos* sites and break the λ-DNA into appropriate segments for packaging into the phage heads. It is this feature of λ-phage's biology which is used by the gene cloner. As long as the *cos* sites are untouched, the λ-packaging enzymes will unconcernedly do their packaging job no matter what lies between the sites (Figure 5.8).

The trick used by the genetic engineer is to cut the *cos* sites from the λ-DNA and clone them in a plasmid. The plasmid is then broken open with an appropriate restriction endonuclease and a stretch of eukaryotic DNA prepared with a similar endonuclease is inserted. It is vital that the eukaryotic DNA is of the correct length: 35–45 kbp. The λ-packaging enzymes are now added and any stretches of eukaryotic DNA of the right length with *cos* sites at each end will be packaged into λ-phage. The phage is now used to infect *E. coli*. When the eukaryotic DNA–*cos* hybrid reaches the interior of the bacterium it exists and replicates as a plasmid (Figure 5.9).

This rather complicated procedure is, as might be expected, rather less efficient than using straightforward phage or plasmid cloning. Nevertheless it is invaluable if long stretches of DNA such as make up many mammalian genes or *a fortiori* two or more linked genes are to be analysed.

5.5.4 Bacterial Artificial Chromosomes (BACs)

Even the 35–45 kbp of DNA accepted by cosmids is small compared with the size of many quite modest genomes. The genome of *Caenorhabditis elegans* is, for instance, about 100 Mbp and that of *Drosophila* nearly 200 Mbp. Compared with the 3.2 Gbp genome of humans 35–45 kbp is almost vanishingly small. In order to sequence complete genomes it has thus been important to find vectors which would accept even larger stretches of DNA than cosmids. In order to clone larger stretches of DNA techniques have been developed using bacterial and yeast artificial chromosomes.

It would be inappropriate in a text of this type, whose main topic is neurobiology, to discuss in detail the methodology involved in this type of cloning. Interested readers should consult one of the molecular genetics texts listed in the Bibliography. Only a brief outline is given.

Bacterial artificial chromosomes make use of the mechanisms involved in bacterial conjugation. In *E. coli* these mechanisms are driven by a circular plasmid known as the **fertility factor (F)**. During conjugation F replicates and a copy is transferred to the conjugation partner. This process is very rapid so F can spread quickly through an *E. coli* population. In some cases the F plasmid incorporates one or more 'insertion sequences'. These are mobile DNA sequences which move between the

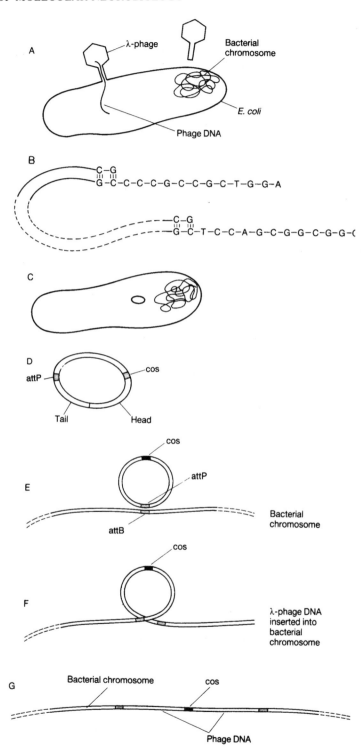

Figure 5.7 Insertion of λ-phage DNA into *E. coli* chromosome. (A) λ-phage is shown injecting its DNA into *E. coli*. (B) λ-phage DNA has two 'sticky' or 'cohesive' (cos) ends. (C) The cos ends ensure that λ-phage DNA circularises in *E. coli*. (D) Enlarged view of plasmid-like λ-phage DNA. The cos site is marked. This is followed by a sequence of genes programming the structure of the head and tail regions of the phage. Lastly there is a region labelled attP which is able to attach to a region (attB) of the bacterial chromosome. (E) attP of the phage DNA finds attB on the bacterial chromosome. (F) Union between the bacterial and phage DNA occurs. (G) The loop straightens out leaving λ-phage incorporated in *E. coli* chromosome as a 'provirus'.

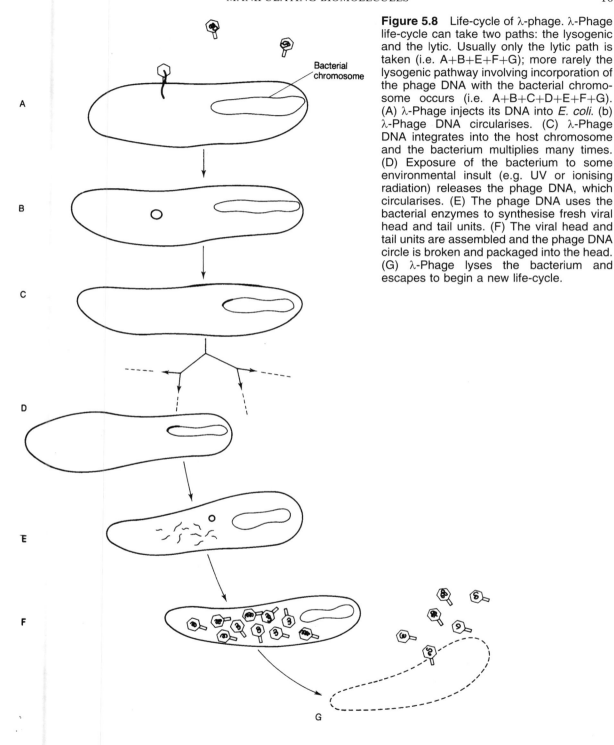

Figure 5.8 Life-cycle of λ-phage. λ-Phage life-cycle can take two paths: the lysogenic and the lytic. Usually only the lytic path is taken (i.e. A+B+E+F+G); more rarely the lysogenic pathway involving incorporation of the phage DNA with the bacterial chromosome occurs (i.e. A+B+C+D+E+F+G). (A) λ-Phage injects its DNA into *E. coli*. (b) λ-Phage DNA circularises. (C) λ-Phage DNA integrates into the host chromosome and the bacterium multiplies many times. (D) Exposure of the bacterium to some environmental insult (e.g. UV or ionising radiation) releases the phage DNA, which circularises. (E) The phage DNA uses the bacterial enzymes to synthesise fresh viral head and tail units. (F) The viral head and tail units are assembled and the phage DNA circle is broken and packaged into the head. (G) λ-Phage lyses the bacterium and escapes to begin a new life-cycle.

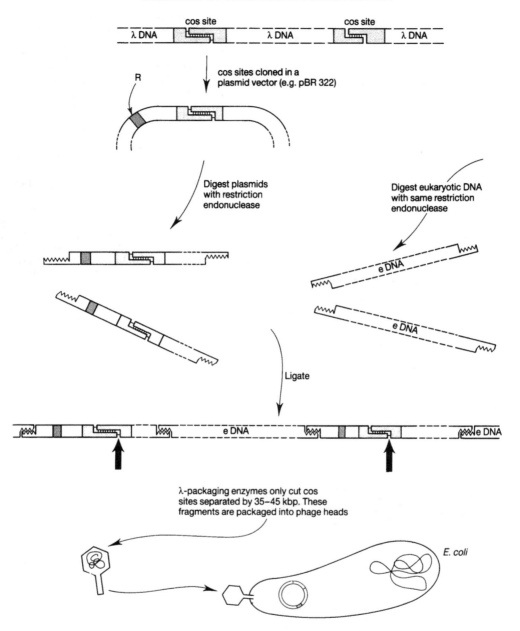

Figure 5.9 Cloning of eukaryotic genes by the cosmid technique. Cos sites are cloned in a plasmid vector which contains an antibiotic resistance gene (R). The plasmid is digested with a restriction enzyme and the same endonuclease is used to cleave the eukaryotic DNA (eDNA) which it is desired to clone. Ligase is added to the mixture and a complex set of fragments results. Some of these fragments will, however, consists of two cos sites separated by 35–45 kbp of eukaryotic DNA. The λ-packaging enzymes recognise such lengths and proceed to package them into λ-phage heads. These then infect *E. coli*. λ-Phage–eukaryote recombinant DNA circularises through its cos sites (as in the lytic phase shown in Figure 5.8) and the bacterium divides many times. The recombinant DNA circle consists mostly of eukaryotic DNA but also possesses the antibiotic resistance gene. This, as we shall see, is vital for selecting the bacteria containing the recombinant DNA.

bacterial chromosome and F plasmid. This enables crossing over to occur between plasmid and chromosome and when this happens the plasmid becomes incorporated into the bacterial chromosome. On subsequent conjugation events the integrated F factor ensures that the whole bacterial chromosome is transferred to the recipient *E. coli.*

It is not difficult to see how the genetic engineer can make use of this system. In place of the insertion sequence the geneticist introduces the DNA sequence he or she is interested in cloning. This sequence is usually about 100 kbp in length but can be up to 300 kbp, nearly an order of magnitude greater than that available with the *cos* system. BACs have been used in the analysis of the human genome (see Chapter 6).

5.5.5 Yeast Artificial Chromosomes (YACs)

To clone even longer stretches of DNA recourse is made to yeast artificial chromosomes. Yeast cells (which are, of course, eukaryotic) contain a large 6.3 kbp plasmid which, as it has a circumference of about 2 μm, is often known simply as the 2-micron plasmid. In forming an artificial chromosome the plasmid is cut open and '**linearised**' by adding yeast **telomere** DNA to its ends. If a DNA stretch containing the **centromere** is now inserted into the linearised plasmid an artificial yeast chromosome (YAC) is synthesised. When mitotic division occurs and the spindle fibres appear, they attach to the centromere and the artificial chromosome divides and segregates along with the yeast cell's other chromosomes.

Once again it is not difficult to see how the gene cloner can make use of such a system. The DNA sequence of interest to the cloner is spliced into the 2-micron plasmid and an artificial chromosome formed as described above. This is allowed to replicate with the yeast and thus yield large quantities of the sequence under investigation. Although there is often some faultiness in the replication, DNA sequences up to 1 Mbp in length can be cloned by this method. This is invaluable not only in sequencing large genomes but also in cloning genes together with their 'upstream' promoter and regulatory regions.

5.6 ISOLATING BACTERIA CONTAINING RECOMBINANT PLASMIDS OR PHAGE

The techniques described in the preceding section allow the molecular biologist to insert a fragment of DNA into a bacterium such as *E. coli* and (remembering that bacteria can divide once every twenty minutes) multiply it a billion-fold in the space of 24 hours. Similar, though not quite so rapid, multiplication can by obtained with YACs. For simplicity, however, we shall restrict ourselves to prokaryocytes in this and the following sections. The processing of YACs is essentially similar.

Not every bacterium, of course, will contain a plasmid or phage with a foreign gene. The insertion step, it will be remembered, is very 'hit and miss'. One simply adds the plasmid or the phage to a population of bacteria in the presence of Ca^{2+}. Some will be infected, others not.

How, then, can we select out only the bacteria containing recombinant plasmids? The trick is to incorporate into the plasmid or phage in addition to the gene of interest a gene or genes conferring **resistance** to one or more antibiotics, e.g. **ampicillin**, **chloramphenicol**, **tetracycline**. Growth of a bacterial population on a medium containing one or more of these antibiotics ensures that only those bacteria containing the plasmid of interest survive (Figure 5.10).

5.7 THE 'SHOTGUN' CONSTRUCTION OF 'GENOMIC' GENE LIBRARIES

If a eukaryotic genome is cleaved by a number of restriction endonucleases and inserted into cloning vectors and these in turn are grown up inside bacteria or yeast as described above, we end up with what has been termed a '**genomic DNA library**' (Figure 5.11). This is often called the 'shotgun approach'. It is as if we attacked a colony of bacteria with a shotgun full of genes and fragments of genes. Very much a hit-and-miss affair. All the eukaryotic genes should be present in the bacterial population. But if we are interested in a particular gene, and in practical circumstances we always are, there is a horrific 'needle-in-the-haystack' problem. How can we possibly find the gene of interest amongst the tens of millions of others? It has been remarked that although we may have a library, we

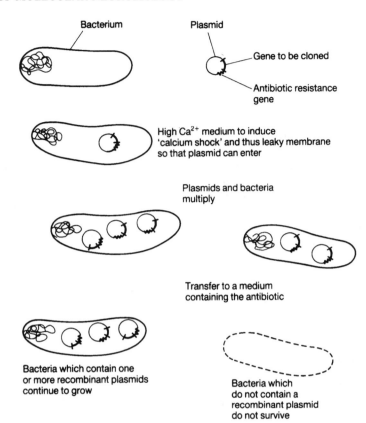

Figure 5.10 Procedure for selecting bacteria containing recombinant plasmids. Only those bacteria which possess plasmids expressing an antibiotic resistance gene survive challenge by a medium containing the antibiotic.

lack an index. Indeed it's much worse – we do not really have a library at all, just a higgledy-piggledy heap of books! Even worse still, the genes (books) may be in fragments, some parts in one plasmid, others in another, and some of the material may be just so much meaningless scribbling, not a book or part of a book at all, in other words it may be an intron or part of an intron.

5.8 A TECHNIQUE FOR FINDING A GENE IN THE LIBRARY

Let us look at a way by which we may fish out the DNA we are interested in. Suppose that we know the **amino acid sequence**, or even a short stretch of the amino acid sequence, of the protein under investigation. It will then be possible, knowing the genetic code, to synthesise a short stretch of RNA which corresponds to that sequence. It is, of course, important to bear in mind the degeneracy of the code (see Chapter 2) and hence to synthesise all the

possible RNA sequences which might correspond to the amino acid chain. Next we make this RNA stretch highly radioactive. We now have a probe with which we can fish for a gene in the library (Figure 5.12).

The next step depends on replicating the exact spatial location of the bacterial colonies in the culture dish on a nitrocellulose filter. This is done by pressing the filter down on to the culture dish. Some members of each of the colonies will become attached (Figure 5.13). When this has been done we subject the replicated colonies to **alkali digestion** which releases their DNA, binds it to the filter and, finally, opens up the double helices thus exposing their complementary surfaces. After further treatment to remove proteins and other contaminants the radioactive RNA probe is added. Given time, the probe will find its complementary surface. The longer the probe the more certain is the complementarity and the firmer the DNA–RNA duplex. The filter is now thoroughly washed to get rid of

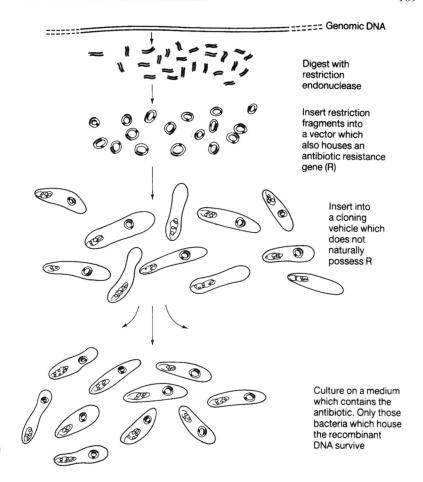

Genomic DNA

Digest with
restriction
endonuclease

Insert restriction
fragments into
a vector which
also houses an
antibiotic resistance
gene (R)

Insert into
a cloning
vehicle which
does not
naturally
possess R

Culture on a medium
which contains the
antibiotic. Only those
bacteria which house
the recombinant
DNA survive

Figure 5.11 'Shotgun' construction
of a 'genomic' gene library.

unbound probe and the radioactivity localised by
placing it on an X-ray film in a dark room. **If the
geography of the colonies on the nitrocellulose filter
exactly copies the geography of the colonies in the
culture dish** we can go back to the latter and pick
out the culture which holds the gene of interest.
This colony is re-plated and grown up, in the
knowledge that it contains the gene of interest. In
practice this procedure is repeated several times to

eliminate 'false positive' results. In the end, however, one can be reasonably certain that one has
found the plasmid containing the gene one is
looking for. This sequence of operations is shown
in Figure 5.13.

5.9 CONSTRUCTION OF A 'cDNA' GENE LIBRARY

An alternative method by which to construct a gene
library involves the use of an enzyme obtained
from some RNA tumour viruses – **reverse transcriptase**. As its name indicates, this enzyme is able
to reverse the transcription step in protein biosynthesis and synthesise DNA alongside an mRNA
template. This process is shown in Figure 5.14. The
result is first a **single-stranded complementary DNA
((ss) cDNA)**, which can be converted by DNA

NH_2--------- Tyr — Trp —Cys — Arg----------COOH   Polypeptide

5'—UA$_C^U$—UGC—UG$_C^U$—AG$_G^A$—3'   mRNA coding sequence
incorporating a
radioisotope label

Figure 5.12 Construction of an oligonucleotide probe
from a short amino acid sequence. The longer the
oligonucleotide sequence the more accurate the probe.

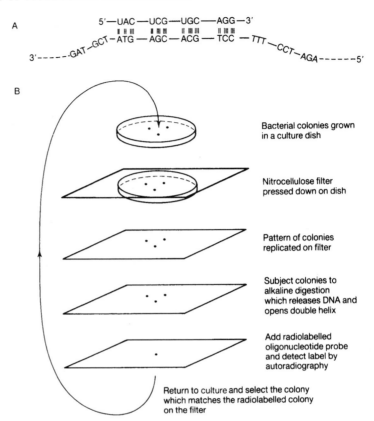

A

5′—UAC—UCG—UGC—AGG—3′

3′------GAT—GCT—ATG—AGC—ACG—TCC—TTT—CCT—AGA------5′

B

Bacterial colonies grown in a culture dish

Nitrocellulose filter pressed down on dish

Pattern of colonies replicated on filter

Subject colonies to alkaline digestion which releases DNA and opens double helix

Add radiolabelled oligonucleotide probe and detect label by autoradiography

Return to culture and select the colony which matches the radiolabelled colony on the filter

Figure 5.13 Using an oligonucleotide probe to 'fish' for a gene in a gene library. (A) The short oligonucleotide probe synthesised in Figure 5.12 is shown hybridising with a complementary stretch of single-stranded DNA. (B) The sequence of steps used to detect which bacterial colony contains plasmids incorporating the gene coding for the polypeptide of interest.

polymerase 1 into **double-stranded cDNA ((ds) cDNA)**, the two strands of which are connected by a hairpin loop; finally an **S1 nuclease** is used to remove this loop and produce a true double-stranded cDNA.

Another and somewhat more efficient means of constructing a cDNA library is to make use of a variant of the nick translation mechanism touched on in Section 4.1.2. This technique eliminates the necessity to use S1 nuclease to digest away the hairpin loop which acted as a primer in the previous technique and consequently does not risk the loss of significant sequences of cDNA. The main steps of the nick translation technique are shown in Figure 5.15. Reverse transcriptase is once again used to synthesise a complementary DNA strand alongside an mRNA template provided with an oligo(dT) primer. Next an RNase (not a DNase as discussed in Chapter 4) is used to nick the RNA in the RNA–DNA hybrid. DNA polymer-

ase 1 is then used to replace the nicked RNA strand with a DNA strand using the RNA fragments as primers.

Whichever technique is used it is clear that cDNA provides an alternative way of cloning. Instead of using restriction endonucleases to break up a genome (Figure 5.11) one can take the **mRNA from a tissue or cell population** and copy it into cDNA. This is obviously of great value if one is looking for tissue-specific proteins – as one frequently is in molecular neurobiology. A **terminal transferase** enzyme is then used to add a homopolymer tail of nucleotides to the 3′ end of each strand of the cDNA double helix and the result inserted into a suitably prepared vector. This sequence of steps is shown in Figure 5.16.

The cDNA library prepared in this way will (in theory) contain all the genes that are being transcribed at the time in the tissue examined. It will, however, be biased towards the genes which are being most actively transcribed at that time.

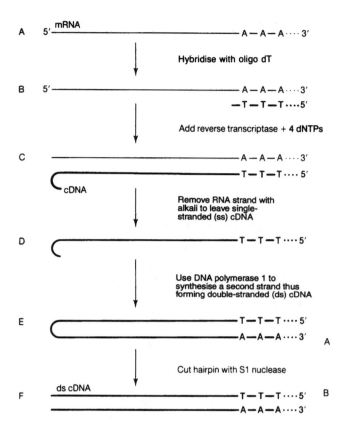

A   5' mRNA ———————————————— A—A—A····3'

Hybridise with oligo dT

B   5' ———————————————— A—A—A····3'
                                    —T—T—T····5'

Add reverse transcriptase + 4 dNTPs

C   ———————————————— A—A—A····3'
                                    —T—T—T····5'
    cDNA

Remove RNA strand with alkali to leave single-stranded (ss) cDNA

D   ———————————————— T—T—T····5'

Use DNA polymerase 1 to synthesise a second strand thus forming double-stranded (ds) cDNA

E   ———————————————— T—T—T····5'
    ———————————————— A—A—A····3'   A

Cut hairpin with S1 nuclease

F   ds cDNA
    ———————————————— T—T—T····5'   B
    ———————————————— A—A—A····3'

Figure 5.14 Action of reverse transcriptase in synthesising cDNA complementary to an mRNA strand. We noted in Chapter 3 that mRNA usually possesses a lengthy polyA tail. This is made use of in synthesising a primer for reverse transcriptase activity. It is base paired with a length of deoxythymidines (dTs). The reverse transcriptase enzyme makes use of this primer to synthesise a DNA strand alongside the mRNA. The latter is removed by alkali digestion. The remaining (ss) cDNA has a hairpin loop at its 3' end. This acts as a primer for the synthesis of a complementary strand by DNA polymerase 1. The hairpin loop is cut away by S1 nuclease, leaving double-stranded (ds) cDNA. RNA is shown light and DNA bold.

The library may include genes whose existence, and whose product, was otherwise unknown.

5.10 FISHING FOR GENES IN A cDNA LIBRARY

There are several procedures for finding the gene of interest in a cDNA library. One much used technique (as we shall see in subsequent chapters) is analogous to that described in Section 5.8. It depends on the amino acid sequence of at least a small section of the gene product being known. If this is the case, then an oligonucleotide probe can be prepared – as described in Section 5.8. The cDNA corresponding to the protein can then be 'fished' out by the hybridisation technique described in Section 5.8.

Another technique depends on the identification of a gene product by immunological, biochemical or patch-clamping techniques (for the last of these

techniques see Chapter 10). In these procedures plasmids containing cDNA are created as described in Section 5.9. It can be arranged that in a colony of bacteria each cell only receives one plasmid. The cells are then separately cloned, lysed, their plasmids extracted, the DNA double helices broken open by heating, the resultant single-stranded DNA bound to nitrocellulose filters, and finally challenged with mixtures of mRNA obtained from the tissue. Only complementary mRNA will be bound. This can later be separated, added to a cell-free protein-synthetic system and protein manufactured. It may then be possible to identify the protein by standard biochemical, immunological or physiological techniques. Having put aliquots of the cDNA clones on one side during the procedure, one can now go back to them and pick out the clone which contains the gene. The cDNA in this clone can then be amplified by recloning and its nucleotide sequence determined (Figure 5.17).

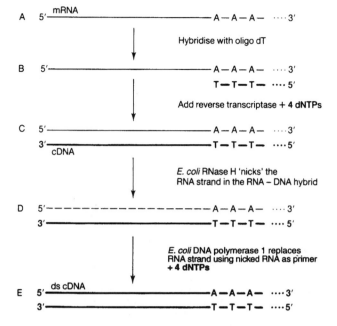

Figure 5.15 Preparation of cDNA using a nick translation technique. The first three steps (A, B and C) in the synthesis are the same as in the previous figure. In step (D) *E. coli* RNase H is used to 'nick' the mRNA strand and in step (E) DNA polymerase 1, making use of the RNA fragments as primers and the (ss) cDNA as template, synthesises (ds) cDNA from the four dNTPs. RNA is shown light and DNA bold.

5.11 POSITIONAL CLONING

This technique is primarily used in the discovery of disease-causing genes. In a sense, it can be seen as a type of 'reverse genetics'. Clinicians describe a disease condition in the human population and positional cloning can be used to run down the gene, clone it, and detect the mutation leading to the disease.

The first requirement is accurate epidemiology. The disease is traced through pedigrees in the population. This, of course, is usually the human population but similar analyses are possible with infra-human animals. Indeed the gene encoding the shaker K^+ channel in *Drosophila* was first identified by this technique (see Chapter 11). In the human instance, the epidemiologist hopes to find families whose members over several generations show the disease in question. Once the inheritance pattern is established the gene can be mapped, using conventional genetic techniques and polymorphic markers (see Chapter 6), to a particular region of a particular chromosome (Figure 6.2). These techniques allow the gene to be located in a region of the chromosome approximately 3 Mbp in length.

This region of the chromosome can then be examined by inserting it into YACs or other cloning agents to produce overlapping DNA sequences which can then by analysed. Since the elucidation of the human genome in 2001 this method is no longer necessary for tracing human genes. Instead the 3 Mbp sequence suspected of including the disease-causing gene can be examined from the genome database. The sequence can be compared with other sequences of known function held on the increasingly sophisticated database and candidate gene(s) discovered.

Once a candidate gene has been proposed it can be cloned from a family member showing the disease, using one of the techniques discussed above, and its base sequence analysed. If the analysis has been successful the gene will show one (or more) mutations compared with sequences obtained from normal family members or from the database.

5.12 THE POLYMERASE CHAIN REACTION (PCR)

It is clear from the preceding sections that amplification of DNA sequences by cloning is a tedious and time-consuming undertaking. In the early 1980s a new technique, the polymerase chain reaction (PCR), was invented by Kary Mullins. It revolutionised the determination of DNA and mRNA sequences by providing a comparatively

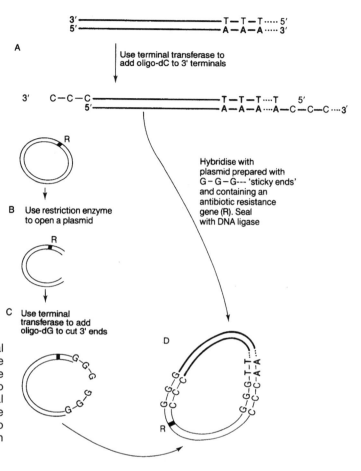

Figure 5.16 Cloning with cDNA. (A) Terminal transferase is used to add an oligo-dC sequence to the 3′ ends of the cDNA. (B) An appropriate plasmid is opened with a restriction enzyme to produce blunt-ended strands. (C) Terminal transferase is used to add an oligo-dG sequence to the 3′ blunt ends. (D) DNA ligase is used to seal the cDNA into the plasmid. Amplification can now be induced in the usual way.

simple method of amplifying nucleic acid strands many million-fold. It is much used nowadays in many areas of molecular biology, not least in molecular neurobiology.

Essentially it depends on the fact that the DNA double helix is a fragile structure only held together by hydrogen bonds between the base pairs. It is consequently easy to denature the molecule by heating it in solution to 90°C or more when the two polynucleotide strands fall apart. On cooling back to 60°C it is possible to attach short oligonucleotide 'primers' to each strand and in the presence of DNA polymerase and the four deoxyribonucleoside triphosphates synthesise a new strand alongside each of them (Figure 5.18). It is, of course, important to use a DNA polymerase that is active at high temperatures. Such an enzyme has been found in the bacterium *Thermus aquaticus*, which

lives in hot springs. The gene for this enzyme (*Taq* polymerase) has, more recently, been spliced into *E. coli* so that there is nowadays no difficulty in obtaining adequate supplies.

In principle it is easy to see that repeating the cycle of heating and cooling (and each cycle lasts about two minutes) and adding appropriate reagents allows an exponential amplification of the nucleic acid: 2, 4, 8, 16, 32, 64, 128 . . . After two dozen cycles the strands have been multiplied several million-fold. Moreover, it is not the entire DNA which is amplified but only the nucleotide sequence between the two primers. This is because the two strands of the double helix are antiparallel (see Section 2.2.1) and because DNA polymerases always act in only one direction: the 5′ to 3′ direction (see Section 3.2). Figures 5.18 and 5.19 show the first four steps of the PCR.

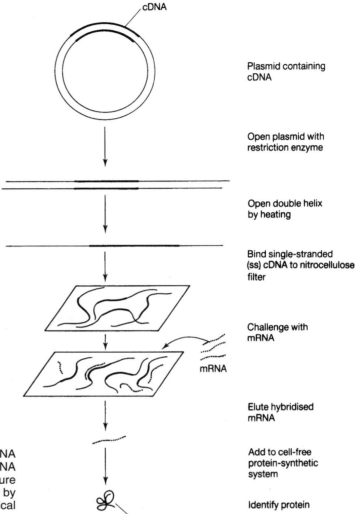

cDNA

Plasmid containing cDNA

Open plasmid with restriction enzyme

Open double helix by heating

Bind single-stranded (ss) cDNA to nitrocellulose filter

Challenge with mRNA

mRNA

Elute hybridised mRNA

Add to cell-free protein-synthetic system

Identify protein

Protein

Figure 5.17 Identification of a gene in a cDNA library by translation of complementary mRNA and identification of the product. This procedure depends on the protein being easily identified by physiological, biochemical or immunological techniques.

It is important to note in Figure 5.19 that it is only the sequence between the two primers which is subject to exponential amplification. Because of this one can also think of the PCR as **extracting** that sequence from the entire molecule, or from a heterogeneous population of molecules. It is also worth bearing in mind that *Taq* polymerase does not possess a 'proof-reading' ability (see Section 4.1.2) and hence an incorrect nucleotide is incorporated about once in 2×10^4 nucleotides. This is not a serious matter for sequence analysis as the same error is not likely to be made in each of the ten million or so strands resulting from a sequence

of PCR cycles. Finally, it is easy to see that the nucleotide sequences in mRNA strands can also be amplified if the mRNA is first transformed into cDNA by using a reverse transcriptase.

Kary Mullins first published the PCR technique in 1985 and in the succeeding five years more than 600 publications involving PCR appeared in the scientific literature. Various modifications and improvements are still being suggested so that investigations using PCR continue to mushroom. Many of these techniques are described in the laboratory guidebook edited by Michael Innis *et al.* (see Bibliography). Some allow the amplification of

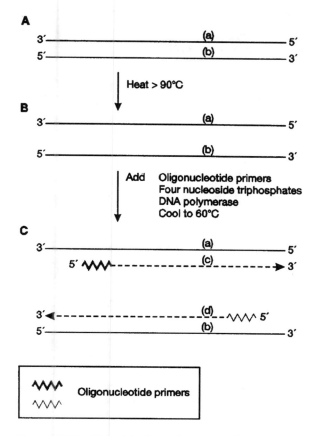

Figure 5.18 Step 1 in the polymerase chain reaction (PCR). (A) The two antiparallel polynucleotide strands of DNA are represented by lines with the direction 5′ to 3′ marked. (B) Heating a solution of DNA causes the two strands to separate. (C) After cooling to 60°C, oligonucleotide primers are attached to the two polynucleotide strands and, in the presence of DNA polymerase and the four nucleoside triphosphates, new strands are laid down (dashed lines). It is vital to note the synthesis is in each case from 5′ to 3′.

nucleotide sequences from minute tissue samples and from wax-embedded microtome sections. Such techniques are admirably suited for study of the nucleic acids in minute regions of the brain.

5.13 SEQUENCE ANALYSIS OF DNA

The most important application of recombinant DNA and PCR techniques in molecular neurobiology has (to date) been in the analysis of the primary sequence (and hence structure and function) of neurobiologically important proteins and polypeptides. We shall meet many examples of this as we proceed through this book. For these analyses to be carried through amplification of the DNA is all-important.

Several powerful techniques have been developed for sequencing DNA but they all depend on the availability of adequate quantities of DNA 'purified' by one or other of the procedures described above. The two best-known sequencing techniques are those developed by Maxam and Gilbert and by Sanger. Gilbert and Sanger shared a Nobel prize (Sanger's second) for this development in 1980. It is now considerably easier to sequence a stretch of DNA than to determine the amino acid sequence of its protein. Indeed, the process has now been automated and the total quantity of sequencing data available doubles every 18 months whilst sequencing cost per base pair halves every 18 months. The development of automated sequencing techniques thus follows a 'law' similar to that which was noted by Moore for computers. Moore's 'law' says that computers double their computing power/$ every 18 months. It follows that far more proteins are nowadays sequenced by prediction from their DNA codes than have ever been worked out by direct analysis.

Let us briefly look at Sanger's technique. The method is shown diagrammatically in Figure 5.20. It depends on the use of **dideoxynucleotides (ddNTPs)** which, lacking a hydroxyl group at the 3′ position, are unable to form phosphodiester bonds with other nucleotides. The DNA strand whose sequence is required is mated with a short primer sequence (radioactively labelled) and incubated with DNA polymerase 1, the four deoxynucleotide triphosphates (dNTPs) and a small amount of one or other of the four dideoxynucleotides. Synthesis occurs, as shown in the figure, until a **ddNTP is incorporated**. This incorporation occurs at random. Hence a random collection of different lengths of DNA are synthesised. These random lengths are then subjected to **polyacrylamide gel electrophoresis (PAGE)**. The latter technique separates DNA fragments **according to their length**. It is able to distinguish between fragments differing in length by just one nucleotide. The position of the fragments on the gel is detected by autoradiography.

As Figure 5.20A shows, four different reactions are run using the four different ddNTPs: ddATP,

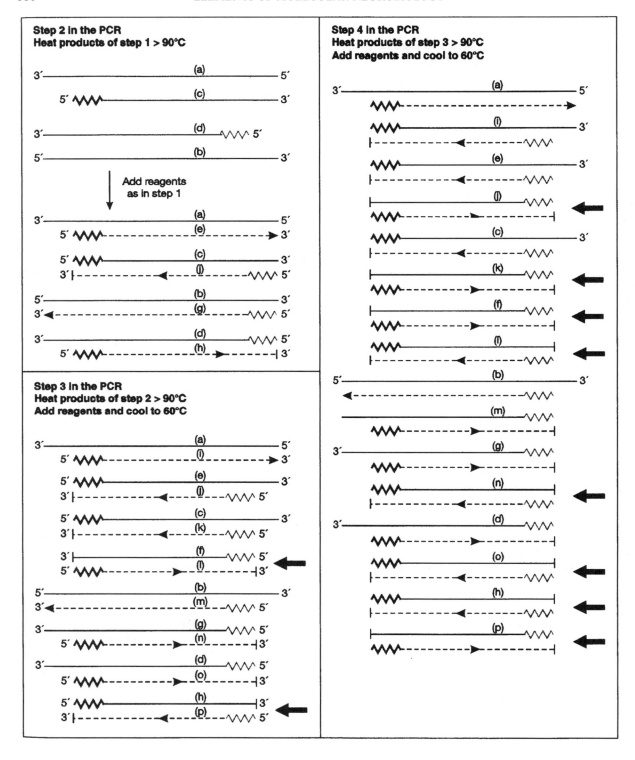

ddCTP, ddGTP and ddTTP. The reaction products are run in columns alongside each other. By carefully examining the position of the polynucleotide fragments in the polyacrylamide columns (Figure 5.20B) the nucleotide sequence can be deduced.

5.14 PROKARYOTIC EXPRESSION VECTORS FOR EUKARYOTIC DNA

In some circumstances it is possible to get a bacterial cell to manufacture the protein specified by eukaryotic cDNA. This is of importance in neurobiology where often only very small quantities of a protein or polypeptide are synthesised. It has already been emphasised that the diversity of neuronal proteins exceeds that of other tissues by anything up to five times.

In order to express the information held in the cDNA it is necessary to insert it into an **expression vector**. An example of such a vector is a plasmid, for instance pBR322, into which has been inserted the **promoter** and **operator** sequences of the *lac* **operon** from *E. coli*. If a eukaryotic cDNA sequence is inserted next to this region, it will sometimes be expressed when the inducer (in this case lactose) is presented (Figure 5.21). Up to 100 mg of pure eukaryotic protein (for instance preproinsulin) have been obtained by this technique. It should be emphasised, however, that it has not proved possible, so far, to express more than a very few eukaryotic genes in bacterial systems. Eukaryotic genes mostly have their own expression signals and also seem to depend on the environment provided by the eukaryotic cell. Indeed attempts to use *E. coli* to manufacture biomedically important proteins on a commercial scale have been fraught with difficulty and disappointment.

5.15 *XENOPUS* OOCYTE AS AN EXPRESSION VECTOR FOR MEMBRANE PROTEINS

One of the best expression vehicles for eukaryotic genes is the large oocyte of *Xenopus*, the clawed frog. This cell may be up to 1 mm in diameter. It is primed, ready to develop into a frog after maturation and fertilisation. It possesses all the appropriate transcriptional and translational machinery in large amounts and 'well-oiled' condition.

mRNA from other cells can be injected into it through a glass micropipette. These mRNAs succeed where endogenous mRNAs – probably due to inhibition by specific binding proteins – fail. The first eukaryotic mRNA to be successfully translated by this system was that coding for rabbit haemoglobin. The oocyte translated this mRNA and manufactured the protein far more efficiently than cell-free systems.

Neurobiologists are particularly interested in membrane proteins – receptors, channels, pumps, etc. The first neurobiological protein to be expressed by the *Xenopus* oocyte was **nAChR**. The mRNA was cloned from the electric organ of the electric ray *Torpedo marmorata* and then injected into the oocyte. It was shown that functional nAChR channels appeared in the oocyte membrane. The oocyte translation machinery was thus able to read the *Torpedo* mRNA, synthesise the protein, and perform the post-translational processing required to glycosylate, assemble the subunits, and insert the whole complex in the membrane.

The presence of ACh-activated channels in the oocyte membrane could be demonstrated by standard physiological techniques. Their properties seem almost (though possibly not quite) identical to their properties in the nervous system.

Figure 5.19 Amplification and 'extraction' by the polymerase chain reaction (PCR). Same symbolism as Figure 5.18. Step 2: The PCR is continued from step 1 of Figure 5.18. The four polynucleotide strands are melted apart once again at 90°C; the PCR reagents added as in Figure 5.18 and four new strands synthesised (dashed lines) alongside the template strands (full lines) in the 5′ to 3′ direction. There are now eight polynucleotide strands labelled (a) to (h). Step 3: This is where the power of the PCR begins to become apparent. After melting and re-annealing as in the previous steps we now have 16 strands (a to p). But, most importantly, two of these newly synthesised double helices (indicated by horizontal arrows) are **end-stopped**. Step 4: The same sequence of melting and re-annealing is repeated. The 16 polynucleotide strands (a to p) from step 3 form the templates for the synthesis of another 16. But note: there are now **eight 'end-stopped'** double helices. It is these end-stopped DNAs that undergo exponential amplification and are, in consequence, said to be 'extracted' from the original DNA.

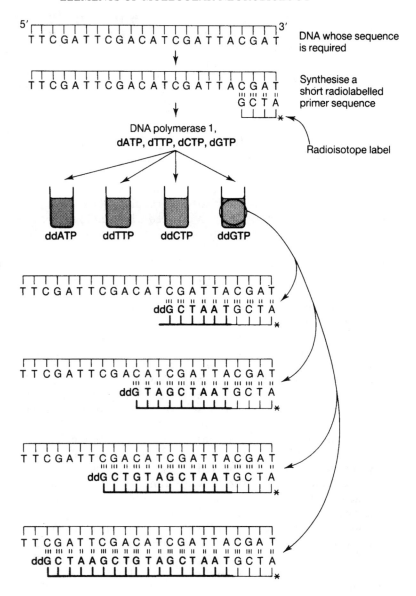

Figure 5.20A

Many other chemical and voltage-activated channels have subsequently been expressed in *Xenopus* oocyte. These include **GABA**-activated channels from the chick brain; **ACh**-activated channels from cat muscle; **serotonin**-, **neurotensin**- and **substance**-P-activated channels from rat brain; **kainate** and **glycine** receptors from bovine retina; **glycine**-, **GABA**- and **serotonin**-activated channels from human fetal cerebrum, and voltage-activated **Na+** and **K+** channels from rat and human fetal brain and cat muscle.

It is clear that the *Xenopus* oocyte is an extremely valuable system for molecular neurobiology. It enables one to transplant, so to speak, molecular entities from regions (brains, retinas, etc.) where they may be exceedingly difficult to study into a

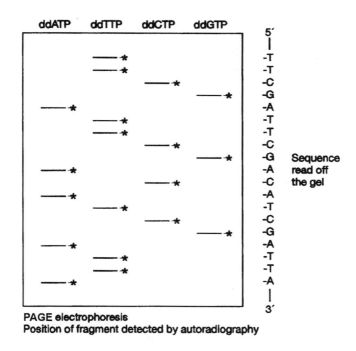

Figure 5.20 Sequence analysis of cloned DNA. (A) The unknown DNA sequence is shown at the top of figure. A short radioactively labelled primer is synthesised against its 3′ end. A reaction mixture of the four deoxyribonucleotide triphosphates and DNA polymerase 1 is prepared in four reaction vessels. A different dideoxynucleotide triphosphate (ddNTP) is introduced into each reaction vessel. The synthesis is allowed to proceed. Only reactions from the right-hand vessel containing ddGTP are shown. A complementary DNA strand is synthesised alongside the template DNA until a ddG is incorporated. This inhibits any further elongation. As ddG is incorporated randomly a series of fragments of different lengths results. Similar reactions are occurring in the other vessels with the other ddNTPs creating different chain lengths. Finally the DNA fragments are eluted and separated by PAGE electrophoresis (see part B). (B) The shortest sequences travel furthest. The positions of the fragments can be detected by autoradiography. The sequence of the original DNA can thus be read off the gel. Modified from Alberts *et al.* (1987), *Molecular Biology of the Gene*, Menlo Park, CA: Academic Press.

robust system where they can be investigated at leisure. But, note, it is not the mature structure which is transplanted, only the genetic blueprint for its manufacture. The manufacturing itself is done by the biochemical machinery within the oocyte.

The genetic blueprint can, furthermore, be slightly altered by **site-directed mutagenesis** (to be outlined in the next section). This allows the neurobiologist to direct the synthesis of slightly different channel proteins. The experimenter can thus investigate structure–function relationships: how much, for instance, does the alteration of one amino acid at a known and specific point in the channel affect its physiological properties? Finally, the oocyte system can be used in a sense back-

wards: that is, it can be used to express an unknown mRNA. If physiological investigation of the oocyte membrane subsequently shows it to possess a rare channel then one or other of the techniques described in the preceding sections allow for the amplification and hence reading of the mRNA base sequence, and determination of the channel protein primary sequence.

5.16 SITE-DIRECTED MUTAGENESIS

Recombinant DNA technology also enables us to carry out another instance of '**reverse genetics**'. Instead of finding a phenotypic change and then looking to see which genes have caused the change

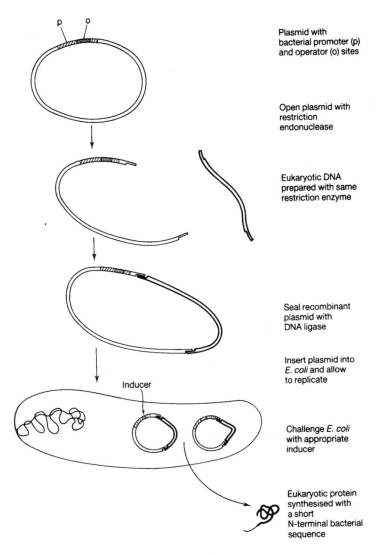

Plasmid with
bacterial promoter (p)
and operator (o) sites

Open plasmid with
restriction
endonuclease

Eukaryotic DNA
prepared with same
restriction enzyme

Seal recombinant
plasmid with
DNA ligase

Insert plasmid into
E. coli and allow
to replicate

Challenge E. coli
with appropriate
inducer

Eukaryotic protein
synthesised with
a short
N-terminal bacterial
sequence

Figure 5.21 Expression of a eukaryotic gene by an *E. coli* plasmid. The promoter and operator sequences of *E. coli*'s *lac* operon (see Figure 3.17) have been inserted into a plasmid. The plasmid is cut with a restriction enzyme and a similarly cleaved fragment of eukaryotic DNA is inserted and ligated. The recombinant plasmid is inserted into *E. coli* and allowed to replicate. In response to the appropriate inducer (in the case of the *lac* operon, lactose) the operon is switched on and the message in the eukaryotic DNA translated into protein. The short length of bacterial polypeptide at the N-terminal end of this protein can usually be removed.

and how, we can proceed in the reverse direction: we can engineer a known change in a gene and then observe the phenotypic result.

One of the most interesting and powerful of these techniques involves the insertion of one or more altered nucleotides into a gene (Figure 5.22). In this way one can arrange matters so that the gene codes for a protein with **one or more unusual amino acids at specific points in its primary sequence**.

In order to achieve this result one has first to determine the base sequence of the amplified DNA molecule. It is then possible to synthesise an oligonucleotide of some 10–15 bases, complementary to the region of interest, but with **a mismatch** where a new codon is to be inserted. This slightly inaccurate oligonucleotide is then mixed with single-stranded DNA from the cloned gene and provided the hybridisation conditions are correct complementary base pairing will take place. The mismatched base pairs will form a tiny loop in the middle of the oligonucleotide sequence. DNA polymerase 1 will then synthesise a complementary strand to the rest of the plasmid using the **mismatched sequence as a primer**. The plasmid is then inserted into *E. coli*, where it will replicate in the usual way. The outcome of the replication is two different types of plasmid: a '**wild-type**' and a **mutant**.

Finally, the mutant and the wild-type have to be separated. This can be done by creating a single-stranded plasmid by alkali denaturation and then using a radiolabelled oligonucleotide identical to the inserted mutant sequence. This will find and label the mutant plasmid. But it is likely (especially if only one or a very few base changes have been made) to find the wild-type as well. However, if one begins to raise the temperature the wild-type begins to lose its slightly mismatched probe sooner than the perfectly matched mutant. Thus the mutant can be detected and its DNA retrieved. It is usually sequenced to make quite sure that it contains the desired mutation. If all is well the mutant DNA can be amplified by cloning in the usual way and then introduced into an expression vector. The structure or (more usually) the function of the slightly altered protein can then be examined.

5.17 GENE TARGETING AND KNOCKOUT GENETICS

Most of the classical work on molecular genetics has been done on microorganisms such as *E. coli* and phage. It is comparatively easy to study genetic defects in these organisms because of their relatively simple genome, rapid generation time and the large populations which can be maintained in the laboratory, indeed in single Petri dishes. Multicellular forms such as *Caenorhabditis elegans* and *Drosophila melanogaster* can also be maintained in reasonably large numbers and with fairly rapid generation times (although nowhere near as large or as fast as microorganisms) and have been basic to much genetics. When, however, we come to the mammals, generation times lengthen and practical

considerations reduce laboratory population sizes dramatically. Yet, in our usual self-regarding way, it is mammalian genetics which is of most interest to us. The 'lower' organisms are all well and good for working out general principles, but the specific processes which operate in ourselves and the other mammals need to be analysed in mammalian systems.

Gene targeting and knockout genetics have provided a means to do just this. They enable the geneticist to insert a **novel gene** in place of the target gene or to '**knock out**' or more subtly alter (by a few base pairs) the target gene and thus determine its function. As both techniques have had a significant impact on studying the genetics of the mammalian nervous system (see Chapter 18 and elsewhere) let us see how they work.

In essence the first technique involves transferring a gene of the geneticist's choice into the mouse's (or other organism's) genome. The transference of DNA from one organism to another has had a long history. It started with the transference of DNA purified from tumour viruses into mammalian cell cultures. Because this process mimicked the normal infection route by which a virus entered a cell it was named **transfection**. There are a number of experimental techniques by which transfection can be accomplished. It is possible, for instance, to use an extremely fine pipette to inject the DNA directly into the nucleus of the recipient cell. In most cases the transfected DNA is found to insert itself into, or recombine with, host DNA at random integration sites distributed throughout the genome. This is known as **heterologous** or **random** recombination.

Because (in these pioneering experiments) tumour virus genes were transferred, the culture cells were transformed into continuously dividing cancer cells. They were, in other words, **immortalised**. This, in fact, has proved very useful in studies of the nervous system. We have already noted that, in general, neurons do not divide in the central nervous system. Introducing tumour viruses by way of these techniques has, however, allowed tissue culture laboratories to establish continuously dividing cultures of neurons *in vitro*.

Transfection is classical work; the trick which allowed the development of gene targeting and knockout genetics was the development of **homologous**, rather than random, recombination. It was

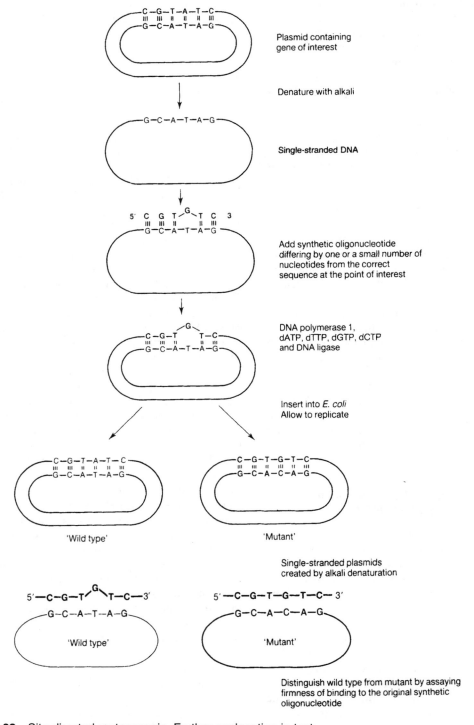

Figure 5.22 Site-directed mutagenesis. Further explanation in text.

found, first of all, that if a number of copies of a DNA sequence were transfected into a cell nucleus they lined up end to end, 5' to 3', in a very regular way, to form lengthy ropes. This immediately suggested that essentially the same thing might happen if the transfected DNA possessed the same terminal sequences as a gene in the cell's own nuclear DNA. This indeed turned out to be the case: in certain rather rare instances a transfected gene (known in these experiments as the targeting vector) lined up alongside its complementary wild-type gene and was inserted into the chromosome (Figure 5.23).

The next question (of course) was to find in which cells this rare event was happening. In order to answer this question markers which could be detected in cell culture were inserted into the targeting vector. These markers can take several forms. In Figure 5.24 the markers are the 'neo-mycin resistance' (neo^r) gene without its promoter so that its expression depended on the targeted gene's promoter and (at the end of the base sequence) the thymidine kinase (tk) gene from the herpes virus. Whilst the neo^r is inserted into the centre of one of the targeting vector's exons, tk is attached at the end (Figure 5.24). It is known that terminal sequences are usually lost in homologous recombination. When the targeting vector finds its homologue on the culture cell's chromosome the integration events slot the homologous DNA into the host chromosome and eliminate or at least inactivate the tk. In the random case the entire sequence, including tk, is inserted into the host chromosome.

All that is now necessary is to use the markers to isolate the homologously transfected cells. All the cultured cells are exposed to solutions containing two lethal agents: a neomycin analogue, G418, lethal to any cell which does not carry a functioning neo^r gene, and gancicloviran, an agent lethal to cells carrying tk. This treatment, known as **positive–negative selection (PNS)**, only allows cells with the homologously recombined gene to survive.

There are various other ways (including PCR) of selecting rare homologous recombinants from populations of cells most of which are hetero-logous. These are explained in the standard texts and need not detain us here. It is clear, however, from the above account that if the target gene's DNA can be isolated and amplified by one of the

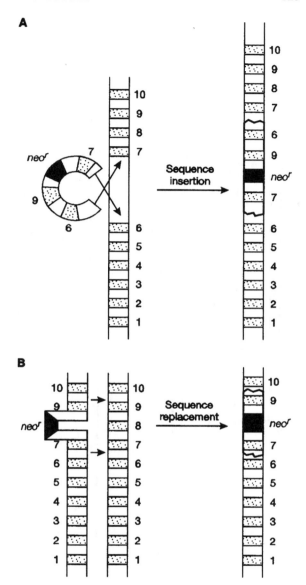

Figure 5.23 Homologous recombination. The gene under investigation is cloned. Generally speaking it will consist of both exons (stippled) and intervening introns (light). The figure shows two out of a number of different homologous recombination techniques. (A) The cloned sequence (exons 6, 7, 8 (containing neo^r) and 9) is inserted. This leads to a partial duplication of the gene. (B) Exons 7, 8 and 9 are replaced by 7, 8 (containing neo^r) and 9. Further explanation in text. After Capecchi (1989), *Science*, **244**, 1288–1292, reproduced by permission of the American Association for the Advancement of Science, 1989.

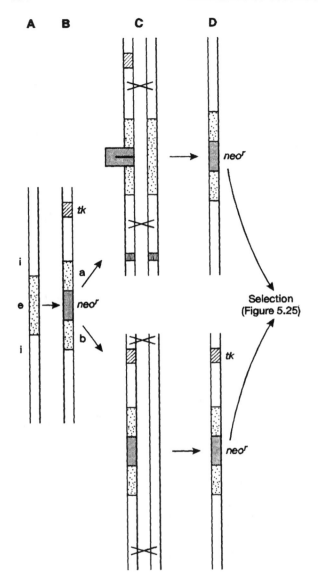

Figure 5.24 Homologous and heterologous recombination. (A) The gene of interest is cloned. It will consist of both exons (e) (stippled) and introns (i). (B) The markers *tk* and *neo*[r] are inserted (see text). (C and D) The gene is now transfected (by insertion) into mouse ES cells. There are two possibilities: (a) Homologous recombination occurs. The *tk* marker falls outside the integration site but the rest of the DNA (including *neo*[r]) is incorporated in the host cell chromosome. (b) Heterologous recombination occurs. The *tk* gene is incorporated along with the gene under investigation. The two cases shown in (D) are exposed to positive–negative selection (Figure 5.25) so that only the homologous

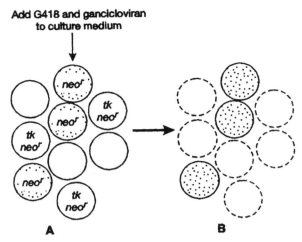

Figure 5.25 Positive–negative selection. (A) G418 and gancicloviran are added to the culture medium of cells containing the two cases of Figure 5.24D. (B) Only the cells with neomycin resistance (*neo*[r]) survive. The others, as indicated by the dashed lines, are eliminated. Further explanation in text.

techniques discussed earlier in this chapter, then homologous recombination provides a means of inserting a modified copy back into the chromosome. In the case described above the insert containing, as it does, *neo*[r], is lethal when homozygous. This is because *neo*[r] disrupts transcription of the exon. The gene, in other words, is '**knocked-out**'. Other techniques allow more subtle alterations to be made to the homologous insert so that its function can be studied in detail.

So far, however, we have only seen how genes may undergo homologous recombination in cells cultured in a Petri dish. Biologists, however, are interested in what these recombined genes do (or, rather, do not do) in the multicellular body, most often that of a mouse. Neuroembryologists, for instance, as we shall see in Chapter 18, wish to know what happens if one or other of the **homeobox** genes is altered or inactivated. Neuropathologists may be interested in the activity (or lack of

recombinants survive. Any transfected DNA which remains uncombined, either homologously or heterologously, is also eliminated. After Capecchi (1989), *Science*, **244**, 1288–1292, reproduced by permission of the American Association for the Advancement of Science, © 1989.

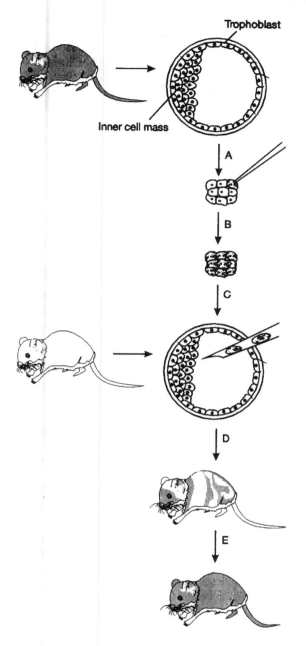

activity) of a mutant gene (Chapter 21). To get the recombined gene into the developing mouse is a tricky and delicate operation (Figure 5.26). First, the cultured cells in which the homologous recombination has been effected must be mouse **embryonic stem cells (ESs)**. Second, these cells must be introduced into the blastocyst stage of the developing embryo. Third, the chimeric mouse pups must be distinguished from those in which the mutated ES cells did not 'take'. Fourth, appropriate mating must be scheduled to ensure that the altered gene is present in all the cells of the mouse (especially the reproductive cells), not just some, as in the chimera. Fifth, a breeding programme must be established to yield animals homozygous for the mutation and this must be confirmed by examination of their DNA. Sixth, and finally, the anatomy, physiology and biochemistry of these '**knockout**' and/or '**transgenic**' mice must be investigated. It may then be tentatively concluded that any defects found are due to a lack of the missing gene's product or to the non-wild-type product of the transgene. It can be seen that the insights obtained from gene targeting do not come lightly. Much hard and careful work is required.

We shall see in Chapter 18 that the ability to alter, insert and knock out genes has been of great help in analysing the significance of the genes controlling early development of the nervous system. The technique has also been applied to create transgenic animals (mostly mice) which mimic human congenital disease. Although the creation of animal models is very promising and in some cases the only way ahead, caution, as ever, is indispensable. Mice and humans are, after all, very different organisms and this difference undoubtedly reaches down into their respective biochemistries and molecular biologies. A neurological example highlights this need for caution.

Lesch–Nyan syndrome is a rare human X-linked neurological disorder characterised by mental

Figure 5.26 Engineering transgenic mice. (A) Blastocyst is obtained from three-day pregnant mouse and cells from the inner cell mass isolated. A cell is isolated, transfected and homologous recombination confirmed by positive–negative selection or other techniques. (B) The transgenic cells are cloned by standard procedures. (C) Cloned cells from the transgenic culture are injected into host mouse blastocyst. It is standard practice to use mice of different coat colours (as indicated in the figure) so chimeric mice can be easily confirmed by observing their striped or patchy coats. (D) Breeding protocols ultimately produce mice homozygous for the transgene. After Zimmer (1992); reproduced with permission from the *Annual Review of Neuroscience*, **15**, 115–137, © 1992, by Annual Reviews. www.AnnualReviews.org.

retardation and compulsive auto-destructive behaviours. The mental retardation is apparent by the
age of three to four months and by the age of two
years all the diagnostic symptoms are present. It is
known that the condition is caused by mutation of
the gene for an enzyme: **hypoxanthine guanine
phosphoribosyltransferase (hprt)**. Because no treatment was available for this devastating complaint it
was thought that the creation of a mouse model
for the disease might allow the development of
successful therapy. Accordingly male mice in which
the *hprt* gene had been knocked out were
engineered. Surprisingly, these transgenic animals
showed none of the symptoms which are so
obvious in humans. Closer biochemical investigation did, however, show that the forebrain levels of
two neurotransmitters, dopamine and serotonin,
had been reduced. This is intriguing as it is found
that in human Lesch–Nyan syndrome the same two
neurotransmitters are affected: but in this case
whilst dopamine is reduced, serotonin is elevated.
This serves to highlight the differences between
mouse and man: clearly the *hprt* gene is affecting
the same synthetic pathways, although in one case
in the opposite way, yet the behavioural outcome
of this influence, when multiplied up through the
two biochemistries and physiologies (especially
neurophysiologies), is very different.

5.18 TARGETED GENE EXPRESSION

The ability to switch genes on and off at different
stages in development or in different tissues and/or
cells is of outstanding value in any attempt to
analyse the development or working of the nervous
system. Techniques which allow this to be done are
just becoming available.

Two major gene-targeting techniques have been
developed. Both depend on controlling promoters
(see Section 3.3.1). In the first case, discussed more
fully in Section 19.3, and called 'enhancer trapping', a transposon is inserted close to the enhancer
sequence of a gene. As we noted in Section 3.3.1
enhancers are frequently tissue- and/or cell-specific.
A 'marker' gene, for example β-galactosidase, is
inserted into promoterless transposons, and the
transposons inserted into the genome of the
organism under investigation. Some of these engineered transposons will integrate themselves into a
chromosome segment influenced by an active

enhancer and their marker genes will consequently
be expressed. The protein product of the marker
gene can then be detected by histochemical or other
techniques. As the whole system, enhancer, promoter, transposon (containing marker gene), will
be inherited each time the cell divides this technique
provides a valuable means of tracing cell lineages.
As such it has been much used in tracing cell
lineages in the CNS, as we shall see in Section 19.3.

In the second case, enhancer trapping is used to
direct a mutated gene to a particular tissue or cell
type. This technique allows activity of specific cells
or cell types in the nervous system to be altered and
thus questions about their normal function to be
answered. This second technique developed from
the discovery in the 1980s that the yeast transcriptional activator GAL4 can be inserted, in the form
of a transposon, into the *Drosophila* genome.
GAL4 acts on a promoter, UAS_G, to transcribe
any attached gene. The trick developed to target
gene expression is to insert UAS_G fused to the gene
of interest into one set of flies and GAL4 into
another. When the flies are crossed GAL4 will
activate UAS_G and the gene of interest will be
transcribed. As transposons insert randomly into
the genome a very large number of GAL4 *Drosophila* lines can be developed. In many of these lines
the GAL4 expression will not be cell- or tissue-
specific but in some cases this will be the case.
When these are found they can be crossed with flies
containing UAS_G fused with the gene under
investigation. The latter gene will then be transcribed. This technique has been used to investigate
the effect of inactivated K^+ channels in motor
neurons, muscles and photoreceptors (see Box 14.2)
and to examine the effect of inactivating the
mushroom body synapses on olfactory learning
and recall (Section 20.4).

5.19 HYBRIDISATION HISTOCHEMISTRY

So far in this chapter we have been considering
some of the ways in which our understanding of
molecular biology is enabling us to manipulate
nucleic acids and the proteins for which they code.
In this section we shall look briefly at a technique
which uses this manipulative ability to investigate
the anatomy of the central nervous system. This is

the technique which has come to be known as hybridisation histochemistry.

We have already noted that the brain is the most heterogeneous of the body's organs. Neuroanatomists, for instance, recognise more than fifty distinct cell types in a structure as comparatively simple as the retina. We have already noted that the brain itself expresses upwards of 125 000 different mRNAs. These messengers specify the multitude of different molecular structures, the neuropeptides, the transmitter-related enzymes, the receptor molecules, etc., which characterise different regions, and different cells, of the brain. Hybridisation histochemistry aims to detect these different mRNAs *in situ*. The technique thus complements other powerful histological techniques such as histo- and cytochemistry, immunohisto- and immunocytochemistry, etc.

In essence, *in situ* hybridisation histochemistry involves the production of RNA probes complementary to the mRNA being expressed in the cell. These mRNA strands are produced by the cloning or PCR techniques discussed in the preceding sections of this chapter. They are then applied to histological sections (frozen or paraffin wax) of the brain or other parts of the nervous system. If the probes find complementary mRNA in the tissue section they attach by Watson–Crick base pairing; if no complementary mRNA is present in the tissue they can by washed out of the section.

The probes must, of course, be attached to an entity which can be visualised in the microscope. There are many ways in which this can be done. One of the first but still one of the most popular techniques is that of nick translation (see Section 4.1.2). In the majority of cases the entity attached is a radioisotope: ^{35}S, ^{32}P or ^{3}H. The location of the probe, after unbound probe has been washed off, can then be located by autoradiography. Radioisotopes are not, however, the only markers available. Various other means of labelling the RNA probe have been tried including enzymes, fluorochromes and mercury. Perhaps the most promising non-radioactive marker is **biotin**. This small molecule can be attached to the RNA probe by nick translation and detected in sections prepared for both the light and the electron microscope. It should be noted, however, that the procedure is complex. Biotin cannot be visualised on its own. The techniques of immunocytochemistry are used to conjugate silver or gold particles to the biotin and it is the latter which is ultimately detected in the section.

This short résumé serves to indicate latent power of the technique. Needless to say it is complicated and time-consuming. But, used with caution and in conjunction with other histo- and cytochemical techniques, *in situ* hybridisation promises the development of a truly functional molecular neuroanatomy. As we shall see in Chapter 18, it has been used to good effect in detecting the mRNA transcripts of genes active in early development. Striking images have been obtained of the expression of gap and pair-rule genes in *Drosophila* embryos and the technique has also been employed to visualise the distribution of a large family of POU-domain transcription factors in mammalian central nervous systems.

5.20 DNA CHIPS

In the preceding section we looked at how hybridisation histochemistry enabled the investigator to determine which mRNAs were being synthesised in cells at different times and in different circumstances. DNA chips complement this technique but are in many ways more powerful. Although they lack the anatomical precision of the hybridisation method they more than compensate by the number of different mRNAs they can detect in a sample.

DNA chips consist of a square of glass or nylon about the size of a microscope cover slip on to which samples of DNA are laid out in geometrical array (Figure 5.27). The formation of these arrays is automated so that the DNA is delivered in microdroplets at specific points on the glass and dried and treated so that each droplet adheres to its specified position. Many thousand droplets can be applied to a chip.

The DNA may be cDNAs from different genes (in principle a series of chips could carry cDNA from all the genes in a genome) or synthetic oligonucleotides. Synthetic oligonucleotides are built up on the slide. In one recent instance, during an investigation of the human genome, 25 000 different 60-unit oligonucleotides were automatically synthesised and deposited on a 1×3 inch glass slide.

The nucleotide sequences of the oligonucleotide or cDNA probes are known. The chips are then exposed to an extract of fluorescently labelled

A

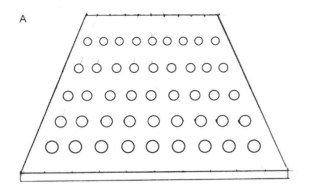

B

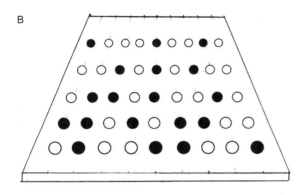

Figure 5.27 Gene chip methodology. (A) An oligo-nucleotide or cDNA is robotically deposited on a glass or nylon 'chip'. Each patch in the array consists of a known oligonucleotide or cDNA. A fluorescently la-belled sample of an unknown mRNA mixture is applied and hybridisation allowed to take place 'overnight'. (B) 'The next morning': mRNAs have attached to known oligonucleotide or cDNA patches (dark patches). Computer-controlled autoscanning detects the patterns of fluorescence and thus the mRNA sequences in the sample. The greater the fluorescence in a patch the more precise the binding.

mRNA derived from the tissue under investigation. These mRNAs stick to complementary oligonu-cleotide or cDNA probes and their positions are determined by laser activation of their fluorescent labels.

Knowing where the cDNA or oligonucleotide probes are located on the chip it is straightforward to read off the mRNAs present in the sample and thus determine which genes are active in the tissue (or cell). This analysis is nowadays mostly done by

computer. The chip is placed in a scanner which automatically detects the positions where fluores-cence is occurring and feeds the data into the computer, which is already programmed with a map of where each probe is positioned in the microarray.

It is clear that gene chips provide a simple and powerful method for monitoring the expression of thousands of genes in a tissue sample. They can show changes in this expression over time (espe-cially in development) and in response to infection or to environmental stress. One of the most interesting uses of this new technology has been in the study of complex diseases states such as schizophrenia. The ability to monitor the expres-sion of thousands of genes during the development of the nervous system combined with sophisticated techniques of computer-aided data analysis pro-mises to discriminate a subset of schizophrenic brains from normal controls. It must, nevertheless, be borne in mind that complex syndromes such as schizophrenia are not 'all in the genes'. Environ-mental influences may and often do trigger the realisation of a potentiality which in other circum-stances would not be expressed.

5.21 CONCLUSION

It will have become apparent from the foregoing pages of this chapter that the techniques emerging from molecular biology are wide-ranging, powerful and advancing with great rapidity. They are revolutionising our understanding of biology, including neurobiology.

We have, of course, only scratched the surface of an enormous subject. The techniques outlined, however, have proved of crucial importance in neurobiology. We shall see, in the next few chapters, that our understanding of the many membrane pumps, receptors and channels which underlie the phenomena of neurophysiology has been revolutionised by their application. Mem-brane proteins, for instance, are notoriously diffi-cult to isolate and analyse. Were it not for recombinant DNA techniques they would have retained their mystery for far longer than now seems likely. Moreover, as we noted in Sections 5.16 and 5.17, molecular biology nowadays pro-mises far more than a 'passive' analysis and understanding. It also promises action and control.

Site-directed mutagenesis holds out the prospect of changing the structure of defined parts of neurobiological proteins. Expression systems such as *Xenopus* oocyte provide means of studying the functioning of these subtly altered proteins in isolation. Homologous recombination allows the engineering of transgenic animals in which the effect of introducing altered or totally inactivated genes into the genome can be studied. Further into the future one can foresee a time when our knowledge of molecular structure and our control over the reverse transcriptases of retroviruses could allow us to attempt gene therapy: that is to replace the defective genes responsible for neuropathologies such as phenylketonuria, Tay–Sachs disease, Huntington's disease and many others (see Chapter 21). If and when such a time comes it will bring a host of problems: financial, legal, ethical. We are already grappling with the outriders of these coming events. It will be essential to have a thorough grasp of the underlying science if rational judgements are to be made.

6

GENOMICS

Publication of draft human genome in 2001 – the prospect of the proteome. **History** – 'big' biology – yeast – *Caenorhabditis* – *Drosophila* – *Homo* – public and private sequencing consortia. **Methodology**. **Chromosomes** – structure – analysis – mapping – VNTRs. Surprisingly small size of human genome – comparison of neurogenes in *Homo*, *Drosophila* and *Caenorhabditis*. **Genes and neuropathology** – channelopathies – neurodegenerations – schizophrenia. **SNPs** – genetic fingerprinting – prospects for a personalised medicine. **Other genomes** – small genomes – skeletal genomes – genomes being sequenced. **Conclusion**: first draft genome only a beginning – the proteome beckons – path from genotype to phenotype open to many external (and internal) influences

The 'final' draft of the human genome, 99.99% accurate (or less than one error in 10 000 base pairs), is due to be published in 2003, exactly fifty years after Watson, Crick and Wilkins published their breakthrough papers on the structure of DNA. Already a first draft of more than 90% of the 3.2 Gbp sequence (3.2 billion base pairs) has been assembled. This event, announced by the US President and UK Prime Minister on 26 July 2000, and published in *Nature* and *Science* in February 2001, attracted widespread media attention. Knowing the full genetic blueprint provides a secure base for studies of human biology, and, in particular, neurobiology. But, as has been frequently pointed out, *only* a foundation. For, as Chapter 3 emphasised, the gene is only the beginning. The processes of transcription, post-transcriptional modification, translation and post-translational manipulation ensure that many subtle influences intervene between DNA code and finished protein product. The route between genome and proteome, between genotype and phenotype, is very far from straightforward, and is open to all manner of modification and moulding. Having pinned down the 3.2 Gbp,

32 000 gene, human genome, the next great task facing biologists is to tease out the yet more complex and multifarious **proteome**.

6.1 SOME HISTORY

Compared to physics and astronomy biology has until recently been a small-scale enterprise. Charles Darwin and Gregor Mendel worked on their own with no outside funding. Watson and Crick solved the structure of DNA using X-ray diffraction data from the basement laboratory of Wilkins and Franklin. The *Drosophila* laboratory of Morgan and the bacteriophage work of Delbrück required very little public funding. Other areas of biology, the ground-breaking neurophysiology of Hodgkin and Huxley, the neuropharmacology of Loewi and Dale, the population genetics of Dobzhansky and the immunological studies of Medawar required similarly modest amounts of financial support. In contrast, it was clear from the beginning that to read the 3.2 Gbp human genome would be no small-scale enterprise. On the contrary, it would require sums of money, state-of-the-art technology and international cross-disciplinary collaborations, more familiar to workers at CERN, Arecibo and Cape Canaveral than to those in university biology departments.

The first move to assemble the requisite resources was made by Robert Sinsheimer at University of California, Santa Cruz, in 1985. At that time the techniques and the technology were scarcely adequate for such a gigantic undertaking and it was not until a few years later that the National Institutes of Health (NIH) in the USA and the Medical Research Council (MRC) with the help of the Imperial Cancer Research Fund (ICRF) in the UK began to pay serious attention to, and release serious quantities of money into, the project. In the early 1990s the Wellcome Trust in the UK agreed to fund development of the Sanger Centre (named for the double Nobel prize-winner) in Cambridge, UK, where about a third of the sequencing ultimately took place. This centre was officially opened in 1993 and was largely responsible for completing the first eukaryotic genome (yeast) in 1996, the first metazoan genome (*Caenorhabditis elegans*) in 1998 and the first human chromosome (chromosome 22) in 1999. But, although the major centres may have been in the UK and in the USA (genome sequencing centres at Washington University (St Louis) and Whitehead Institute for Biomedical Research, MIT) the sequencing project was (and is) international. A useful map showing the extent of the global research effort can be found in *Science*, 2001; **291**, 1204–1205.

The history of the human genome would not, however, be complete without mention of the part played by Craig Venter and the Institute for Genomic Research which he had established in Maryland. In 1998 he announced the formation of a company, Celera Genomics, with the intention of completing a full sequencing of the human genome by 2001. Venter's announcement disturbed the academic dovecotes, not least because in order to fund his work he proposed to patent any sequences he discovered. This was anathema to the academic community who strongly believed that the results of their work, not least because it was publicly and/ or charitably funded, should be placed in the public domain and made available to all, free of charge. Venter also irritated the established investigators in that he was plainly 'piggybacking' on the already published results of their work. Nevertheless, using

what, at bottom, was the cheap and cheerful shotgun technique (see Chapter 5) of smashing the entire genome into fragments and then using banks of powerful computers to search out overlapping sequences and reassemble them, he was able to give the academics a good run for their (and our) money. Indeed Venter's entrance into the field had the useful effect of galvanising the academics to redouble their effort. In the end both approaches ended in near dead-heat and the 2001 'first draft' was published simultaneously, Venter's in *Science*, **291**, 1304–1351 and the International Human Genome Sequencing Consortium's in *Nature*, **409**, 860–921.

6.2 METHODOLOGY

Many of the techniques reviewed in Chapter 5 were used in the analysis of the human genome. In the case of the International Consortium DNA was obtained from 12 anonymous individuals. The final sequence is, consequently, a 'reference' sequence. We all differ. No two of us have precisely the same DNA. We do, however, share 99.9%. But as there are 3.2 Gbp the 0.1% in which we differ constitutes some 3 million base pairs. These, as we shall see later, are called **single nucleotide polymorphisms** (**SNPs** or, colloquially, 'snips') and they crop up every 1000 or so base pairs.

Each of the 23 autosomal and two sex chromosomes was isolated and analysed separately. First, each was broken into roughly 150 kbp fragments and each fragment incorporated into a BAC (see Section 5.5) for cloning. After cloning, each fragment was examined to determine from which part of the chromosome it originated. Having ascertained their place of origin the fragments were 'shotgunned' (Section 5.7) either by enzymes or by physical shearing. This process was repeated several times so that fragments of different break points and length were obtained. This ensures that the base sequences in the fragments overlap. These fragments, usually about 500 bp in length were then cloned using bacterial plasmids, and sequenced (Section 5.13). These sequences were then fed into a computer which searched for overlapping ends. When these were found, longer sequences could be assembled end-to-end so that the sequence of the original BAC insert could be deduced. These lengthy (150 kbp) sequences are known as **contigs**.

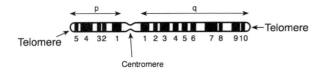

Figure 6.1 Chromosome banding. The figure shows some of the major features of human chromosomes when stained with a stain such as the Giemsa stain and viewed in a microscope. Normally far more bands are distinguishable than those shown in the figure. Further explanation in text.

Knowing from which part of the chromosome the contigs originated (step 2 above), it is now only necessary to assemble the contigs, end to end, to obtain the nucleotide sequence of the chromosome.

The shotgun technique used by Venter and his co-workers at Celera Genomics dispensed with the earlier steps in the International Consortium's procedure. His technique depended on the vast number-crunching power of modern-day computers. The whole genome was broken into fragments that could be sequenced and these fed into banks of computers to find overlapping ends and deduce the entire 3.2 Gbp sequence by 'brute force'.

6.3 SALIENT FEATURES OF THE HUMAN GENOME

The human genome consists of 23 autosomal chromosomes and two sex chromosomes, XX in females, XY in males. Each chromosome constricts to a **centromere** approximately half way along its length. During cell division spindle fibres attach to the centromere to draw the chromatids to opposite poles of the cell before the cytoplasm divides to form two new cells. As the centromere is only *approximately* in the middle it divides the chromosome into a **short arm**, designated 'p', and a **long arm**, designated 'q'. The ends of a chromosome, which also have specialised structure and function, are called **telomeres**.

Chromosomes are not normally visible in a eukaryotic nucleus. They only condense just before cell division. When a chromosome is stained with a dye such as the **Giemsa** stain it exhibits a specific pattern of bands visible in the optical microscope (Figure 6.1). The bands, known as G-bands for the stain, are numbered on each arm from the centromere outwards to the telomere. The rough position of each gene on a chromosome is conventionally given with respect to one of these numbered bands. Thus, as we shall see later, the Na⁺ channel gene (*SNC4A*) is located on the long arm of chromosome 17 somewhere in the region of bands 23–25, i.e. 17q23–25, the gene whose mutation causes Thomsen's disease is at 7q35 and the gene encoding the nicotinic acetylcholine receptor (nAChR) is to be found at 20q13.3.

A more precise mapping of the position of a gene on a chromosome is provided by linkage analysis. In this technique the gene is located with reference to known genetic markers on its chromosome. These markers are distributed along the length of the chromosome at intervals of a few centimorgans (i.e. a few 100 kbp). In many cases they consist of short lengths of tandem repeats (i.e. repetitive sequences of, for instance, C and T: CTCTCTCT). These repetitive sequences, known as VNTRs (**v**ariable **n**umber of **t**andem **r**epeats), are distributed in characteristic ways along human chromosomes.

When a chromosome is broken into fragments and run on an electrophoretic gel, the distance the fragments move depends on their length. When DNA from a family some of whom show a congenital disease and some not is run on an electrophoretic gel, the disease-causing gene is normally seen to co-segregate with a specific VNTR. Figure 6.2 shows the pedigree of a family which inherits an autosomal dominant gene for a disease. The figure shows that all members with the condition also have the same-sized VNTR. It is, in other words, 'linked' to this VNTR. It follows that the gene in question lies very close to the known position of the VNTR. Finally, at a yet more detailed level, the gene is 'run into the ground' in the so-called physical map which shows it as a stretch of nucleotides in the totally sequenced chromosome.

The human genome is presently believed to contain about 32 000 genes spread over 24 chromosomes (see Table 6.1). The small number of genes was a surprise. Table 6.4 shows that the tiny (0.5 mm) and zoologically lowly nematode worm *C. elegans* boasts 18 000 genes and the mustard weed, *Arabidosis thaliana*, has 26 000, whilst rice, *Oryza sativa*, has an astounding 50 000. It had been

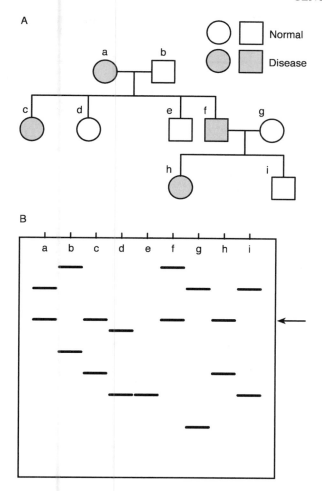

Figure 6.2 Linkage analysis. (A) Family pedigree. The shaded members show the disease condition. (B) Electrophoresis of VNTRs. The arrow shows that a specific VNTR is shown by all members with the disease. Further explanation in text.

thought that the human genome would contain at least 100 000 genes.

Most of the human genome consists of non-coding DNA. Some people refer to this as 'junk' DNA. The actual coding exons make up only about 1.5% of the sequence, only, in other words, about 50 Mbp. Introns make up 24% whilst more than half (53%) of the genome consists of repeat sequences, mostly transposons (see Chapter 4). These transposons are quite ancient; they entered the genome before the appearance of the placental mammals some 100 million years ago. There is also evidence of gene duplication. Some 5% of the sequence is due to the duplication of lengthy segments (10 kb or more). Gene duplication, as we saw in Chapter 4, is a major contributor to genome evolution. Finally, there is evidence that

a couple of hundred genes in the human genome are directly derived from a range of different bacteria. As David Baltimore remarks, the transfer of genes is not restricted to artificially modified organisms (GMOs); it happens quite naturally in the biosphere, indeed it happened to us! Most of these bacterial genes code for enzymes which have been incorporated into the cell's metabolic pathways. One interesting and neurobiologically significant case is **monoamine oxidase (MAO)** which (as we shall see in Chapter 16) is deeply involved in the biochemistry of catecholamine neurotransmitters.

If the human genome is only marginally larger than that of a weed and less than two-thirds that of the rice plant, how is it that we are so much more complex? Complexity is, of course, largely

Table 6.1 Human chromosomes and genes

Chromosome	Size (Mbp)	No. of genes
1	280	2968
2	253	2288
3	233	2032
4	199	1297
5	200	1643
6	189	1963
7	168	1443
8	148	1127
9	142	1299
10	146	1440
11	151	2093
12	145	1652
13	120	748
14	111	1098
15	100	1122
16	102	1098
17	88	1576
18	88	766
19	70	1454
20	70	927
21	45	303
22	48	288
X	160	1184
Y	50	231

Data partly from Gallagher and Dennis (2001), *Wellcome News Supplement*, **Q1**, 13–23.

Table 6.2 Number of proteins assigned to nervous system in humans (H), *Drosophila* (D) and *Caenorhabditis* (C)

	H	D	C
Ion channels			
AChR	17	12	56
CNG/EAG	22	9	9
P2X purinoceptor	7	0	0
Neurotransmitter-gated	61	51	59
Voltage-gated Ca^{2+}	38	9	12
Voltage-gated K^+	55	12	17
Voltage-gated Na^+	11	4	4
Myelin			
Myelin basic protein	1	0	0
Myelin P_0	5	0	0
Myelin proteolipid	3	1	0
Myelin-oligodendrocyte protein	1	0	0
Synapses			
Synaptotagmin	10	3	3
Signalling molecules (morphopoietic)			
Ephrin	8	1	4
Eph receptor	14	1	1
NGF	3	0	0
Neuropilin	2	0	0
Plexin	9	2	0
Semaphorin	22	6	2
Neurotransmitters			
Opioids	3	0	0
Neuropeptide Y	4	0	0
Receptors			
G-coupled	616	146	284
G-proteins	45	15	28
Ras superfamily	141	64	62

CNG=cyclic nucleotide-gated channel; EAG=homologue of *Drosophila* 'ether-a-gogo' (eag) channels. Data mainly extracted from Venter *et al.* (2001), *Science*, **219**, 1343–1345.

an intuitive concept. To date, attempts to devise a quantitative measure have been unsuccessful. Nevertheless, even when all allowance has been made for our ingrained anthropocentricity, there can be little doubt that we are in some sense physiologically more multifarious, anatomically more intricate, behaviourally more various than a weed. In a text devoted to the intricacies of brain and nervous system the point need not be laboured. So if our complexity does not derive from the permutations and combinations of a greater number of genes where *does* it come from?

First of all, it does not derive from different genes. Only 94 of the 1278 protein families encoded by our genomes are specific to vertebrates. All the rest are found in evolutionarily primitive forms such as bacteria and yeasts. These common genes code enzymes for DNA transcription and translation, for the housekeeping duties of intermediary metabolism, protein transport and

chaperoning the three-dimensional conformation of proteins. It is clear that once biochemical solutions to these vital processes were stumbled upon in the early history of the biosphere they spread like wildfire to be universally incorporated, and once incorporated, like the alphabets of languages and the Qwerty keyboards of typewriters, were impossible to shift.

In contrast to these core processes, it appears that major developments in human, compared with other sequenced genomes (*Drosophila*, *Caenorhabditis*, *Saccharomyces* and *Arabidosis*), are proteins concerned with immune response, with the nervous system and its development, with intercellular and intracellular signalling pathways and with haemostasis and apoptosis. Table 6.2 shows the significant increase in the number of proteins assigned to the nervous system in the human genome compared with these other animal genomes.

This increase in the genetic representation of the nervous system is not due, as we noted, to a significant increase in the number of genes but to revolutionary developments in the control of the genes that are there. Generally speaking, human genes have more exons than the genes of other organisms. Moreover, there has been a great deal of 'exon shuffling' (Chapter 4). There have also been significant increases in the amount of cutting and resplicing of the primary transcripts so that, after translation, protein domains are combined and recombined into structures ever more 'rich and strange'.

About 60% of human proteins have two or more alternatively spliced transcripts compared with 22% in *Caenorhabditis*. The large amount of gene duplication has also allowed many variants to evolve (Chapter 4). The olfactory genes are the most striking in this regard. Some 1000 such genes are scattered around the genome. The fact that some 60% of these are non-functional pseudogenes suggests that humans have lost the fine discrimination which the noses of many other mammals possess. Lastly, a combination of these factors – exon shuffling, post-transcriptional processing, etc. – has resulted in the human genome generating many more regulatory proteins than in the other organisms. There are, for instance, twice as many 'zinc fingers' (Chapter 3) as in *Drosophila* and five times as many as in *Caenorhabditis*. Moreover, these regulatory proteins have nearly five times as many zinc finger domains as are found in the homologous proteins of the other known genomes.

In our present state of knowledge it begins to seem that the human genome outstrips other genomes not in size but in the subtlety of the control mechanisms it expresses. It is too early to be sure that this attractive conclusion is correct. Comparisons can at present only be made with a pitifully small number of other organisms: fly, worm, weed and yeast. We need the genomes of other organisms to build up a convincing picture. We shall soon have them.

6.4 THE GENES OF NEUROPATHOLOGY

We noted in the preceding section (Table 6.2) that a significantly larger number of genes are assigned to the nervous system in mammalian than in non-mammalian genomes. We shall see as we go through this book that many of the neurological and psychiatric diseases which afflict us are due to mutations of these genes. For example, a large number of genes code for ion channels in neuronal membranes. Mutations in these genes give rise to at least 20 distressing conditions which have, accordingly, been called **channelopathies**. We shall discuss these conditions in later chapters and especially in Chapter 11 (Section 11.7).

But there are many other disease conditions which derive from the genetics of the nervous system. We shall look at some of the more complex neuropathologies, including the neurodegenerations of old age, in Chapter 21. Table 6.3 shows the distribution of some of the genes responsible for neurological disease. Mutations of these genes give rise to some of the major neuropathologies. The OMIM (*Online Mendelian Inheritance in Man*) web site lists 78 entries for genes responsible for neurodegenerations; 203 for those responsible for different forms of epilepsy; 101 under the heading 'depression'. Only a small selection can be displayed in the table.

Most of the genetic defects listed in Table 6.3 will be discussed as we go through the pages of this book. For ease of reference the page numbers on which they appear are listed in an index of neurological disease at the end of the book. It can be seen from Table 6.3, however, that the genes responsible for mental defect are spread throughout the genome. Furthermore, it is more often than not the case that several genes, often on different chromosomes, are responsible for the disease recognised in the clinic. It may well be that as the genetic basis of mental defect becomes better understood disease syndromes which nowadays go under a common name will be recognised as the complex outcome of many underlying genetic causes.

Table 6.3 Chromosomes, genes and neuropathology

Chromosome	Genes
1	*CPZ2*; *Cx50*; *KCNQ4*; *PARK6*; *PS2*
2	*ALS2*; *DFNA16*; *PARK3*; *SCN2A*; *SPG4*
3	*CMT2B*; *DMT1*
4	*HD*; *SNCA*
5	*GABRG2*; *SCZD1*
6	*PARK2*
7	*CICN1*; *ELN*
8	*CACNB4*; *CMT4A*; *KCNQ3*; *RP1*
9	*DYT1*; *FRDA*
10	*PITX3*
11	*ATM*; *KCNE1*; *NPD1*; *PAX6*
12	*PKU1*
13	*CBJ2*; *Cx46*; *RB1*
14	*PS1*
15	*ALS5*; *TSD*; *UBE3A*
16	*NF1*
17	*CMT1*; *KCNJ2*; *NF1*; *SCN4A*
18	*BPAD1*; *GTS*; *MAFD*
19	*APOE*; *CACNA1A*; *FRDA*; *MSUD1*; *SCN1B*
20	*CHRN4P*; *PRNP*
21	*APP*; *ALS1(SOD1)*; *CRYA1*; Down's syndrome
22	*NF2*
X	*CMT2*; *CMTX1*; *FRMI*; *GJB1*; *KD*; *PMD*

ALS=amyotrophic lateral sclerosis; *ApOE*=apolipoprotein E; *ATM*=ataxia telangiectasia (a devastating progressive degeneration of the nervous system early in life); *BPAD*=bipolar affective disorder; *CACNA1A*=familial hemiplegic migraine (FHM); *CACNB4*=voltage-gated Ca^{2+} channel (idiopathic generalised epilepsy); *CBJ2*=codes for connexin 26 which on mutation causes deafness; *CHRN4*=nAChR subunit which on mutation leads to autosomal nocturnal frontal lobe epilepsy (ADFNL); *CICN1*=Cl^- channel: Thomsen's disease; *CMT*=Charcot–Marie–Tooth; *CPZ2*=cataract, zonular, pulverulent; *CRYA1*: cataract, autosomal dominant; *Cx46*, *Cx50*=connexin genes which in mutation lead to cataract; *DFNA16*=deafness, autosomal dominant; *DMT1*=dementia, familial, non-specific; *DYT1*=dystonia; *ELN*=Williams–Beuren syndrome; *FRDA*=Friedreich's ataxia; *FRM1*=fragile X; *GTS*=Tourette syndrome; *GABRG2*=segregates with generalised epilepsy with febrile seizures (GEFS); *GJB1*=connexin gene which when mutated leads to CMT2; *HD*=Huntington's disease; *KCNJ2*=Andersen's syndrome; *KCNQ3*=voltage-gated K^+ channel: benign familial neonatal convulsions, type 2 (BFNC2); *KCNE1*=K^+ channel, deafness; *KCNQ4*=non-syndromal dominant autosomal deafness (DFNA2); *KD*=Kennedy's disease; *MAFD1*=major affective disorder; *MSUD*=maple syrup urine disease; *NF*=neurofibromatosis; *NMP*=Niemann–Pick disease; *PARK2*=juvenile Parkinsonism; *PAX6*= developmental eye abnormalities; *PITX3*=congenital cataract, corneal opacity, optic nerve defect; *PKU*= phenylketonuria; *PMD*=Pelizaeus–Merzbacher disease; *PRNP*=prion protein; *PS*=presenilin; *RP1*=retinitis pigmentosa; SCDZ1=schizophrenia; *SCN1B*=VGNa+ channel: GEFS; *SCN2A*=VGNa+ channel: generalised epilepsy; *SCN4A*=VGNa+ channel: HyKPP, PC, PAM; *SCNA*=α-synuclein; *SPG4*=spastic paraplegia; *TSD*=Tay–Sachs disease; *UBE*=Angelman syndrome. Data mostly abstracted from OMIM morbid map. Further explanation in text.

Creative use of DNA chip technology (Section 5.20) is likely to prove invaluable in detecting the changed spectrum of gene expression in these complex conditions. Schizophrenia provides an interesting instance. Although only one gene has so far been definitively linked to the condition (Table 6.3), there can be little doubt that schizophrenia is complexly multifactorial. This has been borne out by gene-chip microarray analysis of prefrontal cortex in matched pairs of normal and schizophrenic subjects. It was shown that the levels of a number of mRNA transcripts encoding proteins involved in synaptic function were significantly lower in the schizophrenic group than in a control group. These more profound and finer-grained analyses should allow therapy to become more precisely targeted. Along with advances in detecting and determining single nucleotide polymorphism (to be discussed in the next section) this should allow a far more individually tailored medicine to develop in the twenty-first century. We shall come to look back on twentieth-century diagnosis and therapy as absurdly 'one-size-fit-all', blunt and hit and miss.

6.5 SINGLE NUCLEOTIDE POLYMORPHISMS (SNPS)

As we noted in Section 6.2, no two of us have exactly the same sequence of base pairs in our genomes. Similarly, none of us (except the volunteers for the sequencing projects) have exactly the same sequence as that published (or to be published in 2003) as the **human reference sequence**. We differ from each other and from the reference sequence by about 0.1% which, as there are 3.2 Gbp, works out as about one base pair in every 3000, or, to use round numbers (after all, these are all approximations), about one in every 1000. These differences are termed '**single nucleotide polymorphisms**' or SNPs. The following sequence variation provides an example:

···-C-A-T-A-T-C-G-G-C-T-A-C-G-T-A-···
···-C-A-T-A-T-**G**-G-G-C-T-A-C-G-T-A-···

One polynucleotide sequence shows a C in the place where the other shows a G. In general, as already noted, such variations crop up every 1000 or so base pairs. Alongside the major sequencing project, another group of workers (*the International SNP Map Working Group*) was formed to map the SNPs on the human genome. In the same issue of *Nature* in which the Human Genome Sequencing Consortium published the first draft of the human genome, this group published a paper showing the position of 1.42 million of these SNPs. As SNPs act as landmarks distinguishing one genome from another this 'high-density' map will have many uses. DNA finger-printing techniques in paternity cases and forensic science will become more accurate. An individually tailored medicine, as mentioned in the previous section, becomes a real prospect. Finally, the epidemiological study of disease will be greatly assisted. There have been many studies tracing disease, not least neurological disease, through populations and genealogies. With high density SNP maps, which provide markers every two or three genes along the genome, it will become possible to relate these disease occurrences to particular SNPs and thus to identify 'candidate' genes for these conditions, which can then be subjected to further analysis.

6.6 OTHER GENOMES

In recent years, as automated techniques have become available, a huge scientific effort has been devoted to sequencing. At the time of writing some 599 viruses and viroids, 205 plasmids, 185 organelles, 31 eubacteria, seven archaea, one fungus, three animals and one plant have been completely sequenced. Table 6.4 shows a selection of these genomes. They range from the small (580 kbp, 480 gene) genome of a mycoplasma to the huge (in number of base pairs) genome of a primate, such as that of *Homo sapiens*.

We have already seen (Table 6.2) how even the small number of animal sequences currently known is providing invaluable data with which to compare the human sequence. This small number is destined to grow quickly in the next few years. In our usual self-regarding way it is probably the other chordate sequences which will prove most interesting. Some of these sequences are much smaller than the human, yet the number of genes is not so dramatically different. The smallest genome of all is, unsurprisingly, found in the protochordate ascidian larva. These forms were believed by Charles Darwin and are still believed by most classical zoologists to represent the stem chordates. The larva of the ascidian *Ciona intestinalis*, commonly found in the littoral zone around English shores, consists of only 2600 cells of which some 330 are found in the CNS and of those fewer than 100 are neurons. Its CNS is thus even simpler than that of *C. elegans*. Yet its genome is computed to contain about 15 500 genes, about half the human complement, although about the same as *C. elegans*. Further along the evolutionary ladder the Japanese puffer fish (*Fugu rubripes*) and the zebra fish (*Danio rerio*) are both fully functional vertebrates yet both manage with only a few thousand more genes.

These findings bring out the oddness of human and many other mammalian genomes. Although the puffer fish has about 17 000 genes, its genome consists of only 400 Mbp, compared with the 32 000 genes and 3.2 Gbp of humans. Thus whereas human genomes devote 9.7 Mbp per gene, the puffer fish needs only 24 kbp. This serves to emphasise the large amount of seeming junk in the human DNA sequence. It serves, also, to emphasise that by studying these smaller, 'stripped down', genomes we may be able to tease out the essentials which are lost in the 'noise' of the larger genomes of primates such as ourselves. It is interesting in this regard to note that the bat sequence, at 1.7 Mbp, is even smaller and more 'skeletal'.

The mouse and rat have long been standard laboratory animals and a great deal is known about their biology. Elucidation of their genomes, and comparison with our own, is expected to throw much light on the genetic bases of their heavily researched lives. On the other hand, it is known that both their genomes show high rates of nucleotide substitution compared with those of most other mammals. It may be, therefore, that their genomes will turn out to be so extensively rearranged that they will not form a good comparator with humans.

The genomes of our closest relatives, the other primates, should also be of great interest. The genomes of both the chimpanzee and the rhesus

Table 6.4　Genomes

	Size	No. of genes
Prokaryote		
Mycoplasma genitalium	580 kbp	480
Haemophilus influenzae	1.8 Mbp	1700
Bacillus subtilis	4.2 Mbp	4100
Escherichia coli	4.6 Mbp	4300
Unicellular eukaryote		
Yeast, *Saccharomyces cerevisiae*	13 Mbp	7000
Plants		
Weed, *Arabidopsis thaliana*	117 Mbp	27 000
Rice, *Oryza sativa*	430 Mbp	50 000
Animals		
Nematode worm, *Caenorhabditis elegans*	97 Mbp	18 400
Insect: fruit fly, *Drosophila melanogaster*	180 Mbp	13 600
Protochordate: *Ciona intestinalis*	160 Mbp	15 500
Fish:		
Puffer fish, *Fugu rubripes*	400 Mbp	17 000
Zebra fish, *Danio rerio*	1.7 Gbp	
Mammals:		
Mouse, *Mus musculus*	3.1 Gbp	28 000
Human, *Homo sapiens*	3.2 Gbp	32 000

Other genomes which are being actively sequenced include those of *Xenopus*, the clawed frog, the rat, the pig, the chimpanzee and the rhesus monkey.

macaque are being sequenced. It is common knowledge that the chimpanzee genome differs from the human by only 1 or 2%. This 1 or 2% must be crucial. It will be fascinating to find what it is.

6.7 CONCLUSION

The publication of the 'first draft' of the human genome, along with the other genomes already published and in process of elucidation, is providing biology and, in particular, neurobiology with a firm base. Long ago one of the founders of molecular biology, Max Delbrück, spoke of 'running the map into the ground'. He was referring to the physical nature of the geneticist's 'gene'. Fifty years later the entire human genome has been 'run into the ground', into the 3.2 Gbp of DNA.

The next task, as mentioned above, is working out how the information stored in the 32 000 or so genes of the human genome is expressed in the form of proteins. This task, which is already under way, is referred to as 'functional genomics' or 'proteomics'. It is, perhaps, even more formidable than genomics itself. For we have seen in earlier chapters how the information of the DNA base sequence is chopped and changed and rearranged on the way to being expressed in protein structure. Whereas each of our genomes is comparatively fixed, 99.9% similar to the reference genome published (or to be published in 2003) by the sequencing consortia, our proteomes are very various. The path from gene to protein is open to all sorts of outside influence, from environment, from disease, from culture, from diet: the list is endless. Each of us is provided with an outline, a foundation plan, but the outcome, the proteome, depends on circumstance. And nowhere is this more the case than in the structure and functioning of the brain.

The International Human Genome Sequencing Consortium concluded their 2001 paper by echoing the famous concluding sentence of Watson and Crick's breakthough paper of 1953. '...it has not escaped our notice' write the Consortium 'that the more we learn about the human genome, the more there is to explore.' In the succeeding chapters of this book we shall be reminded of this conclusion time after time. The 1990s were designated 'the decade of the brain' but the great accession of knowledge during that decade has made us yet more aware of how much we do not know. In the years to come genomics is likely to provide a powerful tool and resource base in our task of disentangling the biological bases of our most precious and distinctive possession.

7

BIOMEMBRANES

Huge extent and intricacy of cerebral membrane systems. Biochemical structure: lipids, proteins, carbohydrates. **Lipids**: phospholipids – amphipathic character – bilayers and micelles – artificial bilayers and liposomes – dynamism – lipid packing and membrane fluidity. Sphingosine and its derivatives. Glycolipids: cerebrosides and gangliosides – Tay–Sachs disease. Cholesterol and its function in 'stiffening' membranes. Membrane fluidity – microdomains (rafts) – caveolae. Membrane asymmetry – E- and P-faces – glycolipids and cell–cell recognition. **Proteins**: the 'fluid-mosaic' concept of membrane structure – orientation of proteins in lipid bilayer – 'inside-out' character of lipophilic domains – glycophorin – mobility of proteins in membranes. **Synthesis of membranes** – the *Xenopus* rod outer segment disc – axoplasmic flow carries newly synthesised membrane to appropriate destination. **Myelin and its synthesis** – MS – differences between central and peripheral myelin – Schwann cells and oligodendroglial cells – lipid and protein constitution of myelin – role of proteins in myelin structure – genetic analysis – mutations in mouse (shiverer, jimpy, myelin deficient) and man (Pelizaeus–Merzbacher disease-CMT type 1) – autoimmune attack and MS. **Submembranous cytoskeleton** – complexity in erythrocytes – homologies with neuronal cytoskeletons (fodrins and synapsins). **Cell junctions**: desmosomes, tight junctions, gap junctions – role of tight junctions in blood–brain barrier (BBB) and of gap junctions in metabolic cooperation and as electrical synapses – connexins – open/shut conformations of gap junctions – analogous conformation of gap junction connexins and units of transmitter-gated receptors – role of gap junctions in the retina and in development of CNS – role in synchronised spiking – concept of hyperneuron. **Gap junctions and neuropathology** – deafness – cataract – CMT type 2 – spreading hyper- and hypoexcitability. **Conclusion** – forward look – metabotropic receptors – pumps – ligand-operated ion channels – voltage-operated ion channels

Any electron micrograph will show that the brain is packed full of membranes. It is not difficult to calculate that, flattened out, they would cover several square miles. Much of the physiology of the brain consists of fluxes of ions across them. We shall soon see that membranes are complex and heterogeneous down to the molecular level. Perhaps the brain's computing power becomes more understandable if we think of it as a ten-hectare, ten-nanometre membrane, operating at the molecular level, each part within at most a second's communication time with any other.

Biological membranes are built of three molecular species: always **lipids** and **proteins** and in most cases **carbohydrates** as well. The lipids form a universal matrix whilst the carbohydrates and proteins confer specific biological properties.

7.1 LIPIDS

It has been computed that a small patch of membrane with an area of $1\,\mu m^2$ is built of some 5×10^6 lipid molecules. It can be shown that different membranes consist of slightly different

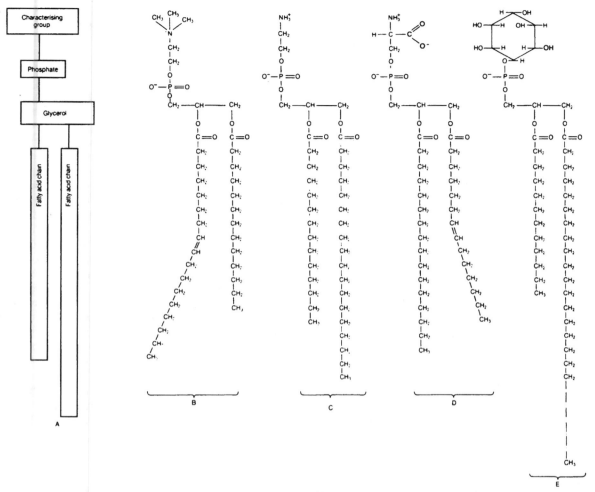

Figure 7.1 Phosphoglycerides: (A) schematic diagram; (B) phosphatidylcholine (lecithin); (C) phosphatidylethan-olamine (cephalin); (D) phosphatidylserine; (E) phosphatidylinositol. Note: the fatty acid chains are very variable in both length and saturation.

lipid mixtures and thus have slightly different properties. The lipids found in biological membranes fall into three major groups: phospholipids, glycolipids and steroids (especially cholesterol). Let us look at each group in turn.

7.1.1 Phospholipids

Figure 7.1 shows the molecular structures of a group of four important membrane phospholipids – the **phosphoglycerides**: phosphatidylcholine (=lecithin), phosphatidylethanolamine (=cephalin), phosphatidylserine and phosphatidylinositol. The figure

shows that they all consist of two long **fatty acid** chains attached through **glycerol** and a **phosphate** group to a hydrophilic 'characterising group': **choline, ethanolamine, serine** or **inositol**.

The important feature which all the molecules shown in Figure 7.1 share, so far as membrane structure is concerned, is that they are all **amphipathic** molecules. Moreover they are amphipathic in an interesting way. One end of the molecule, the nitrogen-containing or (in the case of phosphatidylinositol) the carbohydrate-containing group, is **hydrophilic** whilst the fatty acid chains at the other end of the molecule are **hydrophobic**. This means

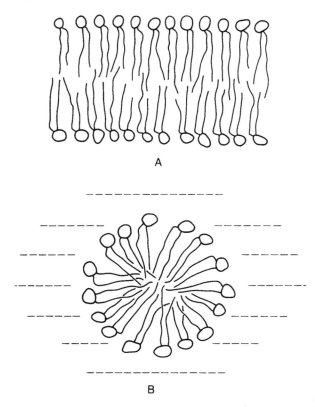

Figure 7.2 (A) Schematic to show a phospholipid bilayer. (B) Schematic to show a phospholipid micelle.

that phospholipids will line up at an air–water or oil–water interface to form a **monomolecular layer**. Cellular interfaces are, however, always between two aqueous solutions. In such circumstances phospholipids form not monolayers but **bilayers** and in some cases **micelles** (Figure. 7.2).

Phospholipid bilayers are, as may be guessed, extremely fragile structures. The forces holding them together are the same hydrophobic forces we met when considering the higher structures of proteins and nucleic acids. On the other hand, if membranes or micelles are disrupted they quickly re-seal. This spontaneous sealing is made use of in the formation of **artificial bilayers** and **liposomes**. Both constructs have been of considerable use in molecular neurobiology. Artificial bilayers may be constructed across a small pore in a partition separating two aqueous solutions whilst liposomes form spontaneously from phospholipids or mixtures of phospholipids introduced into an aqueous phase of appropriate physico-chemical characteristics (Figure 7.3). We shall meet with both types of artificial membrane again in future chapters. We shall see that the study of purified channel proteins has been greatly assisted by inserting them into such structures.

The extremely tenuous nature of phospholipid bilayers (whether natural or artificial) also means that at room temperatures the individual molecules

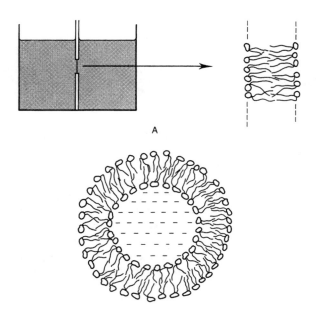

Figure 7.3 (A) Artificial phospholipid bilayer formed across a small hole in a partition between two aqueous solutions. (B) Phospholipid liposome. Note liposomes differ from micelles in that they enclose a volume of aqueous solution.

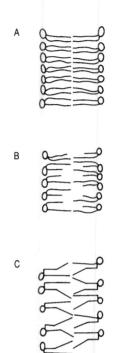

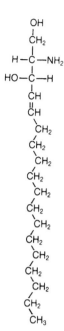

Figure 7.4 Lipid packing and membrane fluidity. (A) The phospholipid fatty acid 'tails' are all fully saturated and of about the same length: they therefore pack compactly. (B) The fatty acid 'tails' differ in length and thus the membrane core is more fluid. (C) The fatty acid 'tails' contain unsaturated bonds and hence, once again, the packing cannot be so tight and the membrane core tends to fluidity.

Figure 7.5 Sphingosine.

are in constant motion. Indeed the hydrophobic fatty acid 'tails' of the molecules have, in this respect, been likened to 'a basket of snakes', bending and twisting about in perpetual motion. The interior of the membrane is thus to all intents and purposes an organic fluid. The individual phospholipid molecules, moreover, continually exchange places with each other. They seldom migrate (or 'flip-flop') from one side of the membrane to another – for this would mean their hydrophilic 'heads' passing through the organic phase in the centre of the membrane – but they move laterally within each lipid monolayer with considerable freedom. Indeed their diffusion coefficient is such that it is calculated that they have a lateral velocity (at 37°C) of about 1 µm/s. This means that the average phospholipid (other things being equal) could travel from one end of an average nerve cell body (diameter, say, 20 µm) to the other in about 30 s. 'Other things', however, as we shall shortly see, are seldom 'equal'!

So far we have distinguished between phospholipids by way of their 'characterising heads': cho-

line, ethanolamine, serine, etc. But phospholipids also differ in the nature of their fatty acid 'tails'. The number of carbon atoms in these tails may vary from 12 to 20. The carbon atoms, moreover, may be linked by **saturated** (single) or **unsaturated** (double) bonds. Both these features have an effect on the nature of the membrane. The shorter the fatty acid tail and the greater the unsaturation, the greater the fluidity of the membrane. This is because both features make the tails more difficult to pack compactly within the membrane's core (Figure 7.4). We shall return to the topic of membrane fluidity when we have considered the nature of some of the other lipids making up the structure of biological membranes.

A second important group of membrane lipids is based not on glycerol but on **sphingosine**. The structure of sphingosine is shown in Figure 7.5. Sphingosine, like the phosphoglycerides discussed above, is an amphipathic molecule. It possesses a hydrophilic 'head' containing an amino group and a long (13-carbon) saturated fatty acid hydrophobic tail. The 'head' bends round, rather like a

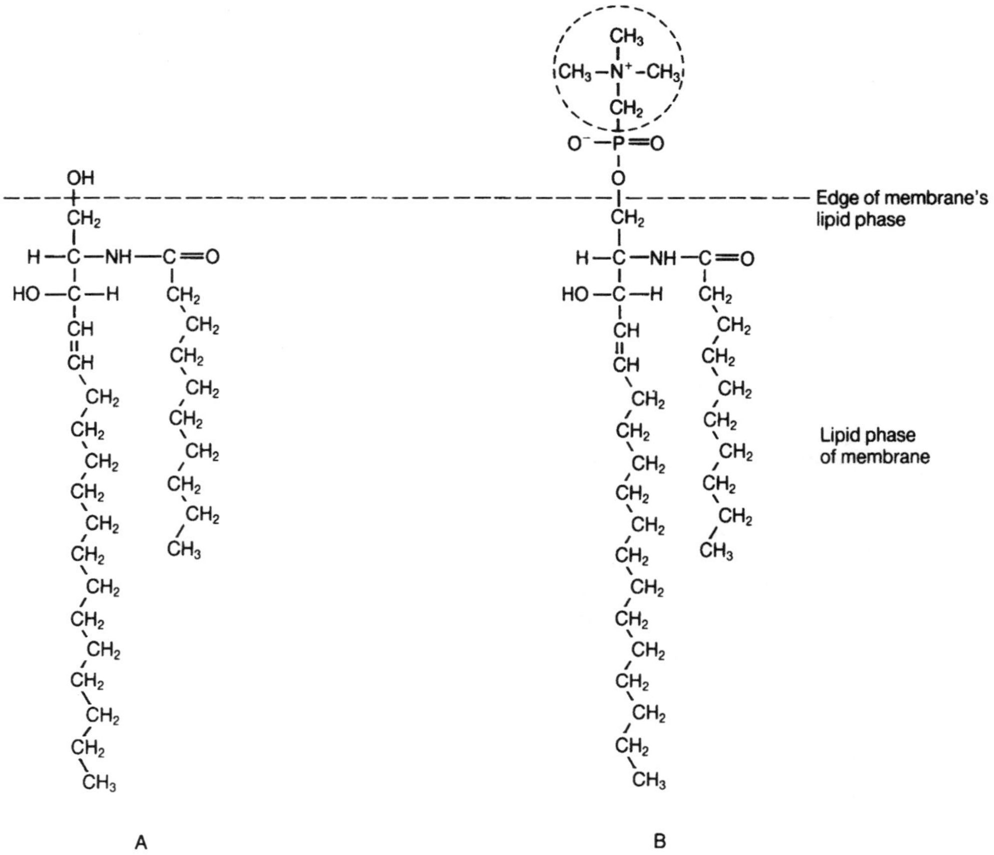

Figure 7.6   (A) Ceramide. (B) Sphingomyelin.

hairpin, and forms a point of attachment for other molecules (Figure 7.6). Thus in cell membranes sphingosine is normally attached to another fatty acid chain to form **ceramide**. Ceramide, in its turn, is often attached through a phosphate group to choline to form a quite common membrane constituent: **sphingomyelin**.

7.1.2 Glycolipids

Ceramide forms the starting point for a number of other important constituents of animal cell membranes – the **glycolipids**. These complicated molecules are shown in Figure 7.7. Instead of being attached through a phosphate group to choline we find that ceramide forms a glycosidic linkage to a monosaccaride – **galactose** or **glucose**.

Galactocerebroside, the simplest of these glycolipids, forms the major glycolipid of the myelin sheath around axons (up to 40% of the outer monolayer). The other glycolipids, as Figure 7.7 shows, are more complicated molecules containing one or more **sialic acid** (*N*-acetylneuraminic acid) groups. There are, in fact, at least thirty different varieties. Collectively they are known as **gangliosides** and they are especially plentiful in the membranes of neurons.

Of particular importance in neurobiology is the ganglioside G_{M2}. This is normally transformed into G_{M3} by the enzyme **hexosaminidase A** (Figure 7.8). Young children who suffer from the inherited **Tay–Sachs disease** lack this enzyme. Hence G_{M2} accumulates in the nervous system. Cytoplasmic bodies begin to fill the neurons. Patients appear unaffected

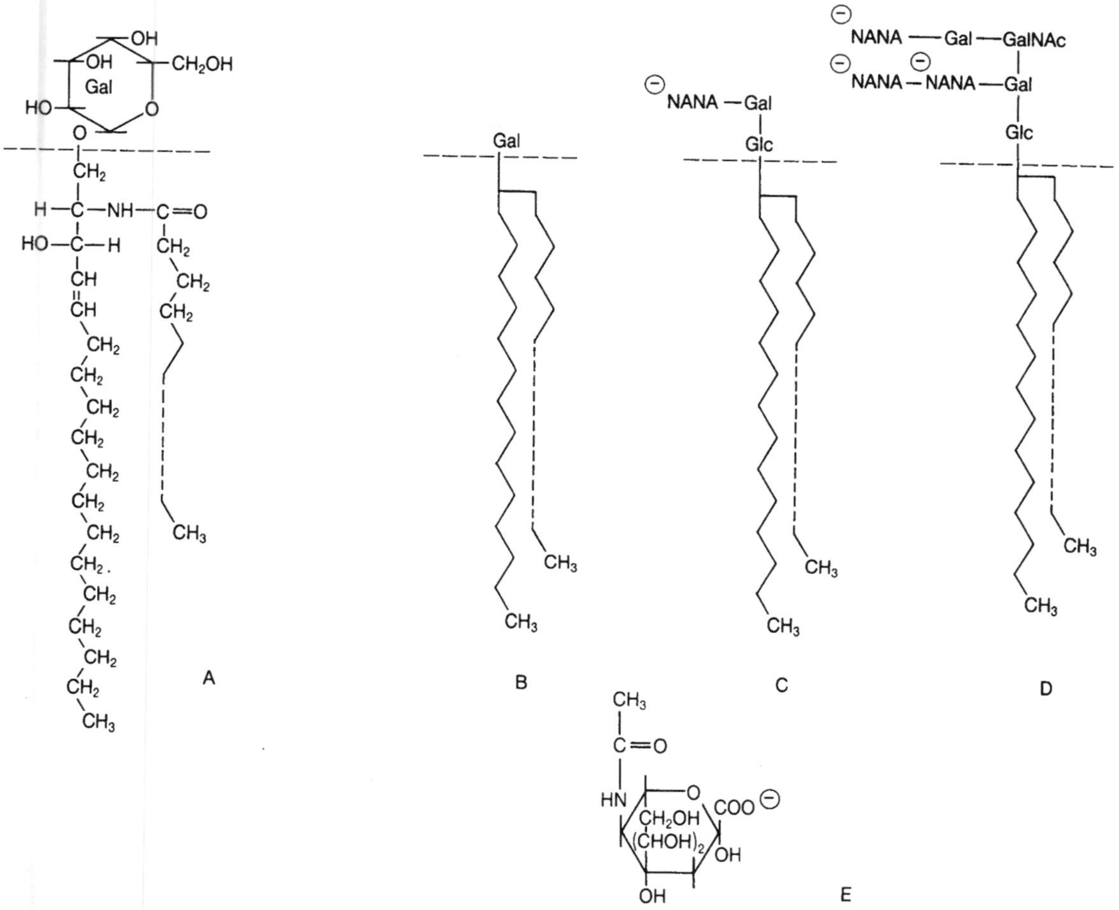

Figure 7.7 Some glycolipids. (A) Galactocerebroside. (B) Schematic of galactocerebroside. (C) Ganglioside G_{M3}. (D) Ganglioside G_{T1}. (E) N-acetylneuraminic acid (NANA) Gangliosides are named according to the number of NANA groups (M=mono, D=di; T=tri, etc.) whilst the number refers to the number of sugar residues subtracted from five. Hence G_{M3} indicates that the ganglioside possesses one NANA and two sugars. Glc=glucose; Gal=galactose; GalNAc=N-acetylgalactosamine; NANA=N-acetylneuraminic acid (=sialic acid).

for the first five or six months of life but they then fail to develop normal mental and motor capacities. Death usually occurs by the third year although, in some cases of late onset, death is delayed until the fifth or sixth year.

7.1.3 Cholesterol

There is one other type of lipid found in most biological membranes: **cholesterol**. This is a very different type of molecule to the phospholipids and

gycolipids we have so far considered. Figure 7.9 shows that the molecule consists of three different regions – a hydrophilic 'head' represented by the hydroxyl group, a flat plate-like steroid ring and a flexible hydrophobic 'tail'.

The amount of cholesterol present in biomembranes is very variable (Table 7.1). Quite large amounts of it are found in some plasma membranes and in myelin, much smaller amounts in intracellular membranes such as endoplasmic reticulum (ER) and mitochondria, and none at all in

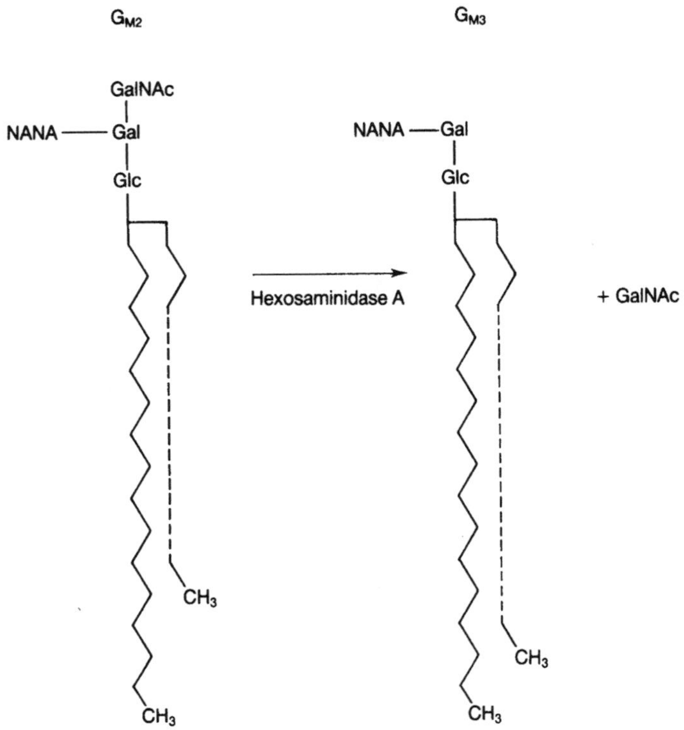

Figure 7.8 Transformation of G_{M2} into G_{M3} by hexosaminidase A.

prokaryotes such as *Escherichia coli*. When it is present it is interpolated between the phospholipid molecules and reduces the fluidity of the membrane (Figure 7.10). We shall see that cholesterol plays an important stabilising role and that when its representation is abnormal serious consequences flow, not least in the disastrous neurodegeneration we know as Alzheimer's disease (Section 21.10).

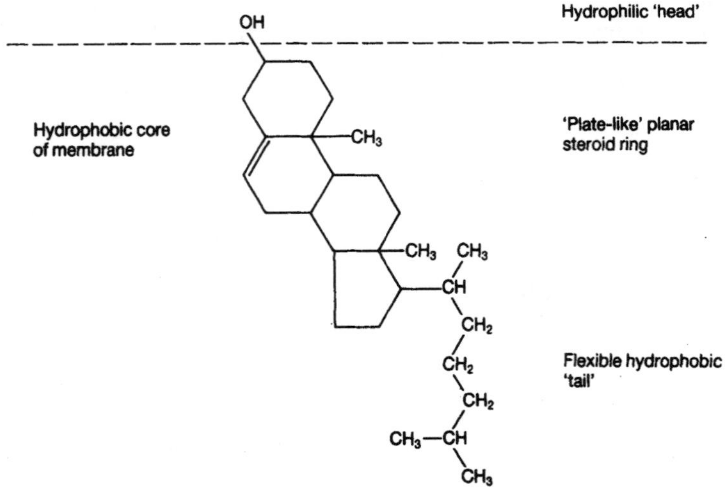

Figure 7.9 Cholesterol.

Table 7.1 Lipid composition of some membranes

Source	Approximate percentages of total extracted lipid								
	CHOL	PC	PE	PS	PI	SP	CE	GA	Other
Rat liver[a]									
Plasma membrane	30	18	11	9	4	14	–	–	1
ER (rough)	6	55	16	3	8	3	–	–	–
ER (smooth)	10	55	21	–	7	12	–	–	2
Mitochondria									
inner	3	45	24	1	6	3	–	–	19
outer	5	50	23	2	13	5	–	–	5
Nuclear	10	55	20	3	7	3	–	–	1
Golgi	8	40	15	4	6	10	–	–	–
Lysosomes	14	25	13	–	7	24	–	–	5
Erythrocytes[a]	24	31	15	7	2	9	–	–	–
Glial plasma membrane[b]	36	25	7	5	–	5	12	–	8
Axon[c]	32	32	23	7	7	–	–	–	–
Synapse[c]	19	34	28	10	2	3	2	–	–
Myelin[b]	42	10	16	5	–	5	16	–	3
Grey matter[d]	22	26	22	8	3	7	5	7	–
E. coli[a]	0	0	80	–	–	–	–	–	20

CHOL=cholesterol; PC=phosphatidylcholine; PE=phosphatidylethanolamine; PS=phosphatidylserine; PI=phosphatidylinositol; SP=sphingomyelin; CE=cerebrosides; GA=gangliosides; Other=other lipids.
Data: [a]Darnell, Lodish and Baltimore (1986), *Molecular Cell Biology*, New York: Scientific American Books; [b]Bradford (1985), *Chemical Neurobiology*, New York: Freeman; [c]Cotman and Levy (1975), in MTP International Review of Science, *Biochemistry Series 1*, vol. 2, 187–205; [d]Siegel *et al.* (1981), *Basic Neurochemistry*, Boston: Little Brown.

7.2 MEMBRANE ORDER AND FLUIDITY

Some workers regard biomembranes as liquid crystals. Liquid crystals (or mesophases) have a lower degree of order than crystalline solids but a higher degree of order than liquids, where the molecules are disordered in position and orientation. The molecules in membranes are disordered in two dimensions (in the membrane plane) but ordered in the third dimension (perpendicular to the plane). There are various types of liquid crystal. Biomembranes are classified as **smectic** liquid crystals. In this type of liquid crystal elongated molecules are aligned roughly parallel to each other, in layers. Research into the properties of liquid crystals is an active field and one which can help us understand the behaviour of biomembranes and their components.

However, the idea that the organisation of lipid molecules is totally random in the plane of the membrane is only partially true. There is much evidence nowadays to support the belief that the membrane lipids are organised into '**microdomains**' or '**rafts**' (<100 nm in diameter). These rafts may be enriched in, for example, cholesterol and sphingolipids, so that specific proteins find easy anchorage in them. The loose organisation shown in Figure 7.10 shows how this can happen. Rafts of specific lipids are important if groups of protein signalling molecules are to be kept close to each other. This rafting is sometimes accentuated by invagination of the membrane to form flask-shaped **caveolae**. These invaginations also help to maintain the 'patchwork quilt' of lipid distribution.

We have already noted that the fluidity of a biomembrane is determined to some extent by the length and saturation of the fatty acid chains forming its core. In artificial bilayers formed of a single phospholipid species there is a sharp transition temperature, characteristic of the particular phospholipid, from a gel state to a fluid state. This **transition temperature** varies in natural membranes, being higher if there is more cholesterol and a

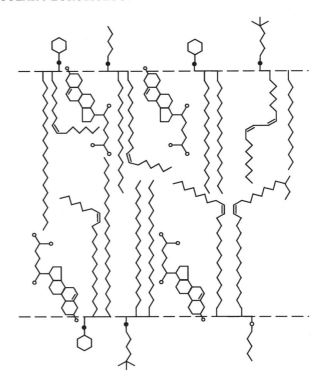

Figure 7.10 Interpolation of cholesterol in a phospholipid bilayer. The comparatively bulky and compact cholesterol stabilises the flexible hydrocarbon 'tails' of the phospholipids.

greater number of saturated fatty acid chains. As the lipid constitution of a membrane such as that which envelops a neuron varies from place to place, the fluidity also varies from place to place. The neuronal membrane thus can be looked at as if it were a patchwork not only of different groups of lipids but also of different fluidities.

7.3 MEMBRANE ASYMMETRY

Membranes are highly asymmetrical. This can be shown by both electron-microscopical and biochemical techniques. The **freeze-fracture** method allows the electron-microscopist to cleave the membrane along its plane of greatest weakness – the centre of the lipid bilayer (Figure 7.11). The outer monolayer (=outer 'leaflet') is defined as having an '**exoplasmic**' or '**E**'-face abutting the extracellular space and an '**exoplasmic fracture**' or '**EF**'-face which is the fracture plane. The inner monolayer (=inner 'leaflet') is, similarly, defined as having a '**protoplasmic fracture**' or '**PF**'-face (again the fracture plane) and a '**protoplasmic**' or '**P**'-face next to the protoplasm.

Biochemistry reveals that, so far as the lipids are concerned, one of the most striking asymmetries involves the **glycolipids**. These are to be found almost exclusively in the outer monolayer and their oligosaccharide moieties extend from the E-face into the extracellular compartment. It has been shown that they have much to do with the processes of **cell–cell recognition**. We shall see the fundamental importance of this process in later chapters, especially in Chapter 19 where we examine how neurons find their way to their correct positions in the brain.

It is also worth noting that significantly more phosphatidylserine and phosphatidylinositol are found in the inner monolayer or 'leaflet' than the outer. Reference to Figure 7.1 will show that these phospholipids bear a preponderating negative electrostatic charge. Hence we find that the P-face is significantly more negative than the E-face.

7.4 PROTEINS

So far we have been discussing the universal scaffolding of biomembranes. Although the

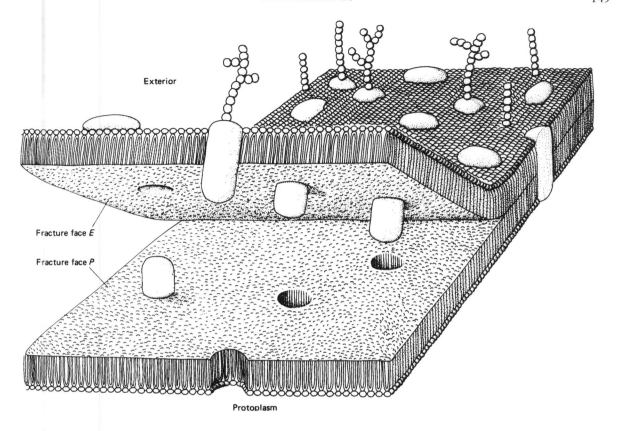

Exterior

Fracture face *E*

Fracture face *P*

Protoplasm

Figure 7.11 Freeze-fractured membrane showing positions of E-, EF-, PF- and P-faces. Reproduced with permission from B. Satir (1975), *Scientific American*, **233**(4), 29–37. Copyright © 1975 by Scientific American Inc.

glycolipids are believed to be very important in intercellular recognition, the most important functional characteristics of most biomembranes is conferred not by their lipid but by their protein constitution. The quantity of protein present in biological membranes varies from about 20% of the mass (myelin) to about 75% of the mass (mitochondrial inner membrane).

For many years it proved extremely difficult to determine exactly where proteins were positioned in the membrane and also the exact nature of the proteins which were there. These problems have yielded to new biochemical and molecular biological techniques. Extraction by the strong detergent sodium dodecyl sulphate (SDS) and subsequent analysis by electrophoresis on SDS–polyacrylamide gel has been used to separate membrane proteins –

more than fifty different types have been distinguished in one membrane (that of *E. coli*) alone. Freeze-fracture etching, as mentioned in the previous section, allows the interior of the phospholipid bilayer to be examined by electron microscopy. Last, but most important of all, the techniques of genetic engineering, as described in Chapter 5, have been much used in recent years to elucidate the primary sequence and conformation of innumerable membrane proteins.

Figure 7.11 has already indicated the position which proteins are nowadays believed to occupy in the phospholipid matrix. They form a mosaic of globular 'icebergs' floating in a lipid sea. It is not surprising that this image has been called the '**fluid-mosaic**' model of the biomembrane. Some of the proteins are confined to one or other monolayer whereas others project all the way through the bilayer and extend into both the extracellular and intracellular compartments. We shall meet many examples of these so-called **transmembrane** proteins in later chapters. We shall see that they are fundamental to the functioning of the nervous system.

In general we can say that transmembrane proteins are so constructed that they possess a **hydrophobic** region or domain which is embedded in the lipid core of the membrane and **hydrophilic** regions which project into the aqueous extracellular and intracellular compartments. In comparison with the globular proteins of the aqueous cytosol, the intramembranous domains of membrane proteins are, in a sense, 'inside-out': their hydrophobic residues project outwards, their hydrophilic residues are tucked inside towards their cores. This ensures that they 'stick' in the lipid bilayer. Membrane proteins, too, whether transmembrane, confined to a single monolayer or merely adhering to a P- or E-face, are always asymmetrical. They are constructed so that one part of their conformation interacts with the lipid bilayer in a specific way. Sometimes they make use of the fact, mentioned above, that the P-face is electrically negative compared with the E-face. Transmembrane proteins, moreover, never rotate or 'flip-flop' across the membrane – this would entail dragging their hydrophilic ends through the membrane's lipid core and this in normal circumstances is not possible.

Studies which involve the incorporation of enzymatic proteins into artificial lipid bilayers show that the activity of such proteins is conditioned by their lipid environment. Features of the lipid bilayer, such as the length of the fatty acid chains, the degree of saturation, the nature of the lipid 'heads', all influence the biological activity of the enzyme. Just as water-soluble enzymes are affected by parameters of the aqueous environment such as pH, so lipid-embedded enzymes are affected by the precise nature of their lipid environment. The significance of the lipid rafts and caveolae mentioned earlier is clear.

We shall meet many examples of transmembrane proteins in the next few chapters – receptors, pumps, ion channels, etc. As an introductory example, however, let us take the very well-known case of **glycophorin**, a protein found in the erythrocyte membrane. Glycophorin, as Figure 7.12 shows, consists of some 131 amino acid residues of which 34 (numbers 62–95) are embedded in the lipid core of the membrane. The great majority of these are hydrophobic residues and one stretch of 23 residues (73–95) consists exclusively of such residues – Phe, Leu, Ile, Val, Try and Thr. It is believed that these residues take the form of an α-**helix** so that their hydrophobic side chains can project into the lipid phase and the hydrogen-bonding potentialities of their amide-groups can be satisfied by the usual α-helical intra-chain linkages. The section of the molecule emerging into the cytosol is composed initially of several positively charged amino acid residues (Arg, Lys) and these presumably are stabilised by electrostatic attraction to the predominantly **negatively charged heads of the phospholipids** in the membrane's P-face. The amino-terminal end of the molecule projects from the E-face of the membrane and **oligosaccharide** groups are attached to many of the Ser, Thr and Asn residues.

7.5 MOBILITY OF MEMBRANE PROTEINS

We have already referred to membrane proteins as floating in a lipid sea. It is not surprising, therefore, to find that they have considerable lateral mobility. We shall see in Chapter 8 that this lateral mobility has been pressed into service, with great effect, in the development of signalling systems based on protein shuttling in the plane of the membrane. Protein diffusion coefficients range from about $10^{-9}\,cm^2/s$ for rhodopsin in retinal rod outer segments (i.e. about $0.1\,\mu m^2/s$) to about $10^{-11}\,cm^2/s$ for proteins in other membranes (i.e. about $0.001\mu m^2/s$). In the first case, rhodopsin could travel across the diameter of an outer-segment disc in about 10 s, whereas at the opposite extreme a protein might require a couple of hours to travel the same distance. There are a number of reasons for these great differences in mobility. It may be, for instance, that the lipid constitution of the membrane makes it more or less fluid, or it may be that the protein is confined to a particularly 'agreeable' lipid 'raft' (see Section 7.2 above). On the other hand it may be, as sometimes happens, that the protein is part of a large quasi-crystalline aggregate of other proteins and thus rendered too bulky to move easily. Or it may be stabilised by structures (cell junctions perhaps) external to the

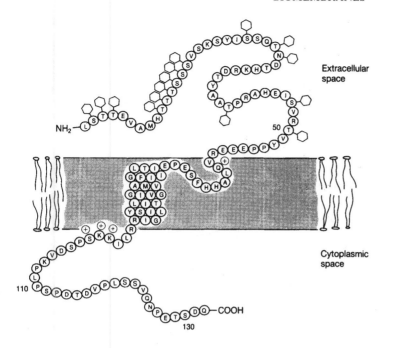

Extracellular
space

50

Cytoplasmic
space

Figure 7.12 Glycophorin in the erythro-cyte membrane. The amino acids are symbolised by their single-letter codes (see Table 2.1). Oligosaccharide groups are represented by hexagons on the appropriate amino acid residues. The positive signs on the K (lysine) residues indicate that this part of the C-terminal polypeptide is believed to be attached to the preponderantly negatively charged P-face of the membrane.

membrane (see Section 7.9). Lastly, it is possible, as we shall see in Section 7.8, that the protein is anchored to one of the elements of the submembranous cytoskeleton.

7.6 SYNTHESIS OF BIOMEMBRANES

We have noted throughout this chapter that membranes are extremely fragile, tenuous structures. It is thus not surprising to find that they are continuously synthesised and broken down throughout the life of the cell. The rate at which this is done is often extraordinarily high. It has been calculated, for instance, that the membrane of *Xenopus* retinal rod discs is synthesised at a rate of $3.2\,\mu m^2/min$. This, no doubt, is an extreme case, but it seems that in many cells an area of membrane equal to the entire surface of the cell is cycled between synthesis and degradation every hour.

The biosynthesis of both membrane proteins and membrane lipids occurs in and on the **rough endoplasmic reticulum (RER)**. We shall discuss the interrelations between ribosomes and endoplasmic reticulum (which together form the 'rough' endoplasmic reticulum) in Chapter 15. Here we can content ourselves by merely stating that proteins destined for incorporation into membrane (whether plasma membrane or the membranes of intracellular organelles) never escape into the lumen of the ER but remain trapped in the ER membrane. This is because the hydrophobic character of the amino acids that follow the signal sequence (see Chapter 15) prevents their squeezing through into the aqueous interior of the ER cisternae. The ER membrane containing the newly formed protein moves toward the Golgi apparatus (Figure 7.13). Glycosylation of the proteins begins in the ER and continues during this movement. Ultimately small 'transport' vesicles bud off the Golgi apparatus and make their way to the plasma and other membranes, where they fuse to form part of the bilayer. Excess membrane, as Figure 7.13 shows, invaginates and makes its way back to the Golgi apparatus in the form of coated vesicles.

This process is by no means haphazard. Radio-labelling shows that both different lipids and different membrane proteins have quite different turnover rates. Once again the processes of molecular recognition must be at work. In the special case of neurons it is not difficult to show that there is a busy traffic of membrane vesicles in the axon (see Chapter 15). Some of these vesicles will, of course, contain secretory materials but others will be involved in the membrane turnover process

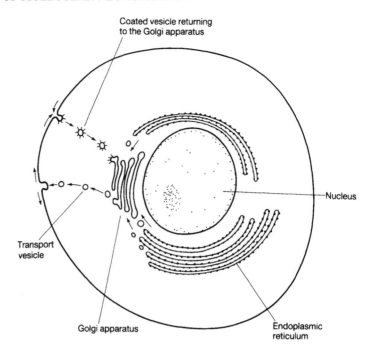

Figure 7.13 Membrane synthesis. The figure shows that not only is fresh membrane being continuously synthesised but also that 'old' membrane is just as continuously resorbed and carried back to the Golgi apparatus.

described above. If the flow of vesicles is interrupted by constricting the axon it can be shown that they mount up on both sides of the constriction. In other words there is a flow of fresh membrane material out to the synaptic ending and a counter-flow of old membrane and debris back to the cell body.

7.7 MYELIN AND MYELINATION

In the vertebrates the axons of both central and peripheral neurons often (not always) become ensheathed in a physiologically very important whorl of membranes: **myelin**. Defect in central myelin leads to the incapacitating condition known as **multiple or disseminated sclerosis (MS)**. The formation and upkeep of myelin is clearly of great importance.

Myelin is formed in both central and peripheral nervous systems by glial cells. In the peripheral nervous system the glial cells responsible are the **Schwann cells**. In the central nervous system (CNS) **oligodendroglial cells** perform the same task. The myelination process is, however, different in the two cases. Peripheral axons become associated with a sequence of single Schwann cells which form a

sort of gutter into which the axon sinks. The opening of this gutter to the extracellular space gradually becomes narrower until its two sides meet to form the **mesaxon**. The mesaxon then begins to grow in length and spirals around the axon, forming first of all a lose spiral of membrane and later a tight whorl. This process is shown diagrammatically in Figure 7.14.

In the CNS the process is different. Instead of a single Schwann cell being associated with a single axon it is found that single oligodendroglial cells myelinate many, sometimes up to fifty, different axons. This is achieved by the oligodendroglial cells sending out huge extensions of their plasma membranes, sometimes as much as ten times the diameter of their own cell bodies, which wrap around neighbouring axons to form their myelin sheaths (Figure 7.15). This is perhaps the most remarkable instance of membrane synthesis known. Oligodendroglial cells may synthesise up to three times their own weight of myelin every day. This huge effort in membrane synthesis has at least one significant advantage. It means that more myelinated axons can be packed into a given volume than if, as in the peripheral nervous system (PNS), a separate glial cell myelinates each axon.

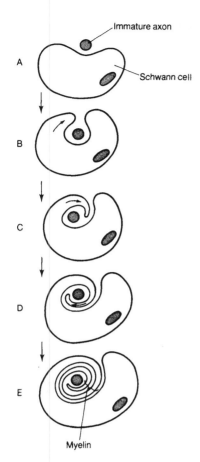

Immature axon

A

Schwann cell

B

C

D

E

Myelin

Figure 7.14 Formation of myelin around a peripheral axon. The process starts at A and proceeds through to E and beyond so that ultimately a tight spiral of perhaps fifty layers of myelin results. Note how the growing myelin whorl tucks under earlier whorls. This process is assisted by P₀ and MAG, which both project into the extracellular space and have similar structures to the cell adhesion molecules (CAMs).

It can be seen from Table 7.1 that the lipid composition of myelin differs markedly from that of its parent glial cell plasma membrane. It contains significantly more cholesterol and phosphatidylethanolamine. Its protein constitution is also very specific (Table 7.2). **Central myelin** contains two major proteins – **myelin basic protein (MBP)** (in fact a family of five proteins derived by differential splicing of the transcript from the MBP gene (14–21.5 kDa)) and **proteolipid protein (PLP)** (30 kDa). **Peripheral myelin** also contains two major proteins,

MBP and **protein zero** (P_0) (25 kDa). Both central and peripheral myelin contain a number of minor proteins, of which the most interesting is **myelin-associated glycoprotein (MAG)**.

The major proteins play an important role in maintaining the structure of both peripheral and central myelin. MBP, rich in lysine and arginine residues, is found on the P-face of developing myelin. The positive charges of the lysine and arginine residues may attach the protein to the generally negatively charged P-faces of the membrane (see Section 7.3 above). It is also possible that the five N-terminal hydrophobic residues may penetrate the lipid monolayer and provide additional help in holding it in place. The presence of positively charged MBP is thought to draw the myelin lamellae together to form a compact structure (Figure 7.16). This compacting is visible in the microscope as myelin's '**major dense lines**'.

The genetic programming of both mouse and human MBP is also becoming known. It has been shown that in both mouse and humans MBP is present in five different forms, having four different molecular weights: 21.5 kDa, 18.5 kDa, 17 kDa (two varieties) and 14 kDa. Careful genetic analysis shows that the MBP gene, located on mouse chromosome 18, is very lengthy: 30–35 kbp. It can also be shown that it consists of seven fairly short exons interrupted by lengthy introns. It turns out that the five different varieties of MBP are formed by differential splicing of the primary transcript. Thus the 21.5 kDa protein is programmed by all seven exons; the 18.5 kDa type is formed by the omission of exon 2 (coding for 26 amino acids); the two 17 kDa varieties by omitting either exon 6 (coding for 41 amino acids) or exons 2 and 5 (which together code for approximately the same number of amino acids); and, finally, the 14 kDa is programmed from a secondary mRNA which lacks exons 2 and 6 (together coding for 67 amino acids).

Mutations of the MBP gene disrupt central myelin (surprisingly, peripheral myelin remains largely unaffected). Mice subject to these mutations (the **shiverer** (*shi*) and **myelin-deficient** (*mld*) mutations) are affected by uncontrollable shivering and convulsions, leading to death at between 50 and 100 days of age. Another pathology in which it is becoming clear that MBP is involved is **multiple sclerosis**. In this disabling condition it is not that the MBP gene has mutated but that MBP becomes

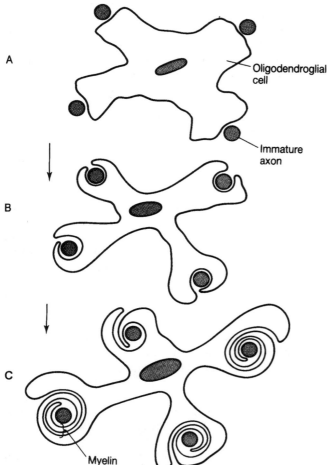

A

Oligodendroglial
cell

Immature
axon

B

C

Myelin

Figure 7.15 Formation of central myelin. Extensions of the oligodendroglial cell wrap around axons to form myelin as in Figure 7.14.

the object of immune attack. It may be that MBP is mistaken for a pathogen; it shares several amino acid sequences with, for example, adenovirus. Or it may be that when the immune system organises an attack on an invading pathogen, for instance measles encephalitis virus, it also attacks homologous regions in MBP. We shall meet other well-established cases of **autoimmune attack** on the nervous system in Section 10.1.4, where we discuss myasthenia.

P_0, the second major protein in **peripheral myelin**, is a member of a very large superfamily of proteins which include the immunoglobulins and the cell adhesion molecules (see Chapter 19). As Figure 7.16 shows, it is believed to have a single transmembrane helix and a very large extracellular

Table 7.2 Myelin proteins

Myelin protein	CNS	PNS	Chromosome (mouse)	Mutation
MBP	+	+	18	Shiverer (shi) Myelin-deficient (mld)
PLP	+	−	X	Jimpy (jp) (mice) Myelin-deficient (md) (rats) PMD (humans)
P_0	−	+	1	
MAG	+	+	7	

PMD = Pelizaeus–Merzbacher disease.
After Lemke (1992), in Z. Hall, ed., *An Introduction to Molecular Biology*, Sunderland, MA: Sinauer; Mikoshiba *et al.* (1991), *Annual Review of Neuroscience*, **14**, 201–217.

domain. This extracellular domain is homologous to an immunoglobulin configuration (see Figure 19.20) and interacts with a similar extracellular configuration formed by a P_0 from the adjacent membrane. Junctions made by P_0s play an important role in both holding the whorls together in mature myelin and ensuring the correct layering of myelin during development. Finally, **type 1** (the most common type) **Charcot–Marie–Tooth** disease (**CMT** disease) is caused by mutation of the P_0 gene. There is a duplication, caused by unequal cross-over during recombination, of a 1.5 Mbp sequence in chromosome 17p12. CMT1 disease has an incidence of about 1/2500 in Western populations. There is peripheral demyelination, axonal degeneration and swelling of Schwann cells. The symptoms include wasting of the lower limbs and sometimes the hands. **Type 2** CMT disease is much rarer and is caused not by defective myelin protein but by defective gap junctions. It is discussed in Section 7.10.

PLP, the second major protein in **central myelin** is, like P_0, a transmembrane protein. Although it plays the same role in central myelin that P_0 does at the periphery, it has no structural homology. It appears that central myelination by oligodendroglia is a comparatively recent development. It is not, for instance, found in cartilaginous fish. We noted the space-saving advantages of oligodendroglia compared with Schwann cell myelination above. PLP, unlike P_0, has three transmembrane helices and its large extracellular domain (between helix 2 and 3) entangles with a partner from an adjacent whorl to provide stability. The PLP gene has been much studied and numerous mutations have been isolated. We have already noted (Section 3.3.2) that the *jimpy (jp)* mutation in mice and **Pelizaeus–Merzbacher (PMD)** disease in humans is due to faulty splicing of the mRNA transcript from the PLP gene. Other PLP mutations are *rumpshaker* in mice, *myelin deficient (md)* in rats and *shaking pup* in dogs.

Finally, it should not be forgotten that there are numerous 'minor' proteins. The best known of these (as mentioned above) is **myelin-associated glycoprotein (MAG)**. MAG is a large protein (100 kDa) which exists in at least two different isoforms, generated by alternative splicing of the primary transcript. It, too, is a member of the immunoglobulin superfamily and has a strong

homology to the cell adhesion molecule, N-CAM (see Figure 19.20). It is interesting to note that it reaches a peak concentration at approximately three weeks after birth (mouse) and this, with the observation that it recognises and binds to axonal membrane, suggests that it is critically involved in myelinogenesis.

7.8 THE SUBMEMBRANOUS CYTOSKELETON

All eukaryotic cells possess some form of cytoskeleton. We shall examine the neuron's cytoskeleton again in Chapter 15. Here we will introduce the subject by a brief description of the best-known submembranous cytoskeleton – that found in **erythrocytes**.

Erythrocyte membranes can be obtained very easily. If blood is subjected to osmotic shock by being placed in hypotonic saline, the erythrocytes (RBCs) burst and appropriate centrifugation will separate the membrane fragments (ghosts) from the haemoglobin. If these fragments are subjected to SDS–polyacrylamide gel electrophoresis at least **12 major bands** can be detected by staining with the dye Coomassie blue (Figure 7.17). Some of the most prominent of these bands are caused by proteins involved in the submembranous cytoskeleton, i.e. '**band 3 protein**', '**band 4.1 protein**', **spectrin**, **actin**, **ankyrin**.

The **band 3** protein is the major transmembrane 'anchorage' protein for the submembrane cytoskeleton. It is also believed to function as a channel for small anions. It is both considerably bigger and considerably more complicated than glycophorin. It is a dimer consisting of two identical chains built of no less than 929 amino acid residues each. Figure 7.18 shows that both the N-terminal and the C-terminal end of each dimer are on the cytoplasmic face of the membrane. The C-terminal appears to be bound firmly to the P-face of the membrane. It is this domain of the protein which acts as the anion exchanger. Chloride ions are exchanged for bicarbonate ions – an important aspect of the erythrocyte's job in respiration. The amino acid chain then traverses the membrane, back and forth, probably as an α-helix, ten times. Some of the loops of this great molecule project out beyond the E-face and one of these forms a point of attachment for oligosaccharide chains. Finally the N-terminal, as

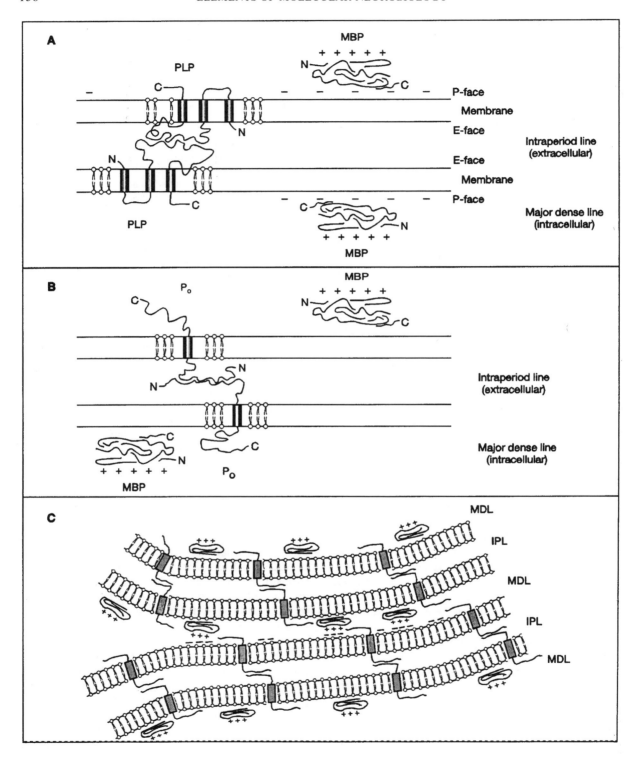

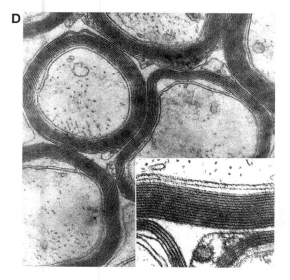

D

Figure 7.16 Involvement of myelin proteins in the formation and stabilisation of the myelin sheath. (A) Central nervous system myelin showing the interaction of PLP in the extracellular space and MBP in the intracellular space. (B) Peripheral myelin. P_0 plays an analogous role to PLP in central myelin. (C) Four whorls of peripheral myelin showing the function of P_0 and MBP in holding myelin together. MDL=major dense line; IPL=intraperiod line. (D) Electron micrograph of transverse section of rat optic nerve ($\times 32,800$). Inset at bottom right a higher magnification ($\times 58,600$) in which the major dense and intraperiod lines are well shown. From T.S. Leeson, C.R. Leeson and A.A. Paparo (1988). *Text/Atlas of Histology*, Philadelphia: Saunders, with permission.

Figure 7.18 shows, projects far into the cytosol and it is this second domain of the molecule which provides firm anchorage for elements of the submembranous cytoskeleton.

The submembranous cytoskeleton of the erythrocyte consists of at least five other proteins: α- and β-**spectrin** (240 kDa and 220 kDa, respectively), **ankyrin**, **band 4.1 protein** and **actin**. Figure 7.19 shows that these proteins form an intricate mesh just beneath the P-face of the membrane. They are responsible for maintaining the biconcave shape of the RBC. **Spectrins** are members of a class of submembranous proteins which includes the **fodrins** of epithelial microvilli and neurons. Both α- and β-spectrins are fibrous proteins and spontaneously assemble to form a two-stranded rope – the αβ-dimer (about 100 nm in length and 5 nm in diameter). Two αβ-dimers join together tail to tail to form an $(\alpha\beta)_2$ tetramer. The spectrin tetramers next form a mesh by interacting with short lengths of actin and these junctions are strengthened by another submembrane protein, the **band 4.1 protein**. Band 4.1 helps to attach the network to the membrane by binding to membrane-embedded **glycophorin** and **band 3**. Last, but very far from least, additional binding to the P-face of the membrane is provided by a fifth submembrane protein, **ankyrin**, which joins the free ends of the spectrin tetramers to other band 3 proteins.

A spectrin-like protein, **fodrin**, is (as mentioned above) also found in neurons. Here again it forms part of the submembranous cytoskeleton. It is both similar to and different from erythrocyte spectrin. It, too, is a tetrameric fibrous protein. Whilst two of the subunits appear to be identical to the α-subunits of spectrin, the other two differ and are named γ-subunits. The fodrin of neurons is thus referred to as an $(\alpha\gamma)_2$ tetramer. There is a surprising amount of fodrin present in neurons, up to 3% of the total protein in some cases. Moreover, erythrocyte-type $(\alpha\beta)_2$-spectrin is also found in lesser quantities. It appears that whilst $(\alpha\beta)_2$-spectrin is restricted to the cell body the $(\alpha\gamma)_2$-fodrin is concentrated in the axon.

In addition to the spectrin-like fodrins, it appears that the other characteristic proteins of the erythrocyte cytoskeleton are also found in neurons. **Synapsin**, which is localised in the termini of central and peripheral axons, where it is a component of the walls of synaptic vesicles, is a homologue of band 4.1 protein. A variant of ankyrin is also present in neural membranes. Actin is well represented. It begins to look as if the submembranous cytoskeleton discovered in erythrocytes has a wider significance. Perhaps something rather like it exists in other cells, and in particular in neurons. We shall see in Section 15.4 that the organisation of the erythrocyte cytoskeleton shows interesting similarities to the interconnections of fodrins, synapsins in **synaptic vesicle membranes**. In addition, the fodrin cytoskeleton of **subsynaptic membranes** is believed to anchor ion channel proteins, holding them in

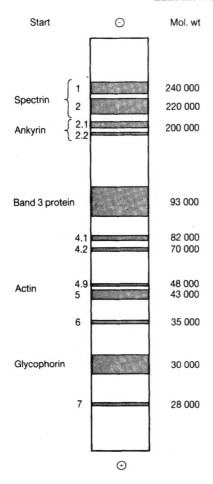

Figure 7.17 SDS–polyacrylamide gel electrophoresis of erythrocyte membrane proteins. The figure shows that if a preparation of erythrocyte membrane is pipetted on to one end of the gel (start) and a voltage applied from one end to the other of the gel (− to +), the various proteins will migrate at velocities related to their molecular weights. After Avers (1986), *Molecular Cell Biology*, Menlo Park, CA: Benjamin/Cummings.

position, just as the spectrin cytoskeleton of the erythrocyte holds the anion channel, band 3 protein, in position (see Section 17.2). Indeed fodrin is a major component of **subsynaptic densities** as well as being concentrated in synaptic endings and at nodes of Ranvier. As these are domains of a neuron's membrane where channels of one sort or another are particularly densely concentrated it seems that, once again, it has an anchoring function.

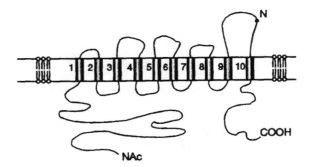

Figure 7.18 Organisation of the band 3 protein in the erythrocyte membrane. The transmembrane α-helices are represented by cylinders, The figure shows a spread-out 'plan' view. In reality the ten α-helical regions are grouped together (the membrane has, of course, a third dimension) to form a compact unit. This compact structure is believed to be associated with a second similar molecule to form a dimer. NAc=*N*-acetylated methionine; N=point of attachment of oligosaccharide chain. After Jay and Cantley (1986), *Annual Review of Biochemistry*, **55**, 511–538.

We have emphasised in the preceding account that we are only considering the submembranous cytoskeleton. In the erythrocyte there is little else, but in most cells there is a great deal more. In neurons the cytoskeleton extends deep into the cytosol and is composed of numerous other protein elements: neurotubules, neurofilaments (of various sorts), actin and a multitude of binding proteins. We shall consider them in detail in Chapter 15.

7.9 JUNCTIONS BETWEEN CELLS

There are three major types of junction between cells, each of which serves a different purpose. **Desmosomes** hold neighbouring cells in a tissue together, **tight junctions** prevent materials diffusing in the intercellular space between two cells and **gap junctions** allow communication between neighbouring cells. Of these three only the latter two play important roles in the brain.

7.9.1 Tight Junctions

Figure 7.20 shows the structure of a typical tight junction. The plasma membranes of the two cells touch each other at intervals, indeed may even fuse, so that all possibility of materials diffusing between

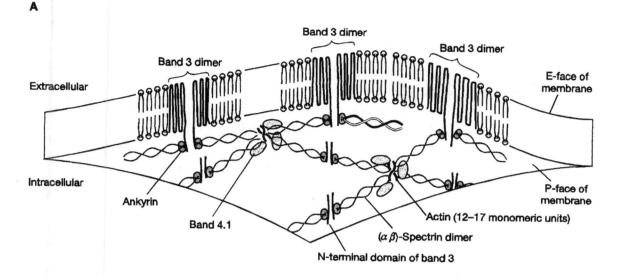

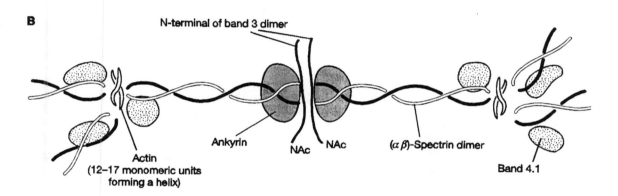

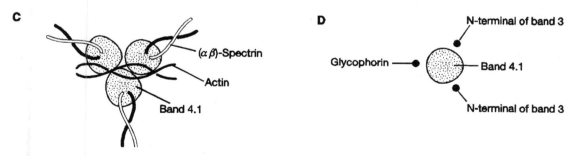

Figure 7.19 Submembranous cytoskeleton of an erythrocyte. (A) Schematic diagram to show disposition of the cytoskeleton on the P-face of the membrane. (B) Enlargement to show the organisation of the major cytoskeletal elements. (C) Plan view of the trimeric junctional complex of band 4.1, actin and α/β-spectrin. (D) Plan view to show the association of band 4.1, glycophorin and the N-terminals of band 3. Further explanation in text.

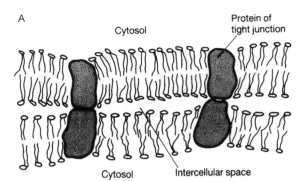

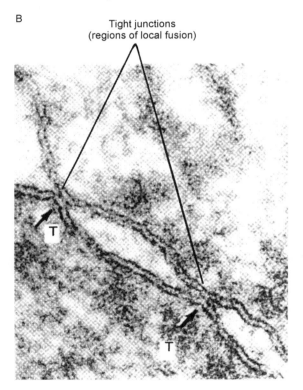

Figure 7.20 Tight junction. (A) Two cell membranes are joined together by the articulation of large globular proteins. (B) Electron micrograph of tight junctions between two epithelial cells in rat intestine (× 326 250). From N.B. Gilula (1974), *Cell Communications* (ed. R.P. Cox), New York; reprinted by permission of John Wiley & Sons, Inc., © 1974.

the two cells is eliminated. Freeze-fracture electron microscopy shows that the junction consists of globular proteins in both adposed membranes which articulate with each other so that the

intercellular space is completely sealed. It is also possible that the junctions are made without proteins but by lipids in the outer leaflets of the adposed membranes interacting with each other to form micelles. Figure 7.20 also shows that one such sealing is usually not enough. Two or more seals are usually made and the number appears to vary with the tissue and the diffusional forces that need to be counteracted. For instance a sequence of six or more tight junctions are made between the epithelial cells that form the wall of the small intestine.

The most important tight junctions in neurobiology are those that are responsible for the '**blood–brain barrier**'. It has been known for over a century that small molecules do not escape from the capillaries of the brain as easily as they do from capillaries in other parts of the anatomy. However, the physical basis of this barrier has only fairly recently been elucidated. It appears that it is due to the presence of well-developed tight junctions between the endothelial cells of brain capillaries. The tight junctions between brain endothelial cells are grouped in pairs, thus forming a double seal and effectively preventing the escape of small molecules from the blood. Tight junctions are also developed between the choroid cells lining the ventricles. The brain is thus protected from unwanted chemical influences. The molecules that reach the neurons have to pass through the endothelial cells of the capillaries and then they are monitored and filtered by the glial cells, especially astrocytes (see Chapter 1).

7.9.2 Gap Junctions

Gap junctions play several important roles in the functioning of the CNS. They were first discovered in the CNS of the crayfish, where they function as **electrical synapses**. But they have since been found in many non-nervous tissues and have been assigned a great variety of functions ranging from metabolic cooperation between neighbouring cells to the signalling involved in growth and development. Figure 7.21 shows that they are formed once again by proteins developed in both adposed membranes. The proteins are called **connexins** and their molecular weights vary from 26 kDa to 50 kDa. This provides the nomenclature. The different connexins are designated

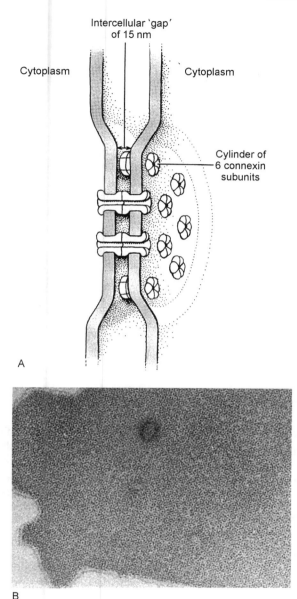

A

B

Figure 7.21 Molecular structure of gap junctions. (A) Diagrammatic representation of gap junction. (B) Electron micrograph of purified suspension of gap junctions. Both A and B show that gap junctions are associated in large, densely packed clusters so that they provide a significant, adjustable, channel of communication between adjacent cells. From *Molecular Cell Biology* by James E. Darnell *et al.* Copyright © 1986 Scientific American Books, Inc. Reprinted with permission.

according to their **molecular weights**: **Cx26** through to **Cx50**.

Six connexins are grouped to form a cylinder surrounding a central hydrophilic canal. This hexagonal structure is called a **connexon** and is adposed to a similar connexon in the adjacent membrane. Thus a hydrophilic pore extends across the intercellular space between one cytosol and the next.

The amino acid sequences of many connexins are now known. The first to be analysed were the human and rat liver connexins. The technique used was similar to that described in Chapter 5. From a purified preparation of liver gap junctions a 19-amino acid primary sequence was determined and a matching oligonucleotide probe synthesised. This probe was then used to screen a library of liver cDNA. It hybridised with a cDNA of 1574 bases. The coding sequence of this DNA was found to specify a polypeptide of 283 amino acids. A very similar sequence (differing in only four amino acid residues) was detected by the same technique in rat liver.

One of the most interesting findings to emerge from this molecular biology is the marked similarity in hydropathic profile between the connexin protein and the subunits of ligand-gated ion channels. These subunits will be considered in Chapter 10. Here, however, it can be said that in each case the primary sequence contains strings of twenty of so hydrophobic amino acids. These hydropathic domains form four membrane-spanning α-helices (see Chapter 2). It should be noted, however, that whereas the connexon of the gap junction is formed by a group of six connexins, ligand-gated ion channels consist of only five (or, in the case of P2X purinoceptors, three) subunits. Thus although the units of which gap junctions and ligand-gated ion channels are built may be similar and evolutionarily related, they are put together in rather different ways in the functionally very different structures. We shall outline further the evolutionary relationships between the various proteins of neuronal membranes in the next four chapters.

Here, however, we should note that **connexons** are often arranged in large clusters. In this they resemble the **nicotinic acetylcholine receptor**. We shall see in Chapter 10 that the latter, especially in electric organs, frequently form huge conglomerations. In the case of connexon clusters the whole

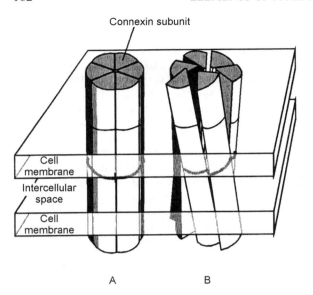

Connexin subunit

Cell membrane

Intercellular space

Cell membrane

A   B

Figure 7.22 Two connexons in a gap junction showing (A) closed and (B) open conformation. Each of the six connexin subunits is believed to rotate so that the channel opens and closes rather like the iris diaphragm of a camera.

structure is believed to provide both a way of transferring metabolites (up to molecular weight of 1.5 kDa) from one cell to another and a pathway of low electrical resistance for electrical signalling. Neighbouring cells are thus both metabolically and electrically coupled. This metabolic coupling can be demonstrated by following the diffusion of dye molecules between cells. The classical studies used the 475 Da **Lucifer yellow**; more recently a smaller molecule, **neurobiotin** (323 Da), has gained favour as it percolates through connections which the larger molecule is unable to penetrate.

The connexon coupling between cells can be controlled. It has been shown that a connexon can exist in various stereochemically different forms (Figure 7.22) – open, partially open and closed – and that these transitions are affected by the concentration of intracellular Ca^{2+}, by pH and by cAMP.

In the CNS, gap junctions were first investigated amongst **glial cells** and in the **retina**. In the nervous systems of **leeches** and **amphibia**, **gap** junctions between neighbouring glia cells have been shown to mediate electrical signalling as well as metabolic

communication. Although a recent report suggests that glial cells may, after all, be excitable and generate action potentials the conventional wisdom is that any electrical communication is by slow 'local-circuit' currents (see Chapter 12). There have also been observations of gap junctions uniting astrocytes, oligodendroglia and neurons in mammalian central nervous systems. This would, at the least, provide a means of metabolic communication between glia and neurons. It may be that, in view of the blood–brain barrier, this provides a vital route for the transport of materials from blood to neuron (see Chapter 1).

Gap junctions also play significant roles in the developing nervous system. Studies of the early differentiation of *Xenopus* neural tube show that initially all the cells are extensively coupled to each other via gap junctions. The cells gradually uncouple as differentiation occurs. This developmentally timed uncoupling has been observed in numerous vertebrate preparations. In particular **neurobiotin** tracing of cells in the developing cerebral cortex shows that the cells are extensively connected. Injecting neurobiotin into a single cortical cell results in clusters of up to eighty neurons showing the tracer. Indeed, the clusters extend through the width of the developing cortex in the form of a column. We shall see in Chapter 20 that an adult cortical column, often regarded as the 'unit' of physiological activity, is made up of cells all originating together from the same small region of the circumventricular neuroepithelium. Neurobiotin tracing shows that these cells not only share complementary cell adhesion molecules but also form an intercommunicating cytoplasmic volume. When synaptic junctions form and synaptic activity commences the cells largely uncouple. This appears to be a general phenomenon. There is, in other words, evidence that the onset of synaptic activity between two cells causes a downregulation of their metabolic coupling via gap junctions.

Nevertheless, far from all gap junctions disappear in the adult nervous system. The adult **retina**, for instance, provides several very interesting cases of gap junctions uniting both sensory cells and neurons together. In all the lower vertebrates so far examined (salamander, toad, fish, turtle), many thousands of **rod cells** are connected by electrically conducting gap junctions (Figure 7.23). This ensures that stimulation of

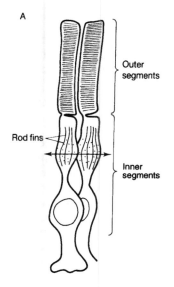

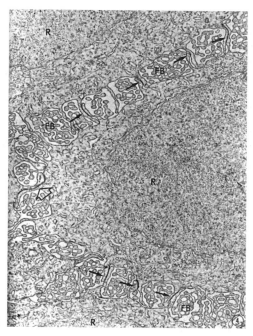

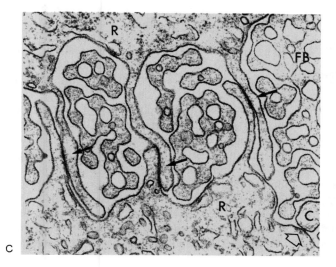

Figure 7.23 Gap junctions in the retina. The figure shows the organisation of the gap junctions which develop between certain types of rod cell in the toad retina. (A) Two rod cells are shown. The outer part of the inner segments develops a series of ridges or 'fins'. (B) Transverse section at the level of the arrow in A. The fins of adjacent rod (and, indeed, cone) cells interdigitate. Gap junctions are made between the fins of adjacent cells. In this way large groups of rod (and cone) cells are electrically interconnected. (C) Electron micrograph of three of these interdigitations (R=rod cell inner segment; C=cone; FB=fibre basket (glial cell process); the arrows point to the gap junctions). From Gold and Dawling (1979), *Journal of Neurophysiology*, **22**, 292–310, with permission.

any one rod cell spreads to a large population. This organisation is not found in mammals but instead it is found that groups of **rods** connect via gap junctions to **cones**. In all vertebrates, both higher and lower, it is found that gap junctions develop between **horizontal** cells and this once again ensures that excitation is spread laterally through the retina, this time to groups of cones. It is clear that gap junctions or electrical synapses play an essential role in the electrophysiology of

the vertebrate retina, especially in increasing its sensitivity. No doubt this is part of the underlying organisation which allows a mammalian retina, under optimal conditions, to respond to a single photon in the visible spectrum.

Finally, it is important to note that in recent years it has become apparent that it is not only the retina which retains gap junctions into adulthood. There have been a number of reports showing that systems of inhibitory interneurons in the neocortex, hippocampus and elsewhere are linked together by gap junctions. These systems are thus electrically (as well as biochemically) united and exhibit **synchronised spiking** (action potentials). Indeed, a recent report shows that the γ-EEG wave is impaired in the Cx36 knockout mouse. Some have seen this gap junction interconnexity as creating a single nervous unit, a '**hyperneuron**', and playing an important role in the functioning of the CNS, perhaps, indeed, forming the physiological correlative of consciousness.

7.10 GAP JUNCTIONS AND NEUROPATHOLOGY

At least 15 connexin genes have been identified in the mammals. These genes are spread over five or six chromosomes: **human** 1, 6, 13, 15 and X; **mouse** 3, 4, 10, 11, 14 and X. Although they are so widespread, the genes (with the exception of that for Cx36) have a common structure: a single intron separating a small non-coding exon from a much larger exon which encodes the whole connexin sequence. Mutations of these genes have been implicated in a number of neuropathologies.

7.10.1 Deafness

Although over a hundred different forms of genetic deafness are known, well over half are due to a G$\Rightarrow$A substitution at codon 70 (G70A) in the *CBJ2* gene coding for **connexin 26 (Cx26)** (chromosome 13 (13p11–12)). This leads to a premature 'stop' codon and a defective connexin protein being expressed in the stria vascularis, basement membrane, limbus and spiral prominence of the cochlea (see Figure 13.25). It is believed that a defective Cx26 prevents the proper recycling of K^+ from synapses at the bases of the hair cells back into the K^+-rich endolymph via the stria vascularis This disrupts the physiology of the organ of Corti and thus causes deafness. We shall see, in Chapter 11, that mutation of a gene encoding a K^+ channel (MinK) has a similar effect. It has been suggested that the prevalence of a mutated Cx26 gene in the population is due to marriage of similarly challenged men and women.

7.10.2 Cataract

As with deafness there are many genetic causes of cataract. One cause, however, is a mutation in the **connexin 50 (Cx50)** gene on chromosome 1. A transition at nucleotide 262 leads to C being replaced by T (C262T) and this, in turn, causes proline$_{88}$ to be replaced by serine (P88S). A second cause of cataract is an A to G transition at position 188 (A188G) in the **connexin 46 (Cx46)** gene on chromosome 13. This leads to the substitution of serine for asparagine at position 63 in the Cx46 protein (N63S). Yet another cause is a frameshift mutation at nucleotide 1137 in the **Cx46** gene. This causes a mistranslation of 56 C-terminal amino acids in the Cx46 protein. All of these mishaps cause defective gap junctions between lens fibres and thus to patchy, dust-like, lens opacities known as **pulverulant cataract**.

7.10.3 Charcot–Marie–Tooth (type 2) disease

This is a rare X-linked disease. Whereas the more common type 1 CMT is caused by mutations affecting the genes coding for myelin proteins (especially P_0) (see Section 7.7), the much rarer type 2 CMT is caused by mutation of the connexin gene, *GJB1*, encoding **connexin 32 (Cx32)**, on the X chromosome (Xq13.1). Numerous amino acid substitutions have been detected as well as frameshift and premature 'stop' codons. It has been found that the majority of these mutations occur in the connexin domain lining the pore or in the domain which forms the surface of attachment with the neighbouring connexin subunit in the adjacent membrane. Type 2 CMT involves loss of neurons in the anterior horn of the grey matter and in the posterior root ganglia, especially in the lumbar and sacral regions.

Both types of CMT disease are peripheral neuropathies. Whilst it is not surprising that type 1, affecting P_0, is confined to the periphery, it is somewhat surprising that type 2 is similarly restricted, for connexin 32 figures prominently in oligodendroglia. It is possible that in the CNS other connexin genes can substitute for defects in connexin 32.

How does defective connexin 32 cause type 2 CMT disease? It has been shown that Schwann cells express Cx32 and concentrate it in the **uncompacted** myelin membranes adjacent to the nodes of Ranvier (paranodal region; see Figure 14.14) and in the incisures of Schmidt–Lanterman. In these regions gap junctions allow communication between the cytoplasm of the Schwann cell body and the cytoplasmic collar of the myelin sheath which wraps around the axon adjacent to the node of Ranvier. Derangement of these channels of communication could have disastrous effects on the well-being of the Schwann cell, its myelin and the enwrapped axon.

7.10.4 Spreading Hyperexcitability (Epilepsy) and Hypoexcitability (Spreading Depression)

Because they function as communication channels between cells, gap junctions have been suspected of being involved in spreading hyperexcitability (epilepsy) and spreading hypoexcitability (spreading depression). It has been shown that epileptic foci show enhanced expression of **connexin 43** when compared with normal controls. Connexin 43 forms the basis of gap junctions between astroglia. Uncoupling gap junctions with agents such as **halothane** blocks the spread of hyperexcitability in experimental preparations.

Vice versa the rate of spread of hypoexcitability (spreading depression) induced by the focal application of K^+ or glutamate is similar to the diffusion rate of Ca^{2+} through glial cell populations (20–50 µm/s). Spreading depression (SD) is characterised by changes in electrical impedance, lengthening of refractory periods, increases in concentration of extracellular K^+ ($[K^+]_o$) and variation in tissue volume. It has been detected in hippocampus, olfactory bulb, spinal cord, superior colliculus and cerebellum. SD may underlie several nervous diseases: seizure discharges, migraine, cerebral ischaemia.

Knockout techniques have created mice lacking connexins 32 and 43. These and other molecular biological techniques will undoubtedly throw much new light on the role of gap junctions in health and disease in the next few years.

Table 7.3 Classification of membrane receptors

Metabotropic	Ionotropic
DA-Rs	
NE-Rs	
mACh-Rs	iACh-Rs
mGlu-Rs	$GABA_A$-R
	Gly-Rs
	iGluR-s
$GABA_B$-Rs	
$5\text{-}HT_{1,2,4,5,6,7}$-Rs	$5\text{-}HT_3$-R
P2Y-Rs	P2TX-Rs
CB-Rs	
Opioid Rs	
Peptide Rs	
NK-Rs	

ACh = acetylcholine; CB = cannabinoid; DA = dopamine; GABA = γ-aminobutyric acid; Glu = glutamate; Gly = glycine; 5-HT = 5-hydroxytryptamine (serotonin); NE = norepinephrine (noradrenaline); NK = neurokinin; P = purine.
The table only shows those receptors described in detail in succeeding chapters. There are many others, falling into one category or the other (especially the metabotropic category), which will be mentioned in passing.

7.11 CONCLUSION AND FORWARD LOOK

This chapter has covered a very wide and rapidly advancing topic. Biomembranes are at the heart of brain physiology. The next four chapters build on the fundamental concepts of membrane structure and function developed here, applying them to the specific characteristics of nerve cell membranes. Thus in Chapter 8 we look at a superfamily of important membrane receptors and note how their action depends on the ability of proteins (so-called G- or N-proteins) to shuttle to and fro in the lipid bilayer of the membrane to influence distant membrane-embedded enzymes and channels. Because the response time of these receptors is comparatively long-lasting (tens of milliseconds) they are known as '**metabotropic**' receptors (Table 7.3). In Chapter 9 we consider the very important role which biological membranes, and especially neuronal membranes, play in separating ionic solutions of different concentrations. We shall see that transmembrane proteins act as '**pumps**' creating these all-important ionic imbalances. In Chapters 10 and 11 we examine the way in which membrane proteins act as '**gates**' or '**valves**' controlling the flows of ions back across neural membranes, down their electrochemical gradients.

We shall see that there are two cases. First, in Chapter 10, we shall look at gates controlled by chemical ligands. In many instances these gates open and close very rapidly (less than a millisecond), and the effect on the membrane potential is consequently also very rapid. Although, like the receptors discussed in Chapter 8, they respond to chemical agonists, the mechanism and rapidity of response is so different that they are clearly distinguishable from the former and known as '**ionotropic**' receptors (Table 7.3). In the second case, discussed in Chapter 11, the gates are controlled by voltage across the membrane. These gates, once again, are many and various, and of great antiquity. Normally they also act very rapidly and are responsible for the ion flows underlying innumerable electrophysiological phenomena, especially the all-important action potential.

8

G-PROTEIN-COUPLED RECEPTORS

Cells communicate by messengers and receptors. Messengers vary from small inorganic molecules to large polypeptides. Pharmacology. Primary and second messengers. Receptors are large membrane-embedded protein complexes. Response depends on the type of receptor presented: either an immediate opening of membrane channels leading to abrupt changes in membrane polarity or a complex, membrane-bound, G-protein-based biochemistry usually leading to the synthesis of second messengers. This chapter discusses the latter response. **The superfamily of 7TM receptors**: membership – general characteristics – structure–function relationships – activation – deactivation (desensitisation) – evolution. **The superfamily of G-proteins**: general characteristics (molecular time-switches) – regulatory proteins (GNRPs and GAPs) – collision coupling – nature of membrane-bound G-proteins – types of effector molecule – types of second messenger – collision coupling in synaptic membranes – interactive networks. **The adrenergic receptors (ARs)**: subtypes – the β_2-AR – isolation – primary structure – interaction with G-proteins – adenylyl cyclase – biological significance – desensitisation. **The muscarinic acetylcholine receptors (mAChRs)**: mAChRs and nAChRs – isolation – primary structure – subtypes – interaction with G-proteins – effectors (adenylyl cyclase, K^+ channels). **mGluRs**: isolation – 7TM disposition – not related to other 7TM receptors – G-protein coupling – variety of effectors (PLC, AC, etc.) – variety of effects. **Neurokinin A receptors (NKARs)** or substance K receptors (SKRs): isolation – primary structure – homologies – structure–function relationships. **Cannabinoid receptors (CBRs)**: isolation – molecular structure – localisation. **Rhodopsin**: ubiquity – isolation – primary structure – X-ray crystallography – variety – response to light – G-protein (transducin) coupling – effector (cGMP phosphodiesterase) – effect on rod cell membrane – desensitisation. **Cone opsins**: structure – function – evolution – biophysics. **Conclusion**: molecular structure reveals evolutionary relationships

At the end of Chapter 7 we noted the various types of junction that hold cells together and, in the case of gap junctions, allow communication between a cell and its nearest neighbours. This of course is an absolute condition of multicellularity. Else, as the poet says, ''Tis all in pieces, all coherence gone/All just supply and all relation'. The nervous system is, of course, the great exemplar of this intercellular signalling and coordination. But on a lesser scale the phenomenon is shown by all the cells of a multicellular body.

8.1 MESSENGERS AND RECEPTORS

In this chapter we begin with a discussion of some of the mechanisms by which one cell can communicate with another. This is usually (not always) accomplished by way of chemical substances. This immediately entails two things: first the production and release of appropriate messenger molecules and second the recognition of these molecules by other cells once released. Details of the synthesis and release of these chemical signals will be considered in Chapters 15 and 16 whilst the biophysical and biochemical response of the **target cell** forms the subject matter of Chapter 17. The initial recognition of the signal, or messenger, depends, however, on the nature of the **receptors** embedded in the target cell membrane.

Receptors are large protein molecules which usually span the membrane – often many times. They possess sites (comparable to the active sites of enzymes) that are stereochemically designed to fit specific messengers. Neurons are far from being the only cells to possess such molecules in their membranes. Probably most cells in multicellular organisms possess membrane receptors. They are especially well represented in cells that respond to circulating hormones.

The messenger or signal molecule (whether it be a hormone or a neurotransmitter) exerts its influence by first binding to the receptor to form a **receptor–ligand** complex. This initial step then sets in train a specific set of responses in the target cell. What these responses are depends on the biochemical nature of the receptor. Thus the same messenger may cause very different results with different receptors.

For many years the study of neural receptors was the province of neuropharmacologists. Their approach was to distinguish and differentiate between receptors in terms of the pharmacological agents which switch them **on** and **off**. These agents are called **agonists** and **antagonists**, respectively. A very large number of such agents has been discovered. Not all of them induce maximal responses in the receptor. Thus pharmacologists distinguish between **agonists**, **partial agonists**, **partial antagonists** and **full antagonists**.

A complex and somewhat confusing nomenclature for receptors has grown up in neuropharmacology. The convention has been to name the receptor after its **full agonist**. Thus we shall meet NMDA receptors, quisqualate receptors, kainate receptors, benzodiazepine receptors, and many more as we proceed through the pages of this book. This is a valuable way of discriminating between receptors and, especially, between receptor subtypes. There are, however, difficulties with this taxonomy. First full agonists at one receptor are sometimes partial agonists at another; second the neuropharmacologist's agonist is frequently not the same molecule as the naturally occurring

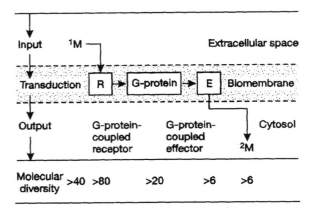

Figure 8.1 Coupling of primary and second messengers by a G-protein system. E=effector molecule; ^{1}M =primary messenger; ^{2}M=second messenger; R= receptor molecule. Partly after Birnbaumer *et al.* (1990), *Biochimica et Biophysica Acta*, **1031**, 163–224; from Smith (1995), in *Biomembranes*, Vol. 1: *General Principles*, ed. A.G. Lee, Greenwich, CT: reproduced by permission of JAI Press Inc.

transmitter for which the receptor is 'designed'. Moves are accordingly afoot to reorganise and clarify the terminology. The **nomenclature committee** of the **International Union of Pharmacology** (**NC-IUPHAR**) periodically publishes recommendations. The advent of molecular neurobiology and the subsequent understanding of the molecular structure and relatedness of receptors is beginning to play a significant role in this clarification.

Let us turn next to the response of the target cell when a receptor–ligand complex is formed. There are two major cases. First, as we shall discuss in the present chapter, the detection of the signal may lead to a complex G-protein-based membrane biochemistry which ultimately leads to biochemical alterations within the target cell or in its plasmalemma. This is the **metabotropic** response and is shown in general terms in Figure 8.1. In other cases, as we shall see in Chapter 10, the chemical signal may open ion channels in the target cell's membrane thus, leading to a change in the electrical voltage across that membrane. This is the **ionotropic** response.

Finally, in this introductory section, it is important to draw a distinction between **primary** messengers and **second** messengers. In the nervous system the primary messenger is the neurotransmitter (or neuromodulator) released from the presynaptic terminal. This crosses the synaptic cleft or gap (or it may diffuse further in the intercellular space) and unites with the receptor in the subsynaptic membrane to form the ligand–receptor complex. This may (though not always) lead to the production of a 'second messenger' on the cytoplasmic face of the subsynaptic membrane (Figure 8.1). This second messenger diffuses into the cytoplasm where it may elicit one or more biochemical effects.

8.2 THE 7TM SERPENTINE RECEPTORS

Neuronal membranes have many different receptors designed to detect many different primary messengers. In this chapter we shall consider six of these receptors: the β_2-**adrenergic receptor** (β_2-**AR**), the **muscarinic acetylcholine receptor** (**mAChR**), the **metabotropic glutamate receptor** (**mGluR**), the **neurokinin receptor** (**NKR**), the **cannabinoid receptor** (**CBR**) and the **visual pigment proteins** (**opsins**) of photoreceptor cells. The molecular structure of these receptors is well understood. With the exception of the mGluRs and CBRs they all turn out to be members of an extremely large superfamily of evolutionarily related proteins (see Figure 8.2). This superfamily is now known to include a family of α-adrenergic receptors (α_{1A}, α_{1B}, α_{1D}; α_{2A}, α_{2B}, α_{2C}); a family of β-adrenergic receptors (β_1, β_2 and β_3); a family of muscarinic acetylcholine receptors (M1, M2, M3, M4, M5); a family of dopaminergic receptors (D_1, D_2, D_3, D_4 and D_5); a family of serotoninergic (5-HT) receptors (5-HT$_{1A}$, 5-HT$_{1B}$, 5-HT$_{1D}$, 5-HT$_{1E}$, 5-HT$_{1F}$; 5-HT$_{2A}$, 5-HT$_{2B}$; 5-HT$_{2C}$; 5-HT$_3$; 5-HT$_4$; 5-HT$_5$; 5-HT$_6$); families of histamine, purine, somatostatin, substance P, vasoactive intestinal peptide (VIP), neuropeptide Y, opioid and tachykinin receptors; the rod, cone and bacterio-opsins; and a large group (over one thousand) of olfactory receptors. It has been pointed out that a significant fraction of the human genome must be concerned in coding these latter receptors (see Section 13.1.2) and an even larger fraction coding all the 7TM membrane-bound proteins.

All these receptors share a common architectural theme. Their polypeptide chain makes seven passes through the membrane (Figure 8.3). This accounts for their being known as the seven transmembrane (7TM) or 'serpentine' receptors. As we have seen in other instances, 'nature' having once stumbled across a valuable theme elaborates innumerable variations. Indeed, the 'stumbling' (in this case) seems to have occurred more than once. As we shall note later, the 7TM mGluRs and P2Y purinoceptors show no sequence relatedness to the main group of 7TM receptors and thus seem to be independently evolved families. To use terms developed first in classical zoology they are analogous rather than homologous to the 'opsin' group. The variations on the 7TM theme in both the main and subsidiary groups have evolved not only to ensure acceptance of different ligands, but also to bring about subtle and not-so-subtle variations in response characteristic from one organism to the next and, indeed, from one part of the same organism to another.

It is believed, as Figure 8.3B shows, that the seven transmembrane sections form the pillars of a hollow column, orientated rather like the iris diaphragm of a camera. Small agonists, such as noradrenaline, acetylcholine and serotonin are thought to occupy binding sites deep within the central cavity of the column. It is easy to see that their presence may well alter the packing of the seven columns. Large agonists such as the polypeptide neuromodulators cannot fit into so small a cavity. Instead their 7TM receptors develop lengthy N-terminal chains projecting into the extracellular space and its is believed that these chains provide the necessary binding sites.

The 7TM receptors share not only a common architectural theme but also a common intra-membranous method of signal amplification. This mechanism, as we shall see, capitalises on the lateral mobility of proteins, in this case G-proteins, in biomembranes, and on the fact that the lipid bilayer holds such proteins in close proximity to each other: they cannot diffuse away into the cytosol. Careful structure–function analysis of serpentine receptors shows that the first, second and third cytoplasmic loops (i-1, i-2, i-3) and the carboxy-terminal tail (Figure 8.3) are crucial to G-protein binding, with the third loop (i-3) particularly concerned with recognising particular G-proteins (Section 8.6).

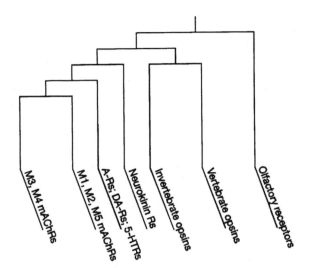

Figure 8.2 Evolutionary relationships of the serpentine receptors. After Findlay *et al.* (1993), *Biochemical Society Transactions*, **21**, 869–873.

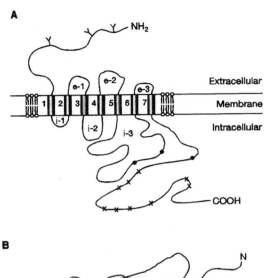

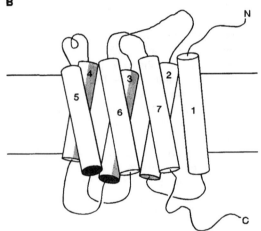

Figure 8.3 The serpentine receptor. (A) The seven transmembrane helices are shown as columns in the membrane, labelled 1–7. The N-terminal sequence is extracellular and is usually glycosylated (Ys). The extracellular loops are labelled e-1,e-2,e-3 and may also sometimes be glycosylated. The intracellular loops i-1, i-2, i-3 provide recognition surfaces for designated G-proteins. The black spots represent phosphorylation sites for PKA and the crosses represent sites for specific desensitising enzymes (e.g. β-ARK in the case of the β-adrenergic receptor). (B) 'Barrel of staves' three-dimensional conformation of a 7TM receptor in a membrane.

Finally, it has long been known that when a receptor has been over-exposed to its agonist it becomes markedly less responsive. This is known as **desensitisation**. There are two cases: **homologous** and **heterologous**. In both cases desensitisation is due to phosphorylation of serine and threonine residues in the carboxy-terminal tail of the 7TM receptor. In homologous desensitisation the phosphorylation is due to an enzyme, **protein kinase A (PKA)**, which is activated by the receptor's own activity (see Section 8.5). In heterologous desensitisation the phosphorylation is due to the activation of PKA as a result of the stimulation of other receptors in the same membrane. Sensitivity is restored by dephosphorylation by the many phosphatase enzymes that populate the cytosol.

8.3 G-PROTEINS

G-proteins are members of a large class of **guanine nucleotide binding** proteins (see Box 8.1), hence their name. They have been likened to 'precision engineered switches' that can turn on and turn off the activity of other molecules. These switches are, moreover, 'time switches'; the duration for which they stay in the 'on' position is precisely determined. It is this 'time switch' capacity that has made them so ubiquitous in the cell's biochemical economy.

BOX 8.1 The GTPase superfamily

Surprising relationships among the proteins are becoming the stock-in-trade of molecular biology. Examples of this phenomenon are found throughout this book. We shall note it particularly strongly in Chapter 18 when we look at the genetics of early development. Many of these relationships have come to light as the result of cancer biology when certain crucial genes mutate. We have already noted, in Box 3.2, the relationship between immediate early genes (IEGs), oncogenes and proto-oncogenes. The superfamily of GTPases tells a similar story.

The superfamily consists of over a hundred members, all of which are GTP-binding proteins with the time-switch facility of catalysing the transformation of GTP to GDP. They exist in two conformational forms, inactive and active. The transition from inactive to active is caused by the exchange of ADP for ATP and the reverse transition, active to inactive, by the hydrolysis of ATP. The superfamily plays many roles in both prokaryotic and eukaryotic cells. Some members (as we see in this chapter) are deeply involved in cell signalling; others play roles in protein synthesis at the ribosome, others in the translocation of nascent proteins in the ER and in the formation and movement of intracellular vesicles (including synaptic vesicles); yet others have a controlling place in cell differentiation and proliferation.

The superfamily gets its name from its universal ability to dephosphorylate GTP. One of the best-known subgroups is the *ras* family consisting of three closely related genes: *H-ras*, *K-ras* and *N-ras*. This family gets its name from the rat sarcoma oncogene carried by two retroviruses (*v-ras*). Later the mammalian proto-oncogene, *c-ras*, was identified. It is concerned with growth-factor signal transduction. Mutations cause neoplastic transformations in mammalian cells. The 21 kDa product of this proto-oncogene (the $p21^{Ras}$ protein) is still structurally the best-known member of the group. Along with the ribosomal elongation factor (EF-Tu) (another member of the superfamily) it has been subjected to detailed X-ray analysis. It can be shown to undergo a major conformational change when it exchanges GDP for GTP. It thus forms a model for understanding the less easily accessible GTPases. For our purposes these are the heterotrimeric G-proteins of membrane signalling systems. The structure of the $p21^{Ras}$ protein resembles that of the α-subunits both in overall conformation and in detailed design. Examination of this structure and its conformational changes, which at first sight seems so far removed from signal transduction at cell membranes, thus provides powerful insight into G-protein biology.

Finally, another powerful reminder of the emerging unity in diversity of molecular biology is the observation that the huge (2818 amino acid) protein encoded by the type 1 neurofibromatosis gene (NF1) has a 360 amino acid run strikingly similar to the GTPase activating proteins (GAPs) (see text) which speed the dephosphorylating action of $p21^{Ras}$. We shall look at the neurofibromatoses in Chapter 21, where we shall see that the symptoms of NF1 include multiple tumours affecting the Schwann cells of peripheral nerves.

G-proteins are switched 'on' by binding to **GTP** and switched 'off' by hydrolysis of GTP to **GDP**. This hydrolysis is catalysed by the GTPase activity of the G-protein itself. The hydrolysis is comparatively slow, having a half-life of a few seconds to a few tens of seconds. Synapses which use this mechanism (see Chapter 16) are accordingly often rather slow acting. A simplified response cycle of a G-protein system is shown in Figure 8.4.

Figure 8.4 shows that in the 'off' or 'inactive' state the G-protein is bound to GDP. The binding is quite firm: the rate constant for GDP release (K_{diss} GDP) is less than 0.03/min. When GDP is released the G-protein enters an 'empty' or 'neutral' state. This state is very transient because as there is normally a higher concentration of GTP than GDP in the cytosol the former quickly enters the empty site. The arrival of GTP causes a conformational

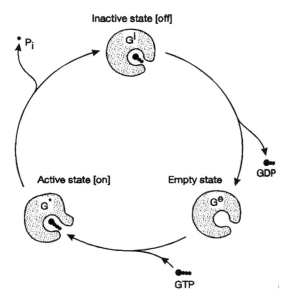

Figure 8.4 Activation–deactivation cycle of G-protein. Explanation in text. From Smith (1995), in *Biomembranes*, Vol. 1: *General Principles*, ed. A.G. Lee, Greenwich, CT: reproduced by permission of JAI Press Inc.

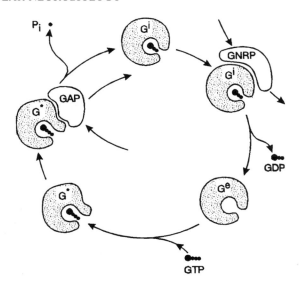

Figure 8.5 Regulation of the G-protein response cycle by GNRP and GAP. Explanation in text. From Smith (1995), in *Biomembranes*, Vol. 1: *General Principles*, ed. A.G. Lee, Greenwich, CT: reproduced by permission of JAI Press Inc.

change in the G-protein. It changes to the '**on**' or '**active**' state (usually represented as **G***). In this 'on' state it is able to activate other biochemicals within the cell. But it is also able to catalyse the hydrolysis of GTP to GDP+P_i and thus return to its original 'off' state. The rate of hydrolysis is crucial: it is this that determines the duration of the G-protein's 'on' state.

To complicate the issue it has been found that in the much-studied archetypal G-protein, p21^{Ras} (see Box 8.1), both the dissociation of GDP from the inactive form and the catalysis of GTP to GDP by the active form are under the control of regulatory proteins: the guanine release proteins (**GNRPs**) and the GTPase-activating proteins (**GAPs**). In the presence of these regulatory factors both processes are markedly speeded up. In Figure 8.5 these regulatory factors have been added to the underlying cycle.

8.4 G-PROTEIN COLLISION-COUPLING SYSTEMS

The G-proteins of membrane signalling systems have a three-part (heterotrimeric) structure

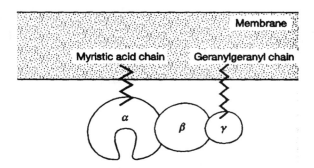

Figure 8.6 Heterotrimeric structure of membrane-bound G-protein. From Smith (1995), in *Biomembranes*, Vol. 1: *General Principles*, ed. A.G. Lee, Greenwich, CT: reproduced by permission of JAI Press Inc.

consisting of a large (c. 45 kDa) α-subunit and smaller β- and γ-subunits (Figure 8.6). All three subunits have evolved a large number of different subtypes. These subtypes are found in different tissues and have different actions on their effector molecules. The α-subunits are the most important (and various) and Table 8.1 gives some of their

Table 8.1 Properties of G-protein α-subunits

Family	Molecular mass (kDa)	Tissue	Receptor	Effector
G_S				
α_s	44.2	Ubiquitous	β_2-AR	AC↑
				Ca^{2+} channels↑
				Na^+ channels↓
α_{olf}	44.7	Olfactory epithelium	Odorant	AC↑
G_i				
α_{i1}	40.3	Nearly ubiquitous	α_2-AR	K^+ channels↑
α_{i2}	40.5	Nearly ubiquitous	M2-ACh	Ca^{2+} channels↓
				K^+ channels↑
α_{i3}	40.5			AC↓ (?)
				PLC↑ (?)
α_0	40.0	Brain	α_2-AR	PLA2↑ (?); Ca^{2+}↓
α_z	40.9	Brain	M2-ACh	AC↓ (?)
α_{t1}	40.0	Rods	Rod opsin	cGMP-PDE↑
α_{t2}	40.1	Cones	Cone opsin	cGMP-PDE↑
α_g	40.5	Gustatory	Gustatory	?
G_q				
α_q	42.0	Nearly ubiquitous	α_1-AR; M_2-ACh	PLC↑
α_{11}	42.0	Nearly ubiquitous	(?)	(?)
G_{12}				
α_{12}	44.0	Ubiquitous	(?)	(?)
α_{13}	44.0	Ubiquitous	(?)	(?)

↑=upregulated; ↓=downregulated; (?)=probably; ?=not known.
Modified from Tang *et al.* (1992), *Cold Spring Harbor Symposia on Quantitative Biology, LVII*, pp. 135–144, New York: Cold Spring Harbor Laboratory Press.

characteristics. Only the α-subunit has a guanine binding site. In the inactive state the three subunits are bound firmly together and both the α- and γ-subunits have fatty-acid 'tails' which attach them to the inner leaflet of the subsynaptic membrane. This attachment (as noted above) ensures that the G-protein is held in the same plane as the receptor and effector molecules and does not diffuse off into the wastes of intracellular cytosol. On the other hand, as we noted in Chapter 7, the lipid bilayer is very fluid and, consequently, lateral movement is usually easy. This allows for continuous shuttling under the influence of thermodynamic forces between comparatively fixed elements, the receptors (Rs) and effectors (Es).

It has been found that when the α-subunit (G-α) binds GDP its affinity for the β- and γ-subunits is much increased. Thus one of the consequences of

GDP binding is to stabilise the heterotrimeric structure. It can also be shown that the tripartite structure has greater affinity for an activated receptor than any of the subunits on their own.

Let us now return to the activation/deactivation cycles of Figures 8.4 and 8.5. We can regard the membrane receptors (Rs) as **ligand-activated GNRPs** and at least some effectors (Es) as **GAPs**. In other words, when a trimeric G-protein collides with a ligand-activated R (R*), GDP will be released from its acceptor site. The empty guanine site soon picks up a GTP molecule and because the affinity of the β- and γ-subunits is less for α-GTP than for α-GDP, the complex tends to dissociate. When the free α-GTP subunit comes into contact with an effector it activates the latter to perform its biochemical function. Very often the effector is an adenylyl cyclase enzyme whose job (when activated) is to catalyse the transformation of ATP into cAMP (a second messenger) plus pyrophosphate. But, as we have noted, the effector can also be regarded as functioning as a GAP. In other words, in addition to carrying out its biochemical function it also speeds the catalysis of GTP to GDP. The α-subunit consequently loses its affinity for the effector and shuttles back along the cytoplasmic leaflet of the lipid bilayer until it contacts the β and γ-complex once again. We have already noted that α-GDP has a strong affinity for the βγ dimer and so the heterotrimeric complex is formed once more. This sequence of events is schematised in Figure 8.7B.

Finally, what role (if any) does the βγ-dimer play? It had, until fairly recently, been thought that this complex (for the two subunits are inseparable in physiological conditions) played no further part in membrane signalling. This is now being questioned. There is evidence to suggest that the complex may in some cases have an inhibitory influence on the α-GTP subunit and/or it may act independently on some effectors. Indeed, quite recently, it has been shown that the βγ-dimer has a direct inhibitory influence on some types of adenylyl cyclase (see below).

8.5 EFFECTORS AND SECOND MESSENGERS

Turning back to Figure 8.1 it is clear that the function of G-proteins is to transmit messages from receptors to effectors. There are various classes of effector molecule: cyclase enzymes, phospholipases, phosphodiesterases, membrane channels. We shall meet instances of all these effectors as we proceed through the pages of this book. Similarly there are various types of second messenger: cAMP, cGMP, inositol triphosphate (IP$_3$ or InsP$_3$), diacylglycerol (DAG) and the ubiquitous Ca^{2+} ion. In this section we examine only two classes of effector, the **adenylyl cyclases** (ACs) and **PIP$_2$-phospholipase (phospholipase C-β)**, both of which engender important 'second messengers'. We shall consider the role of phosphodiesterases and membrane channels and the other second messengers as appropriate in later parts of this book.

8.5.1 Adenylyl Cyclases

These enzymes catalyse the formation of cAMP: a ubiquitous and, perhaps, the most important 'second messenger' in animal cells.

The most important role of cAMP is to activate **cAMP-dependent protein kinase (PKA)**. Once activated this multimeric enzyme phosphorylates (with the help of ATP) one or other of the many biologically active proteins present in the cell – enzymes, receptor and channel proteins, nuclear histones, transcription factors, etc. The phosphorylation is normally of a serine, threonine or tyrosine residue and the effect is either to inhibit (note the desensitisation of G-coupled receptors already mentioned in Section 8.2) or activate the

Formulation of cAMP from ATP catalysed by adenylyl cyclase (AC)

A

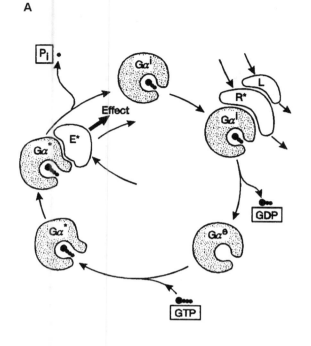

B

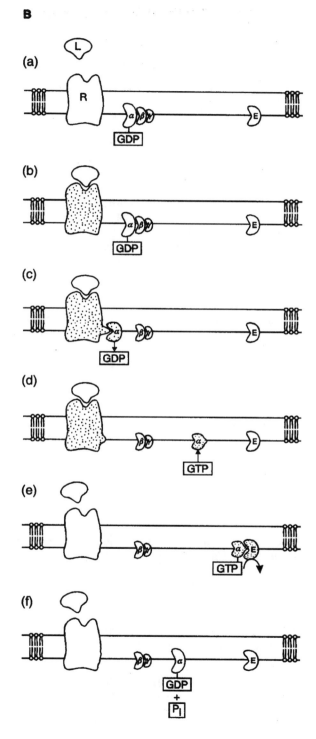

Figure 8.7 Second messenger formation by G-protein collision coupling. (A) Schematic to show how an activated receptor can be seen as a GNRP and an effector as a GAP (compare with Figure 8.5). E*= activated effector; L=ligand; R*=activated receptor. Other symbolism as in Figure 8.4. From Smith (1995), in *Biomembranes*, Vol. 1: *General Principles*, ed. A.G. Lee, Greenwich, CT: reproduced by permission of JAI Press Inc. (B) G-protein collision coupling in a biomembrane. (a) Resting phase. The receptor, G-protein and effector are shown in an unactivated state. (b) A ligand attaches and activates the receptor. (c) The G-protein finds the activated receptor, loses its GDP and dissociates from the βγ-complex. (d) α-subunit is activated by accepting GTP. (e) The Gα-GTP docks with and activates effector causing synthesis of second messenger. (f) Dephosphorylation of GTP leads to α-subunit detaching ready for the cycle to begin again. Stippling indicates activation; E=effector; L=ligand; R= receptor. Further explanation in text.

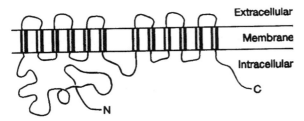

Figure 8.8 Adenylyl cyclase. Plan diagram of adenylyl cyclase showing the 12 transmembrane segments and the extensive intracellular N-terminal domain.

protein. Dephosphorylation back to the original status is by one of the many phosphatase enzymes with which the cytosol abounds.

Molecular biological techniques have shown there to be at least six different adenylyl cyclases in mammalian cells. All have a molecular weight of about 120–130 kDa and examination of hydrophobic sequences indicates that there are 12 transmembrane segments (Figure 8.8). The six cyclases differ in their sensitivity to the $\beta\gamma$-complex of G-proteins and to the calcium binding protein, **calmodulin**. Type 1 AC is, for instance, stimulated by Ca^{2+}/calmodulin and inhibited by the $\beta\gamma$-dimer, whilst type 2 AC does not respond to the first and is stimulated by the second.

It is interesting to find that hybridisation histochemistry locates mRNA for type 1 AC in the neocortex, hippocampus and olfactory system and not (as with the other AC types) generally throughout the brain. This localisation and type 1's calcium sensitivity have suggested that it may be involved in the learning and memory which are features of these parts of the brain. We shall see, in other parts of the book and especially in Chapter 20, that the latter processes are often associated with increased intracellular Ca^{2+}. It can then be argued that this increase stimulates type 1 AC leading to increased quantities of cAMP and that the latter may increase DNA transcription through the cAMP response element (CRE). This, in turn, may lead to the structural changes which are believed to underlie memory (see Section 20.4).

8.5.2 PIP$_2$-phospholipase (Phospholipase C-β)

The activation of this second important effector results in the production of two second messengers:

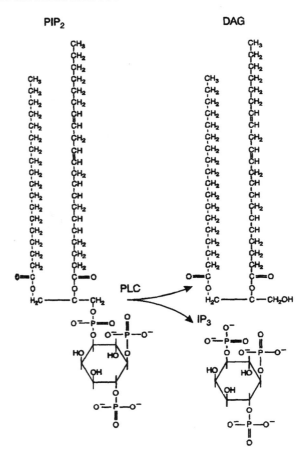

Figure 8.9 Origin of IP$_3$ and DAG from PIP$_2$. Phosphatidylinositol-4,5-biphosphate (PIP$_2$) is converted to diacylglycerol (DAG) and inositol triphosphate (IP$_3$) by the enzyme phospholipase C-β (PLC).

inositol triphosphate (**IP$_3$**) and **diacylglycerol** (**DAG**). Both these second messengers are derived from the phospholipid **phosphatidylinositol** (**PI**) which, as we saw in Chapter 7, is predominantly located in the inner leaflet of the plasmalemma. First phosphatidylinositol is converted to **phosphatidylinositol-4-phosphate** (**PIP**) by the addition of a phosphate group from ATP. This reaction is catalysed by a PI kinase. Next another phosphate group is added, again from ATP and catalysed by a PIP kinase, to yield phosphatidylinositol-4,5-biphosphate (PIP$_2$). Finally (Figure 8.9), **phospholipase C-β**, located in the membrane, cleaves PIP$_2$

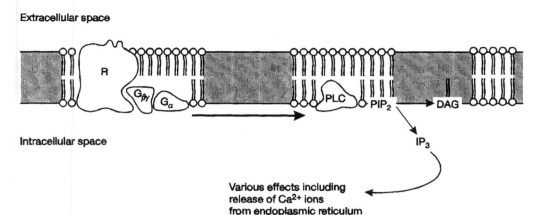

Figure 8.10 IP$_3$ as a second messenger. When the membrane receptor (R) is occupied by a transmitter, a G-protein collision-coupling mechanism activates PLC which cleaves PIP$_2$ into DAG and IP$_3$. DAG remains in the inner leaflet of the plasma membrane while IP$_3$ diffuses into the cytosol as a second messenger. Further explanation in text. PLC=phospholipase C-β; DAG=diacylglycerol; IP$_3$=inositol triphosphate.

into two moieties, **diacylglycerol (DAG)** and **inositol 1,4,5-triphosphate (IP$_3$) or InsP$_3$**.

The production of the two second messengers from PIP$_2$ is shown diagrammatically in Figure 8.10. A receptor in the membrane picks up a signal from a transmitter. A G-protein transmission mechanism activates the membrane-embedded phospholipase. This then reacts with PIP$_2$ to produce IP$_3$ and DAG. IP$_3$ is a water-soluble molecule and hence it readily diffuses away into the cytoplasm. Here it may interact with receptors in the membranes of the endoplasmic reticulum (ER), leading to a release of Ca^{2+} (see Box 10.2). These ions, as we noted above, are known to have many and varied effects on cellular biochemistry. Ultimately IP$_3$ is inactivated by inositol triphosphatase. DAG, on the other hand, is hydrophobic and hence remains behind in the membrane.

We have not finished with the system yet. For the DAG left behind also has a job to do. Figure 8.11 shows that it interacts with a lipid-bound protein kinase, a Ca^{2+}-dependent kinase – **protein kinase C (PKC)**. When the Ca^{2+} concentration of the cytosol rises (an effect, as we have just seen, of IP$_3$) PKC becomes attached to DAG. This interaction requires the presence of phosphatidylserine which, as we saw in Chapter 7, is also concentrated in membrane's inner leaflet. The aroused PKC can now activate proteins which elicit specific biochemical responses. In the case of blood platelets,

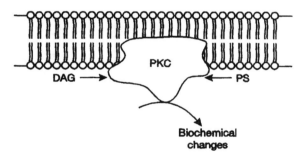

Figure 8.11 DAG as a second messenger. Diacyglycerol (DAG) remains in the membrane and random movement brings it into contact with PKC, which is also in the inner leaflet of the membrane. Activation of PKC also requires the presence of phosphatidylserine (PS), which is also normally present in the inner leaflet. Further explanation in text.

for instance, PKC activates a hydrogen-ion exchange mechanism in the membrane. The consequent alteration in the pH of the cytosol can have profound consequences – not least on the synthesis of RNA. In neurons a number of effects have been demonstrated including synthesis and secretion of neurotransmitters, alterations to the sensitivity of receptors and the functioning of the cytoskeleton. Some recent studies have also suggested that PKC plays a role in determining synaptic plasticity. We shall return to this in Chapter 20 where we consider the molecular basis of memory.

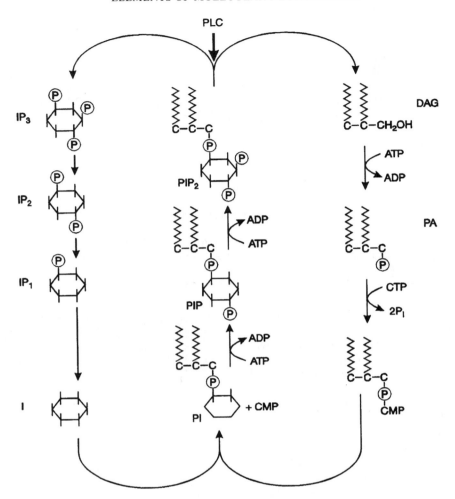

Figure 8.12 Resynthesis of PIP_2 from IP_3 and DAG. The enzyme PLC cleaves PIP_2 into IP_3 and DAG (see text). On the right-hand side of the flow diagram DAG is phosphorylated first to phosphatidic acid (PA) and then interacts with cytosine triphosphate (CTP) to add cytosine monophosphate (CMP). On the left-hand side of the diagram IP_3 is dephosphorylated to inositol diphosphate (IP_2), inositol monophosphate (IP_1) and then to inositol itself (I). The two halves of the flow then come together to yield first phosphatidylinositol (PI) (regenerating CMP), then phosphatidylinositol-4-phosphate (PIP) and lastly back to PIP_2 (see text). After Berridge (1985), *Scientific American*, **253**, 124–136.

Finally, how is the phosphatidylinositol replaced? Evidently the membrane cannot just lose this important constituent each time the second messenger system operates. The resynthesis of PIP_2 is shown in Figure 8.12. The cycle requires the presence of ATP and cytosine triphosphate (CTP) which provide both energy and phosphate bonds. It is interesting to note that **lithium** ions block one of the steps in this resynthesis pathway: the step from inositol monophosphate to inositol. Whether the well-known effects of lithium in controlling the mood swings of bipolar depression (in particular suppressing the manic phase) can be connected to its action in this biochemical pathway is an interesting speculation – but as yet no more than a speculation.

8.6 SYNAPTIC SIGNIFICANCE OF 'COLLISION-COUPLING' SYSTEMS

It is not difficult to recognise the advantages of the G-protein system. It enables a single primary messenger to cause the release of a large number of second messengers into the cytosol. This allows a multifold amplification of the signal. The second messenger may, as we shall shortly see, initiate a variety of effects in the subsynaptic cell. Moreover, the type of second messenger released into the cytosol depends on the effector. The response of a subsynaptic cell to a given primary messenger molecule can thus differ according to the type(s) of effector in its membranes. Finally, there are many different types of α-subunit (Table 8.1). They will have different effects on the same effector, excitatory or inhibitory, more or less rapidly acting, more or less long-lasting. In short, G-protein coupling confers huge amplification and great flexibility of response in the subsynaptic cell.

8.7 NETWORKS OF G-PROTEIN SIGNALLING SYSTEMS

A subsynaptic membrane may be biochemically highly complex. It may present several different 7TM receptors. On to these receptors a single presynaptic terminal may release a mixture of peptide neuromodulators and classical neurotransmitters (Section 16.8). Furthermore, any given stretch of subsynaptic membrane may be served by a number of presynaptic terminals (see Figure 1.19). The G-protein population may also be complex. Experiments with cloned cells have demonstrated the existence of mRNAs for up to seven different α-subunits and three or four different $\beta\gamma$-complexes in the same cell. The number of different heterotrimers in a subsynaptic membrane could thus approach 100. This heterogeneous population of G-proteins may act on a number of different effectors: in addition to various types of adenylyl cyclase and phospholipase C-β, they are also known to act on a variety of channel proteins (see, for instance, Figure 8.20). Biomembranes are fully as complex as, and far more dynamic than, the miniaturised circuit boards of twenty-first century electronics.

If there is indeed a multitude of G-proteins and effectors in a patch of subsynaptic membrane it is obviously necessary that there should be a minimum of 'cross-talk', else the message would be degraded. We are just beginning to understand how this undesirable cross-talk is avoided. Studies of both the 7TM receptors and of effector molecules have indicated that they, too, show subtle variations in their molecular structure. It is now apparent that certain 7TM receptors are tailored to interact with certain Gα-subunits and not others; it is also clear that the effector molecules, the adenylyl cyclases for instance, exist in a number of different isoforms, and consequently that they, also, will only accept certain Gα-subunits. On the other hand, a number of different Gα-subunits do sometimes activate the same effector. At least three Gα-subunits seem to affect certain K^+ channels. Vice versa a single Gα-subunit may have more than one function. A single species of Gα-subunit is, for instance, known to affect both adenylyl cyclase and a Ca^{2+} channel. It is clear, therefore, that without even considering the complications introduced by the activity of the $\beta\gamma$-complexes, the situation in many subsynaptic membranes is of great complexity.

If we look at a 'plan' diagram of a subsynaptic membrane, instead of the usual elevation, we can imagine a very intricate network of activity (Figure 8.13). Moreover, as we noted in Chapter 7, the fluidity of biomembranes, and hence G-protein diffusion constants, is likely to vary from place to place. Indeed the large receptor molecules may well be anchored to the cytoskeleton (Chapter 17) and other barriers may subdivide the membrane into a mosaic of compartments within which, but not between which, G-proteins and their subunits may shuttle. Once again we are struck by the complexity within complexity that characterises the brain.

Having looked at the general features of 7TM receptors and their G-protein signalling systems, it is now time to examine some neurological examples in detail. Of the many instances available we shall confine our attention, first, to two well-known synaptic receptors, the **β_2-adrenergic receptor (β_2-AR)** and the **muscarinic acetylcholine receptor (mAChR)**; then go on to consider the **metabotropic glutamate receptor (mGluR)** and a receptor, or small family of receptors, which responds to a small peptide (the **tachykinin (substance K) receptor**); and, finish by a brief discussion of the **cannabinoid receptors** and the **opsins** of photoreceptor cells.

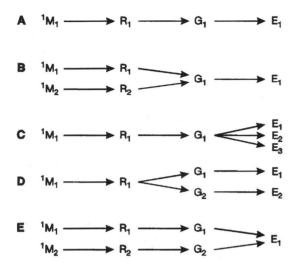

Figure 8.13 G-protein signalling networks in a biomembrane. (A) Primary messenger interacts with receptor, G-protein and effector molecule. (B) Two primary messenger molecules and receptors converge on one G-protein and effector molecule. (C) One primary messenger and receptor affects three different effector molecules. (D) One primary messenger affects two different G-proteins and two different effector molecules. (E) Two different primary messengers, receptors and G-proteins converge on one receptor molecule. There are many other possible networks. ^1M=primary messenger; R=receptor; G=Gα-subunit; E=effector. After Birnbaumer *et al.* (1990), *Biochimica et Biophysica Acta*, **1031**, 163–224; from Smith (1995), in *Biomembranes*, Vol. 1: *General Principles*, ed. A.G. Lee, Greenwich, CT: JAI.

8.8 THE ADRENERGIC RECEPTOR (AR)

The biochemistry and pharmacology of **noradrenaline** (=**norepinephrine**) will be discussed more fully in Chapter 16. Here we shall merely note that it is a member of a class of neuroactive molecules sharing a common six-membered ring structure – **catechol** (Figure 8.14).

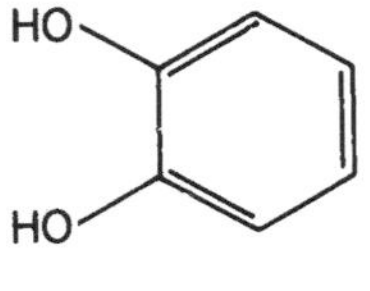

Figure 8.14 Catechol.

As noradrenaline and its congeners (though not adrenaline itself) also possess amine groups, this whole family of important neurotransmitters is commonly referred to as the **catecholamines**.

Ten subtypes of membrane receptor have been identified for adrenaline and noradrenaline, and five receptor subtypes have been identified for the closely related catecholamine, dopamine. These different receptor subtypes have been identified by pharmacological studies using different agonist and antagonist drugs (see Chapter 16, Table 16.7). As we noted in Section 8.2, the adrenaline receptors (which also accept noradrenaline) are classified as α_{1A}, α_{1B}, α_{1D}; α_{2A}, α_{2B}, α_{2C}; β_1, β_2, β_3. All these types of adrenergic receptor have also been found in tissues outside the nervous system. The β_1 receptors, in particular, are known to exist in cardiac muscle membrane. They accept circulating adrenaline from the adrenal medulla and cause the heart to beat more rapidly. Agents such as practolol (the so-called β-blockers), which bind strongly to cardiac β receptors and hence displace adrenaline, are prescribed for angina and other cardiac conditions.

The first adrenergic receptor to have its structure determined was the **β_2 receptor (B_2-AR)**. It is still one of the best known of the G-protein-coupled receptors. In consequence it is to this receptor and its structure–function relations that we turn our attention in this section.

The β_2 receptor was cloned by procedures similar to those described in Chapter 5. First the receptor was isolated and purified by biochemical techniques. After it had been ascertained by immunology that a small peptide fragment of the purified β_2-AR was indeed a fragment of the receptor, its amino acid sequence was determined and a complementary oligonucleotide probe prepared. The probe was then used to fish out the β_2-AR gene from a hamster genome library. This gene was then cloned in a plasmid vector and its nucleotide sequence analysed.

The primary structure deduced from the β_2-AR gene shows that the receptor consists of a protein built of 418 amino acids. Analysis of the sequence for hydrophobic amino acids shows there to be seven hydropathic segments, suggesting that there are seven membrane-spanning helices (Figure 8.15; see Section 7.4). This was the first example of a

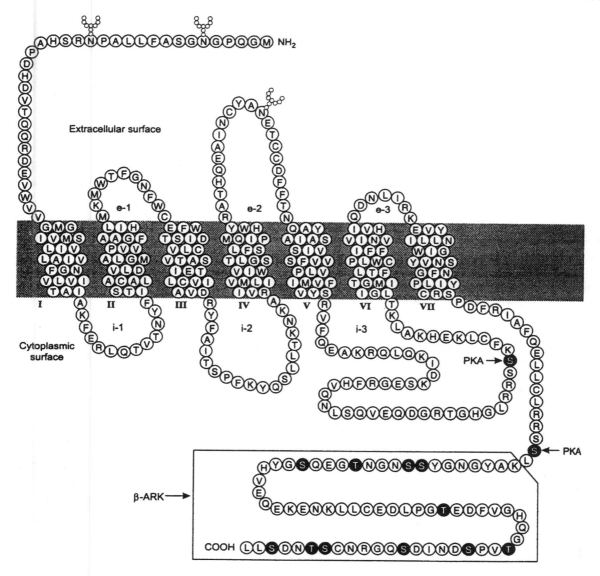

Figure 8.15 Conformation of the human β_2-adrenergic receptor. The N-terminal end of the molecule projects into the extracellular space and two asparagine residues (N) bear oligosacccharide chains (represented by horns). The polypeptide chain then spans the membrane as an α-helix seven times (numbered I to VII in the figure). The molecule ends with a lengthy carboxy-terminal domain which projects into the cytoplasm. Black residues in the intracellular domain indicate sites for phosphorylation by PKA or β-ARK (β-adrenergic receptor kinase), both of which desensitise the receptor. From Kolbika (1992), *Annual Review of Neuroscience*, **15**, 87–114; with permission.

serpentine or 7TM structure and, as we saw in Section 8.2, it proved to be the forerunner of a huge superfamily of such receptors. There is good evidence to show that the catecholamine ligand penetrates deep into the seven transmembrane column and 'docks' in such a way that it interacts with side chains projecting from residues in the third, fourth, fifth and sixth helices.

Next let us examine the interaction of the receptor with its G-protein signalling system. This interaction, schematised in Figure 8.16, provides a good example of the general principles of G-protein biochemistry outlined in Section 8.4. First of all it should be noted that in the absence of an appropriate signal in the form of noradrenaline all three membrane proteins – β_2-AR, G and C – are 'floating' free in the membrane. No doubt they are shuffling to and fro and occasionally bumping into each other, but they do not stick together. When, however, noradrenaline arrives on the outside of the membrane and interacts with the β_2-AR, a whole set of changes occurs. First the presence of noradrenaline changes the **conformation** of the β_2-AR. As we noted above, the catecholamine ligand exerts its influence by intercalating among the transmembrane helices, thus altering the disposition of the membrane-embedded part of the molecule. We shall see later that something analogous occurs when rhodopsin is photo-activated.

The altered conformation of the membrane-spanning segment of the β_2-AR has the effect of rendering it very '**sticky**' to the β- and γ-subunits of the G-protein. On coming into contact with the β_2-AR, the G-protein not only 'sticks' but undergoes a transformation which causes its α-subunit to come free and to release its GDP in exchange for GTP. Not only this, but the freed α-subunit itself undergoes a conformational change which allows it to stick to **adenylyl cyclase**. This, in turn, activates the adenylyl cyclase, which then catalyses the dephosphorylation of ATP to form **cyclic AMP (cAMP)**. cAMP is perhaps the most important of the cell's internal, or 'second', messengers. It diffuses into the cytosol where it may, as we shall see in later chapters, exert a number of biochemical changes. The α-subunit of the G-protein, meanwhile, dephosphorylates its bound GTP to GDP (i.e. in contact with membrane-bound adenylyl cyclase it acts as a GTPase), dissociates from the adenylyl cyclase and assumes its original conformation. When it collides with a $\beta\gamma$ complex it forms the $\alpha\beta\gamma$-complex of the inactive G-protein once again – ready for the whole cycle to start once more.

A further complexity is added to this already rather intricate scenario by the finding that the α-subunit of the G-protein exists in two forms – a form (α_i) which inhibits and a form (α_s) which stimulates the type 1 cyclase enzyme (Table 8.1).

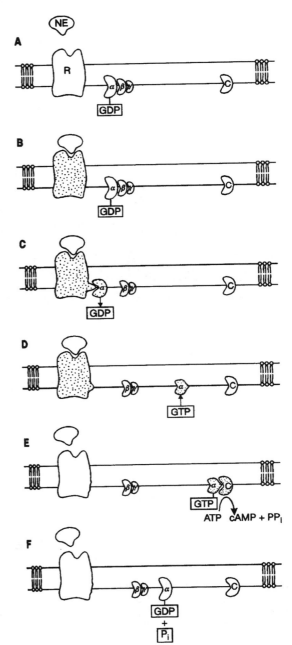

Figure 8.16 Collision-coupling mechanism for the action of noradrenaline at a β-adrenergic synapse. Explanation in text. ATP=adenosine triphosphate; C= adenylyl cyclase; cAMP=cyclic adenosine monophosphate; NE=norepinephrine (noradrenaline). The activated forms of R, G and C are indicated by stippling (compare with Figure 8.7B).

Moreover, as we noted in Section 8.7, the βγ-complex also plays a role: it inhibits some cyclases and potentiates others.

We have already considered the biological significance of this complicated biochemistry (Section 8.6). To recapitulate: the system confers several benefits. First, a single noradrenaline molecule may cause the synthesis of several hundred cAMP second messengers. This is because the activation of a single β_2-AR may lead to the dissociation of tens of G-proteins (remember membranes have a third dimension!). In turn the union of an α_s with a cyclase enzyme may be more or less prolonged and lead to tens or hundreds of cAMPs being synthesised before the α-subunit is dephosphorylated. The system can thus lead to a considerable amplification of the signal. Second, as there are at least six different types of adenylyl cyclase, the response to β_2-AR activation varies from one membrane to another. Third, the α-subunits of G-proteins vary in their effect on the cyclase enzyme. They may, as we saw above, activate or inhibit. Fourth, and finally, the βγ-complexes may exert differential influences on different adenylyl cyclases.

Before leaving the β_2-AR and its associated G-proteins, some further points should be noted. One of the major functions of the cAMP, which as we have seen is the end product of the whole intricate mechanism, is to activate a protein kinase, **protein kinase A** (**PKA**). The function of PKA (in its turn) is to phosphorylate the hydroxyl groups of serine, threonine or tyrosine residues of cellular proteins. These may be channel proteins controlling the flux of K^+ and/or Ca^{2+} ions. In this way comparatively long-lasting effects on membrane potential can be achieved (Figure 8.17). PKA also works back to desensitise the β_2-AR itself. This, as we noted in Section 8.2, is by phosphorylation of serine and threonine residues located in the receptor's carboxy-terminal tail. In fact, this means of desensitisation, although operating on the β_2-AR and other receptors, is not significant on the β_2-AR. Here another kinase, β-adrenergic receptor kinase (β-ARK), phosphorylates the serine and threonine residues in the tail (see Figure 8.15). In both cases these phosphorylations provide the conditions in which yet another inhibitor molecule, β-arrestin, can attach to the receptor. This final attachment is believed to prevent any further interaction with

G-proteins. Ultimately, if the receptor is over-exposed to its agonist for a lengthy period of time, it is removed from the membrane altogether by invagination and sequestration.

8.9 THE MUSCARINIC ACETYLCHOLINE RECEPTOR (mAChR)

Acetylcholine (**ACh**) (Figure 8.18) was the first neurotransmitter to be discovered – by Loewi in 1922 (see Box 16.2). ACh, or **cholinergic**, synapses exist in both the peripheral and central nervous systems. In the peripheral nervous system of vertebrates the neuromuscular junction is always cholinergic and it was here that the pharmacology of ACh was first investigated. We shall look further at this pharmacology in Chapter 16.

Before going any further it is important to make a distinction between acetylcholine's **muscarinic** and **nicotinic** actions. These two completely different effects are due entirely to the interaction of ACh with different receptors. This is a striking example of the importance of receptors in neural membranes. Acetylcholine's nicotinic action is shown at the junction between motor neurons and skeletal muscle, in some central synapses and as we shall see (spectacularly) in the electric organs of electric fish. It is an action which can be mimicked by **nicotine** – hence the nomenclature. Muscarinic synapses, on the other hand, are found on smooth muscle and cardiac muscle, and outnumber nicotinic synapses in the brain by a factor of 10–100. Muscarinic synapses are not affected by nicotine but can be activated by **muscarine**.

We shall discuss the molecular neurobiology of the ionotropic nicotinic acid AChR in Chapter 10. We shall see that its action is very different from that of the **metabotropic** muscarinic receptors we are about to discuss. The pharmacology, biophysics and biochemistry of the two types of synapse are radically different. Instead of the fairly long-lasting and biochemically complex response of muscarinic receptors, we shall see that the nicotinic receptor responds extremely rapidly and then shuts off. Let us, however, turn our attention to the subject of this section: muscarinic acetylcholine receptors (**mAChRs**).

We have just noted that in the brain muscarinic synapses outnumber nicotinic synapses by a factor

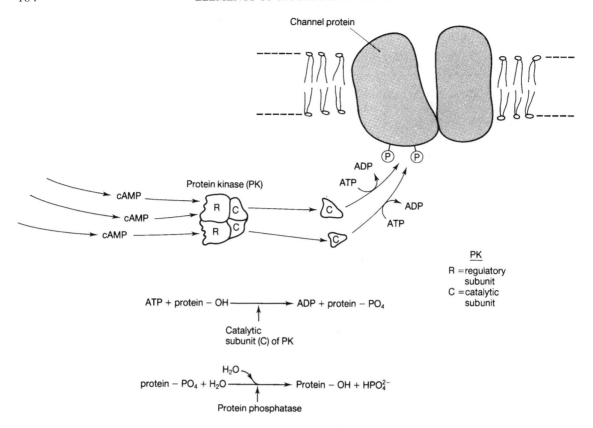

Figure 8.17 Modification of channel conductance by cAMP. Cyclic AMP, shown approaching from the left-hand side of the diagram, interacts with a large tetrameric enzyme – a protein kinase A (PK). Protein kinase A consists of two regulatory (R) and two catalytic (C) subunits. Each regulatory subunit has two binding sites. When more than two cAMPs are bound, the catalytic subunits dissociate. The catalytic subunits catalyse the phosphorylation of specific amino acid side chains in the channel protein, using ATP as both an energy and a phosphate source. The resulting conformational change alters the channel conductance. The lower part of the figure shows that the channel protein is ultimately dephosphorylated by a protein phosphatase.

Figure 8.18 Acetylcholine.

of 10–100. It is thus not surprising that the first muscarinic acetylcholine receptor (mAChR) to be cloned was obtained from brain – the brain of a pig.

Once again the technique involved the use of oligonucleotide hybridisation probes prepared from partial amino acid sequence data. The probes were used to fish for mAChR DNA from a cDNA library prepared from porcine cerebral mRNAs. The cDNA obtained in this way was cloned and injected into the *Xenopus* oocyte. The oocyte expressed the cDNA in the form of receptors in its membrane which showed all the functional and binding characteristics of the M1 subtype of mAChR.

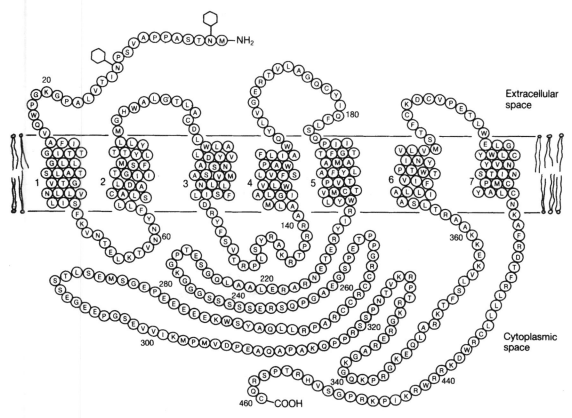

Figure 8.19 Conformation of M1 muscarinic acetylcholine receptor. Explanation in text. Hexagons represent oligosaccharide chains attached to asparagine (N) residues in the amino-terminal end of the molecule. As was the case in the β₂-adrenergic receptor there are seven transmembrane α-helices. The cytoplasmic loop between transmembrane helices 5 and 6 is, however, very much longer than in the adrenergic receptor. After Kubo *et al.* (1986), *FEBS Letters*, **209**, 367–392.

We noted in Section 8.2 that there are various **subtypes** of the mAChR. Four subtypes – M1, M2, M3, M4, – are distinguished, principally by their differential affinities for the anti-ulcer drug **pirenzepine** and its analogues. A more detailed pharmacology is shown in Table 16.2. A fifth subtype, M5, is known only through cloning. The M1 mAChR can be detected through its very high affinity for pirenzepine. It was this antagonist that was used to define the cDNA expressed in *Xenopus* oocyte and confirm that the encoded receptor was indeed the M1 mAChR. It has been shown that the five subtypes are encoded by five distinct genes.

The protein moiety of the M1 subtype of mAChR has a molecular weight of 51.4 kDa and consists of 460 amino acids. In addition there is a large carbohydrate moiety which makes up about 26% of the mass of the entire molecule.

The customary hydropathy analysis indicates that the molecule has seven transmembrane α-helices. The amino-terminal end of the molecule lies on the E-face of the membrane and carries two potential *N*-glycosylation sites. The extensive carbohydrate moiety mentioned above is attached at these positions. The carboxy-terminal tail contains a number of serine and threonine residues which, as was the case with the β-AR, are involved in desensitisation by cytoplasmic phosphorylation. Figure 8.19 shows the amino acid sequence and the molecule's disposition in the membrane.

It has been shown that porcine M1 mAChR activates an α₁ (i.e. inhibitory) G-subunit and hence

switches off adenylyl cyclase. Other types of mAChR may activate α_s (i.e. stimulatory) subunits and thus potentiate the cyclase enzyme. These various effects on the cyclase enzyme control the quantity of cAMP in the cell. We have already noted the effects that this 'second messenger' may have when discussing the β-adrenergic receptor above. An end result of a 'cascade' of biochemical interactions may be to modulate (by phosphorylation) the conductance of, for instance, K^+ channels. Indeed it is known that one of the consequences of activating a muscarinic synapse is down-modulation of the **leak channels** which are largely responsible for the resting potential across nerve cell membranes. This has the extremely important consequence, as we shall see in Chapter 12, of depolarising the membrane.

The M2 muscarinic receptor, present in porcine cardiac muscle, has also been cloned and analysed. It is slightly larger than the M1 subtype, consisting of 466 amino acids. But like M1 it possesses seven transmembrane helices and its amino acid sequence is very similar. Indeed there is 82% homology if both conservative substitutions and identical amino acids are counted.

The M2 mAChR also works through a G-protein system. However, at least in cardiac muscle, instead of using adenylyl cyclase as its effector, Gα acts directly on an ion channel. When ACh is bound to M2 mAChR's ligand site on the E-face of the membrane, the G-protein mechanism on the P-face interacts with an 'inward rectifier' K^+ channel (see Section 11.2.1) so that the latter stays in the open state for longer than normal (Figure 8.20). The influx of K^+ ions which this mechanism allows maintains the membrane at a hyperpolarisation equivalent to the K^+ potential (V_K) (see Chapter 14). The fact that G-protein-coupled K^+ channels operated by a variety of neurotransmitters have been found in the brain suggests that similar mechanisms are at work here also. In atrial cells the 'upmodulation' of the inward rectifier, and the consequent lengthier period of hyperpolarisation, slows the pacemaker and hence the rate at which the heart beats.

The M2 subtype is not confined to cardiac muscle. It is also found in the medulla and pons of the brain. Other subtypes of the muscarinic acetylcholine receptor are also found in the brain. Rapid progress is being made in their analysis. The

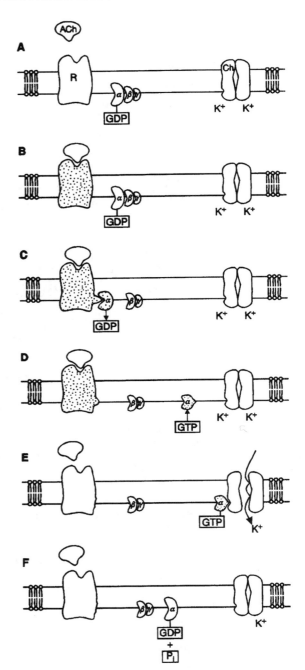

Figure 8.20 Control of a K^+ channel by mAChR-coupled G-protein. Ch=channel protein; other symbols as in Figure 8.7B. Note that the channel is an 'inward rectifier'. This means that K^+ (as shown) flows inward at negative membrane potentials. Further explanation in text.

recognition that the receptors are all members of an evolutionarily related family has been of great value. All of them show considerable amino acid homology in their transmembrane segments. Consequently oligonucleotide probes prepared against the most conserved of these segments (the second) are likely to stick to all five mAChR cDNAs. This turns out to be the case. When a porcine cerebral cortex cDNA library was screened the probes hybridised with five different cDNA sequences. These presumably correspond to the M1, M2, M3, M4 and M5 mAChR subtypes. Indeed there is evidence for even more subtypes. It looks as though each of these mAChR subtypes responds in a (slightly) different way to acetylcholine.

It is clear that mAChR receptors working through a G-protein collision-coupling mechanism can exert very diverse effects. The full range and complexity of these effects are still being elucidated. But the possibilities of long-lasting modulation of various channel conductivities and of biasing metabolic activities in the recipient cell are legion. The muscarinic activity of ACh is very different from the discrete, rapid, effects characteristic of its nicotinic action, which we shall discuss in Section 10.1. It is, therefore, not surprising that mAChRs are so much more prevalent in the brain than nAChRs.

8.10 METABOTROPIC GLUTAMATE RECEPTORS (mGluRs)

We shall see in Section 10.4 that the best-known glutamate receptors control ion channels. They are, in other words, ionotropic. It is only in recent years that G-protein-linked, i.e. metabotropic, GluRs have been discovered. Thus glutamate, an amino acid (see Table 2.1), is analogous to acetylcholine. Both transmitters have two quite different types of receptor: metabotropic and ionotropic. In this section we shall restrict ourselves to mGluRs.

The first mGluR was detected by expression cloning in *Xenopus* oocyte. This is because it bears no sequence homology with other 7TM metabotropic receptors. However, once its sequence was published (in 1991) it became possible (as usual) to use it as a probe to 'fish' in brain cDNA libraries. Five other mGluRs were quickly discovered. At the

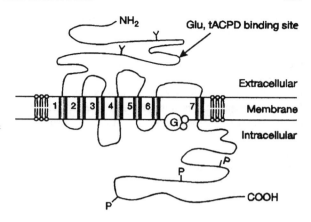

Figure 8.21 mGluR (tACPD receptor). The seven transmembrane helices (M1–M7) are represented by cylinders; glycosylation sites by Ys and phosphorylation sites by Ps. Further explanation in text. G=G-protein. After Hollmann and Heinemann (1994), *Annual Reviews of Neuroscience*, **17**, 31–108.

time of writing, therefore, the mGluRs are seen to constitute a small family, mGluR1–mGluR6.

All the mGluRs are largish proteins (854–1179 amino acids). Each possesses seven hydrophobic sequences which are predicted to form the conventional 7TM structure of G-protein-coupled receptors (Figure 8.21). As usual, the N-terminal is extracellular and the C-terminal, which presents several sites for desensitising phosphorylation, is cytosolic. The cytosolic loop between M5 and M6 is short (compared with the other 7TM receptors considered in this chapter) but highly conserved throughout the mGluR family. It is believed (as in the other 7TM receptors) to be crucial to G-protein coupling. Although the secondary and tertiary structures are similar to that of the other 7TM receptors there is sufficient difference in amino acid sequence and overall conformation to conclude that the mGluRs constitute a separate class, independently evolved. It has been shown that differential splicing of the primary transcript generates a number of subtypes, e.g. mGluR1a, mGluR1b, mGluR1c.

Metabotropic glutamate receptors respond strongly to glutamate and also to quisqualate (QA) and *trans*-1-aminocyclopentane-1,3-decarboxylate (tACPD). The responses, however, vary. The response of mGluR1 and mGluR5 takes the form of large (c. 20 mV), long-lasting (5–10 s), oscillating potentials. This sort of response suggests that the receptors are linked via a G-protein biochemistry to a **phosphatidylinositol** (**PI**) system (see Figure 8.10). Activation of this system leads to the generation of IP_3 and the release of Ca^{2+} ions from the ER and elsewhere (see Box 10.2) and the Ca^{2+} ions can then act on Ca^{2+}-dependent Cl^- channels to induce the oscillating potentials observed.

mGluR2, 3, 4 and 6 work through a different system. They are linked via an **inhibitory** G-protein system ($G\alpha_i$) to adenylyl cyclase. Activation thus tends to shut down the production of cAMP. cAMP may, of course, have many effects. If it is to cause depolarisation of the membrane, an excitatory postsynaptic potential (see Section 17.3.1), then activation of these mGlu receptors will be to inhibit the neuron. The ultimate effect depends, however, on the type of neuron – excitatory or inhibitory – in which the receptor is located. Clearly they will tend to inhibit the activity of excitatory neurons and also that of inhibitory neurons; in other words, in this latter case, release excitation.

mGluRs are widely distributed throughout the mammalian brain. They appear to be mostly located in subsynaptic membranes. It is interesting to note that their distribution overlaps in many parts of the brain with ionotropic GluRs (Section 10.4). This has suggested that the two types may be involved in cooperative activity.

8.11 NEUROKININ RECEPTORS (NKRs)

We saw in Chapters 3 and 4 that a group of small peptides (for sequence data, see Table 2.2), the **tachykinins**, were synthesised in the brain (and, indeed, elsewhere). These molecules, which include **neurokinin A** (**NKA**) (**substance K**), **neurokinin B** (**NKB**) and **substance P** (**SP**), are all synaptically active (see Chapter 16). Their receptors, however, are only sparsely represented in neural membranes. If it were not for the new techniques of molecular biology it is doubtful if their structure and function could have been determined.

In essence the technique adopted was to generate a population of mRNAs from a bovine stomach cDNA library. Neurokinin A, and the other tachykinins, are not, as we noted above, confined

to cerebral tissue. Once having obtained a mixture of mRNAs, it was necessary to select out the strand coding for the neurokinin A receptor (**NK2R**). This was done by using the *Xenopus* oocyte system. The appearance of receptors responsive to NKA and to a lesser extent to the other tachykinins (but not to non-tachykinin peptides) in the oocyte membrane indicated that the appropriate mRNA had been injected. After a number of further fractionations and purifications a single NKA clone was obtained from which the nucleic acid sequence for the receptor could be obtained.

The NK2R was shown to have a molecular weight of 43 066 Da and to consist of 384 amino acid residues. The customary hydropathic analysis of the sequence showed that there were seven transmembrane segments (Figure 8.22). It is thus very obviously a member of the family of receptors we have been considering in this chapter. Unlike the majority of the serpentine receptors the NKR family has been shown to act through a **phosphatidylinositol** second messenger system.

Analysis of the amino acid homology between the NK2R and other members of the superfamily of 7TM receptors confirms their mutual relatedness. If the comparison lumps together identical residues and conservative substitutions then the similarities are as follows:

NK2R compared with opsin: 46% (21% identity)
NK2R compared with β_2-AR: 39% (24% identity)
NK2R compared with M1 mAChR: 38% (24% identity)
NK2R compared with M2 mAChR: 34% (22% identity)

More recently the structures of the other two tachykinin receptors, NK1R and NK3R, have been determined. Reference back to Table 2.2 shows that all three tachykinins share a common C-terminal: -Phe-x-Gly-Leu-Met. This is recognised by all the NKRs. The differing N-terminals of the tachykinins alter the avidity with which they bind to different NKRs and hence confers specificity.

Site-directed mutagenesis has allowed detailed structure–function studies of the NKRs to be undertaken. Specific amino acids could be eliminated or substituted and the consequences on the functioning of the NKR examined. It turned out that the first and second extracellular loops of the

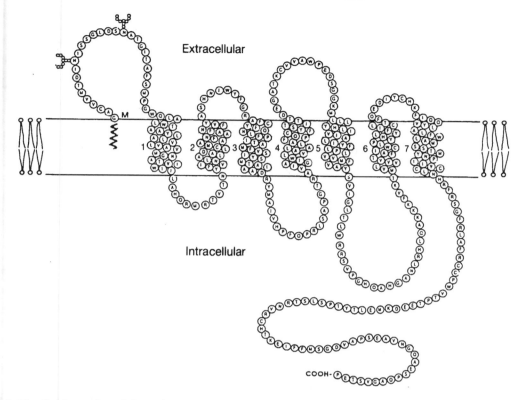

Figure 8.22 Conformation of the substance K receptor. There are seven transmembrane helices. The N-terminal end of the polypeptide chain is believed to be bonded to myristic acid, which is embedded in the membrane. Reprinted with permission from Masu *et al.*, (1987), *Nature*, **329**, 836–838. Copyright 1987, Macmillan Magazines Ltd.

receptors are critical for binding the agonist. It was also found that the second and fifth transmembrane helices of the NKR were also involved. The C-terminal of all the tachykinins appears to penetrate deep into this intramembrane domain and achieve strong binding. This feature reminds us of the binding positions of the catecholamines and other small agonists. The intramembrane helices and the amino acid residues most involved are, however, markedly different in peptide and small molecule binding. Finally, site-directed mutagenesis confirms that the differential affinity of the different tachykinins for the three NKRs is due to differing stereochemical fits between the N-terminal ends of the tachykinins and the differing conformation of the extracellular loops of the three NKRs.

8.12 CANNABINOID RECEPTORS (CBRs)

During the search for NKRs in rat brain cDNA libraries, a novel 7TM G-coupled protein came to light. This 473-amino acid protein responded to no known neurokinin. It was for a while a receptor without an agonist or, in the jargon, an '**orphan receptor**'. The *in situ* mRNA hybridisation technique (see Section 5.19) allowed the distribution of this orphan receptor to be mapped in the brain. When this was done, it was quickly seen to match the distribution of a known radiolabelled cannabinoid ligand: ^{3}HCP-55 940. This led to the suspicion that the orphan was in fact the (until then unknown) cannabinoid receptor. This suspicion was confirmed by further biochemical analysis, especially by the finding that cannabinoid stimulation of the transfected receptor led to inhibition of adenylyl cyclase activity – a known consequence of cannabinoid stimulation. A cannabinoid receptor was soon found in human brain cDNA libraries on the basis of 98% amino acid sequence identity with that of the rat.

A second cannabinoid receptor (CBR$_2$) was subsequently found in the immune system, but not in the brain. In this text it is only the brain receptor (CBR$_1$) that will be discussed. It clearly lies at the heart of the well-known effects that extracts of the plant *Cannabis sativa* (cannabis= marijuana) have on the human brain/mind. The CBR$_1$ receptor is concentrated in the basal ganglia, the hippocampus and the cerebellum, areas that are consistent with behavioural effects of the drug. It is also found in the spinal cord and in the peri-aqueductal grey matter (PAG). This tissue surrounds the central canal of the spinal cord and brain stem and holds the pain pathways as they course up to the brain. Closer examination of the anatomy indicates that CBR$_1$ is located in the presynaptic membrane. We shall discuss its molecular physiology in Section 16.6.

Finally, biochemical and computer searches of DNA databases suggest that cannabinoid receptors are unique to vertebrates. Whilst these investigations have picked up CBR$_1$ orthologues in bony fish, birds and mammals, computer searches of *Drosophila* and *Caenorhabditis* genomes have drawn a blank. It looks as if CB$_1$ receptors and cannabinoid signalling are a vertebrate invention.

8.13 RHODOPSIN

We have already noted (Section 8.2) that opsins belong to the 7TM superfamily. Here, if anywhere, the power of the molecular approach to neurobiology becomes apparent. Not so many years ago visual pigments and adrenergic synapses seemed poles apart. Now their operation at the molecular level is seen to be evolutionarily and functionally closely related. Students of cardiac disease suddenly share a common deep interest with students of ophthalmology. The action of β-blockers may tell us something not only about the cardiac adrenergic receptor but about the means by which a photon stimulates a photoreceptor!

In this section we shall only consider **rhodopsin** and its associated G-protein (often called **transducin (T)**). The organisation of rhodopsin in rod cell outer segments and the further biochemistry of photoreception will be considered in Chapter 13.

Rhodopsin is a ubiquitous photopigment. Not only is it found in the retinal rod cells of all vertebrates but it is also widespread as a photopigment in invertebrates. An analogue, **bacteriorhodopsin**, is found in the 'purple membrane' of the natrophilic archaebacterium *Halobacterium halobium*. As with all other rhodopsins it consists of a membrane-embedded protein, **opsin**, and a light-sensitive pigment group, **retinal**. Although its 7TM tertiary structure is similar to that of animal opsins, its amino acid sequence is very different. It is likely, therefore, that bacterial and animal opsins have separate evolutionary origins and have ultimately converged on the same membrane-bound configuration. This is supported by the fact that retinal is in the **13-*cis*** conformation in bacteriorhodopsin rather than the **11-*cis*** conformation of animal opsins. Because the purple membrane forms crystalline sheets bacteriorhodopsin has been subjected to X-ray crystallography and was, consequently, for long the best-known, structurally, of all the rhodopsins. In *H. halobium* it acts as a 'proton pump' across the membrane.

Retinal absorbs light in the visible range (γ= 400–600 nm). Its structure in animals is shown in Figure 8.23. The figure shows that it can exist in

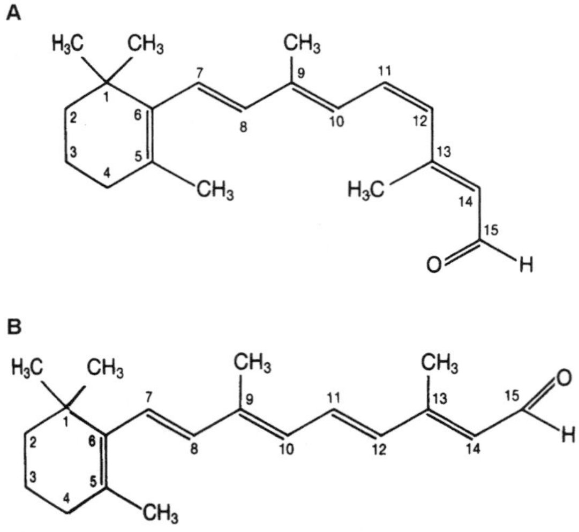

Figure 8.23   (A) 11-*cis*- and (B) all-*trans*-retinal.

two forms – an '11-*cis*' form and a lower-energy, 'all-*trans*' form. The primary event in photoreception is the absorption of a photon of visible light by 11-*cis*-retinal. This provides the activation energy necessary for the transition between the 11-*cis* and all-*trans* form.

It is the 11-*cis* form of retinal that is attached to opsin in the rhodopsin molecule. The attachment is by way of **Schiff base linkage** to a **lysine** residue in the opsin. Transformation into the all-*trans* form breaks this linkage and leads to the dissociation of retinal from opsin (Figure 8.24). This initiates a cascade of biochemical, biophysical and physiological events which ultimately leads to a visual sensation.

Let us turn now to the structure of opsin. The first opsin to have its amino acid sequence analysed was derived from cattle eyes. Subsequently, the usual methods of oligonucleotide hybridisation

have enabled biologists to deduce the amino acid sequences of well over a hundred opsins from a wide variety of different species. All the vertebrate rhodopsins have a chain length of 348 amino acids and show a high degree of sequence homology. The *Drosophila* rhodopsin consists of 373 amino acids and although the overall sequence homology with bovine rhodopsin is only 37%, portions of it are sufficiently similar to be picked out by a bovine oligonucleotide probe. The overall conformation of *Drosophila* rhodopsin and its disposition in the membrane is remarkably similar to that of the mammalian photopigment.

Hydropathic analysis suggests that once again there are seven transmembrane segments – numbered from one to seven from the amino-terminal end (Figure 8.25). The carboxy-terminal projects, as usual, from the P-face of the membrane into the cytoplasmic space and the amino-terminal projects

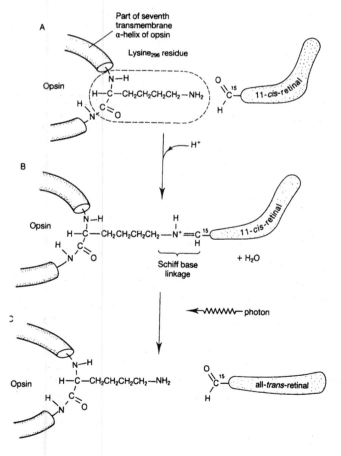

Figure 8.24 Linkage of 11-*cis*-retinal to lysine$_{296}$ of opsin. The linkage, as shown, takes the form of a Schiff base. This linkage is disrupted by a photon of appropriate energy. 11-*cis*-retinal is transformed to all-*trans*-retinal and diffuses out of the opsin molecule.

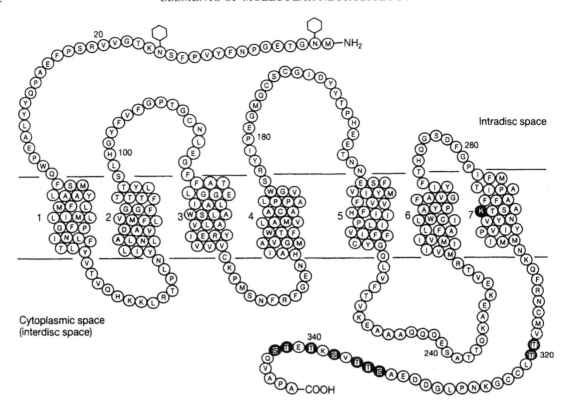

Figure 8.25 Bovine rhodopsin. The structural similarity to both the β_2-adrenergic and the mACh receptor molecules is clear. Once again the amino-terminal end bears oligosaccharide groups; once again there are seven transmembrane α-helices. The heavily circled lysine residue (K) is that to which retinal is attached (see Figure 8.24). The stippled residues in the carboxy-terminal end of the molecule are those that are phosphorylated by rhodopsin kinase (see Figure 8.27 and text). For orientation of the molecule in the outer segment disc see Figures 13.9 and 13.11 and Plate 1. After Lefkowitz *et al.* (1986), *Trends in Pharmacological Sciences*, **7**, 444–448.

into the luminal space within the rod discs (see Chapter 13). The transmembrane helices of both *Drosophila* and bovine rhodopsin and all the loops except that between helices five and six have exactly the same number of residues. The extra amino acid residues in *Drosophila* occur at the amino- and carboxy-terminal ends and in the five/six loop.

Because bovine rhodopsin is so readily available it is not surprising that it was the first animal rhodopsin to be crystallised and have its structure solved by X-ray diffraction. The structure revealed by this technique at the 2.8 Å level largely confirmed the structure worked out by hydropathy analysis as described above. The detail provided by X-ray crystallography is, of course, far greater, and

the three-dimensional disposition of the whole molecule can be viewed from all angles (see Plate 1).

Figure 8.25 shows the seven opsin helices spread out in the usual plan form. X-ray analysis confirms that within the membrane they are grouped to form a somewhat 'dented' cylinder or '**barrel of staves**' (Figure 8.26). The retinal pigment group is located deep down in the centre of this cylinder. The lysine residue to which it is attached through the Schiff base linkage is residue 296 in bovine opsin and residue 319 in *Drosophila*.

The precise chemical environment surrounding retinal, that is the exact nature of the adjacent amino acid side chains from all seven helices, determines the wavelength to which the retinal will

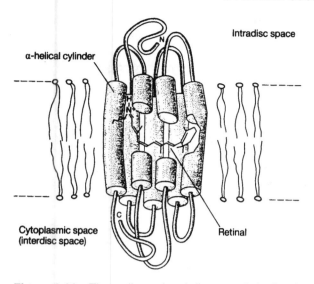

Intradisc space

α-helical cylinder

N

Cytoplasmic space
(interdisc space)

C

Retinal

Figure 8.26 Three-dimensional diagram of rhodopsin molecule in the rod disc membrane.

be maximally responsive. Indeed, investigation of the wide variety of opsins which have evolved in the vertebrates shows that amino acid variation alters λ_{max} from about 350 nm to about 560 nm. On receipt of an appropriate photon 11-*cis* retinal will unbend into the all-*trans* form and this causes a conformational change in the opsin. The X-ray data show that 11-*cis* retinal holds the opsin in a strained configuration which is released once the Schiff base linkage is broken. The change in conformation has far-reaching consequences.

For, as we indicated above, rhodopsin (like β-AR and mAChR) is associated with a system of G-proteins (in the retina, as we noted, often referred to as T-proteins). The association appears to be through the carboxy-terminal sequence, especially a short α-helical region close beneath the P-face of the membrane and designated H8. The other cytoplasmic loops, especially the second, connecting the transmembrane helices on the P-face of the membrane also play a part. Together they form a structure, some 43 Å in length, which may be large enough to interact with both α- and γ-subunits of transducin. A similar collision-coupling mechanism operates to that which we described for the β-adrenergic receptor. However, there is a difference. Figure 8.27 shows that instead

of the α-subunit of the G-protein ($G\alpha_{t1}$ or $T\alpha$) acting on adenylate cyclase to produce cAMP, the α-subunit in rod disc membranes acts on a large tetrameric ($\alpha\beta\gamma_2$) **cGMP phosphodiesterase (PDE)**. When the α-subunit contacts the PDE it binds to and displaces the two γ-subunits. This unveils the catalytic power of the PDE, which is able to open the cyclic GMP ring to form 5′-GMP. The two γ-subunits act as GAPs to speed the dephosphorylation of the $G\alpha_{t1}$ and when this happens the Gα detaches; this allows the γ-subunits to rejoin the α- and β-subunits of the PDE and inhibit its catalytic action.

Figure 8.27 schematises the mechanism described above by which the photoactivation of rhodopsin ultimately results in the formation of 5-GMP. We shall see in Chapter 13 that cGMP concentrations keep the rod cell in a depolarised state. As we noted when discussing collision coupling in the β-AR system, the mechanism allows considerable amplification of the signal to occur. Receipt of a single photon may result in several thousand cGMP rings being opened. This transformation of cGMPs into their straight-chain forms consequently leads to a hyperpolarisation of the membrane and this in turn to a sequence of neurophysiological events ending in the activation of one or other parts of the visual system in the brain.

The conformational change undergone by opsin when retinal's Schiff base linkage is broken has further consequences. It opens the way for inhibition by two cytoplasmic proteins, **rhodopsin kinase** and **arrestin**. Rhodopsin kinase phosphorylates as many as nine serine and threonine residues in opsin's carboxy-terminal tail (see Figure 8.25). This, of course, is analogous to the action of PKA and β-ARK on the β-AR (Section 8.8). But full desensitisation does not occur until **arrestin** attaches. This only occurs when the opsin has been phosphorylated by rhodopsin kinase. When this has happened arrestin attaches and this changes the conformation of opsin in such a way that it strongly inhibits its ability to couple with $G\alpha_{t1}$. The collision-coupling biochemistry is consequently switched off. The phosphodiesterase returns to its inactive state: no further cGMPs are converted to 5-GMPs. The rod cell returns to its resting state. We noted in Section 8.8 that an arrestin homologue, β-arrestin, has a similar effect on β₂-AR. Analysis of its amino acid sequence

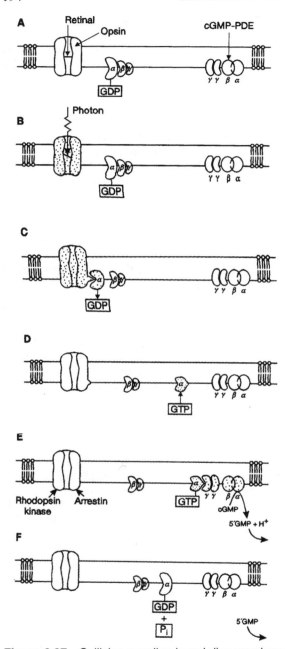

shows that it is 59% identical to that found in rod cells.

8.14 CONE OPSINS

Colour vision is widespread in the animal kingdom. It depends on the presence of pigments which absorb maximally at different wavelengths. In *Drosophila* these pigments are located in different retinula cells of each ommatidium of the compound eye (see Section 19.5). In the vertebrates they are located in **cone outer segments**. In the human retina there are three types of cone pigment having absorption maxima at 419 nm (blue), 531 nm (green) and 559 nm (red). The absorption maximum of rhodopsin (for comparison) is at 496 nm. Differential stimulation of the three categories of cone pigments is the first step in colour vision. It should be noted, however, that so far as mammals are concerned primates are exceptionally well endowed. Alone amongst mammals they have three cone opsins. All the rest develop only two: a long-wave and a short-wave pigment. Perhaps this is due to evolutionary history. It is perhaps significant that a third visual pigment originated only some 40 million years ago (see below). It was at this time that the primates began their long period of arboreal existence. We might speculate that it became important in this habitat to be able to detect the ripeness or otherwise of fruit by their different colours. It is interesting to note, moreover, that although both Old and New World monkeys possess a three-pigment colour system, the originating gene duplication occurred independently in each group: the New World monkeys a little later than in the Old World monkeys. Is it possible that just as the colour vision of phytophilous insects co-evolved with their plant food sources, so the colour vision of frugivorous primates co-evolved with *their* plant resources? The significant advantage conferred by being able to detect the yellow and orange of ripe fruit against a background of green foliage ensured that the mutation was retained in the genome.

The cone pigments, like rhodopsin, consist of a protein (**cone opsin**) attached by a Schiff base linkage to retinal. The differing absorption maxima are believed (as we shall see below) to be due to the three different opsins providing somewhat different chemical environments for retinal.

Figure 8.27 Collision coupling in rod disc membrane. Explanation in text. The figure shows that cGMP-PDE is activated by the α-subunit of the G- (or T-) protein. Once activated, PDE is able to open the ring structure of cGMP. The consequences of this change will be discussed in Chapter 13. The photoactivation of rhodopsin is switched off by rhodopsin kinase and arrestin.

Table 8.2 Homologies of rod and cone opsins

	Percentage sequence homology			
	Rod opsin	Blue opsin	Red opsin	Green opsin
Rod opsin	100	75	73	73
Blue opsin	42	100	79	79
Red opsin	40	43	100	99
Green opsin	41	44	96	100

Data from Nathans (1986). Values below the 100% diagonal are percentage identical residues; values above the diagonal are percentage identical plus conservative substitutions.

Their photoactivation and collision-coupling interaction with cGMP phosphodiesterase is similar to that which we have just outlined for rod opsin.

The three **human** cone opsins have been sequenced by the customary techniques of gene cloning. Table 8.2 shows the sequence similarity with each other and with human rhodopsin. It can be deduced from this table that the cone opsins and rod opsin form a family. It appears that a single ancestral gene duplicated twice at least 500 million years ago. One of the three resulting genes continued to code for rod opsin whilst the others evolved to code for the red and blue cone opsins. A much more recent duplication of the red opsin gene occurred about 40 million years ago. One of the two genes resulting from this final duplication continued to code for red cone opsin whilst the other evolved to code for green cone opsin. The mutability of opsin genes is emphasised by the more than 70 amino acid substitutions which have (to date) been detected in the opsins of patients suffering from **retinitis pigmentosa** (see Box 13.1).

Hydropathy analysis indicates that the cone opsins have retained the seven membrane-spanning helices of rhodopsin (Figure 8.28). All the cone opsins have a lysine corresponding in

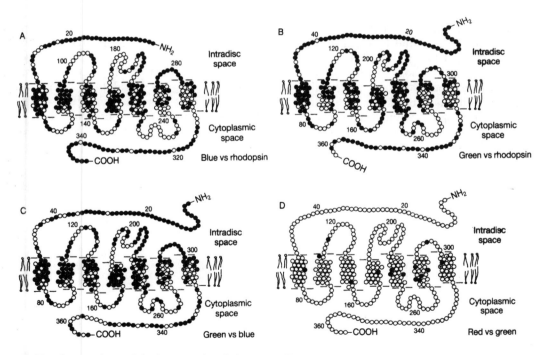

Figure 8.28 Comparison of the human visual pigments. The black residues indicate altered amino acid residues. It can be seen that blue and green cone opsins differ radically from rhodopsin and from each other (see Table 8.2). In contrast the red pigment differs only slightly from the green. The lysine residue to which retinal is attached is shown by a circle with a central dot in each drawing. Further explanation in text. After Nathans *et al.* Reproduced with permission from *Science*, **232**, 193–202, 1986, American Association for the Advancement of Science.

intramembranous position to the all-important lysine in rhodopsin which forms the Schiff base linkage with retinal. There is also considerable amino acid homology in the cytoplasmic loops between the transmembrane helices. We saw in the preceding section that these loops are involved in the vital interaction with the G-protein. Finally, the carboxy-terminal tail, though not strongly homologous to rhodopsin, nevertheless contains a number of threonine and serine residues which, it will be remembered, are phosphorylated in rhodopsin to switch off its photoactivation.

Although each of the cone opsins has, as already indicated, a lysine corresponding to rhodopsin's lysine 296 the distribution of charged amino acid side chains in its vicinity, and the overall charge in its environment, vary. The net intramembranous charge for the blue pigment ($\lambda_{max}=420$ nm) is $+1$, whilst that for the green and red pigments ($\lambda_{max}=530$ and 560 nm) is -1, and that for rhodopsin ($\lambda_{max}=495$ nm) is 0. It is likely that these differences and other subtle biochemical and electrical variations are at root responsible for the different absorption maxima of the different pigments in our retinas, and thus an essential first step in our appreciation of colour. It has, indeed, been shown that three different amino acid substitutions can each transform a green-sensitive to a red-sensitive opsin. These substitutions are alanine$_{180}\Rightarrow$serine, phenylalanine$_{227}\Rightarrow$tryptophan and alanine$_{285}\Rightarrow$ threonine or, in more concise notation, A180S, F277Y and A285T.

It can be seen from this brief account that the molecular analysis of the visual pigments is not only providing a synthesis of biochemistry and neurophysiology, but also beginning to establish a comparative molecular anatomy with all the evolutionary implications that that entails. It is interesting, finally, to note that sequence analysis of *Drosophila* and vertebrate opsins indicates that the common ancestor (no later than 500 million years ago) used an ancestral opsin for photoreception. Colour vision based on opsins sensitive to different wavelengths has evolved separately in the two phylogenies. This provides an impressive case of parallel evolution.

8.15 CONCLUSION

In this chapter we have reviewed some of the intricate biochemistry which the fluid-mosaic structure of biological membranes allows. We have also achieved a first glimpse of how the determination of molecular structure has revealed evolutionary relationships among receptors that had formerly been considered quite distinct. In the next chapter we begin on another vastly important aspect of membrane biology: their ability to separate solutions of different ionic concentration. It does not take very much knowledge of neurophysiology to recognise that upon this ability depends the whole working of the brain.

9

PUMPS

Significance of concentration gradients across neuronal membranes – energetics of establishing a gradient – definition of chemical potential – energy required to pump a Na$^+$ ion against its chemical gradient. **The Na$^+$+K$^+$ pump** – isolation – structure – function – efficiency. **The Ca^{2+} pump** – significance – isolation – structure – function – efficiency – affinity for Ca^{2+} – regulation by calmodulin – homologies between the Ca^{2+} pump and the Na$^+$+K$^+$ pump. **Other pumps and transport mechanisms across membranes**: the Na$^+$/Ca^{2+} exchange pump – Cl$^-$ pump – antiporters and symporters. **Conclusion**: significance of pumps in establishing the ion gradients which create the membrane potentials responsible for basic neurophysiology

So far in this book we have been considering the biochemistry and molecular biology of neurons and neuronal systems. Now it is time to broach a second major theme in molecular neurobiology: **the part that membranes play in separating ionic solutions** and all that follows from this. Ion fluxes, as we remarked at the end of Chapter 8, underlie the functioning of the nervous system. In the nineteenth century the great physicist James Clark Maxwell imagined a demon controlling a trapdoor between one compartment and another of a thermodynamic system. Such a demon, he suggested, could, by judicious opening of the door, allow only atoms above a certain energy to pass from one compartment to the other and hence cause the never-yet-experienced phenomenon of one part of an isolated system warming up whilst the other part cooled. It was later pointed out that for such a thought-experiment to be realised Maxwell's demon would need to be informed about the energy of the atoms in his two compartments – and information (as the twentieth century has discovered) has a thermodynamic cost. This is not the place to discuss the interrelations between thermodynamics and information theory: Maxwell's

demon has only been introduced to emphasise the importance of gates and gate-keepers in membranes.

We shall consider in some detail the gates and channels in neural membranes in Chapters 10 and 11. In this chapter we shall look at some of the pumps that produce the inequality in ionic concentration across membranes in the first place. If such concentration differences did not obtain, membrane gates and channels, no matter what their sophistication, would be of no value; neurophysiology (as we know it) could never have come to be. Table 9.1 shows the concentrations of some important ions inside and outside nerve and muscle cells.

9.1 ENERGETICS

First of all let us look briefly at the energetics of creating and maintaining transmembrane concentration differences. How much energy does it need to pump an ion across a membrane to establish a concentration gradient and to hold that concentration gradient in place? If we neglect the transmembrane voltage the appropriate equation is quite easy to derive.

Table 9.1 Ionic concentrations inside and outside some relevant cells

Ion	Intracellular concentration	Extracellular concentration
1. Squid giant axon	(mM/kg H_2O)	(mM/kg H_2O)
K^+	400	20
Na^+	50	440
Ca^{2+}	0.4	10
Mg^{2+}	10	54
Cl^-	100	560
Organic anions	385	–
2. Mammalian muscle cell	(mM)	(mM)
K^+	155	4
Na^+	12	145
Mg^{2+}	30	1–2
Ca^{2+}	1–2	2.5–5
(Only about 10^{-4} is free)		
Cl^-	4	120
Organic anions	Approx. 150	–
3. Cat motor neuron	(mM)	(mM)
K^+	150	5.5
Na^+	15	150
Cl^-	9	125

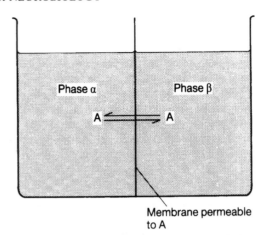

Figure 9.1 Membrane separating two solutions of substance A. A permeable membrane separates two solutions of substance A. When $\mu_A^\alpha \neq \mu_A^\beta$ there will an overall movement of A across the membrane until equality of chemical potential is achieved.

First of all we need to define an important parameter: the **chemical potential**. In essence the notion of a chemical potential is analogous to the better-known electrical and gravitational potentials. It is a familiar idea that electrons will flow from a region of high electrical potential to one of low potential; similarly objects placed in regions of high gravitational potential will fall towards regions of lower potential. The same idea obtains in the world of chemistry. Molecules and/or ions will move from states of high chemical potential to states of low potential.

The chemical potential of a substance, A, in phase α, is symbolised as μ_A^α. Its value is given by the following expression:

$$\mu_A^\alpha = \mu_A^0 + RT \ln x_A^\alpha \qquad (9.1)$$

where μ_A^0 is the chemical potential of A in a pure phase of A, x_A^α is the mole fraction of A in phase α (NB: in a pure phase of A, $x_A^\alpha = 1$) and ln is $\log_e$, the natural logarithm.

Now for all biological calculations x_A^α may be taken as C_A^α – the **concentration** of A in phase α.

Having defined a chemical potential, let us see how it can be used to determine the amount of energy required either to pump materials across membranes against their concentration gradients or, vice versa, the amount of energy that could (theoretically) be tapped when materials flow 'downhill' along their concentration gradients. The situation under consideration is shown in Figure 9.1.

By definition

$$\mu_A^\alpha \neq \mu_A^\beta$$

Let us suppose that

$$\mu_A^\alpha < \mu_A^\beta$$

Then:

$$\Delta\mu_A = \mu_A^\beta - \mu_A^\alpha$$
$$= (\mu_A^0 - \mu_A^0) + RT \ln \frac{C_A^\beta}{C_A^\alpha} \qquad (9.2)$$

i.e.

$$\Delta\mu_A = RT \ln \frac{C_A^\beta}{C_A^\alpha} \qquad (9.3)$$

This expression gives the 'free energy' (symbolised as ΔG) which is available to do work when the chemical substance passes from its high potential to its low potential state. Let us now simplify the symbolism a little. Let us remove the signs indicating the two phases, α and β, in which our solute is supposed to be present, and merely refer to the concentration of the solute on either side of the partition as C_1 and C_2. We can then write:

$$\Delta G = RT \ln \frac{C_1}{C_2} \qquad (9.4)$$

If we now insert the usual values for R and T, i.e. R (the gas constant)$=8.31\,J\,K^{-1}\,mol^{-1}$ and T (temperature in kelvins)$=310\,K$ (i.e. $37°C$), then

$$\Delta G = [(8.31\,J\,K^{-1}\,mol^{-1}) \times 310\,K]\,\ln\frac{C_1}{C_2}$$
$$= 2576\,\ln\frac{C_1}{C_2}\,J\,mol^{-1} \qquad (9.5)$$

Let us make use of equation 9.5 to determine the quantity of energy required to pump a sodium ion from inside to outside a neuron. In this calculation it is important to note that we do not take into account any electrical forces that may be (probably are) acting on the ion. We shall consider these in detail in Chapter 11. There we shall see that the gradient up which an ion is pumped is given by the **Nernst equation**, which takes into account the electrical as well as the concentration differences of ions on the two sides of a membrane. Here, however, let us introduce the subject by substituting in equation 9.5 the values given in Table 9.1 (cat motor neuron) for Na^+ outside ($[Na^+]_o$), i.e. 150 mM and Na^+ inside ($[Na^+]_i$), i.e. 15 mM. The energy required to pump a mole of sodium ions from the inside to the outside against this concentration gradient at $37°C$ is given by:

$$\Delta G = 2576\,\ln\,(150/15)\,J\,mol^{-1}$$
$$= 5932\,J\,mol^{-1}$$

Now a mole of Na^+ consists of 6×10^{23} ions (i.e. Avogadro's number of ions). It follows that 5932 J are required to pump this quantity against the prevailing concentration gradient – and unless the pumping mechanism is 100% efficient (which it is not) considerably more.

To determine the minimum amount of energy required to pump a single ion out of the neuron is a simple matter:

$$\frac{5932}{6 \times 10^{23}} \approx 1 \times 10^{-20}\,J$$

Now the majority of biochemical activities are driven by energy derived from energy-rich phosphate bonds – principally those of ATP. We shall shortly see that membrane pumps are no exception to this rule: the energy required to pump ions against their concentration gradients is also derived from the energy-rich phosphate bonds of ATP. We have just worked out how much energy is required to pump a sodium ion out of a neuron against its concentration gradient. Let us see how many 'energy-rich' phosphate bonds have to be hydrolysed to provide this energy.

The ΔG for the dephosphorylation of ATP to $ADP+P_i$ varies somewhat according to the concentration of ATP. Let us take $-12\,kcal\,mol^{-1}$ as a reasonable value for the situation we are considering. Remembering that $1\,cal=4.18\,J$, it follows that one energy-rich bond is equivalent to about $50\,160\,J\,mol^{-1}$.

Hence the dephosphorylation of a **single ATP** to ADP would yield approximately

$$\frac{5 \times 10^4}{6 \times 10^{23}} \approx 8 \times 10^{-20}\,J$$

It follows that the hydrolysis of one energy-rich phosphate bond will yield sufficient energy to transfer approximately **eight sodium ions**.

These calculations have, as we have already noted, assumed that the pump is 100% efficient and, as we emphasised at the outset, have totally neglected the contribution of electrical forces. They have also totally omitted any consideration of the complex biochemical mechanisms underlying the operation of the pump. It is thus somewhat surprising that they yield an answer so close to the experimentally determined value. It has been shown (as we shall shortly see) that three sodium ions are pumped out in exchange for two potassium ions pumped in for the expenditure of one energy-rich phosphate bond.

This short excursion into the energetics of pumping will, it is hoped, have given the reader some feel for the quantities of energy and the numbers of molecules and ions involved.

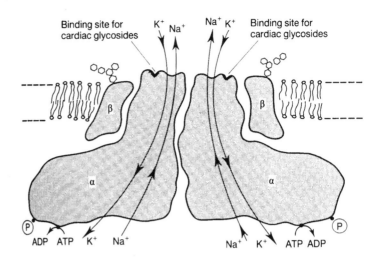

Figure 9.2 Conceptual diagram of the Na$^+$+K$^+$ pump. The schematic diagram shows that the Na$^+$+K$^+$ pump is a dimer each unit of which consists of two subunits: a large catalytic or α-subunit and a smaller glycoprotein β-subunit. The major part of the catalytic subunit lies in the cytoplasmic space. This cytosolic domain has a site for dephosphorylating ATP and another site for the attachment of ATP's γ-phosphate. The extracellular surface of the α-subunit possesses a site for cardiac glycosides such as ouabain and digitalis. The β-subunit is much smaller and bears a short oligosaccharide chain on its external face. It is also believed to be sensitive to cardiac glycosides.

9.2 THE NA$^+$ + K$^+$ PUMP

The Na$^+$+K$^+$ pump is ubiquitous. It is found in the plasma membranes of practically all animal cells. It pumps Na$^+$ ions out of a cell and at the same time pumps K$^+$ ions in the opposite direction. It is thus an example of an '**antiporter**' mechanism. **Three sodium ions** are pumped **out** whilst **two potassium ions** are pumped **in** during one cycle of operation. The pump is energised by the dephosphorylation of ATP, as mentioned above, and one complete cycle is fuelled by the hydrolysis of one molecule of ATP to ADP. The ATPase that catalyses this dephosphorylation is dependent on the presence of both Na$^+$ and K$^+$ ions. The Na$^+$ ions have to be within the cell, the K$^+$ ions outside. The pump is inhibited by cardiac glycosides such as **digitalis** and **ouabain** which affect the K$^+$ site on the external surface. The Na$^+$+K$^+$ ATPase is clearly of fundamental importance in maintaining the ionic concentrations upon which, as indicated above, the entire functioning of a nervous system depends.

The Na$^+$+K$^+$ pump can be isolated from the membranes of several types of cell including mammalian kidney cells and the electric organs of eels and rays where it is highly concentrated. As Figure 9.2 shows, it appears to consist of two identical units each of which, in turn, consists of two subunits. The larger of the two subunits, the α-subunit, has a molecular weight of about 100 kDa, and consists of 1016 amino acids (sheep kidney) or 1022 amino acids (*Torpedo*, the electric ray). The smaller β-subunit, a glycoprotein, has a molecular weight of 55 kDa. The α-polypeptide is the catalytic subunit. The smaller β-polypeptide is almost 20% carbohydrate by mass. Its exact function is not clear. Its close union to the α-subunit is, however, necessary if the α-subunit is to function as an ATPase.

cDNA libraries prepared from sheep kidney and *Torpedo* electroplax have been probed by radiolabelled oligonucleotides prepared from known amino acid fragments of the α-polypeptide. This technique (which was described in outline in Chapter 5) yields DNA clones from which complete amino acid sequences of the α-polypeptide can be deduced. It turns out that there is 85% amino acid homology between the α-polypeptides derived from these two species. This shows remarkable conservatism when it is recalled that a common ancestor of sheep and electric ray can have lived no later than 400 million years ago.

Examination of the polypeptide for hydropathic domains suggests that the amino acid chain spans the membrane no less than eight times, whilst the central part of the molecule forms a large cytosolic domain (Figure 9.3). The cytosolic domain contains the ATP binding and hydrolysis sites. The ouabain binding site, on the other hand, is to be found on the outside of the membrane, possibly between the third and fourth transmembrane helices. It is not yet clear where the ion pore(s) is (are) or, indeed, whether such a pore exists at all. We shall see in the next section that the closely

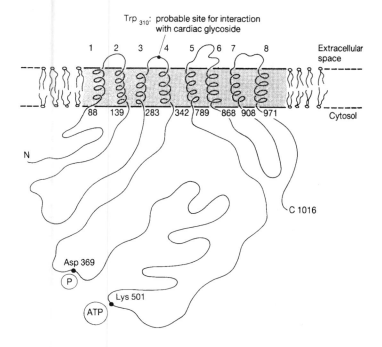

Trp$_{310}$: probable site for interaction with cardiac glycoside

Figure 9.3 Na$^+$+K$^+$ ATPase catalytic subunit. The major part of the α-subunit is located in the cytosol. It is embedded in the membrane by eight transmembrane helices. The largest cytosolic domain (residues 342–789) provides the phosphate binding site (Asp$_{369}$) and the ATP hydrolysis site (Lys$_{501}$). A tryptophan residue (Trp$_{310}$) on the outer face of the membrane between transmembrane helices 3 and 4 is believed to be the site with which cardiac glycosides interact. The figure represents only a two-dimensional plan of the molecule: in reality both the cytosolic and the transmembrane domains would be folded in upon themselves. Data from Shull, Schwartz and Lingrel (1985), *Nature*, **316**, 691–695.

related Ca^{2+} pump has no pore so it is very probable that the Na$^+$+K$^+$ is similarly poreless. A schematic diagram showing the disposition of the polypeptide in the membrane is shown in Figure 9.3.

It is believed that the binding of a Mg^{2+} ATP to the catalytic subunit at **lysine 501** leads to a dephosphorylation with the transference of the cleaved phosphate (i.e. **the γ-phosphate)** to **aspartate 369** (see Figure 9.3). The energy released in the dephosphorylation brings about an extensive conformational change in the molecule involving at least 80 amino acid residues. It is suggested that this conformational change consists in a transition from an α-helical to a β-sheet structure. The conformational change impels three sodium ions out of the cell. The new conformation also allows two potassium ions to bind to the extracellular side of the subunit. This leads, in its turn, to a reverse conformational change which both frees the γ-phosphate from aspartate and impels the two K$^+$ ions into the cell. Thus at the end of this sequence of events one ATP has been hydrolysed to ADP, three sodium ions have been extruded, two potassium ions pumped into the cell, and the catalytic subunit has resumed its original conformation.

This sequence of events is schematised in Figure 9.4.

As mentioned above, we shall see in the next section that the structure of the Na$^+$+K$^+$ catalytic subunit is similar to the Ca^{2+} ATPase of muscle sarcoplasmic reticulum (and indeed also similar to the K$^+$ pump of bacteria). Pumps form yet another group of evolutionarily related proteins.

9.3 THE CALCIUM PUMP

In Chapter 11 and elsewhere we shall note the great importance of the Ca^{2+} ions in neurons, especially at synapses. We shall return to this in more detail in Chapter 15. We can note here, however, that there is a very steep concentration gradient of free calcium ions between the inside and the outside of neurons (see Table 9.1). Thus (in spite of the existence of molecules such as **calmodulin** which mop up intracellular Ca^{2+}) there is just as great a need for a pump to extrude the Ca^{2+} ions, which as we shall see, flow in during a membrane depolarisation as there is for a pump to move K$^+$ and Na$^+$ into and out of an axon after an action potential.

A calcium pump is present in the membranes of most cells. Judged by its immunological reactivity it

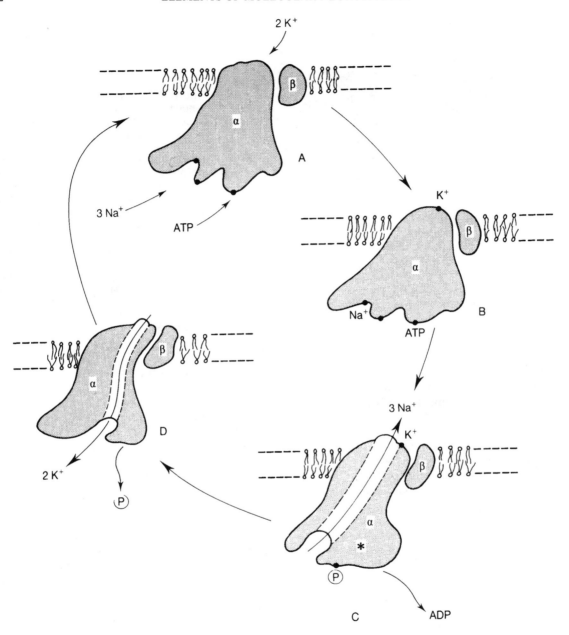

Figure 9.4   Schematic to show the operation of the Na⁺+K⁺ ATPase. (A) Na⁺ ions and ATP approach the pump's cytosolic domain; K⁺ ions approach from the extracellular compartment. (B) Na⁺ ions find their binding sites on the α-subunit and ATP is hydrolysed to ADP. (C) The γ-phosphate from ATP is transferred to the phosphate binding site and the α-subunit undergoes a conformational change so that three Na⁺ ions are discharged to the exterior. (D) K⁺ ions bind to the extracellular surface and a continuation of the conformational change impels two K⁺ ions into the cell and releases the phosphate group. The catalytic subunit returns to its original state. Note that the figure should not be taken to imply that there is a continuous pore through the α-subunit.

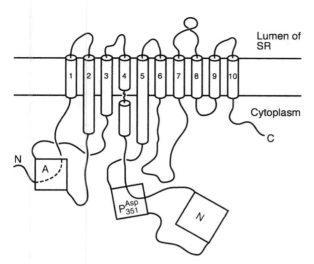

Lumen of SR

Cytoplasm

Figure 9.5 Structure of the rabbit SR Mg^{2+}-dependent Ca^{2+} ATPase. Note the resemblance to the Na$^+$+K$^+$ ATPase catalytic subunit (Figure 9.3). As before, the large cytosolic region can be divided into three major domains. The first domain, A, has been named the 'actuator'; the second domain, N, the 'nucleotide binding domain'; the third the phosphorylation domain, P. Asp$_{351}$ is at the centre of the phosphorylation process. It should be borne in mind that the diagram represents only a two-dimensional plan: in reality both the cytoplasmic and the transmembrane domains are folded in upon themselves (see Plate 2). After Lee and East (2001), *Biochemical Journal*, **356**, 665–683; MacLennan (2000), *Nature*, **405**, 633–634; Toyoshima *et al.* (2000), *Nature*, **405**, 647–654.

takes slightly different forms in different cells. Immunologically distinct forms of the calcium pump have been detected in fast twitch and slow twitch skeletal muscles, cardiac muscle, erythrocyte membranes, etc. The pump has, however, been most intensively studied in the system of internal membranes found in striped muscle fibres – the **sarcoplasmic reticulum (SR)**. The contraction of striped muscles is dependent on triggering by Ca^{2+} ions and hence the concentration of calcium ions in the vicinity of the contractile elements (actin, myosin, etc.) is very critical. It is for this reason that the sarcoplasmic reticulum provides such a rich source for the calcium pump. Indeed it appears that the pump forms up to 80% of the SR membrane.

It comes as no surprise, therefore, that the first gene coding for a calcium pump was isolated from the SR of rabbit skeletal muscle. Essentially the same technique of oligonucleotide hybridisation cloning was used as was employed to isolate the gene coding for the Na$^+$+K$^+$ ATPase. From the nucleotide base sequence obtained in this way the amino acid sequence of the pump was deduced. Again it turned out to be a huge protein consisting of 997 amino acids. Ca^{2+} pumps isolated from other tissues range up to 1220 amino acids in length.

By examining the polypeptide chain for hydropathic domains, the position of the molecule in the membrane can, as usual, be deduced. There are 10 α-helical transmembrane segments. In addition to the membrane-spanning segments the molecule also possesses a large and intricate tripartite cytosolic domain (Figure 9.5). Each domain is believed to have a different function. The first domain is concerned with setting in train the activities of the other two; the second is the **phosphorylation** domain, accepting the γ-phosphate from ATP; the third, the **nucleotide** domain, is concerned with ATP attachment. These three cytosolic domains are believed to be folded together so that they can work interactively as a unit. Evidence from electron microscopy and optical diffractometry support this interesting structure.

These somewhat indirect arguments for tertiary conformation were confirmed when the first X-ray diffraction analyses, at a 2.6 Å resolution, were published at the turn of the century (see Plate 2). Suitable crystals for X-ray diffraction were obtained from the sarcoplasmic reticulum. The molecule measured 100 Å×80 Å×140 Å and had an M_r of 110 000. The ten transmembrane helices were confirmed, as were the three large and quite widely separated cytosolic domains. These domains acted together in much the same way as deduced from the more indirect chemical evidence cited above. Domain A acts as an **actuator**, domain P is the **phosphorylation** domain and domain N the **nucleotide domain**. When ATP attaches to the N-domain mechanical movement occurs so that the actuator domain twists in such a way that an aspartate residue (Asp$_{351}$) is phosphorylated in the P-domain. This causes the P-domain to move so that it

BOX 9.1 Calmodulin

Calmodulin (CaM) is found in all eukaryotic cells. Indeed in typical animal cells it makes up some 1% of the total cell protein (more than a million molecules). Although not itself an enzyme it plays a vital role in many Ca^{2+}-activated processes. As the calcium ion is such a ubiquitous 'second messenger' the significance of CaM can hardly be over-estimated.

Calmodulin is a medium-sized protein (15 kDa) consisting of a single chain of 148 amino acids. Its conformation is rather like a dumbbell with two globular Ca^{2+} binding regions connected by a lengthy α-helical 'bar'. This bar has a central flexible region which allows the molecule to fold around its target proteins. Each globular 'dumbbell' is able to accept two Ca^{2+} ions. This acceptance occurs when the intracellular Ca^{2+} concentration rises above 1 μM. The acceptance is serial: the affinity for Ca^{2+} is ten times greater at the carboxy-terminal dumbbell than at the N-terminal dumbbell, and is dependent on the presence of Mg^{2+} at about 1 μM concentration.

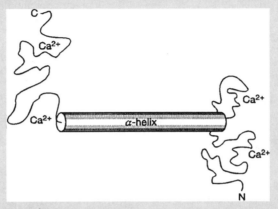

Figure A Calmodulin. For explanation see text.

The acceptance of Ca^{2+} ions alters the conformation of calmodulin. In particular it is believed that hydrophobic surfaces are exposed when Ca^{2+} binds. These hydrophobic surfaces are attracted by similar surfaces on calmodulin-activated enzymes and channels. One important such surface is found at the C-terminal of the Ca^{2+} pump described in this chapter. Other calmodulin-activated proteins are shown in Figure B. Most such proteins have a regulatory subunit to which calmodulin binds. Ca^{2+} binding to calmodulin influences the activity of more than 20 enzymes and structural proteins.

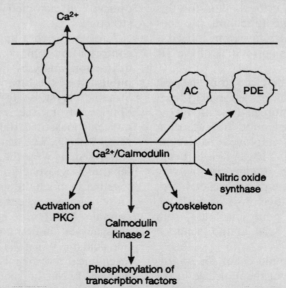

Figure B Some calmodulin activations. AC=adenylyl cyclase; PDE=phosphodiesterase; PKC=protein kinase C.

contacts the loop between transmembrane helices 6 and 7. This, in turn, alters the packing of the transmembrane helices (Figure 9.5; Plate 2). Ca^{2+} is bound to two high-affinity sites within the cage formed by helices M4, M5, M6 and M8 (Figure 9.5) When the α-helical packing is altered the two bound Ca^{2+} ions are displaced and are expelled to the outside. Unlike the ligand- and voltage-gated ion channels which are discussed in the next two chapters there is no central pore or selectivity filter to squeeze through. The expulsion of Ca^{2+} is the result of mechanical rearrangement of the intramembranous helical architecture. A detailed view of this architecture is shown in Plate 2.

For every two Ca^{2+} ions translocated, one ATP molecule is dephosphorylated. The pump also depends on the presence of Mg^{2+} ions; indeed there may be a counter-transport of one Mg^{2+} ion inwards for every two Ca^{2+} ions pumped out. Because free calcium ions are in such very low concentration within a cell the calcium pump is very specific, and has a very high affinity for the ion. The pump is, moreover, controlled by the quantity of Ca^{2+} in the cytosol. If the concentration rises beyond about 10^{-4} mM the pump increases its rate of working. In the erythrocyte membrane this has been shown to be due to the presence of an important regulatory protein, **calmodulin** (see Box 9.1), as part of the pump. This binds excess cytosolic Ca^{2+}, undergoes a conformational change, and causes the work rate of the Ca^{2+} pump to increase. The calmodulin binding domain has been shown to be the C-terminal end of the sequence which in many Ca^{2+} pumps is much lengthier than that developed by the SR pump.

Finally, as mentioned in the preceding section, it is possible to find remarkable similarities between the Mg^{2+}-dependent Ca^{2+} ATPase and the catalytic subunit of the $Na^+ + K^+$ ATPase. Analysis of the hydropathic segments of the molecules indicates striking homologies (Figure 9.6B). The phosphorylation and ATP binding sites are in the same position and, except for three regions where segments of $\geqslant 20$ amino acids have been inserted or deleted from one or other of the proteins, homologous structural features are obvious (Figure 9.6A). There is no doubt that the calcium and sodium/potassium pumps have evolved from a common ancestor.

9.4 OTHER PUMPS AND TRANSPORT MECHANISMS

Plasma membranes possess many other metabolically driven pumps. Some of these are linked to the pumping of Na^+ ions across the membrane. The **sodium–calcium exchange pump** is an example of this linkage. Ca^{2+} ions are pumped out of the neuron against a steep electrochemical gradient by using the energy provided by the Na^+ gradient induced, ultimately, by the $Na^+ + K^+$ pump. The pumping mechanism involves the transport of three Na^+ ions inward for each Ca^{2+} ion extruded. Recently evidence has suggested that defects in *mass 1*, the gene encoding this pump (a single base pair deletion), leads to audiogenic epilepsy in the mouse. It is believed that all non-symptomatic epilepsies (i.e. epilepsies where no structural, metabolic or other neurological abnormalities can be found) are due to mutations of ion channel genes. Future research will show whether a homologue of *mass 1* (**m**onogenic **a**udiogenic **s**eizure **s**usceptible) is also at work in humans. Other cases of ion channel defects leading to epilepsy are reviewed in Section 11.7.

Other pumps are connected to the movement of H^+ ions (protons). An example that is of considerable importance in neurobiology is the **chloride pump**, which extrudes Cl^- ions from the neuron. Rather little is known about this pump. Cl^- ions are not in electrochemical equilibrium across the membranes of most neurons (see Chapter 12). This indicates that a pump ultimately depending on metabolic energy must exist. In some cases it is believed the energy is derived from the ATP–ADP system; in other cases the energy may be derived from the distribution of other ions (for instance H^+ ions) across the membrane.

Two very important neurobiological transport mechanisms, the vesicular neurotransmitter transporters and the neurotransmitter reuptake transporters, will be described in Boxes 15.2 and 16.3. In the first case the energy is supplied by a proton pump. The escape of the H^+ ion from the vesicle energises the uptake of a specific neurotransmitter. Transport mechanisms using this type of reverse coupling are called '**antiporters**'. We have already met examples of antiporters in the form of the $Na^+ + K^+$ pump and, in the paragraph above, the Na^+/Ca^{2+} exchanger. In contrast, the neurotransmitter reuptake pumps make use of the potential

A

```
NKA  INAEEVVVGDLVEVKGGDRIPAOLRIISANGC..KVONSSLTGES 215   233 RNIAFFSTNCVEGTARGIVVYTGDRTVMGRIATLASGLEGGQTP 276
CA   IKAKDIVPGDIVEIAVGDKVPAOIRLTSIKSTTLRVDQSILTGES 184   205 KNMLFSGTNIAAGKAMGVVVATGVNTEIGKIROEMVATEQERTP 248

NKA  VANVPEGLLATVTVCLTLTAKRMARKNCLVKNLEAVETLGSTSTICSDKTGTLTQNRMTVAHM 384   499 VMKGAPERILDRCSSILIHGKEQPLDEELK 528
CA   VAAIPEGLPAVITTCLALGTRRMAKKNAIVRSLPSVETLGCTSVICSDKTGTLTTNQMSVCRM 366   512 FVKGAPEGVIDRCTHIRVGSTKVPMTAGVK 541

NKA  NLCFVGLISMIOPPRAAVPDAVGKCRSAGIKVIMVTGDHPITAKAIAKGVGI 626   683 FARTSPQOKLIIVEGCQROGAIVAVTGOGVNDSPALKKAOI 723
CA   NLTFVGCVGMLDDPPRIEVASSVKLCRQAGIRVIMITGDNKGTAVAICRRIGI 640   675 FARVEPSHKSKIVEFLQSFDEITAMTGDGVNDAPALKKAEI 715

NKA  GVAMGIAGSDVSKQAAOMILLDDNFASIVTGVEEGRLIFDNLKKSIAYTLTSNIPE 779   807 TDMVPAISLAYEOAESDIMKRQPRNPQTDKLYNERLIS 844
CA   GIAMG.SGTAVAKTASEMVLADONFSTIVAAVEEGRAIYNNMKQFIRYLISSNVGE 770   798 TDGLPATALGFNPPOLDIMNKPPRNP...K...EPLIS 829

NKA  PLKPTWWFCAFPYSLLIFVYOEVRKLIIR 1003
CA   PLNVTOWLMVLKISLPVILMOETLKFVAR 988
```

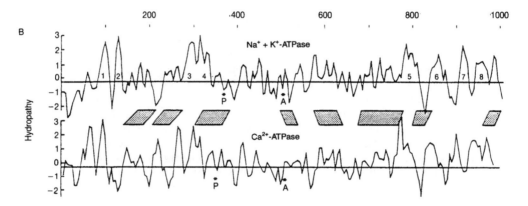

B

Figure 9.6   Comparison between the Mg^{2+}-dependent Ca^{2+} ATPase and the $Na^+ + K^+$ ATPase. (A) The amino acid homology between sheep kidney $Na^+ + K^+$ ATPase (NKA) and rabbit cardiac Ca^{2+} ATPase (CA) is shown (amino acids represented by letter symbols: see Table 2.1). Only sequences with greater than 30% homology are compared and gaps have been added to maximise the homology; homologous residues are indicated by a dot. (B) Hydropathic plots for the two pumps. Hydrophobic regions are drawn above the horizontal axis and hydrophilic regions below it. The parallelograms represent the homologous regions shown in A and their positions along the polypeptide chain. The numbers along the NKA plot indicate the position of the major hydrophobic regions (transmembrane helices); P=the phosphate binding site; A=the ATP binding site. Reprinted by permission from Shull, Schwartz and Lingrel (1985), *Nature*, **316**, 691–695. Copyright © 1985, Macmillan Magazines Ltd.

energy in the Na^+ electrochemical gradient. Na^+ ions and transmitter move in the same direction and in association with each other. Transport mechanisms of this type are called '**symporters**'.

9.5 CONCLUSION

The several metabolically driven pumps we have considered in this chapter are very effective in establishing concentration differences of ions across neuronal and glial cell membranes. In the next chapters of this book we shall examine the ways in which neuronal membranes are able to control the fluxes of ions back down their electro-

chemical gradients. Neurons are, of course, not the only types of cell able to exert this control, but they are certainly the most accomplished. We shall see how ligand- and voltage-controlled gates shape the flows of ions along their gradients and we shall see how these flows, in their turn, affect the electrical polarity of neuronal and neuroglial membranes. It is these variations in electrical polarity that the electrophysiologist picks up with his probing electrode. And, it is these variations, ultimately, that allow neurophysiology, and all that flows from neurophysiology, to happen. Much depends, therefore, on the presence of $Na^+ + K^+$, Ca^{2+} and other pumps in the membranes of neurons.

10

LIGAND-GATED ION CHANNELS

Gated channels are basic elements of neurobiology. Responsible for the phenomena of electrophysiology. Differentiation into voltage- and ligand-gated types. **nAChRs – structure**: isolation – electric fish – snake venom chromatography – polyacrylamide electrophoresis – cDNA cloning – *Xenopus* oocyte expression – determination of primary structure – hydropathy analysis shows 4TM disposition – four different 4TM subunits (α, β, γ and δ) – evolutionary relationships – pentameric structure of complete receptor ($\alpha_2\beta\gamma\delta$); **function**: use of liposomes and artificial bilayers – electron microscopy and patch-clamping – calculation of ion flux – several different 'open' and 'closed' states – location of ACh binding site. **Development** – effects of innervation on distribution in membrane and structure ($\alpha_2\beta\gamma\delta \Rightarrow \alpha_2\beta\varepsilon\delta$). Pathologies. **CNS nAChRs** – permeable to Ca^{2+} – variety of subunit structure. **GABA$_A$Rs**: subclasses – localisation – purification on benzodiazepine columns – 4TM subunits ($\alpha, \beta, \gamma, \delta, \rho$) – expression in *Xenopus* oocyte – pentameric structure of complete receptor ($\alpha, \beta, \gamma, \delta, \rho$). Pathologies. **GlyRs**: purification on 2-aminostrychnine columns – three subunits (48 kDa, 58 kDa, 93 kDa) – 48 kDa subunit has 4TM disposition – two subtypes (α and β) – five copies of 48 kDa subunit form channel – heterogeneity. Homologies of nAChRs, GABA$_A$Rs and GlyRs. **iGluRs**: three major types (AMPA or Q, KA, NMDA) – 3TM subunits – numerous subtypes – pharmacology – biophysics. NMDA-Rs show voltage-sensitive Mg^{2+} blockade of Ca^{2+} channel – act as 'AND' or 'Hebb' gates – LTP and learning – pacemaker activity. **Ionotropic purinoceptors** (P2X-Rs) – 2TM subunits – convergent evolution? **Conclusion**: great heterogeneity of LGICs in neuronal membranes – yet may be grouped into three major families

In Chapter 7 we met with a first example of a gate controlling a transmembrane channel in the variable open/shut states of the gap junction connexon. In this chapter and the next we shall consider some of the many other gates and channels that have been developed in neural cells. Until the advent of genetic engineering and patch-clamping their structure and function were extremely difficult to study. They are sometimes present as only a few molecules per cell and, moreover, are hidden from classical biochemical techniques by insertion in the lipid biomembrane. Nonetheless, as we shall see in the next few chapters, a great deal of neurophysiology ultimately depends on the their presence and operation.

Membrane 'gates' or 'channels' may be divided into two categories: those that are controlled by **chemical molecules (ligands)** and those that are controlled by **transmembrane voltage**. In fact this division is not absolutely clear cut. Ligands have an effect on some voltage-dependent gates, and, vice versa, transmembrane voltage influences at least some ligand-controlled gates. Nevertheless, for the purposes of exposition, we shall consider each in turn.

Both ligand-controlled and voltage-controlled gates depend for their physiological effect on there being a concentration and/or an electrochemical gradient across the membrane. Opening and shutting of these gates allows ions to flow across the membrane in the direction in which the gradient is inclined. This, in turn, results in the various electrical phenomena which the neurophysiologist picks up when investigating the nervous system: action potentials, postsynaptic potentials, electrotonic potentials, receptor potentials, generator potentials, etc. These concentration differences, ultimately due to the membrane pumps which we considered in the last chapter, are shown in Table 9.1.

In this chapter we shall consider ligand-gated ion channels (LGICs) and in the next we shall turn our attention to voltage-controlled gates. We shall proceed as follows. We shall begin by describing the best known of all the ligand-gated channels – the **nicotinic acetylcholine receptor (nAChR)**. We shall then go on to discuss two other ligand-controlled gates, the **GABA$_A$ receptor (GABA$_A$R)** and the **glycine receptor (GlyR)**. We shall see that the molecular structure of these latter two receptors is sufficiently similar to that of the nAChR to indicate that they are all members of an evolutionarily related superfamily. This superfamily has also been shown to include the **5-HT$_3$ receptor (5-HT$_3$R)**. Next, we shall look at another well-known group of ligand-gated channels: those that respond to excitatory amino acids, **glutamate** and **aspartate**, and to **kainate**, **AMPA** (α-amino-3-hydroxy-5-methyl-isoxazole propionate) and **NMDA** (*N*-methyl-D-aspartate). These receptors also form another natural group consisting of protein subunits with three rather than four transmembrane helices. Finally, we shall end with the P2X **purinoceptors**, another natural group, the subunits of which have only two transmembrane domains. The account in this chapter, of course, only scratches the surface. The web site listed in the Bibliography lists 390 LGIC subunits (July 2001).

10.1 THE NICOTINIC ACETYLCHOLINE RECEPTOR

As mentioned above, the type example of a ligand-controlled gate is the nicotinic acetylcholine

receptor (nAChR). We have already met this entity several times in previous chapters. Its structure is nowadays very well known.

We noted in Chapter 8 that cholinergic receptors can be divided into two major classes: **nicotinic** and **muscarinic**. In that chapter we looked at the muscarinic receptor in some depth. We saw that it was coupled to a complex membrane molecular biology involving G-proteins and cyclase enzymes. The response of the nAChR is, as we shall shortly see, simpler and consequently much more rapid.

The molecular biology of nAChRs was first investigated at peripheral sites, especially the **electroplax** of electric eels. The receptor is also well represented in the central nervous system. Different localisation techniques have yielded slightly different results, but most investigations have shown it to be present in the cerebral cortex, hippocampus, colliculi, hypothalamus, interpeduncular nucleus and thalamic nuclei. The two types of nAChR differ (see Box 10.1) in their subunit structure. Whereas both, as we shall see, are pentameric structures, the neuronal type besides being evolutionarily more ancient also has a more heterogeneous set of subunits.

Finally, in this introductory section, it should be noted that it is a little misleading to refer to *the* nicotinic acetylcholine receptor. For, like most of the other receptors we have considered and shall consider, nAChRs vary from one organism to another and from one part of a given organism to another. Brain nAChRs differ from peripheral nAChRs; peripheral nAChRs themselves differ: those on 'fast' muscles have different response times to those on 'slow' muscles. Nicotinic acetylcholine receptors form yet another closely knit evolutionary family (see Box 10.1).

10.1.1 Structure

The most concentrated source of nicotinic acetylcholine receptors is to be found in the **electric organ (electroplax)** of electric fish – the elasmobranch electric ray *Torpedo* and the teleost electric eel *Electrophorus* (=*Gymnotus*). Electroplaxes develop from muscle somites and their innervation is via cholinergic neuromuscular junctions. These are far larger and far more numerous than in normal skeletal muscle. The whole organ is very rich in ACh, nAChRs and the associated enzymes of cholinergic synapses. Indeed electron micrographs of electroplax postsynaptic membranes show extremely dense populations of nAChRs (Figure 10.1). Discharge of the electric organ can generate as much as 600 volts. Finally it is worth noting that cartilaginous and bony fish diverged at least 400 million years ago. Their electric organs, apart from some differences in detailed physiology, are thus remarkable instances of convergent evolution. In spite of the great period of independent evolution their nAChRs are, as we shall see, very similar.

The first molecular-biological studies of nAChRs thus used the very rich source provided by electric organs. Many sophisticated techniques were employed. It is, for instance, possible to obtain purified samples of the nAChR by making use of some snake neurotoxic peptides – the favourites being **najatoxin** (**NajaTX**) from the cobra *Naja naja siamensis* (sometimes called cobratoxin) and α-**bungarotoxin** (α-**BuTX**) from the snake *Bungarus multicinctus*. These neurotoxins bind specifically to the nicotinic acetylcholine receptor. Thus if α-BuTX is bound to a solid chromatography substrate and a solubilised preparation of electroplax membrane run through the substrate, nAChR will be caught by the α-BuTX and held. In fact α-BuTX binds nAChR so firmly that it is almost impossible to wash it off the column. NajaTX binds less firmly and hence provides a better substrate.

After purification of electroplax nAChR by snake venom chromatography it is eluted, denatured with SDS (see Section 7.4), and subjected to SDS–poylacrylamide electrophoresis. Staining with Coomassie blue reveals four bands of material (Figure 10.2). These correspond to four subunits – α, β, γ and δ – in order of increasing molecular weight.

Although the *Torpedo* electroplax is sufficiently rich in nAChR to yield milligrams of the purified receptor, the method of choice for determining primary structure is via the recombinant DNA techniques described in Chapter 5. Small amino acid sequences can be determined from subunit material eluted from the electrophoresis bands. These can be used, as described in Chapter 5, to construct short oligonucleotide probes. The probes, in turn, are used to screen a cDNA library prepared from electroplax mRNA. The cDNA corresponding to the probe can then be cloned.

It is next necessary to ensure that the cDNA isolated by this technique does in fact code for the nAChR. This can be done by transcribing it into mRNA and introducing this into an expression system. Clearly the mRNA for all four subunits must be injected. As indicated in Chapter 5 the expression system is normally *Xenopus* oocyte. There are two ways to ascertain that the protein expressed in the oocyte is indeed nAChR. First, one can examine the oocyte membrane by electro-physiological techniques to establish that it now possesses channels responsive to ACh. Second, one can subject the oocyte membrane to SDS–polyacrylamide gel analysis alongside α-toxin-purified nAChR from *Torpedo* electroplax and establish that the same protein is present.

Finally, having satisfied oneself that the isolated cDNA is indeed the nAChR gene, one can then use one or other of the standard techniques described in Chapter 5 to obtain its base sequence. Having the base sequence it is then possible, knowing the genetic code, to predict the primary structure of the nAChR.

The outcome of this sophisticated molecular biology has been to show that *Torpedo* nAChR is multimeric protein with a total molecular weight of some 268 kDa. It consists of five subunits – two α-subunits (each consisting of 461 amino acids), one β-subunit (493 amino acids), one γ-subunit (506 amino acids) and one δ-subunit (522 amino acids). A pentameric structure has been shown to exist in all the vertebrates so far examined. The subunits making up the pentamer may, however, differ. As we shall see, there is evidence that brain nAChRs are very diverse. The amino acid sequences of human, calf and *Torpedo californica* α-subunits are shown in Figure 10.3.

Inspection of the figure shows that there is very considerable amino acid homology between these three evolutionarily widely separated AChR α-subunits. M1, M2, M3 and M4 in Figure 10.3 indicate the hydrophobic stretches of the α-subunits. The position of these hydrophobic stretches in the primary sequences of all three subunits is very similar. This suggests that all the α-subunits originated by three successive duplications of an ancestral gene. Indeed, as we noted in Chapter 4, the nAChR present in the nervous systems of some insects appears to be a homopentamer ($\psi\alpha_5$) and hence perhaps similar to the ancestral vertebrate complex.

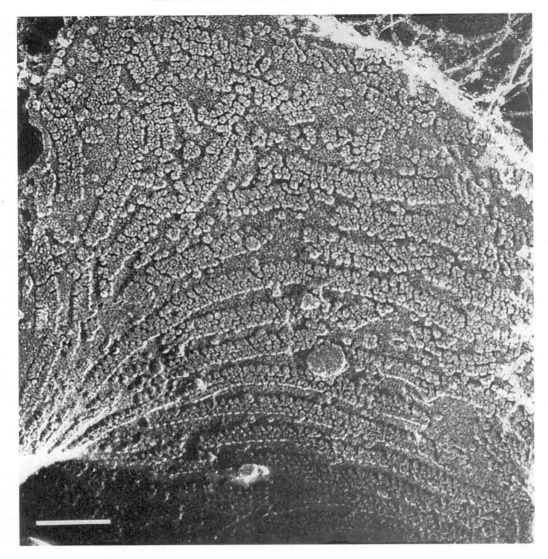

Figure 10.1   Freeze-fracture electron micrograph of the E-face of the postsynaptic membrane of *Torpedo* electric organ. The acetylcholine receptors are arranged in rows. Each AChR can be seen in this surface view to resemble a tiny doughnut. The scale bar represents 0.1 μm. From Hirokawa (1983), *Structure and Function of Excitable Cells*, New York: Plenum Press, pp. 113–141; with permission.

A further question to ask is if, and if so, how, the four different subunits of the *Torpedo* nAChR are related to each other. The answer is given in Figure 10.4. If the four subunits are aligned it is clear that they have far more than chance sequence similarity. Examination of these similarities suggests the evolutionary schematic shown in Figure 10.4B.

The hydrophobic amino acid stretches indicate which parts of the various subunits are embedded in the membrane. The amino-terminal end of the molecule extends into the extracellular space and the amino acid chain then loops through the membrane four times so that all its hydrophobic segments are embedded in the lipid bilayer. The

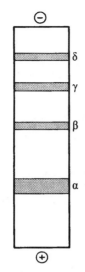

carboxy-terminal end emerges ultimately once again into the extracellular compartment (Figure 10.5A). As all five subunits have similarly placed hydrophobic sequences they are all believed to take up a similar disposition across the membrane.

Lastly how are the five subunits of the pentamer arranged with respect to each other? If sufficient electron microscope images of a symmetrical structure can be obtained the fuzzy individual images can be superimposed and the 'noise' averaged out to give a much clearer representation. This has been done with samples of electroplax membrane to give a clear image, showing that viewed from the surface the nAChR has five-fold symmetry. Indeed this image strongly suggests that the five subunits are grouped around a central pore (Figure 10.6B).

It has already been emphasised that the exact conformation of the transmembrane segments is somewhat speculative. Until fairly recently it was assumed that they each assumed an α-helical conformation: as indicated in Figure 10.5A and B). This is nowadays challenged: there is evidence to suggest that whereas M2 is α-helical, and strongly contributes to the lining of the pore, the other three

Figure 10.2 SDS–polyacrylamide gel electrophoresis of the nicotinic acetylcholine receptor. Electrophoresis separates the four subunits by molecular weight. The α-subunit travels furthest (M_r 40 000), the β-subunit is next (M_r 49 000), then comes the γ-subunit (M_r 57 000) and heaviest of all is the δ-subunit (M_r 65 000). The electrophoresis also shows that there are two copies of the α-subunit for each copy of the β-, γ- and δ-subunits.

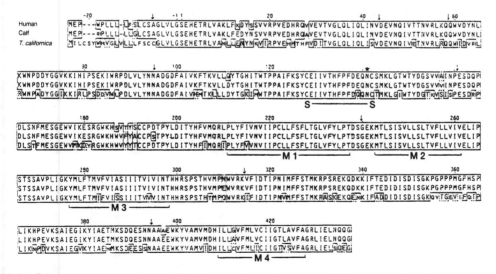

Figure 10.3 Primary sequences of the *Torpedo*, calf and human nAChR α-subunits. The amino acid sequences have been aligned to achieve maximum homology. Amino acids are represented by single-letter code (see Table 2.1). The positive numbering commences at the N-terminal of the mature peptide; negative numbers refer to the signal sequences. Large letters indicate identical residues, dotted lines enclose conservative substitutions. Vertical arrows represent intron splice junctions. S—S shows the disulphide linkage between two cysteine residues in the vicinity of the ACh binding site. M1, M2, M3 and M4 indicate transmembrane segments. Reprinted by permission from Noda *et al.* (1983), *Nature*, **302**, 528–532. Copyright © 1983, Macmillan Magazines Ltd.

A

B

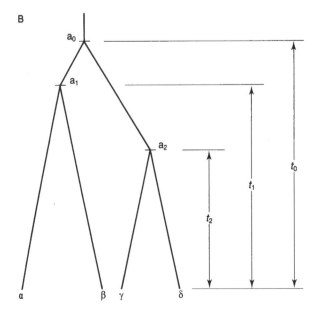

Figure 10.4 Primary sequences of the α-, β-, γ-and δ-subunits of *Torpedo* nAChR and possible evolution of the nAChR gene. (A) Note that the numbering of these sequences commences from the N-terminal methionine residue of the signal peptide. To compare the residue numbers with those in Figure 10.3 it is necessary to count from residue Ser_{25} (arrowed). The four sequences in the figure have been aligned to achieve maximum homology. Gaps (–) have been inserted to achieve this maximum. Identical residues are enclosed by continuous lines. Dashed lines enclose conservative substitutions. Otherwise as for Figure 10.3. Examination of the similarities in the amino acid sequences of the α-, β-, γ-and δ-subunits suggests that they all evolved from a common ancestor by three gene duplications, a_0, a_1 and a_2. Making certain assumptions about the rate of change it is estimated that $t_1 = 0.82t_0$ and $t_2 = 0.65t_0$. From Numa *et al.* (1983), *Cold Spring Harbor Symposia on Quantitative Biology*, **XLVIII**, 57–69; with permission.

A

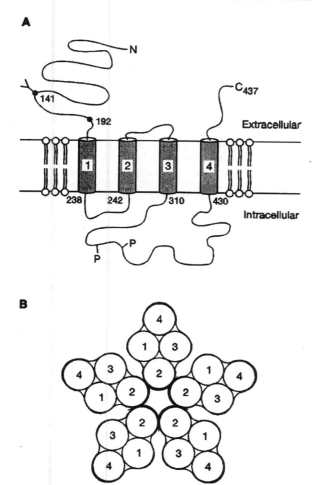

B

Figure 10.5 Disposition of human α-subunit in a membrane. (A) The first 210 residues of the subunit are located in the extracellular space. Glycosylation occurs at residue 141 (Asn). The binding site for acetylcholine is believed to be close to residue 192 and 193 (both Cys). Four segments (represented by dark cylinders) span the membrane. There is a large intracellular loop (residues 310 to 430) between TM3 and TM4. Phosphorylation sites between residues 350 and 375 are indicated by P. Both N- and C-terminals are extracellular. (B) Plan view. The pentameric structure of the complete receptor is seen from above. TM2 of each subunit takes the form of an α-helix and forms the lining of the pore. As indicated in the text, there is evidence that M1, M3 and M4 assume a β-strand conformation. The amino acid sequence of M4 is more variable than M1 or M3 and this may indicate that it is more distant from the pore than the latter two segments.

segments may take the form of β-strands. Each of these outer strands could make one or three passages through the membrane, making tight turns, or β-hairpins (Section 2.1.2). If their backbones were hydrogen-bonded to each other (see Figure 2.2A) a continuous β-sheet would surround the inner ring of α-helical M2 segments. This would help ease the problem presented by interdigitating the side chains of the closely packed α-helical segments of Figure 10.5B. Careful analysis of images such as that shown in Figure 10.6B tends to support the hypothesis that M1, M3 and M4 in each subunit have a β-strand conformation. The ligand binding sites, tailored to accept acetylcholine, are located at the junctions of the α- and β-subunits and the α- and δ-subunits (Figure 10.6C).

Each nAChR pentamer has a diameter of about 9 nm and a central pore of about 2 nm at the synaptic entry narrowing as it passes through the postsynaptic membrane. The pentamer projects 6 nm into the synaptic cleft. This ensures that it stands proud of any extracellular basement membrane which may be present (see Figure 15.16). It will be recalled from Chapter 2 that the collagen-like 'tail' of the important cholinergic enzyme, acetylcholinesterase, is also believed to be inserted into this membrane at neuromuscular junctions. On the other side of the cell membrane the nicotinic acetylcholine receptor protrudes some 2 nm into the cytoplasm. Studies with small cations suggest that the minimum diameter of the channel is no more than 0.80 nm (= 8 Å).

10.1.2 Function

In recent years a number of sophisticated techniques have been developed which enable detailed examination of membrane channels to be undertaken. Many of these have been used to investigate the properties of the nicotinic acetylcholine receptor. Purified nAChR derived from snake-venom affinity chromatography can be inserted into **liposomes** and/or **lipid bilayers** and its physiological properties examined in isolation. Alternatively **patch-clamping** techniques allow the investigation of individual nAChRs expressed in oocyte membranes. **Biochemical** and **electron microscopic** techniques may be used to find the sites on the subunits which bind α-bungarotoxin, thus indicating where ACh normally attaches. Finally the

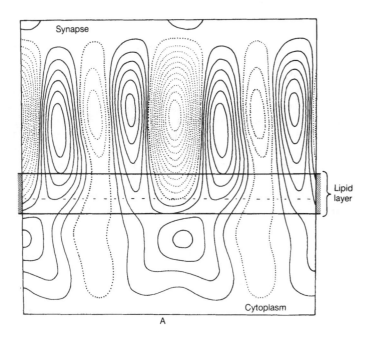

A

Figure 10.6 Pentameric structure of the nicotinic acetylcholine receptor in the post-synaptic membrane. (A) Computer analysed EM representation of the acetylcholine receptor: vertical section, (B) Computer analysed EM of acetylcholine receptors seen end-on. Note five-fold symmetry. (C) Drawing of acetylcholine receptor. The five-fold structure is shown; the position of the ligand and/or neurotoxin binding sites are shown on the α-subunits; the internal cup-shaped channel is shown by dotted lines. Parts A and B reprinted by permission from Brisson and Unwin (1985), *Nature*, **315**, 474–477; copyright © 1985, Macmillan Magazines Ltd. Part C from Stroud (1981), *Proceedings of the Second SUNYA Conversation in the Discipline of Biomolecular Stereodynamics*, ed. by R.H. Shama, Vol. 2, New York: Adenine Press; with permission.

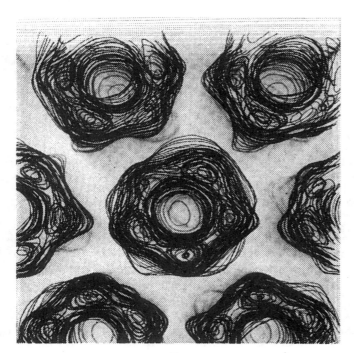

B

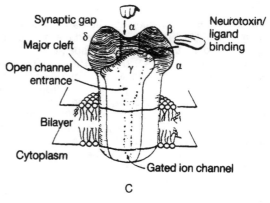

C

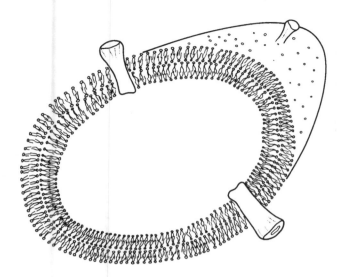

Figure 10.7 Incorporation of nAChRs into a liposome. Three nicotinic acetylcholine receptors have become incorporated into a liposome. Liposomes large enough to hold big multimeric structures such as nicotinic acetylcholine receptors are multilamellar structures formed by a cycle of freeze–thaw steps from suspensions of small liposomes.

techniques of **site-directed mutagenesis**, outlined in Chapter 5, can be used to alter the primary structure of one or more nAChR subunits and the effect on function examined by one of the above techniques. Results from all these approaches enable us to home in on the biophysics of the channel. As these are all techniques of general applicability let us look briefly at each in turn.

Liposomes

In Chapter 7 (Figure 7.3) we noted that because of the amphipathic nature of phospholipids it is not difficult to form liposomes from either pure or mixed solutions of phospholipids. Simple single-layered liposomes are too fragile to hold the comparatively huge nicotinic acetylcholine receptors. It is, however, possible to manufacture large multilayered liposomes by subjecting phospholipid suspensions to repeated cycles of freezing and thawing.

If such multilayered liposomes are created from a suspension of phospholipids and purified nAChR it is found that the latter become incorporated into the structure (Figure 10.7). In most cases the nAChRs are incorporated in such a way that their α-bungarotoxin binding sites face outwards. It can be arranged that the liposomes form around radioactive Na$^+$ and/or K$^+$ ions. It can then be shown that these ions are released when ACh is added to the liposome suspension. Furthermore **tubocurarine**, an antagonist of ACh, prevents the

ACh-activated release of these ions. Finally, the liposomal channels can be examined by the patch-clamp technique described below.

Bilayers

There are several techniques for constructing artificial bilayers. One technique, as shown in Figure 7.3 (Chapter 7), is to form a bilayer across a small aperture (0.2–1 mm diameter) made in a partition separating two aqueous compartments. A defined phospholipid suspension is then introduced into one compartment and into the other a suspension of liposomes containing nAChR. A bilayer containing the nicotinic receptor forms across the aperture (Figure 10.8).

An alternative technique for constructing bilayers is to make use of the spontaneous formation of a monolayer by phospholipids at an air/water interface. It is then possible to allow liposomes containing nAChR to diffuse into the monolayer, thus creating an nAChR-containing bilayer. There are several other techniques for forming artificial bilayers and the interested reader should consult the volume of the *Biophysical Journal* mentioned in the Bibliography.

Patch-clamping

The patch-clamping technique was introduced to neurophysiology in 1976 by Erwin Neher and Bert Sakmann. It has revolutionised the investigation of

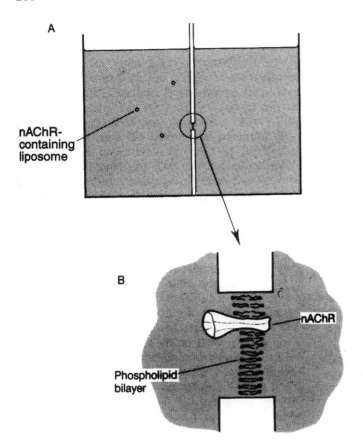

Figure 10.8 Incorporation of nAChRs into an artificial bilayer. (A) An approximately 200 μm hole is made in a Teflon partition separating two aqueous buffer solutions. A defined phospholipid bilayer is formed across the aperture. Liposomes incorporating nAChRs are added to one compartment. (B) Enlarged view. A liposome has fused with the artificial bilayer and its nAChR is now incorporated in the phospholipid bilayer.

the physiological properties of ion channels. Neher and Sakmann were awarded the Nobel prize in 1991. The technique enables the physiologist to examine the fluxes of ions through single channels. The essence of the technique is to place a glass micropipette (tip diameter about 0.5 μm) on to the membrane of interest. A very high resistance (10 GΩ) seal is made between the pipette tip and the membrane. This is essential if currents in the sub-picoampere (pA) range are to be detected. The micropipette is filled with an electrolyte and hooked up to electronics so that the flow of current across the membrane can be measured. The micropipette can also be used to 'clamp' the membrane patch at a predetermined voltage. Further details of 'voltage clamping' are given in Section 14.1 and in Figure 14.2. The membrane may be left *in situ* or by the application of gentle suction it may be detached from the cell and

examined in isolation. The major varieties of this crucial technique are shown in Figure 10.9.

Figure 10.9 shows the various types of preparation that can be obtained: cell-attached patch, whole-cell voltage clamp, inside-out patch, outside-out patch. It is worth noting, especially in the detached patches, that the membrane is sucked into the pipette mouth in an 'omega' form. This means that the area of membrane from which the measurements are made is considerably larger than the area of the pipette tip. This observation is important when calculations of the number of channels per membrane area are made.

Membranes may be obtained from nerve, muscle or other cells, genetically engineered oocytes, liposomes or artificial bilayers. The currents detected by the technique are, of course, minute – to be measured in picoamps (pA). But they are quite sharp. They are due to the opening

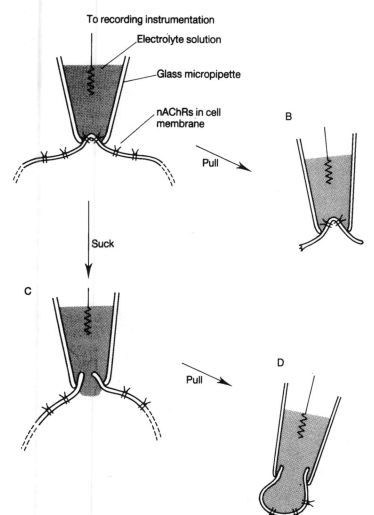

Figure 10.9 Various types of patch-clamping. (A) The recording micropipette is attached to the cell by gentle suction. (B) A sharp pull detaches the cell membrane. The result is an 'inside-out' patch. (C) Alternatively the preparation in A can be subjected to a more vigorous suction. This tears open the membrane. The result is a 'whole-cell voltage clamp'. (D) If C is subjected to a sharp pull the membrane is removed from the cell but spontaneously seals to provide an 'outside-out' patch.

and shutting of **single ion channels**. If and when a second channel opens in the patch the current doubles in magnitude.

In Figure 10.10 it can be seen that in a rat cultured muscle cell, one channel opens for about 30 ms. When the membrane is held at −70 mV a flow of 5 pA of current occurs during that time. The current flow is governed by Ohm's law:

$$I = gV$$

where I is the current, V is the applied voltage and g is the conductance. The magnitude of the current will therefore vary according to the applied voltage and the conductivity of the channel. In the case of the nAChR channel it can be shown that the frequency of opening depends on the quantity of ACh applied.

It is easy to calculate how many ions flow through the channel in the opening event shown on the right-hand side of Figure 10.10. It will be recalled that 1 amp = 1 coulomb per second and that one mole of univalent ions carries a Faraday (= 96 500 coulombs) of electricity. It will also be recalled that a mole is Avogadro's number (N_A), i.e. 6×10^{23} of particles. It follows that:

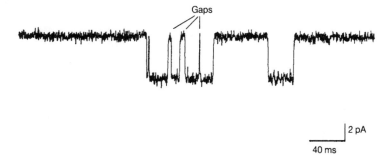

Figure 10.10 Patch-clamp recording of a single nAChR channel. Recording from a rat muscle cell in culture in the presence of 5 μM ACh and the membrane potential at −70 mV. A 'classical' open/shut single channel event is shown on the right. On the left the trace shows several rapid closings of the gate. From Barrantes (1983), *International Review of Neurobiology*, **24**, 259–341; with permission.

$$\frac{[5 \times 10^{-12} \text{ coulombs s}^{-1}] \times [30 \times 10^{-3} \text{ s}] \times [6 \times 10^{23} \text{ ions (g ion)}^{-1}]}{96\,500 \text{ coulombs (g ion)}^{-1}}$$

$$= \frac{9 \times 10^{10} \text{ ions}}{9.6 \times 10^4}$$

$$= \text{approx. } 1 \times 10^6 \text{ univalent ions}$$

This is a considerably larger flow than normally occurs through single nAChR-controlled channels in the subsynaptic membranes of cholinergic neuromuscular junctions. It can be shown that in this position the gates open on average for only about 3 ms instead of the 30 ms of the cultured muscle cell in the example. Furthermore the flow of ions is dependent (as we have just seen) on the driving voltage. For any given ion this is the difference between the membrane potential (V_m) and the Nernst potential of the ion (V_I) (for further analysis see Chapter 17). This difference diminishes as the membrane discharges. Hence the flow of univalent ions in physiological conditions is several orders of magnitude less than that worked out above.

The initial studies with the patch-clamp technique seemed to show that, as indicated on the right-hand side of Figure 10.10, the nAChR channel existed in just two states – open and closed. More recently it has been shown that this is too simple a picture: it seems that there are in fact two different types of open and several different types of closed state. We shall see as we go on through this book that this complexity is a general characteristic of membrane channels. Furthermore it can be shown that (as with the 7TM receptors – see Section 8.2) prolonged binding of agonist molecules to the nAChR leads to **desensitisation**. It is found, in other words, that if the agonist is allowed to remain on

the receptor for a period of seconds to minutes the channel begins to close. This time period is three to four orders of magnitude greater than that required for opening the nAChR gate (microseconds to milliseconds). The desensitisation is due to **phosphorylation** of serine and tyrosine residues in the lengthy cytoplasmic loop between TM3 and 4 (see Figure 10.5). These residues are located in a 25-residue stretch between positions 350 and 375 on the β-, δ- and γ-subunits. It is clear that the patch-clamp technique is beginning to show us something of the molecular complexity that underlies the operation of ligand-gated channels.

Biochemical and Electron Microscopic Techniques

Biochemical and electron microscopic techniques provide a means of determining where the ACh site is located on the receptor. The most usual technique for locating this site is to make use, once again, of the snake-bite venom, α-bungarotoxin. It can be shown that synthetic sequences of amino acids 173–204, and more precisely 185–196, in the α-subunits bind this ACh antagonist. It is consequently believed that it is this part of the pentamer, projecting from the E-face of the postsynaptic membrane (see Figure 10.6C) which acts as the ACh site. More particularly the cysteine residues at positions 192 and 193 have been implicated.

Electron microscopy provides an alternative approach to the localisation of the ACh site (Figure 10.11). In essence what is done is to take electron micrographs of nAChRs in the postsynaptic membrane with and without the addition of α-bungarotoxin. By computerised subtraction of the second image from the first, the position of toxin can be located. Knowing that the toxin occupies the ACh

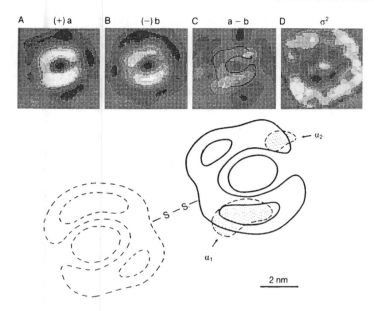

Figure 10.11 Localisation of α-BuTX on the two α-subunits of the nAChR pentamer. (A) An averaged EM image of a negatively stained acetylcholine receptor to which α-BuTX has been bound. (B) The image without the toxin. (C) The image obtained when the computer subtracts B from A. (D) Plot of the standard error of the difference between A and B: the lowest error occurs where there is no toxin. The technique thus allows the localisation of the toxin on the receptor. It is concluded (lower part of figure) that the toxin is located over two parts of the receptor (seen end-on) and these consequently are deduced to be the α-subunits. S—S represents a disulphide bond that is known to link the δ-subunits of adjacent receptor molecules. From Barrantes (1983), *International Review of Neurobiology*, **24**, 259–341, with permission; after Zingsheim *et al.* (1982), *Nature*, **299**, 81–84.

site on the α-subunits, the position of these subunits in the electron micrographic image can be deduced.

Finally, in this section on function, it can be shown that the probability of the nAChR channel opening is greatly increased when two ACh molecules are attached. There is a very small but finite probability of the channel opening even when no ligand is attached at all. When the first ACh binds to its site on the α-subunit it causes a change in the conformation of the nAChR proteins which increases the probability of the channel opening. This probability, however, still remains low. The binding of the first ACh enhances the probability of another ACh binding to the second α-subunit. When both ACh molecules are attached the nAChR pentamer once again changes its conformation and the probability of a brief opening is high. When the channel opens, ions can flow down their concentration gradients. We shall return to the pharmacology of the nicotinic acetylcholine channel in Chapter 16.

10.1.3 Development

Figure 10.12 shows the major steps in the synthesis and assembly of the nAChR complex in muscle fibres. The synthesis occurs in the rough endo-plasmic reticulum (RER) as described in Section 7.6. The assembly of the five subunits into the mature pentamer occurs in the Golgi apparatus and the receptor moves from that position, incorporated in a transport vesicle, to its final home in the folds of the motor end plate. It would seem reasonable to suppose that a similar sequence of events obtains in neurons. In this case the transport vesicles budding off the Golgi apparatus would have to be moved along the dendrites to the appropriate postsynaptic membrane. Alternatively if, as seems likely, many brain nAChRs are presynaptic (i.e. modulating transmitter release) then the transport vesicles would be carried from the perikaryon to the bouton in the axoplasmic flow (see Chapter 15). Throughout life there is a constant turnover. nAChRs have a **half-life** of about a **week**.

A large number of proteins (more than a dozen) are associated with the motor end plate. Many of these form elements of the cytoskeleton. A 43 kDa protein located at the crests of the subsynaptic membrane (see Section 17.2) is the most important of these anchoring proteins and, like the submembranous cytoskeletal elements we discussed in Section 7.8, is believed to hold the nAChR pentamers in position. These proteins are probably attached to the submembranous stretch of the

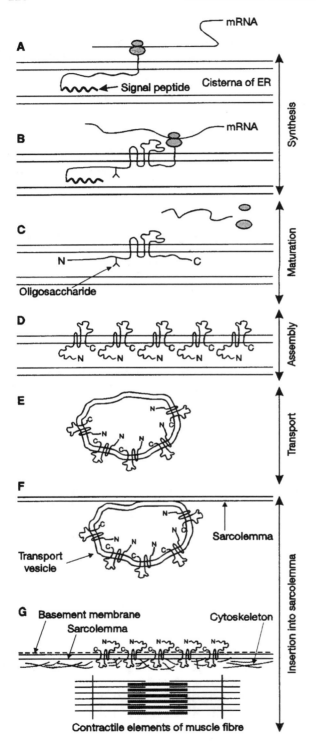

A mRNA

Signal peptide   Cisterna of ER

B mRNA

C

N ———   — C

Oligosaccharide

D

E

F

Sarcolemma

Transport
vesicle

G Basement membrane   Cytoskeleton
      Sarcolemma

Contractile elements of muscle fibre

Synthesis

Maturation

Assembly

Transport

Insertion into sarcolemma

subunit polypeptide chain between M3 and M4 (see Figure 10.5).

During embryology it is found that before the motor neurons reach the muscle, nAChRs are distributed widely (c. $100\,\mu m^{-2}$) and at random in the sarcolemma. Only when a neuromuscular junction has been established are the nAChRs concentrated beneath it. Removal of the junction by sectioning the motor nerve has the opposite effect: the nAChRs are released from the motor end plate and are free to diffuse in the sarcolemma once again. There is evidence to show that the motor neuron secretes a protein, **agrin**, from its terminal on to the sarcolemma and that this exerts a concentrating effect on nAChRs floating, like icebergs, in the muscle fibre membrane.

Yet more interestingly it is found that the character of the nAChR changes when a neuromuscular junction is established. It can be shown that the response time of the acetylcholine receptor to ACh is shortened. The ion channel remains open for only about a quarter of the duration that it stayed open in fetal muscle. The response time of the whole muscle fibre consequently becomes three or four times more rapid.

It has been shown that this change in response time is due to the replacement of the γ-subunit in the nAChR pentamer by a different polypeptide – the ε-subunit. The ε-subunit differs from the γ-subunit in about 50% of its residues. Its hydropathic profile, however, is homologous to that of

Figure 10.12 Synthesis and assembly of nAChRs in muscle fibre. (A) Co-translational insertion of AChR subunit polypeptide into cisternal space of ER. (B) N-glycosylation. (C) Termination of translation. Ribosome and mRNA separate. Subunit polypeptide is inserted in its characteristic position across the membrane. Signal peptide cleaved from subunit polypeptide. (D) Assembly of the receptor from two copies of α-, and one copy of the β-, γ- and δ-subunits. This occurs in the Golgi body. (E) Transport in transport vesicle to sarcolemma. (F) Transport vesicle recognises appropriate place in sarcolemma and the two membranes fuse. (G) The five nAChR subunits cluster to form the mature nACh receptor. Their N- and C-terminals project up beyond the thick basement membrane of the neuromuscular junction. The intracellular domain between M3 and M4 becomes associated with the cytoskeleton. For further discussion of the synthetic process see Chapter 15.

the γ-subunit and hence its disposition in the membrane is believed to be similar.

It seems, therefore, that when a neuromuscular junction is formed on **mammalian** muscle an ε-gene is switched on and a γ-gene switched off. Immediately after innervation both types of subunit are present. Later only the ε-subunit can be detected. It is, once again, concluded that this change is caused by agents released from the motor fibre terminal. Two candidates have been isolated: **CGRP** and **ARIA** (acetylcholine receptor-inducing activator). Both have been shown to be present in terminals and released by impulse activity. Both lead to increased synthesis of nAChR mRNAs. Both would tend to selectively activate nAChR genes held in nuclei close to the motor end plate. It is not yet clear how these neural factors ensure that the ε- rather than the γ-subunit is transcribed. It may be that they alter in some way the activity of the promoter regions exposed to transcription factors.

The acetylcholine receptor at the adult mammalian neuromuscular junction is thus an $\alpha_2\beta\epsilon\delta$ pentamer. This alteration in subunit constitution is very reminiscent of the similar situation in the haemoglobin tetramer which we discussed in Section 4.2.1. It will be recalled that there, too, fetal forms (ε- and γ-) of one of the subunits (the β-subunit) are found. The synthesis of the $\alpha_2\beta\epsilon\delta$ pentamer is confined to nuclei beneath the **subsynaptic** membrane of the neuromuscular junction; nAChR synthesis by nuclei beneath non-junctional sarcolemma tends to be suppressed but the little that is synthesised retains the familiar $\alpha_2\beta\gamma\delta$ structure.

10.1.4 Pathologies

Myasthenia Gravis

Myasthenia gravis is one of the autoimmune diseases. For some as yet unknown reason antibodies are synthesised against the body's own nAChR complexes. The nAChRs are, in consequence, progressively destroyed. This results in increasing muscular weakness. Normally neuromuscular junctions possess a superabundance of nAChRs. ACh released by motor neuron terminals on to the motor end plate is easily taken up by the nAChRs. But as more and more nAChRs are inactivated, ACh becomes less and less able to initiate muscle contraction. Sustained muscular activity becomes progressively more difficult and ultimately impossible.

It is believed that the region which antibodies recognise, the so-called main immunogenic region (MIR), is located on the α-subunit. Segments defined by residues 44–59 and 66–79 are particularly suspected.

Lambert–Eaton Myasthenia

Lambert–Eaton myasthenia (LEM) is believed to be due to an autoimmune response to Ca^{2+} channels on the presynaptic endings of motor nerve terminals. Indeed immunostaining of these endings locates LEM antibodies on particles in the presynaptic membrane which are believed on other grounds to be Ca^{2+} channel proteins (see Section 15.4 and Figure 15.17). We shall see in Section 15.4 that the ingress of Ca^{2+} through Ca^{2+} channels is essential to the release of neurotransmitter (in this case acetylcholine) into the synaptic gap. Once again muscular weakness, flaccidity and ultimately inactivity result.

Congenital Myasthenic Syndromes (CMSs)

In contrast to the autoimmune myasthenias, congenital myasthenic syndromes (CMSs) are caused by **genetic** defects at the presynaptic, synaptic or postsynaptic parts of the synapse. **Presynaptic CMS** is due to insufficient production of acetylcholine in the presynaptic terminal; **synaptic CMS** is caused by insufficiently active **AChE** in the synaptic cleft which leads to a damaging over-stimulation of the muscle fibres; **postsynaptic CMS** results from defective nAChRs. There are three cases of the latter disease: slow channel syndrome, fast channel syndrome and AChR deficiency. **Slow channel syndrome**: mutations on the nAChR genes lead to defects in the channel protein which slow the movements of ions through the channel and to variations in the attraction of ligand and receptor. Patients suffer weakness and fatigue and many show degeneration of muscle fibres. **Fast channel syndrome**: the movement of ions through the channel is too rapid. This again leads to easy fatigability and a diffuse weakness leading ultimately to respiratory failure. **AChR deficiency syndrome**: there are just too few receptors on the postsynaptic membrane. Once again muscular weakness ensues but the course of the condition is more benign than in the other CMSs.

10.1.5 CNS Acetylcholine Receptors

We noted in Section 10.1 that nAChRs are not confined to neuromuscular junctions but are widely (if sparsely) distributed throughout the nervous system, including the central nervous system. Once again oligonucleotide probes prepared from peripheral nAChR data have been used to fish out nAChR sequences from brain cDNA libraries. When these sequences have been cloned and their sequences analysed and translated into protein structure, it has been found that CNS nAChRs are far more diverse than their peripheral cousins. This diversity is shown in their subunit structure and this, of course, translates into diversity of biophysical function.

One feature that distinguishes most CNS nAChRs from those found at the neuromuscular junction is that, in addition to their permeability to the small univalent cations, Na^+ and K^+, they also display a significant permeability to Ca^{2+}. This Ca^{2+} conductance may be significant in providing a 'second messenger' signal in central neurons (see Section 8.5). Another feature that distinguishes the two classes of nAChRs is their response to pharmacological agents. For instance, it has been shown that whereas suberyldicholine is more potent than ACh at the neuromuscular junction, it is less effective than ACh on neuronal nAChRs. Pharmacologists have been able to show a number of other differences in the response of neuronal and muscular nAChRs to synthetic agonists. This, again, implies differences in subunit make-up and structure.

This implication is borne out by structural studies. Molecular biological techniques have shown that there are at least seven α-subunit homologues and three non-α-subunits in vertebrate brains. The α homologues are defined by their ability to bind acetylcholine whilst the non-α have a structural role in the nAChR architecture. The α-subunits have been labelled **α2–α8** (α1 being

---

BOX 10.1 Evolution of nAChRs

Nicotinic acetylcholine receptors constitute a family of great and distinguished antiquity. AChRs originated before the appearance of the Bilateralia, between 2000 and 1500 million years ago. In that remote epoch the most advanced animals were represented by the Cnidaria (Coelenterata): the jellyfish, sea anemones, seashore polyps such as *Obelia*, etc. These animals consist of just two layers of cells, ectoderm and endoderm, separated by a structureless mesoglea. In this mesoglea is to be found a diffuse nerve net (of ectodermal origin) which is responsible for coordinating the tactile responses of polyps and the swimming activity of medusae. It is within the cells of the nerve net that ACh and its nicotinic receptor first evolved.

Neuronal nAChRs. nAChRs are, as we have seen, pentameric structures. The subunit which binds ACh includes a pair of cysteines in the N-terminal domain. This is taken as defining the α-subunit. The ability to bind ACh implies that the α-subunit is ancestral. In the Gnathostomata (the jawed vertebrates) at least seven different α-subunits (α2–α8) have evolved in the nervous system, all sharing this characteristic. Other subunits, the non-α-subunits, having a supporting structural role, have also evolved (nα1–nα3).

Muscle nAChRs. With the appearance of a third germ layer, the mesoderm, between the ectoderm and endoderm (c. 1300 million years ago), the stage was set for the development of muscles, neuromuscular junctions and hence muscle-type nAChRs. These nAChRs retained the α-binding unit of their neuronal progenitors and also evolved a number of different supporting subunits: β, γ, δ and ε.

By careful assessments of the amino acid sequences in these various subunits and by making estimates of the rate of change during evolution (assuming all the while that this is constant (see Section 4.2.1)), a provisional evolutionary tree for the nAChRs has been developed. This is shown in Figure A.

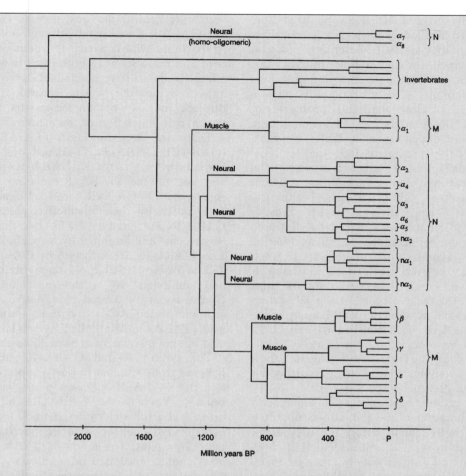

Figure A   Evolution of nAChRs. The chart should be regarded as provisional as receptor subunits are still being isolated and analysed. With the exception of one major branch all the evolutionary lines are of vertebrate subunits. They cover a wide spectrum of forms ranging from humans through other mammals, birds, amphibia and fish. Further explanation in text. Adapted from Le Novère and Changeux (1995), *Journal of Molecular Evolution*, **40**, 155–172; and Ortells and Lunt (1995), *Trends in Neurosciences*, **18**, 121–127, where further details may be found. See also web site listed in Bibliography.

Figure A shows that the present-day richness of the nAChR family is due to a number of gene duplications and subsequent evolutionary development. The chart shows that the $\alpha7$- and $\alpha8$-subunits are the most primitive. These subunits (like the 5-HT$_3$ receptors to which the primitive nAChRs are related (see Figure 11.28)) are capable of forming homo-oligomers in the *Xenopus* oocyte expression system. The other branch of this early bifurcation leads to a number of invertebrate α- or α-like subunits. The earliest muscle subunit, the $\alpha1$-subunit, derived from a bifurcation some 1300 million years ago. The other branch of the bifurcation leads to a large number of neural α- and non-α-subunits and the muscle β-, γ-, δ- and ε-subunits.

It has been argued that the combinatorial possibilities provided by the comparatively recent development of a large number of α- and non-α-subunits underpins the evolution of the highly flexible and heterogeneous vertebrate neurophysiology.

reserved for the muscle nAChR α-subunit) and the non-α-subunits labelled as nα1–nα3 (confusingly another terminology labels the latter as β2–β4 with β1 being reserved for the muscle subunit; this terminology is misleading to the uninitiated as neuronal β-subunits are unrelated to muscle β-subunits). It is clear that this profusion of different nAChR subunits can lead to an even greater profusion of pentameric nAChRs. Neurons, for instance, could present as $(\alpha3)_2/(n\alpha1)_3$, $(\alpha4)_2/(n\alpha1)_3$, $(\alpha2)/(n\alpha1)_3$, etc. As function emerges from molecular structure it is evident that these different nAChRs will have subtly different biophysical characteristics. When expressed in the *Xenopus* oocyte preparation they can indeed be shown to have differing ion conductances as well as differing sensitivities to agonists and antagonists. The heterogeneity may be even greater than indicated. It may be that different α- and nα-subunits come together to form the receptor. In the laboratory nearly all possible pentameric combinations of α- and β-subunits have been constructed and examined in *Xenopus* oocyte: $\alpha3\alpha5/n\alpha1(n\alpha3)_2$; $\alpha3\alpha7/n\alpha1n\alpha2n\alpha3$; $(\alpha5)_2/(n\alpha1)_2n\alpha3$, etc. Again, these heterologous combinations generate distinctive biophysics and pharmacology.

Even if the extreme heterogeneity in which every possible combination of α- and non-α-subunit is allowed does not occur in nature but only in the scientist's laboratory, it is still evident that cholinergic neurons may differ widely amongst themselves in biophysical and pharmacological character. Indeed, this is one more indication that each neuron in the brain is an individual, with its own molecular 'personality', changing from day to day as surrounding circumstances change.

Autosomal Nocturnal Frontal Lobe Epilepsy (ANFLE)

The α4-subunit of neuronal nAChR is coded by the *CHRN4* gene on chromosome 20 (20q13.3). A single nucleotide change in this gene causes an amino acid substitution in the subunit's pore-lining domain. This is one of the causes of ANFLE, a rare condition showing clustered epileptic episodes occurring during sleep.

10.2 THE GABA$_A$ RECEPTOR

Two important inhibitory transmitters (γ-aminobutyric acid (GABA) and glycine) are found in vertebrate central nervous systems (see Chapter 16). GABA is found mainly though not exclusively in the brain whilst glycine is found in the spinal cord and brain stem. In both cases they exert their effect by controlling a channel specific to small anions. When they open, chloride ions course through and, as we shall see in Chapter 17, lead to a hyperpolarisation of the membrane.

There are three subclasses of GABA receptor (GABAR): GABA$_A$R, GABA$_B$R and GABA$_C$R. Some investigators regard GABA$_C$Rs as a specialised set of the GABA$_A$R subtype. The three subclasses of GABAR are distinguished from each other by their distinctive pharmacologies. **GABA$_A$Rs** are activated by GABA, muscimol and isoguvacine and inhibited by bicuculline and gabazine; **GABA$_B$Rs** are activated by GABA, bacloven and 4-amino-3-(5-chloro-2-thienyl)butanoic acid and inhibited by phacloven and sacloven; **GABA$_C$Rs** are activated by GABA, *cis*-4-aminocrotonic acid (CACA) and *trans*-4-aminocrotonic acid (TACA) and inhibited by imidazole-4-acetic acid but are insensitive to bicuculline and bacloven.

The GABA$_A$R and GABA$_C$R subclasses are believed to be situated in postsynaptic membranes and the GABA$_B$R subclass in presynaptic membranes. Whereas the GABA$_A$R and GABA$_C$R directly control a chloride channel, the GABA$_B$R acts through a collision-coupling mechanism involving G-proteins. The action of GABA is thus, like acetylcholine, mediated through two very different types of subsynaptic receptor. In this section only the GABA$_A$R will be considered.

The pharmacology of the GABA$_A$R subclass has been intensively studied. It appears to have binding sites for at least four types of drug. These include GABA itself, the benzodiazepines, picrotoxin and the barbiturates. Just as it is possible to purify the nAChR on an α-toxin column so it is possible to make use of the GABA$_A$ receptor's affinity for benzodiazepine to purify it on a benzodiazepine column. Accordingly it is this subclass of GABAR that was the first to be subjected to detailed molecular analysis.

After purification the GABA$_A$R turned out to consist of five approximately 50 kDa subunits: α, β, γ, δ, ρ. Each subunit consisted of between 400 and 500 amino acids and there was some 30–40% homology in their amino acid sequences. Further research showed that six subtypes of the α-subunit existed (α_1–α_6); three subtypes of β-subunit (β_1–β_3); three subtypes of γ-subunit (γ_1–γ_3); and three subtypes of ρ-subunit (ρ_1–ρ_3). Subsequently two other types of subunit have been identified (ε and π) and further variants of the major five subunits have been found.

To ensure that the entire GABA$_A$R had been obtained the mRNA was expressed in the *Xenopus* oocyte system. The oocyte does not normally possess GABA receptors in its membrane. After injection of the putative GABA$_A$-R mRNA large conductances of chloride across the membrane in response to the external application of GABA could, however, be detected. This proved beyond doubt that the entire channel protein had been synthesised by the oocyte and consequently that the entire gene for GABA$_A$-R had been cloned.

The usual analysis of the amino acid sequence for hydrophobic stretches showed there to be four regions of sufficient length (i.e. about twenty residues) to form α-helical spans of the membrane. This was the case in both subunits. These transmembrane segments are designated M1, M2, M3 and M4 analogously to the similarly named four hydrophobic transmembrane stretches of the nicotinic acetylcholine receptor (Figure 10.13). As with the nAChR, it is considered that M1, M3 and M4 are β-strands and only the pore-lining M2 strand is α-helical. The large, but variable, intracellular loop between M3 and M4 is believed to be concerned in regulatory mechanisms involving phosphorylation.

A proposed structure for the GABA$_A$R is shown in Figure 10.14. It consists of five subunits arranged around a central pore (cf. the nAChR). There are 20 membrane-spanning helices. It is known that the channel diameter can be no more than 5.6 Å at its narrowest point. It is not geometrically possible to pack all 16 α-helices so that each faces the channel's lumen. Some other, as yet unknown, organisation must be adopted. Stereochemical calculations show that five α-helices can be arranged to enclose a channel of pore diameter 5.8 Å: very close to the minimum diameter determined by biophysical measurements (Figure 10.14).

As in the other multimeric receptors there are great possibilities for heterogeneity. Thus the receptor may consist of copies of all the five subunits, or it may consist of three alphas and two deltas, or two betas and three deltas, etc. Furthermore, as we saw above, each subunit comes

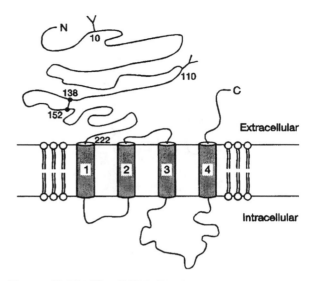

Figure 10.13 The GABA$_A$R subunit. The N-terminal consists of a lengthy (222 residue) sequence in the extracellular compartment. Asn residues at positions 10 and 110 are glycosylated (as shown) and a disulphide bond links Cys residues at positions 138 and 152. The amino acid chain then makes four passes through the membrane. A lengthy intracellular loop connects M3 and M4.

in up to six distinct varieties. A very large number of permutations and combinations is possible. GABA$_A$Rs thus differ, like the neuronal nAChRs, in different parts of the brain, in different neurons and, indeed, in different parts of the same neuron.

10.2.1 Pathology

There is evidence that a mutation (K289M) in the γ_2-subunit gene (*GABRG2* on chromosome 5q34) segregates with a family subject to **generalised epilepsy with febrile seizures (GEFS)**. This condition, as we shall see in Section 11.8, is also associated with defects in the Na$^+$ channel. Clearly the condition is multifactorial. The K289M substitution occurs in a highly conserved stretch of amino acid residues connecting the M2 and M3 transmembrane domains of the γ_2-subunit. Physiological analysis of mutated GABA$_A$Rs inserted into *Xenopus* oocytes showed that GABA-mediated Cl$^-$ currents were reduced to about 10% of normal. This suggests a cause for GEFS symptoms.

Plan view

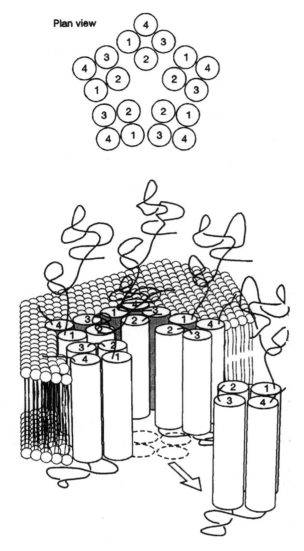

Figure 10.14 Schematic of the GABA$_A$ receptor. Five subunits (cf. Figure 10.13) are packed around a central Cl$^-$ channel. Reproduced with permission from MacDonald and Olsen, *Annual Review of Neuroscience*, **17**, 569–602, 1994 by Annual Reviews Inc.

10.3 THE GLYCINE RECEPTOR

Next let us look at the glycine-activated channel. These channels are found on postsynaptic membranes in the brain stem and spinal cord of mammals. Similar channels are found throughout the vertebrates and in many invertebrates. They are extremely narrow (<5.2 Å) and, like the GABA-activated channels, very selective for anions. Of the ions present on either side of neural membranes they thus allow only Cl$^-$ to pass.

The study of the glycine receptor (GlyR) has been greatly helped by its affinity for **strychnine**. Strychnine thus plays something of the same role for the GlyR that the α-toxins played for the nACh receptor, and benzodiazepine for the GABA$_A$ receptor. The GlyR may, for instance, be purified by chromatography through a column of agarose beads to which has been attached a derivative of strychnine, 2-aminostrychnine. The GlyR so purified consists of three subunits, a 48 kDa α-subunit, a 58 kDa β-subunit and a 93 kDa polypeptide.

The position of these subunits in the GlyR has been clarified by immuno-electron microscopy. Monoclonal antibodies can be prepared against the chromatographically purified GlyR polypeptides. These can then be conjugated to gold and reacted with central synapses. Figure 10.15 shows that the antibody against the 93 kDa polypeptide (subsequently called **gephyrin**) is located on the cytoplasmic side of the subsynaptic membrane.

It is thought that the glycine binding part of the receptor consists of five copies of the 48 kDa and one or two copies of the 58 kDa polypeptide. The five copies of the 48 kDa polypeptide form the ion channel. The 93 kDa polypeptide is associated with this transmembrane complex, but is located on the P-face of the membrane where it is believed to anchor the receptor to the cytoskeleton of the subsynaptic density (Figure 10.16B).

The strychnine binding subunit of the rat spinal cord glycine receptor (the 48 kDa α-subunit) has been successfully cloned and its amino acid sequence determined. It consists of 421 amino acids and a precise molecular weight of 48 383 Da. It turns out that, like the preceding receptors of this chapter, there are a number of variants. At least three varieties of the strychnine binding or α-subunit have been characterised (α_1, α_2 and α_3) as well as a non-strychnine binding β-subunit. A combination of α- and β-subunits form the customary pentamer surrounding the ion channel. Hydropathic analysis reveals that the amino acid sequence of each subunit possesses four transmembrane segments designated M1, M2, M3 and M4 (Figure 10.16A). Once again we become aware of a common theme underlying channel architecture. It

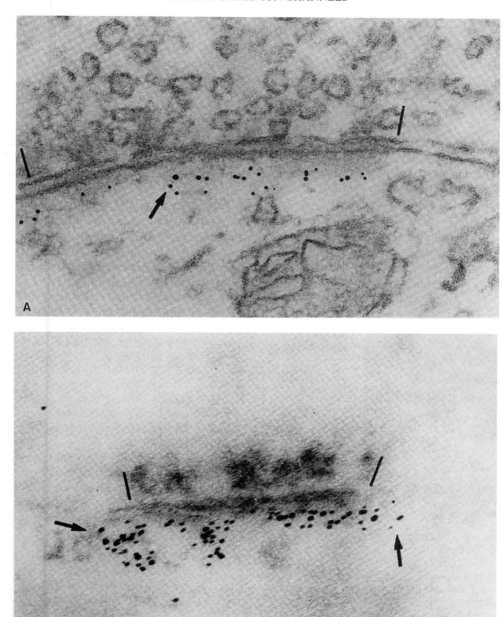

Figure 10.15 Localisation of the 93 kDa polypeptide of the glycine receptor. (a) The 93 kDa GlyR subunit is labelled by an immunogold technique and examined in the electron microscope. The arrow points to one of the labels. The 93 kDa subunit is clearly on the intracellular side of the subsynaptic membrane. The preparation is of rat spinal cord (×100 000). (B) The arrows again point to the immunogold-labelled 93 kDa subunit in the intracellular space. The bars indicate the position of an active zone in the presynaptic terminal (see Chapter 15). Most of the GlyRs lie beneath this active zone although some (arrowed) lie outside it (×100 000). Reproduced from Triller *et al.* (1985), *Journal of Cell Biology*, **101**, 638–688, by copyright permission of The Rockefeller University Press.

A

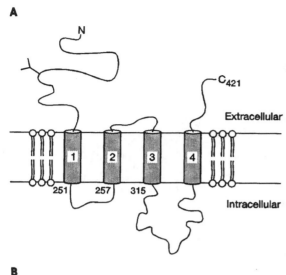

B

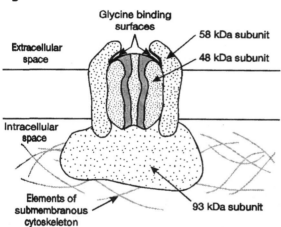

Figure 10.16 The 48 kDa subunit of the glycine receptor. (A) The schematic figure shows the disposition of the 48 kDa subunit in the membrane. The similarity with nAChR and GABA$_A$R subunits is clear. (B) Schematic figure to show the disposition of the 93 kDa, two copies of the 58 kDa and five copies (only three shown) of the 48 kDa subunits. The glycine binding surfaces covering both 58 kDa and 48 kDa subunits are shown.

is clear that the various subtypes of nAChR, GABA$_A$R and GlyR are all members of an evolutionarily related family.

The relatedness of nAChR, GABA$_A$R and GlyR is borne out by a comparison of the amino acid sequences in the nAChR α-subunit, the GABA$_A$R

Table 10.1 Percentage amino acid homologies in the nAChR, GlyR and GABA$_A$R receptor subunits

	GABA$_{A\alpha}$R	GABA$_{A\beta}$R	GlyR
nAChR$_\alpha$	19 (38)	15 (32)	15 (37)
GABA$_{A\alpha}$R	100	35 (57)	34 (56)
GABA$_{A\beta}$R	35 (57)	100	39 (59)

The sequences compared are those of bovine GABA$_A$ receptor (α- and β-subunits), rat 48 kDa glycine receptor and bovine muscle nAChR α-subunit. The first figure in each case represents percentage identical residues, the figure in parentheses the percentage identical plus conservative substitutions. Data from Barnard, Darlison and Seeburg, 1987, *Trends in Neurosciences*, **10**, 502–509.

α- and β-subunits and the α-subunit of the glycine receptor. These homologies are shown in Table 10.1.

There are many suggestive similarities between the three receptor subunits. The regions of greatest homology are, for instance, to be found in the transmembrane segments. The M2 helix, in particular, is remarkably similar in the GABA$_A$ receptor and the glycine receptor. Although the amino acid homology does not extend to nAChR there is good evidence that the M2 helix in that receptor plays a major role in lining the channel wall. It is likely, therefore, that the invariance of the M2 helix in the GlyR and the GABA$_A$ receptor indicates that it plays the same role here also. There are many other fascinating aspects of molecular comparative anatomy which our increasing knowledge of this family of ligand-gated ion channels is bringing to light. The interested reader can pursue them in the references given in the Bibliography.

10.4 IONOTROPIC GLUTAMATE RECEPTORS (iGluRs)

Mammalian brains possess two important excitatory amino acids (EAAs): **glutamate** and **aspartate**. They are widely distributed throughout the brain and spinal cord. Glutamate and its receptors (GluRs) are the best known. Stimulation leads to three major types of response. In the first case glutamate induces rapid (c. 1 ms) membrane depolarisations. In this respect glutamate resembles the action of acetylcholine on the nicotinic acetylcholine receptor. In the second, and perhaps more interesting, case glutamate once again causes a membrane depolarisation but this time of a much longer duration (10–15 ms) often accompanied by other, more complex, events. In the third case (already discussed in Section 8.10) glutamate causes a response which works through a G-protein system.

Now although the membrane responses to glutamate may be classified into these three major classes, pharmacological and molecular biological analyses show there to be at least four different types of receptor: three subclasses of ionotropic receptor and one subclass of metabotropic receptor. They are classified by their preferred ligands:

1. **AMPA** or **Q receptors** (agonist: α-amino-3-hydroxy-5-methyl-4-isoxazole propionic acid (quisqualate); antagonist: 6-cyano-7-nitro-quinoxaline-2,3-dione (CNQX));
2. **KA receptors** (agonist: kainate);
3. **NMDA receptors** (agonist: N-methyl-D-aspartate (an analogue of glutamate); competitive antagonist: D-2-amino-5-phosphonovalerate (D-AP5) non-competitive antagonist: phenylcyclidine (PCP));
4. **tACPD receptors** (agonist: *trans*-1-aminocyclopentane-1,3-decarboxylate (tACPD)).

In fact the pharmacological situation is not quite clear cut: tACPD only activates GluR2, 3 and 4 whilst quisqualate is the preferential agonist for GluR1 and 5.

It is found that the Q, KA and NMDA receptors are responsible for ionotropic responses whilst the tAPCD receptor is (as we noted in Chapter 8) metabotropic. We shall see, however, that the NMDA receptor's response is far more complex than standard ionotropic receptors such as the nAChR or the Q and KA GluRs.

It was not until 1989 that the first glutamate receptor (GluR1) was cloned and its structure determined. This was because there is no high-affinity, high-specificity ligand (such as the snake toxins used for AChR or benzodiazepine for $GABA_A$) which could be used to purify the receptor and thus obtain an initial oligonucleotide probe. Instead the technique of expression cloning (using the *Xenopus* oocyte (Section 5.14)) was used to isolate a glutamate receptor clone. This receptor was labelled GluR1. But once this had been done, and the nucleotide sequence published, it

immediately became possible to synthesise appropriate oligonucleotides and search for other GluRs in brain cDNA libraries. Within five years 28 GluR genes had been characterised coding 22 ionotropic and six metabotropic receptors. These could be grouped into 13 subfamilies: ten ionotropic and three metabotropic receptors. No doubt further genes and perhaps subfamilies will be reported in the years to come.

10.4.1 AMPA Receptors

The first four GluRs to be characterised (GluR1–GluR4) were members of the Q or AMPA group. They are all of a similar size (about 900 amino acids) and their sequences are about 70% identical. Hydropathy analysis implies **three** transmembrane segments. In place of the second transmembrane helix (M2) of the AChR family there is a hairpin-like loop (Figure 10.17). This, as we shall see in Chapter 11, has some similarity to the voltage-gated cation channels although their hairpins are made from the extracellular compartment. Comparison of amino acid sequence with nAChR, $GABA_A R$ and GlyR shows only some 20% identity – hardly above chance – so it is unlikely that they all are members of a single superfamily.

Figure 10.17B shows that the complete GluR consists of five copies (cf. nAChR, etc.) of the 3TM subunit arranged around a central pore. The re-entrant hairpin of each 3TM subunit lines this canal. When activated this central pore allows the passage of both K^+ and Na^+ leading to the membrane depolarisation of the excitatory post-synaptic potential (EPSP) (see Section 17.3.1). In addition it has been shown that the pore (except in GluR2) is also quite permeable to Ca^{2+} and Mg^{2+} ions. Finally, experiments have indicated that the channel currents are inhibited by PKA. Desensitisation by phosphorylation of residues in the cytoplasmic C-terminal is very rapid (GluR1 c. 36 ms and GluR4 c. 8 ms).

10.4.2 KA Receptors

KA receptors are classified according to their affinity for kainate. There are three low-affinity subtypes (Glu5, Glu6, Glu7) and two high-affinity subtypes (KA1, KA2). There are a number of splice variants of all these types of KA receptor. All of these receptors share the tertiary and quaternary structures outlined above and shown in Figure 10.17. Their amino acid sequences set them apart from GluR1–GluR4, as do their pharmacological sensitivities and biophysical responses. Kainate receptors have been detected in a number of brain regions including hippocampus (mossy fibres), cerebellum, amygdala and striatum.

Both AMPA and KA receptors consist of either five identical subunits (homo-oligomeric) or a pentamer of five different subunits (hetero-oligomeric). This provides GluRs with a great variety of subtly different biophysical characteristics.

Finally, we noted in Section 3.3.2 that the mRNA for some GluR subunits is subjected to an editing process. This editing inserts an arginine residue into TM2 of the GluR2, GluR5 and GluR6 subunits. We noted above that homo-oligomeric GluR2 channels synthesised from their DNA code do not conduct Ca^{2+}. In the cell, however, the mRNA strand is edited so that an arginine residue is inserted at the appropriate place. Ca^{2+} conductivity is thus ensured. Arginine is also inserted by a similar editing into the GluR5 and 6 subunits of the KA receptor.

10.4.3 NMDA Receptors

Although NMDA receptors have a rather low amino acid homology with non-NMDA GluRs (25–29%) they once again share the 3TM architecture and pentameric quaternary structure of Figure 10.17. The hydrophobic (presumably transmembrane segments) are, as before, arranged in the $1+2$ pattern allowing a lengthy intracellular domain between M2 and M3. The NMDA receptor differs, however, from the non-NMDA GluRs in possessing a rather more extensive N-terminal, extracellular sequence. It is believed that this latter domain may be involved in ligand binding. As with the AMPA and KA GluRs, there are a number of subtypes: NMDA-R2A, NMDA-R2B, NMDA-R2C, NMDA-R2D. Within this classification there is, again as with the AMPA and KA GluRs, a great deal of diversity created by differential splicing of the various DNA transcripts.

All three of the ionotropic receptors discussed above are found throughout the brain and especially in telencephalic structures. NMDA receptors are particularly heavily represented in the hippocampus. This is especially noteworthy because, as

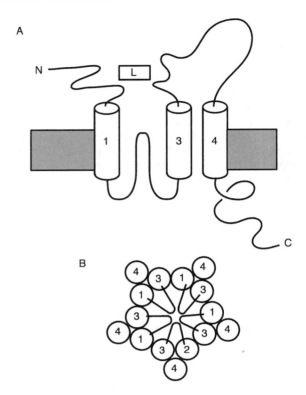

Figure 10.17 Structure of ionotropic GluRs. (A) Several different dispositions of the hydrophobic segments have been suggested. The figure shows the model most favoured at present. Three TM segments are shown. A re-entrant hairpin is shown between TM1 and TM3. (B) Plan view of the pentamer of subunits surrounding the ion pore. Note that the re-entrant loop is believed to form the wall of the channel. L=ligand.

we shall see, NMDA receptors are suspected of being involved in synaptic plasticity and short-term memory. We shall see in Chapter 20 that the hippocampus is believed to be deeply implicated in these processes. It is thus worth devoting a little space to considering its biophysical and pharmacological characteristics.

Biophysics of the NMDA Receptor

The NMDA receptor has been subjected to exhaustive biophysical and pharmacological analysis. It has also been investigated by patch-clamp analysis of cultured cells, especially cerebellar Purkinje cells and hippocampal pyramidal cells,

BOX 10.2   The inositol triphosphate (IP₃ or InsP₃) receptor

We have noted in many places in this book the importance and ubiquity of Ca^{2+} ions in controlling cellular biochemistry. The cytosolic concentration of Ca^{2+} is very low ($<1\,\mu M$). Reserves are, however, held by calmodulin (see Box 9.1) and in the cisternae of the endoplasmic reticulum. We saw in Section 8.5 that one of the major roles of IP₃ as a second messenger is to release Ca^{2+} ions from their storage reservoir in the ER. This is achieved by the interaction of IP₃ with specific receptors, the IP₃ receptors, in the membrane of the ER.

IP₃ receptors were first detected in the Purkinje cells of mouse cerebellum. Cloning revealed a huge 260 kDa protein of some 2749 amino acid residues. No sequence homology could be detected with the Ca^{2+} channels of the plasmalemma. Hydropathy analysis showed that the amino acid sequence contained six hydropathic segments towards its C-terminal end. These hydropathic sections are, as usual, believed to zigzag through the ER membrane (Figure A). Both the C-terminal and the N-terminal are situated in the cytosol. The N-terminal is extremely lengthy and exhibits both Ca^{2+} and ATP binding sites. Evidence from site-directed mutagenesis suggests that a 650-residue sequence towards the N-terminal forms the IP₃ binding site. The IP₃R subunit may thus be considered to be divided into three domains: a ligand binding domain; a regulatory domain; and channel-pore domain (Figure A).

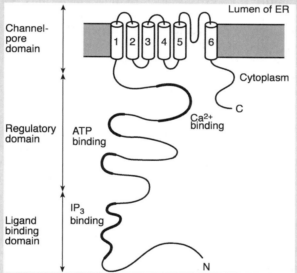

Figure A   Inositol triphosphate receptor subunit. The schematic figure shows the disposition of the subunit in the membrane of the ER. The full IP₃R consists of four of these subunits grouped around a central pore. Further explanation in text.

There is evidence to show that the fully formed receptor consists of four of the subunits shown in Figure A. At present it is believed that these are all identical; in other words the receptor is a homotetramer. Each subunit binds one IP₃ molecule but preliminary work indicates that, unlike the nACh receptor, there is no cooperativity. When IP₃ binds the channel opens and Ca^{2+} passes down its concentration gradient from the lumen of the ER into the cytosol.

The cerebellar IP₃ receptor exists in several isoforms derived by differential splicing of the primary transcript. Furthermore, there at least three genes coding for three different IP₃Rs: IP₃R1, IP₃R2 and

IP$_3$R3. These receptors share 60–70% sequence identity. The original IP$_3$ receptor isolated from mouse Purkinje cells is designated the cerebellar or type 1 receptor (IP$_3$R1). The existence of several types of IP$_3$ receptor opens the possibility of multiple Ca^{2+} signalling pathways. One tissue may contain several different types and splice variants of the receptor. The ubiquity of the channel in eukaryotic cells and its similarity to another major class of Ca^{2+} channels (ryanodine receptors) suggests that it is evolutionarily ancient (Figure 11.28).

The type 1 IP$_3$ receptor is most abundant in the cerebellum but is also found in the hippocampus and the cerebral cortex. Within the cell it can be localised on the membranes of the smooth endoplasmic reticulum (SER) but it is also present in rough endoplasmic reticulum (RER) and on the nuclear membrane. It is also to be found in numerous tissues outside the brain.

Finally, it is worth noting that this is the first receptor we have discussed that is not inserted into the plasmalemma. There seems little doubt that the internal membranes of the cell will contain many different receptors and pumps. We shall meet an instance of one of the latter when we consider the pumps for neurotransmitters developed in the membranes of synaptic vesicles in Box 15.2.

where as we noted the transmitter is known to be present in large quantities.

In striking contrast to other ionotropic receptors it is found that, in addition to its sensitivity to NMDA and other agonists, the NMDA receptor is also sensitive to the **voltage** across the membrane in which it is embedded. This makes it appropriate to draw the present chapter towards a close with a discussion of its characteristics. For, in the next chapter, we shall proceed to a discussion of some very well-known voltage-sensitive channels. The NMDA receptor thus serves as an example of 'hybrid' channel: it is ligand operated, yet sensitive to voltage.

It can be shown that the voltage-dependent opening of the NMDA channel is affected by the presence of **Mg^{2+}** ions. When these ions are present (and they usually are in physiological conditions) the channel is blocked and small membrane depolarisations have little or no effect. We shall encounter other cases of Mg^{2+} blockade when we come to consider the inwardly rectifying K$^+$ channels in Section 11.2.1). However, in the case of NMDA channels, it is found that when the membrane is depolarised by some 30 mV from its resting state the Mg^{2+} blockade is overcome and the channels begin to open (Figure 10.18). As the depolarisation is increased a larger number of channels open. In this respect, as we shall see in Chapter 11, the NMDA channel resembles the Na$^+$ channel: there is positive feedback between membrane depolarisation and opening of the channel.

But which ions pass through the channel? Experiments have demonstrated that Ca^{2+} is the most important ion to flux through although Na$^+$ and K$^+$ ions also pass. Compared with the NMDA channel the AMPA and KA channels are only sparingly permeable to Ca^{2+}. The most significant ions in these latter channels are Na$^+$ and K$^+$. The fact that the NMDA channels are designed to allow Ca^{2+} ions to flow through into the cytosol from the outside is of great physiological and biochemical significance. We shall see in later chapters how important Ca^{2+} ions are in synaptic transmission and as intracellular 'second messengers'.

Once the voltage across the NMDA membrane has been reduced sufficiently to overcome the Mg^{2+} blockade the positive feedback effect mentioned above tends to keep the NMDA channels open. This accounts for the comparatively long-lasting response which, as we noted at the outset, is characteristic of NMDA receptors. The situation is, however, by no means clear cut. It seems likely that the channel can exist in a number of different sub-states: some more long lasting (10–15 ms) than others (1–3 ms), in some cases fully open, in others in various states of 'partial' opening.

The Mg^{2+}-mediated voltage dependency of NMDA channels could have very important consequences for synaptic physiology. It may, for instance, provide a biochemical basis for an '**AND**' or '**Hebb**' gate, and thus for associative learning. The NMDA receptor can only be actuated by an

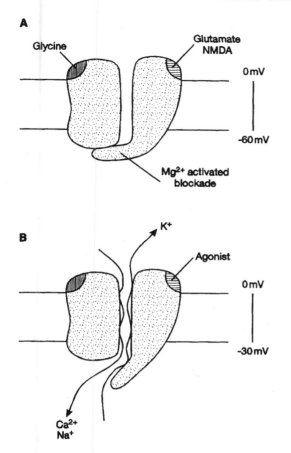

Figure 10.18 Physiological responses of the NMDA-activated channel. (A) The NMDA-controlled ion channel is closed. The figure shows a site for agonist molecules (glutamate, NMDA) and for antagonists (glycine). Agonist molecules will not open the channel at normal membrane resting potentials if Mg^{2+} ions are present. (B) When the membrane is depolarised the Mg^{2+} blockade is removed and agonist molecules can open the channel. Ca^{2+} and Na^+ flow inwards down their electrochemical gradients and K^+ flows out. These ion currents tend to keep the membrane depolarised and hence the channel open.

EAA when the membrane in which it is situated is to some extent depolarised. One can suppose therefore that to fire a subsynaptic cell it is necessary for two presynaptic terminals to be active at once: one releasing an excitatory transmitter on to a non-NMDA receptor (thus depolarising the membrane) and the other releasing an EAA on to a nearby NMDA receptor (Figure 10.19). Perhaps this is why NMDA receptors are so well represented in the hippocampus. We shall return to NMDA receptors in Chapter 20. We shall see that they have been implicated in 'long-term potentiation' (LTP) which is much discussed as a biophysical basis for associative learning and short-term memory. We shall also see (Chapter 19) that NMDA receptors may be implicated in the synaptic plasticity of the developing brain.

Finally, it is worth noting that NMDA receptors may also be involved in pacemaker variations of membrane polarity. We shall see in Chapter 11 that calcium-dependent potassium gates are present in some neural membranes. It is easy to see that if NMDA receptors are activated by an EAA, the inflowing Ca^{2+} ions could open nearby gates of this type. The outflow of K^+ ions would tend to repolarise the membrane. This repolarisation would bring into play the Mg^{2+} blockade of the NMDA receptor. The inflow of Ca^{2+} would be cut off. Any cytosolic Ca^{2+} would be quickly bound. The outward flow of K^+ through the Ca^{2+}-dependent channels would cease. The membrane would then be ready for another cycle of depolarisation and repolarisation: perhaps by release of fresh glutamate on to AMPA, KA and NMDA receptors.

The NMDA channel provides a good example of the developing understanding that ligand-gated channels are not the simple 'open'/'shut' mechanisms that they were once thought to be. Indeed the NMDA receptor shows even more subtlety than we have yet discussed. It is known, for instance, to be potentiated by external glycine down to a concentration of $10\,\mu M$. This tends to pile yet more complexity upon what is already a complex story. For we saw in Section 10.3 that glycine is an inhibitory transmitter in the brain stem and spinal cord. It now appears that in addition to its inhibitory activity it potentiates the excitatory effect of glutamate at the NMDA receptor.

In conclusion, we can see that cerebral glutamate receptors are many and varied, ranging from straightforward iGluRs through the Hebbian NMDAs to the recently revealed mGluRs discussed in Section 8.10. We have seen that there are many members of each type and that each member has many subtypes produced by differential splicing of the DNA transcript. It is salutary to remember that

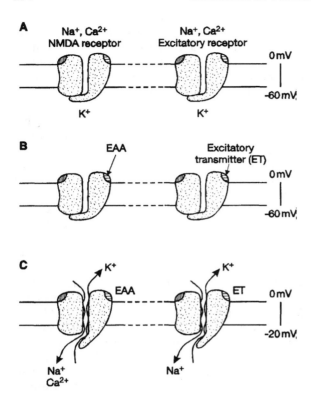

Figure 10.19 The NMDA receptor as a logic gate in a subsynaptic membrane. (A) An NMDA receptor and an excitatory receptor (activated by an EAA or some other excitatory transmitter) are shown in the same subsynaptic membrane. (B) An EAA and (another) excitatory transmitter (ET) is liberated by the pre-synaptic terminal. (C) The ET opens its ion channel and the membrane depolarises. This removes the Mg^{2+} blockade from the NMDA receptor, whose ion channel consequently opens to allow the flow of Na^+, Ca^{2+} and K^+.

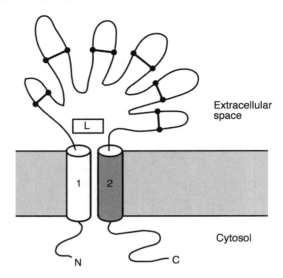

Figure 10.20 Conformation of the P2X purinoceptor subunit. Note that both N- and C-terminals are cytoplasmic and that the major part of the 472 amino acid receptor is deployed in the extracellular compartment. Only two transmembrane segments are present and it is suggested that the large extracellular domain forms six leaflets bound at their bases by disulphide linkages. The stippled segment indicates the pore-lining domain. L=ligand.

this huge variety of excitatory receptors has been characterised in only the last ten years. This gives some impression of the power of modern molecular biological methods and the intensity of the scientific effort devoted to the subject. It is also salutary to recognise that the subtle range of GluRs represents the state of neuronal membranes in the brain far more closely than the classical nicotinic acetylcholine receptor at the neuromuscular junction. It seems likely that many of the synaptic receptors in the brain are, like the GluRs, richly complex in their variety and in their responses to numerous different neurotransmitters, neuromodulators and ions.

10.5 PURINOCEPTORS

We shall see in Chapter 16 that the purine nucleoside **adenosine** and its phosphorylated derivatives (AMP, ADP and ATP) have been accorded full status as neurotransmitters. Purine receptors are divided into two large classes: **P1 receptors** sensitive to adenosine and AMP and **P2 receptors** sensitive to ADP and ATP. Both classes are further subdivided into subclasses. It has been found that whilst the members of one subclass of P2 receptor, the P2Y receptors, are metabotropic, the members of the other subclass, P2X receptors, are ionotropic. We shall discuss their pharmacology in Section 16.5 (Table 16.8). The molecular structure of members of both types of receptor has been elucidated. The structure of the P2X subunit differs from the subunits discussed in the preceding pages of this chapter in that it makes only two passes through the membrane (Figure 10.20). In this it resembles the epithelial Na⁺ channels and the mechanosensitive channels of hair cells which we shall discuss in Section 13.3. The members of the other subclass, the P2Y metobotropic receptors, show the canonical 7TM conformation of other G-protein-linked receptors. Even here, however, the amino acid sequence does not indicate any close relationship to the classical 7TM receptors discussed in Chapter 8. It may be, therefore, that the purinoceptors constitute an evolutionarily primitive and independently evolved group of channels. If this is so it raises fascinating questions of convergent (rather than divergent) evolution at the molecular level.

There are at least seven and probably more different P2X subunits (see Table 16.8) varying from 388 to 595 amino acids in length so that the possible permutations and combinations and, consequently, biophysical characteristics are very large. A major feature of all the subunits is the very large glycosylated extracellular loop which contains the ATP binding site. The subunits are believed to assemble as trimers to form the physiological channel which is lined by M2. The channel is cation selective, allowing the passage of Na⁺ and K⁺ with almost equal ease and to a lesser extent Ca²⁺.

10.6 CONCLUSION

In this chapter we have made a rapid overview of the great variety of ligand-gated channels that modern techniques are bringing to light. We shall return to a further consideration of their pharmacology in Chapter 16. In the present chapter, however, we have seen how the approach through molecular biology has revealed remarkable order in the otherwise overwhelming complexity. We have seen that at least three major ligand-gated channels – nAChR, GABA_AR and GlyR – are members of a single, evolutionarily related superfamily. It is becoming clear that these receptors, and all their many subtypes, are variations on a single architectural theme. This theme, in its essence, consists of units built of four membrane-spanning segments (M1, M2, M3 and M4) joined by stretches of hydrophilic amino acids. The M2 segment is α-helical whilst the other three segments are probably β-strands. Five of these units are

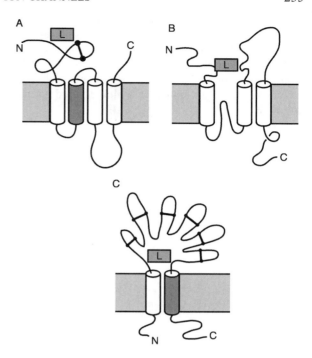

Figure 10.21 Subunit structure of the three major groups of LGIC. (A) nAChR family. (B) GluR family. (C) P2XR family. The pore-lining domain is stippled. L = ligand. Bars indicate disulphide linkages between cysteine residues. After Khakh (2001), *Nature Reviews Neuroscience*, **2**, 165–174.

assembled to form a pentamer surrounding an ion pore. The peripheral nicotinic acetylcholine receptor is constructed from four different subunits, either α₂βγδ or α₂βεδ. The nAChRs of the central nervous system show greater heterogeneity. There are at least seven different α-homologues and three non-α-subunits. The possible permutations of these ten subunits in the overall pentameric architecture is very large. 4TM subunits are also the basis of the GABA_AR architecture. In this case five different subunits are distinguished, α, β, γ, δ and ρ, and, again like nAChR, they are assembled into a pentameric structure. The same architecture is found in the glycine receptors.

When we come to the iGluRs and their subclasses, the AMPA, kainate and NMDA receptors, we find a different architecture. Instead of four transmembrane passes these receptor subunits make only three. In place of the second transmembrane helix, M2, of the AChR superfamily we find a re-entrant hairpin. There is evidence that this, like M2, lines the pore of the pentameric channel. We shall find something similar in the next chapter when we come to consider voltage-gated cation channels.

Finally, the simplest subunits of all are found in the large and evolutionarily ancient family of ionotropic purinoceptors. These receptors, the P2X receptors, are built of subunits making only two passes through the membrane. For convenience all three subunit designs are shown in Figure 10.21.

11

VOLTAGE-GATED ION CHANNELS

Central role of voltage-gated channels in neurophysiology. Classification. **The KcsA bacterial K$^+$ channel** – structure – biophysics. **Neuronal K$^+$ channels** – nomenclature and overview. Kir channels (2TM (1P)) – inward rectification – genetics, distribution and role. K$^+$ 'leak' channels (4TM (2P)) – structure – genetics – role. Kv channels (6TM (1P)) – shaker channels – structure – biophysics – heterogeneity – genetics – accessory subunits – Kvdr (delayed-response K$^+$ channels) – KCNQ channels – MinK – Eag channels – ligand-modulated K$^+$ channels. **Ca^{2+} channels** – diversity – physiology – structure – genetics – pharmacology – biophysics. **Na$^+$ channels** – structure – diversity – genetics – biophysics. **Ion selectivity and voltage sensitivity** – H5 hairpin/P-domain and ion selectivity – gating current – sliding-helix voltage sensor. **Cl$^-$ channels** – ClC channels – Cln channels – phospholemman. **Channelopathies**: K$^+$ channels – Ca^{2+} channels – Na$^+$ channels. Cl$^-$ channels. **Evolution of ion channels. Conclusion and forward look**: themes and variations in the evolution of channel proteins

Excitable cells, as we shall see more fully in Chapter 14, depend on the existence of voltage-controlled gates in their membranes. These gates, once again, are many and various. The most important are those that open to allow the passage of ions such as Na$^+$, K$^+$, Ca^{2+} and Cl$^-$. Classification of the multitude of different Na$^+$, K$^+$, Ca^{2+} and Cl$^-$ channels is complex. There are at least eight types of Na$^+$ channel, six types of voltage-dependent K$^+$ channels (not counting subtypes), six major groups of inwardly-rectifying K$^+$ channels, eight or nine voltage-independent K$^+$ channels, and five types of Ca^{2+} channels. Various types of Cl$^-$ channels, though less important in the electrical phenomena of neurophysiology, also have significant roles to play.

Voltage-sensitive channels are responsible for the electrical excitability of nerve and muscle cells and for the sensitivity of sensory cells. In addition there are numerous ligand-coupled ion channels which modulate the electrical potential across neuronal membranes (detail of this bewildering variety of channels may be found in the 2001 Nomenclature Supplement of *Trends in Pharmacological Sciences*).

With the exception of most ligand-gated K$^+$ channels all of these channels respond to voltage changes across the membrane. As we shall see more fully in Chapter 12, the resting potential across most cell membranes is about 50 or 60 mV (inside negative to outside). This may not seem very much. It must be remembered, however, that membranes are very thin – no more than 6 or 7 nm across. Hence the voltage drop is in fact very steep. A potential gradient of 60 mV in 6 nm works out as 10^5 V cm^{-1}. We must assume that voltage-sensitive proteins are very delicately poised in this intense electric field. Any change in the potential gradient will affect their conformation – and the openness or shutness of any ion channel they may contain.

There are two ways of classifying this still growing multitude of voltage-sensitive channels: first by function, second by structure. Nomenclature is also important if we are to keep clear what it is we are talking about. Thus a channel that responds to a voltage change across the membrane by opening to admit the passage of, say, K$^+$ ions is termed a **K channel**. The current of K$^+$ ions through the channel is designated I_K. If the channel is built in such a way that the passage of K$^+$ ions is modulated by ATP, then it is designated a **K$_{ATP}$** channel and the current is termed the $I_{K(ATP)}$ current.

In this chapter we shall only consider a few of the great variety of channels known to exist in the CNS. We shall take them in roughly evolutionary order. We shall start by examining one of the most spectacular advances of recent years: the solution by X-ray diffraction techniques first at the 3.4 Å level and then at the 2 Å level of a bacterial K$^+$ channel. This is the first and, until the recent determination of bacterial Cl$^-$ channels, the only ion channel to have had its structure elucidated at the atomic level. Although it has **no voltage sensitivity** it is nevertheless believed to represent the core architecture, especially that of the central pore, of the larger voltage-sensitive K$^+$ channels of eukaryocytes.

We shall then go on to consider a selection of the great variety of K$^+$ channels found in animal nervous systems before turning to Ca^{2+} and Na$^+$ channels and then finishing the chapter with a brief account of Cl$^-$ channels. Further details on channel structure, pharmacology and biophysics can be found in the publications listed in the bibliography and in the Receptor and Ion Channel Supplements of *Trends in Pharmacology* or in volumes I–IV of *The Ion Channel Facts Book*.

11.1 THE KcsA CHANNEL

As we shall see in Section 11.2.3, the first K$^+$ channel to be sequenced was the *Drosophila* 'shaker' channel. In the early 1990s the amino acid sequence of the extracellular pore region (P) of this channel was established. With the rapid growth in knowledge of genomes it became possible to search the gene libraries of other organisms for similar sequences. One such organism was the Gram-positive soil bacterium, *Streptomyces lividans*. A 1080 nucleotide sequence between start and stop signals with a segment containing the code for a P-region was found. This is the *KcsA* gene. When translated it coded for a 17.6 kDa protein (**KcsA**) and when this was incorporated into giant liposomes it was shown to be a K$^+$ channel. Because it is possible to culture large number of *S. lividans* over-expressing the KcsA protein, and because the protein has great intrinsic stability, it is possible to generate the large quantities necessary for crystallisation and X-ray diffraction analysis.

The KcsA channel subunit is a 2TM protein (Figure 11.1A). The extracellular connection between the two transmembrane helices is known as the P-region. The complete KcsA channel consists of four of these 2TM subunits grouped around a central canal. The inner helix of each subunit forms the lining of the canal. X-ray diffraction allows us to visualise the three-dimensional structure of the channel, first at a resolution of 3.2 Å and subsequently at 2 Å (Plate 3A). The four pairs of helices lean together like the poles of an inverted tepee (wigwam) with the vertex pointing into the cell (Figure 11.1C). The pore runs along the centre of the tepee with a vestibule situated just beneath the selectivity filter. The amino acid sequence forming the P-region folds into the space at the extracellular base of the tepee to form this filter, discriminating between K$^+$ and Na$^+$ (Figure 11.1C).

The selectivity filter is remarkable. As with all K$^+$ channels the permeability for K$^+$ is very great (almost equal to free diffusion) but that for Na$^+$ is almost immeasurably small, more than 10 000 times less than that for K$^+$. The question for all K$^+$ channels has long been: how can they allow such an easy passage for K$^+$ (radius 1.33 Å) and yet so firmly exclude the smaller Na$^+$ (radius 0.95 Å)? The 2 Å crystallography has allowed this question to be answered.

Figure 11.1 (*Opposite*) The KcsA channel. (A) Subunit structure. (B) Plan of pore region. The amino acid sequences of the KcsA and shaker-type K$^+$ channel are shown below. The similarity of the selectivity filter sequences is obvious. (C) Schematic to show the organisation of two of the four subunits in the membrane. The central cavity where K$^+$ ions wait to pass through the filter is shown by broken lines. The carbonyl oxygens of GYG and also D are shown by black dots. V = vestibule. After MacKinnon *et al.* (1998), *Science*, **280**, 106–109; Zhou *et al.* (2001), *Nature*, **414**, 43–48.

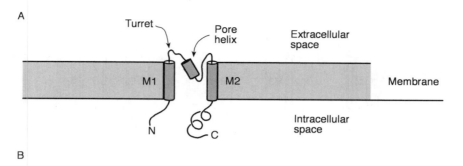

KCsA ———— E R G A P G A Q L I T Y P R A L W W S V E T A T T V **GYG** D L Y P V T L ————

Shaker ——— E A G S E N S F F K S I P D A F W W A V V T M T T V **GYG** D M Y P V G F ———

51   60   70   80   86

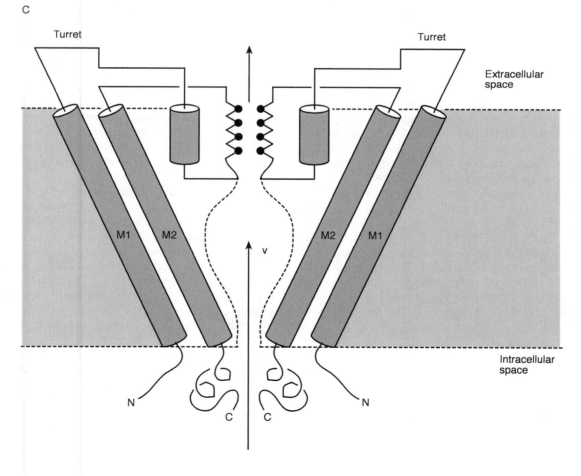

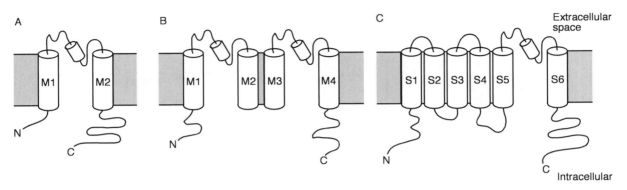

Figure 11.2 Architecture of K+ channels. (A) 2TM(1P); (B) 4TM(2P); (C) 6TM(1P). Note the large C-terminal domain in A and C. This is involved in the 'ball and chain' inactivation of the channel. See text.

It turns out that five of the amino acids in the P-region of each subunit (-T-V-G-Y-G-) (Figure 11.1B) run into the base of the tepee turning their side chains outward, and their carbonyl groups inward towards the pore axis. The four subunits thus create a rigid carbonyl-lined pore 3 Å in diameter. K+ ions form coordinate bonds with the carbonyl oxygens and fill the space with great precision. The ions have about the same energy as they have when surrounded by H_2O molecules in aqueous solution. This is not the case with Na+ ions. These ions do not fit so precisely the carbonyl-lined channel and their energy is consequently considerably higher than when closely surrounded by water molecules in aqueous solution. Hence K+ ions will flow nearly as easily as if free in aqueous solution but Na+ ions will be held back.

The 2 Å structure shows that the K+ ion has four selectivity sites which it has to satisfy (squeeze through) in the P-region. The vestibule (Figure 11.1C; Plate 3) holds a K+ ion surrounded by eight H_2O molecules. Because, as we saw above, the energetics are so similar, the K+ ion slips away from its sheath of water molecules and enters the first site and from there to the next on its journey through the filter. The rejected water molecules also enter the filter. In fact at any one time K+ ions alternate with H_2O molecules on the selectivity sites. In other words only two K+ ions occupy the channel at the same time separated by H_2O. The selectivity sites are energetically so similar that when a K+ ion approaches from the inside it knocks the line of K+ ions and H_2O molecules in the channel so that the one nearest to the extra-

cellular compartment is detached and diffuses to the outside. This process has been vividly pictured as rather like one of the pendulum toys sometimes seen on the desks of stressed executives. When the left-hand metal sphere swings into the line of hanging spheres the impulse is transmitted through the group so that the sphere at the far right swings out.

The 2 Å structure also shows how the conductivity of the KcsA channel varies according to the extracellular K+ concentration. It can be shown that at low extracellular K+ the selectivity filter loses one of its dehydrated K+ ions and undergoes a conformational change rendering it non-conductive. The channel thus switches between open and shut states according to the external concentration of potassium ions.

The permeability characteristics of the KcsA channel have thus found a highly satisfactory solution in the 2 Å structure provided by X-ray diffraction. Does this solution apply to other K+ channels? A strong indication that it does is the finding that the sequence (-T-x-G-Y-G) is conserved in the great variety of K+ channels found in eukaryocytes. It has been dubbed the **K+ selectivity sequence**. Further evidence that the architecture is conserved across K+ channels has been derived from experiments with scorpion toxin. These toxins have a characteristic fold extending over 35–40 amino acid residues and held rigidly in place by three disulphide linkages. The toxins block K+ channels by lying across the **extracellular environs** of the pore. The X-ray diffraction shows that in the KcsA channel this surface is rather like the keep of a castle: a squarish flat area watched over by four corner turrets (Figure 11.1C). The toxin lies diagonally across this area from one turret to another blocking the pore exit. Mutational analysis, inserting different amino acid residues into the toxin, shows that this area is much the same in KcsA as in the ubiquitous shaker-type K⁺ channels of the animal kingdom (see below). The outer keep of the *S. lividans* KcsA channel may thus be conserved in the K⁺ channels of higher organisms. It will be fascinating to discover how the 2TM subunit K⁺ channels of prokaryocytes have been adapted to form the much larger voltage-sensitive K⁺ channels of the animal kingdom: channels that play such important roles in the functioning of nervous systems.

11.2 NEURONAL K⁺ CHANNELS

Biophysical and molecular biological studies have shown there to be a huge variety of neuronal potassium channels. Some are ligand-gated but the majority, to a greater or lesser extent, are sensitive to transmembrane voltage. They have many functions but one of their major roles is to stabilise the membrane potential at approximately the Nernst potassium potential (V_K) (see Chapter 14). When their molecular structure was determined (see below) they were found to fall into three major groups: those that made **two passes (2TM)**, those that made **four passes (4TM)** and those that made **six passes (6TM)** through the membrane (Figure 11.2). Each of these three groups can be further subdivided. The 2TM channels form a large group (at least 12 members) of **inward rectifying K⁺ channels (Kirs)**, the 4TM channels form a group of at least 11 so-called **'leak' channels**, and the very large group of 6TM channels are subdivided into six families (**Kv channels, KCNQ channels, eag-like channels**, and three types of **ligand-activated channel**). All members of this great multitude of K⁺ channels share one thing in common: the pore

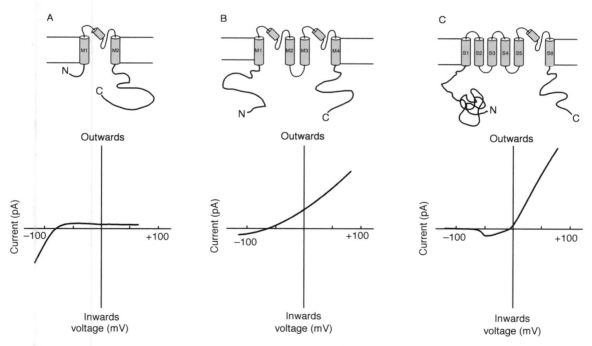

Figure 11.3 Current–voltage relations of 2TM(1P), 4TM(2P) and 6TM(1P) K⁺ channels. (A) Inwardly rectifying channel. Only small K⁺ currents flow as the membrane is depolarised due to Mg²⁺ blockade. In hyperpolarised conditions the blockade is removed and inward current flows. (B) Leak channel. K⁺ ions flow along their electrochemical gradients. (C) Voltage-gated K⁺ channel (KV). In the case shown here there is a delay before the outward current commences when the membrane is progressively depolarised. After Goldstein *et al.* (2001), *Nature Reviews Neuroscience*, **2**, 175–184.

(P) loop we first met in the bacterial KcsA of the preceding section. It is this that confers K^+ selectivity.

We shall note below that the structure of the 6TM K^+ channels was elucidated some five years before that of the 2TM channels and seven years before that of the 4TM channels. Confusingly for twenty-first-century readers the terminology developed for 6TM channels was transferred to the 2TM structures. Thus although, as in Figure 11.2, the two transmembrane helices of the 2TM subunits are labelled M1 and M2, they are sometimes designated as S5 and S6. Similarly the hairpin between S5 and S6 is sometimes referred to as H5 and sometimes (analogously with the KcsA channel) as P. The yet more recent elucidation of 4TM channels has forced a further revision. It can be seen from Figure 11.2 that the TM domains are numbered M1 to M4 and the H5 hairpin is denominated 'P'.

The temporal characteristics and sensitivity to voltage, neurotransmitters and 'second messengers', such as Ca^{2+} etc., of K^+ channels varies widely.

The three major groups vary in response to alteration of membrane polarity. The 2TM channels are blocked when the membrane is depolarised and only allow K^+ current to flow (outwards) when the membrane is hyperpolarised (Figure 11.3A); the 4TM channels allow K^+ to 'leak' down their electrochemical gradients (Figure 11.3B); the 6TM channels open when the membrane is depolarised and allow K^+ to flow through with increasing ease as the voltage across the membrane becomes positive (Figure 11.3C).

In addition to their major structural and physiological classification (2TM(1P), 4TM(2P), 6TM(1P); 'inward rectification' (ir), 'leak', 'outward rectification') K^+ channels have been given a more detailed classification according to their particular characteristics: i.e. the **fast or early K^+ channel**, or **A channel (K_v or K_A)**, the **delayed** or **delayed rectifier K^+ channel (K_{vdr})**, the **serotonin-dependent K^+ channel (K_S or $K_{5\text{-HT}}$)**, the **Ca^{2+}-dependent K^+ channel (K_{vCa})**, etc. The K^+ currents which course through them when they open are termed $I_{K(A)}$, $I_{K(dr)}$, $I_{K(S)}$ and $I_{K(Ca)}$ currents, etc.

All these channel types, and many others, including those whose voltage sensitivity is modulated by second messengers, and those with very weak or absent voltage sensitivity and operated by second messengers and other ligands, exist throughout the animal kingdom although the variety increases with the evolution of advanced forms. Moreover, it transpires that all the major types of K^+ channel can be further classified into subtypes according to their kinetics and single channel conductances. Indeed it is nowadays believed that K^+ channels are the most diversified of all channels. The mix of K^+ channels in its membrane once again confers personality on an individual neuron.

As would be expected, this great diversity of channels influences a huge number of physiological phenomena. In addition to stabilising the resting potential (V_m) across neuronal membranes, K^+ outflow determines the shape and duration of an action potential (Chapter 14). Hence the characteristics and number of these channels in a neuron's membrane are very important. Furthermore, the duration and magnitude of the action potential invading a synaptic terminal governs, as we shall see, the inflow of calcium ions. This influx, in its

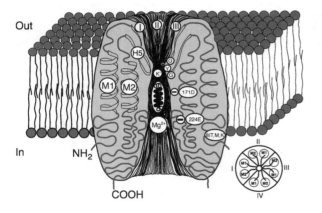

Figure 11.4 Mg^{2+} channel blockade of Kir channels when membrane is depolarised. The main figure shows a vertical section through the tetrameric channel. On the left the M1 and M2 helices are shown and the H5 selectivity filter presenting the G-Y-G residues. A K^+ ion is shown in the filter. Beneath the filter a polyamine is shown blocking the channel and below that a Mg^{2+} ion. The smaller figure on the right represents a cross-section of the channel showing the disposition of the M1 and M2 helices and the H5 filter (black). From Nichols and Lopatin, 1997; with permission, from the *Annual Review of Physiology*, Volume 59 © 1997 by Annual Reviews; www.AnnualReviews.org.

turn, controls the quantity of neurotransmitter released (see Section 16.4). The distribution and mix of potassium channels on synaptic terminals is thus of considerable importance in the large-scale functioning of the nervous system. Finally the K$^+$ 'leak channels' are also ubiquitous in neuronal membranes. These activate slowly with small membrane depolarisations and allow K$^+$ to enter the neuron down its electrochemical gradient, thus ensuring that the resting potential is restored. These channels thus buffer the membrane against small fluctuations in potential, preventing the initiation of accidental action potentials.

11.2.1 2TM(1P) Channels; Kir Channels

2TM channels (Figure 11.2A) have the biophysical character of '**inward rectification**'. Inward rectification was initially called 'anomalous rectification' as it is the opposite of that occurring in 'classical' (6TM) K$^+$ channels (see Section 11.2.3). Instead of an inward flow of K$^+$ ions when the membrane is depolarised, it is found that the 2TM channels shut down at potentials positive to about $-40\,\text{mV}$ and conduct more current when the membrane is hyperpolarised (i.e. negative to the Nernst potassium potential, V_K). This 'anomalous' behaviour is largely due to a voltage-dependent block of the channel pore by cytoplasmic cations, particularly Mg^{2+}, and by polyamines, when the membrane is depolarised (Figure 11.4). Because 2TM K$^+$ channels show this property of inward rectification they are known as **Kir** channels. Their physiological importance is to stabilise the membrane at its resting potential (V_m), near V_K. This is due to their high K$^+$ conductance as the membrane exceeds V_m (further explanation in Chapter 14).

Table 11.1 shows that since the identification of the first Kir channel in 1993 a very large family has been uncovered. They all share the common 2TM subunit structure and, as Figure 11.4 shows, the fully formed channel consists of four of these subunits grouped around a central canal. It is tempting to believe that the structure of the KcsA channel described in the preceding section provides a model for this four subunit architecture. While this temptation should not be resisted too strongly it is nevertheless the case that the amino acid sequence of the KcsA channel bears more resemblance to the 6TM voltage-dependent K$^+$ channels than to the 2TM Kir channels. There are several other pointers from genetic and biophysical studies which suggest that we should be cautious in transferring the bacterial KcsA structure too directly to the Kirs.

Table 11.1 shows that the Kir family is widely distributed in the brain. There is evidence that cerebral Kir channels are controlled (via receptor and G-protein) by a number of transmitters including substance P, GABA, acetylcholine and

Table 11.1 Mammalian Kir genes

Gene	Chromosome	Channel	Expression (human)
KCNJ1	11q24	Kir1.1	Kidney
KCNJ10	1q	Kir1.2	Kidney
KCNJ15	21q22.2	Kir1.3	Kidney
KCNJ2	17	Kir2.1 (IRK1)	Heart, nervous system
KCNJ12	17p11.2–11.1	Kir2.2	Heart, nervous system
KCNJ4	22q13	Kir2.3	Heart, nervous system
KCNJ3	2q24.1	Kir3.1 (GIRK1)	Heart, brain
KCNJ6	21q22.1–22.2	Kir3.2 (GIRK2)	Brain
KCNJ9	1q21–23	Kir3.3 (GIRK3)	Brain
KCNJ5	11q24	Kir3.4 (GIRK4)	Heart
KCNJ16	17q	Kir5	Brain
KCNJ8	12p11.23	Kir6.1	Ubiquitous
KCNJ11	11p.15.1	Kir6.2 (K$_{ATP}$)	Pancreas, brain, muscle, heart

Modified from Nichols and Lopatin (1997), *Annual Review of Physiology*, **59**, 171–191; Riemann and Ashcroft (1999), *Current Opinion in Cell Biology*, **11**, 503–508; Ashcroft (2000), *Ion Channels and Disease*, San Diego: Academic Press.

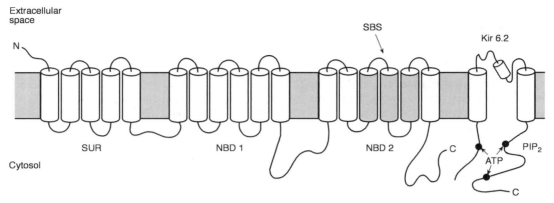

Figure 11.5 Plan diagram to show Kir6.2 with SUR modulator. The 17TM sulphonylurea receptor (SUR) is shown on the left with its sulphonylurea binding site (SBS) shaded. Nucleotide binding domains (NBD) are shown on the intracellular loops of the SUR. The activity of the Kir is modulated by PIP$_2$, which interacts with the C-terminal domain, and inhibited by ATP, which attaches at the three points indicated in the figure. After Riemann and Ashcroft (1999), *Current Opinion in Cell Biology*, **11**, 503–508.

somatostatin. In this way they are crucial in allowing neurotransmitters to influence the electrical excitability of neuronal membranes.

Kir channels also play significant roles in cardiac physiology and elsewhere. In the sarcolemmas of cardiac muscle cells the anomalous voltage sensitivity of Kir3.1 (GIRK1) ensures that the resting potential is prolonged. This extends the period of diastole and is consequently crucial in determining cardiac rhythm. Indeed, it is the ACh-controlled biophysics of GIRK1 that are largely responsible for the deceleration of heart rate in response to ACh which Otto Loewi observed at the origin of pharmacology (see Box 16.2). Its anomalous voltage-sensitivity is also important in other situations. Some glial cells make use of its opening under hyperpolarising conditions to remove K$^+$ ions from the intercellular space.

It would clearly be inappropriate in a book of this nature to review the physiological characteristics of all the Kir channels listed in Table 11.1. Instead only two examples, both of which have neurobiological significance, will be discussed. Both exemplify the fact that Kir channels are modulated by a great variety intracellular agents and G-protein subunits and, in particular, by PIP$_2$. The first, Kir6.2 (an ATP-sensitive channel (K$_{ATP}$)), provides an instance of control by a cytosolic agent (ATP), and the second, Kir3.1 (GIRK1), shows, as noted above, a Kir channel controlled by G-protein membrane biochemistry.

Kir6.2: An ATP-sensitive Potassium Channel (K$_{ATP}$)

ATP-sensitive K$^+$ channels have a weak voltage sensitivity and are controlled by the concentration of an internal agent, in this case ATP. They are closed by high concentrations of ATP. In this way cell metabolism can be linked to the electrical excitability of the membrane. The K$_{ATP}$ channel consists of Kir6.2 and an unrelated modulator subunit (Figure 11.5). The latter subunit is a large (17TM) **sulphonylurea receptor (SUR)**. There are two variants of SUR and these mix and match with the two variants of Kir6.2 to give channels with different biophysical and pharmacological characteristics. Although it has been suggested that the ATP binding sites were located on SUR, they are now believed to be present on the Kir molecule (Figure 11.5). Their precise positions have yet to be identified although mutations in both amino- and carboxy-terminal domains decrease ATP sensitivity.

Kir3.1 (GIRK1): G-protein-linked Muscarinic Potassium Channel (K$_{ACh}$)

We noted in Section 8.9 that the M2 muscarinic AChR controlled a K$^+$ channel through a G-protein system. This G-protein-linked channel, when open, allows a 7–50 pS K$^+$ current to flow inwards. It may be blocked by Cs^{2+} and Ba^{2+} ions and by tetraethyl ammonium chloride (TEA Cl). Its biophysics were first analysed in cardiac muscle. The molecular structure of this channel has now been elucidated. It was determined by screening a rat heart cDNA library and isolating a 4.2 kbp clone. cRNA from this clone, when injected into *Xenopus* oocyte, was found to induce functional K^+ channels. These channels were shown to be electrophysiologically and pharmacologically almost identical to the G-protein-coupled muscarinic K^+ channels of the rat's heart.

The 4.2 kbp cDNA predicts a 501-amino acid protein which, like K_{ATP}, has two transmembrane helices separated by an H5 hairpin. Site-directed mutagenesis has shown that an Asp_{172} on M2 is crucial to voltage sensitivity. It has also been shown that not only $G\alpha$ but also the $\beta\gamma$ complex are significant activators of the channel and that Mg^{2+} ions are required. This was the first member of the now large (and rapidly enlarging) family of Kirs to be identified and it was, consequently, named **GIRK-1**.

11.2.2 4TM(2P) Channels; K^+ 'Leak' Channels

The first 2P channel subunit was cloned from the yeast *Saccharomyces cerevisiae* in 1995. It turned out to be a large 8TM non-voltage-gated K^+ leak channel. The first 2P subunit to be identified in animals was located in neuromuscular tissues of *Drosophila* in 1996 and its gene (which turned out to be the first of a large family) was designated *KCNK0*. *KCNK0* was shown to code for a 4TM membrane-bound 2P K^+ leak channel (Figure 11.2B). Subsequently over 50 *KCNK* genes have been detected in DNA databases and 14 have been cloned to generate KCNK channels.

The large number of *KCNK* genes suggests that many physiologically different KCNK channels exist in neuronal membranes. Some of these differences are coming to light. The KCNK3 channel expressed in hypoglossal and cerebellar granule cells is inhibited by lowered pH and by a number of neurotransmitters acting through G-protein-linked receptors. If K^+ permeability is inhibited (i.e. P_K decreased) in this way (other things being equal) the membrane depolarises (see Section 12.2). Hence the neuron becomes more easily excitable (Chapter 14). Thus the control of the KCNK3 channel may have important implications for levels of responsiveness and/or reaction

time throughout the cortex or other regions of the brain.

Vice versa, KCNK3 and KCNK5 channels increase their K^+ conductivity in response to various anaesthetics. This increased permeability would tend to hyperpolarise neuronal membranes (see Section 12.2) and hence reduce excitability. This has provided a rationale for the action of anaesthetics on the nervous system. However, it is early days yet. If anaesthetics do exert an effect through inhibition of K^+ leak channels, it is probably not their only mode of action. Further research will (hopefully) disentangle a complex situation.

11.2.3 6TM(1P) Channels; K_v Channels

The first 6TM K^+ channels were cloned and analysed in 1988 some five years before the 2TM Kir channels and some seven years before the 4TM leak channels were discovered. Since 1988 a great variety of 6TM K^+ channels have been found and analysed (see Table 11.2). They all share a common architecture of 6TM segments numbered S1 to S6 with a 'hairpin' between S5 and S6 known either as H5 or P which, as we have seen, selects with exquisite sensitivity K^+ from all other ions.

It is probable that the first voltage-dependent K^+ channels to develop in evolution were the so-called fast K^+ channels. There is evidence for the existence of the fast K^+ current, also known as the A current (I_{KA}), in Metazoa from coelenterates, such as *Obelia* and the 'sea-pansy', *Renilla*, through the molluscs, arthropods, to all classes of vertebrates. Indeed the squid giant axon (used by Hodgkin and Huxley in their original work on the action potential) seems to be very much an exception to the rule in that it does not exhibit this current. In this preparation only the delayed K^+ current, I_{Kdr}, can be detected.

In the great majority of nerve fibres the fast K^+ current occurs immediately on depolarisation and switches off within ten to a hundred milliseconds whether the membrane has been repolarised or not (Figure 11.6). Thus this current and (presumably) the responsible channel resemble the Na^+ current and channel far more than they resemble the delayed K^+ current and channel which we shall describe below. One other feature distinguishes the fast current, I_{KA}, from the classical delayed current,

BOX 11.1   Cyclic nucleotide-gated (CNG) channels

In Chapter 13 we shall see that cyclic nucleotide-gated channels play crucial roles in both photoreceptors and olfactory receptors. Although they are only very slightly voltage-dependent their subunits are so similar in structure to the 6TM(1P) K^+ channels that they are considered to belong to the same superfamily and are best considered alongside them in this chapter.

Each subunit of the tetramer which makes up the complete CNG channel resembles the shaker K^+ channel in consisting of six transmembrane domains (Figure A). Between the fifth and sixth domains the characteristic H5 (or P) hairpin is inserted into the membrane. Both N- and C-terminals are intracellular and the latter includes a segment of 80–100 amino acid residues which, being homologous to the CN binding region of cGMP kinases, is regarded as the CN binding area. The N-terminal in olfactory CNG channel subunits contains a calmodulin-binding domain.

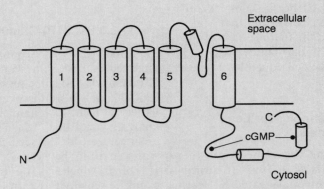

Figure A   α-Subunit of CNG-gated channel. Note the cyclic nucleotide site in the C-terminal with its two α-helical segments. Modified from Zheny and Zagotta (2000), *Neuron*, **28**, 369–394.

Two rather different subunits have so far been identified: α and β. The α-subunit (Figure A) has a molecular weight of 63 kDa whilst the β-subunit is nearly four times as large (240 kDa). This extra weight is almost entirely due to a far more lengthy C-terminal. The 6TM architecture of the β-subunit is very similar to that of the α-subunit. CNG channels may be either heteromeric tetramers of α- and β-subunits or, in the vomeronasal organ, homomeric tetramers of β-subunits. The latter channel differs from the heteromeric structures in that, although it contains cyclic nucleotide binding site, it is not activated by cyclic nucleotides but by nitric oxide (NO).

CNG channels are indiscriminately permeable to monovalent cations such as Na^+ and K^+. This lack of discrimination (compared to the rigorous selection practised by other members of the superfamily) is due to the lack of two amino acids, **tyrosine (T)** and **glycine (G)**, from the K^+ selectivity sequence of the P-region of CNG channels compared with their presence in other members of the superfamily. CNG channels are also permeable to divalent ions, especially Ca^{2+}, but these ions also act to close the channel. This is by way of their combining with calmodulin to form Ca^{2+}/calmodulin which then binds to the calmodulin site and reduces the channel's CN sensitivity.

We shall return to CNG channels in Chapter 13. We shall see that they are near the root of both olfaction (Section 13.1.2) and photoreception (Section 13.2). We shall see that in the first case cAMP opens CNG channels on olfactory epithelia leading to depolarisation and in the second case removal of cGMP leads to the closure of CNG channels resulting in hyperpolarisation.

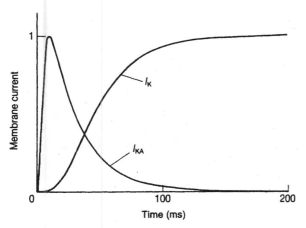

Figure 11.6 I_K and I_{KA}. The membrane is depolarised at time zero. The figure shows the relationship of the fast I_{KA} and delayed I_K potassium currents.

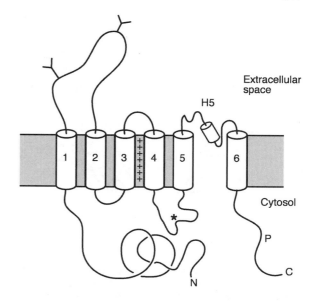

Figure 11.7 Structure of the *Drosophila* fast potassium channel subunit. This schematic shows six transmembrane segments. Between TM5 and TM6 an antiparallel pleated 'hairpin' is inserted into the membrane. The lengthy N-terminal region is coiled to form a globular 'ball'. 'P' indicates the position of the cAMP-mediated phosphorylation site. The star indicates position of receptor site for the N-activation ball. *N*-glycosylation sites are shown in the usual way. Note that transmembrane segment 4 is strongly positively charged. It is believed to be the voltage sensor. Further explanation in text.

I_{Kdr}. It can be shown that the K_A channel opens when the membrane is more hyperpolarised (i.e. less depolarised) than is required for the opening of the K_{vdr} channel.

The function of K_A in excitable cells has been the subject of a great deal of discussion. Several roles have been proposed. First and foremost, the channel is probably important in ensuring that a neuron does not fire when a small 'subthreshold' stimulus is applied. This is due to the fact that the very rapid opening of the fast channel allows a quick outflow of K^+ ions which repolarise the membrane. We shall see the significance of this when we come to consider the generation of action potentials in Chapter 14. Second, it has been suggested that as the channel is active when the membrane is hyperpolarised it may act to increase the duration of the **after-hyperpolarisation (AHP)** following a spike (see Chapter 14). Third, it has been shown that the fast potassium current is affected by a number of modulating agents: Ca^{2+}, serotonin, α-adrenergic agonists and other agents all seem to be able to reduce the flow of potassium ions through this channel. This allows these agents some control over the shape and duration of an action potential.

In the late 1980s it was found that a valuable approach towards an understanding of the fast channel could be made through *Drosophila* genetics. **Shaker** mutants show leg tremor under ether anaesthesia. Voltage-clamp techniques show that the condition is due to a defect in the K_A channel which results in an abnormally broadened spike. This defect shows itself not only in the leg muscles but also in all the other muscles of *Drosophila*'s anatomy. It was concluded that the mutation affects a gene coding for a fast channel protein(s). As *Drosophila* genetics are very well known this at once opened the possibility of a genetic attack on the structure of the fast channel.

The challenge was quickly accepted. The techniques of positional cloning (see Section 5.11) were employed to home in on the *shaker* locus in the *Drosophila* genome. cDNA clones were isolated from this region and found to specify a polypeptide which, when expressed in the *Xenopus* oocyte system, shows all the physiological and pharmacological properties of the K_A channel. The polypeptide has a molecular weight of 80 kDa and consists

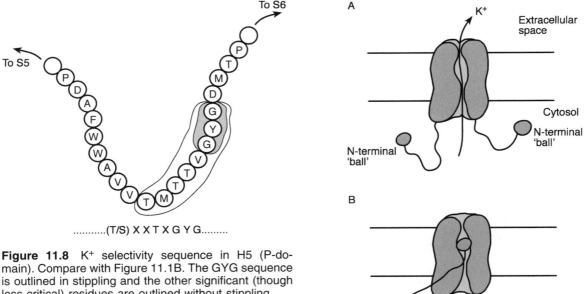

Figure 11.8 K⁺ selectivity sequence in H5 (P-domain). Compare with Figure 11.1B. The GYG sequence is outlined in stippling and the other significant (though less critical) residues are outlined without stippling.

Figure 11.9 Inactivation of K⁺ channel. Sectional view of channel: only two subunits shown. Two conformations are illustrated: (A) open; (B) closed. In B one of the N-terminal 'balls' has worked its way into the central goblet-shaped channel, blocking the exit. This is called N-type inactivation. In addition part B shows that the subunits have closed at the extracellular exit from the pore. This is termed C-type inactivation. Further explanation in text.

of 616 amino acids. Hydropathy analysis shows the molecule to have six transmembrane helices (S1–S6) (Figure 11.7).

The 80 kDa protein turns out to be a subunit of the entire physiological channel. The latter consists of a stack of four of these subunits arranged around a central pore which tapers to about 4 Å in diameter. We shall see in Sections 11.3 and 11.4 that there is a striking similarity to the Ca^{2+} and Na^+ channels. Although these are synthesised as one huge polypeptide they have four homologous domains, each of which shows strong similarities to a single shaker subunit.

Drosophila genetics, site-directed mutagenesis, *Xenopus* oocyte expression, and several other powerful techniques have been used to investigate the shaker channel. Non-conservative amino acid substitutions in S4 strongly affect the voltage sensitivity of the channel. The stretch of amino acids between S5 and S6 forms the customary re-entry 'hairpin'. It shows the K⁺ selectivity signature we have now come to expect (Figure 11.8). It is believed to line the pore formed within the four-unit cluster of the channel protein. On the analogy of the KcsA channel, the S5–H5–S6 elements of the four subunits surrounding the pore are thought to

assume an inverted tepee conformation surrounding a central channel. Surrounding this core the other transmembrane segments, especially the S4 helices, sense the voltage across the membrane (see Section 11.5).

Finally, site-directed mutation has been used to examine the function of the N-terminal sequence. This technique confirmed that the terminal sequence is involved in inactivating the channel 10–100 ms after opening (Figure 11.9). If lysines at positions 18 and 19 and/or the arginines at positions 14 and 17 (both positively charged amino acids) are substituted with neutral residues, the inactivation is reduced. If the region is removed altogether by a protease, inactivation is eliminated.

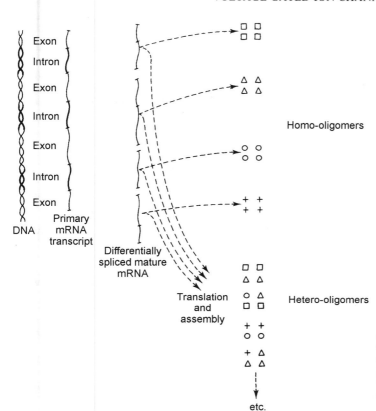

Figure 11.10 Generation of a diversity of A channels. A variety of homo- and hetero-oligomers of the A channel protein may be derived by differential splicing of the primary mRNA transcript of *Drosophila* 'shaker' locus. Exons thin lines; introns thick lines. After Schwarz *et al.* (1988), *Nature*, **331**, 137–142.

It appears, therefore, that this N-terminal sequence acts as a kind of 'ball and chain', inactivating the channel by plugging the entrance when the membrane is depolarised (Figure 11.9).

Further study of the inactivation mechanism has largely confirmed the 'ball and chain' idea. It has been shown, however, that the N-terminal inactivation sequence enters through one of four 'windows' in the cytoplasmic domain of the K⁺ channel and works its way up to block the narrowest part of the pore. Residues deep in the heart of the pore, in its most constricted region, have been identified as the binding site of the N-terminal sequence. Although there are four N-terminals in the homotetrameric K⁺ channel, only one is need for inactivation. The inactivation is accurately mimicked by quaternary amines such as TEA derivatives

In addition to N-terminal 'ball and chain' inactivation, called **N-type inactivation**, another gate located near the outer mouth of the pore causes a slower inactivation, **C-type inactivation**, by collapsing the selectivity filter region. Detailed investigation of how voltage sensing and the two types of inactivation are linked at the molecular level is ongoing. We shall see in Section 11.5 that when the surrounding membrane is depolarised the S4 helix rotates along its long axis, screwing outwards through the membrane. Using the KcsA structure as a model (remember, however, that KcsA is not voltage sensitive) it begins to be possible to see how C-type inactivation may follow mechanically. The rotatory movement of S4 could lead to a cuff of tryptophans just above the exit from the pore rotating into a position which very effectively blocks the opening.

A Multiplicity of Shaker Channels

The *Drosophila* shaker (*sh*) gene is a highly complex unit. It contains at least 23 exons and it has multiple promoter sites so that the initiation of transcription may occur at different places in different tissues. Furthermore, the primary mRNA once transcribed may be spliced at different points, combining different exons, and thus giving rise to different mature mRNA strands. These would then be translated into subtly different channel proteins. These proteins could then assemble to give different homo-oligomers or, even more interestingly, different hetero-oligomers (Figure 11.10).

The mRNAs from the *Drosophila sh* genes can be expressed in *Xenopus* oocyte. It is then possible to show that each of the variant mRNAs can form a physiologically active channel. A variety of homo-oligomers are thus possible. Each homo-oligomer presents subtly different physiological characteristics. All activate and inactivate in response to voltage changes across the oocyte membrane – but the transition rates between these states vary from one homo-oligomer to another. Finally, if variant mRNAs are injected it is possible to express hetero-oligomers in the oocyte membrane and, as suggested in Figure 11.10, these have intermediate physiological characteristics.

Other Members of the Family

Once the sequence of the *Drosophila* shaker gene had been determined it became possible to search cDNA libraries for other K^+ channel genes. *Drosophila* quickly yielded three other genes, *Shab, Shal* and *Shaw*, which coded for three similar K^+ channels. Homologues of all four *Drosophila* K^+ channels have been found throughout the animal kingdom, all of which show the same 6TM 'core' structure. In mammals the homologous channels are called Kv1.x, Kv2.x, Kv3.x and Kv4.x, respectively. The Kv1 channel has since been shown to exist in nine variants, Kv1.1–Kv1.9 (Table 11.2). Subsequent work has led to the cloning and characterisation of more than fifty different voltage-gated shaker-like K^+ channels from animals ranging from *Aplysia* to *Homo*. All seem to be members of one evolutionarily related superfamily and to share a common core structure. The 6TM core has thus been conserved over at least 600 million years.

Shaker Accessory Subunits

The four subunits of the shaker channel are known as α-subunits. In physiological conditions these

Table 11.2 Mammalian shaker and shaker-related channels and genes

Gene	Human chromosome	Channel	Subtype
KCNA1	12p13	Shaker (Kv1)	Kv1.1
KCNA2	1p		Kv1.2
KCNA3	1p13.3		Kv1.3
KCNA4	11p14.1		Kv1.4
KCNA5	12p13		Kv1.5
KCNA6	12p		Kv1.6
KCNA7	19q13.3		Kv1.7
KCNA8	1p		Kv1.8
KCNB1	20q13.2	Shab (Kv2)	Kv2.1
KCNB2	1		Kv2.2
KCNC1	11p14.3–15.2	Shaw (Kv3)	Kv3.1
KCNC2	19q13.3–13.4		Kv3.2
KCNC3	19q13.3–13.4		Kv3.3
KCNC4	1p21		Kv3.4
KCND1	X	Shal (Kv4)	Kv4.1
KCND2	7q		Kv4.2

Data from Ashcroft (2000), *Ion Channels and Disease*, San Diego: Academic Press; and Conley and Bannar (1999), *The Ion Channel Facts Book*, vol. IV: *Voltage-Gated Channels*, London: Academic Press.

subunits are associated with other, smaller (40 kDa) subunits, called β-subunits. A complete shaker or shaker-type channel consists of four α- and four β-subunits. There at least three β-subunit genes and alternative splicing generate a large variety of proteins. Hydrophobicity plots suggest that they have no transmembrane domains and are held on the cytoplasmic face of the membrane. Their presence both increases the K^+ current and dramatically increases the rate of inactivation. As Figure 11.11 shows, there is good evidence that they are associated with the N-terminal inactivation sequence.

We noted above that more than fifty different variants on the 6TM(1P) theme are known in the animal kingdom. One of the most important of these variants in neurophysiology is the delayed rectifier channel, K_{vdr}.

Delayed-response Potassium Channels (Hodgkin–Huxley Channels) (K_{vdr})

The pioneering work of Hodgkin and Huxley in the 1950s showed that the recovery phase of an action potential, the phase in which the axonal membrane

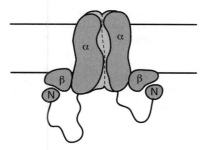

Figure 11.11 Disposition of α- and β-subunits in shaker-type channels. N=N-terminal inactivation ball.

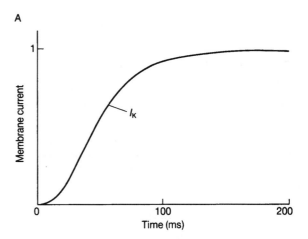

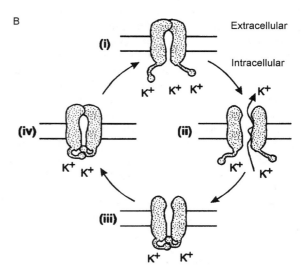

reverts to its normal polarity (see Chapter 14), is due to the efflux of potassium ions. This efflux is through potassium channels in which opening is delayed by about 2 μs after the membrane has been depolarised. Thus, although they are voltage-dependent, they are not as quick to respond to a change in transmembrane voltage as the K_A channel. They also differ from the K_A channel in remaining open much longer whilst the membrane is depolarised. In other words there is no snapping open and shut but a fairly sustained open state until the membrane regains or surpasses its original polarity. These, then, are the 'classical' delayed K^+ channels upon which so much of basic neurophysiology turns (Figure 11.12). The delayed potassium channel is characteristically blocked by TEA Cl.

At the time of writing the gene or genes coding for K_{vdr} activity remains obscure. The channel is believed to be coded by shaker family genes but the precise subunit arrangement has been difficult to establish. It has, however, been shown that the gene for the major delayed rectifier channel in *Drosophila* neurons can be traced to the *Shab* locus.

KCNQ Channels

These channels first came to light because of mutations leading to long QT cardiac syndrome. Three genes have so far been identified: *KCNQ1–3* (Table 11.3). They encode a typical 6TM K^+ channel with a P hairpin between S5 and S6.

Figure 11.12 The 'classical' delayed potassium current. (A) The membrane is depolarised at time zero. The outward potassium current, I_K, commences about 2 μs after time zero. It rises to its full height in about 100 ms. If the membrane is held in a depolarised condition the potassium gates remain open and I_K continues for tens of milliseconds. (B) Schematic to show the functioning of the gates responsible for I_K. (i) When the membrane is at its resting potential the I_K gate is closed. (ii) On depolarisation a voltage sensor causes the gate to open. Potassium ions flow down their electrochemical gradient from the intracellular to the extracellular space. (iii) As the membrane recovers its original polarity, the 'ball and chain' inactivation gate begins to close and shut off I_K. (iv) On achieving the Nernst potassium potential (i.e. $V_m \cong V_K$) the inactivation gate closes. The activation gate then closes (as in (i)) and remains closed until the next depolarisation. The inactivation gate slowly opens. Compare with the Na^+ channel of Figure 11.22. As indicated in text all these operations have a stochastic character.

Table 11.3 KCNQ channels

Gene	Chromosome	Channel	Subtype
KCNE1	11p15.5	KCNQ	KCNQ1
KCNE2	20q13.3		KCNQ2
KCNE3	8q24		KCNQ3

Table 11.4 Ether à go-go channels

Gene	Chromosome	Channel	Subtype
eag		eag	eag
			erg
HERG	7q35–36	EAG	elk

Although KCNQ channels are mostly involved in cardiac physiology, one of them, KCNQ1, is, as we shall see below, important (alongside MinK) in the physiology of the inner ear.

MinK

It has turned out that subunits with only a single transmembrane domain have channel-forming properties. These so-called minimal K$^+$ units, **minK units**, consist of a protein some 130 amino acids in length with a single α-helical transmembrane segment. MinK units were long a source of controversy. Evidently they could hardly form a K$^+$ channel individually. Did they form a multimeric complex? Eventually it was found that they were always associated with the KCNQ1 K$^+$ channel where they acted as a β-subunit. Both KCNQ1 and MinK are expressed in the stria vascularis of cochlea (see Figure 13.25). In this position they are involved in the secretion of the K$^+$-rich endolymph which is essential to the functioning of the cochlear hair cells (Section 13.3.3). Mutations in the *KCNE1* gene lead to reduced secretion of endolymph, collapse of Reissner's membrane and profound deafness.

Eag Channels

When, in the vibrant decade of the 1960s, neurogeneticists observed that under ether anaesthesia some *Drosophila* showed more than usually agitated leg shaking they named the mutation 'ether à go-go'. The term stuck and the mutated gene is now referred to as *eag*. The gene codes for a typical 6TM(1P) K$^+$ channel which proved to be a strong inward rectifier (Table 11.4). Hybridisation probes retrieved two similar genes from *Drosophila* cDNA libraries. The three genes, *eag*, *erg*, *elk*, form a family and each member is now known to be the head of a subfamily coding for channels with differing biophysical characteristics.

When a human hippocampal cDNA library was probed with *eag* a human homologue was identified. This is called *HERG* (human eag-related gene). The sequence similarity is only some 49% so it is not a human form of *Drosophila eag* but a member of the same family. It, too, is a powerful inward rectifier and plays a significant role in cardiac physiology. Although *HERG* was originally found in the hippocampus, and although *eag* was originally detected as a neuromuscular abnormality in *Drosophila,* the role of the HERG in the human nervous system has yet to be clarified.

Ligand-modulated 6TM(1P) Channels

In addition to K$^+$ channels solely sensitive to voltage there are a great variety of K$^+$ channels whose voltage dependence is either modulated by other molecules, especially second messengers, or which are largely voltage-insensitive and controlled by second messengers. Two instances are outlined below.

Serotonin-dependent potassium channel (S channel) (K_{5-HT} or K_S): It is possible by voltage clamp experiments to show that a potassium channel is present in the membranes of *Aplysia* (sea-hare) neurons. This channel, unlike the slow or fast K$^+$ channels, is activated at the membrane's resting potential, remains open when the membrane is depolarised and is not affected by Ca^{2+}. The channel is, however, affected by serotonin, a common neurotransmitter, and by cyclic AMP (cAMP). It can be shown by patch-clamp analysis that serotonin and cAMP exert their effects by reducing the number of I_{KS} channels open at a given time.

Although S channels are few and far between, their cumulative effect is normally to speed the repolarisation of a neuronal membrane after an action potential. If they are inactivated the flow of K$^+$ out of a neuron (responsible for this

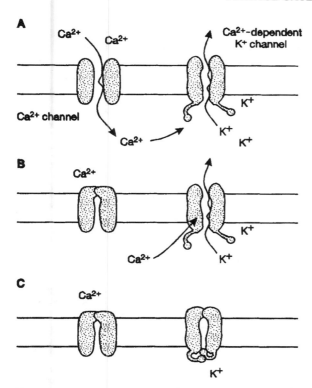

A

Ca²⁺   Ca²⁺   Ca²⁺-dependent K⁺ channel

Ca²⁺ channel

Ca²⁺   K⁺   K⁺

B

Ca²⁺

K⁺

Ca²⁺   K⁺

C

Ca²⁺

K⁺

Figure 11.13 Sensory adaptation and Ca²⁺-dependent K⁺ gates. (A) The membrane is depolarised. Ca²⁺ gates and Ca²⁺-dependent K⁺ gates open. (B) The membrane repolarises but unlike the I_K and I_{KA} gates the I_{KCa} gates remain open in the presence of Ca²⁺. A subsequent depolarising stimulus or transmitter will consequently find it more difficult to depolarise the membrane due to the countervailing efflux of K⁺. (C) Ca²⁺ is removed by intracellular mechanisms and the I_{KCa} channel closes.

repolarisation) is hindered. Hence the period of depolarisation is lengthened. Hence the period during which Ca²⁺ can flow through its channels to the interior is lengthened. We shall see in Chapter 20 that this could have an important role to play in the molecular basis of memory.

Calcium-dependent potassium channel (K_{Ca}): Calcium-dependent potassium channels differ from all the channels we have considered so far in that they are sensitive to the internal concentration of a modulator, in this case calcium ions. As the Ca²⁺ concentration increases in the cytosol these potassium channels open and allow the escape of

K⁺. This makes the membrane more difficult to depolarise – any depolarising voltages applied are counteracted by an outward flow of positively charged potassium ions.

In Chapter 10 we noted the possible interaction of these channels with the NMDA Ca²⁺ channel to produce the oscillating membrane potentials underlying pacemaker activity. Calcium-dependent potassium channels are probably also involved in sensory adaptation. All sensory systems adapt to a constant stimulus by a reduction in impulse frequency. It is not difficult to see that this will occur when it is understood (see next section) that each action potential generated leads to the influx of a small amount of calcium through voltage-dependent Ca²⁺ channels. A rapid volley of impulses due to a steady stimulus being turned on will soon increase the internal concentration of Ca²⁺ ions and thus open the Ca²⁺ gates with the results indicated above (Figure 11.13).

11.3 Ca²⁺ CHANNELS

There are at least six subclasses of Ca²⁺ channel: T-, L-, N-, P-, Q- and R-. These subclasses may be grouped into two major categories: those activated by low and those activated by high voltages. The low-voltage-activated (LVA) channels have only one representative, the T-type channel; all the others are high-voltage-activated (HVA) channels. The five types of HVA channel are distinguished by both pharmacological and biophysical characteristics (Table 11.5).

Table 11.5 Ca²⁺ channels

Type	Distinguishing properties
T	Activates at negative potentials
	Slow inactivation
	Insensitive to micromolar concentrations of Cd²⁺
L	Voltage-dependent block by DHP antagonists
N	Irreversibly blocked by snail toxin, ω-CTx-GU1A
P	Blocked by spider toxin, ω-Aga-IVA (1 nM concentration)
Q	Blocked by ω-Aga-IVA at > 100 nM concentration
R	Insensitive to all above blockers

Data from Randall (1998), *Journal of Membrane Biology*, **161**, 207–213.

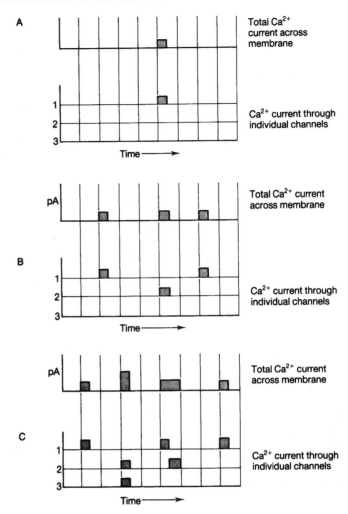

Figure 11.14 Voltage-dependence of Ca^{2+} channels. The schematic shows that when a membrane containing Ca^{2+} channels is progressively depolarised (A→B→C) fresh channels are recruited and the Ca^{2+} current across the membrane increases.

Voltage-gated Ca^{2+} channels are to be found in muscle cells and several other types of cell as well as neurons. In neurons they play (see Chapter 15) a vital role in the release of neurotransmitters from synaptic endings. They are also to be found in the membrane of an axon's initial segment where, as we noted in Section 11.2.4, they can play a part in adaptation. In addition they are found, albeit in smaller numbers, in dendrites, growth cones and cell bodies.

The concentration of calcium ions is very different on the two sides of a neuronal cell membrane. Whereas the extracellular concentration is usually about 3 mM the internal concentration of free Ca^{2+} is seldom more than 10^{-4} mM.

Opening of channels across the membrane allowing Ca^{2+} ions to cascade in can thus result in a very marked change in the concentration of calcium ions in the cytosol.

Calcium channels are very sensitive to transmembrane voltage. Studies using the patch-clamp technique show that the channels open and then rapidly close again. The inactivation rate in fact varies amongst the HVA channels. The P-type is slow compared with the R-type. Furthermore, as was the case with all the other channels we have discussed, the mechanism appears to be more complicated than a simple two-stage open–shut cycle. There seem to be intermediate states. Each L-type channel's open–shut cycle has a time-course of about 0.6 ms. The number of channels going through this cycle on a small patch of membrane depends on the transmembrane voltage (Figure 11.14). As the voltage across the membrane falls, more and more channels are activated. The flow of Ca^{2+} ions consequently increases.

In physiological conditions when the external concentration of Ca^{2+} is about 3 mM each L-type channel allows the passage of about 0.06 pA of current. The number of calcium ions which flow through an L-type channel can be worked out in the same way as we worked out the flow of ions through the nicotinic acetylcholine receptor (nAChR) channel in Section 10.1.2 (remember we are dealing with divalent ions):

$$\frac{2 \times (0.06 \times 10^{-12} \text{ coulombs s}^{-1}) \times (6 \times 10^{23} \text{ ions mol}^{-1})}{96\,500 \text{ coulombs mol}^{-1}}$$

$$= \text{approx. } 7.5 \times 10^5 \text{ ions s}^{-1}$$

Although the number of Ca^{2+} channels per unit area is rather small – ranging from about $4\,\mu m^{-2}$ in heart muscle to $60\,\mu m^{-2}$ in snail neurons – and although the opening time of a channel is to be measured in milliseconds rather than seconds, the flux is evidently great enough to make a considerable difference to the very low free Ca^{2+} concentration normally present in cytosol. We shall return to this in Chapters 15 and 16 where we consider the neuron as a secretory cell and the release of neurotransmitters by presynaptic terminals.

Calcium channels are very selective. This is vital if they are to ensure the influx of calcium ions rather than a mixture of the sodium and potassium ions which are both much more common in the extracellular fluid. Compared to a Ca^{2+} concentration of about 3 mM, Na^+ is present at a concentration of about 145 mM.

In addition to their exquisite sensitivity to transmembrane voltage, calcium channels are also affected by a large number of neurotransmitters and drugs. So far all the agents tested have tended to **inactivate ('downmodulate')** the channel. These agents include GABA, serotonin, somatostatin, noradrenaline (norepinephrine), enkephalin and **1,4-dihydropyridine (DHP)**. This last agent was used to isolate the L-type calcium channel from rabbit skeletal muscle.

11.3.1 Structure

A purified DHP receptor preparation was used to synthesise an oligonucleotide probe and this in turn for extracting and cloning the receptor cDNA. It was found that the protein thus obtained consists of 1873 amino acids with a molecular weight of approximately 170 kDa. It must be borne in mind when considering this Ca^{2+} channel that it has been obtained from skeletal muscle, and neuronal channels may differ in detail.

Examination of this polypeptide shows it to have four homologous domains of about 300 amino acids each. The amino- and carboxy-terminals are both located on the cytoplasmic side of the membrane (Figure 11.15). Hydropathic analysis of the polypeptide shows that each of the homologous domains consists of six hydrophobic segments, S1, S2, S3, S4, S5 and S6. All six segments are believed to span the membrane. It is found that the S4 segment in each homologous domain contains a sequence of positively charged Arg or Lys residues following a regular pattern (Arg/Lys – X – X – Arg/Lys) where X is a nonpolar residue. This pattern is also to be found in all known S4 segments of potassium and sodium channel proteins. It has been concluded (see Section 11.5) that, as with the other ion channels, this regular sequence of positively charged side chains acts as a voltage sensor. Finally, once again in common with the other voltage-dependent channels, the polypeptide chain between S5 and S6 assumes a hairpin formation inserted into the membrane. As with the K^+ channels, it makes an important contribution to lining the pore and conferring ion selectivity.

It is clear that each of the homologous domains resembles a single 6TM shaker-type K^+ channel. We shall see in the next section that the Na^+ channel also consists of four homologous domains. It looks as if these latter channels evolved from shaker-type channels by two gene duplications. This insight is supported by the outcome of a comparison of the amino acid sequences. It can be shown that a stretch of 120 amino acids in the shaker channel protein (304–435) is 27% identical (47% if conservative substitutions are added) with a stretch (1360–1496) in *Gymnotus* sodium channel protein. The centre of homology in the K^+, Na^+ and Ca^{2+} channels is the arginine-rich region

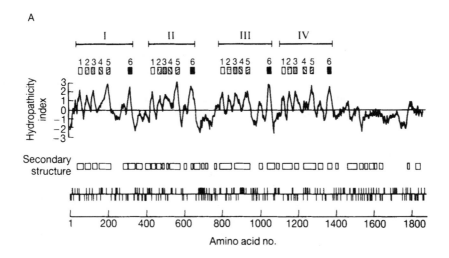

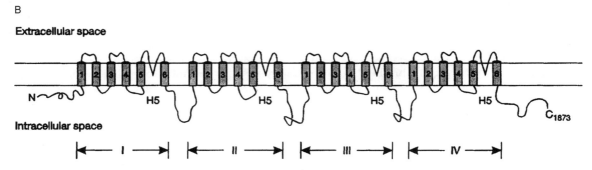

Figure 11.15 Structure of the calcium channel α-subunit. (A) Hydropathy plot of the polypeptide chain. The further above the horizontal line the greater the hydrophobicity. I, II, III and IV indicate the position of the homologous domains while the boxes 1–6 denote the homologous segments within each domain. The white rectangles beneath the hydropathicity plot indicate where the α-helix or β-sheet is predicted and the vertical bars on the line beneath show the position of positively charged residues (Lys, Arg) (upward) and negatively charged residues (Asp, Glu) (downward). (B) Putative disposition in the membrane. In reality (as with the other channel proteins) the four homologous domains make use of the third dimension to cluster together around (in this case) a calcium-selective pore. Reprinted by permission from Tanabe *et al.* (1987), *Nature,* **328,** 313–318. Copyright © 1987 Macmillan Magazines Ltd.

which, as we saw above, is believed to be involved in voltage gating. The general design of this region with alternating hydrophobic and hydrophilic residues is remarkably similar in all three channels.

It has turned out that the Ca^{2+} channel consists of more than just the 170 kDa protein described above. Indeed this 170 kDa channel is now referred to as the α-subunit. It is associated with four other proteins: an intracellular β-subunit (56 kDa); a transmembrane γ-subunit (35 kDa); and a partially extracellular, partially transmembrane, disulphide-linked dimer, the α_2/δ-subunit (155 kDa). The full structure of the Ca^{2+} channel is shown in Figure 11.16. Although the α_1-subunit, when expressed alone, allows the permeation of Ca^{2+} and responds to the various pharmacological blockers and toxins in the same way as the complete channel, its activation and inactivation kinetics are abnormally slow. It is clear that the four subunits work together to provide the pharmacological and biophysical characteristics of the channel. There is evidence, also, that the β-subunit plays a significant

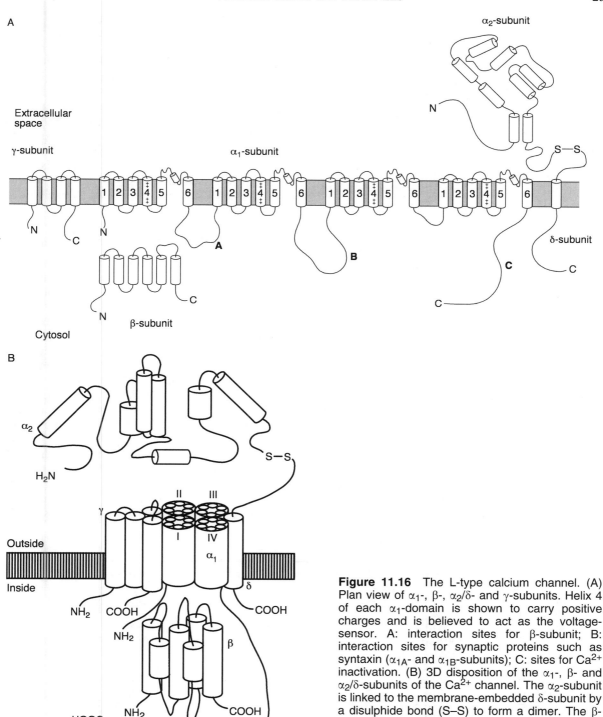

Figure 11.16 The L-type calcium channel. (A) Plan view of α_1-, β-, α_2/δ- and γ-subunits. Helix 4 of each α_1-domain is shown to carry positive charges and is believed to act as the voltage-sensor. A: interaction sites for β-subunit; B: interaction sites for synaptic proteins such as syntaxin (α_{1A}- and α_{1B}-subunits); C: sites for Ca^{2+} inactivation. (B) 3D disposition of the α_1-, β- and α_2/δ-subunits of the Ca^{2+} channel. The α_2-subunit is linked to the membrane-embedded δ-subunit by a disulphide bond (S–S) to form a dimer. The β-subunit is cytoplasmic. Part B from Varadi *et al.* (1999), *Critical Reviews in Biochemistry and Molecular Biology*, **34**, 181–214, with permission.

role in leading (chaperoning) the α-subunit to its place in the membrane. In the absence of the former subunit the density of α_1-subunits falls by a factor of ten.

11.3.2 Diversity

The diversity of Ca^{2+} channels is emphasised by the finding that the α_1-subunits of the HVA channels are encoded by at least seven genes (*CACNA1A–CACNA1F, CACNA1S*) spread over six chromosomes and the α_1-subunits of the LVA channel by three genes (*CACNA1G–CACNA1I*) spread over two chromosomes. Similarly the β- and γ-subunit genes (*CACNB1–CACNB4; CACNA2A–CACNA2C; CACNG1, CACNG2*) are spread over six chromosomes (Table 11.6). Each of the *CACNA* genes encodes a different α_1-subunit (α_1A, α_1B, etc.), varying from 1610 to 2424 amino acid residues, which, when expressed alone, carries a unique Ca^{2+} current ($Ca_v1.1$, $Ca_v1.2$, etc.). The six types of Ca^{2+} channel listed in Table 11.6 are derived by combination of one of the ten α_1-subunits with β- γ- and $\alpha_2\delta$-subunits. When it is recognised that in addition to all this many of the α_1-subunit gene transcripts generate different splice variants, the number of possible pentameric Ca^{2+} channel structures that can be put together is clearly very large.

The pharmacological and other properties of Ca^{2+} channels found in the CNS and elsewhere are due to combinations of one or other of these α-subunits and the β-, γ- and $\alpha_2\delta$-subunits. The L-type channel of sarcoplasmic reticulum is, for instance, a pentameric complex consisting of α_1, α_2, β, γ and δ where the α_1-subunit is 1S, 1C or 1D; the N-type channel, more characteristic of the nervous system, is composed of the same pentamer but in this case the α_1-subunit is 1B. The P/Q-, R- and T-type channels are composed of 1A, 1E and 1G α-subunits respectively. These differences have significant physiological outcomes. In particular, as Figure 11.16 shows, the loop between the second and third homologous domains interacts with synaptic proteins, such as syntaxin, in channels, such as N- or P-type, where the α-subunit is either 1A or 1B, whilst in the L-type channels of skeletal muscle, the 1S subunit is involved in excitation–contraction coupling.

11.3.3 Biophysics

Although the precise biophysical and pharmacological properties of this variety of Ca^{2+} channels varies widely, the Ca^{2+} channel pore through the α_1-subunit remains the same throughout. It is believed to have a comparatively wide extracellular entrance and intracellular exit with a narrow central region where the ion filter is located. There are thought to be two Ca^{2+} binding sites, a low-affinity and a high-affinity site. When a Ca^{2+} ion arrives at the low-affinity site near the extracellular end of the pore, its electrostatic repulsion forces the Ca^{2+} ion on the high-affinity site deeper within the pore to vacate its position and tumble into the cytosol. A very high throughput is calculated to occur: from 10^6–10^7 ions s^{-1}.

The 3.4 Å solution to the KcsA channel (Section 11.1) provided an impulse to re-examine the electivity filter within the pore of the α-subunit.

Table 11.6 Voltage-gated Ca^{2+} channels and genes

Gene (human)	Chromosome	Subunit	Channel	Site of expression
CACNA1A	19p13	α_1A	P/Q-type	Brain, motor neurons
CACNA1B	9q34	α_1B	N-type	Brain
CACNA1C	12p13.3	α_1X	L-type	Heart
CACNA1D	3p14.3	α_1D	L-type	Brain, pancreas
CACNA1E	1q25	α_1E	R-type	Brain, NMJ
CACNA1F	Xp11.23	α_1F		Retina
CACNA1S	1q32	α_1S	L-type	Skeletal muscle
CACNA1G	17q22	α_1G	T-type	Brain
CACNA1H	16p13.3	α_1H	T-type	Brain, heart
CACNA1I	22q12.3	α_1I	?	?

Data mainly from www.neuro.wustl.edu/neuromuscular/mother/chan.html.

The P-loops (H5 or SS1–SS2 regions) of the four homologous domains were examined using biophysical, genetic and modelling techniques. Although they are different from the K^+ channel motifs discussed earlier, they nevertheless show an interesting possibility. In each of the four H5 regions a glutamic acid residue is located towards the bottom of the loop. In the native channel these residues are brought into close proximity with each other, forming a tight collar in the channel lining. There is convincing evidence that these four glutamic acid residues form tight coordinate linkages with Ca^{2+} and thus are a good candidate for the high-affinity site (cf. the coordination linkages of K^+ in the KcsA channel). It will be fascinating if, and when, the structure of a Ca^{2+} channel is solved by X-ray diffraction, to find whether the picture obtained by the more indirect means of genetics and channel biophysics is borne out.

11.4 Na$^+$ CHANNELS

Although the Na^+ channel is the last cation channel to be discussed in this chapter, and although it is probably the last to appear in the evolutionary sequence (see Figure 11.27), it was the first voltage-controlled channel to be sequenced and have its structure solved. It has been remarked that it is an odd quirk of scientific history that the most cumbersome and intricate ion channels (the Na^+ and Ca^{2+} channels) were isolated and elucidated before the far simpler and more ubiquitous shaker and Kir structures.

This is perhaps not so odd when it is recognised that the Na^+ channel is ubiquitous in nervous systems, is highly concentrated in fish electric organs, and that very specific marker molecules can be used to detect it. The marker molecule – **tetrodotoxin (TTX)** – is derived from fish belonging to the family Tetrodontidae – the best known example being the **Japanese puffer fish**. Tetrodotoxin binds very specifically to the Na^+ channel, in a one-to-one fashion. A similar toxin – **saxitoxin** – is synthesised by a dinoflagellate. Once again it binds in a one-to-one fashion with the sodium channel. Both these molecules incapacitate the sodium channel and are hence extremely poisonous.

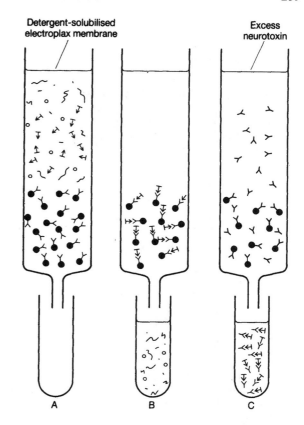

⊻ = Sodium channel protein

Y = Tetrodotoxin

Y = Tetrodotoxin attached to inert bead

Figure 11.17 Purification of the sodium channel on an immobilised neurotoxin chromatography column. (A) Detergent-solubilised electroplax membrane preparation is added to the top of the chromatography column. The column consists of inert beads to which have been attached a neurotoxin such as tetrodotoxin. (B) The electroplax preparation flows through the column and only the sodium channel protein is held by the beads. All the other proteins, polypeptides and membrane fragments flow through unimpeded. (C) A suspension of excess neurotoxin is now passed through the column to elute the sodium channel from the beads.

11.4.1 Structure

Because of the highly specific binding of these neurotoxins the sodium channel can be isolated from detergent-solubilised electroplax membrane

A

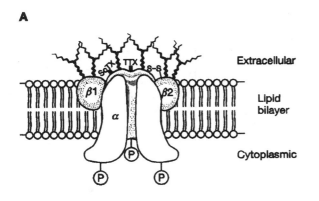

B

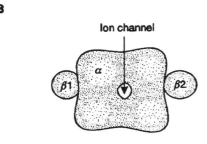

Figure 11.18 Subunit structure of brain Na⁺ channel. (A) Schematic cross-section of channel inserted in membrane. The channel complex consists of the 260 kDa α-subunit and the two smaller β₁- and β₂-subunits. All three subunits are heavily glycosylated on their extracellular surfaces and the α-subunit has binding sites for α-scorpion toxins (ScTX) and tetrodotoxin (TTX). The intracellular surface of the α-subunit has numerous phosphorylation sites. The β₁-subunit is associated non-covalently with α whilst the β₂-subunit is linked via disulphide bonds. (B) Surface view from the extracellular side showing the ion channel in the centre of the α-subunit. From Catterall (1992), *Physiological Reviews*, **72**, S15–S42; with permission.

fragments by affinity chromatography through a column of bead-immobilised tetrodotoxin or saxitoxin (Figure 11.17).

The technique shown in Figure 11.17 isolates a 260 kDa glycosylated protein from eel electroplax. When a similar technique is applied to mammalian brain two other smaller polypeptides (36 kDa and 33 kDa) are found to be associated with the sodium channels (Figure 11.18). The 260 kDa unit is accordingly designated the **α-subunit** and the two smaller polypeptides the **β₁- and β₂-subunits**. A different polypeptide (of approximately 38 kDa) is

associated with sodium channels isolated from mammalian skeletal muscle. It turns out that whereas brain Na⁺ channels are associated with two β-subunits, skeletal muscle Na⁺ channels are associated with a single β-subunit. In both cases the β-subunits modulate the channel properties of the α-subunit.

The Japanese group led by Numa applied recombinant DNA techniques to clone and sequence the **α-subunit** of eel electroplax just as they had done earlier for the acetylcholine receptor. They were able to show that it consists of a single run of 1820 amino acids containing four homologous domains of approximately 300 amino acids each (Figure 11.19A). Unlike K⁺ channels but like Ca²⁺ channels, the sodium channel is not itself built from distinct subunits. The homology of the four domains suggests, as we noted in Section 11.3.1, that they all arose by internal duplication of a single ancestral shaker-type gene.

Hydropathy analysis indicates that within each 300-residue domain there are, as with the Ca²⁺ channel, six membrane-spanning helices termed S1, S2, S3, S4, S5 and S6 (Figure 11.19A and B). Again, as with the Ca²⁺ channel, S4 is found to contain a number of positively charged residues (especially arginine and lysine) in addition to hydrophobic residues. It is consequently believed to be the 'voltage sensor' which detects change of voltage across the membrane and opens the channel. Between each S5 and S6 the polypeptide chain is inserted into the membrane in a hairpin conformation (Figure 11.20). Once again, this formation (designated H5) is believed to line the pore and confer ion selectivity. Finally, the short intracellular section between homologous domains 3 and 4 is responsible for inactivation. When, as we shall see in Section 11.4.3, the pore snaps shut this short length of polypeptide chain occludes the intracellular exit of the pore. This, too, has its analogue in other voltage-sensitive channels. We saw, in Section 11.2.3, that there was good evidence that the fast K⁺ channel is inactivated by an N-terminal 'ball-and-chain' device.

Clearly there are striking structural similarities between Na⁺ and Ca²⁺ α-subunits. There is no doubt that they are evolutionarily related. This relationship is brought out strongly if the amino acid sequences of the homologous domains of the α-subunits are compared:

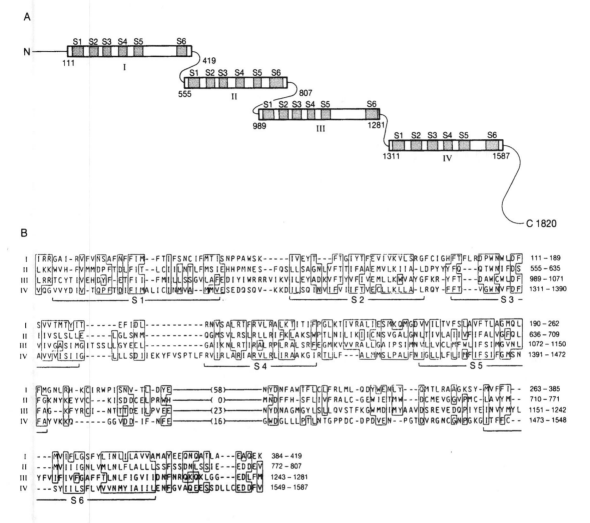

Figure 11.19 Domain structure and amino acid sequences of the four regions of internal homology of the sodium channel α-subunit. (A) The 1820 amino acid sequence contains four homologous domains: I, II, III and IV. The first domain stretches from residue 111 to residue 419, the second from residue 555 to residue 807, the third from residue 989 to residue 1281 and the fourth from residue 1311 to residue 1587. Each homologous domain consists of six transmembrane segments, S1, S2, S3, S4, S5 and S6. (B) Amino acid sequence of the four homologous domains. The six transmembrane segments in each domain are labelled; S4 can be seen to contain a number of positively charged residues (R=arginine; K=lysine) and has been proposed as the voltage sensor. The number of residues in each line of the figure is given on the right; gaps (–) have been inserted to achieve maximum homology; the non-homologous regions of I, III and IV are shown by lines and the number of residues in each indicated in brackets; identical residues and conservative substitutions in all four domains are boxed. Part B reprinted by permission from Noda *et al.* (1984), *Nature*, **312**, 121–127. Copyright © 1984, Macmillan Magazines Ltd.

Domain 1: 32% identical residues; 62% identical + conservative substitutes

Domain 2: 35% identical residues; 59% identical + conservative substitutes

Domain 3: 37% identical residues; 60% identical + conservative substitutes

Domain 4: 32% identical residues; 61% identical + conservative substitutes

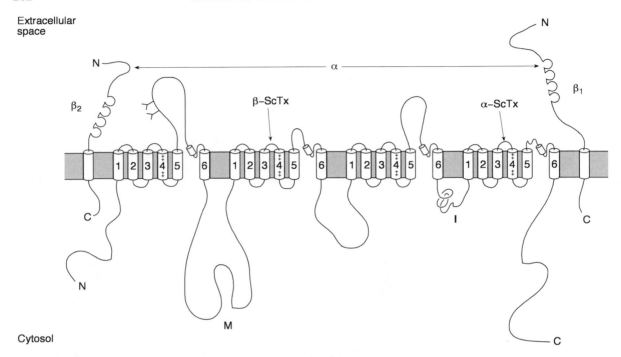

Figure 11.20 Plan of the disposition of the sodium channel in a membrane. The α-subunit consists of four homologous domains. The fourth helix of each of these domains contains many positively charged residues and is believed to be the voltage sensor. Glycolysation sites are indicated by Y. M=modulation region (phosphorylation by PKA and/or PKC); I=inactivation loop with inactivation particle; sites of action by α- and β-scorpiotoxins shown. The β_1- and β_2-subunits have single transmembrane helices and an immunoglobulin-like extracellular domain. The extracellular domain of the β_1-subunit interacts with the extracellular loop leading in to helix 6 of domain 4 of the α-subunit. Modified from Catterall (2000), *Neuron*, **26**, 13–25.

The overall sequence homology is 35% identity or 55% when conservative substitutes are added.

Figure 11.20 shows the proposed architecture of the Na$^+$ channel. Each of the intramembranous segments of each of the four domains of the α-subunit helps form the wall of the channel. The shape of the channel, a tepee with its base facing inwards, resembles that of the KcsA and shaker channels. The diameter of the pore is at least 0.31 nm and that of the entire α-subunit has been computed to be approximately 10 nm. The Na^{2+} α-subunit (like the Ca^{2+} α-subunit) achieves with one lengthy amino acid chain what the gap junctions of Chapter 7, the ligand-gated channels of Chapter 10 and the K$^+$ channels of the earlier part of the present chapter required multiple subunits to accomplish.

In addition to the α-subunit the complete Na$^+$ channel, as noted above, also contains two β-subunits (Figure 11.20). These subunits have an interesting structure, not seen in the β-subunits of other channels. They show the characteristic immunoglobulin folds of cell adhesion molecules (Section 19.8, Figure 19.19). Na$^+$ channel β-subunits thus seem to have two distinct functions. They modulate channel gating, causing increased rates of activation and inactivation, and also bind to elements in the extracellular matrix. It may be that this enables Na$^+$ channels to concentrate in regions such as nodes of Ranvier where they are present in very high densities.

11.4.2 Diversity

As shown in Table 11.7, the α-subunits of the Na$^+$ channels are encoded by a gene family consisting of

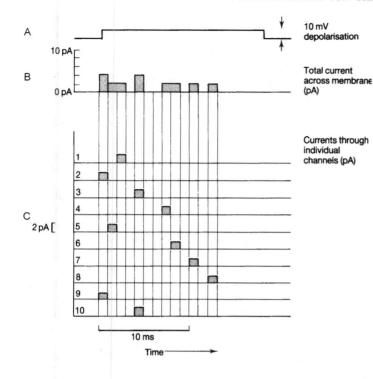

Figure 11.21 Schematic to show current flow through patch of excitable membrane when clamped 10 mV positive to resting potential. (A) The membrane patch is clamped 10 mV positive to resting potential. (B) The current through the membrane measured in picoamps. (C) The current in B is due to the random opening and closing of ten individual sodium channels in the membrane.

at least eight members: *SCN1A–SCN8A*. These genes, as the table shows, are spread over a number of chromosomes: 2, 3, 12, 17. CNS Na^+ channels are coded by a cluster of genes (*SCN1A, SCN2A, SCN3A*) on chromosome 2 with an outlier (*SCN8A*) on chromosome 12. β-subunits are coded by three genes (*SCN1B–SCN3B*) located (human) on chromosomes 19 and 11, respectively.

The pathologies caused by mutation in these genes are discussed in Section 11.7.

Na^+ channels are not so heterogeneous as Ca^{2+} or K^+ channels and they are not classified into subfamilies. Nevertheless, the fact that there are at least eight different α-subunits and three different β-subunits indicates that the biophysical properties of Na^+ channels in different parts of the nervous

Table 11.7 Na^+ channels and genes

Gene (human)	Chromosome	Subunit	Channel	Site of expression
SCN1A	2q24	α1	$Na_v1.1$	CNS
SCN2A	2q23–q24.3	α2	$Na_v1.2$	CNS
SCN3A	2q24–q31	α3	$Na_v1.3$	CNS
SCN4A	17q23–q25	α4	$Na_v1.4$	Skeletal muscle
SCN5A	3p21	α5	$Na_v1.5$	Cardiac muscle
SCN6A	2q21–q23	α6		Neurons/glia
SCN7A		α7		Glia
SCN8A	12q13	α8	$Na_v1.6$	Neurons, glia
SCN1B	19q13.1	β1		Brain, muscle
SCN2B	11q22	β2		Brain muscle
SCN3B	11q23.3	β3		Brain, muscle

Modified from Alexander, S., *et al.* (2001), *Nomenclature Supplement, Trends in Pharmacological Sciences*, Cambridge: Elsevier.

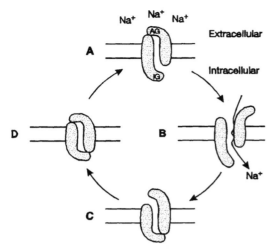

Figure 11.22 Conformational cycle of a sodium channel. (A) In the 'resting' membrane the sodium channel is closed. The activation gate (AG) is shut and the inactivation gate (IG) is open. (B) When the membrane is depolarised the voltage drop is sensed by a 'voltage sensor' and the activation gate opens. Sodium ions flow through down their electrochemical gradient. Because the Na+ ions 'hop' from one site to another in the channel (as indicated) they proceed in single file. (C) After about 1 ms the inactivation gate closes. (D) When the membrane returns to its resting voltage the inactivation gate reopens (but comparatively slowly, i.e. about 1.5 ms) and the channel assumes its original conformation.

system, or in different parts of the same neuron, perhaps at different times in a life history, can vary in subtle ways. The following section describes the biophysics of a typical Na+ channel.

11.4.3 Biophysics

The physiology of sodium channels has been studied by several of the methods by which the nAChR was studied (see Section 10.1.2). In particular the patch-clamp and bilayer techniques have proved important. If a small patch of neural or muscle fibre membrane is clamped some 10 mV positive to its resting potential, sodium channels open and current can be recorded as shown in Figure 11.21.

Figure 11.21 shows that the channels open for about 1 ms and allow about 2 pA of current to pass. A single channel thus has a conductance of about 2×10^{-10} S, i.e. 200 pS. When two channels open at

the same time twice the amount of current is recorded. The exact time of opening or closing of a channel is not predictable. Such systems are called **stochastic**. Once a channel has opened it is inactivated and will not open again whilst the membrane remains depolarised. The channel thus exists in three major conformations: closed, open and inactivated. This cycle is shown diagrammatically in Figure 11.22.

A classical recording from a patch-clamped fragment of myotube membrane is shown in Figure 11.23. The membrane patch (which possessed only two or three active sodium channels) is held at a voltage some 30 mV below its resting potential to ensure that the Na+ gate is closed. It is then depolarised from this holding potential to some 10 mV above its resting potential. This causes Na+ channels to open. Each channel remains open on average for 0.7 ms and then closes again. During the open phase 1.6 pA of current flows.

The number of sodium ions which flux through a single sodium channel in the above experiment during its transient open configuration can be calculated in the same way as for the Ca^{2+} channel (Section 11.3):

$$\frac{(1.6 \times 10^{-12} \text{ coulombs s}^{-1}) \times (0.7 \times 10^{-3} \text{s}) \times (6 \times 10^{23} \text{ ions mol}^{-1})}{96\,500 \text{ coulombs mol}^{-1}}$$

$= $ approx. 7000 sodium ions (about 10 000 ms^{-1})

This is, in fact, a small number compared with the number of sodium ions present. Because tetrodotoxin marks the sodium channel so precisely it is possible to radiolabel the toxin and determine the number of channels present in an area of membrane. We can thus estimate the number of sodium ions cascading in during an action potential if all the channels should happen to open. In rabbit unmyelinated vagus nerve fibres, the concentration of channels turns out to be about 100 channels per square micrometre of membrane whilst in the nodes of Ranvier of myelinated fibres there may be up to 3000 μm^{-2}. Let us consider an unmyelinated axon with a radius of 6 μm. We need to calculate the quantity of sodium ions beneath a square micrometre of neurilemma. We proceed as follows:

The surface area of a cylinder is given by

$$A = 2\pi r l$$

i.e.

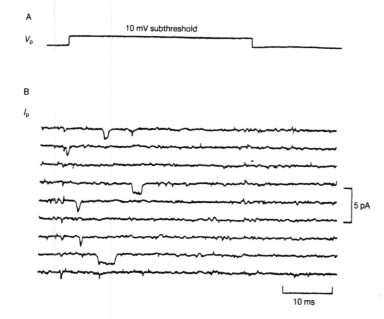

A

V_p

10 mV subthreshold

B

I_p

5 pA

10 ms

Figure 11.23 Current flux across a patch-clamped fragment of sarcolemma. (A) The membrane patch was hyper-polarised by 30 mV to ensure that all sodium gates were closed. 40 mV depolarising pulses were given ensuring that the membrane was depolarised 10 mV above its resting potential and clamped at that value (trace at A). (B) Nine successive records from this patch-clamped membrane. The large downward deflections indicate channel openings. The mean channel current was 1.6 pA and mean lifetime was 0.7 ms. The patch probably contained two or three active channels. In some cases channel openings overlap. The small wiggles represent noise in the recording equipment. Reprinted by permission from Sigworth and Neher (1980), *Nature*, **287**, 447–449. Copyright © 1980 Macmillan Magazines Ltd.

$$l = A/2\pi r$$

The volume of a cylinder is given by

$$V = \pi r^2 l$$

Hence

$$V = \frac{\pi r^2 A}{2\pi r}$$

i.e.

$$V = rA/2$$

But $A = 1\ \mu m^2$, and therefore

$$V = 6/2 = 3\ \mu m^3$$

We now have to calculate how many sodium ions there are in a volume of $3\ \mu m^3$.

The concentration of Na^+ ions in a mammalian motor neuron has been found to be 15 mM (Table 9.1), i.e. 15×10^{-3} M per litre.

Now

$$1\ \mu m^3 = 1 \times 10^{-12}\ ml\ or\ 10^{-15}\ l$$

Therefore $1\mu m^3$ contains $(15 \times 10^{-3}) \times (1 \times 10^{-15})$ moles of Na^+. But one mole of Na^+ ions consists of 6×10^{23} ions (Avogadro's number). Therefore, $1\ \mu m^3$ of mammalian axoplasm contains

$$(15 \times 10^{-18}) \times (6 \times 10^{23}) = 90 \times 10^5\ sodium\ ions$$

Hence $3\ \mu m^3$ of axoplasm will hold 2.7×10^7 sodium ions.

If all 100 channels in the square micrometre of rabbit vagus fibre membrane were to open during the passage of an action potential, then $7000 \times 100 = 700\,000$ ions will flow into the axoplasm. It can be seen that this constitutes about 3% of the sodium ions inside. It is, of course, unlikely that all the sodium channels will open during any single action potential. It is also obvious that the calculation makes some very unphysiological assumptions. Most importantly it overlooks the fact that during an impulse the potential across the membrane rapidly diminishes and indeed reverses. Hence the driving force behind the sodium ions (remember $I = gV$) rapidly diminishes to zero. Nevertheless it is clear that an axon cannot continue conducting impulses for long without the $Na^+ + K^+$ pump described in Chapter 9 working to re-establish the proper internal and external ionic concentrations.

Finally, it should be emphasised that the sodium channel, like the other channels we have considered, is more complex than this first analysis suggests. It appears to be capable of existing in more than just the three states – open, inactivated

Table 11.8 Amino acid sequences in S4 of voltage-dependent channels

K⁺ channels	
Shaker	**R** V I **R** L V **R** V F **R** I F **K** L S **R** H S **K** G L
Shab	Q V F **R** I M **R** I L **R** V L **K** L A **R** H S T G L
Shaw	E F F S I I **R** I M **R** L F **K** V T **R** H S S G L
Shal	F V T **R** V F **R** V F **R** I F **K** F S **R** H S Q G L
Na⁺ channels	S A L **R** T F **R** V L **R** A L **K** T I S V I P G L
Ca²⁺ channels	S V L **R** C I **R** L L **R** L F **K** I T **K** Y W T S L

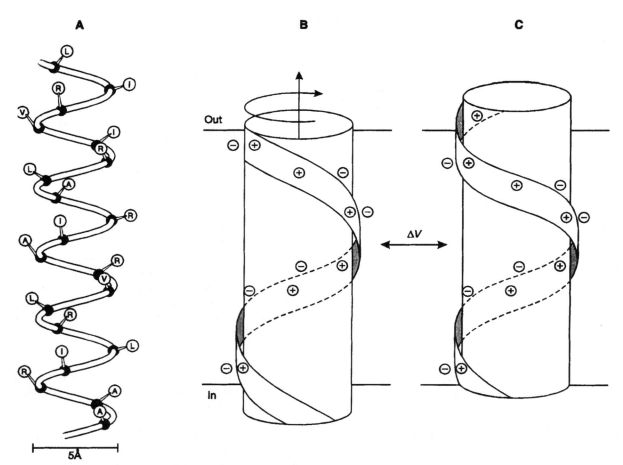

Figure 11.24 S4 'sliding helix' voltage sensor. (A) 'Ball and stick' representation of the S4 helix of domain 4 of the sodium channel. Dark circles represent the α-carbon atoms of each amino acid residue; open circles are labelled with the amino acid single-letter code and show the direction in which the side chain projects. (B) and (C) Movement of the helix in response to voltage change (ΔV) across the membrane. At resting potential (B) all positively charged side chains are paired with fixed negative charges on other transmembrane domains of the channel protein. The helix is held in this position by the negative internal potential of the resting membrane (V_m). On depolarisation this restraint disappears and the helix 'screws' outwards by 5 Å and through 60°. As indicated in C, this leaves an unpaired negative charge on the P-face of the membrane and an unpaired positive charge on the E-face. Further explanation in text. From Catterall (1992), *Physiological Reviews*, **72**, S15–S48; with permission.

and shut – which we outlined above. Careful analysis of the patch-clamp records suggests that there are at least three substrates of the open condition. This untidy complexity turns out to be characteristic of all the channels that have so far been carefully examined. It shows that we are dealing with complex macromolecular structures – not simple mechanical or electronic valves.

11.5 ION SELECTIVITY AND VOLTAGE SENSITIVITY

We have noted, as we reviewed the K^+, Ca^{2+} and Na^+ voltage-dependent channels, that the core region of each was remarkably similar. In particular we saw that the polypeptide chain between S5 and S6 was believed to take the form of a hairpin inserted into the membrane. In the K^+ channels this region is believed to resemble the structure elucidated in the KcsA bacterial channel, while in the Ca^{2+} and Na^+ channels, although its exact structure remains unknown, it is unlikely to differ from the bacterial model very markedly. This region, of about 17 amino acid residues, is, as we have seen, called variously the P-region, H5 or SS1–SS2. We also noted that S4 in each of the three voltage-gated ion channels contained a number of positively charged residues, especially arginine and lysine, arranged -arg/lys-X-X-arg/lys- (i.e. R/K-X-X-R/K) where X is a neutral residue (Table 11.8).

We saw in our review that in each channel there is evidence that H5 is involved in ion selectivity and that S4 acts as a 'voltage sensor'. The evidence is largely derived from site-directed mutagenesis and careful examination of how this affects the biophysics and pharmacology of the channel.

11.5.1 Ion Selectivity

First it must be re-emphasised that ions do not flow through membrane pores like water through a tube. There is biophysical evidence to show that they proceed in single file, from one binding site to the next. This, as we noted, was very satisfactorily confirmed in the 2 Å structure of the KcsA channel derived by X-ray crystallography. It is instructive to note that substitution of certain of the amino acids in this region alters the channel's conductivity. For instance, taking the shaker K^+ channels, it is found that substitution of $Thre_{441}$ and $Thre_{442}$

(closely abutting the selectivity filter (Figure 11.1B)) by serine alters the relative conductivity of the channel to K^+, Rb^+ and NH_4^+. Similarly, substitution of Asp_{431} and $Thre_{449}$ reduces the inhibitory affect of externally applied TEA Cl, whilst substitution of $Thre_{441}$ diminishes the effect of TEA Cl when internally applied.

Analogous results are obtained by mutation studies of the Na^+ channel H5 region. When Lys_{1422} and Ala_{1714} are substituted by negatively charged Glu residues, the ion selectivity of the channel is dramatically changed from Na^+ to Ca^{2+}. Similar experiments indicate that S5 and S6 also contribute to the walls of the pore, at least at its internal and external entrances. The receptor site for tetrodotoxin, for instance, is found to involve acidic residues on the extracellular end of S6 in each domain of the Na^+ channel.

11.5.2 Voltage Sensitivity

First of all we have to distinguish 'gating' from channel 'opening'. The first thing necessary when the voltage across a membrane changes is that the change should be somehow 'sensed'. This 'sensing' has been found to be accompanied by a small but detectable current, the gating current. This current can be detected at the beginning and end of any small depolarisation. The gating 'activates' the channel, which may then open.

We have already noted that S4 is suspected to be the sensor responsible for gating voltage-dependent ion channels. How does it work? Its structure provides a clue. We have seen that every third residue is positively charged. When disposed in an α-helical conformation (remember the repeat is 3.6 residues (Figure 2.6)) all these charged residues will be projecting in approximately the same direction. In the membrane they are neutralised by negatively charged amino acid side chains from other parts of the channel protein. The whole structure, it is hypothesised, is held in position by the voltage across the membrane. However, when a small depolarisation occurs, the forces holding the helix diminish; in particular the negative potential on the internal surface of the membrane is reduced. The reduced electrostatic attraction allows the whole helical structure to 'screw' outwards: a 60° turn and a 5 Å outward displacement (Figure 11.24). This transfers one positive charge to the outside of the membrane, leaving an unpaired negative charge within, a net charge transfer of $+1$. It is believed that the S4 helices of all four domains rotate in rapid sequential order, thus transferring four positive charges from the inside to the outside of the membrane. This is approximately the size of the gating current observed by the electrophysiologist. This conformational change then activates by some as yet unknown means the voltage-gated channel.

An alternative gating mechanism which does not envisage an ion pore at all is given in Appendix 5. The latter mechanism has the advantage that it sees voltage sensing and channel opening as a single phenomenon. It should be said, however, that this is at present very much a minority view. The X-ray crystallographic solution of the KcsA channel and the mutational and biophysical analyses of eukaryotic voltage-gated channels leave little doubt of the reality lying behind the diagrams printed in papers and textbooks. Research in this area, using the combined techniques of molecular biology and biophysics, is fast-moving and exciting. The goal of understanding how voltage-gated cation channels work in atomic detail is beginning to come within reach.

11.6 VOLTAGE-SENSITIVE CHLORIDE CHANNELS

Voltage-gated chloride channels are present in the membranes of all animal cells. We shall see in Chapters 12 and 14 that they play their part in maintaining the resting potential across cell membranes. It is only comparatively recently, however, that we have gained an understanding of the structure of these channels. The two most important members of the class are the **ClC (chloride channel)** and the **Cln (chloride–nucleotide modulated channel)**. A third voltage-sensitive Cl^- channel is remarkable in that it is formed by a surprisingly small 72 amino acid polypeptide, **phospholemman**.

11.6.1 ClC Channels

The first ClC channel (designated ClC-0) to be cloned was derived from the electroplax of *Torpedo californica*. Proteins derived from cDNA clones prepared from this organ expressed a voltage-sensitive Cl^- channel in *Xenopus* oocyte. The protein predicted from the cDNA weighs in at approximately 89 kDa and consists of a single chain of 805 amino acids. The ClC protein did not appear to be related to any other channel protein and turned out to be the first member of a family of voltage-sensitive Cl^- channels.

The complete ClC channel consists of two 89 kDa subunits. Until very recently there was no certain knowledge of where the voltage sensor or the ion selector are located. This changed, however, as prokaryotic homologues of ClC channels were crystallised and subjected to X-ray diffraction. The first analysis, at 6.5 Å resolution, did not allow the detailed structure of the channel to be determined. Subsequently it proved possible to crystallise ClC channels from *Salmonella enterica* (stClC) and *Escherichia coli* (EcClC) and obtain X-ray structures at 3.0 Å and 3.5 Å, respectively. As in the other cases we have reviewed in these chapters the crystallography largely confirmed the structure deduced from biochemistry and molecular genetics, but added detail and dimensions which allow the precise biophysics to be understood.

The X-ray crystallography shows a dimeric model each of whose halves consists of an intricate membrane-bound protein with no less than 18 helices, labelled A to R (Figure 11.25A). Not all of these helices penetrate all the way through the membrane. Furthermore, the helices do not run perpendicular to the membrane but slant through at a diagonal (Figure 11.25B). The direction of the diagonal slant varies so that the helices in one part of the molecule lean toward those in the other part. This complex structure is said to constitute an antiparallel architecture. As we shall see below, this architecture serves a purpose.

Each subunit forms its own channel. In this the ClC channel is markedly unlike the cation channels discussed above where the channel forms at the centre of a group of subunits. It is also unlike the cation channels in being, in a sense, 'double-barrelled', having two pores in the complete dimeric channel. Because of the antiparallel architecture mentioned above it is believed that partially positively charged ends of oppositely slanted helices point inwards towards each other and form a selectivity filter for the Cl^- ion (Figure 11.25B). This filter is thus in the centre of each subunit, rather than near the exit as in the cation

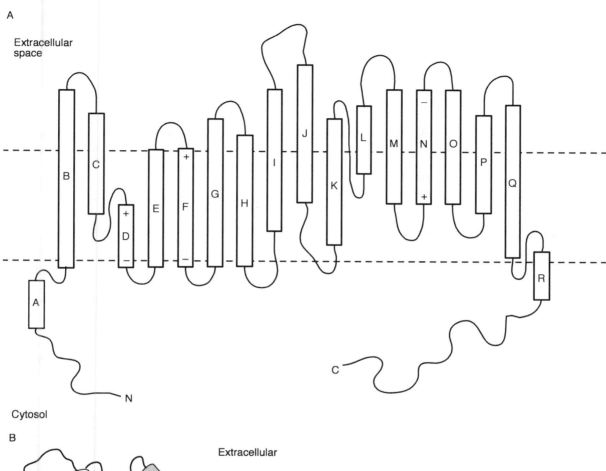

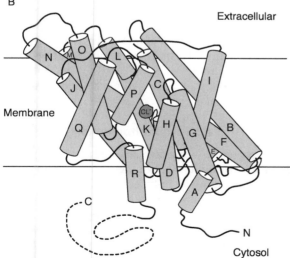

Figure 11.25 Topology of the ClC channel subunit. (A) Plan view to show the 18 α-helices in the subunit. The positions of the charged helices which together form the selectivity filter are shown. (B) Disposition of the α-helices in the membrane. The position of the Cl⁻ ion in the selectivity filter is shown. The figure shows the side of the subunit which interfaces with its partner subunit. Redrawn from Dutzler *et al.* (2002), *Nature*, **415**, 287–294.

channels. The shape of the channel thus resembles more an hourglass than a goblet.

Now that the first detailed structure of a Cl⁻ channel has been determined it will be a fascinating exercise in comparative molecular biology to compare and contrast the different structures, biophysics and evolutionary histories of cation and anion channels.

The best-known eukaryotic ClC channel remains ClC-0. It is strongly anion selective having a permeability sequence $Cl^- > Br^- > I^-$. Even though it is the best-known chloride channel, ClC-0 retains many mysteries. Like the bacterial ClC channel it contains two biophysically identical pores, a feature that has led to its being called the 'double-barrelled' channel. Now that the prokaryote channel structure has been solved it should be possible to gain much greater understanding of the biophysical nature of these pores. It is likely that the StClC and EcClC channel structures will prove just as enlightening for students of anion channels as the KcsA channel proved for devotees of cation channels.

Oligonucleotide probes from the *Torpedo* ClC-0 channel have been used to screen cDNA libraries from various mammalian tissues. The first ClC channel to be isolated in this way was derived from striated muscle and was designated ClC-1. To date ten ClC channels have been identified and their genes designated as *CLCN0* to *ClCN7*, *ClCN-Ka* and *ClCN-Kb*.

We shall see in Chapters 12 and 14 that Cl⁻ ions play an important role in maintaining the resting potential across excitable membranes. If the ClC-1 channel is inactivated the outflow of Cl⁻, which helps to return a membrane to its resting potential after an action potential, is delayed. This has the consequence that sodium channels tend to recover from their inactivation whilst the membrane is still depolarised. This means, as we shall see in Chapter 14, that they reopen. The consequence is repetitive firing, the basis of the muscle impairment known as **myotonia**.

A second member of the ClC family, ClC-2, isolated from most mammalian tissues, including neurons, is activated by strongly hypopolarising conditions ($V_m < -100\,\text{mV}$). When expressed in *Xenopus* oocytes it can also be shown to be activated by hypotonicity. It may, therefore, have a role in controlling cytoplasmic volume, opening when swelling occurs.

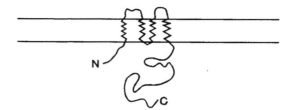

Figure 11.26 Topology of the Cln channel subunit. The four intramembrane β-strands are represented by zigzags.

11.6.2 Cln Channels

This type of channel was first isolated by expression cloning of **epithelial cell** mRNA in *Xenopus* oocyte. Like the MinK channel it is surprisingly small (c. 26 kDa; 235 amino acids). Like the MinK channel it has a single membrane-embedded domain but, unlike MinK, this domain is *not* believed to form an α-helix. Instead it has been suggested that it spans the membrane as two 'hairpins' each containing two β-strands (Figure 11.26). If the complete channel is formed by two of these 26 kDa subunits the pore may be constructed by a cylinder of eight β-strands, i.e. a β- or TIM barrel (see Section 2.2).

When expressed in *Xenopus* oocyte the channel carries a strong outward current of Cl⁻ ions when the membrane is depolarised. The channel is insensitive to internal Ca^{2+} ions but I_{Cln} is inhibited by externally applied nucleotides, including ATP, ADP, GTP, cAMP and cGMP. Hence the 'n' in the channel's name. Perhaps its major location is in internal membranes, such as the endoplasmic reticulum, where its 'external' nucleotide sensitivity would make more sense.

11.6.3 Phospholemman

Our account of chloride channels would not be complete without mentioning the surprising discovery that a small polypeptide (72 amino acids) with just one transmembrane domain can form a voltage-sensitive Cl⁻ channel. Presumably a number of these small units group together around a Cl⁻ pore. Phospholemman obtained its name from the fact that it was first isolated from the sarcolemma of cardiac muscle cells. When expressed in *Xenopus* oocyte it can be shown to carry a Cl⁻ current in strongly hyperpolarising conditions.

11.7 CHANNELOPATHIES

Mutations in the genes coding the channel proteins reviewed in this and previous chapters generally lead to defective function. These defective ion channels cause diseases collectively known as **channelopathies**. Many of these mutations affect the ion channels in the sarcolemmas of muscle cells and fibres rather than those in neurons.

11.7.1 Potassium Channels

Episodic ataxia (EA1) is caused by point mutations of the potassium channel gene *KCNA1*. This disorder, which generally lasts only minutes, is provoked by abrupt postural change and vestibular stimulation.

Benign familial neonatal convulsions, beginning within days of birth but clearing away in weeks to months, are due to mutations in the *KCNQ3* potassium channel gene.

Andersen's syndrome, characterised by periodic muscle paralysis, cardiac arrhythmia and abnormal growth, a very rare condition, is also due to mutation on the *KCNJ2* K$^+$ channel gene.

Weaver mouse: This behavioural syndrome whose nature is graphically expressed by its name is shown by mice holding a mutation in the *KCNJ6* gene which leads to an alteration in the P-region of the Kir3.2 channel. The mutation makes the channel permeable to Na$^+$ and this, in turn, to loss of granule cells in the cerebellum. The mutation will be discussed further in Section 19.1.

11.7.2 Calcium Channels

We saw in Section 11.3.2 that at least 10 different genes encode the α_1-subunits of the calcium channel. Mutations in these CACNA genes cause Ca^{2+} channelopathies. Mutations in the CACNA1A gene (chromosome 19p13) cause a group of conditions including familial hemiplegic migraine (FHM), type 2 episodic ataxia (EA2) and spinocerebellar ataxia, type 6 (SCA6).

FHM is mainly a childhood disorder characterised by intermittent lateral paralysis accompanied by migraine. Most patients (there are exceptions) recover from these conditions and live a normal adult life. The mutation responsible for the condition is a G to A transition at position 850 on the CACNA gene leading to R192Q in the α-subunit of the calcium channel. This, in turn, causes both an increase in channel density (presumably due to increased expression of the gene) and an increase in open probability.

EA2 presents as abnormal eye movements and episodes of vertigo and instability of the trunk usually precipitated by stress and/or fatigue. It, too, is due to mutation of the CACNA1A gene, in this case by deletion of a C at position 4073 which leads to a premature stop codon. The consequences at the ion channel are as yet unknown.

SCA6 patients display cerebellar dysfunctions leading to movement disorders. The late onset form of this disease (SCA6) is due to an expansion of CAG repeats (greater than 21) at the 3′ end of the CACNA1A gene. Again, the exact consequences at the level of the calcium channel remain to be determined.

All of these conditions have variants induced by mutations in slightly different positions in the CACNA1A gene. They are all likely to lead to altered synaptic transmission and/or altered intracellular Ca^{2+} levels and thus to the disease symptoms observed in the patient.

Hypokalaemic periodic paralysis (HoKPP): Mutation of another calcium channel gene, *CACNL1A3*, leads to HoKPP, characterised by recurrent inexcitability of skeletal muscles. Mutations affect the S4 TM segment of the second and fourth domains of the calcium channel protein.

Mouse genetics have also proved valuable in understanding these channelopathies. A number of mutations affecting the mouse α_1 calcium channel subunit have been detected. These cause a variety of behavioural abnormalities including ataxia and epilepsy. Mice, as in other cases, are likely to provide indispensable models for investigating the human condition.

11.7.3 Sodium Channels

The majority of sodium channelopathies presently known affect muscle rather than nerve. The mutations fall into three major classes all of which affect the *SCN4A* gene which encodes the α_4-subunit. The first group of mutations affect the S4 TM segment of domain IV, the second group affect the inactivation loop between domains III and IV and the third group affect the inactivation sequence at the cytoplasmic end of the molecule.

All appear to be single nucleotide substitutions causing single amino acid changes. The majority of mutations are of the third type. They affect the rate of inactivation leading to a persisting inward current Na^+ current causing various types of periodic paralysis. The best known is **hereditary hyperkalaemic periodic paralysis (HyKPP)** characterised by muscle hyperexcitability and delayed relaxation. In addition to HyKPP, mutations of the *SCN4A* gene cause **paramyotonia congenita (PC)** and a diverse group of disorders known as **potassium-aggravated myotonias (PAMs)**.

In addition to muscle disorders, sodium channelopathies are responsible for certain forms of epilepsy. **Generalised epilepsy with febrile seizures (GEFS)** is, for instance, due to a C to G transition at position 387 on the *SCN1B* gene which leads to a $Cys \Rightarrow Try$ substitution in the β_1-subunit of the Na^+ channel. The β-subunit normally accelerates the rate of inactivation and the rate of recovery from inactivation. Mutation thus leads to a persistent inward Na^+ current after activation and this may cause the symptoms of GEFS.

11.7.4 Chloride Channels

Myotonia congenita (Thomsen's disease): This condition is characterised by muscle stiffness, particularly after prolonged rest, It has been shown to be due to $G \Rightarrow A$ transition at position 689 in the *ClCN-1* gene (chromosome 7q35). This leads to $Gly_{180} \Rightarrow Glu$ in the ClC-1 protein in muscle sarcolemma which, in turn, reduces its Cl^- conductance. At least four other point mutations have been detected in the *ClCN1* gene causing congenital myotonias.

11.8 EVOLUTION OF ION CHANNELS

In a frequently cited paper François Jacob characterises evolution as a 'tinkerer'. It takes what it finds on the scrapheap of molecular structures and presses them to new and unexpected uses. Biologists have always known this. Gill arches become jawbones and auditory ossicles; the elasmobranch's spiracle becomes the carotid labyrinth; heart chambers become pacemaker patches, fins become wings, and so on. We have already met many molecular counterparts of this tinkering in the first ten chapters of this book and will meet many more

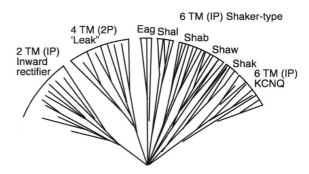

Figure 11.27 Evolutionary bush to show the evolutionary relationships of human K^+ channels. The sizes of the cones are approximately proportional to the number of channels recognised in the summer of 2000, i.e. 13 inward rectifiers; 11 'leak' channels; 17 shaker-type; 5 KCNQ. After Rogawski (2000), *Trends in Neurosciences*, **23**, 394.

in the next ten chapters. In particular we have seen how great families and superfamilies of ion channels have evolved, developing different features and specificities over geological time. In this chapter we have reviewed the great variety of voltage-gated ion channels which lie at the core of neurophysiology. The earliest to make an appearance on the evolutionary scene were the K^+ channels. We noted how they have evolved into three subfamilies: 2TM(1P), 4TM(2P) and 6TM(1P). These subfamilies have themselves diversified to generate a large number of different members. An evolutionary bush is shown in Figure 11.27.

We saw in Box 11.1 that CNG-gated channels show striking homologies with shaker-type K^+ channels. There is little doubt that, as Figure 11.28 shows, they are members of the same superfamily and evolved from shaker-type channels at some time in the evolutionary past. Further to the right of the VGK$^+$ pyramid of Figure 11.28 the voltage-gated Ca^{2+} and Na^+ channels are shown evolving from shaker-type channels by two internal gene duplications. The Ca^{2+} channels were the first to originate and from them came the Na^+ channels.

Figure 11.28 suggests that non-voltage-gated K^+ channels along with Cl^- channels formed the earliest ion channels. It is possible that ion channels controlled by stretch, the earliest mechanoreceptors (see Section 13.3.1), date back to similarly ancient times. Other early sensory receptors (photoreceptors, chemoreceptors) did not control ion channels

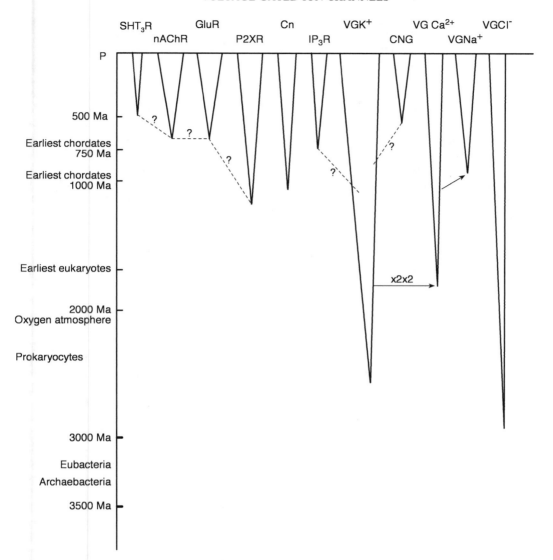

Figure 11.28 Evolutionary relationships of ionic channels. The figure has several question marks: this emphasises that our understanding is at present highly provisional. The chart shows, however, that K^+ and Cl^- channels are at present believed to be the earliest forms of ion channel. On the right-hand side of the chart the Ca^{2+} and Na^+ voltage-gated channels are shown originating from stem K^+ channels by two gene duplications. The Cl^- channel with its very different architecture is shown as a completely independent lineage. The spreading triangles are intended to show that each ion channel has evolved into many varieties. Cn=connexin; CNG=cyclic nucleotide gated channel; VGK^+=voltage-gated K^+ channel; $VGCa^{2+}$=voltage-gated Ca^{2+} channel; $VGNa^+$=voltage-gated Na^+ channel. Other abbreviations as in text.

but used G-protein-linked 7TM mechanisms. Later, as we shall note in Sections 13.1 and 13.2, G-proteins gained control of CNG ion pores in vertebrate olfactory cilia and outer segments and elsewhere.

The dating in Figure 11.28 is, of necessity, only approximate. If it is the case that mechanoreceptors and K^+ channels were the first to originate, it suggests that response to volume changes due to osmotic flows or cell growth were from the first of great importance. Bacterial K^+ channels, as we saw when discussing the KcsA channel of *Streptomyces lividans*, are not voltage controlled. This came later. The ligand-gated receptors do not make their appearance until the Metazoa.

The expanding wedges in the diagram indicate the great structural diversity to be found in each family (for more accurate representation of two such lineages see Box 10.1, the nAChR lineage, and Figure 11.27). The great diversity of shaker-type K^+ channels is due to the fact that (unlike the Na^+ and Ca^{2+} channels) they consists of four independently programmed subunits. This allows, as we noted, great variability: upwards of 50 physiologically different K^+ channels are known. Figure 11.28 indicates that although the subunits of the all the ligand-gated ion channels (LGICs) are highly heterogeneous they nevertheless resemble each other more than they resemble the subunits of any other family. This indicates that they are true families, derived from a common ancestor. In the same way that the K^+ channels show great diversity due to their construction from independent subunits so the intricate multimeric structure of the LGICs have allowed diversification into a great number of subtypes (indicated by the wedge). At the latest count there were nearly 60 different subtypes the nAChR, over 40 subtypes of $GABA_AR$, and 6 subtypes of GlyR.

11.9 CONCLUSION AND FORWARD LOOK

This chapter, like its predecessors, has given us an inkling of the great variety of channels which exist in neuronal membranes. We have seen that there are various physiological subtypes of potassium, calcium, sodium and chloride channels. We have also seen that each of these subtypes can exist in several different conditions of openness or shutness. The more we understand about the brain the more infinitely complex it seems.

Yet at the same time that the complexity increases great unifying principles begin to emerge. At root these unifying principles rest on the major unifying theory of all biology: evolution. We have seen in this chapter that the three major voltage-controlled cation channels – the potassium, sodium and calcium channels – are built according to a common architectural principle. The various types of potassium channel are built around one (or two) homologous K^+-selective P-regions; the ubiquitous shaker-type K^+ channels consist of 6TM regions; the sodium and calcium gates consists of a protein some three times as large, with over 1800 amino acids, and consisting of four domains each of which is rather similar to a single shaker-type channel. Each of these domains, like the shaker channel, consists of six membrane-spanning segments and in each case the fourth segment, like the fourth segment in the shaker channel, is highly polar and suspected of being the voltage sensor. Furthermore, the polypeptide between the S5 and S6 helices is inserted into the membrane as a β-pleated hairpin in a similar way to the P-region in K^+ channels and plays a crucial role in lining the pore and determining which ions to accept. In the sodium and calcium gates the four homologous domains are joined together by stretches of hydrophilic amino acids. These channels thus constitute a superfamily sometimes called the 'shaker superfamily'.

The structures of the first Cl^- channels have only very recently been determined. Their architecture is radically different from the 'shaker'-type architecture of the cation channels. It will be fascinating to learn whether these channels have fanned out into a similar set of variations on a

---

Figure 11.29 Summary of the structures of receptor and channel subunits. (a) Opsin-like G-protein coupled 7TM receptor (note probable disulphide link between e-3 and e-5; i-2 and i-3 are crucial in G-protein coupling; a fatty acid chain embedded in membrane often forms a fourth intracellular loop); (b) Peptide transmitter variant of 7TM receptor (note lengthy extracellular N-terminal probably containing the ligand-binding site); (c) P2X 2TM family; (d) GluR 3TM family; (e) nAChR 4TM family; (f) Kir 2TM(1P) family; (g) K^+ leak 4TM(2P) family; (h) shaker-type V_K 6TM(1P) family; (i) Na^+ and/or Ca^{2+} family; (j) ClC Cl^- family; (k) Cln Cl^- family (note the β-pleated strands in place of α-helices); (l) IP_3 receptor; (m) CNG-gated ion channel family (note structural similarity to (h)). Note that all these subunits are drawn spread out in two dimensions; in reality they form polymeric clusters.

1. G-protein coupled redeptors

2. Ligand-gated ion channels

3. Voltage-gated ion channels

4. Miscellaneous

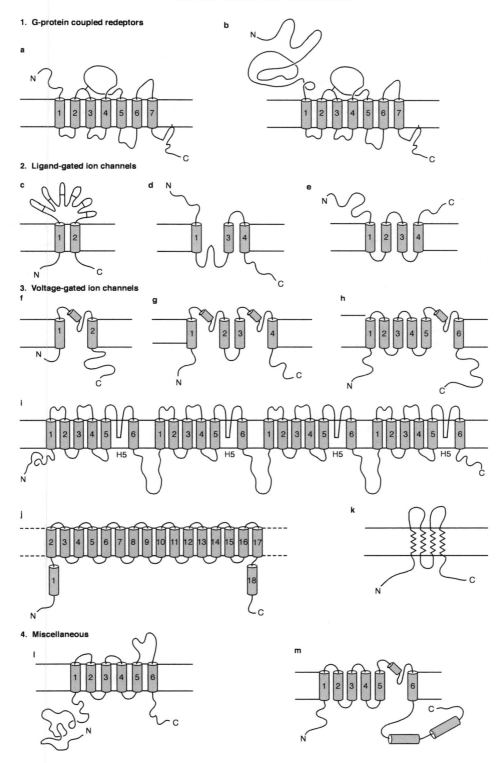

theme. At the time of writing it does not seem that anion channels are anything like as diverse as the cation channels.

Further still, looking back over Chapters 8–11, we can begin to see that the multitudinous variety of channels and pumps that abound in neuronal membranes fall into a much smaller number of natural groups. We saw in Chapter 8 that the great variety of G-coupled receptors, β_1- and β_2-ARs, mAChRs, mGluRs, NKRs and the opsins, all showed a remarkably consistent structure, marked by seven membrane-spanning helices. In Chapter 9 we noted that there is a strong evolutionary relationship between the $Na^+ + K^+$ pump and the Ca^{2+} pump. Next in Chapter 10 we discussed the remarkable family resemblances between nAChRs, $GABA_A$Rs and GlyRs. Finally, in this chapter we have reviewed the striking structural relationships between the various voltage-gated channels.

There are, as usual in biology, many exceptions to this neat classification. The Cln family shows, for instance, four amphipathic β-threads traversing the membrane whilst the ClC family possess 12 transmembrane helices. All of these various designs are displayed for convenience in Figure 11.29.

In the next three chapters we shall see how this complexity of pumps, gates and ion gradients underlies those definitive features of neural biophysics: resting potentials, cable conduction, receptor potentials, and last, but far from least, action potentials. The action potential or nerve impulse, in particular, can be understood as arising from the interaction of many of the gated channels that we have just reviewed. Finally, in Chapter 16, we shall broach the equally central topic of synaptic transmission leading on to a discussion of postsynaptic potentials in Chapter 17. In these two chapters we shall return once again to many of the channels discussed above and in Chapters 8 and 10. Indeed it is in the study of the molecular structure and function of membrane channels that molecular biology is forming a firm union with biophysics to make a major contribution to neurobiology. It might even be said that in some ways these membrane-embedded pumps, channels and receptors are the elements out of which molecular neurobiology is built.

12

RESTING POTENTIALS AND CABLE CONDUCTION

Resting potentials: significance of experimental equipment and appropriate biological preparations; measurement: importance of the squid giant axon preparation; **causes of resting potential**:
definition of electrochemical potentials – derivation of Nernst equation – derivation of V_K, V_{Cl}
and V_{Na} – the Goldman or 'constant field equation' – derivation of resting potential (V_m) –
effect of varying membrane permeability – simplification of Goldman equation. **Electrotonic
potentials and conduction**: dendrites – local circuit neurons – synapses; **biophysics of electrotonic
conduction**: λ, the 'space constant' – significance of length – significance of diameter –
dendrites – giant fibres – significance of myelination; **electrotonic conduction in dendrites**: effects
of activation at different locations and for different durations; the manifold roles of electrotonic
conduction

That the functioning of the nervous system is
accompanied by electrical changes has been known
since the work of Galvani and Volta in the late
eighteenth century. It was, however, only with the
development of mid-twentieth-century electrotech-
nology (especially electronics) and the discovery of
suitable experimental preparations (in particular
the cephalopod giant axon) that a genuine understanding of 'animal electricity' became possible.

In this chapter we shall first consider the origin
of the 'resting potential' (V_m) which exists across all
plasma membranes and in particular across the
plasma membranes of neurons. Second, we shall
turn our attention to the 'passive' flows of electrical
current and changes of electrical potentials across
membranes. We shall deal with 'action potentials' in
Chapter 14. We shall see in these two chapters how
the electrical phenomena of nervous systems are
signs of ion flows controlled by the many gates,
channels and pumps which we reviewed in Chapters
9–11. The theory of the nervous system at this level

begins to take on a coherence and simplicity that
would have delighted earlier investigators.

12.1 MEASUREMENT OF THE RESTING POTENTIAL

We noted above that resting potentials are believed
to be developed across all cell membranes – whatever the type of cell. Most mammalian neurons,
however, have extremely small diameters – seldom
more than about 20 μm. The recognition that certain
large tubular structures (diameter 500–600 μm) in
the squid, *Loligo*, were in fact single giant axons
was thus of enormous value to electrophysiologists.
They were at last able to place fine glass micro-
pipettes filled with an electrolyte inside an axon and
measure the resting and action potentials directly.
They were also able to squeeze out the axoplasm
and subject it to chemical analysis (see Table 9.1).
Most of the pioneering work which established the
physical basis of membrane potentials was done on

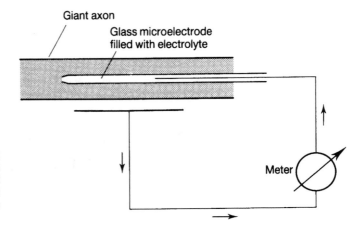

Figure 12.1 Measurement of the resting potential using a squid giant axon preparation. A fine glass microelectrode is inserted into the giant axon. The potentiometer measures the electrical potential across the membrane. The arrows indicate the direction of current flow.

this convenient preparation. Once again we are made aware of the great value of non-mammalian preparations in the prosecution of fundamental neuroscience.

Figure 12.1 shows that when the external electrode is taken to be at 'ground potential' (as is conventional) the internal electrode is found to record a voltage drop of some 50 mV across the membrane. This is the so-called **'resting potential'** (V_m). Our next task is to determine the origin of this ubiquitous potential.

12.2 THE ORIGIN OF THE RESTING POTENTIAL

The giant axon preparation and subsequently other preparations have allowed investigators to determine the ionic concentrations on either side of a neuronal membrane. We have already seen in Chapter 9 how it is possible to calculate the difference in 'free energy' (ΔG) represented by the distribution of a substance on two sides of a membrane. But in that chapter we explicitly put on one side consideration of any electrical forces that might be acting on charged particles such as ions. In this chapter we have to bring these forces into the equation. We have to consider not only the diffusional forces due to concentration differences but also the forces due to differences in electrical potential. For we shall be looking at the distribution of charged ions: Na^+, K^+, Cl^-, Ca^{2+}, etc. The distribution and movement of ions across membranes is responsible for all the multifarious electrical

phenomena that characterise the physiology of nervous systems.

In Chapter 9 we worked out the energetics of a transmembrane system by making use of μ, the chemical potential. In this chapter we introduce another parameter, $\bar{\mu}$, the electrochemical potential. This parameter takes into account the electrical forces which ions feel. In effect it is the chemical potential, μ, with an electrical term added.

The magnitude of the electrical force, P, which an ion, I, will feel depends on its valency, Z, and on the electrical potential, ψ, of its surroundings. We can thus write:

$$P = Z_I \psi^\alpha \qquad (12.1)$$

where Z_I = valency of the ion, I (i.e. +1 for Na and K, −1 for Cl, +2 for Ca, etc.), and ψ^α = electrical potential of phase α in which ion I is located.

Now biologists and biochemists are accustomed to working in moles or grams of ions, not single molecules or ions. A mole of univalent ions (i.e. 6×10^{23} ions) carries 96 500 coulombs of electricity, i.e. one Faraday (F). Hence we can rewrite the preceding equation as:

$$P = Z_I \psi^\alpha F$$

This is the electric force that a mole of ion, I, will experience in phase α.

It will be recalled, from Chapter 9, that the expression for the chemical potential of a substance, A, in phase α, was:

$$\mu_A^\alpha = \mu_A^0 + RT \ln C_A^\alpha$$

It follows that the electrochemical potential for an ion, I, in phase α, is given by

$$\bar{\mu}_I^\alpha = \mu_I^0 + RT \ln C_I^\alpha + Z_I \psi^\alpha F \qquad (12.2)$$

Next let us remind ourselves of the system we are considering. This is shown in Figure 12.2. The ion, I, is present in both phase α and phase β; the two phases are separated by a **permeable** partition or membrane.

Now if, as we emphasised above, the membrane between the two solutions is fully permeable to 'I' then microscopic flows of the ion will occur to ensure that the electrochemical potentials of 'I' on each side of the membrane are identical. The condition for equilibrium is thus:

$$\bar{\mu}_I^\alpha = \bar{\mu}_I^\beta \qquad (12.3)$$

If, next, we expand both sides of equation 12.3 we have:

$$\mu_I^0 + RT \ln C_I^\alpha + Z_I \psi^\alpha F = \mu_I^0 + RT \ln C_I^\beta + Z_I \psi^\beta F$$

Collecting terms we have:

$$Z_I F (\psi^\alpha - \psi^\beta) = RT \ln \frac{C_I^\beta}{C_I^\alpha}$$

Next we take note of two things. First, we can substitute C_o and C_i (the concentrations of I on either side of the membrane) for C_I^α and C_I^β;

second, as any difference in electrical potential between the two phases, α and β, will be felt at the boundary between the phases we can substitute V, the voltage across the membrane, for $(\psi^\alpha - \psi^\beta)$. Hence we may write:

$$V = \frac{RT}{Z_I F} \ln \frac{C_o}{C_i} \qquad (12.4)$$

Equation 12.4 is the **Nernst equation**, one of the most important equations in neurophysiology.

The Nernst equation relates the electrical potential across a **permeable** membrane to the distribution of charged ions that it separates. Note that the membrane must be permeable to the ion under consideration. This, as we have just seen, is one of the premises on which the derivation of the equation is based. If it holds then it is possible to see that when the system is in equilibrium the electrical potential across the membrane exactly counterbalances any concentration differences.

Let us test the equation by supposing the concentrations of ion 'I' on both sides of the membrane to be identical. If we substitute in equation 12.4 we see that the logarithmic term becomes unity. The log of unity is zero. Hence the right-hand side of the equation goes to zero. Hence the Nernst equation predicts no electrical potential should be developed across the membrane. This, of course, is what is observed. When a cell dies the integrity of its membrane and its pumping mechanisms disappear. Ions flow along their concentration gradients until their concentrations equalise inside and outside the cell. The membrane potential vanishes.

Let us test the equation further by substituting the values for $[K^+]_o$ (i.e. 5.5 mM) and $[K^+]_i$ (i.e. 150 mM) which we quoted in Table 9.1. Then,

$$V_K = \frac{RT}{Z_I F} \ln \frac{5.5}{150}$$
$$= 0.027 \ln 0.036$$
$$= -0.089 \text{ V}$$

or $\qquad\qquad -89 \text{ mV}$

This is known as the **Nernst potassium potential**, V_K. Measurement of the actual resting potential across nerve cell membranes, V_m, usually gives values of -50 to -75 mV. The Nernst potassium potential is evidently markedly larger than this, but not too far out. It does suggest that the membrane

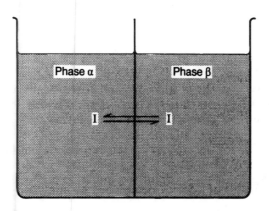

Figure 12.2 Permeable membrane separating two ionic solutions.

potential is caused, as the Nernst equation suggests, by the distribution of ions across it. If, however, values for the concentration of the other ions (Cl^-, Na^+, Ca^{2+}), are substituted in the equation, the predicted values for V_m are very far from what is observed. This is especially the case when the values for Na^+ are substituted.

The reasons for this are very simple. As we have seen in previous chapters cell membranes, especially nerve cell membranes, are very complex structures. Their permeability to different ions varies dramatically. And, as we emphasised above, the Nernst equation only works for ions that can pass unhindered through the membrane in question. It is known, however, that both sodium and chloride ions have very low permeability coefficients in the resting membrane.

Furthermore, V_m does not depend on the transmembrane distribution of a single ion species but on the distribution of several different types of ion: Na^+ and Cl^- (however low their permeability coefficients), as well as K^+. Thus, to gain a fuller understanding of the origin of the electrical potential across nerve cell membranes, we must generalise the Nernst equation. We must derive an equation that takes into account the different permeability of the membrane to different ions and the fact that there is not just one ionic species in play but many.

The equation we are looking for was developed by David Goldman and is consequently known as the **Goldman equation**. The equation is also sometimes known as '**the constant field equation**' because it assumes that the electric field across the membrane (the electrical potential gradient, V_m) remains unchanging. This, of course, is a large assumption to make. However, the Goldman equation provides a useful first approximation to the biophysical situation. It is written as follows:

$$V = \frac{RT}{F} \ln \frac{P_K[K^+]_o + P_{Na}[Na^+]_o + P_{Cl}[Cl^-]_i}{P_K[K^+]_i + P_{Na}[Na^+]_i + P_{Cl}[Cl^-]_o}$$
$$(12.5)$$

where P is the permeability constant of the ion concerned, square brackets indicate concentration of the ion either inside (subscript 'i') or outside (subscript 'o') the cell, and R, T and F have their usual connotations.

Note that whereas the **external** concentrations of the cations, K^+ and Na^+, appear in the numerator

of the equation, the **internal** concentration of the anion, Cl^-, is placed alongside them in this position.

Let us try some test runs on the Goldman equation. First if we make the permeability constants of Na^+ and Cl^- equal to zero, i.e. the membrane is completely impermeable to these ions, then the Goldman equation reduces to the Nernst equation for K^+. Similarly if we make $P_K = P_{Cl} = 0$ then the equation reduces to the Nernst equation for Na^+ and predicts V_{Na} as the potential across the membrane.

Now as we noted at the end of Chapter 11, neuronal membranes are not completely impermeable to any of the small inorganic ions found in the extracellular and intracellular compartments. Although cations and anions, being strongly hydrophilic, would find it next to impossible to make their way through the lipid bilayer of a plasma membrane, we saw in Section 11.2.2 that there are numerous '**leak channels**' through which such ions can pass. Moreover, the 'leakiness' of these channels varies markedly from one cell type to another. Neuroglial cells, for instance, seem to be more permeable to potassium ions than are neurons.

Neurons, as we noted above, resemble other cells in being much more permeable to K^+ than to Cl^- or Na^+:

$$P_K \gg P_{Cl} \cong P_{Na}$$

Let us put some figures to these relative permeabilities. Measurements of the flow of radiolabelled ions across plasma membranes gives the following values:

$$P_K = 1 \times 10^{-7} \text{ cm s}^{-1}$$
$$P_{Cl} = 1 \times 10^{-8} \text{ cm s}^{-1}$$
$$P_{Na} = 1 \times 10^{-8} \text{ cm s}^{-1}$$

Next, let us insert these permeability constants and the appropriate ionic concentrations (Table 9.1: cat motor neuron) into the Goldman equation:

$$V_m = 0.027 \ln \frac{(1 \times 10^{-7}[5.5]) + (1 \times 10^{-8}[150]) + (1 \times 10^{-8}[9])}{(1 \times 10^{-7}[150]) + (1 \times 10^{-8}[15]) + (1 \times 10^{-8}[125])}$$

$$= 0.027 \ln \frac{(55 \times 10^{-8}) + (150 \times 10^{-8}) + (9 \times 10^{-8})}{(1500 \times 10^{-8}) + (15 \times 10^{-8}) + (125 \times 10^{-8})}$$

$$= -0.055 \text{ V}$$

$$= -55 \text{ mV}$$

The value of $-55\,mV$ determined by application of the Goldman equation is quite close to the value of the resting potential across cat motor neurons actually observed by microelectrode recording.

Next, let us see what happens if we increase the potassium permeability by an order of magnitude. If we insert $P_K = 1 \times 10^{-6}\,cm\,s^{-1}$ into the equation, keeping all the other permeability constants unchanged we find:

$$V_m = -83\,mV$$

We have already remarked that the membranes of some glial cells are markedly more permeable to K^+ than the membranes of neurons. Hence we find that the V_m across these membranes is characteristically greater than the customary resting potential of neuronal membranes. Astrocytes, for instance, have V_ms ranging from -70 to $-90\,mV$. This larger than usual K^+ permeability is believed to be of considerable importance in mopping up excess K^+ which diffuses into the brain's intercellular space when neurons are active over appreciable periods of time (see Chapter 14). This mechanism can also be demonstrated to be at work in the retina. Large glial cells, known as Müller cells, take up excess K^+ generated in the nervous part of the retina in response to illumination and discharge it into the vitreous humour. Finally, as we noted in Section 11.2.2, the K^+ leak channels in neuronal membranes are under the control of a number of factors and so, in consequence, is V_m. V_m is, of course, sensitive to variations in the permeability constants of all the ions distributed across it. In Chapter 14 we shall look closely at what happens when P_{Na} suddenly increases.

Before completing this section it is worth noting that as it is much easier to measure the relative rather than the absolute permeabilities of ions the Goldman equation is often written in a slightly different form:

$$V = \frac{RT}{F} \ln \frac{[K]_o + b[Na]_o + c[Cl]_i}{[K]_i + b[Na]_i + c[Cl]_o} \qquad (12.6)$$

where
$$b = P_{Na}/P_K$$
and
$$c = P_{Cl}/P_K$$

We shall see in Chapter 14 that the chloride ion plays little or no part in the generation of action potentials. Hence when considering the ionic bases of action potentials equation 12.6 is often simplified to:

$$V = \frac{RT}{F} \ln \frac{[K]_o + b[Na]_o}{[K]_i + b[Na]_i} \qquad (12.7)$$

We shall see, however, that although the chloride ion is unimportant in action potentials it nevertheless plays a crucial role in the hyperpolarisation of inhibitory synapses. It is important to use the full form of Goldman's equation in these and similar circumstances.

12.3 ELECTROTONIC POTENTIALS AND CABLE CONDUCTION

Not all the electrical signalling in the nervous system is by way of action potentials, or impulses. Indeed it could be argued that some of the most important, if not *the* most important, of the central nervous system's communications depend upon non-impulse signalling. These signals, which are at least one order of magnitude and sometimes two or more orders of magnitude, weaker than action potentials have been termed **electrotonic potentials**. They are small depolarisations of a nerve process's membrane and are caused by the essentially passive spread of electrical current through the conducting fluids inside and outside nerve cells and their processes. Nonetheless, however small electrotonic potentials may be, they can have very considerable effect on the physiology of neuronal membranes and thus on the large-scale functioning of the brain. To see that this is the case we need only recall the sensitivity of some of the ion channels discussed in Chapter 11 to transmembrane voltage. In later chapters we shall see that the influx of, for instance, Ca^{2+} ions by the opening of voltage-dependent gates may lead to all sorts of dramatic consequences.

Consider Figure 12.3. Here the diagram shows a microelectrode inserted into a neural process and a small amount of current injected. This current flows down the process and leaks out through the membrane back into the bathing fluid to complete the circuit. If the process penetrated by the microelectrode is an axon the amount of current injected will, of course, have to be sufficiently minute not to

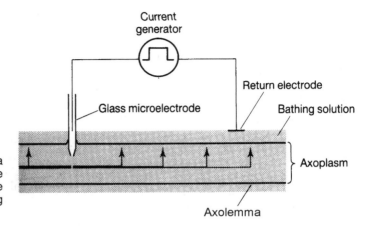

Figure 12.3 Electrotonic conduction in a neuronal process. The current injected by the microelectrode leaks back out through the axolemma to the return electrode in the bathing fluid.

open the voltage-dependent Na^+ gates (see Chapter 11) and thereby precipitate an action potential. If, however, the process is a dendrite or one of the short axons which are to be found in the brain's local circuit neurons (see Chapter 1) or one of the many types of neuron that are developed between the two plexiform layers in the retina then, because there are very few if any voltage-dependent sodium gates present, the current can be larger.

Because of its considerable importance the spread of electrotonic potentials and the **'cable conduction'** upon which it depends has been intensively studied. The conduction is referred to as 'cable conduction' because of its similarity to the transmission of current through long-distance telegraph cables. In both cases the injected current leaks out through the insulating sheath around the conductive core. In consequence some of the mathematical theory developed by telecommunications engineers can be applied to the neurophysiological problem. This theory can be mathematically somewhat fierce. Interested readers will find that titles listed in the Bibliography give exhaustive accounts.

Nevertheless, some brief account of electrotonic or cable conduction is necessary if the functioning of the nervous system is to be understood. We have already seen in Chapters 10 and 11 how ligand- or voltage-controlled gates in the neuronal membrane allow flows of ions into and out of the cell. These flows carry electric current and depolarise or hyperpolarise the membrane. We shall see in the chapters to come that local circuits spreading from these membrane patches are responsible for switching on

or off subsynaptic cells and for the propagation of action potentials.

In order to develop a mathematical framework within which electrotonic conduction can be discussed it is necessary to make some simplifying assumptions at the outset. It is necessary to assume that the following parameters remain constant along the length of the process:

1. r_o: the longitudinal electrical resistance of the extracellular medium;
2. r_i: the longitudinal electrical resistance of the intracellular medium, i.e. the axoplasm;
3. r_m: the transverse electrical resistance of the membrane, i.e. the neurolemma;
4. c_m: the electrical capacity of the membrane.

Next consider Figure 12.4. A current is injected into the neural process at point 'x_0' so that the membrane is depolarised to a value V_0. Two further simplifying assumptions have to be made here. First, that the injected current remains constant for the duration of the experiment. Second, that the nerve process is 'infinitely' long. When these assumptions are made Figure 12.4 shows that the current spreads down the process through the conductive axoplasm or dendroplasm, leaking out through the neurilemma until it attenuates to zero. Our question is: what is the value of the electrotonic potential, V_x, at some point 'x_1'?

It can be shown that the appropriate cable equation (given the above conditions) is:

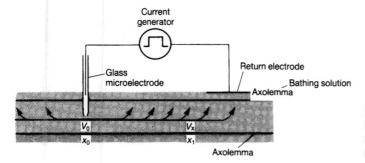

Figure 12.4 Electrotonic conduction. Current is injected at x_0 to induce a voltage across the membrane of V_0. What is the electrotonic potential, V_1, at some point x_1?

$$V = \lambda^2 \frac{d^2 V}{dx^2} \qquad (12.8)$$

where λ is (as we shall see below) a 'space constant'.

Equation 12.8 can be solved to give the following answer to our question:

$$V_x = V_0 e^{-x/\lambda} \qquad (12.9)$$

As usual let us test this equation. When $x \to 0$, $e^{-x/\lambda} \to e^0$, and $e^0 = 1$. Hence $V_x = V_0$, as it should.

Next let us give a little attention to the space constant, λ. If we go back to the derivation of the cable equation (see, for instance, Aidley's text listed in the Bibliography) we shall find that it depends on the three electrical resistances mentioned at the beginning of this discussion:

i.e.

$$\lambda^2 = \frac{r_m}{r_o + r_i} \qquad (12.10)$$

Next let us put $x = \lambda$ in equation 12.9. Then,

$$V_x = V_0 e^{-1} \qquad (12.11)$$

Equation 12.11 shows that λ **is the distance from the point at which current is injected to the point at which the electrotonic depolarisation has fallen to e^{-1}, i.e. 0.37, of its original value (V_0).**

λ is a useful parameter for comparing the spread of electrotonus along neuronal processes of different types and size. Both the length and the diameter of the process are significant. Let us briefly look at each in turn.

12.3.1 Length

The graph in Figure 12.5 shows how the electrotonic potential falls off with distance from the point at which depolarising current is injected. The lower curve represents the case for a process of 'infinite' length (as in the example we have just discussed). In practice this means any process more than three times longer than λ. This is the solution to the cable conduction equation described above. It is clearly applicable to long undersea telegraph cables. But in many neural instances the process under consideration may be comparatively short. The upper curve in Figure 12.5 shows the case where λ is about the same length as the entire process. When this is the

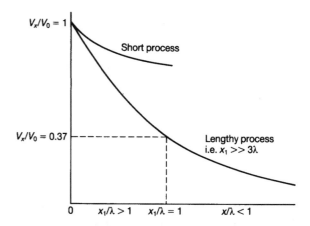

Figure 12.5 Electrotonic conduction in processes of different length. Current is injected at x_0 so that the process is depolarised from a resting potential of $-60\,\text{mV}$ to some value, V_0. The length constant, λ, is the distance from x_0 at which the electrotonic potential (V_x) across the membrane has fallen to $1/e$ (i.e. 37%) of its value at x_0. Further explanation in text.

case, as the figure shows, the curve does not in fact fall away asymptotically to the abscissa (as it does for the case of infinite length) but declines much less steeply, perhaps never diminishing by more than 0.75 of the input voltage. This latter instance is of considerable significance for the spread of receptor potentials in receptor cells, a phenomenon we shall discuss in the next chapter.

12.3.2 Diameter

As the diameter of a process increases, the magnitude of λ also increases. Let us see why this is the case. First of all we can note that while the longitudinal internal resistance, r_i, decreases as the radius of the process increases, the longitudinal external resistance, r_o, remains unaltered. Because r_o is small compared to r_i we can simplify equation 12.10 to:

$$\lambda^2 \cong \frac{r_m}{r_i} \qquad (12.12)$$

Now

$$r_m = \frac{R_m \ \Omega \ \mathrm{cm}^2}{2\pi a \ \mathrm{cm}}$$

where R_m is the resistance of unit area of membrane and a is the radius of the neuronal process

and

$$r_i = \frac{R_i \ \Omega \ \mathrm{cm}}{\pi a^2 \ \mathrm{cm}^2}$$

where R_i is the resistivity of unit cube of cytosol.

Hence

$$\lambda \cong \sqrt{\frac{R_m}{2\pi a} \times \frac{\pi a^2}{R_i}} \ \mathrm{cm}^2$$

$$\cong \sqrt{\frac{R_m}{R_i} \times \frac{a}{2}} \ \mathrm{cm}^2 \qquad (12.13)$$

It is clear from equation 12.13 that, whilst R_m and R_i remain constant, λ will increase as the diameter of the process increases. In Chapter 14 we shall see that action potentials depend upon underlying cable conduction. Hence we can see from the above analysis why it is that so many invertebrates have developed **giant axons**. The larger the diameter the greater the spread of the electrotonic conduction

and hence, other things being equal, the more rapid the impulse propagation. Invertebrates develop giant fibres in order to respond rapidly to an emergency. Conduction along the squid's giant fibre normally leads to the vigorous tail flexure that propels the cephalopod backwards out of danger. In Chapter 14 we shall see that the vertebrates have evolved a different means of increasing impulse propagation rate – **myelination**.

The above analysis is also of particular importance to our understanding of the biophysics of dendrites.

Let us consider an example. Some of the large apical dendrites springing from the pyramidal cells in the cerebral cortex may be up to 10 μm in diameter. If we take

$$R_m = 2500 \ \Omega \ \mathrm{cm}^2$$
$$R_i = 70 \ \Omega \ \mathrm{cm}$$
$$a = 0.0005 \ \mathrm{cm}$$

It follows that $\lambda = \sqrt{\dfrac{2500 \ \Omega \ \mathrm{cm}^2}{70 \ \Omega \ \mathrm{cm}} \times \dfrac{0.0005 \ \mathrm{cm}}{2}}$

$$\approx 0.1 \ \mathrm{cm} \ \mathrm{or} \ 1 \ \mathrm{mm}$$

A similar calculation for an extremely fine dendrite ($d = 0.1$ μm) yields a value for λ of about 100 μm or 0.1 mm. In other words a decrease of two orders of magnitude in diameter leads to a decrease of only one order of magnitude in λ. These values for λ are of considerable interest as they fit well with the magnitudes of dendritic processes that neurohistologists observe in the brain.

Before leaving the topics of electrotonic potentials and cable conduction it is worth noting a further important parameter: time. In our discussion so far we have assumed (as noted at the outset) that the depolarising voltage has been constant throughout the experiment. In neurophysiological reality, however, synaptic depolarisations (or hyperpolarisations) are often brief, transient events. In order to study the consequences of such transient electrotonic potentials we have to take into account the electrical capacitance of the membrane. This determines the time taken to build up electrical charge and/or to evacuate that charge. The time taken to reach e^{-1} of the final voltage is called the **charging time constant** or, alternatively, the **whole neuron**

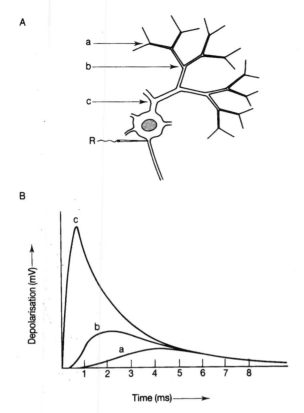

Figure 12.6 Transient electrotonus exhibited at the initial segment of the axon in response to depolarising synaptic events at different points on a dendritic tree. (A) A depolarising current is injected at a, b or c on an idealised dichotomously branching dendritic tree. (B) The graph shows the depolarisation recorded at the initial segment of the axon by recording electrode, R, to each dendritic stimulation. After Shepherd (1979), *The Synaptic Organisation of the Brain*, 2nd edn, Oxford: Oxford University Press.

constant (τ). If the size and branching characteristics of a dendritic tree are taken into account it becomes possible to compute the transient electrotonic potentials exhibited by a patch of membrane in the perikaryon or, more significantly, by a patch on the initial segment of the axon, when a synapse is activated on the dendritic tree. The result of one such computation is shown in Figure 12.6.

Figure 12.6 shows that a synaptic event far out on a dendritic tree has a very different effect on the perikaryon from one close in. Depolarisations close to the perikaryon can have a large enough effect on the initial segment to induce an impulse. Vice versa, events far out on the tree may not depolarise an initial segment sufficiently to initiate an impulse but they may **bias** that membrane towards activation for tens of milliseconds.

The reader will be aware that the foregoing account describes a detailed mathematical theory the full exposition of which must be left to biophysical texts. It will also be clear that the theory is highly abstract. The mathematician's neuron is very different from the sloppy multitude of variegated forms with which the neurobiologist is familiar. Nevertheless the theory provides a background for investigation of the real brain. Indeed, recent research on the dendrites of rat hippocampal neurons using voltage-sensitive dyes has gone far towards providing a 'reality check' of the mathematics in the wetware of the brain. The spread of electrotonic potentials turned out to be very much as the theoreticians predicted. In the final analysis, however, the theory might best be compared to Galileo's ideal world of frictionless pulleys and inclined planes which, although very different from our common experience of the nature of things, nevertheless proved indispensable for the development of a genuine physical science.

12.4 CONCLUSION

In this chapter we have examined the biophysical causes of the resting potentials across membranes which the neurophysiologist's microelectrode detects. We have also looked at the way in which minute electrical currents may spread through neuronal processes in response to tiny membrane depolarisations. In the next chapter we shall consider a selection of sensory cells and note that in all the metazoan cases environmental energies are transduced to alterations in membrane polarity which lead, in turn, to just the sort of electrotonic or cable conduction we have discussed. In Chapter 14 we shall make use of the same biophysical concepts to elucidate the bases of the action potential upon which the nervous system depends for its long-range signalling.

13

SENSORY TRANSDUCTION

Classification: entero- and exteroreceptors – chemo-, mechano-, photoreceptors. General features: stimulus detection and transduction – specific nerve energies – unifying role of molecular biology – sensory and neurosensory cells – receptor and generator potentials. **Chemoreceptors**: **prokaryocytes**: bacterial chemotaxes – flagellar motion – genetic and molecular analysis – binding proteins – R-T proteins – signal transduction – sensory adaptation – prospects for complete molecular understanding of a sensory-motor system; **vertebrates**: **olfaction**: olfactory epithelium – olfactory neurons – olfactory cilia – receptor and binding proteins – specialist or generalist? – olfactory genes encode olfactory receptor proteins (ORPs), 7TM receptors – control CNG ion channels – isolation of CNG channels – resemble voltage-gated channels – patch-clamping – biophysics – generator potentials in olfactory cell – graded responses – adaptation – inherited anosmias; **gustation**: molecular mechanisms – gustducin – receptor potentials. **Photoreceptors**: heterogeneity – vertebrate rods and cones – outer segments – embryology – ultrastructure – location of rhodopsin – ion channels and 'dark current' – hyperpolarisation on illumination – rod-bipolar synapses – sign-inverting and sign-conserving – rhodopsin kinase, arrestin and recoverin induce sensory adaptation – amplification cascade induced by single photon. **Mechanoreceptors**: heterogeneity – evolutionary antiquity – tension detectors; prokaryocytes – mechanosensitive ion channels cloned in *E. coli* MscL and MscS channels – MscL cloning – biophysics – structure–function analyses – crystallography of MscL in *M. tuberculosis* (TbMscL) – atomic structure – function – role of stretch detectors in bacterial life; *C. elegans* – touch sensitivity – mec mutations – touch-detector neurons – genetic analysis of stretch receptor; **Mammals** – membranous labyrinth – cochlea – hair cells – stereocilia and kinocilia – biophysics of stereocilia – receptor potentials – ion channels – analogies with *C. elegans* – adaptation – microphonic potentials – physiology of frequency detection. **Conclusion**: sensory transduction only the beginning of the sensory process

According to the empiricist tradition there is nothing in the mind which was not first in the senses. The detection and transduction of happenings in the environment has always been of great interest to scientists and philosophers. In this chapter we shall look at some representative examples of these processes at the molecular level.

The animal kingdom has developed an overwhelming variety of different sense organs. Let us make a start by looking at two different (but not mutually exclusive) classificatory schemes. Sense organs may be subdivided on the basis of whether they 'look' inside at the internal environment (**enteroreceptors**) or outwards at the external environment (**exteroreceptors**). Or, secondly, they may be classified on the basis of the type of stimulus to which they are most responsive. In this chapter we shall concentrate on the second classification. We shall accordingly look at some examples of

chemoreceptors, **photoreceptors** and **mechano-receptors**.

Before we begin it will be well to outline some general features shared by all sensory systems. Sensory cells have the job not only of detecting but also of **transducing** the impinging stimulus into a signal in the organism's internal 'language'. In multicellular forms that have developed a nervous system this signal is ultimately an action potential in a sensory nerve fibre. As all action potentials are very much alike the central nervous system can only tell the type of external energy (chemical, electromagnetic, mechanical) by taking note of which fibres are carrying impulses.

The initial transduction is not, however, directly into nerve impulses. The first steps remain firmly in the realm of molecular biology. Indeed in the first example we look at in this chapter they never get further than that level. Bacteria (being small) have no need to develop a nervous system. This does not mean that they lack senses. We shall see that many of them have well-developed chemosensory devices. But even in multicellular forms with well-developed nervous systems we shall see that molecular biology still intervenes between signal detection and nerve impulse.

This molecular biology results, as we shall see, in changes in the electrical polarity across the sensory cell's plasma membrane. These changes, which may be hyperpolarisations or, and more usually, depolarisations, are called **receptor potentials**. In the case of olfactory reception (the second example in this chapter) we shall see that the sensory cell also conducts action potentials. It is a **neurosensory** cell, having the dual function of sensory cell and sensory axon. In this case the receptor potential leads directly to an action potential. It is thus better known as a **generator potential**.

In our final three examples, mammalian gustatory cells, photoreceptor cells and hair cells, the receptor potentials developed in response to the appropriate stimulus do not lead directly to action potentials. The sensory nerve fibre, in these cases, is separated by a synapse from the sensory cell. The receptor potentials in these cases thus lead to the release of a neurotransmitter which falling on the underlying dendrite of the sensory fibres leads to a generator potential in this latter process. This in turn initiates an action potential in the sensory fibre.

We have noticed throughout the preceding chapters of this book that molecular neurobiologists are slowly uncovering a remarkable uniformity of design at the molecular level. We shall see in this chapter that this underlying unity is also becoming apparent in the structure and function of the sensitive regions of sensory cells.

13.1 CHEMORECEPTORS

13.1.1 Chemosensitivity in Prokaryocytes

Chemosensitivity evolved very early in the history of life on earth. We may speculate that once having hit upon a satisfactory mechanism there was little pressure to change it. It is thus not absurd for neurobiologists to follow their interest into organisms that very definitely lack all vestige of a nervous system: the bacteria.

There is no doubt that motile bacteria are sensitive to the chemical substances in their environment. It has long been known that they will swim up a concentration gradient of an attractant chemical. They can both sense the material in the environment and act on the sensation. How do they do it?

Motile bacteria are propelled by flagella. Prokaryotic flagella are far simpler than those of the eukaryotes. They consist of a single tubular array of flagellin subunits twisted into a helix (Figure 13.1). They also have a very different mechanism for producing motion. Instead of the sliding-filament system believed to be at work in eukaryotic flagella they have the distinction of being the only organic structures so far known to employ a rotary mechanism. The cellular end of the flagellum is rotated at about 100 revolutions per second by a mechanism energised by a transmembrane hydrogen ion gradient.

Most flagellated bacteria have more than one flagellum. *Escherichia coli* has five to ten. When they all rotate in the anticlockwise direction the bacterium swims forward smoothly. Things, however, are very different if they rotate in the clockwise sense. In this case, because of the helical structure of the flagellum, the flagella all pull outwards, resulting in an irregular 'tumbling' motion (Figure 13.2).

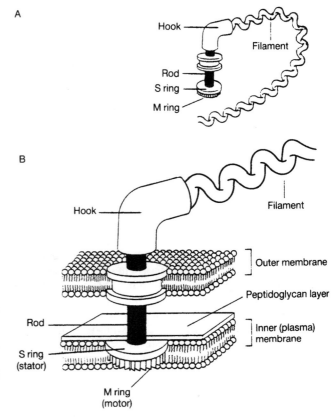

Figure 13.1 Rotary mechanism of a bacterial flagellum. The mechanism penetrates both the outer and inner membranes surrounding the bacterium. Energy derived from a proton gradient causes the 'M ring' (or motor) to rotate relative to the 'S ring' (or stator) at about 100 revolutions per second. The stator is embedded in the peptidoglycan layer. A rod links the M ring to a hook and then to a helical flagellar filament. A 'bearing' in the outer membrane acts as a seal. From Adler (1976), in Goldman, Pollard and Rosenbaum (eds), *Cell Motility*, Cold Spring Harbor, NY: Cold Spring Harbor Laboratory; with permission.

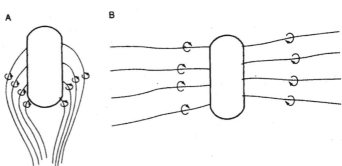

Figure 13.2 Anticlockwise and clockwise rotation of bacterial flagella. (A) Anticlockwise rotation. The flagella stream together as a single bundle which propels the bacterium forwards. (B) Clockwise rotation. The flagella each pull away from the bacterium in the direction of the straight arrows. According to the varying strength of the pull from each flagellum the bacterium veers from side to side and tumbles hither and thither.

If the swimming of a bacterial cell is followed under a microscope it will be seen to consist of a series of smooth 'runs' (several seconds) interspersed with episodes (about 0.1 s) of chaotic tumbling. When the cell comes out of its tumble it will set off smoothly again but in a completely random direction. If, however, the cell is placed in a gradient of chemical attractant it is found that when the swimming is in the direction of the source fewer tumbles occur than when it is moving in any other direction. The net result is that the bacterium migrates up the concentration gradient towards the source of the attractant (Figure 13.3).

Clearly this phenomenon provides a valuable system for the investigation of the general problem of chemoreception. The flagella of a bacterium may

A

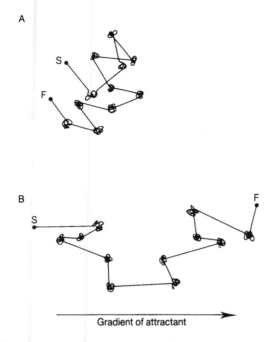

B

Gradient of attractant

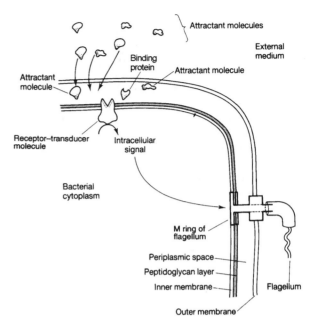

Figure 13.3 Bacterial migration along a concentration gradient of chemical attractant. (A) Bacterial path in the absence of an attractant. The swim alternates between periods when the flagella rotate anticlockwise giving a smooth straight 'run' and short periods when the flagella rotate clockwise causing 'tumbling'. The direction in which the bacterium is facing when tumbling ceases is completely random; hence the next run will be in a random direction. The bacterium makes no progress. S=start; F=finish. (B) Bacterial path in the presence of an attractant. The length of time between tumbles is increased when the bacterium is moving up the gradient and decreased when it is moving in the opposite direction. Hence, although the tumbling episodes still randomise the direction in which the runs occur, the bacterium nevertheless swims up towards the source of the attractant. S=start; F=finish.

Figure 13.4 Interaction of attractants with binding proteins and receptor-transducer molecules in bacterial membranes. Chemical attractants (shown in the external medium) are of two types. Either they fit the active site of the receptor-transducer molecule directly or they are 'adapted' to do so by first fitting a binding protein in the periplasmic space. The receptor-transducer molecule generates a signal (see Figure 13.6) which affects the direction of rotation of the flagellar motor and hence frequency of tumbling. Further explanation in text.

be attached to a glass slide and the rotation of the cell (about 10 Hz) observed in response to various chemicals. The genetics of *E. coli* are, of course, very well known so the sensory system may also be examined by genetic techniques.

It has been shown that detection of attractant (or repellent) molecules is by way of transmembrane proteins. There are two mechanisms. Either an attractant molecule interacts with a **binding protein (BP)** (sometimes known (confusingly) as a receptor molecule) in the bacterium's periplasmic space,

causing a conformational change so that it 'fits' one or other of the transmembrane proteins. This has been shown to be the case with sugars and peptides. **Galactose binding protein (GBP), ribose binding protein (RBP), maltose binding protein (MBP)** and **dipeptide binding protein (DBP)** have been isolated. Or, in the case of amino acids, the attractant interacts with the transmembrane protein directly. The transmembrane proteins have the ability to affect the rotation of the flagella. They are, in other words, able to transduce the signal represented by the attractant molecule as well as detect it. In order to distinguish them from the binding proteins in the periplasmic space they are known as **receptor-transducer (R-T) proteins** (Figure 13.4).

This system has one other very important feature. It shows **sensory adaptation**. This, as we have seen, is a feature of all sensory systems. Indeed, we have already met it in the guise of desensitisation when considering the receptor proteins of Chapters 8, 10 and 11. In the case of *E. coli* it is found that a sudden immersion in attractant suppresses tumbling. But after a time it begins to be shown again and, in a few minutes, returns to its normal frequency. This adaptation, moreover, is restricted to the specific attractant molecule. Addition of a different attractant inhibits the tumbling in the usual way. The biological significance of adaptation or desensitisation is obvious. Because its molecular biology is so well known *E. coli* provides a valuable system for examining all these processes.

Binding Proteins

The detailed molecular structure of a number of binding proteins has been determined. The **galactose binding protein (GBP)** has, for instance, been successfully subjected to high-resolution X-ray diffraction studies and its overall architecture and binding site configuration solved. It has been shown to consist of two 'wings' separated by a crevice which contains the binding site. When the site is occupied by the substrate – galactose or glucose – the wings 'close' by moving towards each other through an angle of 18°. This alteration in three-dimensional conformation enables the binding protein to interact with the appropriate transmembrane R-T molecule. There is evidence that other binding proteins have a similar three-dimensional conformation.

Receptor-transducer (R-T) Proteins

The genes for a number of the R-T proteins in *E. coli* have been isolated by the techniques of molecular genetics and their nucleotide sequences determined. The best known of these genes are the *tsr, tar, trg* and *tap* genes. They encode four R-T proteins: **Tsr**, sensitive to serine (attractant) and leucine (repellent); **Tar**, sensitive to aspartate and maltose (attractants) and Co^{2+} and Ni^{2+} (repellents); **Trg**, sensitive to galactose and ribose attached to receptor proteins (attractants); and **Tap** sensitive to dipeptides attached to receptor proteins.

From the nucleotide sequences the amino acid structure of these R-T proteins can be determined

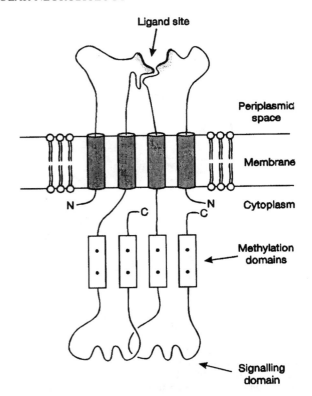

Figure 13.5 Bacterial receptor-transducer molecule. The two subunits of the dimer are shown; each has two transmembrane domains and each contributes to the ligand site. Methylation and signalling domains are also shown.

and hydropathic analysis performed to predict membrane-spanning segments. All four of the proteins consist of over 500 amino acids but appear to have only two transmembrane sequences. There is strong homology between the four, especially in the cytoplasmic domain where the processes of signalling to the flagellar apparatus and sensory adaptation occur.

Figure 13.5 indicates that the R-T molecule is held in the membrane so that an amino acid sequence of about 150 residues projects into the periplasmic space. This forms the attractant and/or repellent binding site. The major part of the molecule from about residue 215 to the carboxy-terminal end lies in the bacterial cytoplasm. This part of the molecule is highly conserved across the four proteins studied, generates the signal to the flagellar apparatus and is involved in the adaptation mechanism. The fully formed R-T protein is a dimer built of two of these 50 kDa polypeptides.

Signal Transduction

Chemotactic signalling is initiated by the binding of the attractant molecule or of the attractant-binding molecule complex to the periplasmic domain of the R-T molecule. There is evidence that this causes a conformational change which is transmitted through the membrane to alter the configuration of the cytoplasmic domain. This altered cytoplasmic domain then initiates a biochemical signalling cascade which ultimately affects the activity of the flagellum.

Genetic analysis has allowed the elements of this signalling pathway to be isolated and analysed. It consists of four significant proteins: **CheA, CheW, CheY** and **CheZ** ('Che' for 'chemotactic'). CheA and CheY transmit the 'tumble' signal from the receptor-transducer to the flagellum (Figure 13.6). When CheY attaches to the flagellar motor it causes the flagellum to rotate in a clockwise direction thus inducing tumbling. This pathway is inhibited when an attractant molecule binds externally to the binding protein and the complex so formed attaches to an appropriate R-T molecule. When, however, a repellent binds to the binding protein, the opposite effect occurs: the receptor-transducer is activated, the pathway to the motor is, in its turn, activated and the flagellum turns in a clockwise sense. Hence in the presence of an attractant the bacterium will move up the concentration gradient and in the presence of a repellent it will continue tumbling. When neither attractant nor repellent is present a balance of activation and deactivation ensures that smooth runs are interspersed with tumbles.

Figure 13.6 shows the biochemistry responsible for activation and deactivation. When a repellent attaches to an R-T protein the cytoplasmic domain of the latter binds CheW and induces the self-phosphorylation of CheA. CheA's phosphate group is rapidly passed on to CheY. It is only the

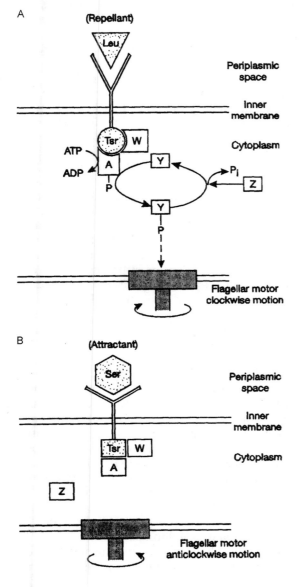

Figure 13.6 Molecular biology of *E. coli* sensorimotor system. (A) The Tsr R-P protein accepts a repellent molecule (Leu). CheW and CheA are activated. CheA accepts phosphate from ATP and passes it on to CheY. CheY diffuses to the flagellar motor and induces a clockwise rotation and hence tumbling. CheY is eventually dephosphorylated by CheZ. (B) The Tsr R-P protein accepts an attractant molecule (Ser). The consequent conformational change inactivates CheA and CheW so that CheY remains unphosphorylated and consequently inactive. The flagellum resumes its anticlockwise rotation and the bacterium moves smoothly forward. Data from Bourrett, Borkovich and Simon (1991), 'Signal transduction pathways involving protein phosphorylation in prokaryotes', *Annual Review of Biochemistry*, **60**, 401–441.

phosphorylated form of CheY that can induce clockwise motion in the flagellum. The final protein in the system, CheZ, eventually dephosphorylates CheY and thus terminates its influence.

Sensory Adaptation

Sensory adaptation occurs when the bacterium finds itself in an unvarying concentration of attractant. This is due to methylation of the R-T protein. For this reason these proteins are sometimes referred to as methyl-adapting chemotaxis proteins (MCPs). That adaptation is due to methylation was shown by the isolation of mutants in which methylation domains of the R-T proteins (Figure 13.5) were **inactivated**. Such mutants did not adapt; in the presence of an attractant they would cease tumbling for hours, even days, instead of for only a minute or so. It appears, therefore, that when an R-T protein is deactivated by binding an attractant it exposes extra methylation sites. An enzyme, methyl transferase, attaches up to eight methyl groups. This increases the activity of the R-T protein again and, by the signalling system described above, causes tumbling to commence once more.

The great virtue of this system lies in the fact that the genetics of *E. coli* are so well known and so easily manipulated. The same system is at work in *Salmonella typhimurium*. The fact that these two bacteria (one a parasite within the gut of mammals, the other in the eggs of birds) have evolved separately for at least 150 million years indicates that similar chemosensitive mechanisms are probably widespread in the bacteria. There are good prospects for following up the successful genetic analysis of the R-T proteins and the cytoplasmic signalling system to provide a complete understanding of bacterial chemotaxis. This would provide the first instance in which a complete sensory-motor system has been run into its molecular ground.

13.1.2 Chemosensitivity in Vertebrates

Let us now turn from chemosensitivity in one of the simplest of living forms to chemoreception in some of the most advanced. In this section we shall look briefly at mammalian olfactory and gustatory reception.

Olfactory reception is carried out by the olfactory epithelium in the nasal cavity and vomeronasal organ. The area that this epithelium covers in the nasal cavity ranges from a few square centimetres (human) to well over a hundred square centimetres (dog). It consists of three types of cell: **supporting** (glia-like) cells, which secrete mucus; **neurosensory** cells or **olfactory neurons**; and **basal** cells, which appear to be stem cells capable of dividing and forming new functional neurons throughout life. Olfactory neurons are, with the exception of small populations of stem cells (see Section 19.2), the only neurons in the mammalian body that renew themselves throughout life.

Let us concentrate our attention on the olfactory neurons. Figure 13.7 shows that they are bipolar cells with a single unbranched dendrite which squeezes up between the supporting cells to end in a small swelling – the olfactory knob. From the knob project up to 20 lengthy cilia. These cilia carry the sensory membrane of the olfactory neuron. Because the olfactory neuron has the dual function of detecting the stimulus and transmitting the nerve impulse into the brain it is in fact a neurosensory cell. Custom and practice, however, ensure that it is normally called a sensory neuron.

The ultrastructure of olfactory cilia is not greatly different from that of other cilia. They contain the usual internal axoneme but are, in mammals at least, non-motile. They are also unusually long and thin – ranging from 30 to 200 μm in length but often only 0.1–0.2 μm in diameter. This adaptation of a motile organelle, the cilium, to serve a sensory function is also found (as we shall see) in photoreceptors (rod and cone outer segments) and mechanoreceptors (kinocilia of hair cells).

The bunch of thin lengthy cilia springing from an olfactory bulb undoubtedly increases the sensory surface area dramatically. Freeze-fracture electron microscopy shows, furthermore, that the membrane of each cilium contains a high density of globular particles. It is believed that these are the **olfactory receptor proteins (ORPs)**. This is another point of analogy with the rod and cone cell outer segments we shall consider in the next section. The membranes of outer segments are also densely populated with receptor molecules – in this case rhodopsins and/or iodopsins.

The olfactory cilia lie in the mucus secreted by the supporting cells where they form a dense mat.

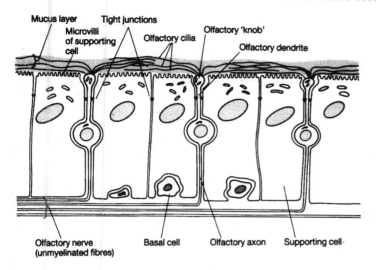

Figure 13.7 Structure of olfactory epithelium. The olfactory cilia are usually very lengthy and lie over the surface of the epithelium embedded in the mucus. Tight junctions between the epithelial cells and between the epithelial cells and the olfactory neurosensory cells prevent any penetration of the intercellular space.

The sensitivity is acute. Humans can detect airborne odorants within the range 10^{-4} M to 10^{-13} M. One of the signs of a bad head cold is a marked falling-away of olfactory sensitivity. This is because the olfactory cilia become engulfed in the extra amount of mucus produced by the supporting cells.

Normally, however, the mucus layer plays a role in olfaction. It contains numerous small (20 kDa) proteins. They have been compared to the **binding proteins** which (as we saw above) play a significant role in bacterial chemoreception. However, in the mammalian case, these proteins are found not only in the olfactory but also in the respiratory epithelium. Hence, they probably do not play quite as significant a role as their analogues in bacteria (and, indeed, in insects) play. It is likely that they have very broad affinities and merely capture and present odorant molecules to the receptor proteins in the olfactory cilia.

The olfactory system is able to detect and discriminate between a huge variety of odorants. For many years it was a matter of controversy whether a cilium is a generalist or a specialist, Does it, in other words, express just one type of ORP or several? The evidence now suggests that the cilia are specialists. This, however, is not to say that a cilium recognises only one type of odorant molecule. On the contrary, it has been shown that a given ORP responds to a number of chemically related odorant molecules. Vice versa, a single type of odorant molecule is recognised by a number of different ORPs. Combinations of receptor activations signal the odour to the brain along the olfactory nerve (via the olfactory bulb). In this way the olfactory system is rather like the words of a language or the notes of music. Combinations of words or of notes create infinite number of different literary or musical works.

A large family of olfactory receptor genes (over 1000) has been identified in the last few years. As the human genome consists only of some 32 000 genes it follows that the olfactory receptor proteins take up about 3% of all human genes. They are spread over all our chromosomes with the exception of chromosome 20 and the Y chromosome. The large number of genes and their wide chromosomal representation indicates the vital importance of the olfactory sense. In fact, this importance relates more to infra-human mammals than to ourselves and the other primates. Our sense of smell is greatly impoverished compared with the majority of mammals. This is shown by the fact that more than 60% of our olfactory genes are in fact **pseudogenes**; only about 350 are functional.

The olfactory proteins are 7TM 'serpentine' receptors and belong to the same superfamily as the 7TM receptors we examined in Chapter 8. This relatedness means that an understanding of the molecular biology and pharmacology of well-known 7TM receptors, for instance the β-adrenergic receptor (see Section 8.2), can give insight into the *modus operandi* of the receptors in the olfactory cilium. It will also be remembered that rod and cone opsins also belong to this superfamily. The molecular biology of the outer segment thus finds parallels in the olfactory cilium. We shall examine it and note the various common features in Section 13.2.

It will be recalled that the three-dimensional structure of the 7TM proteins resembles that of a **'barrel of staves'**. It is believed that odorant molecules come to occupy the interior of the barrel. This interior is designed to fit different odorants in different olfactory receptor proteins. It must be borne in mind, however, that whereas most G-protein-coupled 7TM receptors have rather precise affinities with their ligands, the olfactory receptor–odorant fit must be much looser. For, as mentioned above, a single olfactory receptor recognises, with varying degrees of certainty, a range of odorant molecules.

The 7TM receptor in the olfactory cilium is (as usual) linked via a G-protein system to a membrane-bound effector. This, in most cases, is adenylyl cyclase (AC) which, as usual, catalyses the production of cAMP from ATP. It has recently become apparent that other second messengers are involved to a lesser extent: in particular cGMP and IP₃. The second messenger controls an ion channel (Figure 13.9). Because the channel is controlled by a cyclic nucleotide it is called a **cyclic nucleotide-gated channel (CNG channel)** (see Box 11.1). We shall meet other examples of CNG channels in later parts of this chapter. Although the olfactory CNG channel was not the first to be isolated (that honour goes to rod cell channels), it is probably the best characterised.

Isolation of CNG channel subunits was first achieved by what has come to be traditional methodology. First a 63 kDa protein was purified from bovine rod and cone outer segments. When the protein was inserted into an artificial lipid bilayer, single channel currents could be detected in response to cGMP. Next a bovine retinal cDNA library was probed by oligonucleotides representing sections of this protein. A clone coding for a 690-residue protein was isolated. When this was expressed in *Xenopus* oocyte it formed a functioning cGMP-gated channel. Using sequences derived from this protein, numerous other CNG channels

have been isolated. In particular they have been isolated from olfactory epithelia.

As we saw in Box 11.1, hydropathic analysis of CNG channel subunits shows that they bear a marked resemblance to the voltage-gated channel subunits. They have six transmembrane domains (Figure 13.8). S4 is strongly homologous to the S4 voltage sensor of voltage-gated channels. There is also a 'hairpin' between S5 and S6. It will be recalled that a similar, so-called H5 domain exists in voltage-gated channels where it is believed to control the type of ion allowed through the channel.

Four subunits form the CNG channel. There are at least two different subunit types, α and β, and both exist as a number of subtypes. The β-subunits have a large C-terminal cytoplasmic domain which binds the cyclic nucleotide second messenger. The human CNG channel consists of two α- and two β-subunits. When cAMP attaches to the β-subunits the channel is activated.

The exceedingly delicate task of patch-clamping an olfactory cilium has been accomplished and conductance channels have been demonstrated which open in response to activation of the CNG channels by cAMP. Because there are about 2400 channels μm⁻² on cilia compared with only about 6 channels μm⁻² on the olfactory knob and dendrite, studies of single channels have usually used the latter regions of the olfactory cell. Channels in these regions have been shown to have very similar biophysical characteristics to those on the cilia.

In the absence of Ca²⁺, olfactory CNG channels respond to cAMP with a conductance of about 45 pS. Ca²⁺ and other divalent cations reduce this conductance to about 1.5 pS. These conductances

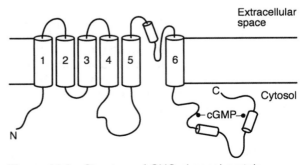

Figure 13.8 Structure of CNG channel protein.

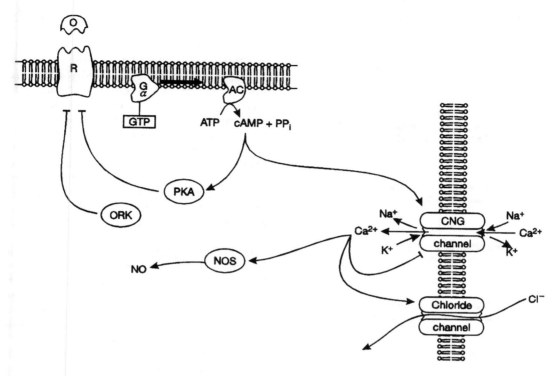

Figure 13.9 Molecular biology of olfactory reception. The figure should be compared with those in Section 8.4. A collision-coupling mechanism results in the synthesis of cAMP. This opens the CNG-channel so that Na$^+$, K$^+$ and Ca^{2+} can flow down their electrochemical gradients. Ca^{2+} has a number of effects as shown. Further explanation in text. AC=adenylyl cyclase; G=G$_{olf}$; NOS=nitric oxide synthase; O=odorant molecule; ORK=olfactory receptor kinase; PKA=protein kinase A; R=receptor protein. Adapted from Zufall *et al.* (1994), *Annual Review of Biophysics and Biomolecular Structure*, **23**, 577–607.

can also be obtained by the application of cGMP. This was at first surprising for it was not thought that cGMP was present in olfactory cilia. More recently, as indicated above, cGMP has been shown to exist as a minor second messenger. It may be, also, that cGMP is the outcome of Ca^{2+}-induced formation of NO from NO synthase (Figure 13.9). We shall see in Section 16.9 that the major target of nitric oxide is soluble guanylyl cyclase, which it activates to catalyse the formation of cGMP from GTP.

The biochemical 'cascade' basic to the response of an olfactory cilium is shown in Figure 13.9. The receipt of an odorant molecule (O) by the olfactory receptor protein (R) leads to a classical collision-coupling mechanism in the membrane. The G-protein, **G$_{olf}$**, induces an adenylyl cyclase enzyme to catalyse the production of cAMP which then opens

a CNG channel. Na$^+$, K$^+$ and Ca^{2+} flow along their electrochemical gradients and the membrane is depolarised. This depolarisation is assisted by the increase in intracellular Ca^{2+} concentration opening a Cl$^-$ channel. This is our first example of a **generator potential**. It initiates an action potential in the olfactory nerve fibre. The action potential propagates without decrement to the olfactory bulb in the forebrain (Figure 13.10).

Sensory adaptation is largely due to Ca^{2+} ions. Figure 13.9 shows that increased intracellular Ca^{2+} tends to **close** the CNG channels. This, of course, is the reason why the electrophysiologists found that the channels only carry significant current flow in the absence of [Ca^{2+}]$_i$. Ca^{2+} ions also desensitise the receptor molecule. In addition to the Ca^{2+} mechanisms, the receptor molecule is also desensitised by the phosphorylating action of **protein kinase A**

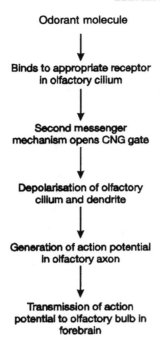

Odorant molecule

↓

Binds to appropriate receptor
in olfactory cilium

↓

Second messenger
mechanism opens CNG gate

↓

Depolarisation of olfactory
cilium and dendrite

↓

Generation of action potential
in olfactory axon

↓

Transmission of action
potential to olfactory bulb in
forebrain

Figure 13.10   Summary of olfactory reception.

(PKA) and **olfactory receptor kinase (ORK)**. It will be remembered from Section 8.8 (Figure 8.17) that cAMP activates PKA. However, the adaptation is very rapidly terminated. It has been shown that isolated olfactory CNG channels retain their normal sensitivity to recurrent bursts of odorant.

The response of the olfactory receptor cell is 'graded'. It varies in direct proportion to the concentration of odorant molecule. This graded response is due to the cAMP second messenger system. In order to discriminate between different odours in real (biological) time the response has to be rapid. Experiment has shown that cAMP production peaks within 40–75 ms of odour application and falls back to zero within 100–500 ms. The amplification provided by the G-protein cascade ensures that numerous channels are activated by one burst of odorant. It has, however, been shown that the channel kinetics are rather slow. Their 'open' states may outlast brief pulses of cAMP by several hundred milliseconds. When the concentration of odorant is high there will be continuous activation of 7TM receptors. This will

ensure that the CNG channel is kept steadily open by burst after burst of cAMP.

Finally, it should be noted that with the isolation and sequencing of the entire human olfactory subgenome it is rapidly becoming possible to trace the source of many olfactory defects. Humans suffer from several dozen specific **anosmias**. They are due to deficiency in one or other olfactory receptor molecules. Many of these anosmias are inherited in a Mendelian fashion. There is a clear analogy here with human colour blindnesses, which are also due to defective receptor proteins – in this case iodopsins. The genetics of human anosomias will soon become as well known as the genetics of human colour blindness.

Before leaving the topic of chemoreception in mammals it is worth observing that recent investigations of mammalian gustatory cells are beginning to reveal that mechanisms analogous to those operating in olfactory neurons are at work here also. Gustatory receptor cells are grouped together in the taste buds which are to be found on the tongue and pharynx. Their sensitive ends take the form of microvilli which project into the lumen of the taste bud. Chemicals diffusing in from the buccal cavity stimulate these endings. The mechanism of stimulation varies from one tastant molecule to another. In the case of Na^+ sensitivity, so important for the internal homeostasis of all animals, the response is directly through an amiloride-sensitive channel. In a number of other cases, however, the receptor molecule in the microvillus membrane is a 7TM protein and acts via a G-protein (known as a **gustducin**) collision-coupling system to generate a second messenger. The second messenger may be either inositol triphosphate or cAMP. In the case of the **sucrose receptor** the gustducin activates an adenylyl cyclase/cAMP system which, via PKA, phosphorylates a K^+ channel protein in the gustatory cell's membrane. The phosphorylation tends to close the K^+ channel and the consequent decrease in K^+ conductivity causes the membrane to depolarise. This depolarisation can be detected by microelectrode techniques. It is a first instance of a **receptor potential**. For gustatory cells, unlike olfactory neurons, are true receptor cells: depolarisation leads to the release of a transmitter on to the dendrite of a sensory nerve fibre. Other gustatory cells use other biochemical cascades to generate receptor potentials.

13.2 PHOTORECEPTORS

The animal kingdom has developed a great variety of different photoreceptors. The most intensively researched of all these different types are the vertebrate retinal rod and cone cells. We have already met aspects of their molecular physiology several times in this book, especially in Chapter 8, where we discussed the structure and biochemical function of the visual pigments.

It has been calculated that rod cells are able to detect a single photon of light. How do they do it? We shall now try to put together the isolated fragments that we have already discussed to form a picture of the way in which rod and cone cells work.

First let us remind ourselves of the structure of these fascinating cells. Figure 13.11B shows that both rod and cone cells have a very similar design. Both consist of an outer segment, an inner segment, a nuclear region and a synaptic foot, or pedicle.

It is the outer segments of rod and cone cells that contain the visual pigment and are the photosensitive parts of the cell. Figure 13.11B shows that the outer segments consist of a stack of discs. Electron microscopists have shown that whereas the intradisc space in cones opens into the extracellular space, rod discs are separated from this space by a continuous boundary membrane.

Figure 13.11A shows that during embryology rod and cone outer segments first appear as cilia springing from their respective cells. This, in itself, is interesting as we have already seen that the sensitive regions of olfactory cells are also modified cilia. It begins to look as if there is a common design principle at this ultrastructural level. We have already seen evidence of common design at the molecular level. It seems, once again, that evolution having 'found' a workable solution has been content to develop variations on a theme.

A vestige of the ciliary origin of rod and cone outer segments remains in the form of the 'connecting cilium' which joins the outer to the inner segment of the mature cell. This retains the characteristic ring of nine peripheral microtubular doublets of motile cilia but lacks the central pair.

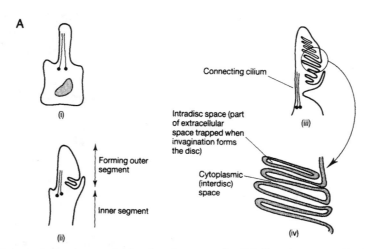

Figure 13.11 Structure and development of rod and cone cells. (A) **Development**. Rods and cones originate as ciliated cells lining the ocular ventricle. (i) Each cell bears one cilium. (ii) The cilium grows and its membrane hypertrophies and begins to invaginate. (iii) Hypertrophy of the membrane continues and invagination proceeds at the base of the forming outer segment. The original ultrastructure of internal filaments characteristic of cilia remains in the 'connecting cilium' which runs between the outer and inner segments. (iv) Detail of (iii) to show how the mode of formation of the discs by invagination traps a small fragment of the extracellular compartment as the intradisc space. The space between the discs (the interdisc space) is cytoplasmic.

(continued on next page)

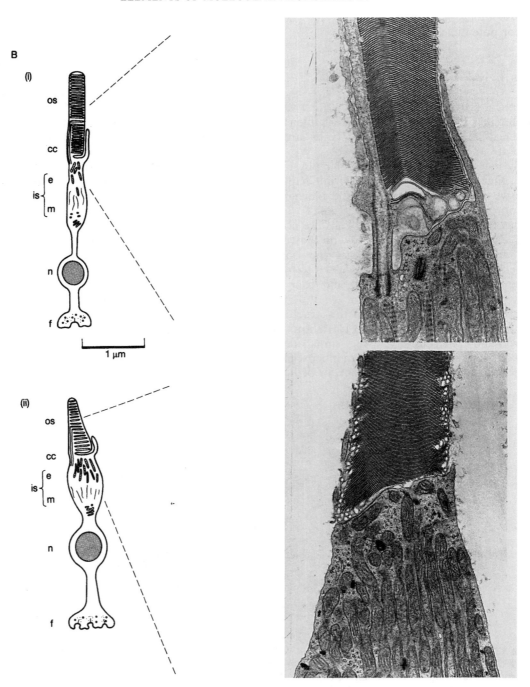

Figure 13.11   *(continued)* (B) **Structure**. (i) Rod. (ii) Cone. cc=connecting cilium; e=ellipsoid; f=foot (or pedicle); m=myoid; n=nuclear region; is=inner segment; os=outer segment. Electron micrographs from Hogan, Alverado and Weddell (1971), *Histology of the Human Eye*, Philadelphia: W.B. Saunders Company; with permission.

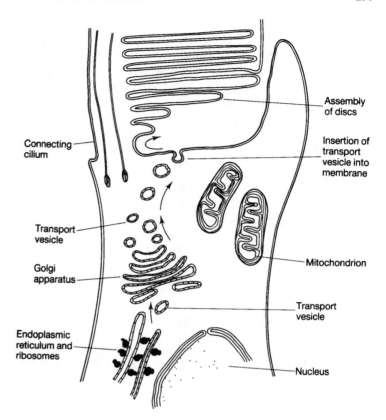

Figure 13.12 Disc membrane synthesis in rod inner segments. Schematic diagram to show the route of synthesis of visual pigment and rod and cone discs. The pigment is synthesised in the endoplasmic reticulum (ER) and (along with the other phototransduction proteins – see Chapter 8) is inserted into the ER membrane. The latter buds off as transport vesicles to form the *cis* face of the Golgi apparatus. Further packaging and glycosylation occurs in the Golgi body and transport vesicles bud from its *trans* face and move towards the plasma membrane at the top of the inner segment. They fuse with the membrane at this point and create new rod discs. Further detail of these processes in Chapter 15. After Besharse (1986), in *The Retina: A Model for Cell Biology Studies*, ed. R. Adler and D. Farber, Orlando, FL: Academic Press, pp. 297–352.

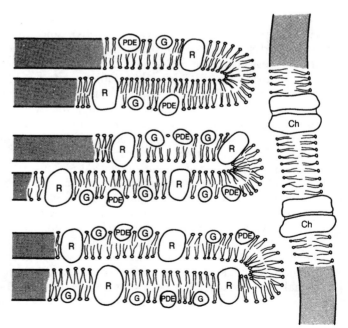

Figure 13.13 Detail of rod discs. Three rod discs are shown abutting the boundary membrane of the outer segment. R=rhodopsin; G=G-protein (=transducin); PDE=phosphodiesterase; Ch=channel protein. Compare with Figure 8.27.

BOX 13.1   Retinitis pigmentosa

The eye (like the ear) is subject to a large number of genetic diseases. The genes responsible are distributed over nearly the whole of the human genome (20 out of the 23 chromosomes). This reflects the complexity of the eye's structure. Its development and maintenance involve the coordinated operation of many genes, any one of which may malfunction. The commonest form of inherited visual defect is **retinitis pigmentosa (RP)**. The name, in fact, covers a very heterogeneous group of pathologies which, together, have an estimated prevalence of about 1/3000. The condition may be inherited as autosomal dominant (adRP), autosomal recessive (acRP), or linked to the X chromosome (xlRP). arRP is by far the most frequent form, being shown by at least half all families suffering from RP. The disease is caused by the (presently) irretrievable degeneration and death of rods, and to a much lesser extent cones. The degenerating rod outer segments are scavenged and digested by the pigment epithelium, giving rise to a heavy pigmentation and hence the name, retinitis pigmentosa. The loss of rods leads to a cascade which ultimately affects cones and the retinal blood supply. Early indications are night blindness and abnormalities in the electroretinogram (ERG). The end result, sooner or later, is blindness.

There is evidence for two varieties of arRP. In the first case there is a general degeneration of rods across the retina, in the second the rods degenerate in patches which gradually coalesce. In both cases cones survive until more than 75% of the neighbouring rod outer segments have degenerated. It may be that these two conditions demand different therapeutic approaches. In both cases it is vital to develop an understanding of the underlying causes.

We have noted in this chapter some of the molecular and cellular elements of rod cells. All of these elements are open to mutational attack leading to misfunction and consequent visual impairment. Genetic analysis has implicated chromosomes 1, 3, 6, 7, 8 and the X chromosome. X-linked RP has been subjected to particularly intensive genetic analysis and the mutations responsible have been traced to two genes on the short arm of the chromosome. Neither gene has as yet been characterised or its biochemical effect determined.

Understanding is more advanced in some cases of autosomal dominant RP. The first mutation responsible for adRP to be tracked was located on chromosome 3. It was shown to lead to an amino acid substitution in rod opsin ($Pro_{23} \rightarrow His$). Since then more than 70 other mutations affecting the amino acid sequence of rod opsin have been detected. In some cases these mutations affect less essential regions of the molecule and hence the onset of RP is delayed and is less severe. The first mutation to lead to adRP (mentioned above) affected a large US population of Irish origin and causes very early onset RP and major visual impairment by the second decade of life. The mutation seems to be restricted to the USA where it accounts for over 15% of adRP patients. The fact that it is not found in Europe suggests that it is a case of 'founder effect' in the early Irish emigrant community. Another US Irish pedigree, with a less severe adRP, causing impairment only in the fourth or later decades of life, was shown to be linked to mutation on chromosome 6. The candidate gene encodes a structural protein in rod outer segment discs. This protein is called **peripherin-RDS** (the RDS suffix indicates that it was first characterised in mice suffering from slow retinal degeneration).

Genetic analysis of arRP is not so far advanced. It has been difficult to establish sufficiently extensive genealogies to undertake decisive genetic analysis. Nevertheless a $G \rightarrow T$ transversion has been discovered in codon 249 of one arRP kinship, leading to GAG being substituted by TAG. This is transcribed as UAG, which is a stop signal in translation at the ribosome. The resulting opsin would thus lose its sixth and seventh transmembrane segments. The homozygous patient had suffered night blindness for as long as she could remember although her heterozygous parents and sibling were normal. Another arRP has been shown to be due to a $G \rightarrow A$ transition at codon 150. This leads to a GAG triplet (glutamate) being changed to AAG (lysine). Once again the heterozygous kinship showed no ill effect whereas the homozygous patient suffered severe night blindness. Other work suggests that arRP is not confined to mutation on chromosome 3; it can be due to mutant genes on a number of different chromosomes. It is rapidly becoming possible to engineer transgenic mice which express some of these defective genes. The molecular biological defects underlying RP are thus slowly being unravelled. In one case of mouse retinal degeneration (rd) it is, for instance, known that the pathology is due to a defect not in opsin but in the β-subunit of cGMP PDE. This defect has also recently been identified in some cases of human arRP.

We have noted in this chapter and in Section 8.13 that rhodopsin and cGMP PDE are present in large quantities in outer segment discs and that rhodopsin (at least) forms part of their structure. It is not surprising, therefore, that genetic errors in the manufacture of these elements (and, of course, peripherin) should lead to loss of outer segment integrity. But it is not only the manufacture and integrity of rod discs that is at risk but also their removal. Rod discs are scavenged and digested by the retinal pigment epithelium (RPE). This process is also under genetic control. Perhaps it is a developing imbalance between synthesis and degradation that is responsible for the slow development of most RPs.

Molecular neurobiology is just beginning to elucidate this complex of interacting processes and, as in other cases, it is leading to crucial insights into the underlying causes of a devastating human disease. Recent success in inserting normal genes into mice homozygous for rds and rd resulting in blockage of photoreceptor degeneration hopefully points the way to rational therapies.

As we noted in Chapter 7, discs are added throughout life to the base of the outer segment and push up to the tip where they are nipped off and digested in the pigment epithelium (Figure 13.12). The visual pigment – **rhodopsin** in the case of rods – is incorporated into the discs as they are manufactured in the inner segment. Rhodopsin is present in very large amounts – about 70% of outer segment protein is rhodopsin. It has been calculated that $0.1\,\mu m^2$ of disc membrane contains some 2000 rhodopsin molecules. We noted in Chapter 7 that in *Xenopus* disc membrane is synthesised at some $3.2\,\mu m^2$ per minute throughout life. This means that approximately 60 000 rhodopsin molecules have to be synthesised every minute – a formidable task.

Rhodopsin molecules are packed very closely together in outer segment discs. Indeed the distance between rhodopsin molecules has been computed to be no more than about 20 nm. It is obvious that the outer segment provides an extremely effective device for detecting light. Incoming illumination (after having passed through the cornea, vitreous fluid, lens, aqueous fluid and the highly transparent nervous elements of the retina) is presented with a stack of perhaps 20 000 discs in each outer segment.

Furthermore, those animals (proverbially cats) that have a **reflective tapetum** behind the outer segments give light a second chance to interact with the piles of rhodopsin molecules.

Let us now focus on single discs. These are shown diagrammatically in Figure 13.13. The rhodopsin molecules (see Section 8.13) are set in the disc membranes so that their carboxy-terminal ends extend into the interdisc (cytoplasmic) space. Although we stated above that the discs are enclosed by a continuous boundary membrane (Figure 13.11), this is not to say that that membrane is totally leak-proof. It is not. It is found to be penetrated by channels that allow the ingress of Ca^{2+} and small cations, especially Na^+. These are the CNG channels which, as we saw in the previous section, show strong homologies with the CNG channels that play so important a role in olfactory cilia. They play a similarly important role in outer segments. However, unlike olfactory CNG channels they are **open** when the outer segment is **unstimulated, in the dark**. Furthermore, instead of acting through a cAMP second messenger system, they are sensitive (as we shall see below) to cGMP.

The fact that the outer segment boundary membrane is more than usually permeable to Na^+

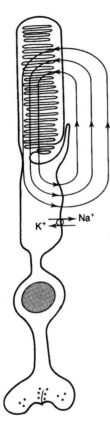

Figure 13.14 Rod cell dark current. Because of its 'leak' channels sodium ions percolate back into the outer segment. They then flow down to the inner segment where an $Na^+ + K^+$ pump extrudes them into the extracellular compartment once again. Hence in the dark a current flows in the direction indicated by the arrows.

percolate down into the inner segment where the sodium pump extrudes them. In the dark, therefore, there is a circuit of sodium ions – pumped out of the inner segment only to enter the outer-segment (Figure 13.14).

All this changes when the cell is illuminated. We saw in Chapter 8 that when rhodopsin receives a photon of light a cascade of biochemical reactions ensues which leads to the transformation of cGMP into the straight-chain 5'-GMP. Now it is found that it is precisely cGMP that keeps the CNG channels open! Patch-clamp experiments on pieces of rod cell outer segment membrane have demonstrated that CNG channel opening significantly increases in the presence of cGMP. Indeed it can be deduced that at least three cGMP molecules are required to keep the pore open. The part played by cyclic nucleotides in opening CNG channels in olfactory cilia should be recalled.

Thus illumination causes the outer segment sodium ion pores to close. The dark current is broken. The boundary membrane of the rod cell begins to hyperpolarise. The same events occur in cone cells when they receive photons of the appropriate wavelength (see Chapter 8). Figure 13.15A shows that microelectrode recording allows this hyperpolarisation to be observed directly. Alternatively, as Figure 13.15B shows, rod cell outer segments can be sucked into a micropipette from a piece of detached retina. If a pencil-beam of light is shone through the pipette and the outer segment illuminated, the interruption of the dark current can be detected. The greater the intensity of the illuminating beam the more the current is reduced.

We shall see in Chapter 16 that synapses are activated by membrane depolarisations. Hyperpolarisations tend to switch synapses off. Hence in the case of rod cells illumination will reduce any synaptic activity in their feet. In the dark the comparatively depolarised rod cell releases

has important consequences for the resting potential (V_m) of the rod cell. Suppose that P_{Na} is not about $1 \times 10^{-8}\,cm\,s^{-1}$ (as we quoted in Chapter 12) but five times greater, i.e. $5 \times 10^{-8}\,cm\,s^{-1}$. If we insert this value into the Goldman equation keeping the permeability coefficients of K^+ and Cl^- unchanged we find that V_m works out at $-22\,mV$.

That the rod cell is depolarised in this way is indeed found when the cell is unilluminated. Sodium ions leak into the outer segment and

Figure 13.15 *(opposite)* Effect of light on rod cells. (A) When a photon of light is absorbed by an outer segment the sodium channels close. The rod cell hyperpolarises. This may be recorded by a microelectrode inserted in the inner segment. The amount of hyperpolarisation recorded depends on the intensity of the light flash. This is shown in the graph to the right. After Penn and Hagins (1969), *Nature*, **223**, 201–205. (B) (i) A toad outer segment (much larger than mammalian outer segments) is sucked into a glass pipette. As it blocks the end of the pipette all the current which flows into or out of it is supplied from the interior of the pipette and can thus be measured. In the dark this current (as the graph at (iii) shows) is a little over 20 pA. (ii) A pencil beam of light is flashed through the pipette. (iii) The graph shows the effect on the dark current of the flash of light starting at the point marked by the vertical

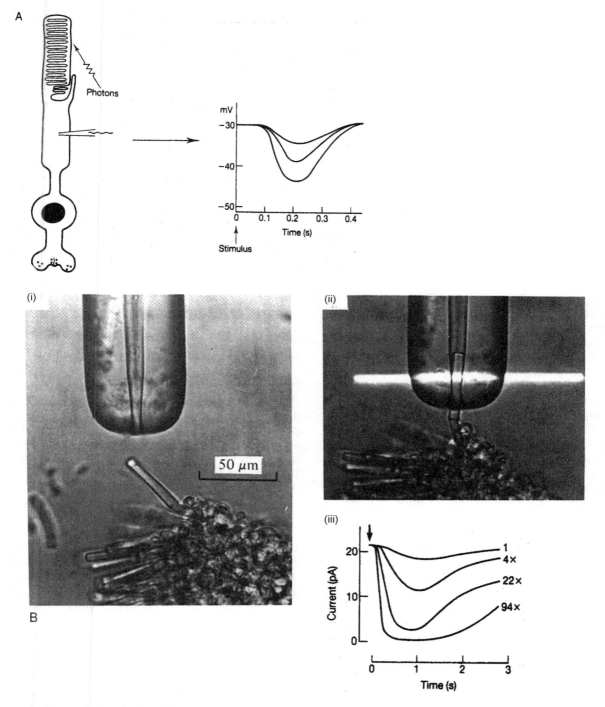

arrow. The reduction in the dark current is dependent on the intensity of the flash stimulus. The bottom line of the graph results from a stimulus 94 × the intensity of flash responsible for the top line. From Baylor, Lamb and Yau (1979), *Journal of Physiology*, **288**, 589–611; reproduced by permission of the Physiological Society.

transmitter on to the dendritic ending of the underlying bipolar cell. Illumination inhibits this release. The transmitter is, in most cases, glutamate. This, as we shall see in Chapter 16, is an excitatory transmitter.

What happens next depends on the nature of the subsynaptic bipolar cell. In the case of rod bipolars the reduction in glutamate due to illumination of the rod cell leads to depolarisation of the bipolar (Figure 13.16). Because hyperpolarisation of the rod cell leads to depolarisation of the underlying bipolar the synapse is said to be 'sign inverting'. The biochemical mechanism responsible for this inversion has similarities to the mechanism keeping CNG channels open in photoreceptor cells. It appears that glutamate actuates a biochemical cascade which removes cGMP that otherwise keeps these CNG channels open. Hence in the dark, when glutamate is present in the rod-bipolar synapse, the CNG channels are closed and the bipolar cell relatively hyperpolarised. When light is switched on and the rod cell hyperpolarises, glutamate is no longer released into the synaptic cleft, the CNG channels in the bipolar cell open and the cell **depolarises**.

The cone-bipolar synapses are of two types: **sign inverting** and **sign conserving**. The sign inverting synapses are similar to those just described for rod cells. The **sign conserving** synapses ensure that when the cone cell hyperpolarises, so does the bipolar cell (Figure 13.16). In this case glutamate in the cone-bipolar synaptic cleft opens cation channels in the bipolar cell, thus increasing the permeability of its membrane and thus ensuring that the cell is somewhat depolarised in the dark. When the rod cell is illuminated and hyperpolarises the supply of glutamate is cut off, the cation channels close and the bipolar cell **hyperpolarises**.

If we turn back to Section 8.13 we shall remind ourselves of how first **rhodopsin kinase** and then **arrestin** inactivate opsin and thus desensitise the rod in response to continuous, steady illumination. In addition, the rod disc also contains a biochemical mechanism ensuring adaptation.

This mechanism consists in a means for replenishing cGMP during sustained illumination (Figure 13.17). The discs contain a Ca^{2+}-sensitive protein called **recoverin**. Recoverin is in some ways the inverse of the ubiquitous calmodulin. Instead of being activated by increased Ca^{2+} concentrations it is activated by decreased concentrations. In other words, it is only active when it is not bound to Ca^{2+}. But on activation it stimulates **guanylyl cyclase** to resynthesise cGMP from 5'-GMP.

But why should the concentration of Ca^{2+} fall when light is switched on? This is a direct consequence of the closing of the CNG channels. For these channels, as we noted above, are permeable not only to Na^+ but also to Ca^{2+}. Thus, one of the consequences of light-induced closing of the CNG channels is a disinhibition of the guanylyl cyclase enzyme and increased quantities of cGMP.

Increased quantities of cGMP, in balancing out the phosphodiesterase (PDE) removal of cGMP, assist the outer segment to adapt to illumination. They assist, in other words, the maintenance of the cGMP concentrations on which the open state of the CNG channels depends. The longer illumination persists the more cGMP is synthesised by this route, the more the CNG channels open, and the less hyperpolarised is the rod cell.

Once again we are struck by the neat feedback and feedforward loops which ensure that the system is tuned to an optimal fitness for purpose.

It is now possible to answer the question with which we started this section. How is it possible that rod cells can detect single photons? The G-protein → second messenger → ion channel → receptor potential cascade enables the minute energy of a single photon to lead to a 100 mV change in the membrane of an optic nerve fibre and (who knows) to illimitable action in the world (Figure 13.18).

13.3 MECHANORECEPTORS

Once again an enormous variety of mechanoreceptors have evolved in the biological world. Mechanoreceptors predate not only the animal kingdom but also the eukaryocytes. This huge antiquity emphasises that they play many important roles in the life of cells. In particular, detection of tension in cell membranes during growth and volume regulation has always been crucial. There have, however, until quite recently been seemingly insuperable obstacles to establishing their molecular biology. This is because there are no highly enriched sources of mechanoreceptors (compare the significance of electric organs for early nAChR characterisations) nor are there suitable toxins that can be used for chromatographic isolation.

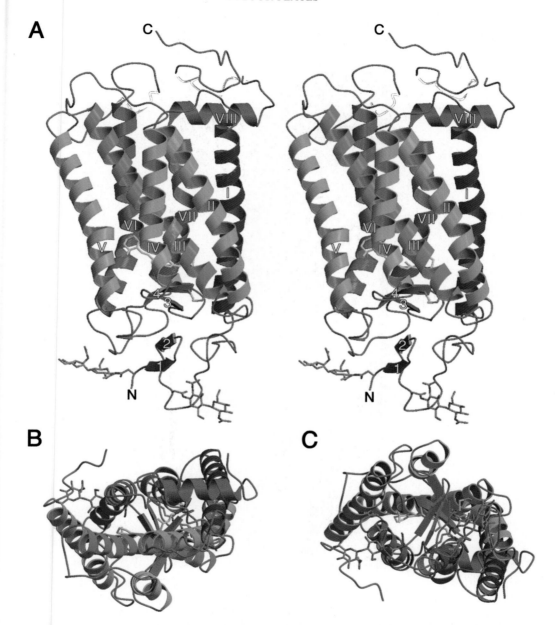

Plate 1   Rhodopsin. (A) Ribbon diagrams orthogonal to plane of membrane (stereopair). Defocus eyes to get 3D effect. (B) View from cytoplasmic (interdisc) side of membrane. (C) View from extracellular (intradisc) side of membrane. The ribbons represent alpha-helices and are numbered I–VIII. Note that helix VIII does not traverse the membrane but runs parallel to the cytoplasmic surface (see also B). Anti-parallel beta-strands on the extracellular (intradisc) end of the molecule are labelled 1, 2, 3 and 4 and are shown as arrows (see also C). 11-cis retinene (not shown) nestles in the centre of the seven TM helices and holds the whole structure in its inactive state. Note that the molecule in (A) is the other way up from its representation in figure 8.25. Reprinted with permission from Pazewski, K. et al., 2000, 'Crystal structure of Rhodopsin: A G-Protein-Coupled Receptor', *Science*, **289**, 740. Copyright (2000) American Association for the Advancement of Science.

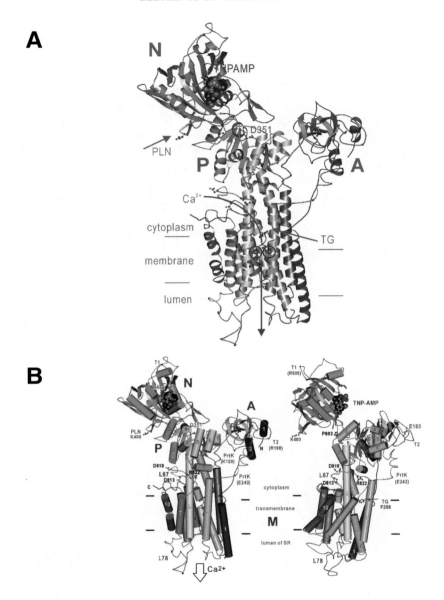

Plate 2 Ca^{2+} pump. (A) Ribbon diagram. (B) Cylinder diagrams (alpha-helices represented by cylinders). The transmembrane helices are numbered 1–10. The model is orientated so that the longest helix, M5, is vertical and parallel to the plane of the paper. It is 60 Å in length and hence provides a scale bar. The right hand diagram is rotated 50° around M5. The three cytoplasmic domains are labelled A, N and P (see text pp. 203–5) and helices in A and P are also numbered. Beta-strands are represented by arrows. D351 (Asp$_{351}$) is the residue at which phosphorylation occurs and TNP-AMP shows where the adenosine of ATP attaches to the nucleotide domain. PLN and TG indicate the binding sites for phospholamban and thapsigargin and a purple sphere represents one of the two Ca^{2+} ions on its transmembrane binding site. For other detail consult reference cited below. Note, finally, that the models are the other way up to the figures in chapter 9. Reprinted with permission from Toyoshima, C. et al., 2000, 'Crystal structure of the calcium pump of sarcoplasmic reticulum at 2.6 Å resolution', *Nature*, **405**, 648. Copyright (2000) Macmillan Magazines Ltd.

EXTRACELLULAR

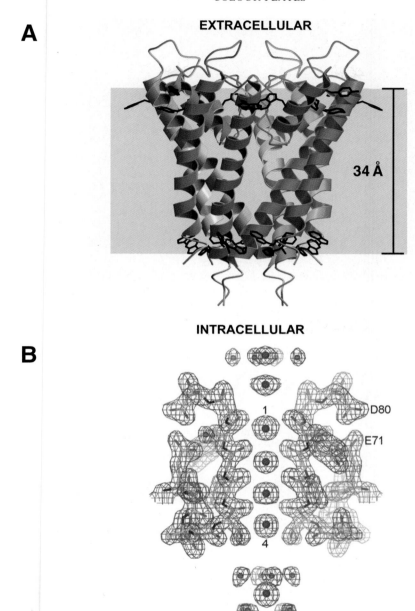

Plate 3 The KcsA channel. (A) Ribbon diagram of the tetrameric complex embedded in the membrane. The selectivity filter is at the top of the figure surrounded by the four 'turrets'. (B) Electron density diagram of the K^+-selectivity filter at 2.0 Å resolution. Four K^+ ions are caught in the filter (green spheres) and water molecules (red spheres) associated with K^+ ions can be seen outside (top of figure) and inside (bottom of figure) the membrane. D80 (Asp_{80}) and E71 (Glu_{71}) identify amino acid residues. Part A reprinted with permission from Doyle, D. A. *et al.*, 1998, 'The structure of the potassium channel: molecular basis of K^+ conduction and selectivity', *Science*, **280**, 73. Copyright (1998) American Association for the Advancement of Science. Part B reprinted with permission from Zhou, Y, et al., 2001, 'Chemistry of ion coordination and hydration revealed by a K^+ channel-Fab complex at 2.0 Å resolution', *Nature*, **414**, 45. Copyright (2001) Macmillan Magazines Ltd.

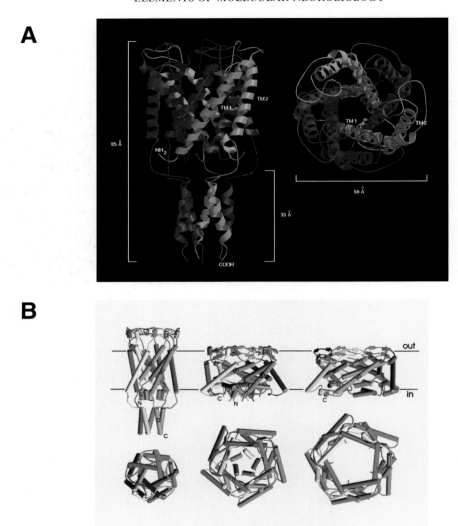

Plate 4 The MscL channel. (A) Ribbon diagram of the mechanosensitive channel from *M. tuberculosis*. (TbMscL). Side view on left; extracellular view on right. the five subunits are individually coloured and the N and C terminals of one of the subunits (cyan coloured) and its transmembrane helices (TM1, TM2) are labelled. Note that only the upper part of the molecule is embedded in the membrane. (B) Cylinder models of the MscL channel of *E. coli* (EcoMscL). Upper row shows the molecule from the side, lower row looking upward from the periplasm. The figure shows (from left to right) closed/resting conformation; closed/expanded conformation; open conformation. The five subunits are (as in (A) above) differently coloured and only one (blue) is labelled. The TM helices are labelled M1 and M2 and the other helices S1, S2, S3. The C and N terminals of the blue subunit are also indicated. Horizontal lines show the approximate position of the membrane. When the membrane is stretched the S1 helices are, at first, dragged over to plug the incipient pore (middle figures), if stretching continues the S1 helices are ultimately pulled away to open a large passageway (right hand figures). Part A reprinted with permission from Chang, G. et al., 1998, 'Structure of the MscL homolog from *Mycobacterium tuberculosis*: A gated mechanosensitive channel', *Science,* **282**, 2224. Copyright (1998) American Association for the Advancement of Science. Part B reprinted with permission from Sukharev, S. et al., 2001, 'The gating mechanism of the large mechanosensitive channel MscL', *Nature,* **409**, 721. Copyright (2001) Macmillan Magazines Ltd.

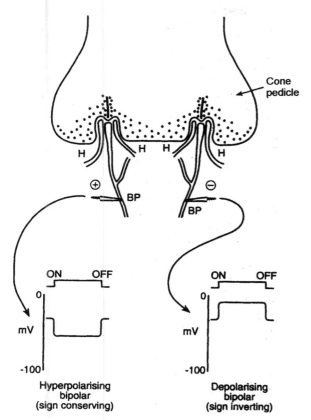

Figure 13.16 Sign-conserving and sign-inverting cone-bipolar synapses. The lower part of the figure shows the response of the bipolar cell when the photoreceptor is illuminated. The cone-bipolar synapse is complex. The cone pedicle makes synaptic 'contact' with horizontal cells (H) and bipolar cells (BP). The bipolars are either hyperpolarising, +, (sign conserving), or depolarising, −, (sign inverting). The lower part of the figure shows the responses of the bipolar cells when the cone is illuminated. The response is greater the more intense the illumination. Further explanation in text.

In spite of these seemingly overwhelming difficulties, molecular structures for a number of mechanosensitive ion channels have recently been determined. Amongst the prokaryocytes the gene for an *E. coli* mechanosensitive channel has been cloned and the molecular structure of the channel protein established. More recently, as we see below, the mechanosensitive channel in *Mycobacterium tuberculosis* has been analysed to the 3 Å level. Considerable progress has also been made in

elucidating the structure of mechanosensitive channels in *Caenorhabditis elegans*. It turns out that the latter channels show strong homologies with a vertebrate epithelial Na⁺ channel. This channel must not be confused with the voltage-dependent Na⁺ channels of Section 11.1. Epithelial channels show very little voltage dependence and permit the flow of large numbers of Na⁺ ions accompanied by water. It turns out that the *C. elegans* mechanosensitive and epithelial Na⁺ channels form the first members of a large new family of channel proteins.

We shall proceed as follows. As in Section 13.1, we shall begin with a brief account of prokaryote receptors and then pass via a brief discussion of mechanosensitive channels in *C. elegans* to the immeasurably more complex mechanoreceptive systems that have evolved in mammals, in particular the ear.

13.3.1 A Prokaryote Mechanoreceptor

Although bacteria are clearly sensitive to mechanical stimuli, and although, as mentioned above, sensitivity to membrane tension is of importance in volume regulation, especially osmoregulation, and cell growth, it is not obvious how the receptors can be isolated and characterised. The small size of most prokaryotes ($\leqslant 1 \mu m$) precludes patch-clamping. However, it is possible to inhibit cell division and thus generate large '**spheroplasts**' of up to 6 μm in diameter. These spheroplasts are large enough to patch-clamp and detect single channel conductances.

Patch-clamping was discussed in Section 10.1.2 (Figure 10.9). A fragment of membrane is sucked into the tip of a micropipette. In the case of the mechanosensitive channels in *E. coli* spheroplasts the technique is doubly useful. Not only does it allow the detection of ion channels but it also allows the tension on the spheroplast membrane to be varied. Carefully increasing (or decreasing) the suction on the pipette increases (or decreases) the tension on the membrane. Using this technique two types of mechanosensitive channel were found – a large conductance (3000 pS) and a small conductance (1000 pS). The large conductance channel (**MscL: mechanosensitive channel (large)**) was shown to be non-selective whilst the small conductance channel (**MscS**) was anion specific. That they were distinct channels and not two states of the

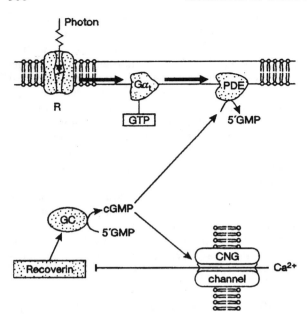

Figure 13.17 Biochemical adaptation in the rod disc. Arrows symbolise activation; —| symbolises inhibition. $G_{\alpha t}$ = G-protein (transducin) α-subunit (see Table 8.1); GC = guanylyl cyclase; PDE = phosphodiesterase; R = rhodopsin. Explanation in text.

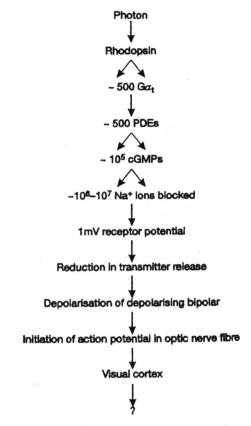

Figure 13.18 Amplification cascade in the visual system. Adapted from Alberts *et al.* (1964), *Molecular Biology of the Cell* (3rd edn), New York: Garland; and Lamb and Pugh (1992), *Trends in Neurosciences*, **15**, 291–298.

same channel was later confirmed by 'knocking out' the gene for MscL. The conductance due to MscS can still be detected in these 'knock out' *E. coli* spheroplasts.

The analysis of the MscL channel was completed by using some of the sophisticated techniques of modern molecular biology. These techniques involve the incorporation of *E. coli* membrane fragments into liposomes composed of foreign phospholipids and testing again with the patch-clamp technique to confirm the presence of the mechanosensitive channels. The proteins can be extracted from these liposomes by the use of mild detergents and refined by biochemical techniques, all the time patch-clamp testing for MscL activity, until a pure preparation of channel protein is obtained. This turns out to have M_r of about 17. The N-terminal amino acid sequence of this 17 kDa protein was sequenced and then the very full database of the *E. coli* genome searched for a gene containing a corresponding nucleotide sequence. Such a sequence was found in a gene

whose function was previously unknown. Through complex molecular biological procedures the gene was caused to express itself and the resulting protein incorporated into phospholipid liposomes which, on being patch-clamped, showed the presence of the mechanosensitive channels.

Once the MscL channel had been isolated and incorporated into artificial phospholipid liposomes, its biophysical characteristics could be examined in detail. Like most membrane channels it shows multiple conducting states (Figure 13.19). It also shows strong dependency on lateral tension in the membrane. There is, in fact, a steep sigmoid curve relating the probability of being open with the tension. There is no evidence of ion selectivity and

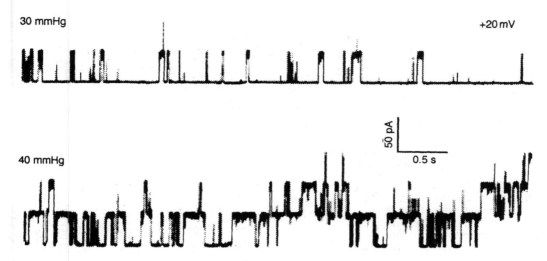

Figure 13.19 Biophysical characteristics of the MscL channel. The channel protein was purified and incorporated into phosphatidylcholine/phosphatidylserine liposomes. The liposomes were patch-clamped and the membrane subjected to differing degrees of suction by the patch-clamp pipette (30 mmHg and 40 mmHg). The upward excursions represent ion flows carrying a depolarising current. It can be seen that the open probability is significantly increased when the suction is increased. From Blount *et al.* (1996), 'Membrane topology and multimeric structure of a mechanosensitive channel protein in *Escherichia coli*', *EMBO Journal*, **15**, 4801. Reproduced by permission of the European Molecular Biology Organisation.

this, along with its large conductivity, suggests a wide water-filled pore.

The isolation of the MscL protein and the identification of its gene allows the molecular biologist to determine its molecular structure (Figure 13.20A). It turns out to be a relatively small protein consisting of some 136 amino acid residues. Examination of the sequence shows that the N-terminal is followed by a section of hydrophilic residues and then by segment of strongly hydrophobic amino acids (19–49) followed by another hydrophilic region (50–69), then a second hydrophobic segment (72–100) and, finally, a hydrophilic sequence to the C-terminal. Careful examination of the physico-chemical characteristics of the polypeptide indicates that the two hydrophobic segments form α-helices (M1 and M2) spanning the membrane. Both the N-terminal and C-terminal are located inside the cell and there is reason to believe that there is another short helical region (S3) in the extracellular segment between M1 and M2 (Figure 13.20A). On the analogy of other voltage-gated channels, such as the Na$^+$ channel outlined in Chapter 11, it has been

suggested that the S3 region forms part of the lining of the channel. Indeed there is, as we shall see below, accumulating evidence from mutagenesis studies to support this hypothesis.

A single MscL protein could not form a gated channel with the biophysical characteristics of the native MscL channel. The operational channel must be a multimeric complex. Although there is still some debate, the consensus at present is that the MscL channel consists of five MscL subunits grouped around a central water-filled channel (Figure 13.20B). This, as we shall see below, has been confirmed by X-ray crystallography. When the membrane is subjected to mechanical stretch the subunits are pulled somewhat apart and the channel between them is opened.

The isolation of the MscL channel protein and its gene, *mscl*, opens the possibility for genetic investigation of structure–function relationships. This investigation is, of course, greatly helped by the intimate understanding of *E. coli* genetics built up over the years. Research is ongoing at the time of writing, but already, as mentioned above, several interesting relationships have been teased out. One

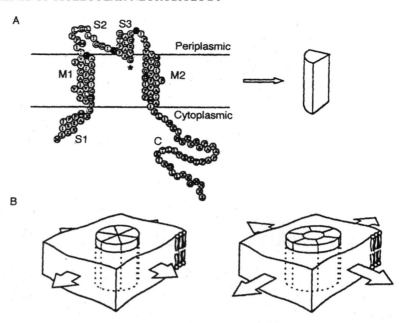

Figure 13.20 Structure and membrane topology of *E. coli* MscL channel. (A) The disposition of the MscL subunit in the membrane. The asterisk shows the position of a critical glutamine residue. Mutation of this residue affects the sensitivity of the channel to stretching forces. (B) Six MscL subunits are grouped to form a compact cylinder through the membrane. When the membrane is stretched a hydrophilic pore opens in the centre of the cylinder. From Sukhaerev *et al.* (1997), *Annual Review of Physiology*, **59**, 633–657. With permission from the *Annual Review of Physiology*, Volume 59, © 1997, by Annual Reviews, www.annualreviews. org

amino acid residue at the 'bottom' of the S3 periplasmic helix seems particularly critical. This is the glutamine residue in position 56 (Q56) (see Figure 13.20A). A number of amino acid substitutions at this point significantly lengthen the time for which the channel stays open under a given stretch. Indeed, some of these substitutions, in particular the substitution of proline for glutamine at this position (in the short-hand of molecular biologists: Q56P), make the channel sensitive to stretching forces far below those normally required, indeed below those necessary to open the MscS channel. Substitutions in other parts of the channel protein, though not so dramatic in their effects, nevertheless alter function in interesting and thought-provoking ways. Clearly this ongoing genetic research into structure–function relationships combined with X-ray crystallography (see below) will ultimately lead to an understanding of this mechanosensitive channel in full molecular, indeed atomic, detail.

Do similar channels develop in the plasma membranes of other bacteria? Once the nucleotide sequence for the *mscl* gene had been determined it became possible to prepare matching oligonucleotide probes to search for homologous genes in other bacteria. Such genes have been shown to exist in a number of other bacteria, for instance,

Mycobacterium tuberculosis, *Haemophilus influenzae*, *Pseudomonas fluorescens*, *Bacillus subtilis*, *Clostridium perfringens* and *Staphylococcus aureus*. Although MscL channels are clearly widespread in prokaryotes, searches for homologues in eukaryotic cells have so far proved disappointingly negative.

The MscL channel in *M. tuberculosis* (TbMscL) has proved particularly valuable. It shows a 37% sequence identity with *E. coli* MscL (EcoMscL) and of all the MscL channels cloned it proved the easiest to crystallise. When the crystals were subjected to X-ray analysis a fascinating structure was revealed. TbMscL was shown to consist of five identical subunits arranged around a central pore. Each subunit consisted, as expected, of two transmembrane helices designated **M1** and **M2**, and a continuation into the cytosol at the C-terminal end. The total pentameric channel complex surrounds a vase-shaped canal which in the closed conformation (as seen in the crystals used for X-ray diffraction) has a minimum radius of 1 Å (Plate 4). The C-terminal domains of the five subunits surround a continuation of the pore, at its narrowest diameter, into the cytosol. This continuation takes the form of another five α-helices, labelled **S** helices (Plate 4). The MscL canal is thus a double structure. There is evidence to show that when the membrane is stretched the M1 and M2 helices rotate rather like an iris diaphragm and take up more diagonal positions, but the cytosolic ring, comparatively free from stretching forces, remains unchanged and plugs the exit from the pore. Further stretching pulls the S-helices apart from each other thus opening a wide canal of up to 30 Å in diameter (see Plate 4). This evidence comes from mutational substitution of amino acids in critical parts of the S and M helices in *E. coli*: direct structural analysis of the open TbMscL has yet to be achieved.

Finally, what role do these mechanosensitive channels play in the life of bacteria? Perhaps somewhat surprisingly there is no generally accepted answer to this question. The most likely reason for their existence has to do with osmoregulation. When exposed to fresh water, perhaps downpours from the sky, bacteria experience considerable osmotic stress. Water molecules flow down their concentration gradients into the cell. The consequent swelling would certainly be sensed by the mechanosensitive channels in the plasma membrane. It may be that in letting solutes (other than macromolecules) flow out they counteract the osmotic stress by making the water concentration gradient less steep. In addition to a function in detecting osmotic swelling, stretch receptors have also been thought to play a part in detecting swelling due to growth thus signalling the onset of a period when cell division should occur.

13.3.2 Mechanosensitivity in *Caenorhabditis elegans*

Like all animals *C. elegans* is sensitive to touch. Two touch stimuli have been investigated. A gentle touch on the body with an eyelash glued to a cocktail stick and a more vigorous prod with a thin wire. The most interesting results have come from investigations of the response to the gentle stimulation.

When *C. elegans* is placed in a Petri dish it moves forward with a sinusoidal motion until it encounters a tactile stimulus. If it is given a gentle touch on the anterior part of its body the movement reverses and the worm moves backward across the dish; if the gentle touch is applied to the posterior it will move forward. A large number of mutations (over

440) affecting 15 genes have been shown to interfere with these responses. That the worms were still capable of movement was confirmed by using the vigorous wire prod. Most of these mutations were classified as *mec* **mutants** because the phenotype they generate is described as mechanosensitive abnormal (Mec). In consequence the genes are called *mec* **genes** and the proteins which they designate **MEC proteins**.

It can be shown that touch is detected by six touch detector neurons: two anterior lateral microtubule cells (left and right) – ALML and ALMR; two posterior lateral microtubule cells (left and right) – PLML and PLMR; one anterior ventral microtubule cell – AVM. One further neuron seems to be involved although, unlike the others, it is unable to mediate touch avoidance on its own; this is the posterior ventral microtubule cell – PVM. The anatomical location of these touch-detector neurons is shown in Figure 13.21.

The touch receptor neurons, as their names indicate, are characterised by a bundle of extra-large microtubules. Electron microscopy suggests that the individual microtubules in the bundle are linked together and they all have the peculiarity of consisting of 15 protofilaments (rather than the 11-protofilament ultrastructure of other *C. elegans*

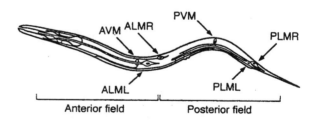

Figure 13.21 *C. elegans* touch receptor neurons. There are two fields of touch sensitivity in the nematode's body: an anterior and a posterior field. The position of these fields is determined by the positions of the touch detector neurons. ALML= anterior lateral microtubule cell left; ALMR=anterior lateral microtubule cell right; PLML=posterior lateral microtubule cell left; PLMR=posterior lateral microtubule cell right; AVM=anterior ventral microtubule cell. From Tavernarakis and Driscoll (1997), *Annual Review of Physiology*, **59**, 662. With permission from *Annual Reviews of Physiology*, Volume 59, © 1997, by Annual Reviews www.annualreviews.org

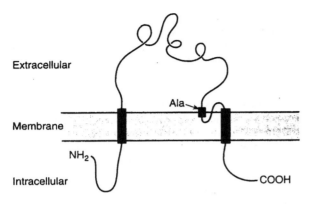

Figure 13.22 The MEC-4 subunit. There are two transmembrane domains and a small membrane insertion before the second transmembrane helix. When alanine$_{713}$ is replaced by a bulkier amino acid death ensues. Mutations affecting the second transmembrane domain lead to neurodegeneration. The *C. elegans* mechanosensitive channel is believed to consist of six or more subunits so that the second transmembrane domains of all the subunits form the lining of the channel. The channel may thus consist of various combinations of MEC-4, MEC-6, MEC-10 and perhaps other subunits.

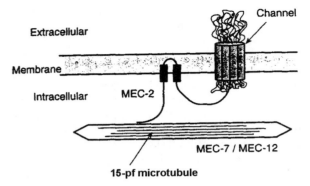

Figure 13.23 Proposed linkage provided by MEC-2 between channel to one of the 15-protofilament (pf) microtubules of a touch-sensitive neuron in *C. elegans*. It is proposed that the N-terminal of MEC-2 interacts with a microtubule and the C-terminal with the channel. After Tavernarakis and Driscoll (1997), 'Molecular modelling of mechanotransduction in the nematode *Caenorhabditis elegans*', *Annual Review of Physiology*, **59**, 662.

microtubules or the 13 protofilaments found in microtubules elsewhere in the animal kingdom; see Section 15.3, Figure 15.12). These extra-large and ultrastructurally unusual microtubules are much shorter than the neuron (no more than about 5% of the latter's length) and seem to run obliquely to the latter's longitudinal axis. Their distal ends are believed to be embedded in the boundary membrane of the neuron, suggesting some mechanical linkage. The microtubules are essential to the functioning of the neuron. If they are disrupted by colchicine or by mutation, touch sensitivity is lost.

The touch-receptor neurons themselves run just beneath the cuticle ('skin') of the worm to which they are attached by an extracellular 'mantle' of fibrous material. The ultrastructure suggests that there is mechanical continuity between the cuticle and the microtubules within the touch neuron's processes.

Genetic analysis proceeded by generating mutations in the worm and examining their progeny for loss of response to gentle touch. When such a loss was detected molecular biological techniques were employed to determine what structure the mutated gene normally designated. In some cases the genes were found to be responsible for factors necessary for the development of the touch cells themselves or for controlling the action of other *mec* genes. In more interesting cases genes were discovered that designated channel proteins and accessory elements connecting the channels to the cuticle on the one hand and to the neuronal microtubules on the other.

Mec-4 and *mec-10* are the two genes that designate the channel proteins – MEC-4 and MEC-10. These genes have also been called '**degenerins**' because mutations in them, leading to defective MEC-4 and MEC-10, cause the cells in which they are expressed to swell and die. Both MEC-4 and MEC-10 belong to a superfamily of proteins which includes the three subunits of the mammalian epithelial Na$^+$ channel protein (ENaC). There are two transmembrane segments, both C-terminal and N-terminal are intracellular and there is a large extracellular loop (Figure 13.22). Although the precise structure of the MEC channel has yet to be determined, analogy with other channels, especially the mammalian ENa channel, suggests an assembly of six subunits, possibly a mixture of MEC-4s and MEC-10s.

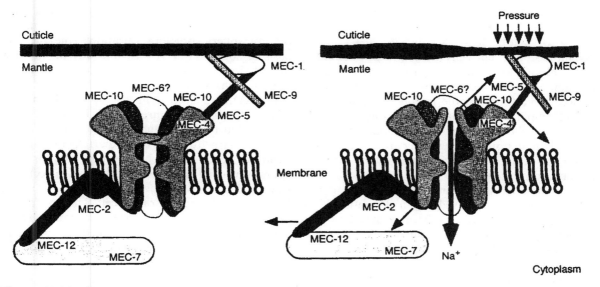

Figure 13.24 Conceptual model of *C. elegans* touch receptor. Explanation and nomenclature in text. From N. Tavernarakis and M. Driscoll, 1997, 'Molecular modelling of mechanotransduction in the nematode *Caenorhabditis elegans*', *Annual Review of Physiology*, **59**, 679. With permission from *Annual Reviews of Physiology*, Volume 59, © 1997, by Annual Reviews, www.annualreviews.org.

There is some evidence that *mec-6* encodes another channel subunit, MEC-6, in which case the channel would consist of a heteromultimer of MEC-4, MEC-6 and MEC-10 subunits.

As mentioned above far more than two *mec* genes are known. The roles played by the proteins that these other *mec* genes encode has in many cases been elucidated. Mutations in *mec-7* and *mec-12* disrupt the 15-protofilament microtubules that are so characteristic of the touch-receptor neurons. Another gene, *mec-2*, has been shown to designate a 481 amino acid protein (similar to the stomatin of mammalian red blood cell cytoskeleton) which is believed to act as a link between the MEC channel and the microtubules (Figure 13.23). Yet other genes, *mec-1, mec-5, mec-9*, code for proteins (MEC-1, MEC-5, MEC-9) that are found outside the plasma membrane in the mantle.

An interesting, though so far somewhat speculative, model has been proposed to link all these elements together into a functioning tactile receptor. This model is shown in Figure 13.24. MEC-1, MEC-5 and MEC-9 are organised to transmit pressure on the cuticle to the MEC-4 subunit and open the channel. When the channel opens Na+ ions flow down their concentration gradients and the neuron is depolarised. This mechanical transmission works if the channel is stabilised by attachment to an internal microtubule. This attachment, as we have seen, is provided by MEC-2. Alternatively, the mechanical transmission might displace the microtubule network within, which, as it is attached to the membrane (see above), might increase tension in the plasma membrane and thus open the channel. The number of 'mights' in the preceding sentence shows that we are here at the frontier of research. Only further careful experiment will eliminate false hypotheses where they exist.

Although, as indicated, much remains speculative, the *C. elegans* touch receptor is much the best-known animal mechanoreceptor at the molecular level. It may, indeed, have more than local interest to specialists in nematode neuromuscular physiology. It has been suggested that the molecular mechanism worked out for the *C. elegans* mechanoreceptor may help elucidate the operation of vertebrate hair cells. It is to these ubiquitous mechanoreceptors that we turn next.

13.3.3 Mechanosensitivity in Vertebrates: Hair Cells

Mechanoreceptive hair cells are found throughout the animal kingdom. They are particularly ubiquitous amongst the vertebrates. They are, for instance, to be found in the lateral line canals of fish and amphibian tadpoles, where they have a function in echolocation, but more importantly, from the human point of view, they are involved in balance and in the detection of sound by the ear, in the human case the inner ear.

The mammalian inner ear, like the mammalian eye, is a very highly evolved organ. Essentially it consists of a complicated membranous structure – the **membranous labyrinth** – held by connective tissues threads within a fluid-filled cavity in the skull's tympanic bone. It consists of three major functional compartments: three **semi-circular canals**, which detect the movement of the head; the **utriculus** and **sacculus**, which detect the position of the head with respect to gravity and are sensitive to linear acceleration; and lastly (but only in the mammals) a **cochlea**, coiled like a snail's shell, which detects sound (Figure 13.25A). Not only is the membranous labyrinth suspended in an aqueous fluid, the **perilymph**, but it also contains an aqueous fluid, the **endolymph**. Thus, in Figure 13.25, the **scala media** contains endolymph whilst the scala vestibuli and scala tympani are filled with **perilymph**. These two fluids differ radically in their ionic constitution. Whereas perilymph resembles other extracellular fluids in having a high Na^+ concentration ($140 \, meq \, l^{-1}$) and a low K^+ concentration ($5 \, meq \, l^{-1}$), the endolymph is much more like an intracellular fluid in being rich in K^+ ($150 \, meq \, l^{-1}$) and poor in Na^+ (1–$2.5 \, meq \, l^{-1}$). We shall see that these ionic constitutions are important in the molecular physiology of the inner ear.

At the heart of this intricate structure are to be found groups of hair cells. The tops of these cells and their sensory hairs project into the endolymph (Figure 13.25B) whereas the remainder of their boundary membranes are bathed in extracellular fluid which, in turn, is in equilibrium with the perilymph. Because they are interposed in this way between lymphs of such very different ionic constitutions a very large electrical potential (about $150 \, mV$) is developed across their membranes.

Textbooks of sensory physiology explain how these cells are arranged so that movement, gravity, or atmospheric pressure variations (the physical basis of sound), will distort, one way or another, their 'hairs'. Figure 13.26 shows the structure of a typical labyrinthine hair cell.

If a scanning electron micrograph of the surface of a single hair cell (Figure 13.27) is examined it can be seen that the hairs tend to stack together into a 'wigwam' structure (preparative artefact cannot be excluded). Each hair cell has up to fifty or sixty 'hairs' springing from its surface. As the sectional view of Figure 13.26 shows, they increase in length from across the surface of the cell. The majority of these 'hairs' are in fact modified (and greatly enlarged) microvilli. They are called **stereocilia**. In addition to the stereocilia a single true cilium, the **kinocilium**, a little taller than the majority of stereocilia and with a bulbous tip, is found at the 'tall' end of the group.

If we increase the magnification yet further and examine a single stereocilium we note first of all that it is much larger than a standard microvillus. Instead of being a micrometer or so in length (as is a microvillus) it may be up to $5 \, \mu m$ and, in some lizards, up to $30 \, \mu m$ in extent. Its diameter is also decidedly greater (100–$900 \, nm$) although it tapers at its basal insertion to less than $100 \, nm$. The interior contains a complicated meshwork of actin filaments cross-linked by fibrin. This meshwork is connected to the membrane by numerous fibrous connectives. Whether this actin network is directly involved in signal detection is not yet known.

It is, however, known that movement of the stereocilia bundle activates the hair cell. Movement occurs by hingeing at the tapered base, the major part of a stereocilium remains rigid. The movement necessary to generate a receptor potential and thence an action potential in an auditory nerve fibre is minute. It can be shown that the lower threshold is a movement of about $0.3 \, nm$, a movement which, as Hudspeth remarks, corresponds (if the stereocilium were scaled up) to the movement of the top of the Eiffel Tower by a 'thumb's breadth'. At the end of Section 13.2 we noted how a single photon of light was detectable by a rod cell and hence could engender huge effects through the visual cortex. We find another instance of this amplification through multiple orders of magnitude here. A movement of $0.3 \, nm$ is at or below the

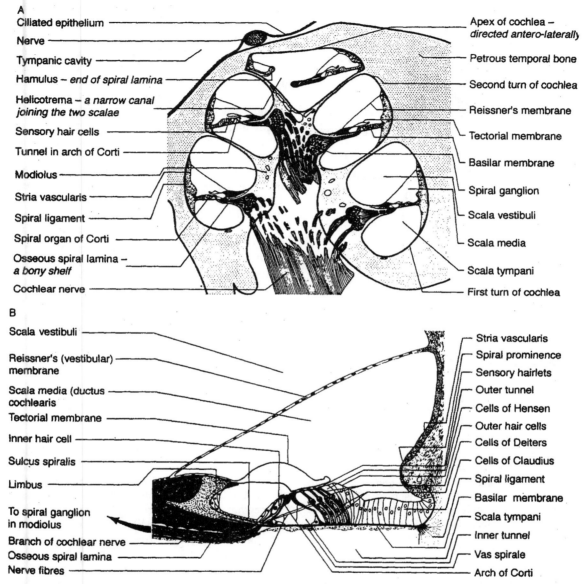

A

Ciliated epithelium

Nerve

Tympanic cavity

Hamulus – *end of spiral lamina*

Helicotrema – *a narrow canal joining the two scalae*

Sensory hair cells

Tunnel in arch of Corti

Modiolus

Stria vascularis

Spiral ligament

Spiral organ of Corti

Osseous spiral lamina – *a bony shelf*

Cochlear nerve

Apex of cochlea – *directed antero-laterally*

Petrous temporal bone

Second turn of cochlea

Reissner's membrane

Tectorial membrane

Basilar membrane

Spiral ganglion

Scala vestibuli

Scala media

Scala tympani

First turn of cochlea

B

Scala vestibuli

Reissner's (vestibular) membrane

Scala media (ductus cochlearis)

Tectorial membrane

Inner hair cell

Sulcus spiralis

Limbus

To spiral ganglion in modiolus

Branch of cochlear nerve

Osseous spiral lamina

Nerve fibres

Stria vascularis

Spiral prominence

Sensory hairlets

Outer tunnel

Cells of Hensen

Outer hair cells

Cells of Deiters

Cells of Claudius

Spiral ligament

Basilar membrane

Scala tympani

Inner tunnel

Vas spirale

Arch of Corti

Figure 13.25 The cochlea. (A) The cochlea is wound in the form of a snail's shell. The sectional view in the figure is taken vertically through the 'snail shell' conformation. (B) Detail of the scala media and organ of Corti. From Freeman and Bracegirdle (1976), *An Atlas of Histology*, London: Heinemann. Reproduced by permission of Butterworth Heinemann.

movement induced by Brownian motion. At its limit our auditory sense is constrained by the uproar of random molecular motion.

Delicate experiments using a glass microprobe to bend stereocilia this way and that have demonstrated that mechanical distortion leads to changes in the electrical potential across the hair cell's plasma membrane (Figure 13.28). Threshold movements induce receptor potentials of about 100 μV. These experiments, moreover, show that movement of the stereocilia in one direction produces a different result from movement in the other. Indeed

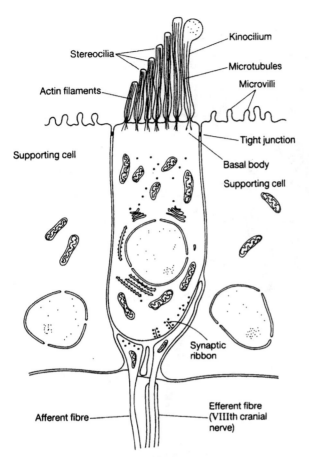

Figure 13.26 Labyrinthine hair cell to show stereocilia and kinocilium. Both stereocilia and kinocilia are nipped in at their bases: they are thus both stiff and able to bend at their junction with the cell body. There are normally many more stereocilia than shown (up to 60). At the base of the cell, above the efferent fibre, are numerous synaptic vesicles many of which are associated (as in rod and cone feet) with synaptic ribbons.

frequencies up to at least 100 kHz. There seems to be no time for the elaborate biochemistry of 'collision coupling' and second messengers which we saw to be at work in chemo- and photoreception. Gated channels in the hair cell membrane must open and shut very rapidly indeed. The electrical response must be due to near-instantaneous flows of ions down their concentration gradients. Where are these gates located?

Again careful experiments have gone far to answer this question. It is probable that the maximum ionic flows in response to movement occur toward the tips of the stereocilia (Figure 13.29). The response, moreover, is found to be extremely rapid, beginning within a few microseconds of stimulus onset and saturating within about 100 microseconds. The currents (about 50 pS at 30°C) are mostly carried by K^+ ions which, as we noted, are present in such high concentration in the endolymph into which the hairs project. The fact that experiment shows that other cations including Ca^{2+} and even some small organic cations, e.g. choline, can pass shows that the channel is fairly wide – perhaps 0.7 nm in diameter. It is estimated that there are about four such channels in each stereocilium.

A model for the biophysics of stereocilia has been put forward by Hudspeth and others. We have already noted that stereocilia are connected near their tips by a molecular thread (tip link) to the next tallest neighbour (Figure 13.27). Evidence has accumulated to show that these threads are in fact attached to ion channels. It is proposed that when the stereocilium is at rest the channels are somewhat leaky. Figure 13.30 represents this by giving an 'open probability' of about 0.1. When a mechanical stimulus moves the assembly of stereocilia towards the kinocilium at the taller end of the group, the 'open probability' of the channels is shifted towards unity. Because, as we have seen, the hairs project into a K^+-rich endolymph, K^+ is the principal ion to flow into the cilium. This, combined with a parallel influx of Ca^{2+} ions, induces a membrane depolarisation. When the movement is in the opposite direction the 'open probability' is decreased towards zero. The K^+/Ca^{2+} flux is shut off and the membrane hyperpolarises.

Identification of the ion channels has proved difficult as there are only three or four in each stereocilium. The amount of protein is thus

if the stereocilia bundle is moved towards the kinocilium, a marked **depolarisation** (up to 20 mV) occurs whilst movement of the stereocilia bundle in the opposite direction elicits a **hyperpolarisation** of some 5 mV (Figure 13.28). Movement at right angles to this axis induces no change in membrane polarity at all. These receptor potentials saturate at movements above 100 nm.

The transduction process is extremely rapid. Many mammals, members of the Cetacea and the Chiroptera, for example, respond to sound

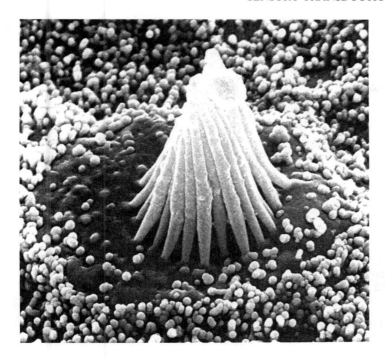

Figure 13.27 Scanning electron micrograph of hair cells in the labyrinth of toad. The bulbous-ended kinocilium can be clearly seen at the top of the 'wigwam' of stereocilia ($\times 16\,000$). Courtesy of Dr R.A. Jacobs.

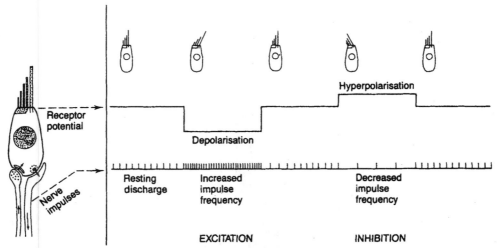

Figure 13.28 Electrical response of hair cell to movement of stereocilia. The figure shows that movement of the bundle of stereocilia in one direction causes a depolarisation of the hair cell whilst movement in the opposite direction causes a hyperpolarisation. This is translated into increased and decreased impulse frequency in the sensory fibre. From Flock (1965), *Cold Spring Harbor Symposia on Quantitative Biology*, 133–145; with permission.

vanishingly small (a few attomoles). Is it possible, however, to use a comparative approach? One of the striking outcomes of modern molecular biology has been the demonstration of a remarkable unity across the living world at the molecular level. In the previous section we looked at the touch receptors in *C. elegans*. We saw that there was evidence that they were related to vertebrate epithelial Na$^+$

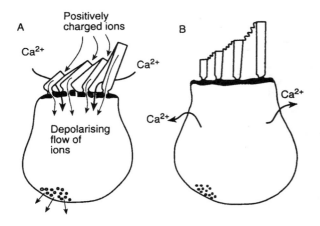

Figure 13.29 Mechanoelectrical transduction in hair cells. (A) The stereocilia are all pushed towards the left and this causes K^+ and Ca^{2+} channels to open in their tips. The inward flow of positively charged ions causes a depolarisation which opens further Ca^{2+} channels, thus leading to release of transmitter on to the dendritic ending of the underlying sensory neuron. (B) When the stereocilia return to their resting position the ion channels at their tips are closed. The Ca^{2+} which has flowed in during the depolarising stage is sequestrated or pumped out of the cell, the membrane repolarises, and the release of transmitter ceases.

channels (ENaCs). Could the ion channels in stereocilia belong to the same family? Suggestive evidence in support comes from studies of channel blockers. ENaCs are very sensitive to amiloride. The renal Na^+ channel is blocked by nanomolar concentrations of this agent. It is found that amiloride also blocks the channels in stereocilia. It may be, therefore, that the *C. elegans* channel and that of vertebrate hair cells are related. If this is so some of the elegant molecular biology carried out on the nematode may have relevance to the operation of the human ear.

The mechanotransductive mechanism outlined above ensures that the response to mechanical stimulation is almost instantaneous (as it is observed to be) but it is also known that most hair cells adapt very rapidly (at most a few tens of milliseconds). In other words, when the deflection of the stereocilia is maintained for more than about a millisecond the influx of K^+/Ca^{2+} and the consequent depolarisation ceases. How is this brought about? Recent investigations have revealed a remarkable mechanism within the stereocilium. A schematic of the ultrastructure is shown in Figure 13.31. In addition to external tip links the figure shows that channel protein is attached via a myosin

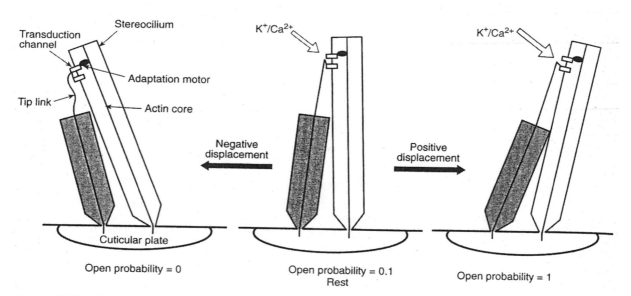

Figure 13.30 Gating-spring model of hair cell stimulation. The gate is pulled open when the cilium moves to the right. Explanation in text. From Gillespie (1995), 'Molecular machinery of auditory and vestibular transduction', *Current Opinion in Neurobiology*, **5**, 449–455; with permission.

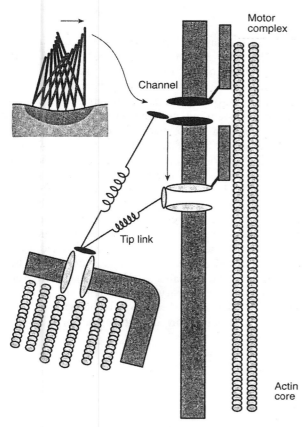

The mechanism responsible for adaptation depends on the influx of Ca^{2+} when the ion channel is opened. One theory suggests that this leads to a loosening of the linkage between the myosin 'motor' and the actin microfilaments within the stereocilium. This, in turn, allows the ion channel to slip down the stereocilium's membrane thus releasing tension on the tip link. The 'open probability' of the channel then decreases and the membrane polarity returns to its resting value. When the stereocilium returns to its normal, near-upright position, the channel protein, now under no internal tension, moves toward an 'open probability' of zero. In the absence of Ca^{2+} influx the motor climbs back up the actin filament pulling the channel with it, thus returning the tension on the tip link to its original value and the open probability to about 0.1. A word of caution: although the hypothetical mechanism of adaptation is very appealing much more research is needed to establish its reality beyond a peradventure.

In addition to Ca^{2+} influx through the stereociliary transduction channels further Ca^{2+} ions will enter through voltage-dependent channels located in the cuticular plate. These channels open when the membrane depolarises which, as we have seen, occurs as a consequence of stereociliary displacement. The influx of Ca^{2+} ions leads in turn to the release of transmitter substance from the base of the hair cell. Depending on the nature of the transmitter and the nature of the underlying dendrite this will either initiate or inhibit the initiation of a generator potential. In the first case an action potential will be triggered which will be propagated along a sensory nerve into the central nervous system.

Although the ground-breaking investigations of the biophysics of the vertebrate hair cell have been carried out on the hair cells of the inner ear there is little doubt that the mechanisms there discovered can be generalised to apply to the working of hair cells in other parts of the subphylum Vertebrata. Hair cells are, in fact, evolutionarily very ancient and develop in all classes of the vertebrates from the ancient agnathan fish to the mammals. In the Cyclostomata, which include the only living representatives of the Agnatha (the hagfish, *Myxine* and the lamprey, *Lampetra*), hair cells are developed in both the lateral line canals and the ears.

Figure 13.31 Adaptation in hair cells. When the hair is deflected the channel is opened and the 'motor complex' detaches from the actin filaments. The channel slips down the stereocilium (as shown), the tension on the gate is reduced and it closes. Later, when the stereocilium returns to its upright position, the 'motor complex' climbs back up the actin filament, dragging the channel back to its resting position. Further explanation in text. From Gillespie and Corey (1997), 'Myosin and adaptation by hair cells', *Neuron*, **19**, 955–958; with permission.

'motor' to the internal actin microfilaments. This attachment provides a platform to stabilise the position of the channel in the membrane so that increased tension on the tip links does not simply pull the whole channel protein down the stereocilium in the plane of the membrane. Again an interesting comparison can be made with the way in which the *C. elegans* touch receptor is attached to tubulin molecules with the neurosensory cell (Figures 13.23 and 13.24).

It should be borne in mind that although this model of the hair cell's mechanism accounts for the **microphonic potentials** which can be detected in the cochlea it does not, on its own, provide an explanation of the ear's ability to discriminate between different frequencies of sound. The hair cell, as we have seen, may be able to respond in a matter of microseconds; synapses and nerve fibres are much more sluggish – generally having response times measured in milliseconds. Nerve fibres, furthermore, are restricted by their refractory periods (see next chapter) to conduction rates of less than a thousand impulses per second. Hence it is impossible for single auditory fibres to keep up with the response of a hair cell detecting sound of, say, 15 kHz (human) or still less 100 kHz (bat). The explanation of the cochlea's remarkable frequency discrimination ability lies in the differential resonance of different regions of the basilar membrane. This explanation, known as 'place theory', originally proposed by Helmholtz in the mid-nineteenth century, is set out in textbooks of physiology.

13.4 CONCLUSION

In this chapter we have looked at just a few examples of the innumerable sensory cells found in the living world. More extensive accounts may be found in the texts on sensory systems listed in the Bibliography. We have progressed from a bacterial system which connected sensory detection to action in a very direct way, through neurosensory cells of the olfactory system which transduce sensory stimuli into action potentials in the same cell, to sensory cells which detect a stimulus and via complicated biochemistry and biophysics produce receptor potentials to, finally, receptor cells which respond directly to a stimulus by the opening and closing of ion gates. The last two types of sensory cell are incorporated into highly evolved sense organs: eyes and ears. The detection of the stimulus in these cases is only the beginning of the sensory process. It forms the raw material which is worked upon in deeper parts of the sense organs and *a fortiori* in the sensory cortices.

14

THE ACTION POTENTIAL

Centrality of the action potential in neuroscience. **Analysis by the voltage-clamp technique**: ion flows responsible for changes in membrane potential during action potential – changes in membrane permeability responsible for these flows – number of ions involved. **Patch-clamping**: single-channel analyses – positive feedback between Na^+ channel opening and membrane depolarisation – number of Na^+ channels in unit area of axonal membrane – K^+ channels – role of delayed K^+ current (I_K) – role of fast K^+ current (I_{KA}). **Propagation of action potential**: orthodromic and antidromic propagation – role of local circuits – role of AHP in ensuring unidirectional propagation – role of $Na^+ + K^+$ pump in re-establishing the ion distributions across membrane. **Initiation of impulse**: initial segment of axon – role of Ca^{2+}-dependent K^+ channel in sensory adaptation. **Rate of propagation**: significance of λ (the space constant or electrotonic length) – giant fibres versus myelination – saltatory conduction – multiple sclerosis (MS) – role of node of Ranvier

The **action potential** (=**nerve impulse or 'spike'**) lies at the heart of neurophysiology. It is the element with which the brain scientist builds a theory of brain activity – whether it be the EEG of the association areas, the visual system's 'primal sketch' or the cortical computation that precedes a conscious act. It is the action potential that Sherrington had in mind when he wrote his famous passage describing the brain stirring from slumber:

Swiftly the head-mass becomes an enchanted loom where millions of flashing shuttles weave a dissolving pattern, always a meaningful pattern though never an abiding one; a shifting harmony of sub-patterns. Now as the waking body rouses, sub-patterns of this great harmony of activity stretch down into the unlit tracts of the stalk piece of the scheme. Strings of flashing and travelling sparks engage the lengths of it. This means that the body is up and rises to meet its waking day. (Sherrington, 1951, p. 187)

But for the molecular neurobiologist the action potential is itself a point of synthesis. A synthesis of numerous underlying elements – different voltage- and ligand-gated channels, differential ion flows, the biophysics of cable conduction and electrotonic potentials.

14.1 VOLTAGE-CLAMP ANALYSES

The modern phase of our understanding of the action potential began with the experiments of Hodgkin and Huxley in the early 1950s. Working with the squid giant axon preparation they were able to model closely the shape of the action potential and determine the nature of the ion flows responsible for that shape. Figure 14.1 shows a classical recording from this preparation.

Figure 14.1 shows that when a microelectrode is placed within a squid axon a **resting potential** (V_m) of about $-60\,mV$ can be recorded. When an impulse or **action potential** passes over this patch of membrane the potential is reversed so that the internal electrode becomes about $+35\,mV$ compared to the external.

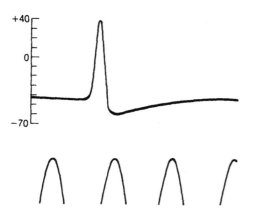

Figure 14.1 Action potential recorded from a squid giant axon. The vertical scale indicates the potential of the internal electrode (mV), the sea-water outside being taken to be at 0 or ground potential. Time marker 500 Hz. This is one of the first pictures of a complete action potential to be published. Reprinted by permission from Hodgkin and Huxley (1939), *Nature*, **144**, 710–711. Copyright © 1939, Macmillan Magazines Ltd.

This reversal of polarity is very quickly over. The peak of the action potential is reached about 0.5 ms after onset and the membrane has returned to its resting potential value in about 1.5 ms. The figure shows that the recovery process in fact overshoots the resting potential value. The **after-hyperpolarisation (AHP)** persists for 1–2 ms before the normal resting potential (V_m) is re-established.

In order to study this extremely rapid voltage change Hodgkin and Huxley used a device first developed by Marmont and Cole to clamp the voltage across the membrane at any desired value and hold it at that value. Figure 14.2 shows a schematic diagram of a voltage-clamp circuit.

In essence a voltage-clamp circuit consists of four electrodes – two inside (1 and 2 in Figure 14.2) and two outside (3 and 4 in Figure 14.2). Electrodes 1 and 3 measure the electrical potential across the axonal membrane. Electrode 2 is used to inject electrons (i.e. current) into or remove them from the axon. If it is required to **depolarise** the membrane then electrons can be removed from the inside of the axon. This removal of negative charge from inside the axonal membrane is monitored by electrodes 1 and 3. The electronic circuitry ensures that just enough current is withdrawn to hold the membrane at the required polarity. This current is obviously identical to the current which ions are carrying across the membrane at that particular depolarisation. Vice versa, if it is required to study transmembrane currents when the membrane is hyperpolarised the current flow in electrode 2 may be reversed. Electrons injected into the axon will **hyperpolarise** the membrane.

As Hodgkin and Huxley remarked, the voltage-clamp technique enables one to slow down the millisecond events of the action potential and study them at leisure. It is comparable to taking a movie

Figure 14.2 Voltage-clamp circuit. In this schematic figure two electrodes are inserted into the axon and two are placed outside. Electrodes 1 and 3 measure the voltage across the membrane. When a command voltage is applied to the feedback amplifier current is injected (or removed) from the axon by electrode 2. The current flow across the membrane required to maintain it at any specified voltage can then be measured by electrode 4. After Katz (1966), *Nerve Muscle and Synapse*, New York: McGraw Hill.

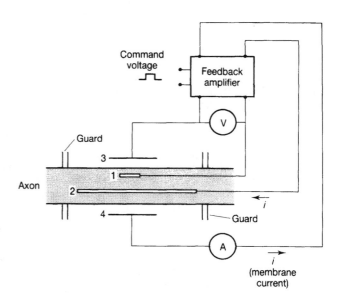

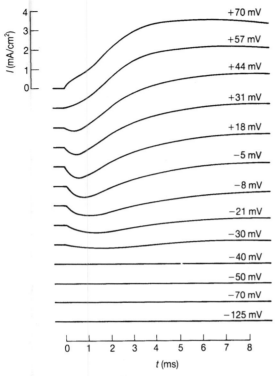

Figure 14.3 Voltage-clamp measurements of currents across a squid giant axon. Currents (mA/cm²) across the membrane of a squid giant axon clamped at various potentials. Downward deflection indicates inward current. Further explanation in text. After Hodgkin, Huxley and Katz (1952), *Journal of Physiology*, **116**, 424–448.

of a fast-moving event and then studying the action frame by frame under a magnifying glass or, better, using the freeze-frame control on a video recorder.

Figure 14.3 shows the effect of clamping the transmembrane potential at a number of different values. The resting potential is $-60\,\mathrm{mV}$ and the voltage is stepped upward at discrete intervals. Inward current is represented by a downward deflection of the curve. The figure shows that no transmembrane current can be detected until the membrane is set at $-30\,\mathrm{mV}$ (a depolarisation of $30\,\mathrm{mV}$). An inward current then begins to show itself followed by a sustained outward current. The initial inward current becomes more definite as the membrane is stepped to $-21\,\mathrm{mV}$, $-8\,\mathrm{mV}$, $-5\,\mathrm{mV}$, $+18\,\mathrm{mV}$ but then begins to diminish until it disappears altogether at $+57\,\mathrm{mV}$. The outward current, however, grows more and more marked.

The next question, of course, is: exactly which ions are carrying these observed electric currents? This question is also easily tackled with the voltage-clamp technique. There are two complementary approaches. First, one can remove the suspected ionic carrier and see if that has the expected result on the transmembrane current, or, second, one can block the membrane channels used by the suspected ions and, once again, observe the effect on the current across the voltage-clamped membrane. Both these approaches are shown in Figure 14.4. If the external Na^+ ions are replaced with the impermeant cation choline, the initial inward current is eliminated. If the potassium channels are blocked by tetraethylammonium chloride (TEA), then the sustained outward-going current disappears.

The voltage-clamp experiments of the 1950s were thus able to dissect the action potential into two phases. A sudden initial ingress of Na^+ ions followed by a longer-lasting egress of K^+ ions. It is clear that the electrical phenomena of the action potential (Figure 14.1) are ultimately based on abrupt changes in the permeability of the membrane to these ions. Let us make use of the Goldman equation (Chapter 12) to assure ourselves that such permeability changes could account for the shape of the impulse. Let us suppose, first, that the membrane becomes suddenly very much more permeable to sodium ions. Let us suppose, in other words, that instead of $P_{Na} \approx 1 \times 10^{-8}\,\mathrm{cm\,s^{-1}}$ it becomes $1 \times 10^{-6}\,\mathrm{cm\,s^{-1}}$ – an increase by a factor of 100. The Goldman equation would then predict a transmembrane potential of about $+43\,\mathrm{mV}$ – very much what is observed at the peak of the action potential. Similarly if we increase the permeability constant of the potassium ion, P_K, by a factor of about five to $5 \times 10^{-7}\,\mathrm{cm\,s^{-1}}$ then a transmembrane potential of about $-77\,\mathrm{mV}$ is predicted – as observed during the after-hyperpolarisation.

It is also possible to work out the number of ions required to flow through unit area of membrane to elicit these voltage changes. If we take the capacitance of the membrane to be about $2\,\mu\mathrm{F\,cm^{-2}}$ (see Appendix 3) and the change in potential during the rising phase of the action potential to be about $100\,\mathrm{mV}$ (i.e. from $-60\,\mathrm{mV}$ to $+40\,\mathrm{mV}$), then remembering from elementary physics that

$$Q = VC$$

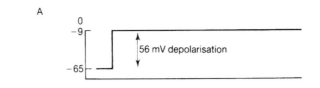

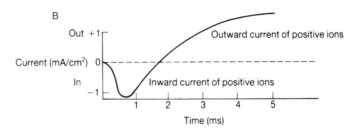

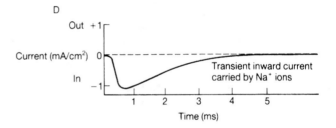

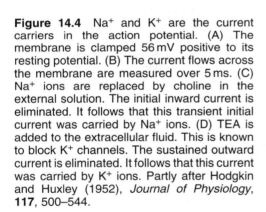

Figure 14.4 Na$^+$ and K$^+$ are the current carriers in the action potential. (A) The membrane is clamped 56 mV positive to its resting potential. (B) The current flows across the membrane are measured over 5 ms. (C) Na$^+$ ions are replaced by choline in the external solution. The initial inward current is eliminated. It follows that this transient initial current was carried by Na$^+$ ions. (D) TEA is added to the extracellular fluid. This is known to block K$^+$ channels. The sustained outward current is eliminated. It follows that this current was carried by K$^+$ ions. Partly after Hodgkin and Huxley (1952), *Journal of Physiology*, **117**, 500–544.

where Q=quantity of electricity in coulombs (C), V=potential difference across the membrane in volts (V) and C=capacity of membrane in farads (F), we have

$$Q = 100\,\text{mV} \times 2\,\mu\text{F cm}^{-2}$$
$$= (1 \times 10^{-1})\text{V} \times (2 \times 10^{-6})\text{F cm}^{-2}$$
$$= 2 \times 10^{-7}\text{C cm}^{-2}$$

Next we note that $1\,\text{cm}^2=1\times10^8\,\mu\text{m}^2$ so that the quantity of electricity carried across $1\,\mu\text{m}^2$ is

$$Q = (2 \times 10^{-7}) \times (1 \times 10^{-8})$$
$$= 2 \times 10^{-15}\text{ C}$$

Then we recall that 96 500 (i.e. approx. 1×10^5) C are carried by 1 g ion and it therefore follows that

2×10^{-15} C are carried by $2 \times 10^{-15}/1 \times 10^5$
$$= 2 \times 10^{-20}\text{ g ions}$$

Furthermore as 1 g ion consists of 6×10^{23} ions (Avogadro's number)

2×10^{-15} C are carried by $(2 \times 10^{-20}) \times (6 \times 10^{23})$

$$= 12 \times 10^3 \text{ ions}$$

This calculation emphasises how very delicately balanced the membrane is. It requires only the influx of twelve thousand or so Na^+ ions to reverse its polarity. Reference back to Chapter 11 will show that when the membrane is clamped at 10 mV positive to its resting potential about 10 000 Na^+ ions course through a single channel each millisecond. This flow, of course, diminishes $(I = gV)$ as the voltage across the membrane moves towards V_{Na}. It is also important to bear in mind that the neurilemma possesses several different types of channel. A few microseconds after the opening of the sodium gates the 'delayed' K^+ gates open. Rather little Na^+ will have flowed through a single channel (e.g. 100 ions in 10 μs) before these I_K channels open. The outward flow of K^+ ions tends to repolarise the membrane. Hence the very few Na^+ ion gates implied by the above calculation must be multiplied many-fold to achieve the conductivity needed to drive the membrane to its action potential peak (see Figure 14.7).

14.2 PATCH-CLAMP ANALYSES

We looked at the techniques of patch-clamping in Section 10.1.2. We saw that by this means single channels could be studied. We saw, furthermore, in Chapters 10 and 11 that patch-clamping had been extensively used to investigate ion channels. In particular we noted the characteristics of single Na^+ and K^+ channels. It is the functioning of multiplicities of these voltage-gated channels that underlies the ion flows detected by the Hodgkin–Huxley voltage-clamp analyses.

It will be recalled from Chapter 11 that sodium channels open in response to a depolarising voltage, stay open for about 0.7 ms, **and then close again**. It will also be recalled that the number of Na^+ channels varies widely in different neurons – from about 10 μm^{-2} in rat sarcolemma, to 100 μm^{-2} or so in rabbit unmyelinated vagus fibres, up to about 300 μm^{-2} in squid giant axon and several thousand per square micrometre in nodes of Ranvier.

Individual channels in any given membrane population will have slightly different trigger

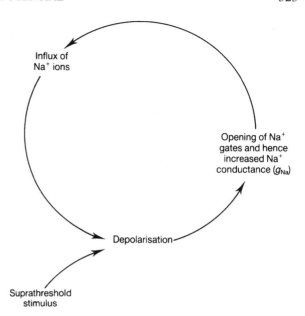

Figure 14.5 Positive feedback loop responsible for the initial phase of the action potential.

voltages – but, as already noted, once having opened they shut after about 0.7 ms and do not open again. It is easy to see that the affect of this sudden leakiness to sodium ions will be precisely the Na^+ current that Hodgkin and Huxley observed in their voltage-clamp experiments.

Once a depolarising stimulus reaches the threshold value Na^+ channels open and sodium ions begin to flood in, thus depolarising the membrane yet further. This has the effect of opening yet more sodium gates and yet more sodium ions cascade in, and so on. This positive feedback loop ensures that the rising phase of the action potential shoots up to its peak in half a millisecond or so (Figure 14.5).

The patch-clamp technique allows one to examine the channel events underlying an action potential. If a small patch of membrane containing only two or three channels is repeatedly depolarised to threshold at regular intervals, the sodium channels can be exercised again and again. Figure 14.6B shows three such exercises of the channels in response to 60 mV depolarisations from a 'holding potential' of −110 mV. If the experiment is repeated 144 times and the results summed it is equivalent to depolarising a patch of membrane

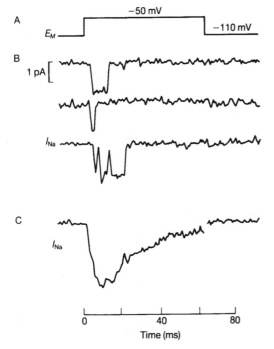

Figure 14.6 Single and summed responses of Na⁺ channels in a patch of myotube membrane to a 60 mV depolarisation from a holding potential of −110 mV. (A) The membrane potential is stepped from a holding potential of −110 mV to −50 mV. (B) Three records showing the opening of sodium gates in the membrane and the inward flux of current carried by Na⁺ ions. (c) The sum of 144 trials. From Hills (1984), *Ionic Channels of Excitable Membranes*, Sunderland, MA: Sinauer; after Patlak and Horn (1982), *Journal of General Physiology*, **79**, 333–351; with permission.

containing several hundred channels once. The result, shown in the lower part of Figure 14.6, shows a marked similarity in profile to the sodium current curve of Figure 14.4.

We noted in Chapter 11 that each sodium channel allows about 1.6 pA of current to flow. Reference to Figure 14.4, however, shows that the sodium current measured by voltage-clamping entire axons is to be measured not in picoamps but in milliamps. In order to achieve currents in the mA range approximately 10^9 channels must be opening. However, it can be easily seen that this is not an unreasonable number. For if we suppose that there are 100 channels μm^{-2} then there are 100×10^8 channels per square centimetre – which is

an order of magnitude more than the number required. We can also work the calculation the other way and get a feel for the density of sodium channels in the membrane. If we suppose that the diameter of the channel protein is about 10 nm (see Chapter 11) then a square micrometre of membrane (i.e. 1×10^6 nm²) could contain 10 000 such channels if they were packed tightly edge to edge. But it is found that $1 \mu m^2$ of membrane only contains about 100 such channels. The packing density is thus only about 1%. In reality, as we have already noted, neurotoxin labelling shows that the packing density is rather higher: several hundred per square micrometre of membrane in squid giant axon. This implies that not all the channels in a patch of membrane open at once.

Let us now turn our attention to the potassium channels. In Chapter 11 we noted that excitable membranes possess a number of different K⁺ channels. The channels that are responsible for the sustained outward flux of K⁺ ions (I_K) shown in the voltage-clamp experiments are the so-called delayed K⁺ channels. These open rather sluggishly to depolarisation of the membrane – by sluggish is meant about 2 μs! They remain open, moreover, as long as the membrane is depolarised. The greatly increased K⁺ conductance they confer ensures that the membrane is brought back to the resting potential and beyond, to the AHP. We shall see, later, that this AHP is not a mere detail but has an important neurophysiological consequence.

Figure 14.7 shows how the combined action of all these biophysically different channels affects the electrical potential across the axonal membrane. The outcome is the central phenomenon of neurophysiology: the action potential. The huge complexity, both anatomical and physiological, of human and infra-human nervous systems is built around this rapid electrical fluctuation and yet, as mentioned at the beginning of this chapter, it is itself the resultant of the cooperative activity of large numbers of different, exquisitely 'designed', molecular channels and machines. Molecular neurobiology derives directly from the Watson–Crick reduction of Darwinism to the molecular: yet few can remain unastonished at the way all the molecular elements fit together to form such a unified whole.

In addition to the delayed K⁺ channel practically all neurons (except squid giant axons) possess fast

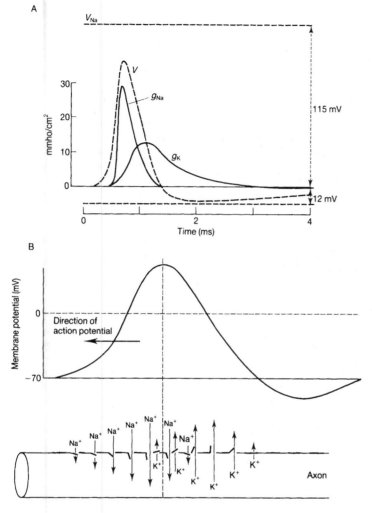

Figure 14.7 Major ion channels and conductances responsible for the action potential. (A) Changes in the sodium and potassium conductances (g_{Na} and g_K) predicted by the Hodgkin–Huxley theory. The left-hand ordinate shows the conductance of the membrane for Na^+ and K^+ ions (mmho/cm^2); the right-hand ordinate shows the voltage across the membrane. The broken line shows the total voltage change across the membrane as the action potential develops. The Nernst sodium (V_{Na}) and potassium (V_K) are shown by horizontal broken lines at top and bottom of the figure. From Hodgkin and Huxley (1952), *Journal of Physiology*, **117**, 500–544; reproduced by permission of the Physiological Society. (B) The upper part of the figure shows the change in voltage across the membrane of an axon (compare with broken line in A above). The lower part of the figure shows the flows of sodium and potassium ions which are the major cause of this voltage change. These flows, in turn, are due to the opening and closing of Na^+ and K^+ channels in the axonal membrane.

K^+ channels (responsible for the fast K^+ current (I_{KA})). We noted in Figure 14.3 that small depolarisations of the membrane did not initiate an action potential. Only when the depolarisation had reached a threshold 10–20 mV below the resting potential did the sodium gates open and an action potential occur. The neuronal membrane is thus, in a sense, elastic. It 'bounces' back from small, subthreshold, depolarisations. This may be helped by the presence of the fast K^+ channels. Although the channels are present in very small numbers their presence allows K^+ ions to flow out rapidly in response to small depolarisations and thus quickly repolarise the membrane. Not all workers, it should however be said, accept this mechanism. It has been

pointed out that most of the fast K^+ channels require the membrane to be hyperpolarised before they open. The significance of this observation will become apparent in the next section of this chapter where we consider the role of the AHP in impulse propagation.

14.3 PROPAGATION OF THE ACTION POTENTIAL

So far we have been considering the ionic movements occurring in a single small patch of axonal membrane. The whole essence of the action potential, however, is that it is a signalling device for tying together the far-flung operations of the brain,

BOX 14.1 Early history of the impulse

Homer spoke of 'wind swift thought'. He marvelled at the rapidity of the human mental process. Shakespeare, too, in Hamlet's soliloquy on man refers to his rapidity of movement and celerity of thought. The speed with which messages are transmitted from mind to hand and with which decisions are made have long been a source of wonderment.

It was not until the late eighteenth century that a beginning was made in understanding the nature of the impulse and hence the rate at which it was propagated. It was then that the science of electricity was emerging and much interest was shown (not only in scientific circles) in the possibility that nerve transmission might also be an electrical phenomenon. This idea gained wide popularity by the success of demonstrations such as those presented by Stephen Gray in the mid-eighteenth century. In these a boy, often a pupil at Charterhouse School, was suspended by silken threads from the rafters of the hall. The boy's feet were touched by an electrically charged body and his face immediately became capable of attracting feathers. There is no record of how the boy was chosen, what he felt, or of the hilarity amongst his classmates.

Nevertheless, whatever the schoolboy may have thought, the experiment was repeated again and again, and not only in England. It was shown at the French court by Nollet and in Leipzig by Hausen. It had one serious implication. It showed that the human body conducted electricity. But it was not until the 1780s that enlightenment began to dawn. Galvani carried out the experiments that set in motion two great sciences: neurophysiology and electromagnetism. Here is his description of what happened:

> I dissected and prepared a frog and laid it on a table, on which, at some distance to the frog was an electric machine. It happened by chance that one of my assistants touched the inner crural nerve of the frog with the point of a scalpel: whereupon at once the muscles of the limbs were violently convulsed.
>
> Another of those who used to help me in electrical experiments thought he had noticed that at this instant a spark was drawn from the conductor of the machine. I myself was at the time occupied with a totally different matter; but when he drew my attention to this, I greatly desired to try it myself and discover the hidden principle.

Serendipity plays its part in science – a surprisingly large part. But it demands a prepared mind. Galvani might so easily have dismissed his assistant's interruption as a nuisance, an irrelevance. Instead he pursued his assistant's observation with a single-minded intensity which, in a century so fascinated by electrical phenomena, gained widespread publicity.

There is no space here to follow the immediate history of Galvani's experiments or to deal adequately with his controversy with his fellow electrician, Volta: suffice it to say that, as he writes in his 1791 treatise, he believed he had at last solved one of the major problems of life: '. . . perhaps at last the nature of animal spirits, which has been hidden and vainly sought for so long will be brought to light with clarity'.

But what form did the electrical message in the nerves take? The measuring instruments at the end of the eighteenth and beginning of the nineteenth century were too slow to pick up the rapid fluctuations that we now know to constitute the action potential. These instruments did not really become available until the development of electronics in the middle of the twentieth century. Nevertheless the middle of the nineteenth century saw considerable progress in establishing the nature of the impulse and its rate of propagation. Carlo Matteuci, Emil Du Bois-Reymond and most significantly Hermann von Helmholtz all made important contributions. Helmholtz by means of a brilliantly simple experiment succeeded in establishing the rate of impulse propagation in a frog's sciatic nerve. Estimates in the eighteenth and early nineteenth centuries had ranged from 9000 feet/min to 57 600 million feet/min and most believed, often on metaphysical grounds, that the actual velocity could never be ascertained. Helmholtz's experiment gave a value of about 90 feet/second (i.e. 30 m/s), a value still accepted today.

Neurophysiology was at last getting to grips with Homer's 'wind swift thought'. Nerve fibres did not conduct impulses at the speed of light. The brain could not compute instantaneously. The cerebrum was being brought into the realm of 'common things', of physical science. Emil Du Bois-Reymond, in particular, was a particularly convinced advocate of this new understanding. He conceived the destiny of physiology, and within physiology, neurophysiology, to be a closer and closer union with the physical sciences. The account of molecular neurobiology in the pages of the present book show just how far that programme has taken us. Yet Du Bois-Reymond was concerned to draw limits around present and future physical interpretations. In a famous lecture which he gave in Berlin in 1872 he set out his position with great force and clarity. In this lecture *'Uber die Grenzen des Naturkennes'* ('On the Boundaries of Science') he distinguished between questions the answers to which we do not know but which are in principle answerable, and questions to which again we do not know the answers but which are in principle unanswerable. This distinction between questions to which we could reply *ignoramus* from those to which we had to reply *ignoramibus* formed something of a *cause célèbre* at the end of the nineteenth century. Questions such as the origin of life and the neurophysiology of the brain were for Du Bois-Reymond 'world enigmas' for which the correct answer was ignoramus, while questions about the origin of sensation and consciousness, the relation of mind to matter, were questions to which the only correct answer was ignoramibus, we do not and can never know the answer. We shall return to this in Appendix 1.

for sending messages to the muscles and/or for carrying information from the sense organs to the CNS.

If a depolarising stimulus is delivered at a point midway along an axon it will be found that an action potential starts off in both directions from that point (Figure 14.8). The normal physiological direction (from the perikaryon to the synaptic bouton) is called the **orthodromic** direction, the opposite non-physiological direction is termed the **antidromic** direction.

The fact that both antidromic and orthodromic propagation occurs from a stimulus midway down an axon is easily explained in terms of the cable conduction theory we looked at in Chapter 12. It will be recalled that local circuits spread through the axoplasm from a point of depolarisation. In the cases we considered in Chapter 12 these circuits

caused small depolarisations (electrotonic depolarisations) at small distances from the point at which the depolarising voltage had been applied. But in an excitable membrane, such as an axon, the small depolarisations are likely to bring the membrane to the threshold required to activate the sodium gates. The positive feedback mechanism described in the preceding section then takes over. An action potential develops on that section of membrane rather than an electrotonic potential.

Once an action potential develops on that segment of membrane local circuits will *a fortiori* spread out and depolarise new segments – thus the action potential runs in both directions away from the point of stimulation.

The reader will no doubt have noted a difficulty in the scenario depicted in Figure 14.9. If local circuits spread out in each direction from the region

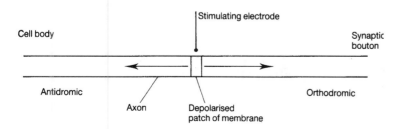

Figure 14.8 Orthodromic and antidromic propagation of an action potential. A membrane is given a suprathreshold stimulus by a stimulating electrode. The action potential propagates in both directions from the point of stimulation.

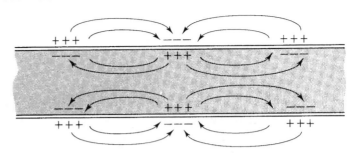

Figure 14.9 Local circuits underlying the propagation of an action potential. Enlargement of Figure 14.8 to show the local circuits set up by a patch of depolarised membrane.

where an action potential has developed how is it that the action potential does not perpetually 'spark back'? Indeed it looks as if the proposed mechanism is altogether a non-starter.

The answer to the conundrum is simple. It has to do with the AHP tacked on to the end of an action potential and with the fact that the sodium channels after having opened during the action potential snap shut and remain shut until the resting potential is re-established. Let us do a little more arithmetic. Figure 14.1 shows that the AHP lasts for at least a millisecond, usually more. Let us suppose that the action potential travels at $20\,\mathrm{m\,s^{-1}}$ – an average value for frog motor nerve fibres. What is the length of fibre left behind the action potential in the AHP state?

$$V = 20\,\mathrm{m\,s^{-1}}$$
$$= 20 \times (1 \times 10^{6})\,\mathrm{\mu m\,s^{-1}}$$

If the AHP state lasts for 1 ms (as we supposed) then the length of axonal membrane left behind the speeding action potential in this state is

$$l = [(2 \times 10^{7})\mathrm{\mu m\,s^{-1}}] \times [1 \times 10^{-3}\mathrm{s}]$$
$$= 2 \times 10^{4}\mathrm{\mu m}\ \text{or 2 cm}$$

This stretch of fibre is populated by inactivated sodium gates. The depolarising influence of local circuits cannot open them. Moreover, the fast potassium channels which are especially sensitive when the membrane is hyperpolarised ensure that any incipient depolarisation is quickly reversed by an outflow of K^{+} ions. Thus the local circuits spreading out from the membrane patch carrying the action potential can only affect fresh membrane. They will be too attenuated to affect membrane 2 cm behind the action potential. Hence the action potential propagates itself without decrement in one direction only (Figure 14.10).

A much favoured analogy for the propagation of an action potential is that of a train of gunpowder or a line of dominoes. In both cases transmission uses the stored energy of the system. A spark running down the gunpowder trail leaves only ash behind, the dominoes fall flat in transmitting the impulse to the next in line and have to be stood up on edge again for a further transmission. And so it is with the nerve impulse. It depends on the energy locked in the ion gradients across the membrane. However, unlike the gunpowder trail or the line of dominoes nerve impulses can be repeated many times before all the stored energy is used up. The properties of the ion channel proteins ensure that the transmembrane ionic gradients are not eliminated by one or a few action potentials. But ultimately the ion gradients must be re-established by means of the $Na^{+}+K^{+}$ ion pumps in the

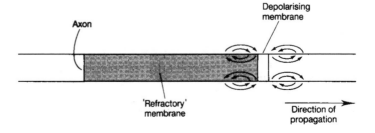

Figure 14.10 Unidirectional propagation of an action potential. Local circuits spread out from the region of depolarised membrane but can only affect 'fresh' membrane. The membrane over which the impulse has passed is refractory. Further explanation in text.

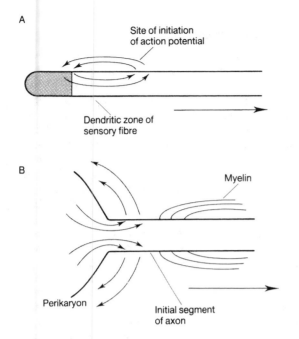

A

Site of initiation
of action potential

Dendritic zone of
sensory fibre

B

Myelin

Perikaryon

Initial segment
of axon

Figure 14.11 Local circuits initiate action potentials at initial segments or dendritic zones. (A) Dendritic ending of a sensory fibres. The ending is depolarised, either directly by environmental energy or by transmitter from an adjacent sensory cell. This depolarisation is termed a generator potential. Local circuits spread from this depolarised region to initiate an action potential in the sensory fibre. (B) Initial segment of an axon. Local circuits spreading from an excitatory postsynaptic potential (EPSP; see Chapter 16) on the perikaryon or dendritic tree of a neuron open the sodium gates in the membrane of the axon's initial segment and initiate an action potential.

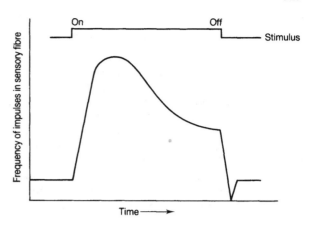

On   Off

Stimulus

Frequency of impulses in sensory fibre

Time ⟶

Figure 14.12 Adaptation in a sensory nerve fibre. After an initial burst of impulses in the sensory fibre the frequency falls off to a much lower value – in some cases almost to the pre-stimulus value. This is sensory adaptation.

membrane. These pumps are at work throughout the neuron's life. Their efforts are swamped when an action potential travels down the axon. They remain hard at work, however, during the quiescent periods between episodes of impulse transmission.

14.4 INITIATION OF THE IMPULSE

Action potentials never normally start up half way along an axon as was suggested in Figure 14.8. When the neurophysiologist is not playing his tricks, action potentials start either at the **initial segment** of the axon (motor or internuncial neurons) or at a **dendritic zone** (sensory neurons). In both

cases the initiation is due to local circuits depolarising excitable sections of membrane below their thresholds. The Na^+ gates begin to open and the action potential takes off.

It is likely that at both regions shown in Figure 14.11 some form of adaptation occurs. Adaptation is best known at sensory endings and, indeed, we have already encountered it when discussing sensory cells in Chapter 13. The majority of sensory fibres respond to a stimulus by a rapid burst of impulses after which the frequency decreases. The intensity of the stimulus is conveyed to the central nervous system by the frequency of the initial burst. The subsequent plateau of low frequency signalling keeps the CNS informed that the stimulus is still present but is not designed to allow good discrimination between different stimulus intensities. The phenomenon of sensory adaptation is shown in Figure 14.12.

It will be recalled from Chapter 11 that the Ca^{2+}-dependent K^+ channel (I_{KCa}) is believed to be involved in sensory adaptation. It is now possible to see how this may work. Consider Figure 14.13. Local circuits from a depolarised sensory ending act on a patch of excitable membrane to initiate an action potential. It will also be recalled from Chapter 11 that Ca^{2+} gates are strongly voltage-dependent and open when the membrane is

BOX 14.2   Switching off neurons by manipulating K$^+$ channels

We saw in Chapter 11 that neurons express a wide variety of voltage-dependent K$^+$ channels and we have seen in subsequent chapters something of the role these channels play in setting the resting potential and bringing about recovery from the repolarisation of the action potential. The rapid advance in our understanding of the molecular biology of these channels is beginning to allow scientists to manipulate them in various ways in the living brain.

It will be recalled that Kir channels are principally concerned in maintaining the resting potential. Their K$^+$ conductivity decreases as depolarisation of the membrane increases. Consequently if these channels are over-expressed the neuronal membrane becomes less excitable. This has been shown to be the case by using an adenovirus to insert an ecdysone-inducible promoter to drive the expression of Kir2.1 in cultured rat superior cervical ganglion neurons In the presence of ecdysone the electrical activity of these neurons was effectively suppressed.

In contrast to Kir channels, shaker-type K$^+$ channels normally only open during depolarisation, If they were to remain open permanently, even when the membrane was in its 'resting' state, the neuron would, once again, be very difficult to excite. This too has been shown to be the case by clever experimentation. Using the GAL4-UAS$_G$ system described in Section 5.18 it has proved possible to insert inactivated K$^+$ shaker channels into **specific cells** in the *Drosophila* nervous system. The gene fused with the UAS$_G$ promoter encoded an α-subunit with point mutations (D316N and K374Q) in the M3 and M4 transmembrane helices. These mutations shifted the voltage sensitivity of the channel toward V_K, thus ensuring that it was almost fully open at resting potentials. To ensure that it stayed open the N-terminal inactivation domain was deleted. These modifications ensured that the cells in which it was expressed were rendered inexcitable. Their electrical activity, in other words, was 'knocked out' and the investigators consequently designated them EKO cells.

The ability to form EKO cells in specific regions of the nervous, sensory and muscular systems of *Drosophila* (and, in the future, other organisms) clearly points the way to a powerful means of investigating the development and functioning of these systems. Specific neurons or systems of neurons can be 'shut down' and their role in the normal functioning of the nervous system elucidated. The next steps in this exciting technique for analysing the functioning of nervous systems is to find means of controlling the expression of the mutant shaker gene not only spatially but also temporally. These steps are already being taken. The GAL-UAS system has been connected to a drug-inducible promoter. This promoter, known as GeneSwitch, can be turned on by an activator (RU486 or mifespristone). By this means EKO activity has been turned on and off in specific cells of *Drosophila* larval muscles. The lag between applying the activator and the appearance of EKO activity was no more than five hours. It is, of course, a giant leap from the larva of a fly to the brain of a human, but these giant leaps are achieved here, as elsewhere, by a series of small steps.

It is clear, then, that we are at the beginning of an exciting period in which the increasingly sophisticated techniques of molecular biology will be used to switch on and off ion channels in selected parts of the nervous system and thus switch on and off specific systems of neurons. These specific systems can, moreover, be labelled by coupling a visible marker, such as green fluorescent protein (GFP), to the inserted channel.

depolarised. The membrane is strongly depolarised during an action potential. Hence Ca^{2+} cascades into the fibre. Internal Ca^{2+}, however, tends to open the Ca^{2+}-dependent K$^+$ gates. K$^+$, in consequence, flows out. This leads to an increased duration for the AHP. But during the AHP the Na$^+$ channels are tightly closed – unresponsive to local circuits – and the fast K$^+$ channels are primed to open in response to depolarising currents. Hence the longer the I_{KCa} current persists the longer the fibre remains unresponsive to generator potentials. Hence sensory adaptation.

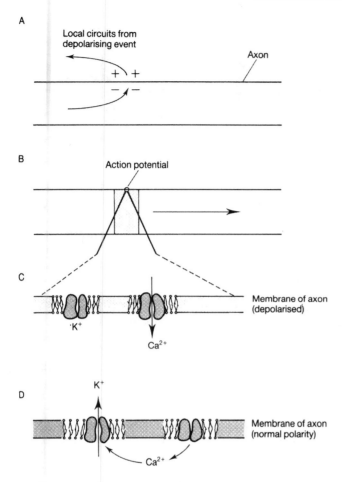

Figure 14.13 An ionic mechanism for sensory adaptation. (A) Local circuits from some depolarising event (either synaptic or sensory) depolarise the axon below threshold. (B) An action potential is initiated and propagates away down the axon. (C) Enlargement of the membrane at which the action potential is initiated. Ca^{2+} gates open in response to depolarisation and Ca^{2+} ions flow into the axon. (D) The Ca^{2+} ions affect neighbouring Ca^{2+}-dependent K^+ gates and K^+ ions flow out, thus enhancing the AHP. Further explanation in text.

14.5 RATE OF PROPAGATION

We noted in Chapter 12 (equation 12.13) that, other things being equal, the greater the diameter of a process the greater the value of λ, the **space constant** or **electrotonic length**. In that chapter we were particularly concerned with cable conduction in non-impulse conducting processes. But the analysis applies equally well of, course, to axons. The greater the diameter, the further the local circuits spread down an axon. We saw in that chapter that this physical fact accounts for the evolution of giant fibres in all large invertebrate animals. The rapid signalling that these fibres allow enables the mollusc or annelid or crustacean to respond rapidly to emergencies.

Vertebrate animals have, however, evolved an alternative means for speeding an impulse –

myelination. Myelinated nerve fibres are seldom above $20 \, \mu m$ in diameter; yet, in mammals, they conduct action potentials at well over twice the rate that the $600 \, \mu m$ fibres of the squid can manage. Whilst squid giant fibres seldom conduct at rates much above $50 \, m \, s^{-1}$, mammalian alpha fibres can sustain conduction rates of over $120 \, m \, s^{-1}$. The ability to conduct with great rapidity whilst remaining small is, of course, of great advantage: not only is the message delivered promptly but a large number of small fibres can provide a great deal more information than one large fibre.

We discussed the nature and formation of the myelin sheath around nerve fibres in Section 7.7. We saw there that myelin is formed by glial cells in both the central and peripheral nervous systems. In both cases it is formed by the wrapping of the glial membrane tightly around the axon. In both cases

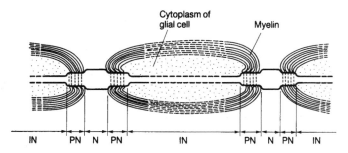

Figure 14.14 Schematic longitudinal section through a myelinated fibre to show the node of Ranvier. N=node of Ranvier; IN=internode; PN=paranode. The stippling on the paranodal membrane represents rows of globular particles which form a firm junction between this part of the axolemma and the myelin. After Livingstone *et al.* (1973), *Brain Research*, **58**, 1–24.

gaps are left between one glial cell and the next in the chain where the axolemma is exposed to the intercellular space. These gaps are called **nodes of Ranvier**.

Figure 14.14 shows that the region of myelin sheath abutting the node is particularly complex. As the figure shows, it is called the **paranode** to distinguish it from the internode which forms the major extent of the myelin sheath. It will be recalled from Chapter 7 that the glial cell membrane is particularly firmly anchored to the axolemma in this position. This anchorage is believed not only to weld the myelin firmly to the nerve fibre but also to make a high resistance seal to electrical current. It also, and importantly, obstructs the movement of membrane-embedded proteins from the node into the internodal region. If myelin is removed from a myelinated axon by diphtheria toxin, or by enzyme attack, the sodium channel and sodium/potassium pump proteins spread out more evenly throughout the axolemma. As there is no evidence that new channels have been synthesised it appears that the paranodal regions of the myelin sheath normally exercise a restraining influence on the movement of these and other membrane proteins. The reader may recall that an analogous situation obtained with sarcolemmal nAChRs after denervation.

Now how does this organisation assist the transmission of action potentials? An answer to this question was proposed many years ago. The tightly wound spirals of the myelin sheath act as a very effective electrical insulator. External current flows cannot penetrate it to affect the ensheathed axolemma. But, of course, the local circuit mechanisms underlying impulse propagation depend on such external current flows. It follows that it is only at the nodes, where the axolemma is exposed, that the local circuits can exert their depolarising effect. Hence the action potential 'jumps' from one node to the next. Technically this is known as **saltatory conduction** (Figure 14.15). The profound significance of the myelin sheath for the propagation of impulses explains the devastating effects of demyelinating diseases such as multiple sclerosis (see also Section 7.7). It may be that the redistribution of Na^+ channels along demyelinated axon (mentioned above) could account for the uncertain course of the affliction: remissions and exacerbations are a feature of the disease.

It might be objected, however, that it is not obvious why this mechanism should increase conduction rate. Surely, it might be said, local circuits would anyway spread down axons and depolarise membrane equally far away whether the membrane was insulated by myelin or not? This

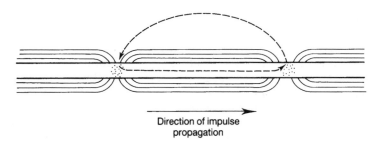

Figure 14.15 Saltatory conduction in myelinated axon. The stippled regions at the nodes indicate where the voltage-dependent gates are located. The broken arrows show the direction of the local circuits when an action potential is travelling down the axon.

objection has force. The answer, however, is inherent in equation 12.12:

$$\lambda^2 \approx \sqrt{\frac{r_m}{r_i}}$$

It will be recalled from Chapter 12 that r_m is the transverse resistance of the membrane and r_i the longitudinal resistance of the axoplasm. It is clear that r_m is far greater in a myelinated axon: the electrical resistance of the tightly wound myelin whorls is up to four orders of magnitude greater than neurilemma on its own (see Appendix 3). Hence λ, the space constant, is greatly increased. Hence the influence of local circuits is felt far further down a myelinated than a comparable unmyelinated fibre.

It is also worth remembering that one of the features of the paranodal region of the myelin sheath is that it obstructs the movement of membrane-embedded protein. It is found, in consequence, that there is a very high density of sodium channels at the node – two or three orders of magnitude greater than in the internodal axolemma. This is also the case for the $Na^+ + K^+$ pump proteins.

Now we calculated in Chapter 11 that the quantity of Na^+ flowing into a 12 μm axon during one action potential constituted up to 3% of the Na^+ present beneath that area of membrane. This would place an intolerable strain on the $Na^+ + K^+$ pump if it were to occur all along the membrane and a rapid tattoo of impulses was transmitted. For it has been calculated that a single pump only expels about 200 Na^+ ions per second. In myelinated axons the ion fluxes only occur at the nodes. Intra-axonal diffusion will buffer the concentration build-up of sodium ions. The high concentration of pumps at the node can get rid of the excess during periods of quiescence.

It appears, therefore, that the node allows extra-large current fluxes to occur through the numerous sodium channels so that extra-large local circuits are initiated. This also increases the distance at which electrotonic conduction can exert its effect.

Thus the development of myelin whorls around vertebrate axons increases the rate of impulse propagation in several ways: partly by increasing the reach of local circuits; partly by increasing the magnitude of local circuits; partly by buffering the influx and efflux of ions.

The giant fibre systems of invertebrates tackle the buffering problem in a different way. They make use of the mathematical fact that whilst surface area increases as the first power of the radius the volume increases as the second power. Hence, although the 600 μm squid axon may develop as many as 500 sodium channels per square micrometre of axolemma, the Na^+ flux when an action potential occurs will only increase the internal sodium concentration by less than 0.01%

Finally it is worth noting that the high densities of pumps and channels which characterise the axolemma at the nodes are also to be found at the axon's initial segment. This is to be expected for we saw in the last section that this is where impulses are initiated.

14.6 CONCLUSION

In this chapter we have seen how the major features of the action potential – its initiation, propagation and velocity – all find an explanation at the molecular level. As our understanding of ion channels and pumps increases so will our appreciation of the finer points of impulse transmission. It has already been emphasised that the heterogeneity of their membranes confers different 'personalities' on different neurons. These personalities are expressed in subtly different stimulabilities, adaptabilities and propagation rates. Moreover, like human personalities, neuronal personalities are not fixed and immutable. Neuronal membranes change over time and in response to 'experience'. Although the action potential remains an all-or-nothing event, propagated without decrement from initial segment to telodendria, the comparison of neurons to the units of silicon-based computers becomes ever more strained.

15

THE NEURON AS A SECRETORY CELL

Metaphors and analogies in neuroscience. Dale's 'law' and its exceptions. Synthesis in perikaryon and secretion at telodendria necessitates axoplasmic transport. Neurosecretion, volume transmission. Ultrastructure of perikarya indicates intense synthetic activity: protein manufacture – co-translational insertion – anchorage of ribosomes to ER – proteins injected into lumen of ER or retained in ER membrane – N- and/or O-linked core glycosylation – clathrin-coated transport vesicles – Golgi body – terminal glycosylation and other modifications – clathrin-coated secretory vesicles – axoplasmic transport. Fast and slow, antero- and retro-transport. **Axonal cytoskeleton**: actin microfilaments – heterogeneous IFs – dynamic MTs – various MAPs and tau proteins. Biophysics investigated by immuno- and electron microscopy – video microscopy – preparations of single MTs. Fast axoplasmic movements energised by kinesins and cytoplasmic dyneins – links between kinesins and transport vesicles – slow axoplasmic flow uses different mechanisms. **Exocytosis at terminal**: initiated by Ca^{2+} influx – vesicles often contain a variety of molecules. **NMJ an 'honorary' synapse**: accessible and relatively easy to study – structure – physiology – biophysics. Clathrin-coated vesicles attached to presynaptic cytoskeleton by synapsin 1 – Ca^{2+}-dependent release – vesicle docking mechanism at P-face of presynaptic membrane – Q-SNAREs, R-SNAREs and other molecules – vesicle attachment sites and active zones – possible mechanisms of exocytosis – recycling of vesicle membrane – dynamin – neurotransmitter transporters refill vesicles. Axo-axonic synapses often control the operation of presynaptic terminal

Science progresses by way of metaphor and analogy. Electricity is likened to a fluid, valency bonds to hooks and eyes, atoms to billiard balls. The science of the brain is no exception. Descartes in the seventeenth century likened the brain to the intricate hydraulic mechanisms of his day; the nineteenth and early twentieth centuries saw a powerful analogy in the telegraph cable and the telephone exchange; nowadays, at the beginning of the twenty-first century, the computer metaphor is all-pervasive. Metaphors and analogies both help and bias our understanding. The power of the computer analogy perhaps prevents us seeing the brain from other equally significant viewpoints. In particular it prevents us seeing the brain as if it were an immense gland. Yet this view has much to commend it. Neurons can be seen not so much as relays or on/off valves but as secretory cells. The electrical phenomena of electrotonic and action potentials can, on this analogy, be seen as epiphenomena or, at best, merely as triggers for the release of secretions – the neurotransmitters and modulators.

In this chapter we shall begin to consider this aspect of the nervous system. We shall consider the neuron as a secretory cell. In the next chapter we shall look at the pharmacology of some of the neurotransmitters and neuromodulators that are secreted and in the chapter after that we shall consider the action of these secretions on underlying (subsynaptic) membranes and cells.

15.1 NEURONS AND SECRETIONS

Neurons synthesise many different transmitters and modulators. We noted in Chapter 1 that in the 1930s Dale proposed that any given neuron only synthesised one type of transmitter. It followed that all the terminations of that neuron released that same substance. The action of that substance depends, of course, on the nature of the receptors in the subsynaptic membrane. The nACh receptor, as we have already noted, responds to acetylcholine in a very different way from the mACh receptor. **Dale's 'law'** or 'principle' has stood the test of time remarkably well. But like practically all generalisations in biology it is now found to be honoured more in the breach than in the observance. Numerous exceptions, both in the vertebrates and in the invertebrates, are now known. Indeed it begins to seem that a quite common pattern is to find a peptide neuromodulator accompanying a monoamine in a single presynaptic terminal (see Chapter 16).

We shall look in detail at the various transmitter and modulator molecules in Chapter 16. Here, however, we can note that they range from small molecules such as nitric oxide, acetylcholine, the catecholamines and individual amino acids, to quite large peptides. We have already considered the structure and evolutionary relationships of some of the latter in Chapters 2, 3 and 4 (see also Table 2.2). The site at which these molecules are synthesised differs. Small transmitters such as acetylcholine are put together (as we shall see) in the terminal. Peptide modulators, in contrast, are synthesised in the perikaryon. However, even the synthesis of acetylcholine requires appropriate enzymes – in particular choline acetyltransferase (CAT) – and these have to be manufactured in the perikaryon. For it is only in the perikaryon that the necessary protein-synthetic equipment – the DNA information tape, the ribosome machine-tools, the endoplasmic reticulum, Golgi apparatus, etc. – exists.

Neurons, thus, are very special types of secretory cell. The secretory membrane (i.e. the presynaptic membrane) can be more than a metre distant from the region where the secretion is synthesised. The connection between the two regions is, of course, provided by the axon. Up till now we have been considering the axon only as if it were a telegraph

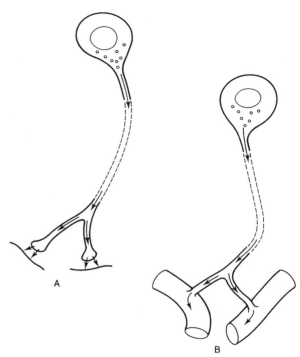

Figure 15.1 Neurons as secretory cells and neurosecretory cells. (A) Typical neuron. The transmitter is released across the synaptic gap and affects the membrane of the cell immediately subjacent. (B) Typical neurosecretory cell. The neurosecretion is released into a blood vessel and carried by the blood flow to exert an effect on a distant cell.

wire – for conducting nerve impulses. Now we begin to see that it has another and equally important function. It has to act as a conduit for the passage of secretory materials manufactured in the perikaryon and destined for release at the synaptic terminal. Moreover, as we shall see, the transport is not in one direction only. Materials also flow back from the terminal to the perikaryon. Axoplasmic flows play crucial roles in the functioning of the nervous system.

It should be mentioned at this point that some neurons have developed their secretory function to the extent of releasing their secretion not across a synaptic gap on to the dendritic membrane of another neuron, but directly into the vascular system. These cells are known straightforwardly as **neurosecretory cells** (Figure 15.1). Such cells are found in the hypothalamus and pituitary of

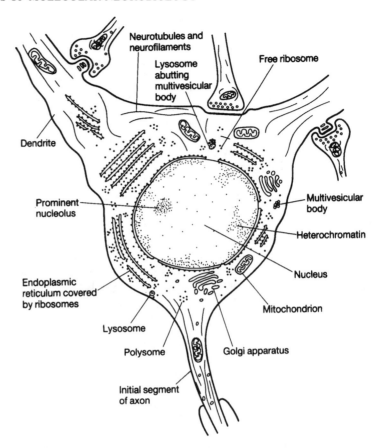

Figure 15.2 Perikaryon of a cortical neuron. Drawing from an electron micrograph of a neuron in the cerebral cortex. Note the massive nucleus, the prominent nucleolus and the complex structure of the cytoplasm.

vertebrates. They are also well developed in many invertebrates. Finally it should be mentioned that, as usual, there are intermediate situations. There is much interest nowadays in the diffusion of neurotransmitters, particularly small molecules, in the intercellular space. This diffusion may take such molecules up to 100 μm from their site of origin until they find an appropriate receptor. This type of transmission can affect whole populations of cells in contrast to the one-to-one interactions of conventional 'punctate' transmission. It is consequently known as 'volume' or 'diffusional' neurotransmission.

Readers interested in the topic of neurosecretion, which underlies the vast subject of neuroendocrinology, should consult one of the texts listed in the Bibliography. In this chapter we shall deal only with the secretory aspect of 'conventional' neurons.

15.2 SYNTHESIS IN THE PERIKARYON

In Chapter 3 we looked at the molecular biology of protein biosynthesis. We saw that the genetic message held in the structure of DNA is first of all transcribed into mRNA and then translated at the ribosome into the specified amino acid sequence of a protein. This process, we noted, was highly complex, involving many different enzymes and both tRNA and rRNA as well as mRNA. But the molecular biology of protein manufacture does not end there. In secretory cells the protein has to be packaged and transported to its release site.

Electron micrographs of the perikarya of typical nerve cells show all the structural features of a cell employed in intense protein biosynthesis. They possess large nuclei with one or more prominent **nucleoli**. Their cytoplasm is richly endowed with

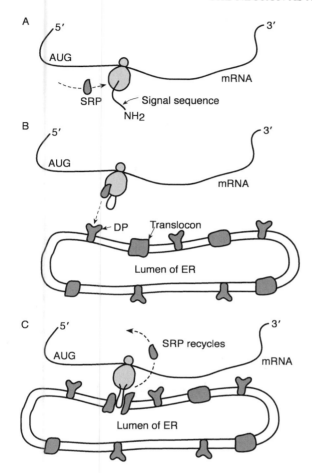

Figure 15.3 Anchorage of a ribosome to the ER. (A) Ribosome attaches to the 3′ end of mRNA and finding a start signal (AUG) commences translation. The initial amino acid sequence is termed the 'signal sequence'. (B) The signal sequence is recognised by a signal recognition protein (SRP) which attaches to it and to the ribosome. (C) The SRP and signal sequence then recognise a 'docking protein' (DP) or 'signal recognition particle receptor' in the ER membrane. The complex then unites with a translocon in the membrane. This causes the SRP to detach and recycle and a channel to open in the translocon to receive the signal sequence. Translation then continues at the ribosome and the growing polypeptide chain passes through into the lumen of the ER.

rough endoplasmic reticulum (RER) and there is always at least one and often a large number of Golgi bodies (Figure 15.2).

Rough endoplasmic reticulum consists of complex stacks of membranes covered by innumerable ribosomes. The latter cause the protuberances or 'roughness' visible in EM pictures. It is these ribosomes that synthesise protein 'for export'.

15.2.1 Co-translational Insertion

It can be shown that the polypeptide chain manufactured at the ribosome is inserted directly into the lumen of the ER. But how does the ribosome find the ER? Not all the proteins that a neuron manufactures are for export. Some of them will be for the usual 'housekeeping' activities of the perikaryon itself. It would be disastrous if these

were to be delivered into the lumen of the ER and sent off for export!

The answer to this question seems to be that the first thirty or so amino acids in the polypeptide chain growing from the ribosome form a signal sequence which is recognised by a **'signal recognition particle' (SRP)** – see Figure 15.3. This particle is in fact a very complex nucleoprotein consisting of six separate polypeptides and a 300-nucleotide RNA moiety. It appears to have two functions. First, it prevents any further translation occurring, thus ensuring that the export protein is not simply dumped in the perikaryal cytoplasm. Second, it is recognised by a 650-residue **'signal recognition particle receptor'** or **'docking protein' (DP)** embedded in the ER membrane. The ribosome, nascent polypeptide chain, and SRP become anchored to the ER.

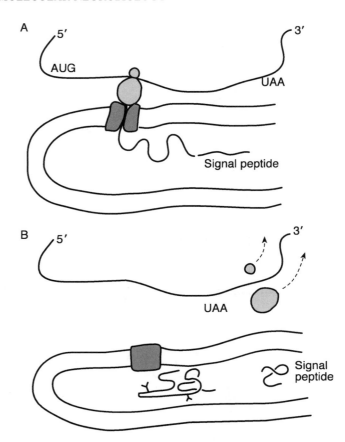

Figure 15.4 Co-translational insertion of polypeptides at the ER. (A) The polypeptide chain is synthesised at the ribosome and inserted into the lumen of the ER and the signal sequence removed. (B) On reaching a stop signal (e.g. UAA) translation ceases and the two parts of the ribosome separate to recycle. Enzymes within the lumen of the ER proceed with any appropriate post-translational processing and the translocon returns to its original state.

The story is far from finished at this point. It is found that if the SRP nucleoprotein remains attached to the signal sequence no further translation occurs. Normally, however, once the ribosome has become anchored to the membrane the signal recognition protein detaches to recycle. This allows the signal sequence of amino acids, which are generally hydrophobic, to attach themselves to a large protein complex in the ER membrane known as a **translocon**. The translocon has a central hydrophobic pore which is usually closed. The signal sequence is shuffled through the complex and opens the gate of the central channel. The ribosome is held in position above the pore by the polypeptide chain whilst the N-terminal protrudes through the pore into the lumen of the ER. Translation then continues as described in Chapter 3. The polypeptide is thus synthesised and inserted into the lumen of the ER at one and the same time. It is never exposed to the cytosol and does not fold until it is within the ER. This ensures that it does not get stuck outside the translocon. The whole process is termed **co-translational insertion**. Finally the N-terminal signal sequence is cut off by enzymes within the lumen of the ER and the ribosome, encountering a stop signal (UAA) on the mRNA strand, dissociates and recycles. The process is shown diagrammatically in Figure 15.4.

There are two possible end results to the process of co-translation (Figure 15.5). In one case the polypeptide is secreted entirely into the lumen of the ER. This is the case with proteins destined for

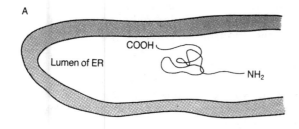

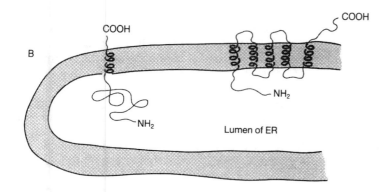

Figure 15.5 Two end results of co-translation. (A) The polypeptide is entirely within the lumen of the ER. (B) Hydrophobic segments of the polypeptide ensure that it is caught in the ER membrane. In many cases several such hydrophobic segments are trapped in the membrane. In these cases the polypeptide chain is inserted into the membrane as a series of hairpin loops.

export such as, in neurobiological instances, peptide neurotransmitters. In the second case the polypeptide never gets through into the lumen at all. This is the case with membrane-embedded proteins.

We have noted in the earlier chapters of this book the great importance of membrane-embedded proteins in molecular neurobiology. In these cases a hydrophobic sequence of amino acids, usually in the form of an α-helix, somewhere along the length of the forming polypeptide chain prevents it passing through the translocon. The resulting polypeptide has a single membrane-spanning segment with its N-terminal in the cytosol and its C-terminal in the lumen of the ER. The membrane-spanning sequence moves (in several steps) laterally through the translocon to take up its final position in the membrane of the ER.

In other cases the forming polypeptide chain consists of alternating segments of hydrophobic and hydrophilic amino acids. The first two hydrophobic segments cling together in the aqueous environment as a 'hairpin' and are inserted into the translocon as such. As the polypeptide continues to grow the first hairpin moves out of the translocon

into the membrane and subsequent hydrophobic hairpins are inserted. In this way multipass proteins with up to two dozen transmembrane sequences are synthesised.

15.2.2 The Golgi Body and Post-translational Modification

We noted in Chapter 3 that post-translational modification of polypeptides and proteins is very common. In particular we saw how some of the peptide neuromodulators – the opioids and enkephalins – are known to be derived from large precursor proteins or polyproteins by post-translational processing. In Chapter 4 we considered the evolutionary implications of this phenomenon. In this section we shall look at where this post-translational processing occurs.

The first steps in post-translational processing occur in the lumen of the ER. Here we find that disulphide bonds are made between cysteine residues and the first steps in glycosylation are taken. The formation of disulphide bonds is generally regarded as stabilising the three-dimensional structure of a protein, and the process of glycosylation – the adding of carbohydrate chains – is very important for many proteins – as we have noted in preceding chapters and shall note again, especially when we come to consider cell–cell recognition in neuroembryology (Chapter 19).

The initial steps in glycosylation are often called **core glycosylation** to distinguish them from the finishing touches that are carried out later in the Golgi body (Figure 15.6). These initial steps often involve the addition of a 'core' oligosaccharide to the amide side chain of **asparagine**. Such glycoproteins, as we noted in Chapter 2, are consequently termed **N-linked**. The core oligosaccharide almost invariably consists of two units of an acetylated monosaccharide, *N*-**acetylglucosamine (NAG)**, and a number of mannose and glucose units.

Alternatively the carbohydrate moiety may be linked to the hydroxyl group in the side chains of **serine** or **threonine** or, more rarely, **hydroxylysine** and **hydroxyproline**. Such glycoproteins, as we noted, are consequently termed **O-linked**. O-linked oligosaccharides are much smaller than the N-linked type. They usually consist of only two, three or perhaps four monosaccharide units – *N*-acetylglucosamine (again) linked perhaps to galactose and very commonly to *N*-**acetylneuraminic acid (=sialic acid)**.

The next phases of post-translational processing occur in the **Golgi body**. Cells active in the synthesis of proteins for export usually have several, sometimes a very large number of, such bodies. This is the case with many neurons. They are often very well endowed with Golgi complexes. Indeed neurons were the cells in which they were initially discovered – by Camillo Golgi in 1898. An understanding of their structure and function awaited, however, the advent of modern techniques in cell biology. Figure 15.7 shows that Golgi bodies have an intricate structure. They consist of a stack of membranous saccules or cisternae. These saccules are believed to be formed at one face by the fusion of vesicles budding off the ER and to be lost at the other face by their tips budding off to form secretory vesicles. The Golgi body is thus said to have a *cis* or forming face and a *trans* or maturing face.

As implied in the previous paragraph, the precursor proteins in the ER reach the Golgi body in transport vesicles. This also applies to membrane proteins. These travel in the membranes of the transport vesicles and ultimately fuse with the *cis* face of the Golgi body. The static picture that the electron microscope gives of cellular ultrastructure is evidently highly misleading. The 'structural' elements of a cell – its internal membranes and (as we shall see) fibres – are in ceaseless turmoil. Nowhere is this more the case than with the ER, transport vesicles, Golgi complex and secretory vesicles.

It is found that transport and secretory vesicles share a common structural feature. They are caged in a lattice work of **clathrin** molecules. Each clathrin molecule is a three-armed ('Isle-of-Man') structure (Figure 15.8) – called a **triskelion** – with a molecular weight of about 180 kDa. A number of triskelions form around the lipoprotein membrane of a vesicle and with the aid of several smaller proteins hold the vesicle together. Such clathrin-gripped vesicles are called **coated vesicles**. Clathrin can very easily assemble and disassemble. Indeed it can be induced to assemble even in the absence of vesicles, when it forms empty cage-like structures.

Once the transport vesicles have fused with the *cis* face of the Golgi body their contained proteins undergo further post-translational processing (see Figure 15.6). The Golgi saccules contain many enzymes which continue the processes of glycosylation – often called **terminal glycosylation** to distinguish it from the core glycosylation of the ER. Other enzymes cut large 'polyproteins' into something closer to their ultimate forms. Yet other enzymes add a signal or tag, perhaps by phosphorylation or by a specific glycosylation, which serves to direct a protein – especially a membrane-bound protein – to its correct destination. Finally, the Golgi body carries out an important role in condensing proteins destined for export into highly concentrated packages.

The intricate molecular biology of the perikaryon is now at an end. The proteins are now prepared, packaged and ready for export (Figure 15.9). In the next section we shall look at what is known about the transport of this material to its final destination. We shall consider, in particular, how it is believed to be moved down the axon. It must, however, be borne in mind that materials, membrane receptors, enzymes, etc. must also be moved out along those other neural processes – the dendrites. This transport, though far less is known

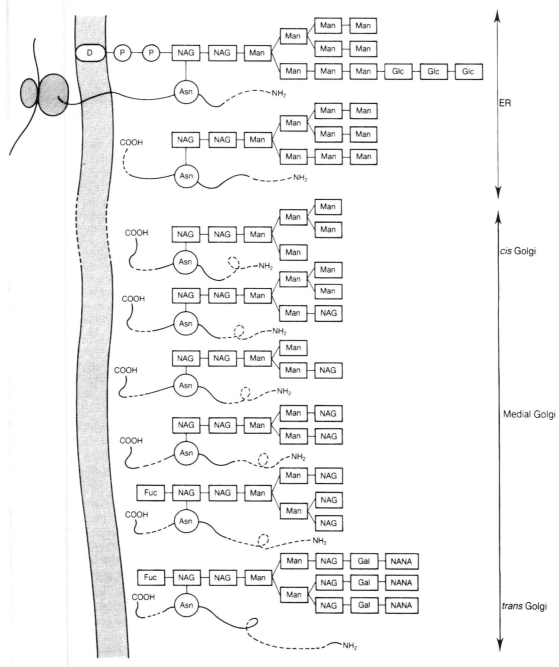

Figure 15.6 Glycosylation in the ER and Golgi body. The figure shows *N*-glycosylation. The initial step is the transfer of an oligosaccharide from dolichol (D) to an asparagine residue in the forming polypeptide chain. Dolichol is a large (75–90 carbon) unsaturated lipid. It is firmly embedded in the ER membrane, which it probably spans three or four times. The subsequent processing, as the figure shows, occurs partly in the ER and partly in the Golgi body. D=dolichol; Asn=asparagine; NAG=*N*-acetylglucosamine; Fuc=fucose; Man=mannose; Glc=glucose; Gal=galactose; NANA=*N*-acetylneuraminic acid (also known as sialic acid).

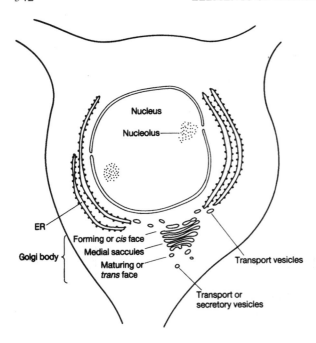

Figure 15.7 Structure of the Golgi body. The perikarya of neurons frequently contain numerous Golgi bodies. The figure shows just one. The flow is from ER via transport vesicles to the *cis* face of the Golgi. The vesicles of the *cis* face coalesce to form saccules and move outwards towards the *trans* face. Ultimately vesicles bud off the *trans* face and carry the fully processed protein to its correct destination.

about it, is just as important for the biology of the neuron as is the transport of transmitters and modulators down the axon (see Box 15.1). The recent introduction of new fluorescent markers (especially green fluorescent protein) has markedly improved our understanding of these 'dendroplasmic' flows.

15.3 TRANSPORT ALONG THE AXON

The secretory vesicles budding off the *trans* face of the Golgi body have far to go. The recurrent laryngeal nerve fibres of a giraffe stretching all the way down the neck, under the aortic arch, and back up again to the larynx may be several metres in length. Whilst this is exceptional, axons of over a metre in length are quite common in large animals.

The topic of axoplasmic transport has thus been one of considerable interest.

Conventional microscopical techniques have revealed, as we shall shortly see, a complex cytoskeleton within the axon. But as we noted in the preceding section these pictures tend to give the wrong impression. The axonal cytoskeleton, like the membranous structures within a perikaryon, is not static like the endoskeleton of the vertebrate body, the exoskeleton of an arthropod or the scaffolding around a building. In vivid contrast it constitutes a scene of dynamic activity.

This dynamism has been revealed both by 'pulse labelling' and by ingenious microscopical techniques. Pulse labelling consists in the introduction of a brief pulse of radioactive label and then finding where the radioactivity has got to in successive intervals of time. When this technique is applied to an axon it reveals waves of radioactive label moving through the axoplasm. Three distinct transport systems appear to be at work. A **fast** system (2–$4\,\mu m\,s^{-1}$, i.e. 20–$40\,cm\,day^{-1}$) and a **slow** system (0.01–$0.04\,\mu m\,s^{-1}$, i.e. 1–$3.5\,mm\,day^{-1}$) transport material from the perikaryon towards the synaptic terminal (**anterograde**), while a system working in the **reverse direction** at about the same rate as the fast system transports material from the terminal towards the perikaryon (**retrograde**).

The fast axoplasmic transport system appears to be responsible for transporting mitochondria and the secretory vesicles which we have followed budding off the Golgi apparatus. These vesicles contain neurotransmitters and/or modulators, glycoproteins, enzymes required for neurotransmitter metabolism in the terminal, etc. The slow system, on the other hand, is believed to carry elements of the cytoskeleton (tubulin, neurofilaments, actin) and to be involved in axonal growth processes. Finally, the retrograde system carries effete membrane and other used materials, neurotrophic factors (e.g. nerve growth factor (NGF)), materials taken up from the synaptic cleft, etc., back from the synaptic terminal to the perikaryon.

Biochemical, immuno- and electron microscopical techniques have helped identify the fibrous elements responsible for these flows. Such cytoskeletal elements seem pretty generally developed throughout metazoan cells. They have become especially well developed in neurons to carry out what is, at the cellular level, an extraordinary task.

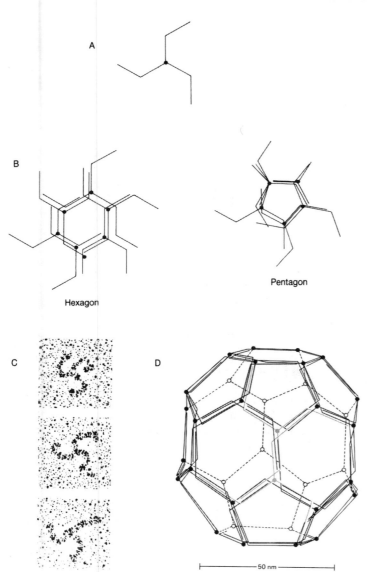

Figure 15.8 Clathrin triskelions and coated vesicles. (A) Clathrin triskelion. (B) Clathrin triskelions spontaneously assemble to form hexagons and pentagons. (C) Electron micrographs of triskelions. (D) Thirty-six triskelions organised as a network of twelve pentagons and eight hexagons form the 'coat' surrounding a coated vesicle. The overlapping arms of the triskelions provide strength with flexibility. Part C reprinted by permission from Ungewickell and Branton (1981), *Nature*, **189**, 420–422; copyright © 1981 Macmillan Magazines Ltd; and from Alberts *et al.* (1983), *Molecular Biology of the Cell*, New York: Garland Publishing, with permission.

It has already been mentioned that many mammalian axons are well over a metre in length. An average diameter for such an axon might be 15 μm. Many are considerably narrower. If we scale things up to a human dimension a water main of diameter 1 metre would have to stretch uninterrupted over 66 km to be comparable. It would have to be prepared to bend and crinkle, carry materials at different speeds and in both directions. Axons are indeed remarkable structures.

Cell biologists recognise three major classes of cytoskeletal fibres: **microfilaments** (diameter about 8 nm); **intermediate filaments (IFs)** (diameter between 7 and 11 nm); **microtubules** (diameter about 25 nm). All three types of fibre are found in neurons, especially in axons.

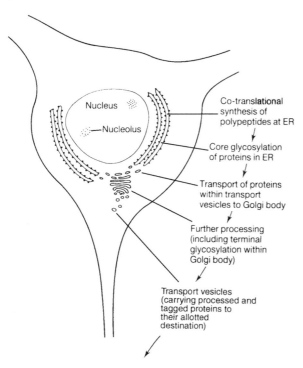

Figure 15.9 Summarising diagram of the pathway of 'export' proteins in neuronal perikarya.

15.3.1 Microfilaments

Microfilaments are mostly composed of actin. We met this cytoskeletal component in Chapter 7 when we were considering the nature of the sub-membranous cytoskeleton. There it was believed to play a role in maintaining the shape of the cell. Actin is also, of course, well known in muscle, where it is crucially involved in the contractile mechanism. It is likely that it plays a part in cytoplasmic movements of all types. It is thus not surprising to find microfilaments well represented in axons. **Actin fibres (F-actin)** are composed of **globular subunits (G-actin)**. Two F-actin strands twist around each other to form a two-stranded rope (Figure 15.10). Numerous cytoplasmic factors are involved in the polymerisation process whereby G-actin units string together to form F-actin. Other factors form cross-links between actin fibres, bundle them together into parallel skeins and/or act as 'spacers' keeping parallel bundles apart by distances of about 200 nm. The molecular biology

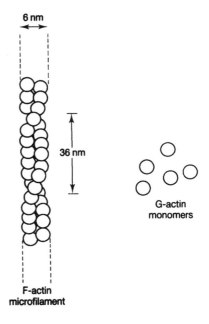

Figure 15.10 Structure of microfilaments. G-actin subunits are in fact more pear-shaped than spherical. Polymerisation into F-actin is usually accompanied by the hydrolysis of ATP to ADP.

of microfilaments is very intricate and is still being worked out. In the axon the microfilaments are usually quite short – seldom more than about 0.5 μm in length.

15.3.2 Intermediate Filaments (IFs)

Unlike either microfilaments or microtubules, intermediate filaments are biochemically extremely heterogeneous. Five major classes are recognised. Each class is found in a specific cell type. In the nervous system glial cells contain glial filaments (a single protein of 51 kDa) whilst in neurons three different types of IF protein are recognised: NF-L (63 kDa), NF-M (160 kDa) and NF-H (200 kDa) (i.e. low, middle and high molecular weight).

Although IFs are extremely various they all seem to share a central, highly conserved 310 amino acid rod-like domain (Figure 15.11). This domain consists of two α-helices wound around each other to form a two-stranded rope. Each α-helix has the characteristic non-polar heptad repeat we discussed in Chapter 2 (Section 2.1.2) which ensures that the two-stranded rope structure is

BOX 15.1   Subcellular geography of protein biosynthesis in neurons

The classical story, as retold in this chapter, is that protein biosynthesis occurs in the perikaryon. There is no doubt that this is, in the main, true. But it has recently become apparent that mRNA also finds its way to other parts of the neuron and in particular into the dendrites. It has been known for some time that polyribosomes and membranous cisternae are associated with the dendritic spines of cerebral neurons (see Box 20.1). More recently, *in situ* hybridisation histochemistry localised mRNA for MAP-2 and the α-subunit of CaM kinase 2 to the neuropil layers of the cerebral cortex and hippocampus. These layers contain dendrites and axons but few cell bodies. Further studies revealed the presence of mRNAs for a number of other significant molecules, including the IP$_3$ receptor, possibly glutamate receptors, and an activity-related cytoskeletal-associated protein (Arc) in these layers. Finally, there is growing evidence that different mRNAs are associated with different neurons. mRNA for Arc is, for instance, most prominent in the dendrites of hippocampal granule cells whilst mRNA levels for IP$_3$ are highest in Purkinje cell dendrites.

Investigators have also shown that whilst some of these mRNA species are constitutive, others are inducible by neuronal activity. The levels of Arc and α-CaM kinase 2 mRNA in the hippocampal dentate gyrus increased dramatically after high-frequency stimulation of the perforant path (for diagram of hippocampus showing this pathway see Figure 20.22). CaM kinase 2 levels were also markedly increased after long-term potentiation. There is also evidence to show that NMDA-R activation is required, at least in the dentate gyrus, to target Arc-mRNA to *active* synapses.

Whilst the induction of fresh mRNA necessarily takes place in the perikaryon there is also fragmentary evidence to show that translation of mRNA in the dendrites may also be inducible by synaptic activity. All of these investigations have obvious implications (as we shall see in Section 20.5) for the physical basis of memory.

Most of these studies of the subcellular distribution of mRNA and its translation machinery are very new and all are very near the edge of what is presently possible. It is by no means certain, for instance, whether the activity-induced increase in specific mRNA species detected in dendrites is due to increased nuclear transcription or increased rate of transport within the dendrite. Nor is it clear whether activity-induced increases in translation are due to increases in ribosomal docking on cisternal membranes or to some facilitation of the translation process itself. Many other intriguing problems remain. It makes good sense, however, that in the lengthy dendritic processes of cerebral and hippocampal pyramidals and in cerebellar Purkinje cells, synthesis of synaptically significant proteins should occur close to their point of use. It makes good sense, and there is evidence that it does indeed take place. It is not only in human bureaucracies that distribution of power to the periphery, subsidiarity, provides the best and most effective solution to problems of control.

stable. At each end of the rod domain there are hypervariable head and tail regions. In the IFs of neurofilaments the tail regions are unusually extensive. It is believed that they may form the cross-linkages that the electron microscopist can see projecting out orthogonally to the neurofilament's long axis. The neurofilament itself is often very lengthy – up to 200 μm. It consists of a number of neural IFs (all three types) lying parallel to each other and end to end in a staggered fashion.

15.3.3 Microtubules (MTs)

Microtubules consist of **tubulin**. Like actin, tubulin is built of **globular** subunits. But here the similarity ends. The tubulin subunits are not all alike (as were the G-actins) but are of two sorts, α- and β-tubulin. Each subunit (or monomer) has three domains, one binds to nucleotides, one to drugs such as **taxol**, and a third which probably forms the binding surface for motor proteins. The α- and β-subunits associate together to form dimers and these in turn join end to end to form a 'protofilament'. Finally, 13 protofilaments line up in parallel and circle to form a hollow tube (Figure 15.12) about 200 μm in length (about the same length as neurofilaments). The tube is polarised: it has head (+) and tail (−) ends. Polymerisation occurs by adding fresh sub-units at the (+) end. 'Treadmilling' has also been observed to occur. In this process units are removed from the (−) end and added to the (+) end.

In axons microtubules are aligned so that their (−) ends point to the perikaryon. In dendrites, however, they are arranged in both directions. It is suggested that the reverse orientation of some dendritic microtubules allows dendrite-specific materials to be recognised and transported to subsynaptic areas.

Polymerisation requires a number of accessory factors and **GTP**. Two of the most important accessory factors are the **microtubule accessory proteins (MAPs)** (200–300 kDa) and the **tau proteins** (60 kDa). These accessories appear to be involved both in the polymerisation process and in the stabilisation of tubulin once it has polymerised. Both MAPs and taus are also involved in cross-linking microtubules to other cellular structures. There are indications that MAPs differ markedly from cell type to cell type. There is also evidence that the MAPs in dendrites are different from those found in axons. We shall return to MAPs and tau proteins in Section 21.10. We shall find that hyperphosphorylated tau forms one of the 'hall-mark' pathologies underlying **Alzheimer's disease**.

Both actin microfilaments and microtubules are well known to be involved in cell movement. Microtubules form the major protein component of the axonemes of eukaryotic cilia and flagella and in association with another protein, **dynein** (of which more below), cause their characteristic beating movements.

15.3.4 The Axonal Cytoskeleton

The distribution of the fibrous 'skeletal' elements within cells can be studied by immuno- and electron microscopy. In the first technique the cytoskeletal element of interest is extracted and purified. An antibody is raised against it and the latter either reinjected into a cell or introduced through an enzymatically weakened plasma membrane. After the antibody has found its cytoskeletal antigen and attached itself another antibody, prepared so that it binds to the first but this time coupled to a fluorescent dye, is introduced. When the cell is exposed to an appropriate wavelength of light the cytoskeletal system under investigation fluoresces.

Electron microscopy has also proved a valuable technique for investigating the cytoskeleton. There are several possible techniques. We shall look at two of the most important. Axons may be prepared either by conventional techniques of chemical fixation and thin sectioning or by the more modern technique of freeze-fracture etching. Electron micrographs prepared by the first technique make the filamentous nature of the axoplasm very evident. It is not difficult to see that neurofilaments and microtubules are present throughout the axon.

An electron micrograph prepared by the second technique is shown in Figure 15.11C. The technique involves freezing fresh nerve fibres extremely rapidly by bringing them into contact with a copper block cooled to the temperature of liquid nitrogen (−196°C) or liquid helium (−269°C). The biochemical nature of the filamentous elements can be revealed by a previous reaction of the tissue with specific antibodies. Antibodies, for instance, can be used to pick out both the neurofilament protein and the cross-bridge protein. These antibodies can then be seen decorating the filamentous elements and their cross-linkages. This provides good evidence that the electron microscope image is indeed of neurofilaments. Similar techniques are used to identify microtubules.

These microscopical and biochemical techniques are beginning to reveal the full complexity of axoplasmic transport. It is believed that the axo-plasm can be subdivided into a core microtubular domain and a more peripheral neurofilamentous domain. Both microtubules and neurofilaments are extensively cross-linked. Actin microfilaments are to be found both beneath the axolemma, where they may form part of a submembranous cytoskeleton (similar to that described in Chapter 7), and also in the core, where they are involved in bringing about the axoplasmic flow of neurofilaments and micro-tubules. Freeze-etched electron micrographs show secretory vesicles attached by cross-links to microtubules. An interpretation of this intricate two-domain system is shown in Figure 15.13.

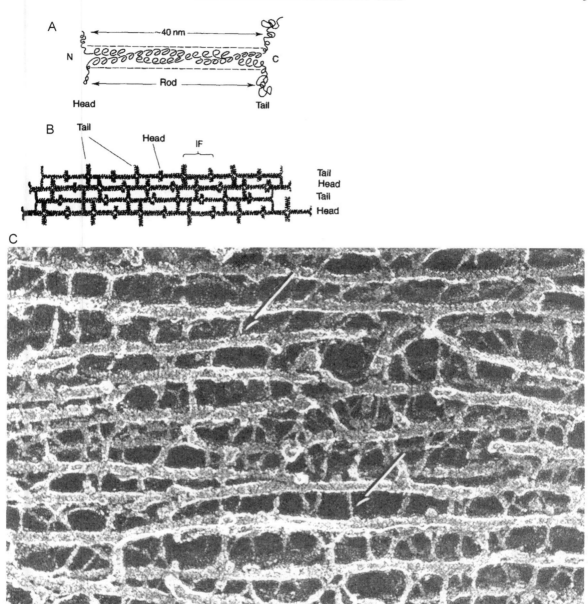

Figure 15.11 Structure of IFs and neurofilaments. (A) An IF consists of a rod-like segment containing of two α-helices wound around each other to form a two-stranded rope. Each rod-like segment ends in a variable region. The C-terminal end (the tail) has a more extensive variable region than the N-terminal end (the head). The whole molecule is about 40 nm in length and 7–11 nm in diameter. (B) The variable entanglements of the heads tend to attach to each other as do the larger entanglements of the tails. The sideways-spreading tail entanglements also become attached to neighbouring IFs so that bundles, 10 nm or so in diameter, are formed. It is probable (not yet certain) that the IFs in these bundles are arranged in an antiparallel manner. There are no (+) or (–) ends as found in microtubules. (C) Neurofilaments observed in the electron microscope using a quick-freeze, deep-etch technique. The arrows indicate cross-bridges between the neurofilaments (×275 000.) Part C from Hirokawa (1986), *Trends in Neurosciences*, **9**, 67–71; with permission.

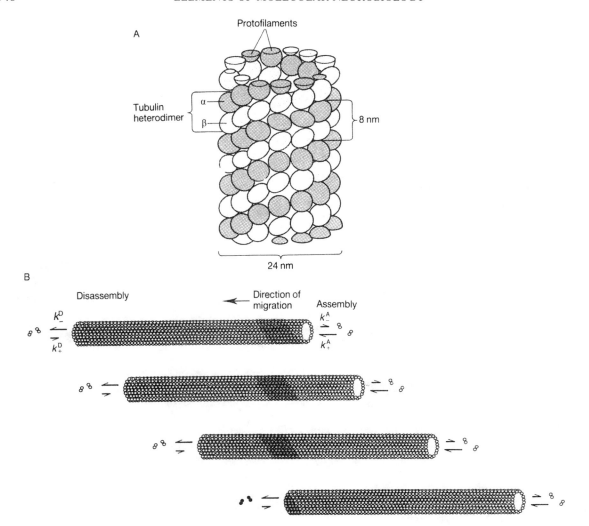

Figure 15.12 Molecular structure of tubulin. (A) Molecular structure of a microtubule. The tube consists of 13 longitudinal rows of α- and β-tubulin subunits. These rows are called protofilaments. (B) 'Treadmilling'. In the presence of GTP, α- and β-heterodimers are added to the (+) end and lost from the (−) end. Incorporation of a radioisotope allows one to follow a patch of radiolabel (black) moving along the microtubule during treadmilling. k_+^A and k_-^A are the rate constants for the addition and removal (respectively) of dimers from the (+) or assembly end whilst k_+^D and k_-^D are the rate constants for addition and removal at the (−) or disassembly end. Reprinted by permission from Margolis and Wilson (1981), *Nature*, **293**, 705–711. Copyright © 1981, Macmillan Journals Ltd.

Dynamism of the Cytoskeleton

Immuno- and electron microscopy, of necessity, provide static images. The biologist has to use his or her imagination to understand the molecular turmoil that these images have caught and frozen. This, however, is not always easy. Indeed the very term

'cytoskeleton' can be deeply misleading. We have already noted that it tends to suggest something stable and enduring, as we regard the skeletons of our own bodies. This analogy may be appropriate for the submembranous cytoskeleton we considered in Chapter 7, but it can easily bias our view of the molecular agitation in the cores of axons.

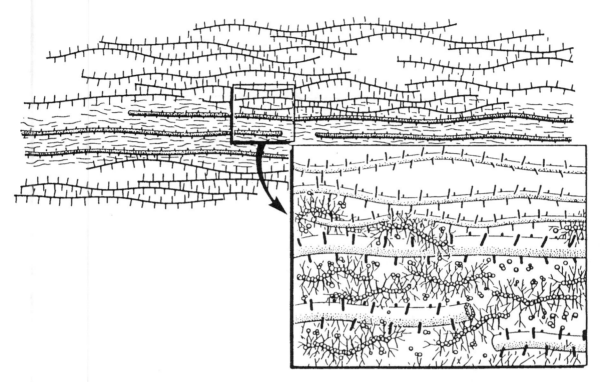

Figure 15.13 Central axonal cytoskeleton. This figure summarises a number of studies on squid giant fibre axoplasm. The major part of the figure shows a microtubule domain surrounded by neurofilaments. The microtubules are associated with numerous short microfilaments. In the higher-magnification inset the intricate interweaving of the microtubules and microfilaments is shown. The microfilaments are interwoven with accessory proteins, both globular and filamentous. Reprinted with permission from Lasek (1986), *Journal of Cell Science*, Suppl. 5, 161–179. Copyright © 1986, the Company of Biologists Ltd.

Fortunately an ingenious new technique has been developed which allows us to glimpse the true molecular dynamism of the 'cytoskeleton'. This technique is called video-enhanced microscopy or, more fully (after its inventor), Allen video-enhanced contrast-differential interference contrast microscopy – **AVEC-DIC**. It not only allows the observation of cytoskeletal elements in unfixed living cytoplasm but also, through clever computer techniques, allows the observation of elements far below the limit of resolution of the best optical microscopes. The theoretical limit of resolution of optical microscopy is about 0.25 µm; the AVEC system allows the observation of fibres with a diameter of only 25 nm – an order of magnitude improvement.

In the early 1980s this new methodology was applied to the study of axoplasmic transport. The axon studied was our old friend the giant axon of the squid. Immediately it became apparent that the axoplasm was teeming with activity. Mitochondria were moving along the axon in both directions with a jerky motion; multivesicular bodies (carrying used membrane) and multilamellar bodies (probably mitochondrial remains) were moving toward the perikaryon; dense shoals of transport vesicles were moving in the opposite direction, from the perikaryon to the synaptic terminal.

These observations were followed up by examining the system *in vitro*. We noted in Chapter 11 that one of the great virtues of the squid giant axon was that the axoplasm could be extracted and analysed. It proved possible to sandwich extruded squid axoplasm between two cover glasses and examine it by AVEC microscopy (Figure 15.14). It was found that organelle movement in the centre of the

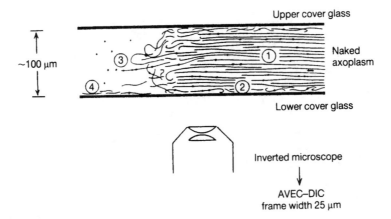

Figure 15.14 *In vitro* examination of extruded squid axoplasm. The axoplasm is sandwiched between two cover glasses and observed by AVEC-DIC microscopy. Organelle movement in the central domain (1) of the axoplasm resembles that seen *in vivo* and is particularly well seen where filaments protrude from the axon surface (3). Microtubules can be clearly observed at the periphery (2) and where they have fallen free from the axoplasm (4). Small particles which become attached to MTs in this last position are transported along their lengths. Reprinted with permission from Weiss (1986), *Journal of Cell Science*, Suppl. 5, 1–15. Copyright © 1986, the Company of Biologists Ltd.

sandwiched axoplasm was not noticeably different from that in the squid axon itself. The *in vitro* system allows numerous experimental manipulations to be undertaken with a view to establishing the nature of the observed movement, the chemical environment required, the energy source, etc.

In addition to the 'cover glass sandwich' system of Figure 15.14 it has also proved possible to extract single microtubules and show that in appropriate conditions they will not only transport organelles and indeed carboxylated latex beads, but also glide across glass cover slips themselves. AVEC microscopy shows that organelles (and latex beads) move in both directions on single microtubules and pass each other without apparent difficulty.

The molecular cause of these fascinating movements is a topic of intense research at the present time. The *in vitro* systems described above allowed the isolation and assay of two classes of motor protein: **kinesins** (500 kDa) and **cytoplasmic dyneins** (1200 kDa).

Kinesin

Kinesins (KIFs) are also well known in other parts of cell biology, where they are involved in the transport of vesicles and other material from one part of a cell to another and in the formation of the

spindle fibres at cell division. They consist of two large globular domains which have ATPase activity connected by stalks to a carrier complex (Figure 15.15A). Kinesins are members of a large superfamily of microtubule-associated motor proteins: KIF1A, KIF1B, KIF2, etc. Altogether 45 different kinesins have been identified in the human genome database and 38 have been detected in the brain. They all share common 320–340 amino acid globular motor domains but their stalks and carrier complexes are very various.

Cytoplasmic Dynein

Cytoplasmic dyneins are closely related to the dyneins of cilia and flagella mentioned above. Like the kinesins they are involved in the formation of the spindle fibres at cell division and like the kinesins they consist of two large globular domains. They are, however, very much larger molecules and far less varied. The globular domains weigh in at about 530 kDa each and they are connected via two 74 kDa stalks to a 60 kDa binding complex (Figure 15.15B). The two globular domains each contain four ATP binding sites and provide the energy for movement along microtubules.

A

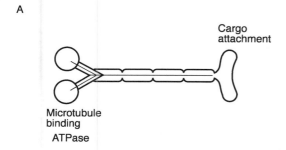

B

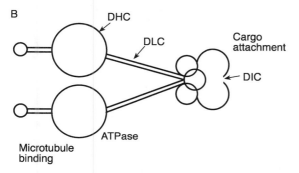

Figure 15.15 Structure of kinesin and cytoplasmic dynein. (A) Kinesin. There are several forms of kinesin which vary in the structure of their 'cargo attachment' domains. (B) Cytoplasmic dynein. These are much larger molecules than kinesins with a more complex structure. DHC=dynein heavy chain; DIC=dynein intermediate chain; DLC=dynein light chain. After Almenar-Queralt and Goldstein (2001), *Current Opinion in Neurobiology*, **11**, 550–557.

'Brachiation'

Cytoplasmic dyneins and kinesins are involved in organelle transport within most eukaryotic cells as well as with the movements associated with mitotic and meiotic cell division. Both are microtubule-dependent ATPases with this activity located in their two large globular 'heads' (Figure 15.15). The heads attach to a microtubule and the tail complex attaches to the organelle or vesicle being translocated.

But which is the more permanent attachment? Does the head attach to the microtubule and the tail waft cargo from one position to the next – like a ciliated epithelium? Or, does the tail attach to the cargo and both head and tail move along the microtubule? It appears that the latter is the case. Glass beads coated with kinesin or cytoplasmic

dynein can be shown to move along purified microtubule guide lines. It can also be shown that whereas beads coated with kinesin move towards the **plus** end of the MT, that is in the anterograde direction, those coated with cytoplasmic dynein move toward the **minus** end, that is in the retrograde direction (Figure 15.16A). The means by which the motor proteins transfer the energy tapped from ATP hydrolysis into mechanical movement is also beginning to be understood.

Essentially kinesin and cytoplasmic dynein discovered bipedalism several billion years before the hominids. The production of relative motion by 'rowing' is very familiar to molecular biologists in other contexts: it lies at the root of both muscle movement (**myosin**) and the movement of eukaryotic flagella and cilia (**dynein**). But this would not work for axoplasmic movements. Every oarsman knows that after the power stroke the oar is lifted from the water and swung back for the next stroke. If this were the case with kinesin or cytoplasmic dynein they would likely drift away from the MT track. Instead, then, of rowing a different technique has been adopted. Kinesin heads have been shown to 'walk' down the tubulin track, or better 'brachiate', 'hand over hand'. In other words, while one of the two heads attaches to the microtubule the other swings through to the next site and attaches before the first lets go (a useful animation can be viewed at the web site listed in the Bibliography).

Vesicle Attachment

The tail complexes of kinesin and cytoplasmic dynein are of great importance. Because the two motor proteins move in opposite directions along the microtubule the tail complexes have to attach to the appropriate cargo. Indeed cytoplasmic dynein itself must be transported in an inactive state by kinesin to the far end of the axon before it can pick up its cargo of used material, NGF etc., for transport back to the perikaryon. As both motor proteins are present in numerous subtypes it is concluded that different 'tails' confer different attachment specificities. Much has still to be learnt about this and about the attachment and release factors which must also be involved.

Some progress has, however, been made. Two proteins have been identified which link membrane elements to the kinesin light chains. These are the

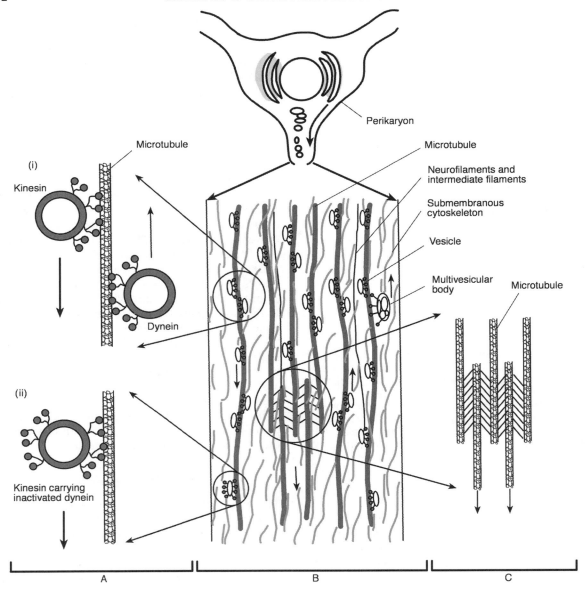

Figure 15.16 Axoplasmic transport summarised. The figure schematises the transport mechanisms which are believed to be operating within an axon. (A) (i) Vesicles coated with kinesin and dynein are shown moving along microtubular guides in the anterograde and retrograde directions. (ii) Kinesin motivated vesicle carrying inactivated dynein towards the axon terminal. After Vallee and Bloom (1991), *Annual Review of Neuroscience*, **14**, 59–92. (B) Schematic diagram showing secretory vesicles (containing neurotransmitter or membrane components) being moved outwards in the anterograde direction to the axon terminal. Other material, perhaps using membrane or breakdown products from neurotransmitters, are moved towards the perikaryon (retrograde direction), mostly in the form of multivesicular bodies. These ultimately empty their contents into perikaryal lysosomes for further digestion. Mitochondria are also moved by the microtubular transport system. Partly after Allen (1987), *Scientific American*, **156**, 26–33.(C) Possible presence of dynein and/or kinesin cross-links 'rowing' tubulin and attached cytoskeletal and other elements in the 'slow' anterograde stream. After Vallee and Bloom (1991), *Annual Review of Neuroscience*, **14**, 59–92.

amusingly named 'sunday driver' (Syd) and, interestingly, the amyloid precursor protein (APP). We shall return to the latter protein in Section 21.10, where we shall see that it is deeply involved in the aetiology of Alzheimer's disease. Both proteins are believed to have an important role in attaching cargo wrapped inside membranous vesicles to kinesin motors. Other kinesins, such as KIF17 which is restricted to dendrites, appear to make attachments to specific receptor subunits, in this case the NMDA 2B subunit. It begins to seem possible that signalling and kinesin-based axonal transport are closely linked.

Cytoplasmic dynein is, as we have noted, a much larger molecule. It is believed to attach to its cargo by a number of polypeptides in its light chains which interact with *trans*-vesicle membrane proteins. In particular the retrograde transport of neurotrophins is thought to occur in this way and this, of course, is of great importance (see Section 19.9).

Slow Axoplasmic Transport

The mechanisms responsible for slow axoplasmic transport are less well understood than those for the fast stream described above. There is much evidence to show that the cargo carried in this stream consists of α- and β-tubulin monomers, of the subunits of neurofilaments, clathrin and other elements of the cytoskeleton, as well as some enzymes and calmodulin. Two proposals have been put forward to account for this slow stream. On the one hand, it has been suggested that complete modules of these cytoskeletal elements are synthesised in the perikaryon and move in the slow stream as cohesive units. There is evidence to show that movement of such cytoskeletal complexes is powered by cytoplasmic dynein although a member of the kinesin superfamily may also be involved. These, it has been suggested, form cross-links between adjacent microtubules. In this scenario, multiple 'oars' generate a 'rowing' or 'millipede crawling' mechanism. The cross-links 'row' microtubules bearing cytoskeletal complexes through the stationary MTs of the axoplasmic skeleton (Figure 15.16C). Alternatively, a second proposal visualises the elements of the cytoskeleton being synthesised and packaged as monomers in the perikaryon and transported down the axon as such to be assembled *in situ*, either in the axon or at

the synaptic bouton. The motive force is provided, once again, by members of the kinesin superfamily.

15.3.5 Axoplasmic Transport Summarised

Molecular neurobiologists still have far to go before they achieve a full understanding. Nevertheless progress in the last few years has been remarkable. The development of the new microscopical techniques and the use of the *in vitro* model systems described above have revolutionised the prospects of success. An understanding of the molecular nature of axoplasmic transport is of first rate importance not only for the science of the brain but also in that it allows us to gain some insight into the causes of several distressing neuropathologies, especially those of the ageing brain. We shall return to some of these, especially Huntington's and Alzheimer's diseases, in Chapter 21. It is interesting to note here, however, that a recent report, using mouse knockout technology, suggests that one form of Charcot–Marie–Tooth disease (CMT2A) is due to a loss of function mutation in a kinesin gene.

Finally, it should be noted that most of the analyses of axoplasmic transport described above have been carried out on peripheral fibres. Although we have no particular reason to suspect that central neurons do it differently, the possibility that they do should be borne in mind. In particular it has been suggested that not only secretory granules, cytoskeletal elements, enzymes, etc. are carried along central axons but also fragments of smooth endoplasmic reticulum (SER). If this is the case the synaptic terminal would be a metabolically more complex place than is usually allowed.

15.4 EXOCYTOSIS AND ENDOCYTOSIS AT THE SYNAPTIC TERMINAL

After days, perhaps weeks, of 'brachiating' or 'walking' down the microtubular tracks within an axon the secretory vesicle finally reaches its destination in the synaptic terminal. It now awaits its one brief moment of action. This depends on the arrival of an action potential or, if it is a local circuit neuron, an electrotonic potential, at the terminal. When this happens the membrane depolarisation (as we noted in Section 11.3) opens calcium gates. The influx of Ca^{2+} triggers release of the vesicle's contents into the synaptic gap.

Let us look in a little more detail at this vital process. Examination of synaptic endings in the electron microscope shows that they contain large numbers of synaptic vesicles. These vesicles vary in shape and size according to their contents. Small spherical translucent vesicles (diameter about 50 nm) contain transmitters such as acetylcholine, glutamate, etc. Other small translucent vesicles assume a more ellipsoidal form and these are believed to contain inhibitory transmitters such as glycine. Larger vesicles (diameter >60 nm), often with dense cores, contain catecholamines whilst yet larger vesicles (diameter c. 175 nm) contain peptides.

It should be borne in mind that many synaptic vesicles contain much more than just neurotransmitter. Even the small translucent vesicles containing neurotransmitters such as ACh or glycine contain several soluble proteins and ATP. The larger dense-cored vesicles contain much more besides catecholamines. In particular they can be shown to include ATP, ATPase and dopamine-β-hydroxylase. The large vesicles of peptidergic neurons contain ATP, ATPase, adenyl cyclase, Ca^{2+}, and perhaps some of the enzymes required for the final stages of post-translational processing.

A great deal of synaptic molecular biology has used the 'honorary' synapse of the neuromuscular junction (Figure 15.17). This junction is so conveniently placed for experimental manipulation that it has been called the '*E. coli* of molecular neurobiology'. We shall meet it frequently in the following chapters of this book. It uses ACh as its transmitter.

Figure 15.17C shows an electron micrograph of the frog neuromuscular junction. The transmitter vesicles are very well shown. Each contains 5000–10 000 molecules of ACh and there may be up to and sometimes more than a million vesicles in a presynaptic terminal. The axoplasm behind the cluster of vesicles can be seen to contain dense accumulations of cytoskeletal elements. This is called the **presynaptic network**.

Transmitter vesicles are continuously bumping into the presynaptic membrane and voiding their contents into the synaptic gap. These packets of 5000–10 000 molecules of ACh are called 'quanta'. They cause small depolarisations in the subsynaptic membrane. These small, random depolarisations are called **miniature end-plate potentials (MEPPs)**. This terminology applies only to the neuromuscular junction where the subsynaptic membrane is called the motor end plate. The same phenomenon probably also occurs at central synapses. However, it is only when the presynaptic terminal is depolarised by an action potential that a large population of vesicles fuse with the presynaptic membrane and release their contents. It is only when this occurs that a major effect is exerted on the subsynaptic cell.

Figure 15.18 shows freeze-fracture preparations of the presynaptic membrane of a neuromuscular junction. The specimen has been prepared in such a way that we are looking up at the outer surface of the inner leaflet (PF-face) of the terminal as if we were sitting on the postsynaptic membrane. Figure 15.18A shows the membrane in the resting state, Figure 15.18B 5 ms after stimulation and Figure 15.18C several tens of milliseconds later. The interpretation of these remarkable images is as follows. The rows and scatterings of granules in Figure 15.18A represent intramembrane proteins – probably Ca^{2+} channel proteins or, alternatively, evidence of internal ridges which guide vesicles to appropriate regions of membrane (see below). The large craters in Figure 15.18B indicate where synaptic vesicles have fused and voided their contents into the synaptic gap. The somewhat smaller depressions in Figure 15.18C represent regions where membrane is being retrieved after exocytosis has terminated.

These electron micrographs help us visualise the mechanisms at work in the presynaptic terminal. Let us consider them in sequence.

15.4.1 Vesicle Mustering

In Section 15.2 we noted that when vesicles were packaged in the perikaryon they usually became caged in clathrin basket-works. It looks as if at least some vesicles are attached to the kinesin 'rowing arms' through their clathrin coats. Clathrin, of course, is not neuron-specific. It is found in most secretory cells. Another 'coat' protein, however, does seem to be restricted to neurons. This is **synapsin 1**. This protein (a dimer of synapsin

A

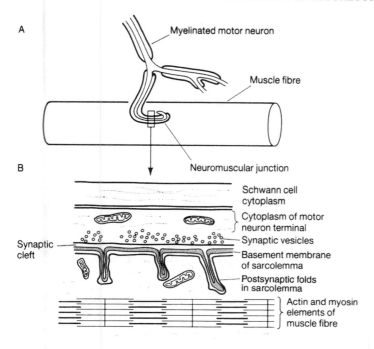

Myelinated motor neuron

Muscle fibre

Neuromuscular junction

B

Schwann cell cytoplasm

Cytoplasm of motor neuron terminal

Synaptic vesicles

Synaptic cleft

Basement membrane of sarcolemma

Postsynaptic folds in sarcolemma

Actin and myosin elements of muscle fibre

C

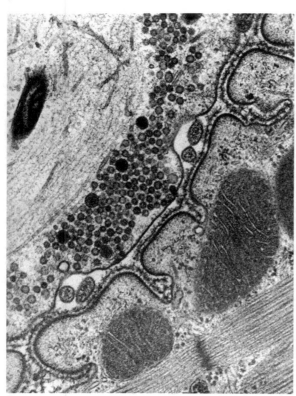

Figure 15.17 The neuromuscular junction. (A) Motor nerve fibres normally branch to innervate a small group of muscle fibres (the 'motor unit'). Each branch ends on the sarcolemma of a muscle fibre in a tiny spiral. (B) Magnified diagram of portion enclosed in rectangle. The sarcolemma of the muscle fibre immediately beneath the nerve terminal (the motor end plate) is invaginated into a number of post-junctional folds (this is not seen at other synapses). A thick basement membrane is secreted by the sarcolemma. The synaptic cleft between the neuronal membrane and the sarcolemma's basement membrane measures about 4 nm. (C) Electron micrograph of frog neuromuscular junction. The presynaptic terminal of the motor neuron is in the upper left part of the micrograph and the muscle fibre in the lower right. The sarcolemma of the muscle fibre is thrown into junctional folds and the basement membrane is well shown, as are the synaptic vesicles in the presynaptic ending. Micrograph produced by Dr John E. Heuser of Washington University School of Medicine, St Louis, MO; reprinted with permission.

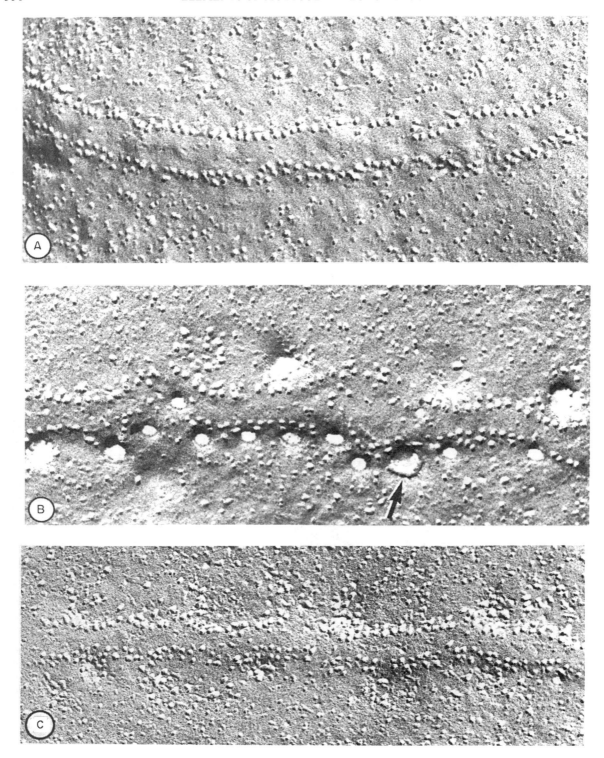

1a (86 kDa) and synapsin 1b (80 kDa)) forms a protein coat specifically around the small (50 nm) vesicles. It is present in large quantities in synaptic terminals (Figure 15.19).

Now synapsin 1 is biochemically almost identical to the band 4.1 protein of the erythrocyte submembranous cytoskeleton (see Figure 7.19). It will be recalled from Chapter 7 that the band 4.1 protein binds both to spectrin and to actin. We also saw that it is anchored to the erythrocyte membrane by binding to the multipass transmembrane protein, band 3. Presynaptic terminals contain high concentrations of **fodrin**, a spectrin-like protein. It appears, therefore, that synapsins are not only structurally very similar to but also play a similar role to that of the erythrocyte band 4.1 protein. In other words they bridge between the lipoprotein wall of the vesicle and one of the cytoskeletal elements which, as we noted above, are so well represented in the terminal.

15.4.2 The Ca²⁺ Trigger

We noted in Chapter 11 that free Ca^{2+} is in very short supply within cells. Any influx of Ca^{2+} from outside can thus have a dramatic effect. One of these effects is to activate several kinase enzymes within the terminal. Two of these Ca^{2+}-dependent kinases (protein kinase A (PKA) and Ca^{2+}-calmodulin-dependent protein kinase 2 (CaM kinase 2)) are known to phosphorylate synapsin 1. CaM kinase 2 is, indeed, bound to synapsin 1. In the presence of Ca^{2+} it phosphorylates synapsin 1 in such a way that its affinity for both vesicle membrane and actin is greatly reduced (Figure 15.20A).

This sequence of biochemical reactions ensures that Ca^{2+} influx into a synaptic terminal causes release of vesicles from their synapsin anchorages. The 'desynapsinated' vesicle is then free to migrate toward the P-face of the presynaptic membrane. But Ca^{2+} has more immediate effects as well. We shall see below that it also triggers the fusion of synaptic vesicles to the presynaptic membrane and the release of neurotransmitter into the synaptic cleft.

15.4.3 Vesicle Docking

The next events are still far from fully worked out. A very complicated piece of molecular machinery is involved. Much of it was invented billions of years ago and has been conserved ever since. Many of the proteins involved at the synapse can also be studied in yeasts. The complexity and the conservation is largely due to the fact that membranes do not fuse spontaneously (see Chapter 7). The repulsive forces between two phospholipid membranes at close quarters (1–2 nm) in aqueous solution is very high. Specialised fusion proteins are required to unite the membranes of exocytotic vesicles, in this case synaptic vesicles, with the cell membrane, in this case the presynaptic membrane.

Several classes of proteins play complicated yet coordinated roles. The most prominent are the so-called **SNARE** proteins. **SNARE** is an acronym derived from 'soluble NSF attachment protein receptors'. The SNAREs form a large superfamily of proteins related by a common c. 60 amino acid sequence known as the **SNARE motif**. They also share flanking sequences which attach them to membranes and/or enable protein–protein interactions to occur.

Before the large number of SNAREs presently known became apparent they were classified into two groups: vSNAREs associated with **v**esicles and tSNAREs associated with **t**arget membranes

---

Figure 15.18 *(opposite)* Freeze-fracture replicas of the presynaptic membrane of frog neuromuscular junction. (A) The electron micrograph shows the membrane 3 ms after the nerve had been stimulated. The double row of particles may represent the undersurface of the dense bars which delimit an 'active zone'. Further explanation in text. (B) The electron micrograph shows the presynaptic membrane 5 ms after stimulation. Large pits (one of which is arrowed) adjacent to the parallel rows of particles are visible. These are believed to represent the sites where synaptic vesicles have fused with the membrane and voided their contents. (C) The electron micrograph shows the membrane 50 ms after stimulation. Shallow depressions with large numbers of intramembranous particles are common. These are believed to represent the final stages in the retrieval of the membrane after exocytosis (see Figure 15.22). The freeze-fracture electron micrographs are of the PF faces of the membrane (see Figure 7.11). From Heuser and Reese (1979), in *The Neurosciences: Fourth Study Program*, ed. F.O. Schmitt and F.G. Worden, Cambridge, MA: MIT Press, pp. 573–600; with permission.

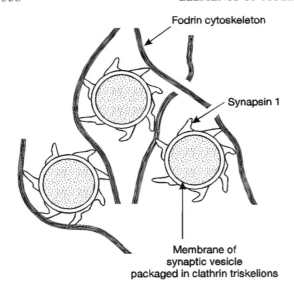

Fodrin cytoskeleton

Synapsin 1

Membrane of
synaptic vesicle
packaged in clathrin triskelions

Figure 15.19 Attachment of vesicles to the fodrin cytoskeleton within a presynaptic terminal. Explanation in text.

(in synapses the presynaptic membrane). This classification has now been replaced by one based on the biochemical structure of the molecules. Once again two major groups are recognised: **Q-SNAREs** have a central glutamine (Q) residue and **R-SNAREs** have a central arginine (R) residue in the SNARE motif. Nevertheless the older classification retains some value in describing the process of exocytosis at the synapse. Thus vSNAREs include **synaptobrevins 1** and **2** (=vesicle associated membrane proteins: **VAMPs 1** and **2**) whilst tSNAREs are represented by **SNAP-25s** (25 kDa synaptosomal associated proteins) and **syntaxins 1A** and **1B**.

In addition to the SNAREs a number of other proteins play significant roles at the synapse. These include **synaptotagmins** and **synaptophysins** in the vesicle membrane; **neurexins** and **calcium channels**, in the P-face of the presynaptic membrane. Most of these proteins exist in multiple isoforms, which allows specific matching of vesicle and synaptic membrane. In addition several non-membrane-bound proteins play essential roles: **n-sec1**, **rab3A**, **α-SNAP**. This exceedingly complex piece of molecular machinery (so far as it is presently understood) is shown in Figure 15.20B.

It must not be forgotten that the trigger for assembling all the actors in this intricate scene is the arrival of an action potential and the consequent opening of Ca^{2+} channels. There is good evidence that **synaptotagmin** is the Ca^{2+} sensor. It is a 421 amino acid protein possessing Ca^{2+} binding domains similar to those of PKC and makes just one pass through the vesicle membrane. Knockout genetics have been used to create mice with defective synaptotagmin. It is interesting to find that hippocampal cells from these mice appeared quite normal in the optical microscope but the fast response to a Ca^{2+} trigger had been eliminated. There was no synaptic activity.

Synaptophysin, as shown in Figure 15.20A, is a 4TM protein and it may, with other similar subunits, form an ion pore (compare the 4TM structures in Chapter 10). When the fusion complex forms (Figure 15.20B) the synaptophysin pore may align with a twin channel (physophylin) in the presynaptic membrane. This alignment may be compared with the gap junction structure of Section 7.9. It will be recalled that the connexin subunits of gap junctions also have a 4TM conformation. There is evidence from capacitance studies that pores *do* form during the early stages of vesicle fusion and similar fusion pores have been detected in other cells. If such a pore does develop it would allow the extrusion of a small amount of transmitter just before exocytosis releases the major part into the synaptic gap.

The exact nature of the interactions between the membrane-bound proteins is, however, still far from being fully understood. At the time of writing the vSNAREs and tSNAREs are believed to be crucial to the **fusion** of the vesicle and presynaptic membranes. They form a 'binding complex'. The non-membrane-bound proteins are also essential. **rab3A** is a member of the *ras* superfamily of GTPases (see Box 8.1). As is the case with other *ras* GTPases, it is believed to play a significant part in timing the process. It is believed to attach to the binding complex and if this remains stable (i.e. docked) for a sufficient length of time for rab3 to hydrolyse its GTP, it locks the complex into position.

The time allowed before rab3 locks the complex ensures that only specific matches between vesicle and presynaptic membrane are made. **NSF** (*N*-ethylmaleimide-sensitive fusion protein) and **α-SNAP** (soluble NSF-attachment protein) are

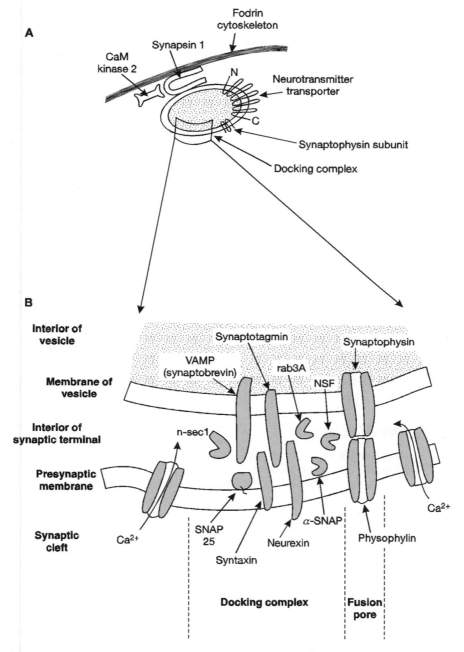

Figure 15.20 Vesicle docking at the presynaptic membrane. (A) A synaptic vesicle is shown attached through one of its synapsins to an element of the cytoskeleton. The docking complex is located within the rectangle. (B) Enlarged cartoon of the docking complex. Further explanation in text. CaM kinase 2 = Ca^{2+}-calmodulin-dependent protein kinase 2; n-sec1 = member of a class of secretory proteins; NSF = N-ethylmaleimide-sensitive fusion protein; rab3A = a *ras*-type GTPase; SNAP = synaptosomal associated protein; VAMP = vesicle associated membrane protein. After Pevsner and Scheller (1994), *Current Opinion in Cell Biology*, **6**, 555–560; and Jessel and Kandel (1993), *Cell*, **72**, *Neuron*, **10**, 1–29.

required to complete the fusion. Lastly **n-sec1** is required in the initial steps of the reaction. Its dissociation from syntaxin (perhaps with the assistance of a Rab) allows syntaptobrevin (a vSNARE) to bind with syntaxin and SNAP25 (tSNAREs). It will be remembered from Section 11.3.1 that the Ca^{2+} channels in synaptic membranes have binding sites for syntaxin. The Ca^{2+}-triggered molecular machinery is clearly very closely interlocked. Ultimately the three SNAREs settle into a stable ternary complex known as the 'core complex' and form a strong link between the vesicle and presynaptic membranes.

The involvement of n-sec1 at this crucial point in the formation of the fusion complex is of considerable interest. n-sec1 is the neural (hence the 'n' prefix) homologue of proteins required for secretion in yeast cells (*Saccharomyces cerevisiae*). Indeed homologous secretory proteins (secs) are found not only in yeast but throughout the animal kingdom, from *Caenorhabditis elegans* through *Drosophila* to the mammals. This suggests that the actors in this intricate secretory drama have been conserved from very early times, the times of the first eukaryocytes. This has allowed their molecular biology to be dissected in forms as distant as *Drosophila* and *C. elegans*.

15.4.4 Transmitter Release

Figure 15.17C shows the synaptic vesicles thronging close to the P-face of the presynaptic membrane. If instead of a neuromuscular junction we were to examine a typical central synapse we would find that the vesicles were ordered by a **presynaptic grid**. This grid is formed of dense projections springing from the presynaptic membrane and pointing inwards into the synaptic interior (Figure 15.21). The projections are bound together by fine filaments and form a highly ordered network on the P-face of the presynaptic membrane. The 'design element' seems to be a hexagon. Six triangular spaces surround each projection and these are probably membrane areas specialised for the attachment of synaptic vesicles. Indeed fracture-freeze preparations show hexagonal arrays of intramembranous particles in the P-face of the presynaptic membrane. It is tempting to conclude that these are Ca^{2+} channels surrounding one of the dense projections of the presynaptic grid and

provide points of attachment for transmitter vesicles. In consequence they have been termed **vesicle attachment sites (VASs)**.

The influx of Ca^{2+} has been followed *in vivo* in the presynaptic terminals of the squid giant axon. The method used involved fast-frame video recording of the Ca^{2+}-induced luminescence of aqueorin. Aqueorin responds to variations of Ca^{2+} concentrations of plus or minus a few hundred micromoles. The technique confirms that in synaptic transmission a rapid influx of Ca^{2+} (less than 16.6 ms) occurs leading to short-lived 'nanodomains' or 'quantum emission domains' (QEDs) (it will be remembered that synaptic vesicles are said to contain a 'quantum' of neurotransmitter molecules). The QEDs are about 0.5 μm in diameter and are located in register with the vesicle attachment sites. Groups of these transient QEDs are defined as constituting a microdomain.

Neuromuscular junctions show a different though analogous ultrastructural organisation. Instead of hexagonal VASs they show dense bars opposite the junctional folds of the subsynaptic membrane. The membrane adjacent to these bars is called an **active zone**. It is here that the vesicles fuse to the membrane. By histochemical staining using ω-conotoxin, which binds to Ca^{2+} channels, it can be shown that the lines of intramembranous particles in Figure 15.18A are Ca^{2+} channels. It seems, therefore, that calcium channels are aligned in the active zones adjacent to the bars. Figure 15.18B shows that the exocytotic craters lie alongside these particle lines – as would be expected if the theory of active zones is correct.

Since the release of neurotransmitter from the small translucent vesicles is very rapid (about 200 μs after Ca^{2+} influx) it is argued that they must already be in contact with the membrane, next to the Ca^{2+} channels, when the action potential invades the terminal. There would not be sufficient time for vesicles to move from internal cytoskeletal attachment sites. It is concluded, therefore, that there are two populations of these small vesicles: a **releasable** group which are already docked on the P-face of the presynaptic membrane and a **reserve** group held within the terminal attached by synapsin to the cytoskeleton. This organisation into two populations does not seem to apply to the larger dense-cored and peptide-containing vesicles. Their release times are more

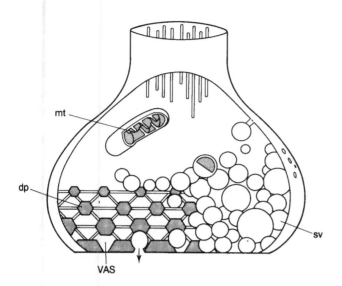

Figure 15.21 Organisation of presynaptic grid within a central synapse. The regular hexagonal structure of the presynaptic grid is shown in this diagram. Each triangular space constitutes a 'vesicle attachment site' (VAS) upon which the synaptic vesicles can dock and through which they can void their contents into the synaptic cleft. The triangular spaces surround a regular array of dense projections (dp) in a hexagonal pattern. sv = synaptic vesicle; mt = mitochondrion. From Akert *et al.* (1972), in *Structure and Function of Synapses*, ed. G.D. Pappas and G.D. Purpura, New York: Raven Press, pp. 67–86; with permission.

sluggish (some 50 ms) and this suggests that they are all attached to the cytoskeleton at some distance from the presynaptic membrane.

The release of the vesicle's contents is by exocytosis. It is a special case of the secretory process shown by many other cells. In consequence the study of this process in other cells, not only in other vertebrates but also in *Drosophila* and *C. elegans*, is very relevant. Exocytosis, like the release of vesicles from their attachment via synapsin to the cytoskeleton, is strongly Ca^{2+}-dependent. It may be that the fusion pore formed at the initial union of vesicle and presynaptic membrane dramatically widens. It may be (according to the attractively named 'kiss-and-run' hypothesis) that the transmitter is released through a transient widening of the synaptophysin fusion pore. As yet we do not know. But whatever does happen happens extremely quickly. The contents of the vesicle are voided into the synaptic gap in microseconds.

15.4.5 Dissociation of Fusion Complex and Retrieval and Reconstitution of Vesicle Membrane

How is the SNARE fusion complex holding the vesicle membrane to the P-face of the presynaptic membrane disassembled? We saw above that NSF and α-SNAP are required to form the SNARE complex. The SNARE complex forms between two opposing membranes (vesicular and presynaptic). This (following a nomenclature derived from organic chemistry) is called the '*trans*' complex. But once the membranes have fused, the SNARE complex is all in one membrane, the '*cis*' position. In this position NSF and α-SNAP switch to the opposite function: instead of assembling the complex they disassemble it. The various proteins of the complex are then free to recycle ready to form another fusion complex in response to a fresh pulse of Ca^{2+} influx.

Meanwhile the lipoprotein elements of the exocytosing vesicle's membrane mix intimately with the lipoprotein of the presynaptic membrane. Indeed the vesicle's proteins can be shown to diffuse rapidly (diffusion constant c. $2 \times 10^{-10}\, cm^2\, s^{-1}$) away into the presynaptic membrane. However, they are not lost. Within about 10 seconds after exocytosis coated pits begin to appear. This means that the presynaptic membrane begins to invaginate in association with a coat protein – usually clathrin. The vesicle proteins diffused in the presynaptic membrane are recognised by clathrin and this induces the endocytosis. Another protein, **dynamin**, is also thought to be involved at this stage. There is evidence to show that it is instrumental in cutting the forming vesicle free from the presynaptic membrane and 'pinching' together the cut ends to

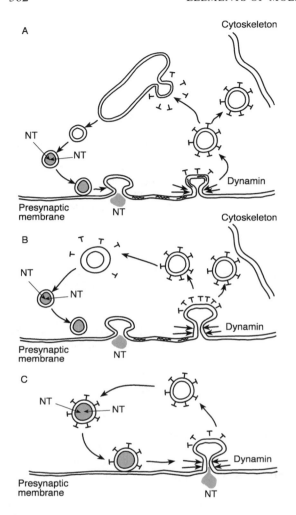

Figure 15.22 Three views of synaptic vesicle cycling. (A) Synaptic vesicles fuse with presynaptic membrane and release their contents into the synaptic cleft. Constituents of the vesicle membrane diffuse into the presynaptic membrane and are ultimately recognised by clathrin or synapsin 1. In both cases vesicles are formed by invagination. Dynamin molecules (arrows) nip the vesicles away from the presynaptic membrane. Synapsin-coated vesicles attach themselves to the cytoskeleton and clathrin-coated vesicles lose their clathrin and fuse with an endosome. Vesicles bud from the endosome and are refilled with neurotransmitter (NT). (B) Cycle as in part A but without passage through endosome. (C) 'Kiss-and-go' hypothesis. Transmitter released through fusion pore; shape of vesicle maintained by a membrane scaffolding. Partly after Mundigi and de Camilli (1994), *Current Opinion in Cell Biology*, **6**, 561–567.

form a complete endocytotic vesicle. We shall see, in Chapter 20, that the gene encoding an insect homologue of dynamin, *shibshire (shi)*, has been used to selectively inactivate synapses in *Drosophila* and thus to gain insight into processes by which memories are stored and retrieved.

Once the endocytotic vesicle is cut away from the underlying presynaptic membrane it is free to recycle (Figure 15.22B), or it may be (as in other secretory cells) that it fuses with an endosome within the presynaptic terminal (Figure 15.22A). Here the membrane components are reassembled and fresh vesicles bud off into the terminal. Alternatively, if the kiss-and-run hypothesis is accepted, the vesicle never loses its integrity and is refilled with transmitter and used again (Figure 15.22C). Whichever mechanism obtains the process is extremely rapid and proceeds on a massive scale within the brain. It has been calculated that a quantity of membrane having an area the size of an American football field is retrieved and reused in the human brain every 24 hours.

15.4.6 Refilling of Vesicle

If the vesicle contains small neurotransmitters it is refilled with neurotransmitter within the terminal. In other cases, where the transmitter is not synthesised in the terminal but in the perikaryon, the vesicle is moved back along the axon in the retrograde transport stream. Such vesicles normally fuse together to form a **multivesicular body**. Multivesicular bodies can often be seen in the perikarya of active neurons. We shall consider the syntheses of transmitter and modulator molecules in the next chapter.

In those cases where the transmitter is synthesised within the terminal it is passed back into the vesicle by means of specific transporter molecules in the vesicle's membrane. An account of the discovery and nature of these transporters is given in Box 15.2. A proton (H^+ ion) pump provides the energy for the transport. It either produces a potential across the membrane (Figure 15.23) or, if the membrane is permeable to a diffusible anion such as Cl^-, an electrochemical gradient is established. In either case the stored energy is used to drive the transport of the appropriate neurotransmitter into the vesicle. The transporter molecules are large 12TM proteins. There are four classes: one for ACh, one for catecholamines, one for glutamate and one

BOX 15.2 Vesicular neurotransmitter transporters

The nature of some of the 'transporters' which allow the passage of neurotransmitters from nerve terminal cytoplasm into synaptic vesicles has quite recently been clarified. The story is very interesting as it not only involves the elucidation of new channel structures but also shows once again how work on evolutionarily very distant forms has proved basic to understanding our own neurobiology. In addition, it throws an intriguing new light on the possible co-evolution of functionally related structures.

In a sense the story begins in the 1950s. When the essential nature of the genetic code had been established several of those involved felt it time to leave the gene for 'fresh woods and pastures new'. Many felt that the most important biological problem after DNA was the brain. Prominent amongst these emigrants was Sydney Brenner. He looked for an organism that would do for neurobiology what *E. coli* had done for genetics. He decided that the small nematode *C. elegans*, with its rapid generation time, simple nervous system and behaviour, would meet the specification. The rest, as they say, is history.

Some of the first behavioural mutants to be isolated displayed uncoordinated movements. These were termed unc mutants and some 120 *unc* genes have subsequently been defined. In particular, mutations at the *unc-17* locus caused jerky, coiling locomotion. *Unc-17* worms could on the one hand be shown to be synthesising ACh and on the other not to be affected by drugs poisonous to

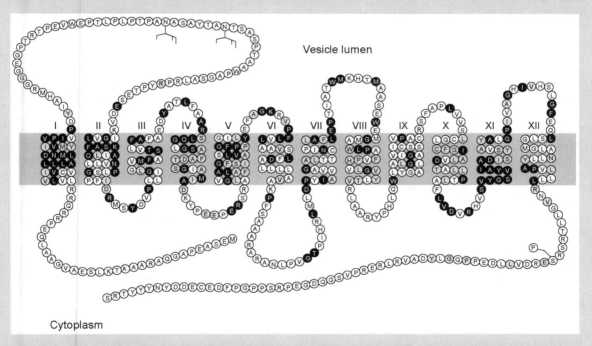

Figure A Vesicular ACh transporter (VAChT). The amino acid sequence and proposed topology of the VAChT is shown. Grey circles represent residues identical in all presently known VAChTs. Black circles indicate identity with residues in VMATs. Intravesicular glycosylation sites are shown and a phosphorylation site for PKC (P) on the C-terminal is indicated. From Usdin *et al.* (1995), *Trends in Neurosciences*, **18**, 218–224; with permission.

AChE. When the *unc-17* gene was cloned and translated it was shown to code for a protein with 40% sequence homology to two vesicular monoamine transporters (VMAT1 and VMAT2). Furthermore, immunostaining showed that the UNC-17 protein co-localised with synaptotagmin. These and other techniques convincingly showed that UNC-17 was a protein associated with the vesicle membrane.

The story passes next to that other favourite of the molecular neurobiologist: the electric organ of *Torpedo*. Using *unc-17* probes in cDNA libraries prepared from the exceedingly rich concentration of ACh vesicles in this organ, it was possible to isolate and clone a vertebrate homologue. It was shown, by expression in a kidney fibroblast cell line, that this homologue was an acetylcholine transporter. The *Torpedo* homologues were in turn used to isolate rat and human vesicular ACh transporters (VAChT). The molecular reason for the *unc-17* behavioural defect in *C. elegans* thus became clear. Although ACh was synthesised it was not sequestrated into vesicles and hence could not be voided into the synaptic cleft. Hence drugs poisonous to AChE were ineffective. Total loss of *unc-17* is fatal.

Hydrophobicity analysis of the amino acid sequences of VAChTs from *C. elegans*, *Torpedo*, rat and human and VMATs from rat, cow and human shows that they all share a common 12TM structure (Figure A). Transport of neurotransmitters into the vesicle is linked to the movement of a hydrogen ion (proton) in the opposite direction (see Figure 15.23). For this reason these channels are defined as neurotransmitter-H^+ antiporters.

The story has a final twist. *C. elegans* genetics had shown that *unc-17*, long before its product had been determined, was closely linked to the choline acetyltransferase (ChAT) gene (*cha*). The discovery that *unc-17* encoded an Ach transporter was thus thought-provoking. It prompted complete sequencing of the *cha–unc-17* complex. It turned out that *unc-17* is contained within the *cha* gene. This work with *C. elegans* suggested a close look at the human genome. The genes for both ChAT and VAChT were mapped to chromosome 10. Once again it was shown that the human VAChT gene is contained within the ChAT gene. This is also the case in all the other mammals that have been analysed. The functional neatness of this nested arrangement is clear. There is no point in ChAT synthesising ACh if it cannot be sequestrated; there is no point in VAChT in vesicular membranes if there is no ACh to transport.

The details of transcription have still to be understood. In some ways the *cha–unc* complex is like the *E. coli*'s lac operon which we discussed in Section 3.4.2. Operons, whilst not unknown, are, however, not common in eukaryotic cells. Moreover, the two gene products of the *cha–unc* complex arise from one common and then several different exons. It looks as if the complex shares characteristics of both the prokaryotic operon and the eukaryotic alternative splicing mechanisms. In other words, it may be that there is a common regulatory element upstream of the complex and the latter is then transcribed by separate transcriptional starts. The transcripts are spliced together in the cytoplasm.

Finally, the fact that VAChT and the VMATs show a 40% sequence homology suggests that the first members of a large family of vesicular transporters have been found. This family may include transporters for ATP, glutamate, GABA and glycine. They may all operate by a similar proton antiport mechanism and, just conceivably, they may have a similar genetic representation.

for GABA and glycine. They ensure that the vesicle is filled with the correct neurotransmitter. In many respects, not least in their 12TM conformation, they resemble the synaptic reuptake transporters discussed in Box 16.3.

15.4.7 Termination of Transmitter Release

Although it has taken a long time to describe, the response of a synaptic terminal is in fact extremely rapid. There are two cases. The small transmitter

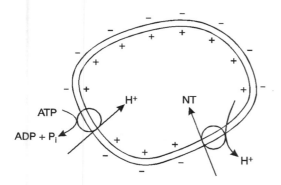

Figure 15.23 Filling synaptic vesicles with a specific transmitter. A proton pump energised by the hydrolysis of ATP drives H^+ into the vesicle. The potential or electrochemical gradient this creates is used to drive the transport of neurotransmitter (NT) into the vesicle.

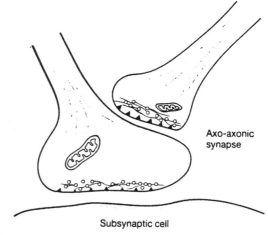

Figure 15.24 Axo-axonic synapse.

molecules in the translucent vesicles are released in less than a millisecond; the larger transmitters and modulators in the cored and larger vesicles are voided in less than 10 ms. The release is terminated by removal of the triggering Ca^{2+} ions. There are extremely efficient mechanisms within the synaptic terminal for mopping up any free calcium ions. These mechanisms include sequestration in mitochondria, in cisternae of ER (if this is present), and perhaps most importantly by the ubiquitous calcium binding protein, **calmodulin**. Once free calcium ions are removed the trigger to release vesicles from the cytoskeleton and for fusion and exocytosis is eliminated. The synaptic terminal returns to its resting state.

15.4.8 Modulation of Release

Finally it should be borne in mind that the release of transmitters and modulators from synaptic terminals is quite commonly under the control of so-called axo-axonic synapses. These are synapses made by other neurons directly on a synaptic terminal such as we have been considering (Figure 15.24). The operation of such synapses can modulate the release of transmitter by the 'subsynaptic synapse'. This

modulation may be brought about through second messenger systems (cAMP, etc.) or directly by depolarising or hyperpolarising the subsynaptic membrane. We shall consider these complexities more fully in Chapters 16 and 17, where we examine the structure and functioning of synapses.

15.5 CONCLUSION

In this chapter we have seen how material elaborated in neuronal perikarya is moved along axons to their terminal boutons and finally secreted into the synaptic cleft. It should be borne in mind that material will also be moving out along the dendrites. This latter transport is not nearly so well known or so intense (but see Box 15.1). For the material arriving at the axon terminal has the all-important function of communicating with the subsynaptic cell – neuron or muscle. Our next task, therefore, is to look at the neurotransmitters, the neuromodulators, their associated enzymes, inhibitors, reuptake mechanisms, etc. which make the presynaptic terminal such a busy and interesting place.

16

NEUROTRANSMITTERS AND NEUROMODULATORS

Complexity of synaptic structure of the brain. Infelicity of 'telephone exchange' image. Fuzzy distinction between neurotransmitters and neuromodulators. Ionotropic and metabotropic responses, punctate and diffusional transmission. Criteria for neurotransmitters. **Neurotransmitters: Acetylcholine**: synthesis – localisation – pharmacology – reuptake – feedback control – pre- and postsynaptic AChRs. **Amino acids**: excitatory (glutamate, aspartate) – synthesis – wide distribution – NMDA – reuptake; inhibitory (GABA, glycine) – synthesis – ubiquity – reuptake. **Serotonin (5-HT)**: synthesis – localisation – varicosities – classes of receptor – feedback control – reuptake – inactivation – pharmacology. **Catecholamines**: dopamine: synthesis – localisation – wide distribution (varicosities) – mood control; **dopamine** – classes of receptor – feedback control – reuptake – inactivation – pharmacology – Parkinsonism; **noradrenaline** – classes of receptor – feedback control – reuptake – inactivation – pharmacology – bioamine theory of depression. **Purines**: NANC fibres – classification – pharmacology – pathologies. **Cannabinoids**: location – molecular biology – endocannabinoids – effect on brain and behaviour. **Peptides**: not recycled in terminal – heterogeneity – distribution; **substance P**: distribution – C-fibres and pain; **enkephalins**: wide distribution – HD – types of opioid receptor – analgesic effect – denervation supersensitivity and 'withdrawal symptoms'. Cohabitation of peptides and non-peptides – significance. **Nitric oxide** – not sequestered into vesicles – no receptors – no reuptake mechanisms – diffusion in intercellular space – affects guanylyl cyclase – synthesised by NO synthase (NOS) – synthesis of NOS induced by many agents – NOS activated by Ca^{2+} – NOS found in many parts of the brain. The brain operates by fluxes of neurochemicals affecting diverse receptors, changing the biochemistry of underlying cells: it is a chemical computer

In Chapter 15 we noted how secretory and other molecules were synthesised in the neuron's perikaryon and then transported to be released from or incorporated into the synaptic ending. In this chapter we shall look at the nature of some of the secretory molecules that are released into the synaptic gap. It should be noted that far from all such molecules are elaborated in the perikaryon and transported down the axon. Many, as we shall see, are synthesised and resynthesised in the

terminal. The enzymes required for this synthesis and the complex protein machinery responsible for their release and reuptake do, however, require the protein synthetic apparatus of the perikaryon. Axoplasmic transport provides a vital communication channel between perikaryon and synapse.

Only a couple of decades ago no more than about half a dozen synaptic transmitters were recognised. The old telephone-exchange metaphor of the brain predominated (see Chapter 15). The brain was 'seen' as a huge number of 'hard' wires and junctions. The wires were the axons, the junctions the synapses. We noted in Chapter 1 that there are some 10^{11} of the former and some 10^{14} of the latter. It was believed that there was quite sufficient complexity in these numbers to account for the activity of the brain. The synapses only needed to be on/off switches, like the on/off units of the digital computer.

In the last twenty or so years there has been a dramatic shift in our perceptions. The telephone exchange/computer analogy is now seen to be only (at best) a first approximation. We saw in Chapter 1 how many different types of synaptic apposition are developed in the brain: not only axo-dendritic and axo-somatic, but also axo-axonic, dendro-dendritic and, perhaps most significant of all, by the development of axonal 'varicosities' which allow 'en passant' release of neuroactive molecules. The latter substances may diffuse through the brain's intercellular space (perhaps for several millimetres) until they reach an appropriate membrane receptor through which to exert their action. Indeed, as we shall see in Box 19.1, there is good evidence that neurotransmitters play important roles as growth factors in the developing brain. Synapses themselves are, moreover, very far from being simple yes/no logic elements. This recognition has been one of the most important outcomes of work in molecular neurobiology. Finally instead of the half-dozen or so transmitters of a decade ago neurobiologists now recognise fifty or so different molecules and the number is still rising (Table 16.1).

These molecules are often divided into two categories: **neurotransmitters** and **neuromodulators**. The distinction has to do with their synaptic activity. Neurotransmitters have a direct effect on a subsynaptic membrane (this, in the case of varicosities, may be at some distance from the presynaptic membrane) whilst neuromodulators 'modulate' or 'regulate' the action of the transmitter. This they may do by affecting the transmitter sensitivity of the subsynaptic membrane or by influencing the release of transmitter from the presynaptic terminal. The actions of neurotransmitters may (as we have already noted) be divided into two categories: ionotropic and metabotropic depending on whether they act directly to open ion pores (Chapter 10) or operate through a collision-

Table 16.1 Neurotransmitters and neuromodulators

Nitric oxide (NO)

Acetylcholine (ACh)

Cannabinoids

Amino acids
Glutamate; aspartate; glycine

Amino acid derivatives
(a) Derived from tryptophan: indoleamines – serotonin
(b) Derived from phenylalanine: catecholamines – dopamine; noradrenaline; adrenaline; octopamine
(c) Derived from glycine: taurine
(d) Derived from glutamate: γ-aminobutyric acid (GABA)
(e) Derived from histidine: histamine

Purines
Adenosine
Adenosine monophosphate (AMP), diphosphate (ADP) and triphosphate (ATP)

Peptides
(a) Opioid: enkephalins; β-endorphin; dynorphin
(b) Neurohypophyseal: vasopressin; oxytocin; neuro-physins
(c) Tachykinins: substance P (SP); neurokinin A (NKA) (=substance K (SK)); neurokinin B (NKB); kassinin; eledoisin
(d) Gastrins: gastrin; cholecystokinins (CCKs)
(e) Somatostatins: somatostatin 14; somatostatin 28
(f) Glucagon-related: vasoactive intestinal polypeptide (VIP)
(g) Pancreatic-polypeptide related: NPY
(h) Miscellaneous: bombesin; neurotensin (NT); brady-kinin; angiotensin; calcitonin gene-related peptide (CGRP)

For formulae of neuropeptides see Table 2.2.

coupling system in the subsynaptic membrane (Chapter 8).

It must be noted straight away, however, that the distinction between neurotransmitters and neuromodulators is far from sharp. The same synaptically active molecule can act at some synapses as a transmitter and at others as a modulator. Similarly a transmitter may have an ionotropic action on one subsynaptic membrane and a metabotropic action on another. We have already met many examples of this. The type example is acetylcholine. We saw

---

BOX 16.1   Criteria for neurotransmitters

1. The molecule must be synthesised within the neuron from which it is released. Enzymes and substrates for its synthesis must be found in that neuron.
2. The molecule must be stored within the neuron from which it is released.
3. Presynaptic stimulation (usually, but not necessarily, electrical) must lead to the release of the molecule.
4. Controlled application of the molecule at the appropriate site should elicit the same postsynaptic response as presynaptic stimulation.
5. Agents which block the postsynaptic response to presynaptic stimulation should also block the response to exogenously applied putative transmitter.
6. The postsynaptic response to the putative transmitter molecule when exogenously applied must be terminated rapidly.
7. The suspected molecule must behave identically to the endogenous transmitter with respect to pharmacological potentiation, inhibition, inactivation, etc.

---

that its nicotinic action is an extremely rapid (c. 10 µs) opening of an ion channel leading to a membrane depolarisation, whilst its muscarinic action is much more long-lasting and involves a collision-coupling and cAMP second messenger system. It is not just the presynaptic ending and its contents that define the nature of the synapse. The subsynaptic membrane is equally, if not more, important.

We shall look at the molecular neurobiology of subsynaptic membranes in Chapter 17. In the present chapter we shall focus our attention on the pharmacology of the more important neuro-transmitters/neuromodulators and their receptors. We shall find that they fall into a number of natural groups. **Acetylcholine** (the first to be discovered) and the **cannabinoids** (the last) still stand, to an extent, in classes of their own; next there is a group of **amino acid** transmitters some of which are excitatory whilst others are inhibitory; biochemically related to the amino acids are the **indoleamines** such as **serotonin** and the **catecholamines** of which noradrenaline is perhaps the best known example; in recent years considerable interest has been focused on the **purines**, especially adenosine and adenosine triphosphate; then there is a large (and growing) group of neuroactive **peptides**; finally there is the important though unconventional transmitter, **nitric oxide (NO)**. In addition to these established transmitters and modulators there are a number of molecules that are found in

the brain and probably have some synaptic activity. Such molecules include **histamine**, some **phenethyl-amine derivatives**, for example octamine and tyramine and so on. We shall not discuss this last group of putative transmitters further.

Before discussing the groups of established transmitters and modulators it is important to list some of the criteria that should be met by any molecule suspected of synaptic activity. For, after all, synaptic endings contain a great variety of substances and there is no reason to suspect that they all play a role at the synaptic junction. A list of appropriate criteria is given in Box 16.1.

16.1 ACETYLCHOLINE

Acetylcholine is synthesised in synaptic terminals from **choline** and **acetate**. The latter is derived from **acetyl coenzyme A**. Choline, on the other hand, is partly obtained by reuptake from the synaptic gap

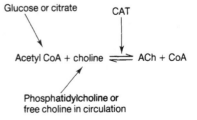

Figure 16.1   Synthesis of acetylcholine. Explanation in text.

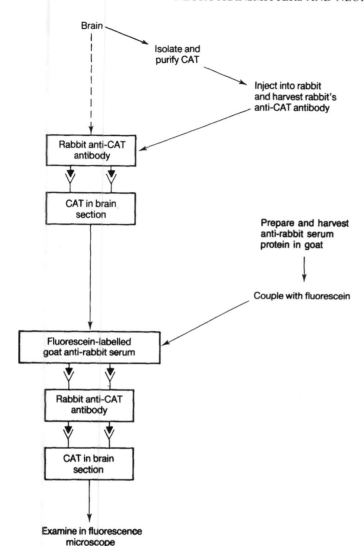

Figure 16.2 An immunohistochemical technique for locating CAT in the CNS. This 'sandwich technique' is just one of many ways of localising CAT in the brain. First an anti-CAT antibody is raised in, for example, a rabbit and reacted with the CAT in the tissue section. This complex, in turn, is reacted with a fluorescein-labelled anti-rabbit serum from a goat. The resultant complex can be visualised in the fluorescence microscope.

and partly from the blood where some of it is transported partly as free choline and partly as the phospholipid phosphatidylcholine. Acetyl coenzyme A is derived from glucose through glycolysis or from citrate.

The synthesis (as shown in Figure 16.1) is catalysed by **choline acetyltransferase (CAT or ChAT)**. This enzyme provides a much used marker for cholinergic synapses. One of the ways of detecting CAT within the brain involves the techniques of immunohistochemistry. First CAT is extracted and purified. Next antibodies are raised against it in another organism such as a rabbit.

This anti-CAT antibody is reacted with CAT in the tissue section. Next it is necessary to produce a marker that can be identified in the microscope. This is usually done by coupling a fluorescent molecule, such as **fluorescein**, to yet another antibody. This antibody is prepared by injecting rabbit serum proteins into, say, a goat. The anti-rabbit antibody is then coupled with fluorescein. Finally this fluorescently labelled anti-rabbit antibody is reacted with the tissue section. It attaches itself to the only rabbit protein present in the section which happens to be the anti-CAT antibody. The location of CAT in the CNS can then be determined by the

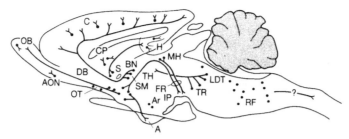

Figure 16.3 Cholinergic pathways in the rat brain (parasagittal section). A=amygdala; AON=anterior olfactory nucleus; Ar=arcuate nucleus; BN=nucleus basalis of Meynert; C=cerebral cortex; CP=caudate putamen; DB=diagonal band nucleus of Broca; FR=fasciculus retroflexus; H=hippocampus; IP=nucleus interpeduncularis; LDT=lateral dorsal tegmental nucleus; MH=medial habenula; OB=olfactory bulb; OT=olfactory tubercle; RF=reticular formation; SM=stria medullaris; TH=thalamus; TR=tegmental reticular formation. From Cuello and Sofroniew (1984), *Trends in Neurosciences*, **7**, 74–78; with permission from Elsevier Science, © 1984.

Figure 16.4 Pharmacology of the cholinergic synapse. Acetyl-CoA is derived from mitochondrial metabolism. Choline acetyltransferase (CAT) catalyses the formation of ACh. β-Bungarotoxin promotes and botulinum toxin and cytochalasin-B block the release of ACh into the synaptic cleft. α-Bungarotoxin, curare and hexamethonium block nicotinic receptors while atropine blocks muscarinic receptors. Acetylcholinesterase (AChE) breaks ACh into choline and acetate. AChE is inhibited by physostigmine. Whereas acetate escapes ultimately into the circulation, choline is taken back into the terminal. This reuptake is blocked by hemicholinium.

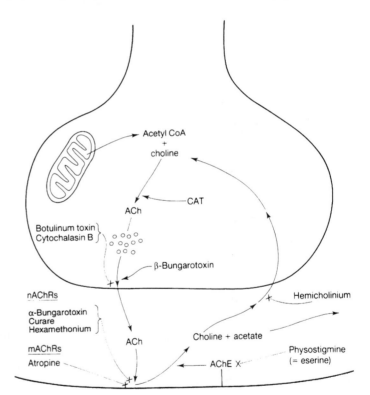

fluorescence microscope. Figure 16.2 shows the sequence of steps in this immunohistochemical technique diagrammatically.

There are a number of other ways of determining the presence of cholinergic synapses in the brain. These range from Koelle and Friedenwald's histochemical technique for localising the enzyme acetylcholinesterase to electrical recording from suspected cells after iontophoretic release of acetylcholine. The cholinergic pathways revealed by these techniques are shown in Figure 16.3. The majority of the cell bodies are located in midbrain nuclei: the basal **nucleus of Meynert**, the **diagonal band nucleus** of Broca, and the **nucleus preopticus**

Table 16.2 Acetylcholine receptors

Name	Structure	Agonist	Antagonist	Channel blocker
1. Muscarinic				
M1	460 aa; 7TM	Muscarine	MT7	
M2	466 aa; 7TM	Muscarine	Tripitramine	
M3	590 aa; 7TM	Muscarine	4-DAMP	
M4	479 aa; 7TM	Muscarine	MT3	
M5	532 aa; 7TM	Muscarine	4-DAMP	
2. Nicotinic				
Neuronal	4TM subunits	Nicotine	α-Conotoxin	Hexamethonium

The muscarinic AChRs are G-protein-linked; the nicotinic AChRs control an ion channel. The neuronal nAChR subunits are designated α2, α3, α4, α6, α7. In the CNS these form homo- or heteropentamers. The pharmacology is accordingly complex. Pharmacologists have established agonists, antagonists and channel blockers for a large number of the possible permutations and combinations of subunits. This pharmacology is, however, too complex to be displayed in this chapter and the interested reader is referred to the 2001 Nomenclature Supplement of *Trends in Pharmacological Sciences*. Adapted from Alexander, S. *et al.*, 2001, Nomenclature Supplement, *Trends in Pharmacological Sciences*, Cambridge, Elsevier.

magnocellularis. Axons course up from these basal regions to innervate the neocortex and especially the hippocampus. We shall see in Chapter 21 that Alzheimer's disease seems especially to affect cholinergic neurons. The fact that the hippocampus receives such a strong cholinergic input and that this region of the brain has long been regarded as involved in short-term memory may account for one of the most obvious symptoms of Alzheimer's disease: loss of memory.

We have already considered aspects of the cholinergic synapse in previous chapters. In particular we looked in depth at the nature and operation of both nicotinic and muscarinic acetylcholine receptors. We noted that there were many different subtypes of both types of ACh receptor. In Section 2.1.2 we considered the molecular nature of the acetylcholinesterase enzyme which plays so vital a role in inactivating ACh once it has exerted its effect on the subsynaptic membrane. It but remains to put these various elements together into an overall picture of the synapse. This is done in Figure 16.4.

Figure 16.4 indicates some of the more important drugs that affect the operation of this synapse. It must be emphasised that very far from all the pharmacological agents that influence cholinergic synapses are shown in the figure. Readers interested in neuropharmacology should consult one of the texts listed in the Bibliography for more complete information. The figure shows, however, that drugs can affect

- the release of ACh from the presynaptic terminal,
- the subsynaptic ACh receptor (because nicotinic and muscarinic receptors are radically different they have a very different pharmacology),
- the action of acetylcholinesterase,
- the reuptake of choline by the presynaptic terminal.

Figure 16.4 shows the nAChRs and mAChRs on the subsynaptic membrane. In central cholinergic synapses there is evidence that both these receptors are also present on the presynaptic membrane. In particular there is evidence that presynaptic muscarinic receptors work through their G-protein collision-coupling mechanism (see Chapter 8) to downregulate the release of ACh from the presynaptic terminal. We shall see that feedback control of the release of transmitter is a very general feature of synaptic physiology. Table 16.2 summarises the pharmacology of cholinergic receptors.

Finally, it should be borne in mind that the most intensively investigated cholinergic synapses lie outside the central nervous system either in the autonomic nervous system or at the neuromuscular junction of skeletal muscle.

16.2 AMINO ACIDS

A number of amino acids found in the central nervous system satisfy at least a majority of the criteria for synaptic activity listed in Box 16.1. In most cases their actions are ionotropic rather than metabotropic. Although some fifteen amino acids have been proposed as neurotransmitters only four have so far met with general acceptance. There are two acknowledged excitatory amino acids (EAAs): **glutamic acid** and **aspartic acid**. Both have two carboxylic acid groups (Figure 16.5). There are two accepted inhibitory amino acids (IAAs): γ-**amino-butyric acid (GABA)**, and **glycine**. Table 16.3 shows the other amino acids and amino acid derivatives that have been proposed as neurotransmitters.

16.2.1 Excitatory Amino Acids (EAAs): Glutamic Acid and Aspartic Acid

Both glutamic and aspartic acids are non-essential amino acids which are synthesised from glucose and other precursors in the Krebs cycle (Figure 16.6). This cycle takes place in synaptic mitochondria. It is from one of the intermediates in the cycle – either oxaloacetate or α-ketoglutarate – that aspartate and glutamate are derived.

Both aspartate and glutamate are widely distributed in the central nervous system. Iontophoretic application of very small amounts of either amino acid (10^{-15} M in the case of glutamate; aspartate is somewhat less potent) leads to an almost instantaneous depolarisation of the subsynaptic membranes of virtually all neurons. Reference to Section 10.4 shows that the membrane receptors for these ionotropic responses are classified by means of their pharmacological agonists and antagonists. We saw that AMPA or Q (quisqualate) and KA (kainate) receptors were the two most important.

We also noted in Chapter 10 that another type of glutamate receptor existed in the brain: the NMDA receptor. We saw that this, although an ionotropic receptor, has a slower and more complicated response to activation by EAAs. NMDA receptors are particularly strongly represented in the hippocampus and the cerebellum. We shall consider their putative significance in these regions in Chapter 20. Finally, we noted that a number of metabotropic

Excitatory amino acids (EAAs)

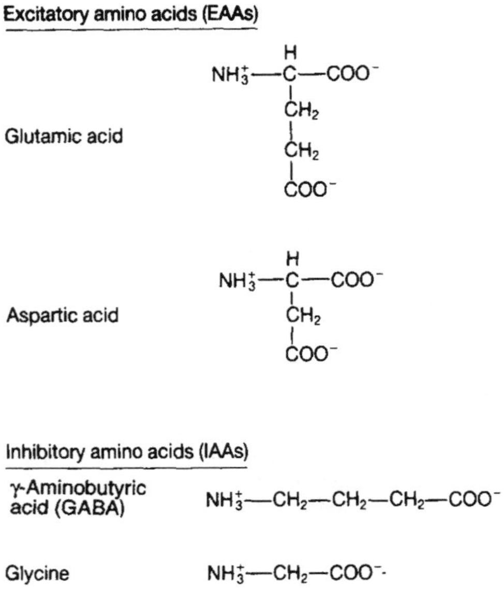

Figure 16.5 Structures of amino acid neurotransmitters. Note that EAAs are dicarboxylic whereas IAAs are monocarboxylic amino acids.

Table 16.3 Amino acid neurotransmitters

Excitatory	Inhibitory
Glutamic acid	γ-aminobutyric acid (GABA)
Aspartic acid	Glycine
Cysteine	Taurine
Cysteine-sulphinic acid	Serine
	β-Alanine

Amino acids below the dashed line have yet to satisfy all the criteria of Box 16.1 Nevertheless there is some evidence that they have a transmitter activity.

receptors for glutamic acid have also been characterised.

It appears that aspartate and glutamate, once released into the synapse, are rapidly taken up by a high-affinity transport system in the aspartatergic and glutamatergic nerve endings and also by adjacent glial cells. Within glial cells glutamate is transformed into glutamine by glutamine synthetase and then transferred back into the glutaminergic nerve ending where it is deaminated once

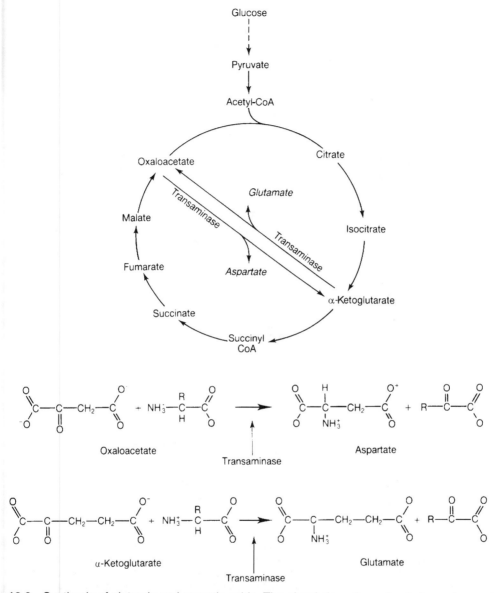

Figure 16.6 Synthesis of glutamic and aspartic acids. The glycolytic pathway leads from glucose to pyruvate and thence via acetyl-CoA into the Krebs cycle. Transaminases are able to convert oxaloacetate and α-ketoglutarate to aspartate and glutamate, respectively (see lower part of figure).

more to glutamate. This pathway is shown in Figure 16.7. A similar biochemistry obtains for aspartate, which appears to share the same uptake system.

Because of their seemingly ubiquitous distribution in the brain it has been difficult to locate specifically glutamatergic or aspartatergic pathways. It is, however, possible to home in on aspartatergic neurons by making use of the fact that its high-affinity uptake system will carry **D-aspartate**, which once in the cell is not further metabolised. Radiolabelled D-aspartate can thus

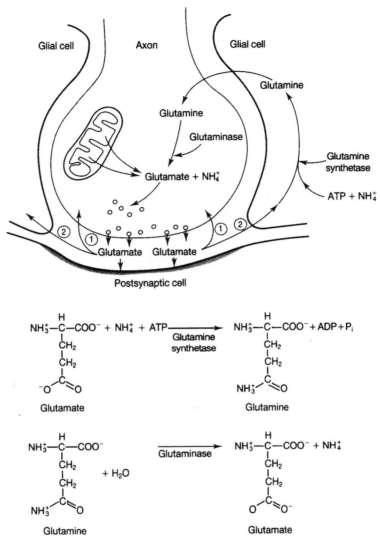

Figure 16.7 Glutaminergic synapse. Glutamate is released from synaptic vesicles into the synaptic cleft. Two reuptake paths are shown: (1) back into the synaptic terminal; (2) into neighbouring glial cells. In glial cells glutamine synthetase forms glutamine from the glutamate and this then passes back into the synaptic terminal. Here glutaminase re-forms glutamate and this forms a pool of free glutamate with fresh glutamate derived from Krebs cycle activity in mitochondria. The free glutamate is sequestrated into synaptic vesicles to await the arrival of the next action potential. The reactions catalysed by glutamine synthetase and glutaminase are shown in the lower part of the figure.

be used to locate aspartatergic synapses. Sectioning of nerve pathways coupled with observations of the loss of this high-affinity uptake system and the diminution of aspartate and/or glutamate has thus been used to trace EAA pathways.

16.2.2 Inhibitory Amino Acids (IAAs): γ-Aminobutyric Acid and Glycine

γ-Aminobutyric acid (GABA)

GABA is found throughout the CNS of both vertebrates and invertebrates. In the vertebrates its concentration is greater than the perhaps better-known neurotransmitters acetylcholine and noradrenaline. Perhaps 25–45% of all nerve terminals contain this transmitter.

GABA is formed by the decarboxylation of glutamate by **glutamic acid decarboxylase (GAD)**. The enzyme is present in the cytosolic phase of GABA-ergic terminals. The reaction is shown in Figure 16.8.

GAD allows the synaptic terminals of GABA-ergic neurons to be identified histologically. The technique is, in essence, the same as that described above for ChAT-containing terminals. Antibodies raised against the enzyme can be coupled with markers which can then be located in thin or ultra-thin sections of the CNS either with the light or the electron microscope. Again, as with cholinergic synapses, iontophoretic release of GABA and electrophysiological recording of the response have also been employed to complement the immunohistochemistry.

GABA-ergic synapses have been identified in many regions of the brain. They are found in the retina, cerebellum, cerebral cortex, hippocampus, thalamus, olfactory bulb and in the basal ganglia. Indeed there is evidence that loss of GABA synapses in the brain's basal ganglia is one of the factors in **Huntington's disease**. We shall see later that other transmitters may also be involved. The symptoms of the condition – sudden uncontrollable movements of the body, especially the limbs – are likely to be due to the development of imbalances between excitatory and inhibitory synapses in some of the basal ganglia. For it is known that in some instances (see later) these ganglia contain the motor programs that control bodily movements.

Turning to the GABA-ergic synapse itself, it can be demonstrated that release of GABA into the synaptic cleft results in a hyperpolarisation of the receptor membrane. Pharmacological evidence shows that there are two major types of GABA receptor: **GABA$_A$** and **GABA$_B$** receptors. The pharmacology of these receptors is given in Table 16.4. We discussed the GABA$_A$ receptor in detail in Chapter 10. We noted that it had a strong affinity for the **benzodiazepines**. These so-called 'anxiolytic' (i.e. anxiety-reducing) drugs enhance the activity of GABA-ergic synapses in some as-yet unknown way, increasing the ability of GABA to open the

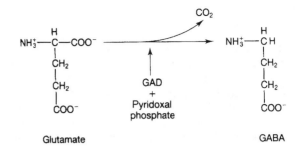

Figure 16.8 Synthesis of GABA from glutamate. GAD (glutamic acid decarboxylase) requires pyridoxal phosphate as a co-enzyme.

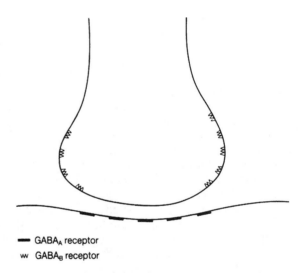

— GABA$_A$ receptor
w GABA$_B$ receptor

Figure 16.9 Differential location of GABA$_A$ and GABA$_B$ receptors.

receptor's chloride channel. It is believed that the GABA$_A$ receptors are found on the subsynaptic membranes of 'classical' inhibitory synapses. In contrast there is evidence that in many cases (not all) GABA$_B$ receptors are located on presynaptic membranes (Figure 16.9).

The presynaptic location of the GABA$_B$ receptor suggests that it is involved in presynaptic inhibition. There is evidence (as we saw in Section 10.2) that unlike the GABA$_A$ receptor it is metabotropic in action. It is believed to work through a G-protein collision-coupling mechanism to **inhibit** membrane-bound adenylyl cyclase. The consequent reduction in intraterminal cAMP is in turn believed

BOX 16.2   Otto Loewi and vagusstoff

The sources of scientific creativity, like those of other forms of creativity, lie deeply hidden. Indispensable, however, is a profound and continuing obsession with the subject matter. Isaac Newton, when asked to account for his genius in setting the universe to mathematics, replied 'by always thinking unto it'. Henri Poincaré who, with Einstein, was the foremost mathematical physicist at the beginning of the twentieth century, found the solution to a deep problem in mathematics which had been troubling him, suddenly, as he stepped aboard a bus. Similarly the young Kekulé saw the 'atoms gamboling before his eyes' whilst travelling on the top deck of a London omnibus, and, later, the essentials of aromatic chemistry came to him as he gazed, sleepily, into the embers of his sitting room fire. Coming to our own subject we can recall Linus Pauling working out the structure of the α-helix with sheets of paper as he lay in a hospital bed; and, finally, Otto Loewi is credited with establishing the foundations of neurotransmission in a waking dream.

Otto Loewi was born at Frankfurt am Main in 1873 and trained in medicine at Strasbourg and Munich. After a short period practising clinical medicine he turned to pharmacology and held professorial positions first at Marburg and then in Vienna. In 1938, after the Nazi occupation of Austria, he fled to Britain and finally to New York, where he died in 1961. He described how he made his seminal discovery to another early worker, Walter Cannon. 'One night, having fallen asleep reading a light novel he awoke', says Cannon, 'suddenly and completely with the idea fully formed that if the vagus nerves inhibit the heart by liberating a muscarine-like substance, the substance might diffuse out into a salt solution left in contact with the heart while it was subjected to vagal inhibition, and that then the presence of this substance might be demonstrated by inhibiting another heart through the influence of the altered solution. He scribbled the plan on a piece of paper and went to sleep again. Next morning, however, he could not decipher what he had written! Yet he felt it was important. All day he went about in a distracted manner, occasionally looking at the paper, but wholly mystified as to its meaning. That night he again awoke, with a vivid revival of the incidents of the previous illumination, and after this he remembered in his waking state both occasions.'

Loewi's seminal experiment is shown in Figure A. He was able to show that when the parasympathetic nerve to the heart (the vagus nerve) was stimulated a substance was released (he called it 'vagusstoff') which when carried in solution to a second heart slowed that heart also. Loewi also showed that the sympathetic nerve to the heart, which causes cardio-acceleration, also released a diffusible substance into the Ringer solution.

It remained to establish the chemical nature of the substances released by the parasympathetic and sympathetic nerves. After a certain hesitation he accepted that 'vagusstoff' was identical to acetylcholine and that the substance released from the sympathetic fibres was similar to, if not identical with, adrenaline. We now know it is noradrenaline. In 1935 he was asked to give the prestigious Ferrier Lecture to the Royal Society. By this time he had accepted that vagusstoff and acetylcholine were one and the same and that 'the substance released by the cardio-accelerator nerves shares many of the properties of adrenaline'.

In the same lecture Loewi generalises his discovery at the periphery to the centre. He writes that 'the transmission of excitation from afferent neurons to efferent ones, an essential feature of activity in the central nervous system, may also be effected by humoral means'. This conclusion chimed in very well with the work of Sherrington and other contemporary neurophysiologists on the functioning of the synapse (see Box 17.1). Finally, Loewi emphasises that 'in all cases where neurohumoral transmission occurs, the dominion of the nerve extends only as far as the liberation of the transmitters: the spot where the transmitter is released is, in that sense, the effective organ of the nerve'.

One can see in these words written in the mid-1930s a prefiguring of the vast subject matter of neuropharmacology which we know today. Loewi, in his tentative and careful way, points to the essence of the matter. His waking dreams were truly prophetic. Experts in efficiency, gurus of Total Quality Management, seldom understand the crucial part 'unallocated thinking time', as it was once somewhat gracelessly phrased, plays in the academic enterprise.

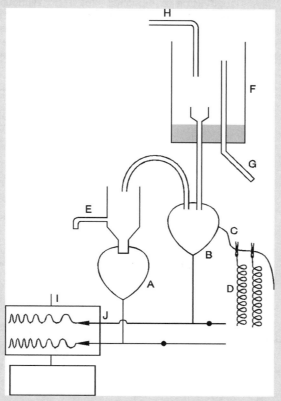

Figure A Loewi's experiment. A=denervated frog heart; B=innervated frog heart; C=vagus nerve; D=stimulating electrodes; E=cannula maintaining constant hydrostatic pressure in heart A; F=cylinder of Ringer's solution; G=overflow maintaining constant hydrostatic pressure in heart B; H=inflow of fresh Ringer's solution; I=kymograph; J=writing stylus. Note how stylus from heart A records a similar deceleration to that recorded from heart B.

to reduce the amount of GABA released when the terminal is invaded by an action potential. We have already noted that an analogous **feedback control** of transmitter release is found in central cholinergic synapses and we shall see that a similar mechanism is at work in many other well-known types of synapse.

Reuptake of GABA is similar to that of the excitatory amino acids mentioned above. There are powerful uptake systems (Na$^+$-dependent) in both GABA nerve terminals and surrounding glial cells. Much of the recycled GABA is probably directly reusable. A proportion, however, is converted to succinic semialdehyde by the mitochondrial enzyme **GABA aminotransferase (GABA-T)**. As Figure 16.10 shows, this is believed to occur in both synaptic terminals and glia. The succinic semialdehyde enters the Krebs cycle, from which GABA can once again be obtained via oxaloacetic and glutamic acid.

Table 16.4 Classification of GABA receptors

Name	Structure	Agonist	Antagonist
GABA$_A$	4TM subunits	Isoguvacine (competitive site)	Bicuculline (competitive site)
		Flunitrazepam (modulatory site)	Flumazenil (modulatory site)
GABA$_B$	7TM heterodimer	L-Baclofen	Saclofen

Data from Alexander, S. *et al.*, 2001, Nomenclature Supplement, *Trends in Pharmacological Sciences*, Cambridge, Elsevier.

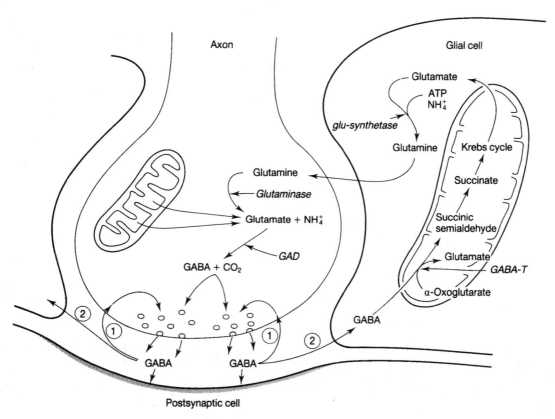

Figure 16.10 GABA-ergic synapse. GABA is released from synaptic vesicles into the synaptic cleft. Two reuptake paths are shown: (1) back into the synaptic terminal; (2) into neighbouring glial cells. In the mitochondria of glial cells GABA is converted by GABA-aminotransferase (GABA-T) to succinic semialdehyde. This enters the Krebs cycle from which glutamate (Figure 16.6) emerges. Glutamine synthetase (glu-synthetase) convertes glutamate to glutamine and this is taken up by the synaptic terminal. Under the influence of glutaminase and GAD, glutamine is converted first to glutamate and then to GABA which is sequestered once again in vesicles. Further supplies of glutamate are available from Krebs cycle intermediates in presynaptic mitochondria.

Glycine

Glycine is the other well-established inhibitory amino acid transmitter. Unlike GABA, glycine is mostly restricted to the spinal cord and brain stem.

Glycine is the smallest and one of the commonest amino acids. It is carried across the blood–brain barrier by a transport protein. In the CNS it is derived from serine by the action of serine transhydroxymethylase (Figure 16.11).

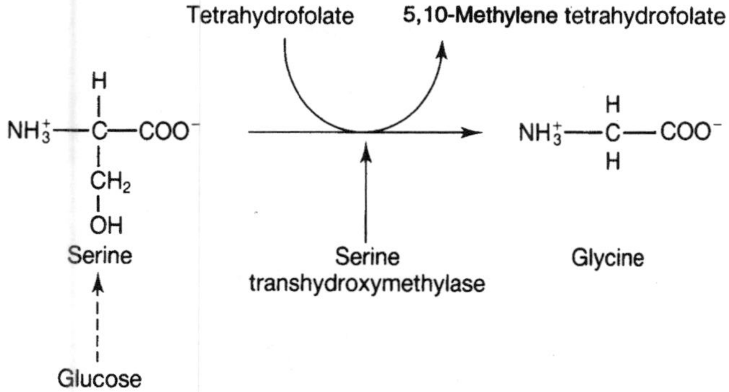

Figure 16.11 Synthesis of glycine. Most neural glycine is formed from serine by a folate-dependent reaction catalysed by serine transhydroxymethylase. Serine itself is derived from glucose via 3-phospho-*d*-glycerate and 3-phosphoserine; glycine may, in addition, be formed from neural peptides and proteins.

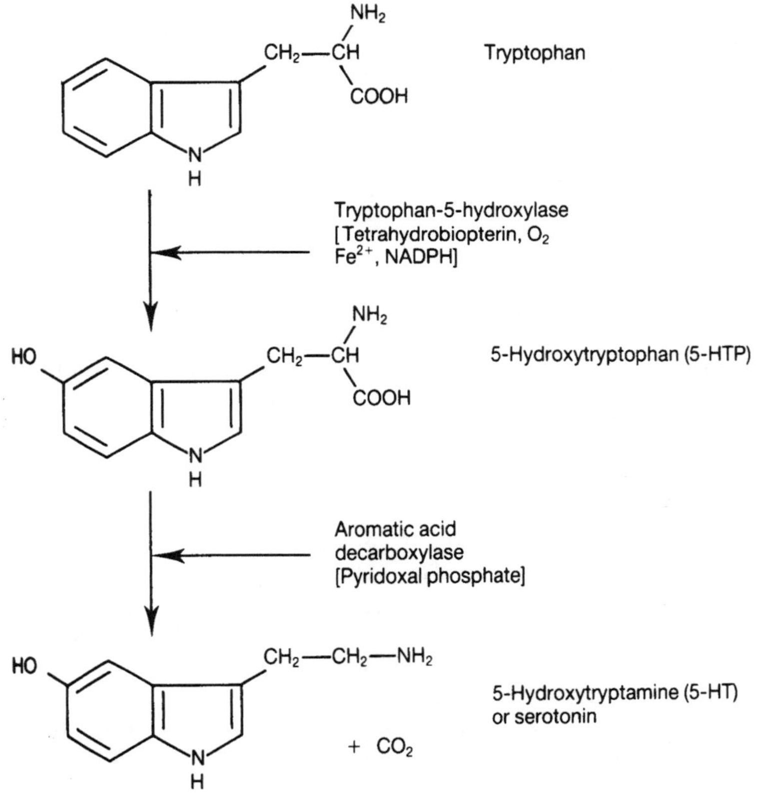

Figure 16.12 Synthesis of serotonin. Serotonin is synthesised from tryptophan – an essential amino acid – in two steps. The first step is catalysed by tryptophan-5-hydroxylase and requires a number of cofactors (in brackets); the second step is catalysed by aromatic acid decarboxylase and requires pyridoxal phosphate as co-factor.

Glycine is probably stored (like GABA) in small elliptical vesicles in presynaptic terminals. Release of the transmitter induces rapid hyperpolarisation of the subsynaptic membrane. A high-affinity Na^+-dependent reuptake system exists to pump glycine back from the synaptic cleft into nerve terminals and glial cells.

We discussed the molecular structure of glycine receptors in Section 10.3 and noted that it is competitively blocked by **strychnine**. Small doses of strychnine thus cause enhanced reflex responses whilst large doses result in generalised motor disinhibition and general convulsions. Death is caused by disinhibition of the respiratory muscles and hence asphyxia.

Figure 16.13 Reaction of serotonin with formaldehyde vapour. This technique for detecting the presence of serotoninergic neurons was first developed by Falck and his co-workers in the early 1960s. To prevent the diffusion and enhance the reaction the tissue is normally freeze-dried before exposing it to dry formaldehyde vapour.

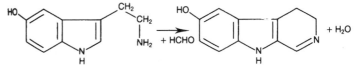

Serotonin + Formaldehyde → 3,4-Dihydro-β-carboline + Water

(Yellow fluorescence)

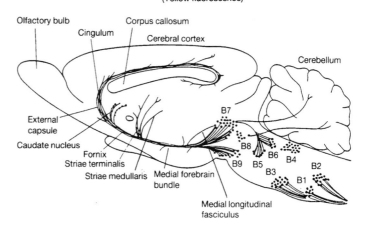

Figure 16.14 Serotoninergic pathways in rat brain (parasagittal section). Serotoninergic neurons are grouped in nine or so nuclei in the brain stem, pons and midbrain. The nuclei B6 to B9 project forward to the diencephalon and telencephalon, whilst the more caudal groups project to the medulla and spinal cord. After Cooper, Bloom and Roth (1991), *The Biochemical Basis of Neuropharmacology* (6th edn), New York: Oxford University Press.

16.3 SEROTONIN (=5-HYDROXYTRYPTAMINE, 5-HT)

Figure 16.12 shows that serotonin is synthesised from tryptophan via the intermediate 5-hydroxytryptophan. The synthesis requires two enzymes – **tryptophan hydroxylase** and **aromatic amino acid decarboxylase**. The first step in the sequence is rate-limiting.

Serotonin can be localised in the brain by reacting the tissue with formaldehyde vapour. Serotonin is converted into 6-hydroxy-3,4-dihydro-carboline which when illuminated in the ultraviolet ($\lambda = 390$ nm) gives a strong yellow-green fluorescence ($\lambda = 520$ nm). The reaction is shown in Figure 16.13. The technique can be made yet more specific by eliminating any fluorescence from catecholaminergic neurons by the prior application of the catecholaminergic-specific neurotoxin 6-hydroxydopamine (6-OHDA).

Subjection of brain sections to this technique reveals that serotonin is localised in the **midbrain**, the **pineal gland**, the **substantia nigra** and **raphe nuclei** of brain stem, and the **hypothalamus**. It can be detected in the varicosities (see Chapter 1) of fibres coursing up to the cerebral

cortex from cell bodies located in the raphe nuclei (Figure 16.14).

Serotonin is stored in vesicles in at least some serotoninergic terminals. Release of serotonin does not result in the rapid response (depolarisation or hyperpolarisation) of the subsynaptic membrane which is characteristic of the amino acids and nicotinic acetylcholine. We shall see in Chapter 17 that its action is more like the muscarinic action of acetylcholine or the action of noradrenaline at the β-adrenergic receptor: in other words it acts through a collision-coupling second messenger system. The response, moreover, may be quite widespread as release from varicosities results in diffusion of the transmitter over many subsynaptic cells. In this instance there is no sharp localisation of pre- and subsynaptic membranes, one 'above' the other. We shall see below that the type of response elicited may be either inhibitory or excitatory.

Seven classes of serotonin receptors are nowadays recognised: 5-HT$_1$, 5-HT$_2$, 5-HT$_3$, 5-HT$_4$, 5-HT$_5$, 5-HT$_6$ and 5-HT$_7$. The classification depends on differential response to a number of agonists and antagonists. The classes have been further subdivided into subclasses as shown in Table 16.5. All, except the 5-HT$_3$ receptor, are

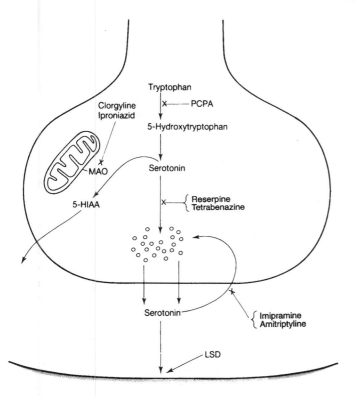

Figure 16.15 Pharmacology of the serotoninergic synapse. The synthetic step from tryptophan to 5-hydroxytryptophan is blocked by *p*-chlorophenylalanine (PCPA); the sequestration of serotonin into vesicles is blocked by both reserpine and tetrabenazine. Any free serotonin in the terminal is in danger of being deaminated by monoamine oxidase (MAO) on the outside of mitochondria to 5-hydroxyindoleacetic acid (5-HIAA). This deamination may be blocked by MAO inhibitors such as clorgyline and iproniazid. Powerful reuptake mechanisms in the presynaptic membrane are inhibited by imipramine and amitryptiline whilst lysergic acid diethylamide (LSD) partially potentiates subsynaptic serotonin receptors.

Table 16.5 Classification of 5-HT receptors

Receptor	Structure	Agonist	Antagonist	Effector
5-HT$_{1A}$	421 aa; 7TM	8-OH DPAT	WAY100635	cAMP↓
				K$^+$ channel↑
5-HT$_{1B}$	386 aa; 7TM	Sumatriptan	SDZ21009	cAMP↓
5-HT$_{1D}$	377 aa; 7TM	Sumatriptan	BRL15572	cAMP↓
5-HT$_{1E}$	365 aa; 7TM	–	–	cAMP↓
5-HT$_{1F}$	366 aa; 7TM	LY334370	–	cAMP↓
5-HT$_{2A}$	471 aa; 7TM	DOI	Ketanserin	IP$_3$/DAG
5-HT$_{2B}$	479 aa; 7TM	BW723C86	SB200646	IP$_3$/DAG
5-HT$_{2C}$	458 aa; 7TM	Ro600175	Mesulergine	IP$_3$/DAG
5-HT$_{3A,3B,3C}$	487 aa; ion channel	SR57227	Granisetron	Cation channel
5-HT$_4$	387aa; 7TM	BIMU8	GR113808	cAMP↑
[a]5-HT$_{5A,5B,6,7}$	357–448aa; 7TM	–	–	–

[a]5-HT$_{5A,5B,6,7}$ are as yet poorly known.
Chemical names of agonists and antagonists in Alexander, S. *et al.*, 2001, Nomenclature Supplement, *Trends in Pharmacological Sciences*, Cambridge, Elsevier.

metabotropic. The pharmacological agents used to make this intricate classification are shown in Table 16.5.

Whereas all the subclasses of 5-HT$_1$, 5-HT$_2$ and 5-HT$_4$ receptors have been found in the brain (as well as elsewhere), the three subclasses of the 5-HT$_3$ receptor have so far only been found in the peripheral nervous system. **5-HT$_1$** receptors are usually located in the **presynaptic** membrane. They are coupled to a G-protein system leading to

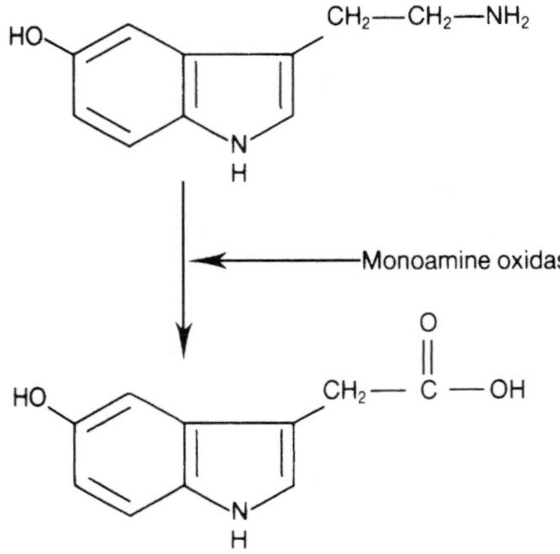

Figure 16.16 Deamination of serotonin by MAO.

decreased adenylyl cyclase activity. Activation of the system by 5-HT leads to **presynaptic inhibition**. As in the GABA system described above, there is evidently a negative feedback loop controlling the release of serotonin from serotoninergic terminals. In contrast **5-HT$_2$** receptors are located postsynaptically and, although they are similarly coupled to a G-protein mechanism, activation by 5-HT leads to depolarisation and thus **excitation** of the subsynaptic cell.

5-HT is quickly removed from serotinergic synapses by a high-affinity reuptake system. This reuptake is blocked by the well-known antidepressant drug **Prozac** (fluoxetine). Once back inside the terminal it is sequestered once again into vesicles. Any serotonin not in vesicles tends to be deaminated by **monoamine oxidase (MAO)** into **5-hydroxyindoleacetic acid (5-HIAA)** (Figure 16.16).

16.4 CATECHOLAMINES

We noted in Chapter 8 that a group of important neurotransmitters share a common ring structure – **catechol**. The two most important neuroactive catecholamines are **dopamine (DA)** and **noradrenaline (=norepinephrine, NE)**. Figure 16.17 shows the synthetic pathway from phenylalanine.

Tyrosine hydroxylase is the rate-limiting enzyme. As Figure 16.17 indicates, it requires tetrahydrobiopterin (BH4) as a co-factor and is sensitive to oxygen concentration. α-Methyl-*p*-tyrosine (AMPT) competitively inhibits tyrosine hydroxylase and consequently blocks the synthesis of all catecholamines. The figure also shows that

noradrenaline can be converted into adrenaline by methylation of the amino group. This reaction occurs in a few nerve cells in the brain stem. In general, however, noradrenaline is the transmitter at adrenergic synapses.

The location of dopamine and noradrenaline in the brain can be determined in much the same way as we noted for serotonin. Formaldehyde vapour transforms both noradrenaline and dopamine into quinonoids which fluoresce with a green colour ($\lambda = 470$ nm) when illuminated with light of $\lambda = 405$ nm. This technique shows that both dopaminergic and noradrenergic neurons have their cell bodies in the brain stem. Dopaminergic cell bodies are located principally in **substantia nigra** and **ventral tegmentum**. Noradrenergic cell bodies are found in various nuclei in the medulla and in the **locus coeruleus** beneath the cerebellum. We noted in the previous section that a neurotoxin, 6-hydroxydopamine (6-OHDA), selectively destroys catecholaminergic neurons. This selectivity can be made specific for dopaminergic neurons by using an agent such as **desipramine** to protect noradrenergic cells from the attentions of 6-OHDA. Thus we can establish the pathways of noradrenergic and dopaminergic neurons by a process of elimination. Figure 16.18 shows that catecholaminergic brain fibres sweep up from brain stem nuclei to innervate large parts of the cortex.

Both dopaminergic and noradrenergic fibres resemble serotoninergic fibres in developing **varicosities** along their lengths. These varicosities are of considerable importance in setting the level of activity over large regions of cortex. Figure 16.19

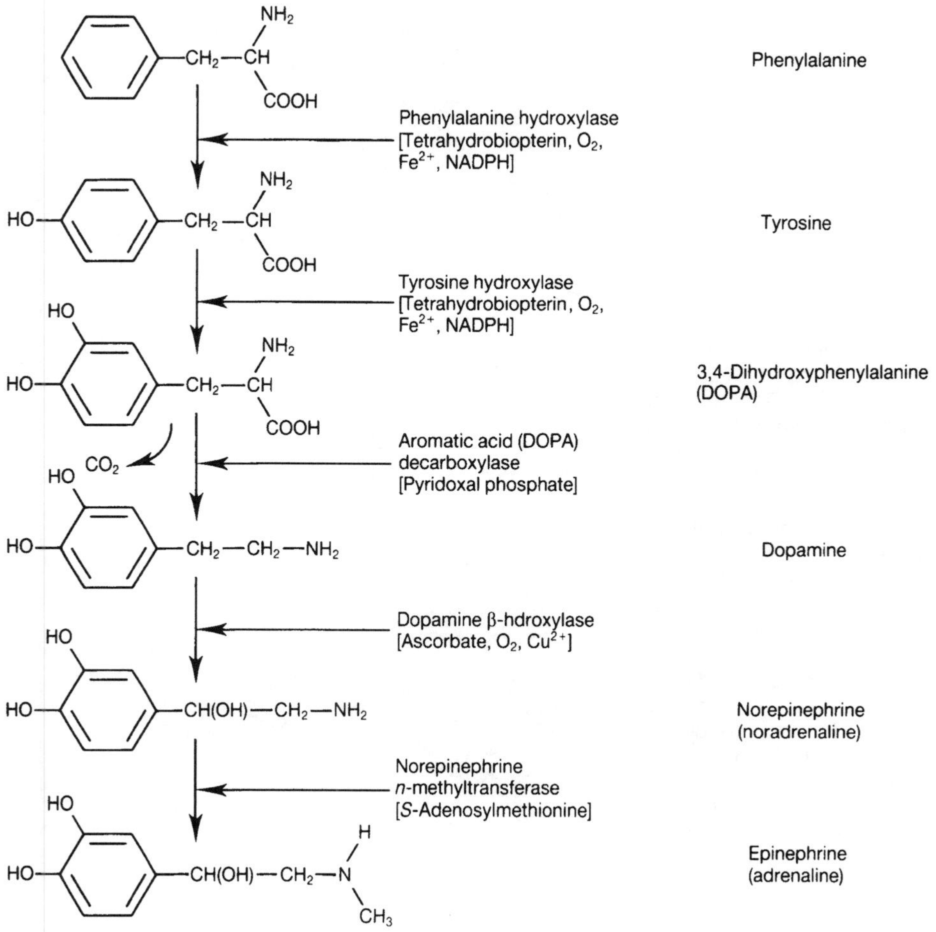

Figure 16.17 Synthesis of catecholamines from phenylalanine. The enzymes catalysing each step of the synthetic pathway are shown and their co-factors are bracketed.

shows the pathway of noradrenergic fibres from the **locus coeruleus**. The fibres enter the neocortex anteriorly and course backwards in layer 6. They send up branches into the upper layers of the cortex at regular intervals and develop many varicosities. It has been computed that in the rat cerebrum some 6000 varicosities are present in each cubic millimetre of cortex and that in consequence no neuron in the cortex is more than 30 μm distant from this source of noradrenaline. Activity in coerulean fibres is thus considered to release a 'mist' of transmitter which percolates through the extra-cellular space of the cortex until it is either sequestrated or finds an appropriate receptor.

Let us next look briefly at the pharmacology of each of the two catecholamines.

16.4.1 Dopamine (DA)

This catecholamine is stored in large, dense-cored vesicles in dopaminergic terminals. It is released in the usual calcium-dependent manner. Amphetamine (and some similar molecules) also bring about the release of dopamine into the synaptic

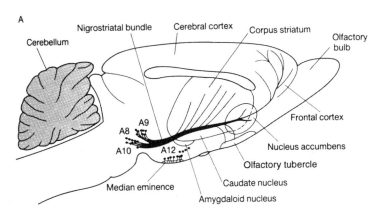

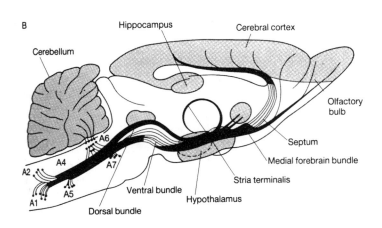

Figure 16.18 Dopaminergic and noradrenergic pathways in rat brain (parasagittal section). (A) Dopaminergic pathways. The major pathway (the nigrostriatal bundle) originates in the substantia nigra (A8, A9) and courses forwards to innervate the corpus striatum. (B) Noradrenergic pathways. A major pathway originates in the locus coeruleus (A6) and projects forwards as a number of distinct fibre bundles giving off branches to many brain regions. Other noradrenergic nuclei are in the ventral part of the brain stem (A1, A2, A5 and A7) and these send a few fibres back down the cord although most mingle with coerulean fibres ascending to the forebrain. After Bradford (1986), *Chemical Neurobiology*. Copyright © 1986 by Henry F. Bradford. Used with permission of W. H. Freeman and Company.

Table 16.6 Classification of dopamine receptors

Name	Structure	Agonist	Antagonist
D1	446 aa; 7TM	SKF81297	SCH23390
D2	443 aa; 7TM	PHNO	Raclopride
D3	400 aa; 7TM	PD128907	S33084
D4	387 aa; 7TM	PD168077	L745870
D5	477 aa; 7TM	–	–

D1 exists in a number of different isoforms derived by alternative splicing the primary transcript of the D1 gene. Chemical names of agonists and antagonists in Alexander, S. *et al.*, 2001, Nomenclature Supplement, *Trends in Pharmacological Sciences*, Cambridge, Elsevier.

cleft. There appear to be at least three pharmacologically distinguishable types of dopamine receptor and two others known from clones (Table 16.6). The best-known type is the D1 receptor. This operates by a collision-coupling mechanism leading ultimately to the synthesis of cAMP.

The activity of dopamine in the synaptic cleft is terminated by a powerful reuptake system in the presynaptic membranes of dopaminergic neurons. Dopamine is then sequestered into vesicles along with ATP and some inorganic ions such as Mg^{2+} and Ca^{2+}. Any dopamine not so sequestered is liable to attack by **monoamine oxidase (MAO)** and **catechol-*O*-methyl transferase (COMT)**. These enzymes convert dopamine to 3,4-dihydroxyphenylacetic acid (DOPAC) and 3-methoxytyramine, respectively (Figure 16.20). Both the latter compounds lack synaptic activity. It should be noted, however, that the enzymic degradation of catecholamines is significantly slower than the degradation of acetylcholine by acetylcholinesterase. It should also be noted that there are at least

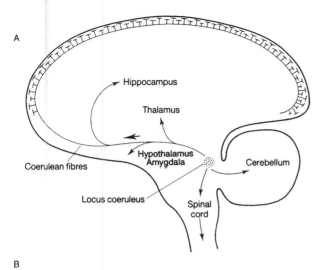

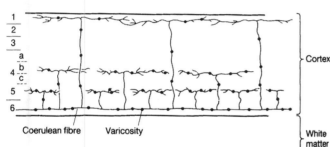

Figure 16.19 Schematic to show the major route of coerulean fibres in the human brain. (A) The locus coeruleus is a small nucleus (c. 20 000 neurons) in the central grey matter of the metencephalon. A major tract of fibres from cells in this nucleus courses forwards and enters the bottom layer of the cerebral cortex anteriorly and then runs towards the posterior. (B) Section through the cerebral cortex to show the disposition of branches from coerulean fibres. The major ramifications are in layers 5, 4(b) and 1. Varicosities are well developed in all regions.

two distinct forms of MAO. These forms are distinguishable by their substrate specificity and by their sensitivity to inhibitors. **MAO-A** has a substrate preference for catecholamines and serotonin and is inhibited by clorgyline, whilst **MAO-B** prefers β-phenylethylamine and benzylamine and is inhibited by deprenyl.

Finally, it should be mentioned in this section that **parkinsonism**, a pathology that generally only affects the elderly, is due to a deterioration of the dopaminergic pathways from the substantia nigra to the corpus striatum. The symptoms can be ameliorated by oral administration of L-dopa. This molecule, unlike dopamine, is carried across the blood–brain barrier by an amino acid transport protein. Once in the brain it is decarboxylated to dopamine and goes some way to making up for the lack of the endogenous molecule. We shall return to parkinsonism in Chapter 21.

The pharmacology of the dopaminergic synapse is schematised in Figure 16.21.

16.4.2 Noradrenaline (=Norepinephrine, NE)

Noradrenaline, like dopamine, is stored in large, dense-cored vesicles. These vesicles contain ATP and a number of proteins and divalent metal ions in addition to noradrenaline. Release of noradrenaline into the synaptic cleft occurs in a calcium-dependent manner – as described in Chapter 15. Once again a number of drugs, including amphetamine, facilitate the release, whereas others, for example guanethidine, have the opposite effect and block it.

The subsynaptic membrane possesses two main types of adrenergic receptor: **α-adrenoceptors** and **β-adrenoceptors** (both of which have numerous subtypes). The pharmacological distinction between these different receptor types is shown in Table 16.7. We discussed the β₂-adrenoceptor in depth in Chapter 8. It will be recalled that it is linked via a collision-coupling mechanism to the

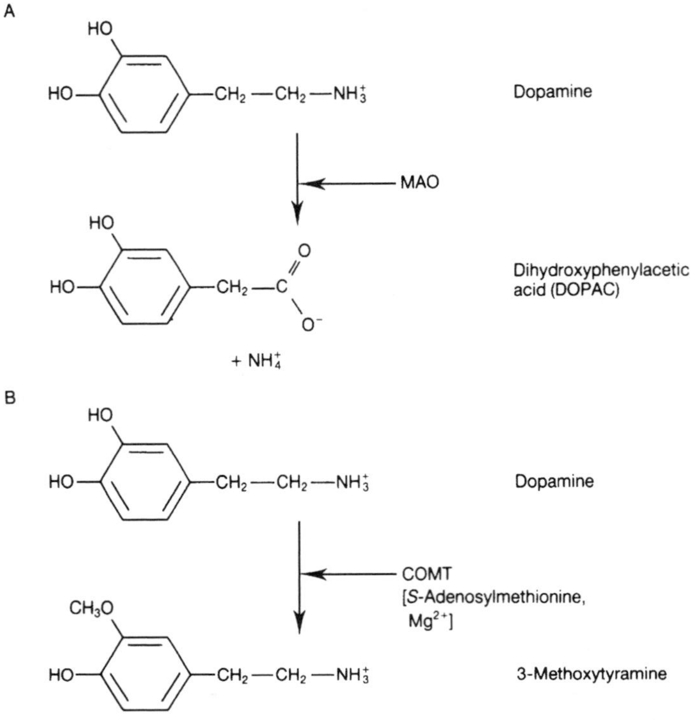

Figure 16.20 Inactivation of dopamine by MAO and COMT. (A) Monoamine oxidase (MAO) removes the terminal amino group from dopamine to form dihydroxyphenylacetic acid (DOPAC). (B) Catechol-*O*-methyl transferase (COMT) adds a methyl group from *S*-adenosylmethionine to dopamine to form 3-methoxytyramine. Mg^{2+} is required as a co-factor. 3-Methoxytyramine may be further transformed to homovanillic acid (HVA) by the action of MAO and this, in turn, is acted on by COMT to form DOPAC.

Table 16.7 Classification of adrenoceptors

Receptor	Structure	Agonist	Antagonist
α_{1A}-Adrenoceptor	466 aa; 7TM	A61603	KMD3213
α_{1B}-Adrenoceptor	519 aa; 7TM	–	Spiperone
α_{1D}-Adrenoceptor	572 aa; 7TM	–	BMY7378
α_{2A}-Adrenoceptor	450 aa; 7TM	Oxymetazoline	BRL44408
α_{2B}-Adrenoceptor	450 aa; 7TM	–	ARC239
α_{2C}-Adrenoceptor	461 aa; 7TM	–	Prazosin
β_1-Adrenoceptor	477 aa; 7TM	Xamoterol	CGP20712A
β_2-Adrenoceptor	413 aa; 7TM	Procaterol	ICI118551
β_3-Adrenoceptor	408 aa; 7TM	BRL37344	SR59230A

Chemical names of agonists and antagonists in Alexander, S. *et al.*, 2001, Nomenclature Supplement, *Trends in Pharmacological Sciences*, Cambridge, Elsevier.

synthesis of cAMP. This also seems to be the case with the α_1-receptor. The α_2-receptor, in contrast, does not seem to be linked to adenylyl cyclase and is believed to be located on the presynaptic membrane, where it causes an inhibitory response in that membrane. Once again we meet the same negative-feedback design principle: release of the transmitter acts as its own shut-off.

This feedback control is more complex than has yet been demonstrated in other types of synapse. A build-up of noradrenaline in the synaptic cleft is detected by the β_2-receptor and the consequent

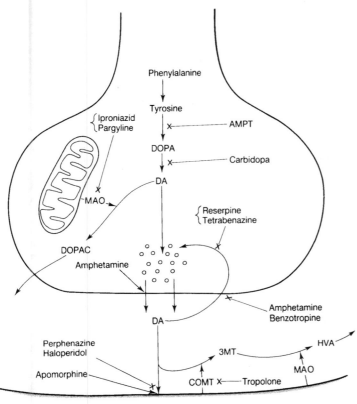

Figure 16.21 Dopaminergic synapse. The synthetic step from tyrosine to DOPA is blocked by α-methyl-*p*-tyrosine (AMPT) – a competitive inhibitor of tyrosine hydroxylase. The step from DOPA to dopamine (DA) is blocked by carbidopa, which inhibits DOPA decarboxylase. The sequestration of DA into vesicles is blocked by reserpine and tetrabenazine. Monoamine oxidase (MAO) (believed to be bound to the exterior of motochondria) converts free DA to dihydroxyphenylacetic acid (DOPAC). MAO is inhibited by pargyline (a B-type MAO inhibitor) and iproniazid (both an A- and a B-type MAO inhibitor). The release of DA into the synaptic cleft is potentiated by amphetamine (probably mainly because of its effect on reuptake). Once in the cleft DA is exposed to both COMT and MAO. COMT converts DA to 3-methoxytyramine (MT) and MAO converts MT to homovanillic acid (HVA). Amphetamine and benzotropine block the reuptake of DA back into the presynaptic terminal; reserpine and tetrabenazine block resequestration once DA gets back into the terminal. Apomorphine is a DA agonist at both post- and presynaptic sites. Perphenazine and haloperidol are antagonists at the postsynaptic membrane. Tropolone blocks the postsynaptic action of COMT.

membrane hyperpolarisation (i.e. inhibition) effectively prevents the release of any further transmitter. But there is more. For the β_2-receptor, which is also located in the presynaptic membrane, is activated only when low concentrations of noradrenaline are present in the cleft and leads to increased transmitter release. Thus there is a two-way, a negative and a positive feedback, control of the release of noradrenaline from adrenergic terminals. The **'push–pull'** action of these two types of presynaptic receptor modulates the amount of noradrenaline present in the synaptic cleft.

Noradrenaline is removed from the synaptic cleft by an Na⁺-dependent ATPase. This pump is very efficient and removes all but a very low concentration of noradrenaline from the cleft. Once back in the terminal the noradrenaline (like dopamine) is sequestered into dense-cored vesicles. This sequestration process depends on Mg^{2+} ions. Agents that chelate Mg^{2+}, such as **reserpine**, inhibit this storage process. As any free noradrenaline is inactivated by

MAO or COMT, adrenergic terminals subjected to reserpine become depleted in the transmitter. Reserpine acts in the same way at the dopaminergic synapse. The breakdown of noradrenaline in the presynaptic terminal is shown in Figure 16.22.

It was found many years ago that MAO inhibitors (MAOIs) (e.g. iproniazid, nialamide, pargyline) lifted psychological states of depression. This observation led to what has been called the **biogenic amine theory of depression**. It was suggested that endogenous depression was caused by a deficiency in biogenic, especially catechol, amines. This idea received support from the finding that reserpine tends to cause sedation in experimental animals and depression in humans. Further support was provided by the finding that **tricyclic** compounds such as desipramine, imipramine and amitriptyline (Figure 16.23) which are known to block reuptake of noradrenaline and serotonin into presynaptic terminals are powerful antidepressants. The biogenic amine theory thus argues that

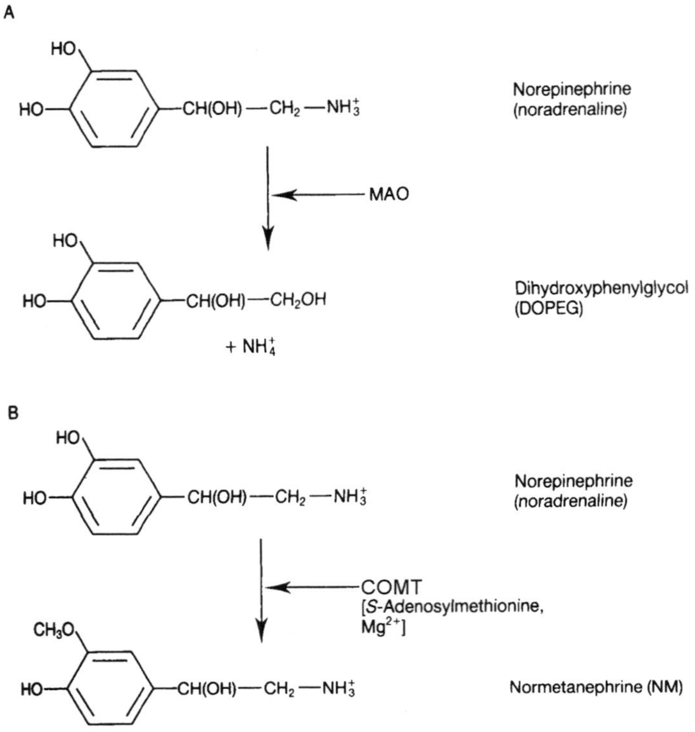

Figure 16.22 Inactivation of noradrenaline by MAO and COMT. (A) MAO deaminates nor drenaline to 3,4-dihydroxyphenylglycol (DOPEG). (B) COMT methylates noradrenaline to noremetanephrine (NM). *S*-Adenosylmethionine and Mg^{2+} are required as co-factors. NM may be further metabolised by MAO and aldehyde reductase to 3-methoxy-4-phenylglycol (MHPG).

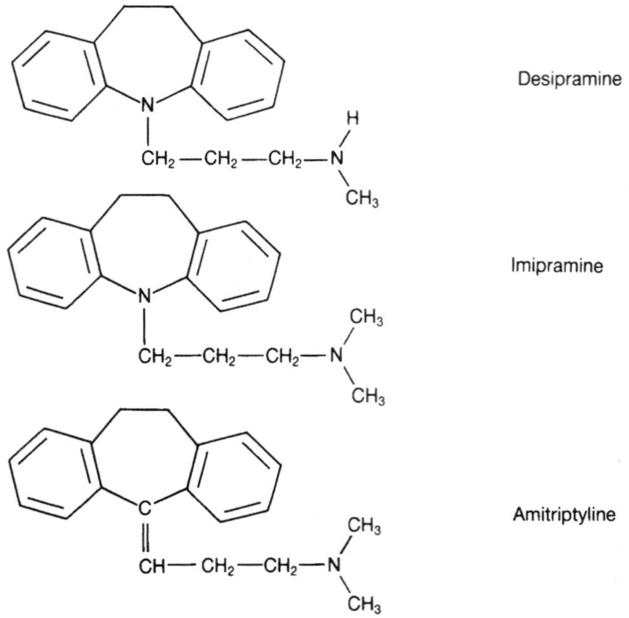

Figure 16.23 Structures of tricyclic antidepressants.

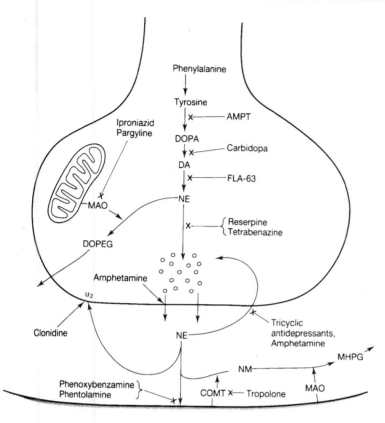

Figure 16.24 The noradrenergic synapse. The pharmacology of the adrenergic synapse is similar to the dopaminergic synapse. Only additional features will be described here. The synthetic step from dopamine (DA) to noradrenaline (norepinephrine: NE) can be blocked by Fla-63-bis-(1-methyl-4-homopiperazinyl-thiocarbonyl)-disulphide (FLA-63). Mitochondrion-bound MAO converts unsequestered NE to 3,4-dihydroxyphenylglycol (DOPEG). The tricyclic antidepressants and amphetamine block reuptake of NE released into the synaptic cleft. Clonidine is a powerful agonist at presynaptic α_2-receptors and has a lesser effect on postsynaptic receptors. Phenoxybenzamine and phentolamine are powerful blockers of postsynaptic receptors. COMT in the synaptic cleft converts NE to normetanephrine (NM) (blocked by tropolone) and MAO converts NM to 3-methoxy-4-phenylglycol (MHPG).

depressive states result from a paucity of catecholamines and/or serotonin in the synaptic cleft. In this simple form the theory is, no doubt, simplistic. The brain is a very complicated place. The biogenic amine theory of depression can, however, be developed, and is being developed, into a more sophisticated theory which has much to commend it. We shall return to consider depressive illness further in Chapter 21.

The pharmacology of the noradrenergic synapse is extremely intricate. Large numbers of drugs have been shown to affect various aspects of its operation. Once again the interested reader must consult one of the neuropharmacology texts listed in the Bibliography for full details. Figure 16.24, however, shows some of the major features of the synapse.

It should be borne in mind, when considering Figure 16.24, that the various subtypes of adrenergic receptor have different pharmacologies (Table 16.7). This is consistent with the situation obtaining with the other receptors considered in this chapter (with the possible exception of the glycine receptor). It is clear that synapses are bewilderingly complex places. Nonetheless some broad principles are beginning to emerge from the thicket of detail – principles that have to do with release and reuptake (see Box 16.3), with sequestration and enzymic degradation, with feedback control of transmitter release. We shall now leave our consideration of 'classical' transmitters and venture into the comparatively new and rapidly developing topics of purines, cannabinoids and neuroactive peptides.

16.5 PURINES

The transmitter activity of purine nucleosides and nucleotides was first recognised at the periphery. It was shown that externally applied ATP caused vasodilation, especially coronary vasodilation. Later it was shown that the nerve plexuses of the gastrointestinal tract contained many non-adrenergic, non-cholinergic (NANC) fibres. The transmitter used by these NANC fibres turned out to be ATP. It was accordingly proposed that these fibres should be re-christened purinergic fibres. It is possible that ATP released by purinergic fibres is degraded to adenosine by enzymes in the synaptic cleft.

In the early 1970s purinergic receptors were identified on the presynaptic membrane of the neuromuscular junction (NMJ). It was shown that release of adenine nucleotides reduced the release of ACh. It was also found that adenine nucleotides similarly reduced the release of noradrenaline from adrenergic terminals. It is now known that purine transmitters are not restricted to the periphery. Indeed of all the tissues that have been investigated the mammalian brain turns out to have the highest levels of adenosine and the richest variety of receptors. Purine transmitters are now known to be associated with many cerebral systems and to be widely involved in cerebral physiology.

We noted in Section 10.5 that purine receptors have been divided into two large classes: **P_1 receptors** sensitive to adenosine and AMP; **P_2 receptors** sensitive to ATP and ADP. The P_2 purinoceptors are further subdivided into subclasses. The 2X subclass is ionotropic whilst the 2Y subclass is metabotropic (Table 16.8).

Although the brain is well stocked with purines, there remains some debate whether purines are released from neurons, glia, or both cell types. In all cases, however, the action is principally neuromodulatory. This conclusion is supported by the observation that ATP is contained in many types of synaptic vesicle. In particular, there is good evidence that it cohabits with noradrenaline and acetylcholine and is released along with them into the synaptic cleft, where it can modulate their activity. Cohabitation will be discussed again in Section 16.8. P_{2X2} subunits are widely expressed in the CNS and are known, along with P_{2X3} subunits, to be strongly expressed in the postsynaptic membrane of dorsal horn neurons and other neurons belonging to the 'pain pathway'.

Because of their widespread distribution in the brain purines are implicated in numerous functions and pathologies. We have already noted their role in pain perception but perhaps their most global function is to promote homeostasis. We saw above

Table 16.8 Pharmacology of the P_2 purinoceptors

Name	Structure	Agonist	Antagonist
1. Metabotropic			
P_{2Y1}	373 aa; 7TM	ATP > ADP	MRS2179
P_{2Y2}	377 aa; 7TM	UTP = ATP	–
P_{2Y4}	365 aa; 7TM	UTP > ATP	ATP
P_{2Y6}	328 aa; 7TM	UDP >> UTP > ATP	–
2. Ionotropic			
P_{2X1}	399 aa, 2TM	ADP	NF023
P_{2X2}	472 aa; 2TM	–	–
P_{2X3}	397 aa; 2TM	ATP	–
P_{2X4}	388 aa; 2TM	–	–
P_{2X5}	422 aa; 2TM	–	–
P_{2X6}	431 aa; 2TM	–	–
P_{2X7}	595 aa; 2TM	ATP	–

Data from Alexander, S. *et al.*, 2001, Nomenclature Supplement, *Trends in Pharmacological Sciences*, Cambridge, Elsevier.

that they reduce release of ACh and noradrenaline from cholinergic and adrenergic terminals. They have similar effects on many other transmitter systems. Because of this, considerable interest has focused on a possible use in the treatment of stroke. In stroke a reduction of blood flow to the affected region leads to the release of excitatory amino acids and a consequent fatal increase in Ca^{2+} flow into neighbouring neurons. It is suggested that the application of purines (or purine agonists) could both diminish the release of EAAs and by causing vasodilation also rapidly reduce blood pressure.

Research into the normal and pathological activity of purines in the brain continues to show how they modulate the activity of other transmitters. In addition to stroke they have been implicated in conditions ranging from epilepsy, through anxiety, depression, to substance abuse.

16.6 CANNABINOIDS

Cannabis (also known as marijuana or hashish) is perhaps the best-known and most controversial of illicit drugs. As we noted in Section 8.12, it is an extract of the female flowering tops of *Cannabis sativa*. This extract contains a large number of psychoactive compounds of which the most active is Δ^9-tetrahydrocannabinol (Δ^9THC). Ingestion or inhalation of cannabis induces a number of responses in humans. These include feelings of euphoria, sedation, impaired memory, analgesia, appetite stimulation and altered perception.

Δ9-Tetrahydrocannabinol

We saw in Section 8.12 that in screening rat brain cDNA libraries for neurokinin receptors an 'orphan' receptor came to light which was ultimately shown to be a cannabinoid receptor and designated CB_1. It turned out to be a 7TM receptor some 473 amino acids in length and found principally in the hippocampus, the basal ganglia and cerebellum, and the periaqueductal grey matter (PAG) of the spinal cord. We also noted in Section 8.12 that closer examination of these regions showed that the receptor was mostly located in presynaptic membranes. How then does it induce the well-known effects of cannabis/marijuana/ hashish on the brain/mind?

There are two major possibilities although, as always in the brain, there are probably other mechanisms at work as well. The first possibility flows from the fact that when cannabis binds to its receptor the resulting inhibition of adenylyl cyclase lowers the level of cAMP in the presynaptic terminal. cAMP, as we saw in Section 8.8, activates PKA and PKA, in turn, phosphorylates, and thus inactivates, K^+ channels, especially K_A channels. It follows that activation of cannabinoid receptors reduces cAMP levels and thus disinhibits K_A channels, consequently shortening the duration during which the presynaptic membrane is depolarised when an action potential invades the terminal. In addition, there is evidence that a direct G-protein inhibition of N-type Ca^{2+} channels occurs. In both cases neurotransmitter release would be diminished (see Section 15.4).

It is unlikely that cannabinoid receptors would exist in the brain if there were no endogenous ligand with which to react. Two candidate endo-

cannabinoids have been detected: arachidonylethanolamide ('**anandamide**') and 2-arachidonylglycerol (**2-AG**). It is likely that anandamide is the endocannabinoid released in the brain. How it exerts its inhibitory action is still a matter of intensive research but it is suggested that it is released into the synaptic gap by the **postsynaptic** cell and diffuses across the gap to bind to CB_1 receptors in the presynaptic membrane, thus inhibiting (or at least moderating) the release of neurotransmitter (Figure 16.25). There is evidence to show that this occurs at GABA-ergic inhibitory synapses in the hippocampus and Purkinje cells in the cerebellum when the postsynaptic cell is depolarised. CB_1 receptors are widespread in the brain and the transmitter whose release is inhibited varies from one part of the brain to another. In the rat

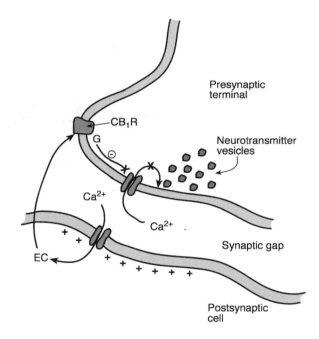

Figure 16.25 Possible organisation of cannabinoid signalling at a synapse. The postsynaptic membrane is depolarised and the resulting influx of Ca^{2+} stimulates the production of endogenous cannabinoid (EC). This diffuses out towards the presynaptic terminal and activates a CB_1 receptor (CB_1R). An inhibitory G-protein system shuts down Ca^{2+} channels in the presynaptic membrane, thus inhibiting the release of neurotransmitter when an impulse invades the terminal. After Montgomery and Madison (2001), *Neuron*, **29**, 567–570.

BOX 16.3   Reuptake neurotransmitter transporters

We saw in Box 15.2 that considerable progress had been made in understanding the molecular nature and genetic coding of vesicular neurotransmitter transporters. In this box we shall see that similar advances have been made in understanding the nature of the transporters which play an equally important role in retrieving neurotransmitters from the synaptic cleft.

As we have seen in this chapter, reuptake of neurotransmitters from the synaptic cleft into the presynaptic terminal plays a vital role both in conserving neurotransmitters and/or their constituents and in regulating the activity of the synapse. Reuptake transporters thus play a key role in synaptic physiology. Because of this role they are also the site of action of many pharmacological agents which affect the operation of synapses.

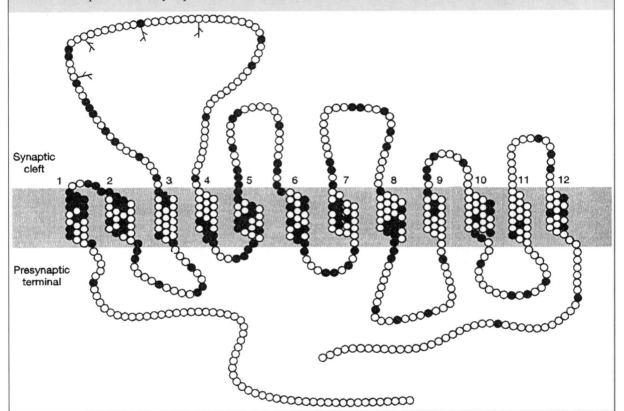

Figure A   The dopamine reuptake transporter. The filled circles represent residues common to rat dopamine, human noradrenaline, rat serotonin, rat GABA and canine betaine transporters. Reproduced with permission from Amara and Kuhar (1993), *Annual Review of Neuroscience*, **16**, 73–93. Copyright © 1993 by Annual Reviews Inc.

The first transporters to be cloned were the rat and human GABA transporters. Oligonucleotide probes derived from these transporters were used to isolate noradrenaline, dopamine and 5-HT transporters from cDNA libraries. They turn out to be glycoproteins with M_rs ranging from 60 to 85 with about 50% sequence identity. They form, in other words, a reasonably close-knit molecular family. Hydropathy analysis suggests that (as with the vesicular transporters) they have 12 transmembrane helices. A large polypeptide loop between M3 and M4 bears a number of N-glycosylation sites. This loop projects into the synaptic cleft and may be involved in selecting the appropriate transmitter to transport. It may also, consequently, be the site of attachment for synaptically active pharmacological agents.

Hybridisation histochemistry has been used to localise the transporter mRNAs within the brain. cDNA of the DA transporter, in particular, shows intense hybridisation in the substantia nigra, olfactory bulb and specific regions of the hypothalamus: all regions known to possess dopaminergic neurons. The DA transporter is also interesting in that it binds **cocaine** and related drugs. It thus appears to be the major brain '**cocaine receptor**' and consequently a crucial focus in treating addiction to this and related substances. The transporter has also been shown to bind the neurotoxin 1-methyl-4-phenylpyridinium (MPP+) which, as we shall see in Section 21.9, has been implicated in that well-known DA pathology, Parkinson's disease.

Unlike the vesicular neurotransmitter transporters, the reuptake transporters are **symporters** (or co-transporters). In other words, the transport of the neurotransmitter is coupled with the movement of Na^+ ions down their concentration gradient. This coupling implies that the transport is, at least potentially, electrogenic. In other words, transport into the synaptic terminal should be accompanied by an electric current. These currents have, indeed, been observed in patch-clamping experiments on glutamate uptake in salamander Müller cells and skate horizontal cells.

prefrontal cortex it is glutamate, in the hippocampus ACh and GABA, in striatum dopamine, and in the PAG GABA and glutamate.

Cannabinoids have numerous effects in the brain, but especially in the hippocampus (memory), the basal ganglia and cerebellum (movement) and the periaqueductal grey matter of the spinal cord and brain stem (analgesia). Experiments with CB_1 'knockout mice' show that short-term memory is significantly **improved**. This suggests that endocannabinoids assist in memory loss. Memory, as we shall see in Chapter 20, is a highly complex phenomenon and the endocannabinoids almost certainly play only a bit part in the process. Nevertheless, the role of cannabinoids in deadening memory is consistent with its effect on human substance-abusers. The effects on the basal ganglia are also complex. In experimental animals behavioural movement is decreased at low concentrations of cannabis (hypomotility) but movement progressively increases as dose concentration increases, ending in catalepsy at $2.5\,mg\,kg^{-1}$ (rats). The cerebellum is, of course, also crucially involved in the control of movement and it presents (as noted above) high concentrations of CB_1 receptors. However, whether or how activation of cerebellar receptors by cannabinoids affects movement is not at present known. Finally, there is

much evidence that the well-known analgesic effects of cannabinoids are due to interaction with CB_1 receptors in pain-pathway interneurons of the periaqueductal grey matter of the spinal cord. Here the similarity with the endogenous opiates is clear and we turn to these and the pain pathway in the next section.

16.7 PEPTIDES

We have already considered the structure, synthesis, post-transcriptional and post-translational modification of the better-known neuroactive peptides in earlier parts of this book (Section 2.1.1; Section 3.4.3; Section 3.4.5; Section 4.2.3). We noted that over fifty such molecules are nowadays recognised and we saw that in many cases they form families, derived by the post-transcriptional processing of a single mother protein (pre-protein or polyprotein).

In this chapter we shall briefly look at their synaptic activity. First, it should be noted that neuropeptides differ from most of the smaller molecules we have discussed in earlier parts of this chapter in that they are not recycled and/or resynthesised in the presynaptic terminal. Neuropeptides are synthesised by processes described in Chapter 15 in the neuronal perikaryon and are transported to the presynaptic terminal in the

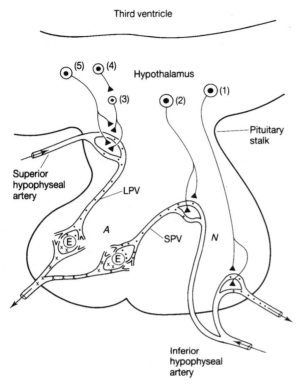

Figure 16.26 Hypothalamo-pituitary system. The figure schematises the major pathways in the peptidergic hypothalamo-pituitary system. Peptidergic neuron (1), located in the hypothalamus, releases oxytocin or vasopressin directly into the general circulation via the capillary bed in the neurohypophysis (*N*) (posterior pituitary). Peptidergic neurons (2) and (3) synthesise releasing hormones (e.g. corticotrophin-releasing hormone (CRH), thyrotrophin-releasing hormone (TRH), etc.) and secrete them into the capillary bed of the hypophyseal portal system. Neurons (4) and (5) link the rest of the brain to the peptidergic neurons – they act with 'conventional' transmitters either on the cell body (4) or the presynaptic endings (5) of peptidergic neurons. The cells (E) which the releasing factors effect are located in another capillary bed of the hypophyseal portal system in the adenohypophysis (*A*). LPV=long portal vessel; SPV=short portal vessel. After Kandel and Schwartz (1985), *Principles of Neural Science*, New York: Elsevier.

axoplasmic flow. As axons can be lengthy it is consequently possible for an overworked neuron to run out of presynaptic neuropeptide.

Second, it should be borne in mind that neuropeptides function in other ways than as neuro-

Table 16.9 Distribution of some neuroactive peptides

Peptide	H	B–H	SC	N	GI	Pa	Sk	Pi
Enkephalin	+	+	+	+	+			
Substance P	+	+	+	+	+	+	+	+
Neurotensin	+	+	+		+		+	
VIP	+	+			+	+		
Gastrin	+	+			+	+		
NPY	+	+	+	+	+	+	−	−

H=Hypothalamus; B–H=brain minus hypothalamus; SC=spinal cord; N=peripheral nerve; GI=gastrointestinal tract; Pa=pancreas; Sk=skin; Pi=pituitary.

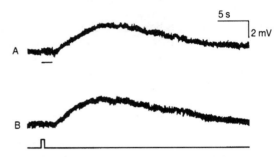

Figure 16.27 SP-induced 'slow' EPSP. (A) Response of a neuron in the inferior mesenteric ganglion to stimulation in the presence of cholinergic blockers hexamethonium (400 μM) and atropine (4 μM). The response was induced by stimulation of the lumbar splanchnic nerves (10 Hz, 2 s) shown by horizontal bar. (B) Response of the same neuron in the presence of the same cholinergic blockers to application of SP (5 μM) by a micropipette (shown by the blip in the lower horizontal line). Reprinted from *Neuroscience*, **7**, Tsunoo, Konishi and Otsuka, with permission from Elsevier Science, © 1982.

transmitters/modulators. They may, for instance, make use of the vascular system for distribution. This is particularly the case in the **hypothalamo-pituitary** system (Figure 16.26). In this respect they are closer to hormones than neurotransmitters and should indeed be regarded as 'local hormones' In yet other cases the vascular system may distribute them to distant parts of the body, outside the nervous system altogether. We noted in Chapter 15 that this leads into the large and diverse subject of neuroendocrinology, which we shall not attempt to enter. Biology, unfortunately, has scant regard for the tidy classificatory schemes of the scientist!

It would clearly be impossible to discuss fifty or more different neuroactive peptides in one short

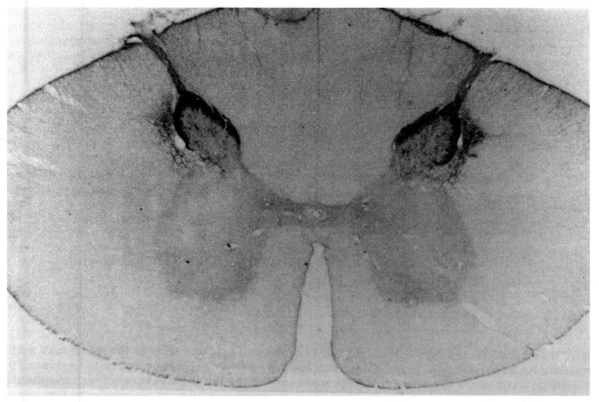

Figure 16.28 Location of SP synapses in the substantia gelatinosa of the spinal cord. The darkly stained regions in the dorsal horn of this transverse section of monkey spinal cord indicate where substance P is concentrated. Courtesy of Dr S. Hunt.

section. We shall therefore concentrate on just a few comparatively well-known examples. We shall look at a family of peptides that are found both in the central nervous system and in the gastrointestinal (GI) tract, the so-called 'gut–brain axis'. We shall, however, only consider their actions in the central nervous system. This family, as Table 16.9 shows, includes the opioids (especially the enkephalins) and substance P (SP), as well as vasoactive intestinal peptide (VIP), cholecystokinin (CCK), neurotensin (NT), gastrin, bombesin and a few others. Of this large and diverse group of peptides we shall only consider the best-known members: substance P and the enkephalins.

16.7.1 Substance P

We saw in Table 2.2 that SP is an 11-residue peptide related to two decapeptides (substance K

(=neurokinin A) and neurokinin B). SP is found in the **dorsal horn** of the mammalian spinal cord and in a number of regions of the brain, especially the **substantia nigra** of the midbrain. It has also been shown to exist in significant quantities in the **inferior mesenteric ganglia**. This ganglion has provided a convenient experimental preparation.

Stimulation of preganglionic fibres elicits long-lasting depolarisation in the ganglion (Figure 16.27). This slow excitatory postsynaptic potential (EPSP) is not blocked by cholinergic antagonists but can be mimicked by the application of SP. Furthermore if **capsaicin**, derived from red pepper, is applied to the ganglion, SP is known to be released. When preganglionic stimulation is applied to these SP-depleted ganglia, the slow EPSP is not developed. These various approaches suggest, therefore, that the slow EPSP, which may last from 20 s to 4 min, is due to substance P.

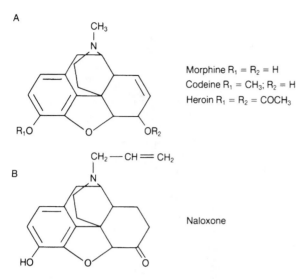

A

Morphine $R_1 = R_2 = H$
Codeine $R_1 = CH_3$; $R_2 = H$
Heroin $R_1 = R_2 = COCH_3$

B

Naloxone

Figure 16.29 Structures of morphine and related drugs.

In the spinal cord substance P has been found to be associated with small-diameter primary afferent fibres (Figure 16.28). These fibres are the **C fibres** which are implicated in pain sensations. This observation does not, however, exclude SP from other afferent pathways – it may also, for instance, be involved in pathways carrying information from **baroreceptors**.

16.7.2 Enkephalins

Table 2.2 showed that two pentapeptide enkephalins – **met-enkephalin** and **leu-enkephalin** – have been identified. Another endogenous opioid has been detected in the pituitary and in extracts of the duodenum. This is **dynorphin** and is leu-enkephalin extended at the C-terminal end by a further 17 amino acid residues: -Tyr-Gly-Gly-Phe-Leu-Arg-Arg-Ile-Arg-Pro-Lys-Leu-Lys-Trp-Asp-Asn-Glu.

Enkephalins are distributed widely throughout the brain. On average met-enkephalin is about three times more prevalent than leu-enkephalin. The heaviest concentrations are to be found in the **dorsal horn** of the spinal cord, the **periaqueductal grey matter**, the **limbic system**, and the **basal ganglia** – especially the **globus pallidus**. This distribution is interesting as the dorsal horn of grey matter and periaqueductal grey matter contain the multisynaptic spinothalamic tract. This tract is believed to be responsible for our sensation of dull aching pains (in contrast to sharp pricking pains). It will be recalled from Section 16.6 that the CB_1 receptors for endocannabinoids are also located in these regions. The limbic system has been implicated in emotional response whilst the globus pallidus is part of the system that controls motor output (see Section 16.2.2 above). It is worth noting that in cases of **Huntington's disease** the concentration of met-enkephalin in the globus pallidus diminishes to a third of normal – from $1.5\,ng\,g^{-1}$ to $0.5\,ng\,g^{-1}$.

The discovery in the early 1970s that enkephalins bind to the same sites as opium, morphine, codeine, etc. (Figure 16.29) was one of the most exciting events in the history of neuropharmacology. It promised that at long last we should have some insight into both the neural events underlying pain and the way in which various exogenous opiates can relieve pain; it promised also some insight into the mechanisms of addiction that have for millennia provided such anguish for addicts and problems for society.

Opioid receptors, again like the CB_1 cannabinoid receptors, are all 7TM membrane-embedded proteins. They have been subdivided into three types: δ, κ and μ. Their pharmacological characteristics are shown in Table 16.10.

The three types of opioid receptor are distributed somewhat differently in the brain. The **μ-receptors**

Table 16.10 Opioid receptors

Name	Structure	Agonist	Antagonist
Delta	372 aa; 7TM	β-end=leu=met>dynA	Nalytindole
Kappa	380 aa; 7TM	dynA>>β-end>leu>met	Nor-binaltrophine
Mu	400 aa; 7TM	β-end>dynA>met=leu	CTOP

β-end=β-endorphin; dynA=dynorphin A.
Data from Alexander, S. *et al.*, 2001, Nomenclature Supplement, *Trends in Pharmacological Sciences*, Cambridge, Elsevier.

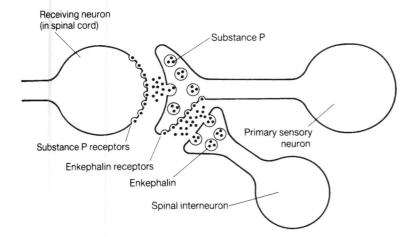

Figure 16.30 Enkephalinergic control of primary pain fibre in substantia gelatinosa. A primary sensory neuron using SP as its transmitter synapses with a neuron of the spinothalamic tract. Its input into the 'pain-pathway' is controlled by the enkephalinergic interneuron which is shown acting presynaptically to inhibit the release of SP. After Iverson (1979), *Scientific American*, **241**, 118–129. Copyright © by Scientific American Inc. All rights reserved.

have a widespread distribution but are especially dense in the **thalamus** and **amygdala**. It is believed that they may be involved in pain regulation and in sensorimotor coordination. The **δ-receptors**, in contrast, have a more restricted distribution. They are especially heavily concentrated in olfactory regions of the brain such as the **caudate putamen** and the **nucleus accumbens**. Finally the **κ-receptors** are once again rather sparsely distributed but are found especially in the **hypothalamus** and **preoptic area**. It is likely that they are involved in water balance and food intake as well as in pain perception.

It is believed that enkephalinergic neurons involved in the regulation of the sensation of pain exert their effects by releasing their transmitter on to the presynaptic terminals of pain-pathway neurons. It was indicated above that the spinal cord and brain stem pain-pathways are multisynaptic routes. There is thus much opportunity for enkephalinergic interneurons to modulate the flow of messages. Indeed some have suggested that this is one of the ways in which the CNS keeps the pathway damped down during the heat of the battle or game. Only afterwards do we feel our bumps and bruises, or worse. The putative interneuronal action of enkephalinergic neurons is shown in Figure 16.30.

The molecular mechanism by which enkephalins are able to damp down activity in spinothalamic tract neurons remains to be discovered. There is circumstantial evidence that enkephalins occupying μ-receptors lead to membrane hyperpolarisation. If

the μ-receptors are on the presynaptic terminals of spinothalamic tract neurons or, as in Figure 16.30, primary 'pain' neurons, it is clear that the depolarisation required for transmitter release would be less likely to occur. Other investigations have shown that in some tissues enkephalins inhibit the action of adenylyl cyclase. It will be recalled that a similar mechanism has been proposed for the endocannabinoids.

It is likely that the painful symptoms experienced by addicts when an exogenous opiate is withdrawn are due to a type of 'denervation supersensitivity'. It can be argued that the continuous presence of the opiate in the addict's pain pathways leads to a compensatory increase in synaptic sensitivity 'downstream' in the pathway. This is known in other cases of diminished impulse traffic in a multisynaptic pathway. It seems that the number of receptors on subsynaptic membranes increases to balance the low level of activity. Hence (returning to the pain pathway) when the drug is removed, downstream synapses will respond to activity which would not normally be transmitted. The addict's 'drying-out' phase can thus be accompanied by extremely distressing withdrawal pains.

16.8 COHABITATION OF PEPTIDES AND NON-PEPTIDES

We have already noted that 'Dale's principle' (one neuron – one transmitter) is now known to have many exceptions. It has recently become apparent that neuroactive peptides frequently share a neuron

Table 16.11 Cohabitation of neuroactive peptides and non-peptides

Non-peptide transmitter	Peptide	Tissue (species)
Dopamine	Enkephalin	Carotid body (cat)
	CCK	Brain (rat, human)
	Neurotensin	Ventrotegmental neurons
Noradrenaline	Neuropeptide Y	Cerebral cortex
		Hypothalamus
		Brain stem
	Somatostatin	Sympathetic ganglia (guinea pig)
	Enkephalin	Sympathetic ganglia (rat, ox)
		Adrenal medulla (several species)
		Locus coeruleus (cat)
	Neurotensin	Adrenal medulla (cat)
Serotonin	Neuropeptide Y	Brain stem
	Substance P	Medulla (rat)
	Enkephalin	Medulla (rat, cat)
Acetylcholine	VIP	Autonomic ganglia (cat)
	Enkephalin	Preganglionic nerves (cat)
		Cochlear nerve (guinea pig)
	Neurotensin	Preganglionic nerves (cat)
	Somatostatin	Heart (toad)
	Substance P	Ciliary ganglion (bird)
GABA	Somatostatin	Thalamus (cat)
	Substance P	Medulla (rat)
	Enkephalin	Medulla (rat)
	Neuropeptide Y	Basal ganglia (rat)

Adapted from Lundberg and Hökfelt, 1983, *Trends in Neurosciences*, **6**, 325–332.

with a non-peptide transmitter. Table 16.11 shows that practically all combinations of peptide and non-peptide transmitters have been found. As these cohabitations have been largely established by immunohistochemical techniques which are sometimes not absolutely specific for neuropeptides, there remain some uncertainties in the identifications. Table 16.11, nevertheless, shows the very widespread nature of this phenomenon.

The exact significance of cohabitation remains to be determined. It may be that the peptide and the non-peptide act synergistically or, in contrast, it could be that the peptide or the non-peptide act presynaptically to inhibit the release of their companion. One might imagine, for example, that because of the different geography of synthesis the neuropeptide in a hard-working neuron might run out before the non-peptide. If the former acted to inhibit or potentiate the release of

the latter, interesting control circuits may be possible.

Alternatively the two transmitters may modulate each other's subsynaptic effects. The peptide, for instance, may work through a second messenger system on the postsynaptic cell's biochemistry. This may result in the ion channels in the **postsynaptic** membrane being 'up' or 'down' modulated. The subsynaptic response to the cohabiting non-peptide transmitter (such as ACh or one of the amino acids) could thus be controlled. Finally, it is always possible that the two cohabitants have quite distinct and independent roles.

These are just a few of the possibilities that our contemporary understanding suggests. It is likely that different possibilities, and some not yet thought of, are realised in different systems. The molecular complexity at the synapse is evidently far greater than was suspected even a few years ago.

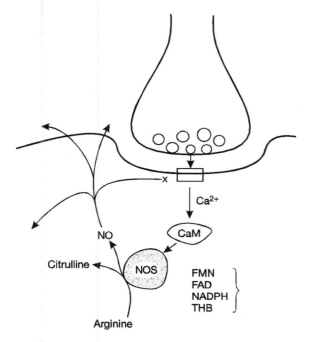

Figure 16.31 Synthesis of NO in a subsynaptic cell. The presynaptic terminal releases appropriate transmitters to open Ca^{2+} gates. These might be NMDA receptors, neuronal nAChRs, etc. Alternatively Ca^{2+} inflow may be due to one of the voltage-dependent Ca^{2+} channels described in Section 11.3. The Ca^{2+} ions first interact with calmodulin (CaM) so that it can activate nitric oxide synthase (NOS). NOS can then (in the presence of co-factors FMN, FAD, NADPH and THB) catalyse the reaction arginine→citrulline by removing NO. NO then diffuses both in the intercellular space and within the cytosol of the subsynaptic cell. It also downregulates NMDA channels, thus switching off the flow of Ca^{2+} ions on which its synthesis depends.

16.9 NITRIC OXIDE (NO)

The first demonstration that nitric oxide (NO) was a neuroactive molecule was not made until 1988. Since that time there has been something approaching an avalanche of papers reporting the involvement of NO in a wide variety of physiological and neurophysiological processes. This work is subtly altering our perspectives on synaptic signalling. NO is a very small molecule (30 Da) compared with classical transmitters and even smaller when compared with the peptides reviewed in the previous section. It may be that other small molecules, for instance CO and even the free radical OH, also have synaptic activity.

Not only is NO unusually small but its synaptic physiology is also unconventional. It is not sequestered into vesicles, it has no subsynaptic receptors, there is no reuptake system into the presynaptic ending. NO diffuses away from its site of production and being such a small molecule moves freely through the intercellular space affecting neurons up to 100 μm distant. This novel mode of transmission is sometimes called 'volume' transmission in contrast to the 'wiring' transmission of more conventional synapses; the networks established by such transmission are similarly often referred to as **'gasnets'** to contrast them with the more punctate 'telephone exchange' networks made by the majority of synapses. Nitric oxide's targets are still being investigated but it is already known that it can affect **guanylyl cyclases** and thus exert control on cGMP production. It has also been shown that it downregulates **NMDA** channels and thus reduces Ca^{2+} flow into the cytosol.

Although NO is the smallest neuroactive molecule, it is synthesised by one of the largest enzymes, the approximately 300 kDa **NO synthase (NOS)**. In fact, NOS comes in at least three different forms: an inducible NOS **(iNOS)** possibly present in glial cells, and two constitutive NOSs, one of which is found in neurons (**ncNOS**) and another in endothelial cells (**ecNOS**). It is likely that other variations on the theme will be discovered in the near future. All the NOSs require co-factors, such as FAD, FMN, NADPH and THB, to provide electrons with which to oxidise one of the terminal nitrogens of arginine to generate citrulline and thus release NO (Figure 16.31).

The factors inducing the expression of the iNOS gene include many microbes and microbial products, tumour cell products, etc. It is, in other words, induced by the by-products of **inflammation**. Once induced the synthesis of iNOS is high and continued (measured in hours). iNOS is not itself sensitive to Ca^{2+} concentrations. In contrast the two cNOS enzymes are highly sensitive to Ca^{2+} levels and produce a transient burst (measured in minutes) of nitric oxide when these levels rise. The Ca^{2+} ions appear to work through the calcium binding protein, **calmodulin**. In the presence of Ca^{2+} calmodulin binds to NOS and releases its catalytic activity. As NO downregulates NMDA channels, there is a strong negative feedback ensuring that NO production is only transient. The relationships of Ca^{2+}, ncNOS and NO are shown in Figure 16.31.

Antibodies raised against NOS and hybridisation histochemistry against its mRNAs have enabled workers to localise NO production sites in the brain. These studies have shown high densities of NOS in the **cerebellum**, **olfactory bulb**, **hippocampus (dentate gyrus)**, **superior** and **inferior colliculi**, some **tegmental nuclei** and also a few isolated cells in the cerebral cortex. Some of these areas, especially the hippocampus, cerebellum and olfactory bulb, are believed to be involved in the plastic changes underlying memory. We shall return to the possible involvement of NO in this process in Chapter 20.

16.10 CONCLUSION

At the outset of Chapter 15 we noted that the brain might be seen as an immense gland rather than as an intricate electronic computer. We are now in a position to feel the force of this analogy and also to recognise the import of Freeman's proposition (Chapter 1) that the flow of activity in the brain could best be likened to the 'continuum of a chemical reaction'. Unlike the majority of chemical reactions, however, the brain's activity occurs not in just one or at most a few phase-spaces but in billions of interconnected compartments.

It is often asked: why are there so many neurotransmitters and neuromodulators? The question betrays a misconstrual of the brain. An inappropriate analogy is at work. The brain is, once again, being likened, probably unconsciously, to a telephone exchange or electronic computer. But we have seen that this is not the way things are in the brain. Its operation is far more subtle. At a certain generality of definition no doubt the brain is a computer. But, so far as its hardware, its operating system, is concerned, it is a chemical computer.

Thus the answer to the question posed above is easy. The brain does not operate by 'yes/no' gates, 'on/off' switches, but by fluxes of chemicals affecting diverse receptors which in turn often modulate intracellular biochemical processes. Only at a very first analysis can synapses by classified straightforwardly into 'excitatory' and 'inhibitory'; deeper investigation almost always shows a host of subtle variations, mostly at the molecular level. Only at a very first analysis does the brain consist of discrete synaptic appositions; further research shows that many synapses are made 'en passant'; many transmitters and modulators are released from 'varicosities' to percolate through the extracellular space to exert 'long-range' influence. The biochemical environment in which neurons work is subtle and ever-changing. In the next chapter we shall see that subtlety is also very much a feature of the subsynaptic cell and its responses.

17

THE POSTSYNAPTIC CELL

Receptor complexity of the postsynaptic membrane. **Synaptosomes** – preparation – physiology – contents. **Postsynaptic density** – preparation – biochemistry – cadherins, PSD-95-attachment of receptors to cytoskeleton – intricate multimolecular architecture. **Electrophysiology: excitatory synapses**: ionic bases of EPSPs – local circuits to IS – initiation of AP – significance of size and site of synapses – graded characteristic – spatial and temporal summation; **inhibitory synapses**: ionic bases of IPSPs – local circuits to IS – inhibition of AP – significance of size and site of synapses – graded characteristic – spatial and temporal summation. Algebraic summation of EPSPs and IPSPs at IS. **Biophysics of channels in postsynaptic membranes**: single-channel biophysics – operational complexity – populations of channels – comparison with AP – subtlety of response of postsynaptic membrane. **Second messenger control of ion channels**: Ca^{2+} channels – K^+ channels – physiological significance. **Second messenger control of gene expression**: induction of third messengers – IEGs (*c-fos* and *c-jun*) – products detected in active neurons – may control neuropeptide synthesis – role of Fos in enkephalinergic neurons – direct action of second messengers on promoters – importance of Ca^{2+} – Ca^{2+} and NMDA channels – the Ca^{2+}, cAMP, PKA, CREB cascade. **The pinealocyte**: Descartes – control of reproduction – photoperiod – NE, NPY, VIP release on to pinealocyte membrane – G-protein networks – control of IEGs by second messenger combinations – release of variety of third messengers – SNAT gene transcription – transcript translated into SNAT – synthesis and secretion of melatonin – release into vascular system. Can the presynaptic terminal, synaptic cleft and postsynaptic membrane be seen as a single neurobiological unit?

In Chapter 16 we reviewed some of the transmitters and modulators that the presynaptic terminal releases into the synaptic gap. In this chapter we shall look at what happens next. What effect do these substances have on the postsynaptic cell? We shall see that, just as action potentials result from the activity of a variety of membrane-embedded voltage-gated channels, so the response of sub- or postsynaptic membranes is the outcome of the underlying activity of a multitude of G-protein-coupled and channel-coupled receptors (see Chapters 8 and 10).

We saw something of the morphological variety of synapses in Section 1.2 and we also noted the major features of the 'classical' chemical synapse (Figure 1.16). We saw that the presynaptic and the postsynaptic membranes were separated from each other by a gap or cleft of some 30–40 nm. That this gap is not 'empty' is shown by the structure of synaptosomes.

17.1 SYNAPTOSOMES

Synaptosomes can be obtained from the central nervous system by homogenisation in buffered sucrose solutions and subsequent density gradient centrifugation. One of the fractions in the centrifuge tube contains broken-off nerve endings or

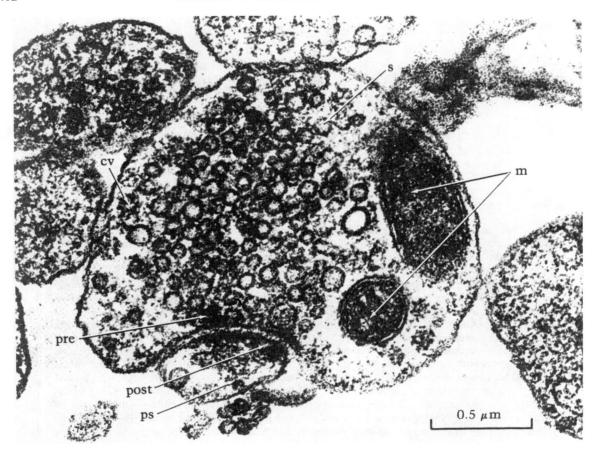

Figure 17.1 Synaptosome. Synaptosome from rat cerebral cortex after incubation in physiological saline for 10 minutes. cv=coated vesicle; m=mitochondria; post=post (i.e. sub-) synaptic density; pre=presynaptic grid element; ps=postsynaptic 'bag'. From Csillag and Hajós (1980), *Journal of Neurochemistry*, **34**, 495–503, by permission of Blackwell Publishing.

synaptosomes. Microscopical examination shows that they consist of not only the presynaptic terminal but also a portion of adhering postsynaptic membrane. This provides good evidence that there is some adhesive material between the two membranes.

Synaptosomes have proved of considerable importance in brain research. Although they are broken-off boutons, their boundary membrane reseals to form a tiny (1–2 μm) sac containing mitochondria, vesicles, neurotransmitters, enzymes and the other elements of synaptic terminals. When incubated in appropriate physiological media they

can be shown to respire and an active $Na^+ + K^+$ pump ensures that a resting potential (-30 to -60 mV) is maintained across their membrane. Finally, when electrically depolarised or treated with depolarising agents (e.g. veratridine), synaptosomes can be shown to release neurotransmitters into the incubating medium. Because synaptosome preparations are obtained from the homogenisation of a large number of terminals a cocktail of neurotransmitters is released. The nature of this cocktail varies according to the brain region from which the synaptosomes were obtained. Thus cerebral-cortical synaptosomes release GABA,

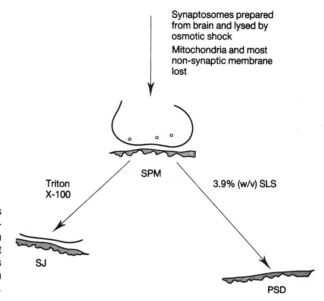

Synaptosomes prepared from brain and lysed by osmotic shock

Mitochondria and most non-synaptic membrane lost

SPM

Triton X-100

3.9% (w/v) SLS

SJ

PSD

Figure 17.2 Preparation of SPMs, SJs and PSDs from synaptosomes. Treatment of lysed synaptosomes (synaptic plasma membranes: SPMs) with Triton X-100 yields synaptic junctions (SJs); treatment of SPMs with sodium lauryl sarcosinate (SLS) yields postsynaptic densities (PSDs). After Kelly and Cotman (1977), *Journal of Biological Chemistry*, **252**, 786–793.

glycine, glutamate, aspartate; synaptosomes derived from the corpus striatum are enriched in dopamine; those derived from the spinal cord are enriched in glycine.

Figure 17.1 shows not only that a portion of postsynaptic membrane adheres to the presynaptic terminal but also that it carries with it the postsynaptic density. We noted in Chapter 1 that this density was a characteristic of chemically conducting synapses. It has proved possible to isolate this density from the synaptosome. We shall consider its constitution in the next section.

17.2 THE POSTSYNAPTIC DENSITY

If synaptosomes are exposed to strong detergents such as Triton X the pre- and postsynaptic membranes are released from the rest of the complex (Figure 17.2). This, the so-called **synaptic plasma membrane** (SPM), may then be isolated from the other elements of the synaptosome by density gradient centrifugation. The fact that the pre- and postsynaptic membranes stay together when the rest of the synaptosome has been liquidated suggests that they are held together by some electron-translucent matrix. The analysis of this region can be carried a step further by subjecting the SPMs to 3.9% w/v sodium lauryl

sarcosinate. This dissolves away all but the subsynaptic or **postsynaptic density (PSD)**.

The proteins and polypeptides of the SPM and PSD can be analysed by polyacrylamide gel electrophoresis. In both cases the gel shows a large number of different bands indicating a variety of molecules. These are likely to differ in different synapses. One major group of molecules within the synaptic cleft have molecular weights of more than 65 kDa and are glycoproteins. There is good evidence that these molecules are **cadherins** (Section 19.8). These molecules, as we shall see in Chapter 19, are crucial to the pathfinding mechanisms by which neurons find their 'correct' positions in the brain. It is likely, therefore, that they are involved in locking appropriate pre- and postsynaptic membranes together and stabilising the complex. Another very significant family of molecules, found particularly in the PSDs of glutaminergic excitatory synapses, belong to the **PSD-95** family. This family of proteins contain multiple socalled 'PDZ' domains which enable them to bind to other proteins and polypeptides to form supramolecular complexes. There is evidence that these complexes form scaffolds or rafts beneath the membrane to hold together receptor proteins. PSD-95s are particularly well represented in spine synapses and we shall meet them again in Section 19.4 and Box 20.1 The other major proteins in the postsynaptic density appear to be smaller, having molecular weights ranging from 40 kDa to 55 kDa. Carrying the analysis a little further, it can be deduced that the proteins in the postsynaptic density are extensively cross-linked, perhaps forming a network. There is also evidence that fibrous proteins, perhaps actin, tubulin, fodrin and (possibly) neurofilaments, are constituents of the network. It is easy to speculate that this intricate network interacts with the PSD-95s to anchor receptor molecules in the postsynaptic membrane.

Evidence to support this speculation has come from studies of the neuromuscular junction (Section 10.1.3). Electron microscopy shows that the postsynaptic membrane of the junction is folded into a large number of invaginations and that there is a filamentous meshwork beneath it (see Figure 15.16). It can be shown, furthermore, that the acetylcholine receptors are clustered on the crests of the folds. The density of nAChRs on the crests may reach $10\,000\,\mu m^{-2}$ compared with about $20\,\mu m^{-2}$ on the rest of the sarcolemma.

We also noted in Chapter 10 that the electric organs of electric fish provide a spectacular source of cholinergic neuromuscular junctions. The electrocytes of the electric ray, *Torpedo*, have thus been used in investigations of the structure and function of the neuromuscular postsynaptic meshwork. It has been found that the postsynaptic material can be removed by alkaline solutions or low concentrations of lithium di-iodosalicylate. Analysis of the extract by gel electrophoresis reveals a number of proteins and in particular a group with a molecular weight of about 43 kDa. If antibodies are raised against this protein, tagged with colloidal gold particles, and then reacted with the *Torpedo* electrocyte, the cytoplasmic faces of the crests of the postsynaptic membrane are labelled (Figure 17.3). This suggests that it is indeed part of the postsynaptic apparatus.

There is suggestive evidence that the 43 kDa protein plays a role in anchoring nAChRs into the postsynaptic membrane. A number of techniques for determining the mobility of large proteins such as the nAChR in biomembranes are available. For instance, electrocyte membranes may be fused with liposomes. If this is done it is found that nAChRs do not diffuse freely throughout the hybrid membrane. If, however, the electrocyte membranes have their postsynaptic network removed by alkali extraction, the lateral diffusion is much enhanced.

Evidence is thus accumulating from various different sources to support the view that postsynaptic densities represent anchorages for neurotransmitter receptors. These anchorages are associated with elements of the cytoskeleton, especially actin and fodrin (Chapter 7). It is likely that the rather amorphous densifications seen by the electron microscopist (Chapter 1) will turn out to have an intricate multimolecular architecture.

17.3 ELECTROPHYSIOLOGY OF THE POSTSYNAPTIC MEMBRANE

A couple of decades ago, when things were simpler, neurophysiologists divided synapses into two functional types: excitatory and inhibitory. We have seen that, in the twenty-first century, the plot has thickened. Nevertheless it is sensible to start with the simple picture and add complications later.

17.3.1 The Excitatory Synapse

The type example of the excitatory synapse is the vertebrate cholinergic neuromuscular junction. We have already looked at the nature of the nicotinic acetylcholine receptor in depth (Chapter 10) and we considered its pharmacology in Chapter 16. Here we can note, first of all, that when ACh is released from the presynaptic terminal and lands on the nAChR, a flux of both Na^+ and K^+ ions traverses the postsynaptic membrane (Figure 17.4). The two ions pass through the nAChR channel with approximately equal ease.

Figure 17.4 shows that both Na^+ and K^+ flow down their electrochemical gradients. The force behind both ions is simply the difference between the membrane potential (V_m) and their respective Nernst potentials (V_{Na} and V_K). If we suppose that the permeability constant of K^+ increases tenfold and that the permeability constant of Na^+ increases until it approaches 0.75 of that of K^+ (i.e. much more than tenfold) and insert these values into Goldman's equation (equation 12.5) we obtain $V_m \approx -7\,mV$. And this is very much what is observed. In the case of the neuromuscular junction this depolarisation is called an **end plate potential (EPP)**. In the central nervous synapses it is known as a **postsynaptic potential (PSP)**. Moreover as it

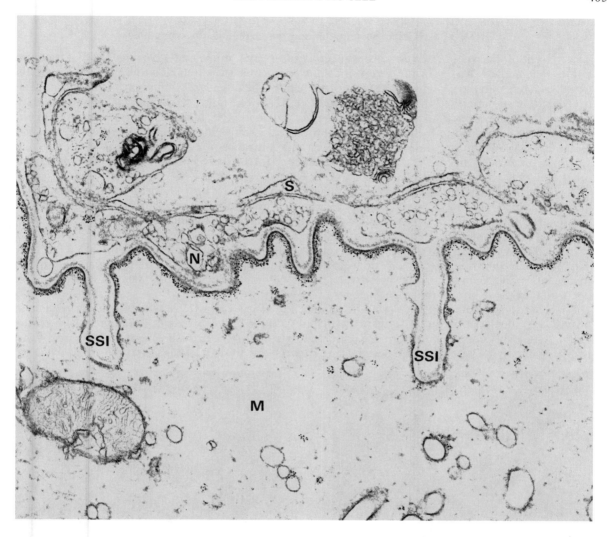

Figure 17.3 *Torpedo* electrocyte membrane labelled with anti-43 kDa protein antibody. The black dots are 5 nm gold particles attached to the anti-43 kDa protein antibody. It can be seen that these particles are concentrated on the cytoplasmic side of the crests of the subsynaptic folds and do not extend deep into the invaginations. Further explanation in text. N=nerve terminal; S=Schwann cytoplasm; M=sarcoplasm of muscle; SSI=subsynaptic invagination. Magnification: ×938 000. Reproduced from Sealock, Wray and Froehner (1984), *Journal of Cell Biology*, **98**, 2239–2244 by copyright permission of the Rockefeller University Press.

leads, as we shall see, to excitation of the post-synaptic cell it is called an **excitatory postsynaptic potential (EPSP)**.

If we now recall our discussion of electrotonic potentials and cable conduction in Chapter 12 we can see that we have the classical situation for the spread of local circuits. In the neuromuscular junction these spread from the EPP to other parts of the sarcolemma where, if they find voltage-dependent Na$^+$ gates, they will initiate an action potential. Similarly, in central synapses local circuits spreading from the EPSP will open any sodium channels embedded in the neuronal membrane. Now we have seen that, in general, sodium

BOX 17.1   Cajal, Sherrington and the beginnings of synaptology

The concept of the synapse was established over a century ago, in the mid-1890s. Before that time the notion that the grey matter of the cortex and spinal cord was a vast reticulum of anastomosing cells was dominant. Camillo Golgi was the most prominent of the reticularists, but many other prominent neuroanatomists supported the position. Meynert (1872), for instance, wrote that 'anastomoses occur between (all) the processes of the cells of the cortex...(we are) led to admit the existence of a nerve fibre network embedded in the grey matrix and forming its third diffused morphological constituent'. Golgi, in his 1906 Nobel oration, says 'I cannot abandon the idea of a unitarian action of the nervous system without being uneasy that by doing so I shall be forced to accept old beliefs' (presumably the organology of a long discredited phrenology).

It is one of the ironies of the history of neuroscience that the histological technique that Golgi invented, which nowadays bears his name, was instrumental in destroying the very theory he passionately supported. Forel writes (in his autobiography) of how, meditating on Golgi images of the CNS, and on the results of degeneration experiments, whilst on holiday at Fisibach, 'it was as though scales had fallen from my eyes...the more I reflected, the clearer it seemed that we had been fathoms sunk in preconceived opinion...why do we always look for anastomoses?...all the data suggests the theory of simple contact...' Beware preconceived opinion! Beware seeing the data through the distorting lenses of a too firmly held theory.

It was Santiago Ramón y Cajal who in the late 1880s and early 1890s finally won the argument against the reticularists. Ironically, again, he shared the 1906 Nobel prize with his antagonist Camillo Golgi. In his 1894 Croonian Lecture to London's Royal Society he provides chapter and verse for the neuronist theory: 'The connections established between the fibres and the nerve cells take place by

Figure A   (i) Santiago Ramón y Cajal. (ii) Charles Scott Sherrington. From Roger Grant (1966), *Charles Scott Sherrington: An Appraisal*, London: Nelson; with permission.

means of contact, that is with the help of genuine articulations... the cells are polarised, that is, the nerve current always enters by way of the protoplasmic apparatus of the cellular body (our dendrite), and that it leaves by the axis cylinder which transmits it to a new protoplasmic apparatus.'

It remained only for the term 'synapse' itself to be coined. This fell to another of the founding fathers of our subject, the English neurophysiologist, Charles Sherrington. The story runs as follows: by the 1890s Foster's *Textbook of Physiology*, the premier text in English, had gone through six editions. Foster planned a seventh edition of which part 3 would be devoted to neurophysiology, and he asked Sherrington to undertake it. Sherrington agreed but found a need for a term for the junction between neurons. In 1943, in a personal communication to a younger physiologist, John Fulton, Sherrington writes: 'You enquire about the introduction of the term "synapse"; it happened thus. M. Foster had asked me to get on with the Nervous System part (part iii) of a new edition of "Text of Physiol." for him. I had begun it, and had not got far with it before I felt the need of some name to call the junction between nerve cell and nerve cell (because the place of junction now entered physiology as carrying functional importance). I wrote to him of my difficulty, and my wish to introduce a specific name. I suggested using "syndesm". He consulted his Trinity friend Verral, the Euripidean scholar, about it, and Verral suggested "synapse" (from Gk. [clasp]) and as that yields a better adjectival form, it was adopted for the book.'

In Sherrington's mind the synapse was a functional concept. It accounted for the phenomenon then known as 'central reflex time', now known as 'synaptic delay'. He says the synapse is 'a surface of separation at the nexus between neuron and neuron' and he goes on to suggest some of the physico-chemical characteristics such a surface of separation might show. We need not follow him into these interesting but now outdated speculations. We can, however, observe that although a victory for the neuronists had been won in the 1890s, pockets of resistance remained and the reticularists were not finally routed until the advent of the electron microscope in the 1950s.

The history of the synapse has many interesting facets and many useful lessons for us. Most importantly it emphasises how important it is to keep an open mind, to remain, however difficult it may be, in a condition of 'suspended disbelief'. It also shows how much advance in neurobiology depends on the development of technique. This, of course, hardly needs emphasis to the readers of this book. Lastly it emphasises that science is generated by human beings, and the story of their lives, their struggles, victories and defeats, helps to humanise the material which is often too aridly summarised in textbooks.

channels are not found in dendritic membrane but only in axolemma. We saw in Chapter 14, moreover, that voltage-dependent sodium gates are particularly densely distributed on the initial segment of the axon (see also Appendix 3). This, then, is where the local circuits spreading out from the EPSP make their influence felt (Figure 17.5).

Referring back to Chapter 12 again, we can see that the position at which the EPSP occurs on the neuron's soma (i.e. perikaryon+dendritic tree) is very important. EPSPs occurring far out on the dendritic tree may only exert a 'biasing' effect on the initial segment, never bringing it below the threshold necessary to initiate an action potential. This biasing effect may, however, be quite long-lasting. Although the EPSP may endure for only a few milliseconds, the bias on the initial segment may persist for tens of milliseconds. In contrast EPSPs occurring close to or actually on the perikaryon will exert a rapid and decisive effect. Local circuits will open sodium gates on the initial segment and set off the action potential. Finally it is worth noting that the area of the synaptic contact is also very significant. Clearly a large synapse will depolarise a comparatively large patch of post-synaptic membrane, resulting in comparatively large current fluxes along the local circuits.

We noted above that the EPSP seldom endures for more than a few milliseconds. This is because, as we saw in Chapter 16, there are always

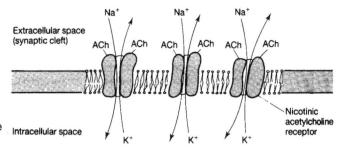

Figure 17.4 Fluxes of Na$^+$ and K$^+$ ions across the postsynaptic membrane of a cholinergic synapse.

mechanisms in the synaptic cleft that inactivate the transmitter. In the case of the cholinergic synapse, we saw that this mechanism takes the form of an enzyme, **acetylcholinesterase (AChE)**, which splits ACh into two parts – acetate and choline. The choline is pumped back into the presynaptic ending and used again for the synthesis of fresh ACh. The phenomenon is quite general. Excitatory transmitters are not allowed to hang around in the synaptic

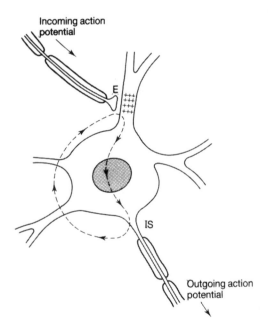

Figure 17.5 Spread of local circuits from an EPSP. The activation of an excitatory synapse (E) leads to the depolarisation of the postsynaptic membrane. Local circuits spread from this region (arrows) and affect the initial segment (IS) of the axon. If the circuits are sufficiently strong then the IS will be depolarised to threshold and an action potential will be initiated which will be propagated without decrement down the axon.

cleft. They are either inactivated by a specific enzyme (or enzymes) or pumped back into the presynaptic terminal.

Finally, it is important to recognise that EPSPs (unlike action potentials) are not 'all-or-nothing' events. The magnitude of an EPSP (i.e. the magnitude of the depolarisation) depends on the quantity of excitatory transmitter released into the synaptic cleft for this determines the number of nAChRs opened. This fact makes possible the phenomenon of **temporal summation**. If a rapid 'tattoo' of impulses arrives at the presynaptic terminal, fresh ACh is released into the cleft before ACh from the earlier impulses has been inactivated by AChE. More nAChRs are opened, more Na$^+$ and K$^+$ flux down their electrochemical gradients, the greater is the membrane depolarisation.

In addition to temporal summation excitatory synapses also show **spatial summation**. If two or more EPSPs occur at the same time on different parts of a neuron's soma the local circuits will sum together. It could thus well be that while one EPSP on its own would not be sufficient to initiate an action potential, two acting together would be adequate to depolarise the initial segment membrane beyond the required threshold.

17.3.2 The Inhibitory Synapse

There are as many, if not more, inhibitory synapses in the CNS as there are excitatory synapses. We noted in Chapter 16 that the best-known inhibitory transmitters are γ-amino-butyric acid (GABA) and glycine. These transmitters are sequestered in small translucent vesicles in the presynaptic ending. There is some evidence that the vesicles are elliptical rather than spherical.

When an action potential arrives at an inhibitory terminal the usual molecular events occur which result in the transmitter being voided into the

Table 17.1 Hydrated radii of some physiological ions

Ion	Non-hydrated radius (Å)	Hydrated radius (Å)	Molecules of water 'carried'
Cl^-	1.81	3.6	4.0
K^+	1.33	3.8	5.4
Na^+	0.95	5.6	8.0
Ca^{2+}	0.99	9.6	16.6
Mg^{2+}	0.65	10.8	22.2

The concept of a 'hydrated radius' is, as Hille puts it, somewhat 'fuzzy'. It should not be thought that a given ion is associated with a specific set of water molecules. Because no covalent bonds are involved the time for which any given water molecule is associated with a specific ion can be measured in nanoseconds. The 'hydrated radius' is best thought of as giving an indication of the strength of attraction between an ion and the surrounding aqueous solvent. Hille, B. (1992), *Ionic Channels of Excitable Membranes* (2nd edn), Sunderland, MA: Sinauer.

synaptic cleft. Once it reaches the synaptic membrane, however, a hyperpolarisation rather than a depolarisation ensues. This is known as an **inhibitory postsynaptic potential (IPSP)**. Let us examine the ionic mechanisms underlying this potential.

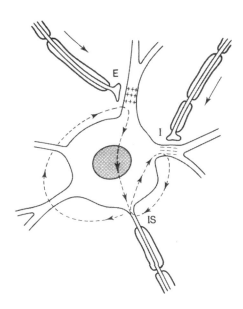

Figure 17.7 Algebraic summation at the initial segment of local circuits engendered by excitatory and inhibitory synapses. E=excitatory synapse; I=inhibitory synapse; IS=initial segment.

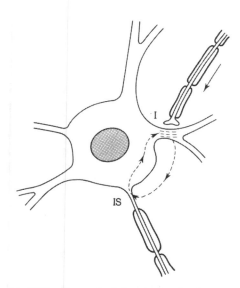

Figure 17.6 Spread of local circuits from an IPSP. The inhibitory synapse (I) hyperpolarises the dendritic membrane immediately beneath. Local circuits run in the opposite sense to those in Figure 17.5 and the initial segment (IS) is consequently hyperpolarised (inhibited).

We saw in Chapter 10 that inhibitory transmitters open very narrow channels in the membrane – channels that will only allow ions with a hydrated radius less than or equal to that of the chloride ion to flux through. Table 17.1 shows that all the naturally occurring ions in and around neurons in fact have greater hydrated radii than Cl^-. It can be shown that both glycine-activated and GABA-activated channels have the following relative conductance sequence: $Cl^- > Br^- > I^- > F^-$. Not only do the channels select ions with a small hydrated radius but they also select anions.

We discussed the molecular nature of the glycine receptor (GlyR) in Section 10. 3. We saw that it was an oligomeric transmembrane complex of which a 48 kDa transmembrane protein is believed to form the channel for chloride ions. An associated 98 kDa submembrane protein anchors the receptor to the cytoskeletal meshwork of the postsynaptic density. We saw that the GlyR could be located at central postsynaptic sites by both immunofluorescence light microscopy and immuno-gold electron microscopy.

Next let us look at the consequence for the postsynaptic membrane when this channel is activated. We can use the same procedure as in the previous section. We can test the effect of opening a Cl^- channel by inserting a greatly increased value for chloride's permeability constant in the Goldman equation. Let us suppose that instead of P_{Cl} being about 1×10^{-8} cm s^{-1} it increases a hundred-fold to 1×10^{-6} cm s^{-1}. Keeping the values for all the other permeability constants unchanged, the Goldman equation predicts $V_m = -69$ mV. This is a **hyperpolarisation** of some 14 mV compared to the resting potential of -55 mV worked out in Section 12.2.

The membrane patch carrying the IPSP clearly acts as a source of local circuits. The direction of the currents is, however, in the opposite direction to that associated with the EPSP. Instead of tending to depolarise the initial segment, these circuits tend to hyperpolarise it (Figure 17.6). They tend, in other words, to make it more difficult to bring the initial segment membrane to the threshold potential required to spark an action potential.

Other aspects of the IPSP are analogous to the EPSP. There is **temporal** and **spatial** summation. The position of the synapse on the postsynaptic neuron's soma and the area of synaptic contact has the same significance for inhibitory as it had for excitatory synapses.

17.3.3 Interaction of EPSPs and IPSPs

We have already noted that central neurons customarily have thousands (sometimes hundreds of thousands) of synaptic contacts dotted over their perikarya and dendritic arborisations. In the general case we can assume that half of these will be excitatory and half inhibitory. It is likely, also, that some members of both types will be active at the same time. It follows that local circuits will spread from IPSPs and EPSPs through the cell ultimately to the axon's initial segment. Here they will summate in an algebraic fashion (Figure 17.7). Whether the axon fires off an action potential depends on which type of postsynaptic potential predominates.

Clearly the situation in the neuron is richly complex. Unfortunately for the neuroscientist this is only the beginning. We shall see in the next sections that further intricacies are built upon what may already seem intricate enough.

17.4 ION CHANNELS IN THE POSTSYNAPTIC MEMBRANE

We have noted in several earlier chapters the importance of the patch-clamp technique for studying the operation of single membrane channels. We also noted that much evidence has accumulated which suggests that channels can exist in more than just 'open' and 'shut' states – that, in other words, there are a number of intermediate conditions. Finally we saw that to account for the millivolt shifts in membrane potential which constitute action potentials, EPSPs and IPSPs, populations of channels must be at work. The individuals in these populations open, generally speaking, in a random fashion.

In earlier chapters of this book we have used a straightforward formulation of Ohm's law ($I = gV$) to describe the movement of ions through channels. Let us develop this formulation a little to take into account the complexity of the situation obtaining in biological membranes. McBurney (1983) has provided a useful analysis of a population of channels. Their operation can be described by equation 17.1:

$$nT + nR(\gamma_c) \underset{\beta}{\overset{\alpha}{\rightleftharpoons}} nTR(\gamma_o) \qquad (17.1)$$

where T = transmitter molecules, R = receptor sites, γ_c = closed channel conductance, γ_o = open channel conductance, and n is the number of transmitter molecules that must bind to the receptor sites to open the channel with a rate constant β.

When the channel is opened the conductance across the membrane increases by $(\gamma_o - \gamma_c)$. Channel closing occurs with a rate constant of α when the transmitter detaches or is otherwise inactivated.

The magnitude of the current carried through the channel will, as usual, be the product of the conductivity of the channel, γ (usually measured in pS), and the driving force across the membrane. The latter is the electrochemical potential gradient for the ion (I) across the membrane. This is given by the difference between the membrane potential (V_m) and the null (or reversal) potential of the ion, i.e. the Nernst potential (V_I) for the ion. The driving force may thus be expressed as ($V_m - V_I$).

The time for which each channel stays open depends on α, the rate constant for transmitter

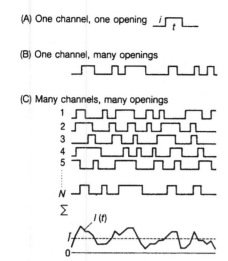

Figure 17.8 Summated effect of a population of ion channels in a postsynaptic membrane. (A) The opening and 'open time' (t) of a single channel depends on numerous factors in its biochemical environment (see text). When it opens a brief pulse of current (i) is carried through. (B) In the presence of a constant concentration of agonist the channel will open many times. This opening will be random depending, as it does, on random collisions of the agonist with the receptor. (C) A postsynaptic membrane will usually contain many activatable ion channels. The summated effect of the random openings of these numerous channels gives the fluctuating current ($I(t)$) shown at the bottom of the figure. This current will fluctuate about a mean level I. Further explanation in text. After McBurney (1983), *Trends in Neurosciences*, **6**, 298–302; with permission from Elsevier Science, 1983.

these gates and the consequent flow of current depolarises the membrane and hence increases the probability that neighbouring channels will open.

The diagram in Figure 17.8 still assumes, however, that a membrane gate exists in just two states – open and closed – although the duration of the open state varies about a mean. This simple picture has now been complicated by the recognition that single gates can exist in several different conductance states (Figure 17.9). It has also been found that a gate may oscillate or 'flicker' between open and shut states. Finally there is evidence that in some conditions a single channel may open in 'bursts' and 'clusters' of bursts. These findings imply that the comparatively simple scheme of Figure 17.8 does not represent the full complexity of ion fluxes in the postsynaptic membrane.

unbinding, in equation 17.1. This will depend on numerous factors in the immediate environment of the channel protein. The greater the value of α the smaller will be the value of τ, the average channel opening constant. Figure 17.8 shows in (A) the opening of one channel over a time of some tens of milliseconds, in (B) the opening of a population of such channels and in (C) the summated effect of such a population on a membrane such as the postsynaptic membrane.

Note the difference between this analysis and the response of populations of voltage-gated channels in excitable membranes (Section 14.2). The difference is, of course, due to the fact that voltage gates are, in a sense, **co-operative**. The opening of one of

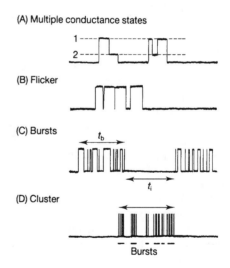

Figure 17.9 Conductance characteristics of single ACh-activated channels. The sketches show some of the features exhibited by single ACh-activated channels. Time base for records (A) and (B) about 50 ms, for records (C) and (D) about 2 and 20 s, respectively. (A) A single channel can exist in more than one conductance state. (B) A channel may flicker rapidly between open and closed states. (C) A channel may open in 'bursts' lasting for up to a second (t_b) followed by a quiescent interval (t_i) before the next burst. (D) The bursts may be grouped in 'clusters'. From McBurney (1983), *Trends in Neurosciences*, **6**, 298–302; with permission from Elsevier Science, 1983.

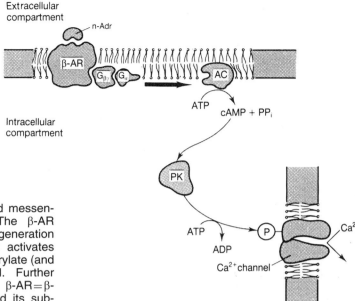

Figure 17.10 Adrenergic-activated second messenger system on presynaptic terminals. The β-AR collision-coupling mechanism results in the generation of cAMP in the presynaptic ending; this activates cAMP-dependent protein kinase to phosphorylate (and thus downmodulate) a calcium channel. Further explanation in text. n-Adr=noradrenaline; β-AR=β-adrenergic receptor; G(βγα)=G-protein and its subunits; AC=adenylyl cyclase; PK=protein kinase.

Returning, however, to our analysis of equation 17.1 and Figure 17.8 we can see that different transmitters, different agonists, different antagonists, may exert their effects in different ways. It may be that an effect is produced by increasing the number of channels open over any given time, or it may be that the synaptically active agent affects τ, the average open time or, finally, it may be that γ, the conductivity of the channel, is affected. Evidently there is great scope for transmitters to elicit subtly different postsynaptic events.

Perhaps this is one of the reasons for the nervous system developing such a bewildering variety of transmitters. It can, for instance, be shown that whereas GABA opens channels in cultured spinal neurons with $\tau=30.4$ ms, taurine (which as we saw in Chapter 16 is a 'doubtful' transmitter) gives $\tau=2.3$ ms; glutamate on crab muscle gives $\tau=1.4$ ms, whilst aspartate on the same preparation causes channel opening with $\tau=0.8$ ms.

17.5 SECOND MESSENGER CONTROL OF ION CHANNELS

So far in this chapter we have been considering very rapid events (usually less than 5 ms) in the postsynaptic membrane. We have seen how they may 'switch on' or 'switch off' the postsynaptic neuron. The next thing we have to do is to remind ourselves that many transmitters do not directly open ion gates at all but work through collision-coupled second messenger systems. The outcome of these second messenger systems may be to affect the conductivity of ion gates elsewhere in the postsynaptic membrane or, indeed, in other parts of the postsynaptic cell's neurilemma. Thus we add another layer of complexity on top of an already intricate enough situation.

We have already looked at some of these modulated channels in Chapter 11. One of the best-known instances is the Ca^{2+} channel (I_{Ca} channel). It can be shown that this is modulated by a second messenger system initiated by noradrenaline. The sequence of events is shown in Figure 17.10. The increased synthesis of cAMP following activation of the β-adrenergic receptor (β-AR) (see Section 8.8) leads to activation of a cAMP-dependent protein kinase, protein kinase A (PKA). The latter phosphorylates the I_{Ca} channel. Phosphorylation (as we have seen in previous chapters) affects a channel's conductivity. In this case there appears to be 'downmodulation'. The flux of Ca^{2+} through the channel is decreased. Although these channels are probably rather few and far between, the effect is nonetheless important. For instance, it will be remembered from Chapter 15 that Ca^{2+} plays a crucial role in the exocytosis of transmitters and other material form presynaptic terminals. It follows that if the postsynaptic membrane happens to be that of a synaptic bouton (see Figure 15.24) then the release of noradrenaline on to β-AR receptors can have very significant consequences.

Another interesting second messenger system is that connected to the muscarinic cholinergic receptor (mAChR). We discussed the molecular biology of this receptor in Section 8.9. We saw that it was coupled to a G-protein signalling system. In many cases this system acts on adenylyl cyclase to generate cAMP. In others, it operates through phospholipase C to produce InsP$_3$ and DAG. This latter mechanism is of considerable interest as there is evidence that InsP$_3$ affects a particular set of K^+ channels in the neuronal membrane. Because they are affected (switched off) by activation of the muscarinic receptor they have been called **M channels**.

The M channels normally allow K^+ to flux out of the cell. They are strongly voltage-dependent. As the membrane depolarises during an EPSP they progressively open. This increases the K^+ conductance across the membrane. If this increased conductance is inserted into the Goldman equation it can be seen that it has the effect of bringing the membrane back towards its resting potential. If, however, these channels are inactivated the EPSP lasts considerably longer. This is exactly what happens when the mAChR is activated (see Figure 17.11). Hence by this indirect route the muscarinic receptor can exert a very noticeable effect on spike initiation at the initial segment.

Let us look at one final instance of this receptor-mediated modulation of ion conductivity. This is the serotonin-dependent potassium channel (K_S). We saw in Chapter 11 that this channel had been studied in the convenient preparation provided by the sea-hare *Aplysia*. It is also to be found in the less readily accessible vertebrate nervous system. The K_S channel in *Aplysia* enables K^+ ions to flux through the membrane to the outside. It will be recalled that the resting potential (V_m) is very sensitive to the concentrations of K^+ inside and outside. If this flow is blocked the external concentration will slowly fall and the internal

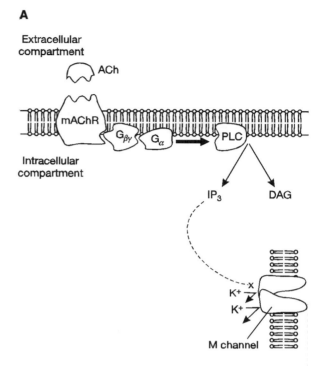

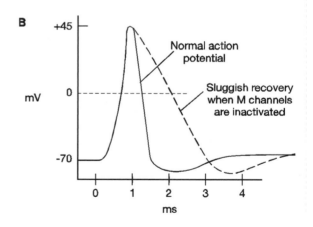

Figure 17.11 The influence of mAChR activation on the impulse initiation. (A) The collision-coupling mechanism generates IP$_3$, which inactivates the M channel. (B) Because the M channel has been blocked the K^+-induced recovery phase of the action potential is retarded. The membrane is thus refractory for longer than normal. When this occurs on an initial segment or dendritic zone, the initiation of subsequent impulses is delayed. Further explanation in text.

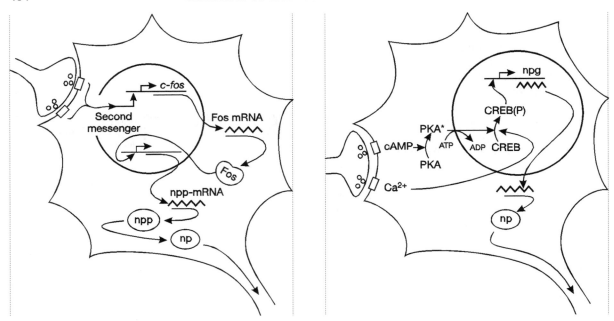

Figure 17.12 Fos as a 'third messenger' in neuropeptide synthesis. The second messenger opens the transcription site on the *c-fos* gene. Fos mRNA is synthesised and translated in the cytoplasm of the postsynaptic cell into Fos. This acts as a transcription factor in the synthesis of neuropeptide precursor mRNA (npp-mRNA). The neuropeptide precursor (npp) is synthesised (e.g. preprodynorphin or preproenkephalin) and the neuropeptide derived by post-translational processing (see Chapter 4). The neuropeptide is then transported along the axon to its synaptic release site. Adapted from Dubner and Ruda (1992), *Trends in Neurosciences*, **15**, 96–103; with permission from Elsevier Science, © 1992.

Figure 17.13 Control of neuropeptide gene expression by Ca^{2+} and cAMP. cAMP activates PKA (PKA PKA*). PKA* phosphorylates CREB protein (CREB CREB(P)). This step is Ca^{2+}-dependent. The concentration of Ca^{2+} ions is often critically dependent on synaptic activity (as shown in the figure). Phosphorylation releases the transcription activity of the CREB protein and neuropeptide mRNA is transcribed. This is translated into the encoded neuropeptide (np) in the cytosol. The neuropeptide is transported along the axon to its destined release site at the synaptic terminal. CREB=cAMP response element binding protein; PKA=protein kinase A; np=neuropeptide. Further explanation in text.

slowly rise. If this occurs the potential across the membrane will slowly fall. In other words a long, slowly increasing EPSP develops.

It can be shown that activation of serotoninergic synapses has just this result. Patch-clamp recording, moreover, shows that the effect is mediated through a cAMP second messenger mechanism which results in the **number** of K_S channels in the open state being reduced. Note that there is no reduction in conductivity of those that are open. This is the general case.

Serotoninergic synapses are also, as we shall see in Chapter 20, found on some presynaptic terminals in *Aplysia*. Here, as mentioned in Chapter 11,

the outflow of K^+ ions through the K_S channels helps to establish the AHP phase of an action potential. If these channels are reduced by activation of the serotonin receptor the duration of the AHP will be increased. We shall see that the increased length of time which allows Ca^{2+} channels to remain open may have very important consequences. It is possible that similar mechanisms are at work in mammalian nervous systems.

In mammalian systems serotonin can be shown to downmodulate yet another channel: the calcium-dependent potassium channel (K_{Ca}). The precise mechanism at work here remains to be discovered. Reference back to Chapter 11 will, however, show that modulation of this channel could have significant neurophysiological consequences. The K_{Ca} channel is, we saw, deeply implicated in adaptation. Reduction in the conductivity of these channels could thus prevent adaptation occurring and thus ensure that a neuron kept on firing at a rapid rate.

We have only discussed the tip of an iceberg. Well over two dozen modulators of ion channels have been discovered and their properties analysed. The old picture of discrete, punctate, synaptic events – either IPSPs or EPSPs – is beginning to recede into history. Nevertheless, it must be borne in mind that however complicated the neurochemical interactions become, the end result, the action potential, remains an all-or-nothing event. The outcome of all the interactions, chemical, spatial, temporal, is 'felt' at the initial segment. **If threshold is reached or exceeded the axon 'fires'; if it is not reached it remains quiescent.**

17.6 SECOND MESSENGER CONTROL OF GENE EXPRESSION

We have so far been considering only the effects that second messenger systems have on the magnitude and duration of EPSPs and IPSPs and on the initiation and adaptation of action potentials. There are, however, many other possibilities. We have been emphasising in this book that the subject of molecular neurobiology brings together the strands of biophysics, biochemistry and molecular biology into a new, very powerful synthesis. This synthesis is nowhere more apparent than in the action of many second messenger systems at deeper levels in the cell. In this section we shall look at the control of gene expression by synaptically derived second messengers. The products of genes switched on in this way are often referred to as '**third messengers**'.

We have already touched on the synaptic control of gene expression in Chapter 3 when discussing the way in which **immediate early genes (IEGs)**, especially c-fos and c-jun, can be very rapidly switched on (induced) by cytosolic second messengers (see Figure 3.18B). In particular, IEGs are believed to encode transcription factors which control secondary genes coding for neuropeptide modulators. That such mechanisms were at work in the CNS first became apparent when c-fos

transcripts were detected in brains subjected to metrazole-induced convulsions. IEG transcripts could not be detected in brains protected from such global discharges by anticonvulsive agents. Subsequently Fos (the induced protein) has been detected in brains subjected to a wide variety of insults: cerebral ischaemia, trauma, elevated concentrations of excitatory amino acids, etc. Later work has refined this somewhat 'sledgehammer' approach. Active neurons can be distinguished from their inactive neighbours by 2-deoxyglucose and hybridisation histochemistry. In some (but by no means all) cases it has been found that c-fos transcription can be correlated with these active cells.

The presence of Fos (and thus of c-fos activation) can be detected in cerebellar cells activated by stimulating motor/sensory pathways in the cerebral cortex. There is also evidence to show that Fos is present in neurons belonging to the pain pathways. Opioid-secreting neurons are stimulated by excitatory amino acids and substance P. Second messengers generated by this stimulation switch on c-fos genes in the enkephalinergic cells. These, in turn, lead to the synthesis of preprodynorphin (PPD) and preproenkephalin (PPE) which, ultimately, after further processing, are released as the opioids dynorphin and enkephalin. This latter sequence of events can also be initiated by dopamine acting through D1 receptors on enkephalinergic neurons. Fos and similar transcription factors can thus be regarded as 'third messengers' which unlock the transcription of the neuropeptide genes (Figure 17.12).

In other cases the second messenger induction of neuropeptides bypasses IEGs altogether. In other words, the second messenger acts directly on the promoter region of the gene concerned and induces the transcription of neuropeptide mRNA. There is, for example, considerable evidence that the cytosolic concentration of Ca^{2+} affects the transcription of these genes. Thus it can be shown that Ca^{2+} influxes due to the opening of NMDA-gated channels can initiate the transcription; it can also be shown that depolarisation due to the opening of nAChR gates also leads to transcription. This is because depolarisation tends to open Ca^{2+} channels in the postsynaptic membrane.

Ca^{2+} ions usually do not act alone. In general they 'assist' other second messengers, especially cAMP. We saw above (Section 17.5) that cAMP exerts its effects (on membrane channels) by activating **protein kinase A (PKA)**. It comes as no surprise, therefore, to find that cAMP acts on the genome by a similar mechanism. It is found that activated PKA catalyses the phosphorylation of a protein which binds to a short consensus site on nuclear DNA: 5TGACGTCA3 (see also Section 3.3.1). This sequence has been termed the **cAMP response site (CRE site)**. The protein which binds to this site is called the CRE binding protein (**CREB** protein) and it is this whose phosphorylation is catalysed by PKA. It is the latter reaction that is Ca^{2+} dependent. This rather complicated cascade of biochemical reactions is schematised in Figure 17.13.

The CREB protein resembles the Fos and Jun transcription factors in possessing a 'leucine zipper' structure (see Figure 3.7C). In solution it exists as a monomer but when activated it attaches to the CRE site and 'zips up' with a partner to form a dimer. Investigations, especially in cultured cells, have shown that genes encoding **somatostatin**, **VIP** and **tyrosine hydroxylase** are expressed through CREB-dependent transcription.

17.7 THE PINEALOCYTE

Finally, before turning in the next two chapters to the molecular biology of brain development, let us consider the second and third messenger systems at work in pineal cells (**pinealocytes**). Analysis of this system brings home, once again, the complexity of molecular neurophysiology. Apart from its inherent interest, the pinealocyte also forms relevant bridging into the next chapters as the mammalian pineal is deeply implicated in controlling reproductive cycles and sexual maturation.

Our understanding of the pineal has come a long way since René Descartes apostrophised it as 'the seat of the soul'. In the lower vertebrates (the fish, amphibians and reptiles) it indeed often functions as a 'third eye'. In the birds and mammals it has, however, lost all direct contact with light. Instead it elaborates a hormone, **melatonin**, which is important in regulating reproductive cycles. Indeed there is a biological argument for supporting Descartes in his conviction that the pineal played a central role in the human constitution. For it is possible that a mutation in the gene controlling the pineal's

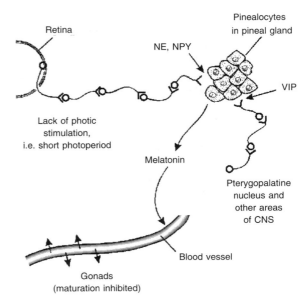

Figure 17.14 Neural control of pinealocytes. Multisynaptic pathways lead from the retina and the pterygopalatine nucleus, etc. to the pineal. The final neurons in the chains release noradrenaline (NE), neuropeptide Y (NPY) and vasoactive intestinal peptide (VIP) on to the pinealocyte membranes. This leads to the release of melatonin into the vascular system, which in turn inhibits the maturation of the gonads.

development is responsible for the delayed onset of reproductive maturity in *Homo sapiens*. The melatonin which the gland secretes inhibits the maturation of the gonads. In humans, unlike other primates, the pineal remains continuously active for the first 14 or 15 years of life, after which time it deteriorates, indeed calcifies. The prolonged childhood which this ensures is undoubtedly one of the most significant developments in the evolutionary path to humankind.

Returning, however, from these broad evolutionary perspectives to the neurochemical detail, we find that pinealocytes have input from both the retina and the pterygopalatine ganglion of the facial nerve (Figure 17.14). Fibres from the retina take a multisynaptic pathway to the pineal and release **noradrenaline (norepinephrine; NE)** and **neuropeptide Y (NPY)**. Fibres from the pterygopalatine ganglion take a similarly multisynaptic pathway to release **vasoactive intestinal peptide (VIP)** on to underlying pinealocyte membrane.

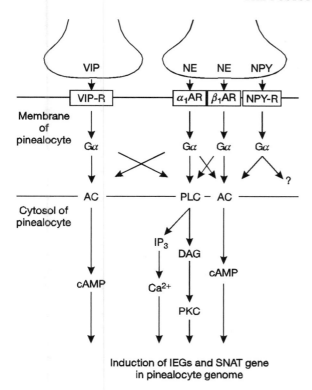

Induction of IEGs and SNAT gene
in pinealocyte genome

Figure 17.15 G-protein network and second messenger cascades in pinealocyte. An intricate G-protein collision-coupling network is shown in the pinealocyte membrane (cf. Figure 8.13). A second messenger cascade is shown in the cytosol of the pinealocyte. **Transmitter substances**: VIP (vasoactive intestinal peptide); NE (noradrenaline); NPY (neuropeptide Y). **Receptors** : VIP-R; α_1-AR; β_1-AR; NPY-R. **Effectors**: AC (adenylyl cyclase); PLC (phospholipase C). **Second messengers**: cAMP; IP$_3$; DAG; Ca^{2+}.

The pathway from the retina is by far the more important of the two. It is inhibited by light. Activation of the pathway only occurs in the dark or low light intensities and short photoperiods. The noradrenaline released operates both α- and β-adrenergic receptors in the pinealocyte membrane. Activation of the α-receptor appears to potentiate the action of the β-receptor. Release of NPY has a similarly potentiating effect on β activity. Finally, it has been shown that release of VIP also acts to increase cAMP levels. These complex synergies at the pinealocyte membrane (Figure 17.15) provide a 'real-life' instance of the G-protein signalling networks discussed in Section 8.7.

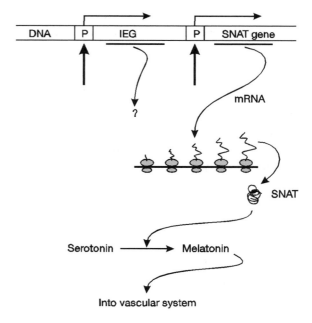

Figure 17.16 Genomic induction and tertiary messengers. Elements of the second messenger cascade of Figure 17.15 induce the transcription of IEGs and the SNAT gene. The transcript from the latter is translated by the ribosomal machinery into SNAT (serotonin-*N*-acetyltransferase). The newly synthesised SNAT catalyses the production of melatonin from serotonin. Melatonin is released into the vascular system and circulates to the gonads. Note that this mechanism is not important in *Homo sapiens* as the pineal is largely inactive in adults of this species. Further explanation in text.

The biochemistry within the pinealocyte is also interestingly complex. It can be shown that both *c-fos* and *c-jun* IEGs are affected. Activation of the α-AR leads via DAG and then PKC activation to transcription of *c-fos*. Stimulation of the β-AR leads via cAMP (but without the involvement of PKA) to transcription of *c-jun*. Homo- and heterodimers of Fos and Jun are thus available to affect the AP-1 site of secondary (presumably structural) genes.

The pinealocyte IEGs thus form a valuable experimental system. Ever since the significance of IEGs was recognised some ten or fifteen years ago it has been puzzling that the great variety of receptors which cell membranes present, the great variety of neurotransmitters and the great variety of second messengers, should be focused down to the comparative poverty of just two or three IEG transcripts. The pinealocyte preparation begins to show that subtle controls of IEGs may be affected by different combinations and quantities of second messengers. The consequent variation in transcription products (third messengers) may well lead to differential expression of secondary genes. Indeed we shall see in the next chapter (Section 18.9) an example of just this: how quantitatively different combinations and permutations of tertiary messengers differentially affect the induction or suppression of dopa-decarboxylase gene expression. We are just lifting a corner of the veil concealing the intricacies of cellular control systems.

It is disappointing, however, to note that (at the time of writing) the relation between the activation of pinealocyte IEGs and the synthesis of melatonin is obscure. It is likely that the Fos and Jun transcription modulators play some subtle role in the pinealocyte's metabolism which has yet to be understood. It has, however, been shown that the PKC activated after α_1-AR stimulation phosphorylates a nuclear histone. Histones are closely associated with DNA in eukaryotic chromosomes. Phosphorylation of the histone allows the **serotonin-*N*-acetyl-transferase (SNAT)** gene to be transcribed. Fresh mRNA from this gene can soon be detected in the cytosol. Indeed it can be detected appreciably before either Fos or Jun can be detected. The SNAT translated from this mRNA is the rate-limiting enzyme in the synthesis of melatonin from serotonin (Figure 17.16). cAMP may also be involved in the translation of SNAT from its mRNA. The melatonin is secreted into the vascular system and on reaching the gonads exerts the effects mentioned above. The upshot of this complex sequence of events is thus to ensure that in many birds and infra-human mammals the gonads only become active when photoperiod increases.

17.8 CONCLUSION AND FORWARD LOOK

This chapter has emphasised the developing recognition that the postsynaptic membrane is of equal importance to the presynaptic terminal in the functioning of synapses. Indeed, some argue that the presynaptic membrane, synaptic cleft and postsynaptic membrane should be treated as a single immensely complicated unit. This region is, moreover, not only of importance in the neurophysiology of information processing in the adult brain. It is also of profound significance in the morphogenesis of the brain, in memory and, when it is in some way disabled, in causing numerous distressing neuro- and psychopathologies. It is to these matters that we turn in our final group of chapters. We shall see that the transcriptional control of gene expression which we began to examine in the last few sections of this chapter plays an all-important role in brain development. Indeed, we shall see in Chapter 20 that it may also play a crucial role in memory. Finally, we shall see in Chapter 21 that subtle variations in molecular structure are multiplied up through the intricacies of neuroanatomy to end in massive effects on brain, behaviour and human well-being.

18

DEVELOPMENTAL GENETICS OF THE BRAIN

The biogenetic 'law': 'ontogeny recapitulates phylogeny' – heterochrony – evo-devo. Morphopoietic gradients – early establishment of longitudinal axis. *Drosophila*: polarity – segmentation. **Vertebrates**: anteroposterior axis in CNS – neurulation – neural crest and neural crest cells (NCCs) – NCCs form ANS and visceral skeleton – dorsoventral and mediolateral axes established – segmentation genes in mouse. **Homeosis and homeotic mutations**. **Homeotic genes**: the *BX* and *ANT* clusters – together form the *HOM-C* – *HOM* genes contain highly conserved 180 bp 'homeobox' – encodes 'homeodomain' – folds into HTH transcription factor motif – control transcription of early onset genes. *HOM* cluster ubiquitous in animal functional homologies – unity of early embryological processes – evolution of *HOM-C*s. **HOM genes and early development of brain**: segmentation of skull – segmentation of hindbrain – rhombomeres – cranial nerves, visceral skeleton – forebrain segmentation – prosomeres – homologies with *Drosophila*. **The POU genes** – neuronal differentiation – homologues of POU in *Caenorhabditis elegans* – encode transcription factors – homologies with *HOM* genes – POU proteins contain homeodomain – POU proteins early localised in proliferative zone of vertebrate forebrain – later restricted to certain cell types, certain brain regions – subtle control systems – evolutionary relationships. Sequential expression of transcription factors. Pax-6: early development of eyes and olfactory systems – unity in diversity. **Other morphopoietic genes**: *Drosophila AS-C* and mammalian homologues: *MASH1*, *MASH2*. Negative control, negative sculpting, allowing a wide variety of cell types to emerge from originally pluripotent, generalist, stem cells

18.1 INTRODUCTION: 'ONTOGENY RECAPITULATES PHYLOGENY'

In Chapter 1, when introducing nervous systems, we noted Haeckel's well-known **biogenetic law**: 'ontogeny recapitulates phylogeny'. Legend has it that the origins of this 'law' lie in an instance of very un-Teutonic serendipity. It is said that the labels fell off von Baer's specimen jars and he suddenly found it impossible to tell which early embryo was which. K.E. von Baer (1792–1876) was one of the founding fathers of modern embryology. Figure 18.1 illustrates his difficulty. Although this story of the mixed-up embryos is nowadays believed to be apocryphal it still serves to make

the point: the early embryos of a wide variety of vertebrates look very alike.

Von Baer was, in fact, quite cautious in his interpretation of this early similarity. He merely noted, in his famous 1828 text on animal embryology, that 'the more general characters of the large group of animals to which an embryo belongs appear earlier than the more special characters', 'from the most general forms, the less general are developed until the most special appear' and 'instead of passing through the adult stages of other animals the embryo departs more and more from them'. Few would dissent from these propositions today.

Later in the nineteenth century Fritz Müller (1864) and Ernst Haeckel (1866) popularised von

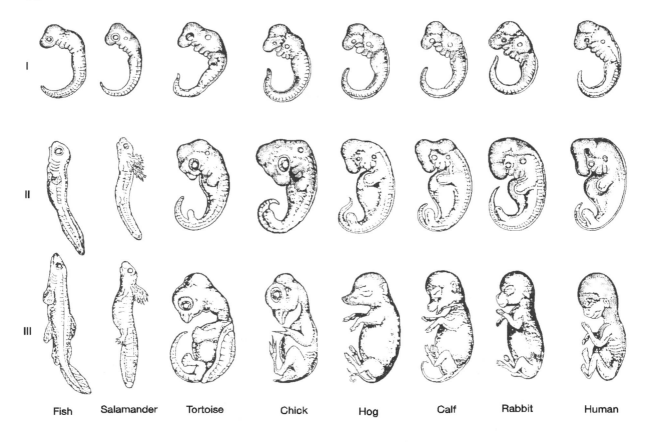

| Fish | Salamander | Tortoise | Chick | Hog | Calf | Rabbit | Human |

Figure 18.1 Vertebrate embryos illustrate the biogenetic 'law'. Modified from Ernst Haeckel (1874), *Anthropogenie*, Leipzig: Engelmann.

Baer's ideas in a more extreme form: the so-called 'biogenetic law'. Müller writes that 'Descendants...reach a new goal by deviating sooner or later whilst still on the way towards the form of their parents, or by passing along this course without deviation, but then instead of standing still advancing still further.' Haeckel, in his 1866 *Generelle Morphologie*, says that the organism 'repeats during the rapid and short course of its individual development (ontogeny) the most important of the form changes which its ancestors traversed during the long and slow course of their palaeontological evolution (phylogeny)'.

It is easy to see that all these formulations encapsulate the same perception: progression from general to particular. It is also easy to see that

damage or genetic mutation at an early stage in development is likely to be far more serious than damage at a later stage. The Müller–Haeckel 'law' was taken to an extreme in the last decades of the nineteenth century, and the twentieth century saw something of a reaction against it. Much more emphasis has been placed on the adaptation of the embryo to its surrounding conditions. We no longer see ontogeny as a straight-line recapitulation of evolutionary history. Indeed, the expression of ancestral characters is often shifted with respect to each other during development, a process known as **heterochrony**. We met an instance of this in Section 17.7 when discussing the pineal's role in the delayed onset of reproductive maturity in humans.

It is salutary to find that the developmental genetics of the 1990s are beginning to come full circle back to confirm von Baer's formulations at the beginning of the nineteenth century, though at a far greater depth. Profound similarities in the molecular genetics of early embryos as widely separated as nematodes, flies and mammals are beginning to provide insight into the early evolutionary history of the Metazoa. Indeed, von Baer's interpretation is being generalised. It is not merely that individuals in the same phylum, for instance the Chordata of Figure 18.1, share common beginnings but that at the level of the gene and the protein all multicellular animals retain signs of a common origin. The early onset genes are to biology as the microwave background is to astronomy: whispers from the earliest of times. And, some of the quietest of these whispers tell of striking molecular homologies in the early development of animal brains.

18.2 ESTABLISHING AN ANTEROPOSTERIOR (A-P) AXIS IN *DROSOPHILA*

Embryology has a long and fascinating history dating back at least to the times of Aristotle. One thing became clear as experimental studies of development began in the nineteenth and twentieth centuries: position was all-important. Grafting experiments in the early embryo showed that grafted organs or tissue were strongly influenced by just where in the body it was placed. Gradients seemed to be everywhere and nowhere more significant than along the anteroposterior axis. A great deal of experimental work was directed to understanding how the vertebrate anteroposterior and limb axes were established. It was not, however, until the techniques of molecular genetics became available in the last few decades of the twentieth century that real progress could be made in establishing the nature of these gradients and subjecting them to experimental manipulation.

It is perhaps not altogether surprising, considering the long history of *Drosophila* genetics, that it is in this organism that the major breakthroughs occurred. Many of the discoveries in the fruit fly have subsequently been shown to apply to a large number of other animals. *Drosophila* does, however, have some peculiarities not shown in the

vertebrates. One of the most striking is that the first 13 nuclear divisions of the zygote take place without the formation of cell membranes. This leads to well over a thousand nuclei coming to inhabit a unitary cytoplasm. These nuclei eventually move to the periphery and form a syncytial blastoderm.

But even before fertilisation of the oocyte occurs an **anteroposterior (or A-P) gradient** has been established in the cytoplasm. Figure 18.2 shows that material flows into the developing oocyte from attached nurse cells. This material includes cytoplasm and a number of genes. One of these, *bicoid (bcd)*, establishes the anterior end of the oocyte. It becomes trapped in the cytoskeleton and its transcript mRNA (or translated protein) establishes a diffusion gradient within the oocyte (Figure 18.3). Mutations of *bicoid* lead to an embryo with two rear ends and no head or thorax. Another gene or, in this case, group of genes, which is introduced into the oocyte by the nurse cells is called *oskar (osk)*. These are trapped at the posterior end and create an inverse gradient to *bcd*. Mutations of the *osk* group lead to embryos with no abdomen. Finally, another gene, or rather group of genes, including *dorsal*, *cactus* and *toll*, are secreted into the oocyte from the nurse cells. These establish dorsal–ventral polarity, in other words, which way up the embryo is.

It is, of course, a pressing question whether analysis of the morphogenetic gradients which are so important in the acellular blastula of *Drosophila* can have any relevance to an understanding of the multicellular blastulae of most other animals. The jury is still out, but the working hypothesis of most embryologists is, yes, *Drosophila* does indeed at the very least point to the way the A-P axis is laid down in other forms. It may be that the morphopoietic gradients develop in the intercellular compartment but, more likely, influence is passed from cell to cell in a cascade or 'bucket brigade' fashion. It is known that the gap junctions (see Section 7.9) which allow communication between a cell and its neighbours form early and are more generally open in embryonic tissue. Thus a morphopoietic signal may diffuse from cell A to cell B, switching on the transcription of the signal in Cell B which then diffuses to cell C, etc.

It is germane, in this regard, to note recent work on the developing neocortex in which high-speed

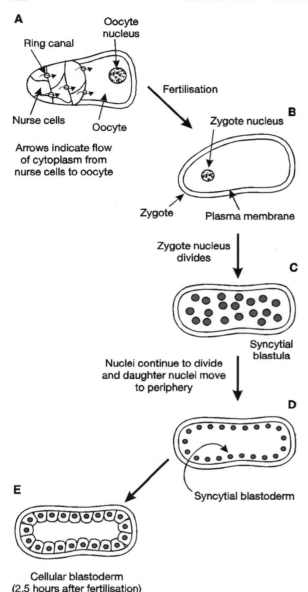

Figure 18.2 Early stages in *Drosophila* embryology. (A) The oocyte with its nurse cells. The arrows indicate the flow of cytoplasm from the nurse cells into the oocyte. (B) After fertilisation the zygote nucleus lies at one end of the cytoplast. (C) The zygote nucleus divides to give a syncytial blastula inhabited by well over a thousand nuclei. (D) Nuclei continue to divide and move to the periphery to form a syncytial blastoderm. (E) 2.5 hours after fertilisation cell walls develop between the blastodermal nuclei. After Mahajanmiklos and Cooley (1994), *Developmental Biology*, **165**, 336–351.

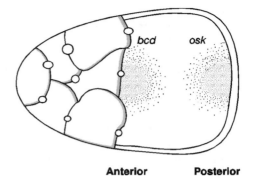

Figure 18.3 Concentration gradients of *bcd* and *osk* transcripts (or coded protein) in *Drosophila* oocyte define anterior and posterior poles.

optical recording techniques show the spreading of Ca^{2+} waves at a velocity of about $100\,\mu m\,s^{-1}$ over distances of $100\,\mu m$ and more. This velocity is far greater than the diffusion rate of intracellular Ca^{2+} (c. $13\,\mu m^2\,s^{-1}$). Furthermore, the intracellular Ca^{2+} levels do not fall off with increasing distance from the source. They are, however, abolished by gap junction blockers such as octanol and halothane. It is suggested, therefore, that a signal molecule, probably IP_3, diffuses through gap junctions causing the release of Ca^{2+} from successive SERs.

18.3 INITIAL SUBDIVISION OF THE *DROSOPHILA* EMBRYO

The first genes to express themselves in the zygote nucleus after fertilisation are members of the *gap* class. Mutations of these genes lead to flies lacking a particular region along the pre-established A-P axis. The most intensively studied members of this class are named *hunchback (hb)*, *Krupple (Kr)* and *Knirps (kni)*. These genes delimit the domains of expression of the homeotic genes which we shall discuss below. They also define the position at which the so-called **pair rule** genes are expressed. The expression of the latter genes leads to a pattern of seven bands or stripes across the embryo (Figure 18.4). These stripes can be visualised in frozen sections by *in situ* hybridisation of their mRNA transcripts with radiolabelled probes (see Section 5.18). The seven stripes define the 'parasegments' which, although out of register with the final

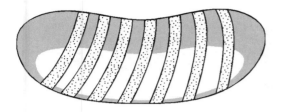

Figure 18.4 Seven parasegments are induced by pair rule genes in the *Drosophila* embryo.

segmentation, coincide with the domains of activity of the homeotic genes (see below).

The pair rule genes also define the realms of expression of the **segment polarity genes:** *engrailed (en)*, *wingless (wg)* and *hedgehog (hh)*. The transcripts of all three of these genes accumulate in 14 narrow stripes along the A-P axis. They establish the A-P polarity of individual segments. The protein product of *hh* appears to act in a very local manner, diffusing only a few cell diameters from its site of origin. Whereas the vertebrate homologues of the *gap* genes which play so important a role in the fly are relatively unknown, some of the pair rule genes, as we shall see in the following sections, do have such homologues.

18.4 THE A-P AXIS IN VERTEBRATE CENTRAL NERVOUS SYSTEMS

In Chapter 1 we saw how the vertebrate central nervous system forms by the invagination of a primordial strip or plate of neurectoderm. This process is called **neurulation** and leads to the formation of a neural tube (see Figure 18.5). In vertebrates (but not in protochordates), along the line where the invaginating neurectoderm pinches off to form the tube, some cells are left behind. These cells constitute the **neural crest**. In other species, especially mammals, the neural crest cells (**NCCs**) form while the neural folds are elevating, before closure. The existence of NCCs is one of the major differences between vertebrate and invertebrate nervous systems. Indeed, it has been argued that they form a fourth germ layer additional to the classic ectoderm, mesoderm and endoderm. The vertebrates, according to this view, are tetrablastic.

NCCs are destined to form the peripheral nervous system (including Schwann cells) and also the adrenergic cells of the adrenal medulla and some skin pigment cells. One neural crest lineage which has received much attention is that which goes to form sympathetic neurons and the chromaffin cells of the adrenal gland. Investigation of this so-called **sympatho-adrenal (SA)** lineage has provided insight into the processes of neuronal differentiation. In the head region, some neural crest cells differentiate into cartilage and bone, especially the visceral skeleton of the branchial arches. We shall see the significance of this when we come to consider the segmentation of the hindbrain (rhombencephalon) in Section 18.8.

The A-P axis of the vertebrate central nervous system is established very early, before neurulation has commenced. Neurectoderm or neural plate is induced to differentiate from surrounding ectoderm by signals from underlying mesoderm, including notochord. The 'primary' organiser (Spemann's organiser after its nineteenth-century German discoverer) is found at the anterior dorsal end of this plate. It sends signals, such as **chordin**, which inhibit a pre-existing gradient of bone morphopoietic proteins (BMPs). Without this inhibition the BMP gradient (a species of TGF-β (see Section 19.9)) will cause mesodermal structures to develop. It is at this very early stage that an A-P axis and, indeed, dorsoventral (D-V) and mediolateral (M-L) gradients can be detected. The mesoderm in front of the notochord (pre-notochordal mesoderm) may continue sending signals into the neural tube, thus influencing brain development, for some time after neurulation has been completed. The molecular nature of these mesodermal signals is still being worked out but may include two proteins, **noggin** and **follistatin**.

In addition to signals inducing an A-P axis from the underlying mesoderm there is also evidence (in *Xenopus* embryos) for signals diffusing through the plane of the neural plate. These signals (as in *Drosophila* oocyte) may spread from both anterior and posterior poles. It is likely that both vertical signals from the underlying mesoderm and planar signals cooperate to establish the A-P axis of the neural plate.

Once the neural tube has formed, D-V and M-L gradients become important. Vertical signals from the notochord induce the floor plate of the neural

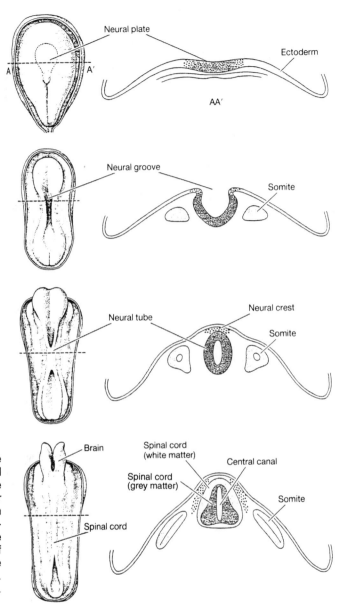

Figure 18.5 Formation of the neural tube (neurulation) and the neural crest. The left-hand column of figures shows the development of the human embryo (third to fourth weeks after conception) from above. The right-hand column shows transverse sections through these embryonic stages at the position marked by the broken line. The origin and development of neural crest cells are clearly shown. From 'The development of the brain' by W.M. Cowan. Copyright © 1979 by Scientific American Inc. All rights reserved.

tube and signals from the floor plate diffuse upwards through the cellular walls of the tube to induce differentiations into motor and commissural neurons and the roof plate (Figure 18.6). It is probable that gene expression at any point in the early neural tube depends on both A-L and D-V gradients. Thus 5-HT cells first appear in the ventral region of rhombomere 1 (see Figure 18.9), suggest-

ing that vertical and longitudinal gradients intersect at this point to release expression of the 5-HT gene.

The vertebrate homologue of *Drosophila hh*, vertebrate *hh1 (vhh1)* or *sonic hedgehog (shh)*, is expressed in midline structures of the early embryo including the notochord and floor plate of the neural tube. The fact that *shh* from zebra fish can replace *hh* in *Drosophila* increases the likelihood

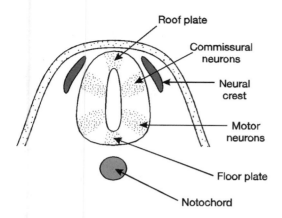

Figure 18.6 D-V differentiation of neural tube. Further explanation in text. After Riji *et al.* (1993), *Cell*, **75**, 1333–1349.

that early morphopoietic mechanisms in insects and vertebrates are conserved. This early 'Baerian' conservation gives way to specialisation as embryology proceeds. It is important to recognise that not all the molecular biology worked out in *Xenopus* applies to the chick or the mammal, or vice versa. Standard texts in embryology emphasise how different the later stages of development are in amphibian, bird and mammal.

18.5 SEGMENTATION GENES IN *MUS MUSCULUS*

Genes controlling segmentation in the vertebrates are not as well known as those carrying out this function in *Drosophila*. Early on in development, before neurulation is complete, an important area dividing the prospective hindbrain from anterior regions appears. This is known as the **isthmic organiser**. It sends signals forward to induce the midbrain and posteriorly to induce the hindbrain.

Anteriorly these signals switch on a homologue of the *Drosophila en (engrailed)* gene, *en-1*, which encodes a transcription factor, Engrailed-1. This is essential to the development of the midbrain. Mouse *en-1* knockouts lack a mesencephalon. Posteriorly the isthmic organiser switches on gradients of *wnt-1* and *en-2*. *Wnt-1* (previously known as *int-1*) encodes a protein similar to that of the *Drosophila wg* gene. Examination of homozygous transgenic mice in which *wnt-1* had been knocked out showed a catastrophic failure to develop large parts of the metencephalon, especially the cerebellum. Such animals, it hardly needs adding, rarely survived more than a few days after birth.

Knocking out the second mouse segmentation gene, *en-2*, did not have quite so disastrous an effect. Transgenic animals normally survived to adulthood. However, examination of their brains showed changes in the folding pattern of the cerebellar cortices and a reduction in size of the Purkinje cells. These seemingly rather insignificant anatomical changes do not seriously disrupt the animal's behaviour. This is surprising when the crucial nature of the *en* gene in *Drosophila* development is recalled. It does, however, remind us that we must be cautious (see Section 5.17) in assuming the preservation of common biochemical pathways and common molecular biological environments across the wide phylogenetic gap which yawns between fruit flies and mice. Although *wg* and *en* act as segmentation genes in *Drosophila* their homologues have evidently assumed other functions in the mammal.

18.6 HOMEOSIS AND HOMEOTIC MUTATIONS

The stage is now set for the switching on of genes that control segmental morphology. These, as we shall see, show fascinating homologies across the whole metazoan subkingdom. The term '**homeosis**' was introduced by William Bateson a century ago in his 1894 *Materials for the Study of Variation*: 'something has been changed into the likeness of something else'. The first case of a homeotic mutation was observed by Bridges in 1915 when he observed that in some rather rare cases the third thoracic segment of *Drosophila melanogaster* was transformed into the likeness of the second segment, bearing a small pair of wings instead of the usual halteres. For obvious reasons he named the mutated gene *bithorax (bx)*. Many homeotic mutations are even more dramatic than *bx*. One of the most striking is the homeotic mutation of the antennapedia gene (*antp*) which transforms one of the head segments into a thoracic segment: instead of antennae a pair of legs is developed. It is clear that mutations in these early genes are normally lethal: the fly does not survive to adulthood.

18.7 HOMEOBOX GENES

Further analysis showed that both *bx* and *antp* were in fact members of complexes consisting of several genes; these complexes have been labelled *BX-C* and *ANT-C*. In the late 1980s and early 1990s both complexes were cloned and their DNA sequenced. When the DNA sequences of the two complexes were compared it was found that they both contain a very similar 180 bp sequence. This region was consequently named the **homeobox**. *BX-C* and *ANT-C* were thus seen to constitute a so-called homeobox cluster, *HOM-C* (Figure 18.7). The sequence of 60 amino acids in the proteins specified by the 180 bp homeobox regions was called the **homeodomain**. The homeodomain is found to twist into a helix–turn–helix (HTH) conformation characteristic of many transcriptional factors, both prokaryotic and eukaryotic (see Chapter 3 and Figure 3.7).

The implication that homeotic genes exerted their effects by coding for transcription factors, thus controlling the activity of early onset genes, has since been confirmed by numerous biochemical and genetic investigations. Once the *Drosophila* embryo develops from its syncytial to a multicellular form its homeobox genes are controlled by segmental polarity genes: the *hedgehog* and *wingless* systems.

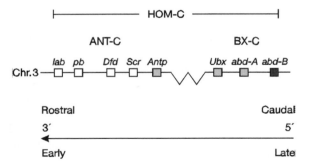

Figure 18.7 The homeobox cluster (*HOM-C*) on *Drosophila*'s third chromosome. *ANT-C* consists of the five genes at the 3′ end of the sequence and *BX-C* the three genes at the 5′ end. The genes are switched on in the 3′ to 5′ sense. The shading of the genes indicates the general rule that the more caudal genes (darker shading) suppress phenotypic expression of the more rostral genes. Further explanation in text. *lab*=labial; *pb*=proboscipedia; *Dfd*=deformed; *Scr*=sex combs reduced; *Antp*=antennapedia; *Ubx*=ultrabithorax; *abd-A* =abdominal-A; *abd-B*=abdominal-B.

It is found that all of the genes of the *HOM-C* cluster are located on *Drosophila*'s third chromosome (Figure 18.7). But, more than this, they express themselves in the same linear order as they assume on the chromosome. Not only are they switched on in order, from the 3′ to the 5′ end of the cluster, but it is also found that the genes that control rostral development are arranged at the 3′ end of the complex and those that control caudal development at the 5′ end. This has been termed **colinearity**.

But, even more interesting than this insight into the means of action of the homeotic and other early onset genes was the finding that the homeobox sequence was strongly conserved throughout the animal kingdom. Using the *Drosophila* homeobox sequence as a probe, similar base sequences were picked out in many organisms, including *Hydra, Caenorhabditis elegans*, frogs, chicken, mice, humans, indeed in all animals that have been investigated. The vertebrate homologues detected by this hybridisation technique are termed ***Hox*** genes. As Figure 18.8 shows, there are 38 of these genes in four different chromosome complexes. Comparison of their sequences shows that they fall into 13 **paralogous** groups. The homeodomains of the proteins which they express are remarkably similar to those of *Drosophila*. For instance, the homeodomain of the protein expressed by one of the *Hox* genes (*Hox 2.2* or, to use more modern nomenclature, *Hoxb-6*) differs by only a single amino acid from that expressed by *Drosophila*'s *Antennapedia* gene. As Haeckel saw long ago, it is evidently very dangerous to alter the genes controlling early development: such alterations are usually fatal.

But more even than this. It is found that not only do *Hox* genes incorporate homologous homeobox sequences but they are arranged in a similar linear order to those of the *Drosophila HOM* complexes. Thus *Hox* genes (paralogue groups 1–4) which specify structures in the anterior part of vertebrate embryos are located anteriorly (i.e. 3′) to those specifying posterior structures (paralogue groups 5–13): in other words the same colinearity is found in fly and mouse. Gene targeting leading to the knockout (see Section 5.17) of *Hox* genes leads to many of the same effects that were first observed in *Drosophila*: principally alteration of segment phenotype. Knockout of *Hoxc-8* leads, for instance, to transformation of the first lumbar into a thoracic vertebra; knockout of *Hoxb-4* to the transformation of the second cervical vertebra into an additional atlas vertebra. Note the so-called '**anteriorisation**'. Knockout of a more caudal gene allows a more rostral gene (otherwise suppressed (see Figure 18.8)) to invade its territory and switch on its (the rostral gene's) structural genes. Finally, the new techniques of molecular genetics have allowed geneticists to transfer mouse *Hoxb-6* (the homologue of *Drosophila ant*) to the fly. When it is switched on the mouse gene shows the same homeotic control as its *Drosophila* homologue. These remarkable discoveries show the astonishing stability of the early events in embryology.

The common ancestor of insects and vertebrates lived far back in the Precambrian period, more than 500 million years ago. Indeed the 180 bp homeobox sequence, as mentioned above, is also found in *Hydra* and *C. elegans*. The sequence seems to have originated with the Metazoa and persisted ever since. The development of clusters of homeobox genes in *C. elegans* and *Drosophila* points to several duplications and subsequent diversifications (as discussed in Chapter 4). Indeed, the rather more primitive insect, the flour beetle, *Tribolium*, possesses only one cluster of homeobox genes rather than the two found in the fruit fly. When we come to the Chordata we find a single *Hox* cluster in *Balanoglossus*, a hemichordate, and *Amphioxus*, a cephalochordate, but by the time we reach the higher vertebrates, such as the mouse, this single cluster has duplicated several times and diversified to form the four *Hox* clusters: A, B, C and D. Evolutionary forces have worked on these clusters to generate the great variety of body plans which exist today. Figure 18.8 shows a suggested relationship between the homeobox clusters of *C. elegans*, *Drosophila* and *Mus musculus*.

18.8 HOMEOBOX GENES AND THE EARLY DEVELOPMENT OF THE BRAIN

The segmentation of the vertebrate skull was something of a *cause célèbre* in the early and mid-nineteenth century. Anatomists from Goethe through Richard Owen to Thomas Henry Huxley made significant contributions. In general it was believed that the skull developed by modification of a number of vertebrae at the anterior end of the animal. It was Huxley in his 1858 Croonian Lecture 'On the Theory of the Vertebrate Skull' who first showed the superiority of the embryological method over the older techniques of comparative anatomy. It is interesting, therefore, to find a century and a half later the techniques of molecular genetics being devoted to the same end: establishing the segmentation of (in this case) the embryonic brain.

Although a sequence of eight macroscopic bulges in the developing rhombencephalon had been recognised for over a century it had been far from clear that these had anything to do with the segmentation which is so clear in the rest of the vertebrate body. Closer analysis, however, showed that each pair of these bulges, referred to as **rhombomeres (r)**, was related to an adjacent branchial arch (Figure 18.9). The trigeminal nerve (cranial nerve V), for example, originates from r2 and r3 and innervates all the muscles associated with the first branchial arch; similarly the facial nerve (cranial nerve 7) originates from r4 and r5 innervates second branchial arch musculature; finally, the glossopharyngeal nerve (cranial nerve IX) originates in r6 and r7 and innervates the third arch musculature. Figure 18.9 shows that a similar rhombomeric organisation can be detected on the sensory side. This strongly implies that rhombomeres are truly segmental structures.

The segmental nature of the rhombomeres is supported by the observation that the most active proliferation of neuroepithelium occurs at the centre of the rhombomere and falls to a low level at the boundary between one rhombomere and the next; developing neurons within a rhombomere synthesise common and distinctive surface molecules; cells within but not in different rhombomeres are able to communicate by way of gap junctions. There is also evidence to show that the sulci between alternate rhombomeres restrict the migration of populations of developing motor neurons and that neurogenesis, itself, is restricted to alternate rhombomeres (r2, r4, r6, r8). It has not, of course, escaped the notice of molecular embryologists that double structure of the rhombomeres is strikingly analogous to the *Drosophila* 'pair rule' organisation.

Thus, once again, the genetic analysis of *Drosophila* has been fundamental to understanding the system. Whilst, as we noted in Section 18.5, the

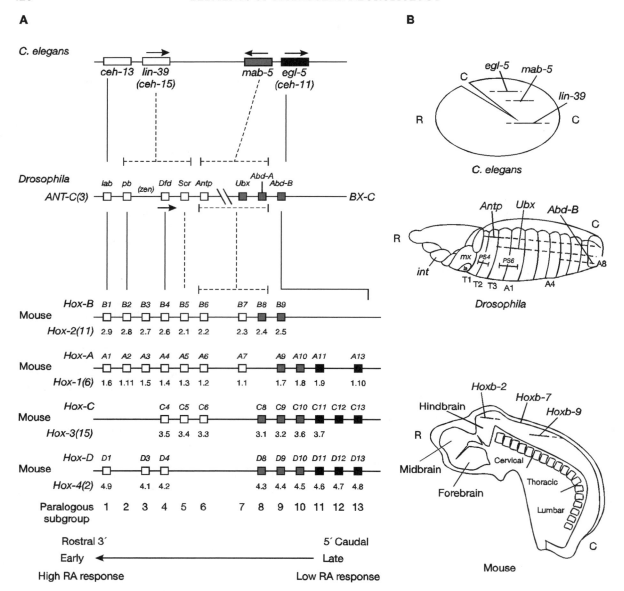

Figure 18.8 Alignment of homeobox genes in *C. elegans, Drosophila* and *Mus.* (A) Alignment of homeobox complexes. The solid lines indicate homologous genes, the dashed lines indicate weaker homology. It can be seen that *C. elegans egl-5, Drosophila Abd-B* and *Mus B9* are homologues; in the *Hox A, Hox C* and *Hox D* complexes there has been massive duplication. The four mouse complexes have the revised names above and older nomenclature below each gene. The 13 paralogous subgroups of the mouse are numbered towards the bottom of the diagram. The long horizontal arrow beneath the paralogous numbering indicates both the colinear relationship between gene order and expression domain in anatomy. It also indicates the order in which the genes are switched on. Small arrows indicate exceptions to this uniform direction of transcription. Finally, at the bottom of the figure, the responsiveness of the gene to retinoic acid (RA) is indicated. The depth of shading of the gene indicates that genes with more caudal expression boundaries suppress the activity of more rostral genes. (B) Indicates the expression territories of the *HOM/Hox* genes. Mouse paralogous groups 1–4 express themselves in the hindbrain, groups 5–13 in the trunk. R=rostral; C=caudal. Modified from Botas (1993), *Current Opinion in Cell Biology,* **5**, 1015–1022; with permission.

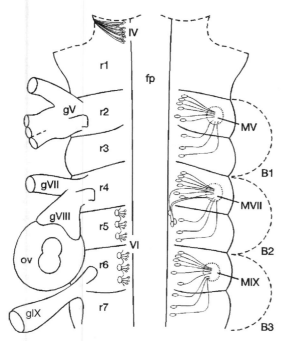

Figure 18.9 Segmental organisation of the hindbrain. The schematic diagram shows the first seven rhombomeres (r1–r7) in the vertebrate (chick) hindbrain and their association with motor nerves (right) and sensory nerves (left). B1, B2, B3=branchial arches; fp=floor plate; gV–gIX=cranial sensory ganglia (trigeminal; facial; vestibulocochlear; glossopharyngeal); MV, MVII, MIX=origins of branchiomotor nerve fibres (trigeminal; facial; glossopharyngeal); IV=trochlear nerve; VI=abducens nerve; ov=otic vesicle. From Lumsden and Keynes (1989), *Nature*, **337**, 424–428; Keynes and Krumlauf (1994), *Annual Review of Neuroscience*, **17**, 109–132, © 1994 by Annual Reviews Inc.; with permission.

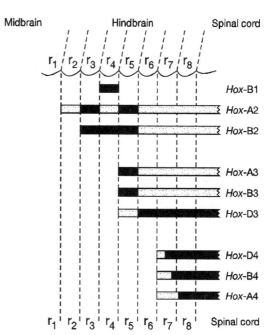

Figure 18.10 Expression of *Hox* genes in the vertebrate hindbrain. At the top of the diagram the rhombomeres of the embryonic mouse hindbrain are represented (about 9.5 days postcoitus). The bars represent the expression domains of the *Hox* genes at this time. The dark regions indicate highest levels of expression. Note that with the exception of *Hox A2* and *Hox B2*, each paralogous group has a common anterior boundary and the higher the paralogue number the more posterior is that boundary. After Krumlauf (1993), *Trends in Genetics*, **9**, 106–112.

segmentation of the brain into neuromeres is not under the control of the *Hox* clusters the latter genes (as the *HOM* genes of *Drosophila*) do determine the specialised nature of the rhombomere. This is emphasised by the homeotic transformations elicited by gene knockout (Section 18.7). In the normal course of events the most 3′ members of all the mouse *Hox* clusters express themselves by day 9.5 postcoitum (pc) in controlling rhombomere phenotype. Figure 18.10 shows that the anterior expression boundaries of paralogous groups 2, 3 and 4 coincide with the two-segment repeat organisation met with above. It can be seen, in other words, that the anterior

expression boundary of paralogous group 2 coincides with the r2/r3 boundary; group 3 coincides with the r4/r5 boundary and group 4 with the r6/r7 boundary. The figure also shows that the expression times and duration of the various *Hox* genes vary widely. It must also be recalled that expression is from the 3′ to 5′ end of the cluster so the more posterior rhombomeres will have a more complex combination of *Hox* genes at work than the anterior segments. The mode of expression is, moreover, complex and is still under investigation; indeed the *Hox* genes affect not only neurogenesis in the hindbrain but also cells in neural crest and branchial arches.

The best interpretation at present is that the *Hox* genes determine the nature of the rhombomere, i.e.

whether it is r2 or r4, etc. The cells within the rhombomere are thereafter restricted to an r2 or an r4 fate. But within this large categorisation they remain pluripotent with respect to specific cell type. Rhombomeres may thus be regarded as 'specification units'. Other signals, probably from the notochord, influence differentiation in the dorsoventral axis and, in the early embryo, there is no differentiation in the third dimension, across the width of the neuroepithelium.

It is not only the hindbrain that shows a segmental organisation; in recent years evidence has begun to accumulate showing that the forebrain also has a segmental origin. Once again studies in *Drosophila* have helped elucidate this segmentation. It is believed that *Drosophila* head consists of six or seven segments. The development of these segments is under the control of the early maternal genes described in Section 18.2. Different concentrations of the *bicoid (bcd)* gradient activate different downstream genes: genes important for head development are *orthodenticle (otd)*, *empty spiracles (ems)* and *buttonhead (btd)*. Genetic analysis shows that mutation of any of these genes leads to loss of a region of the anterior head rather than its transformation into something else; furthermore the domains in which the genes are active overlap. This indicates that these genes shared the properties of both the gap genes (see above) and the homeotic genes which (as we saw above) are able to switch on genes determining segmental characteristics. Finally it has been shown that both the *otd* and the *ems* genes contain a homeobox segment and that the *btd* gene codes for another well-known transcription motif, the zinc finger conformation.

Can we relate these *Drosophila* genetics to vertebrates, especially mammals? Recent work on mouse development has shown that two *otd*-related genes (*Otx1* and *Otx2*) and two *ems*-related genes (*Emx1* and *Emx2*) exist. The homeodomains of the proteins coded by these murine genes differ by only two or three amino acids from their *Drosophila* homologues. Similarly the *Drosophila btd* gene product may be related to one of the mammalian Sp1 zinc finger transcription factors. Furthermore it is found that the *Otx* genes are highly conserved throughout the vertebrates: the homeodomains of Otx proteins differ by only one or two amino acids between mouse, human and *Xenopus*.

The *Otx* and the *Emx* genes are both expressed in the developing forebrain of the mouse. Figure 18.11 shows that both *Emx1* and *Emx2* are expressed in the roof of the telencephalon. The rostral territorial boundary of *Emx2* is slightly anterior to that of *Emx1* but both overlap posteriorly so that in both cases their domains end at the boundary of the telencephalon and diencephalon. In addition *Emx2* is also expressed in the floor of the diencephalon. The two *Otx* genes are expressed somewhat caudal to the *Emx* territories. Although transcripts of *Otx2* can be detected throughout the early brain, its territory recedes by day 7.5 to the anterior region which includes telencephalon, diencephalon and mesencephalon. *Otx1* is also widely expressed, but falls within the *Otx2* domain.

What role do these *Emx* and *Otx* domains play in the embryogenesis of the brain? It is suggested that, as in the case of *Drosophila*, they are involved in segmentation. In other words, just as the hindbrain is subdivided into a series of rhombomeres so the forebrain is subdivided into prosomeres. It is suggested that the embryonic forebrain is built of three diencephalic and three telencephalic segments. The *Emx* and *Otx* system could be involved in the development of these neuromeres in a stepwise fashion, beginning with transcripts from

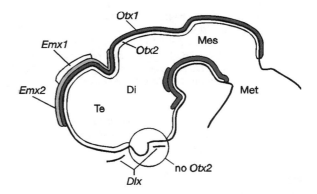

Figure 18.11 Expression of *Otx* and *Emx* in the forebrain of 10 dpc mouse. Di=diencephalon; Mes =mesencephalon; Met=metencephalon; Te= telencephalon. *Otx* is not expressed in the circle at the bottom of the diagram. Instead a number of genes related to the distalless-related family (*Dlx*) are active in this region, which includes the optic chiasma. From Finkelstein and Boncinelli (1994), *Trends in Genetics*, **10**, 310–315; with permission.

Otx2 and followed in sequence by transcripts from *Otx1*, *Emx2* and *Emx1*.

Here we shall have to leave these fascinating neurogenetics. There seems little doubt that strong homologies run between the early embryology of brains in animals as diverse and as evolutionarily distant as *Drosophila* and *Mus musculus*. There is much reason to believe that the vertebrate *Hox* system plays a similar role to the *HOM* system in *Drosophila* in determining segmental phenotype (including as we have seen the neuromeres) in relation to the A-P axis. But we shall have to leave to a future edition the full to and fro analysis of the roles of these early genes in laying down the ground plans of brains. We can now pass to genes further down the developmental cascade and, in particular, to the *POU* and *PAX-6* genes.

18.9 *POU* GENES AND NEURONAL DIFFERENTIATION

After the *Hox* genes have established the neuromere phenotype with respect to the A-P axis the next step in neuroembryology involves the differentiation of its neurons. We have already seen that the brain, especially the brains of mammals, has

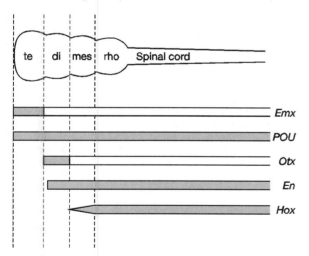

Figure 18.12 Expression domains of some morphopoietic genes in the central nervous system. The dark segments of the bars indicate where transcripts are found. Only five of the several hundred genes active in neuroembryology are shown. te=telencephalon; di=diencephalon; mes=mesencephalon; rho=rhombencephalon.

extremely heterogeneous populations of neurons. Indeed this is one feature which is often quoted as establishing that 'higher' brains are more 'complex' than their ancestral progenitors. The genetic bases of neuronal differentiation are only just beginning to be worked out. Developmental biologists are beginning to uncover intricate networks of regulators, expressed in overlapping temporal and spatial patterns, some early, some late in development, which modulate the regulatory genes responsible for neuronal phenotype. In this section we shall look at work which shows that a large family of so-called *POU* genes plays a central role. Moreover, this role, in contrast to the *Hox* gene family, is played out in mammalian forebrains.

Once again we are aware of the evolutionary lineages that connect the nervous systems of mammals including ourselves to extremely remote forms. One of the genes in the *POU* family, indeed the gene that provides the 'U' in the acronym, encodes the *Unc-86* transcription factor of *C. elegans*. Mutations in the *Unc* gene lead to uncoordinated movements in the nematode. This clumsiness is due to defective development of touch-sensitive cells in the worm's nervous system. The other two letters in the acronym refer to mammalian *Pit-1* (expressed exclusively in the pituitary gland) and *Oct-2* and *Oct-3* (expressed in lymphoid tissue). All four of these genes encode transcription regulators which share two homologous regions: a 60 amino acid homeodomain (called the **POU-homoedomain**) related to the homeodomains of the homeobox proteins discussed above and a 76–78 amino acid sequence which is unique to this protein family and is named the **POU-specific domain**. These two domains are connected by a variable amino acid sequence and the whole unit constitutes the POU domain (Figure 18.13).

Using probes derived from the POU genes described above, numerous other genes encoding POU domains have been detected in mammals, in *Drosophila* and in *C. elegans*. It turns out that there is a large family of *POU* genes with at least twelve members, falling into five major classes. Class 1 includes just one member, *pit-1*; class 2 contains two members, *oct-1*, *oct-2*; class 3 consists of five members, *cf1a*, *brn-1*, *brn-2*, *tst-1*, *ceh-6*; class 4 contains two members, *unc-86*, *brn-3/4*; class 5 just the *oct-3/4* gene; and class 5-i just the *i-pou*

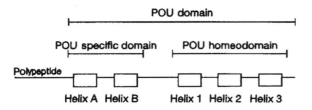

Figure 18.13 Structure of POU domain transcription factors. The POU domain consists of two parts: the POU-specific domain (POU$_S$) and the POU homeo-domain (POU$_{HD}$). The POU$_S$ region contains two helices and the POU$_{HD}$ region three (HTH). The POU$_{HD}$ on its own only provides low-affinity, low-specificity binding to DNA. To achieve high-affinity, high-specificity site binding the two domains have to combine. After Rosenfeld (1991), *Genes and Development*, **5**, 897–907.

(inhibitory-POU) gene. The identification of new POU domain proteins is being continually reported. Furthermore, post-transcriptional modifications generate even greater diversity. These gene products are expressed selectively throughout embryogenesis. Indeed hybridisation histochemistry has detected distinct spatial and temporal patterns of *POU* transcripts at all levels within the nervous system and throughout its development.

We shall see in Chapter 19 that vertebrate neurons originate in a proliferative zone around the brain's ventricles. All known *POU* transcripts are found at one time or another in this proliferative zone. Certain transcripts then become restricted to certain neurons. For instance, *Brn-3* is restricted to sensory ganglion cells derived from the neural crest, whilst transcripts from *Brn-1* and *Brn-2* have been detected in all parts of the cerebrum (layers II–IV) and cerebellum (Purkinje cells). A recently identified POU domain protein, Brn-5, is most strongly expressed in layer IV of the cerebral cortex whilst Brn-3B is found in a subset of retinal ganglion cells. The *Tst-1* gene is expressed in cerebral layers V and VI, cerebellar granule cells, and transiently in myelinating glia. The transcription factor coded by the latter gene binds to the promoter of the cell surface adhesion molecule P$_0$, and in Schwann cells it is believed to influence, by repression, the gene for P$_0$, which, as we saw in Section 7.7, is one of myelin's major structural proteins.

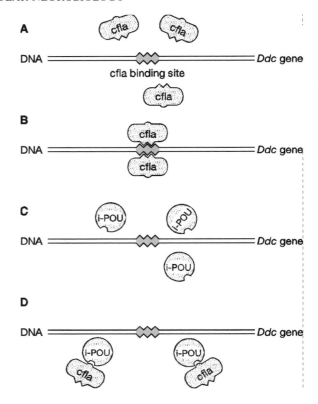

Figure 18.14 Interaction between cf1a, i-POU and the Ddc promoter. (A) The cf1a transcript is stable in solution in the monomeric form. (B) Dimers of the transcript bind to the cf1a binding site and control the transcription of the *Ddc* gene. (C) The i-POU transcript also exists free in the monomeric state and neither binds to each other or to the cf1a binding site on the DNA. (D) Although the i-POU transcript cannot itself bind to the cf1a site, it can (dog-in-the-manger) prevent the cf1a transcription factor binding. It combines with the latter factor, forming stable dimers which remain in solution. After Treacy *et al.* (1991), *Nature*, **350**, 577–584.

The POU domain proteins encoded by the *POU* genes resemble the Hox proteins in acting as transcription factors. In this way they can control the differentiation of specific neuronal phenotypes. The situation promises to be highly complex. It is possible that the transcription proteins encoded by the 12 or more *POU* genes interact in different ways to allow a very large number of neuronal phenotypes to be expressed. For instance, in *Drosophila* the *cf1a* gene product binds to the promoter segment of the gene coding for **dopa-decarboxylase (Ddc)**, an essential enzyme in the pathway leading to dopamine, noradrenaline and serotonin (Chapter 16). It is thus crucial in the differentiation of serotoninergic and catecholaminergic neurons. It has been found, however, that (at least in *Drosophila*) the cf1a POU domain protein does not act alone but in concert with another POU protein, i-pou. It has been found (see Figure 18.14) that the cf1a protein is stable in solution in its monomeric form and requires to dimerise before it can bind to the Ddc promoter. i-pou monomers are also stable in solution but (lacking two basic amino acids) cannot dimerise and bind to the Ddc promoter. However, if i-pou cannot bind to the Ddc promoter it can bind to cf1a. But dimers of cf1a and i-pou no longer recognise the Ddc binding site and hence cannot influence the synthesis of the Ddc enzyme (Figure 18.14). It can be seen that this subtle control system can damp down or switch off the action of cf1a and thus influence the outcome of the neuron's differentiation. The modulation depends on the relative concentrations of the two POU domain proteins and this may well vary during development and from one part of the developing brain to another. This subtle modulation of Ddc expression reminds us of the rather similar subtlety controlling the expression of the pinealocyte SNAT gene (see Section 17.7).

Numerous other interactions between *POU* genes and between *POU* genes and *Hox* genes are coming to light. In *C. elegans*, for instance, it has been shown that the homeobox *mec-3* gene, deeply involved in determining the nematode's neuronal phenotypes, depends on the transcript of *unc-86* (a *POU* gene) for its expression. Once again careful analyses, genetic, biochemical, molecular biological, crossing wide phylogenetic boundaries, are beginning to elucidate the hidden warp and woof of early embryogenesis.

This hidden warp and woof has, of course, an ancestry deep in evolutionary time. We have already noted the fascinating relationships amongst the homeobox genes in widely different animal forms. We have seen that the mechanisms of duplication and subsequent drift (not only of single genes but also of entire clusters) can account for clear relationships between the early genes of creatures as remote as *Hydra* and *Homo sapiens*. Indeed the *Hox* genes are regarded as all members

of one superclass: the '**complex superclass**'. Whereas the members of this superclass have been held together so that they act in a colinear way, the other early developmental genes have become dispersed in the genome. Nevertheless, close examination of the amino acid sequences of the proteins they encode once again shows up unsuspected relationships. It can be shown that many of the genes discussed in this chapter, *en*, *ems*, *POU*, *MASH*, etc., have a more than chance resemblance to each other and to members of the complex superclass. They are said to constitute another homeobox superclass and, because they are dispersed in the genome, this group is called the '**dispersed superclass**'.

18.10 SEQUENTIAL EXPRESSION OF TRANSCRIPTION FACTORS IN *DROSOPHILA* CNS

We shall discuss in some detail the differentiation of cells in mammalian central nervous systems in the next chapter. There, as we shall see, the processes are complex and as yet poorly understood. The *Drosophila* CNS, while still highly complex, is orders of magnitude simpler and provides here, as elsewhere, a valuable experimental preparation. The precursor cells in the CNS (= neuroblasts) can be easily identified and their offspring followed to an end point as a specific type of neuron or glial cell.

Neuroblasts divide asymmetrically. The products of the division are another neuroblast and a small '**ganglion mother cell (GMC)**' (Figure 18.15). The GMCs typically undergo a further mitosis to form two fully mature, post-mitotic, differentiated cells. A large number of different experimental approaches have shown that it is the birth order of the GMC that determines which type of cell it ultimately forms.

Probing more deeply into the molecular biology of this temporal sequencing, it has been shown that as time progresses genes controlling different transcription factors are switched on and off, one after the other. The sequence starts with *hunchback (hb)* and proceeds through *Krüpple (Kr)*, and *POU domain 1 and 2 (pdm)* to end with *castor (cas)* although other genes (e.g. *grainyhead*) may switch on and off later.

$$hb \Rightarrow Kr \Rightarrow pdm \Rightarrow cas$$

It has been shown that whilst the switching on of one of the transcription factor genes is only transient in the neuroblast, the GMC generated at that time by asymmetrical division is marked for life: it will retain the transcription factor with which it is born and will not proceed to switch on the next gene in the sequence. Figure 18.15 shows the outcome of a sequence of neuroblast divisions.

The cellular mechanism ensuring the switching on and off of the genes in correct sequence is still under investigation although some cell-cycle-dependent clock is thought to be implicated. The transcription factor profile of each GMC determines the type of cell into which it ultimately differentiates. Many other factors are involved including, importantly, the lineage of the neuroblast from which the GMC originated. Thus first-born GMCs, expressing significant quantities of *hb* transcription factor, may develop into motor neurons, interneurons or glia.

It will be interesting to discover whether homologues of these *Drosophila* genes are at work in mammalian systems. It is already known that a homologue of *pdm*, the POU domain gene *SCIP* (=*Oct6*), is expressed at high levels in pyramidal cells of layer V of adult rat cerebral cortex. We shall see in Section 19.1 that the subtle control of

homeodomain transcription factors is also responsible for the differentiation of neurons and glia arising from the proliferative zone of embryonic vertebrate neural tube.

18.11 PAX-6: DEVELOPMENTAL GENETICS OF EYES AND OLFACTORY SYSTEMS

This chapter has shown us the remarkable molecular commonalities underlying forms that were previously felt to be widely divergent. We should not finish without citing the fascinating work on the early development of eyes and olfactory systems which emphasises once again this unexpected commonality.

The question has often been asked, when reviewing the vast variety of complex eyes developed in the animal kingdom: did all these different designs arise independently, or do they share a single ancestry? Arthropod compound eyes are, after all, seemingly poles apart from vertebrate vesicular eyes. A common ancestry seems on the face of it unlikely. Yet it is beginning to look as if this is a superficial view and that beneath the surface lurk surprising uniformities. It begins to look as if the answer will be provided by study of the molecular biology of early development. It begins to look as if a common ancestor did exist, but way back before the proterostome/deuterostome divergence some 670 million years ago. The evidence for this remarkable conclusion comes from study of a gene known as **pax-6**.

Pax-6 is one of a number of genes involved in the early development of eyes. As its suffix number indicates, it is a member of a large family of genes. The members of this family encode helix–turn–helix (HTH) (Figure 3.7) transcription factors. The nomenclature, *pax*, refers to the fact that these genes all possess a 384 bp sequence which encodes a 128 amino acid sequence which forms the HTH DNA binding domain. The 384 bp sequence is known as the '*pa*ired bo*x*' (hence *pax*) motif since it was first found in the pair rule genes which, as we saw in Section 18.3, determine the early segmentation of *Drosophila* embryos.

The gene was first detected in *Drosophila* where, because mutation prevents the development of the compound eyes, it is known as *eyeless (ey)*. *pax-6* has subsequently been found in many different

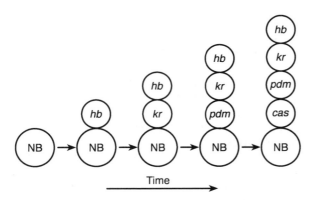

Figure 18.15 Differentiation of neuroblasts in *Drosophila* CNS. The neuroblast (NB) buds off a sequence of ganglion mother cells (shown from left to right). These latter cells retain the transcription factor profile of their date of birth. Further explanation in text. Modified from Isshiki *et al.* (2001), *Cell*, **106**, 511–521.

species of animal from humans through insects and molluscs to round worms (nematodes) and ribbon worms (nemertines). In the early vertebrate embryo *pax-6* expression is at first widely distributed in the central nervous system and it can still be detected in the adult brain, especially in some of the nuclei of the forebrain, the substantia nigra of the midbrain and the granule cell layer of the cerebellum. So far as the forebrain is concerned its expression becomes largely confined to the regions destined to develop into **olfactory** (olfactory epithelium, olfactory bulb) and **optic** (lens, cornea and optic vesicle) structures. Largely, but not entirely, for psychiatrists have shown that individuals of a family carrying a PAX-6 mutation show a distinctive cognitive and psychiatric phenotype traceable to frontal lobe abnormalities.

Mutations of *pax-6* have disastrous effects. We have already noted the effect in *Drosophila*. In the mouse they cause a condition known as small eye (*Sey*). In the heterozygous condition *sey* mice eyes have significantly smaller lenses and optic cups. Homozygous *sey* is lethal. Early mice embryos (9.5 days) show no lens placode and the neural layers of the retina are highly abnormal. The olfactory placode, similarly, fails to develop, and the embryo shows no sign of nasal pits or an olfactory bulb. Clearly *pax-6* plays a crucial role in the early development of eye and nose. Humans heterozygous for *PAX-6* suffer from a condition known as **aniridia**. In this condition there is a complete or partial failure of the iris to develop along with other defects including cataract, corneal opacity, glaucoma, etc. The incidence lies between 1/64 000 and 1/96 000. In addition mutations of *PAX-6* can also lead to defects in the anterior chamber of the eye, including opacity of the central cornea, known as Peter's anomaly. In fact, differences in the degree of inactivation of *PAX-6* lead to a large number of ocular defects affecting numerous parts of the eye, especially the anterior chamber.

Drosophila geneticists have discovered a second *pax-6* gene upstream of *ey*. This gene, *twin of ey* (*toy*), is in fact more like vertebrate *pax-6* than *ey*. It appears that since the common ancestor of vertebrate and insect *pax-6*, a duplication has occurred in the insect line but not in the vertebrate line. Vertebrates only have a single *pax-6* gene. It is fascinating to find that mutations of *ey* and *toy* not only eliminate eye development in *Drosophila* but

also disrupt the neuropil in the mushroom bodies. We shall see in Section 20.4.2 that insect **mushroom bodies** play important roles in olfaction and are vital for olfactory learning. Thus we see that *pax-6* genes play similar roles in insects and vertebrates (including ourselves) in controlling the development of both optic and olfactory structures.

Drosophila geneticists have shown that it is possible to transplant *pax-6* into other parts of the embryo insect's anatomy where it will induce 'ectopic' eyes. Eyes can be induced in legs, antennae, wings, etc. The insect can be covered with eyes! These eyes, moreover, are no mere approximations to the real thing. They consist of a full complement of different cells and structures: primary, secondary and tertiary pigment cells, cornea, cone and pseudocone cells, retinula cells with fully developed rhabdomeres, etc. The cells are organised to form ommatidia and are electrically active. It is not yet known, however, whether the optic nerve fibres project to the correct regions of the insect's brain. Here, then, is a striking instance of latent morphopoietic mechanisms waiting to be triggered or 'uncovered' in insect tissues.

Further work has shown that *pax-6* extracted from embryo mice will also induce ectopic eyes (not, of course, mouse eyes) in *Drosophila*. Finally, it has been found that *pax-6* from squid, normally essential in the early development of cephalopod's highly evolved eye, can also induce ectopic eyes in *Drosophila*. These eyes, again, are anatomically almost normal. Both squid and mouse eyes are, of course, vesicular eyes bearing no obvious resemblance to the compound eyes of arthropods.

All of these findings appear to point in one direction. They suggest that very early in the evolution of the animal kingdom, perhaps at the stage represented by the 'urbilateralia', a genetic system evolved to program the development of complex eyes. This system has remained, essentially unchanged, at the basis of all the huge variety of different complex animal eyes. When *pax-6* is introduced the appropriate biochemical cascades are triggered and appropriate eyes result. Instead of a polyphyletic origin it may be that all eyes, above the level of simple eyes (eyespots), or ocelli, have a unitary origin. The work on insect mushroom bodies noted above may lead to similar conclusions respecting olfactory systems. At the time of writing the jury is still out. But, taken together with the other evidences from molecular biology discussed above, the story of the *pax-6* gene reinforces the gathering vision of a remarkable unity in diversity throughout the animal kingdom.

18.12 OTHER GENES INVOLVED IN NEURONAL DIFFERENTIATION

The genetics of the nervous system is a 'hot' area at the time of writing and there is little doubt that many other genes coding for transcription and other factors will be detected in the next few years. Already a class of genes encoding proteins showing about 80% amino acid homology with *Drosophila* achaete-scute proteins has been detected early in the embryogenesis of mammalian brains. The *Drosophila achaete-scute* gene complex (*AS-C*) is well known to encode transcription factors which play an important role in the early development of both the peripheral and central nervous systems of the fly. There is, accordingly, every likelihood that the so-called *MASH* (mammalian achaete-scute homologue) genes (*MASH-1* and *MASH-2*) play a similar role in controlling the expression of early neuronal genes in mammals.

18.13 CONCLUSION

It needs no further emphasising that the most striking finding in the neurogenetics outlined above has been the almost incredible conservation of the genes for early development throughout the billion-year history of the Metazoa. Other genes that play significant roles in early development, for instance the genes for **transforming growth factors-β (TGF-β)**, which we shall discuss in Section 19.9, also appear early in evolution and are widely distributed throughout the animal kingdom, from sea-urchin to human. It is not only biochemical pathways and cycles that are conserved but also early developmental mechanisms. These twenty-first-century insights go far to vindicate von Baer's early-nineteenth-century theory. Another notable feature of the neural transcription factors detected by contemporary molecular biology is that the majority act in a **negative** way. They tend to shut off the expression of genes rather than turn them on. Thus, from originally pluripotent neuronal stem cells the great variety of neurons in the adult mammalian brain crystallise out. It has been suggested that the genetic control of the brain's development is of necessity negative rather than positive because of the comparatively enormous diversity of its cellular elements and the need to get this heterogeneity and its interconnections correct. We shall see, in the next chapter, that something rather similar happens in the '**negative sculpting**' of synaptic connexity later in embryology and, indeed, is perhaps carried on into the adult brain in the laying down of long-term memory traces.

19

EPIGENETICS OF THE BRAIN

Disparity between gene and synaptic numbers – preformation ruled out – the concept of morphopoietic fields. **The origins of neurons and glia**: neurulation and neural crest cells – proliferative zone – mitosis and cytoplasmic movement – involvement of developmental genes – migration along glial processes – symbioses between neurons and glia – mouse weaver and reeler mutations – formation of forebrain cell layers – cerebral cortex – cell–cell adhesion and the formation of cortical columns. **Neural stem cells (NSCs)**: definitions – location – response to stimuli – presence in song birds – *in vitro* culture – possible therapeutic use. **Lineage tracing**: retrovirus tagging – enhancer trapping. **Morphogenesis of neurons**: genetic determination – over-production of dendritic spines and neurites – negative sculpting. Morphogenesis of *Drosophila* eye – tyrosine kinases (trks) – biochemical pathways. **Pathfinding**: growth cones – movement and its inhibition by cytochalasin – growth of neurites by growth of neurofilaments and MTs – growth along chemical gradients – NGF and other neurotropic substances – contact guidance – semaphorins – eph receptors and ephrins – cell adhesion molecules (CAMs) – structure of CAMs – E to A conversion in N-CAMs – differential splicing of N-CAM mRNA engenders multiplicity of N-CAMs – destination markers (trophic factors, NGF, BDNF, NTs, TGF-βs) – NT receptors – mostly trks. **Morphological sculpting**: optic tecta of frog – experimental analysis – chemoaffinity theory – concept of sensitive periods – transplant and other surgical procedures – morphopoietic fields. **Functional sculpting**: mammalian visual system – columnar structure of visual cortex – receptive fields – mapping of sensory surface – orientation columns, ocular dominance columns – effects of early visual experience – experimental analysis – inactive synapses lose 'struggle for survival' – synapses which fire together survive together – support from the 'three-eyed' frog – possible molecular mechanisms – the NMJ and the search for the retrograde messenger – the role of NMDA-Rs. The practical importance of a knowledge of neuroembryology

Human brains consist of at least 10^{11} cells and upwards of 10^{14} synapses. We have noted the heterogeneity of neuronal membranes and in Chapter 16 we reviewed the ever increasing number of different synaptically active substances. Yet it is a truism that this vast number of units, this immense heterogeneity, is organised into a harmonious functioning unity. How? The brain is perhaps the ultimate challenge for the developmental biologist! In this chapter we shall look at a few promising molecular approaches.

First of all let us clear out of the way any idea that the structure of the brain (and thus its functioning) might be completely **preformed** – directly specified in the 'blueprint' of the genes. It is easy to eliminate this notion. We only have to look at the numbers. It has been calculated that the human genome consists of no more than 32 000 genes (Chapter 6): yet we have just seen that there are at least 10^{11} neurons and 10^{14} synapses. Clearly the numbers do not add up: even with all the differential splicing, polyprotein subdivision, mRNA editing, etc. which we have reviewed in previous chapters the genome is just not large enough to specify the coordinates of each and every neuron, let alone each and every synapse. Preformationism is not an option. Instead we have to consider some mechanism of **epigenesis** – some mechanism of gradual differentiation and elaboration.

The brain is very different from other tissues in its immense interconnexity. It is not enough that its unit cells should differentiate. They must also form appropriate connections with each other. This compounds the problem set out in the previous paragraph. To use the computer analogy – it is not only that transistors must differentiate from resistors and resistors from capacitors but also that they must all somehow come to be interconnected in the proper way. The genome has to specify not only the position and nature of each cell but also its interconnections. Preformationism seems even more a huge impossibility. It seems that the genes could only specify a broad outline plan. We saw in Chapter 18 how they can direct the laying down of criss-crossing chemical gradients through which the growing neurons have to find their way. This crisscross has sometimes been called a **morphopoietic field** or, alternatively, an **epigenetic landscape**. The intersection of comparatively few morphopoietic gradients could lead to an immensely varied epigenetic landscape – see Figure 19.1. Neuronal genes could be very sensitive to the varying concentrations of morphopoietic biochemicals at different points in this field and, as we noted at the end of the last chapter, switch on or off accordingly.

Before considering the neurochemical possibilities in this concept more closely let us consider the origins of the neurons and glia within the central nervous system. We shall, as usual, confine ourselves principally to the vertebrate central nervous system.

19.1 THE ORIGINS OF NEURONS AND GLIA

We noted in Chapter 18 that the vertebrate central nervous system originates as a longitudinal strip of cells on the dorsal surface of the early embryo. We

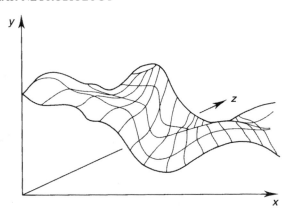

Figure 19.1 An epigenetic 'landscape'. Three chemical gradients laid down along the x, y and z axes of the diagram would create a complex chemical 'landscape' through which a developing neuron would have to find its way. After Waddington (1957), *The Strategy of the Genes*, London: Allen & Unwin.

saw that this strip of cells soon sinks inwards and rolls up to form the neural tube. We also noted that a strip of ectoderm left outside when neurulation is complete, the neural crest, is the precursor of all the sensory nerve fibres, the peripheral autonomic nervous fibres and the Schwann cells. All the other cells of the nervous system originate in the neural tube.

Initially the number of cells in the neural tube is quite small. In amphibian embryos it is believed to be about 125 000. Once the neural tube is complete, however, a rapid proliferative stage begins. The single layer of cells which originally formed the neural plate quickly becomes transformed into a multilayered wall. This **proliferative** stage is, however, interestingly complex. Figure 19.2 shows that the stages of mitotic division are coordinated with movement up and down through the cell layers.

During the **G1 phase** of the mitotic cycle (where the cell is actively metabolising but otherwise quiescent) the cell is situated in the middle of the neuroepithelium. At the onset of DNA synthesis (the **S phase**) the cell nucleus begins to move down toward the ventricular surface. During the **G2 phase** the nucleus is close to the ventricular surface of the neuroepithelium and the lengthy cytoplasmic processes are withdrawn so that the cell rounds up ready for mitotic division. Mitotic division then occurs (**M phase**) so that the cell is duplicated. The

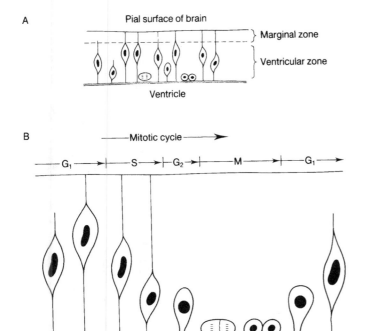

A

Pial surface of brain

} Marginal zone

} Ventricular zone

Ventricle

B

——Mitotic cycle——→

—G₁—→ |—S—→|—G₂—→| —M—→ | —G₁—→

Ventricle

Figure 19.2 Proliferation of cells in the wall of the neural tube. (A) The wall of the neural tube at an early stage consists of two layers: a marginal zone consisting of processes and a ventricular zone where the cell bodies are located. (B) The mitotic cycle of the proliferating neurons is associated with movements up and down through the ventricular zone. Further explanation in text.

two daughter cells then enter the G1 phase once more, their cytoplasmic processes reform and the nuclei return to their original position in the centre of the epithelium.

This sequence of events may be repeated many times. The number of repetitions varies from one brain region to another and appears to be highly characteristic of a given region. What determines the number of repetitions is as yet unknown. But it is crucially important. For once the cell's DNA synthesis has been switched off the cell migrates out of the ventricular, **proliferative**, zone into an **intermediate** zone. Not only this but the cell's 'fate' is also to an extent determined. In other words if its 'date of birth' is known (i.e. the time when it loses its capacity for DNA synthesis) then its final position in the brain, its type and to a large extent its interconnections can be predicted. It will be interesting to learn whether a sequencing of transcription profile is occurring during this proliferative stage as was the case in *Drosophila* neuroblasts (see Section 18.10).

It is found that both neurons and neuroglia originate at much the same time. However, the reproductive rate of neurons is at first far greater than that of glia. In the human fetus the proliferation of neurons occurs during the first 18 weeks after conception. After this period neurons cease proliferating (though there are exceptions in some parts of the brain as we shall see when discussing neural stem cells in Section 19.2). Glial cells, on the other hand, although slow starters, continue dividing for a much longer period – indeed well into postnatal life.

Another feature of the proliferative phase which is worth noting is that the **larger** ('principal' or 'projection' (see Chapter 1)) neurons are the first to develop; the **smaller** ('local circuit' or 'inter-') neurons appear later.

It is not only time of origin that is of significance in determining the fate of the cells streaming away from the proliferative zone but also position in the proliferative zone. This has been worked out most fully for that part of the neural tube destined to form the spinal cord but there is every reason to believe that similar molecular biology operates anteriorly, in the forming brain. We noted in Section 18.4 that *Shh* is expressed in both the notochord and the early neural tube. It is synthesised most actively in the floor of the tube and

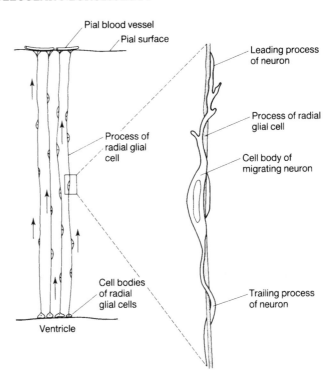

Figure 19.3 Migration of neurons along processes of radial glia. The processes of the radial glial cells extend from the ventricular to the pial surface of the developing cerebrum. On the right of the diagram is an enlargement to show a neuron migrating up a glial process towards the pial surface. After Rakic (1979), in *The Neurosciences: Fourth Study Program*, eds. F.O. Schmitt and F.G. Worden, Cambridge, MA: MIT Press, pp. 109–127; with permission.

diffuses dorsally. This gradient influences the expression of a number of genes encoding transcription factors. Some of these factors are **repressed** by Shh (class 1 homeodomain proteins) and others **induced** (class 2 homeodomain proteins). The Shh gradient determines the levels of these class 1 and class 2 proteins within a progenitor cell and this, in turn, determines its ultimate position in the adult spinal cord and which type of cell it becomes: motor neuron or one of a variety of types of interneuron. This interaction of homeodomain proteins cannot but remind us of the subtle interplay between POU domain proteins in the control of Ddc expression in *Drosophila* (Section 18.9).

The next stage in the development of the brain (though not of the spinal cord where neurons remain central) is the migration of neurons from around the ventricle to the outer surface. This surface because it abuts the innermost of the meninges, the pia mater, is called the **pial surface**. Neurons appear to migrate 'upwards' towards the pial surface along the lengthy processes of **radial glia** (see Chapter 1). The cell bodies of these glia line the ventricular surface of the neural tube and a

lengthy process stretches right across the wall of the tube to terminate against the endothelial wall of one of the blood vessels in the pia mater (Figure 19.3).

The movement of neurons along glial processes seems to be a type of amoeboid locomotion. A leading process entwines itself around the glial fibre and the nucleus is drawn up behind it. In some cases the trailing process behind the neuron merely elongates as the nucleus moves up towards the pial surface. This process may then form the axon. In other cases the trailing process is drawn up behind the nucleus as it makes its slow journey; in yet other cases it becomes detached and is broken down and removed by neighbouring (non-radial) glial cells.

There is evidence for a strong symbiotic relationship between radial glia and migrating neurons. Not only do radial glia act as guidelines for postmitotic neurons but, vice versa, postmitotic neurons appear to be essential to prevent embryonic astroglia continuing their proliferative phase (Figure 19.4). Without such an association astroglia fail to assume their characteristic fibrous form. This is a first example of the great importance of

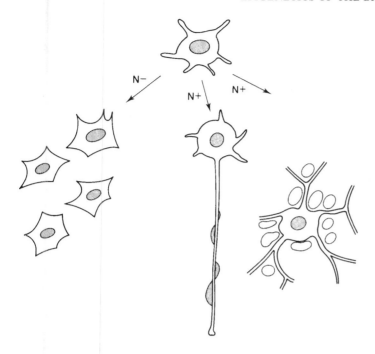

Figure 19.4 Effects of neurons on embryonic astroglia. Neurons and astroglia were rapidly separated from mouse cerebella and either mixed or kept separate. When neurons and glia were mixed (+ N) the glia assumed their characteristic morphology, either developing a long process along which neurons could be seen to 'crawl' or developing radial arms which seemed to 'compartmentalise' the neurons. In the absence of neurons (−N) the glia remained undifferentiated and continued to proliferate. After Hattan and Mason (1986), *Trends in Neurosciences*, **9**, 168–174; with permission from Elsevier Science, © 1986.

cell–cell contact in the development of the nervous system.

The significant role that the fibres of the radial glia play in guiding neurons to their final destination has been examined in mice homozygous for the **weaver** mutation (**wv/wv**). This mutation affects the Kir3.2 K$^+$ channel (see Section 11.8) in granule cell membrane. It results in the granule cells of the cerebellum failing to reach their correct position (Figure 19.5), which leads to the disastrous behavioural upset summed up in the name given to the mutation – weaver.

Although the primary action of the mutation affects the granule cell membrane, there is also evidence that this defective membrane affects the radial glia (in this case Bergmann glia). It appears

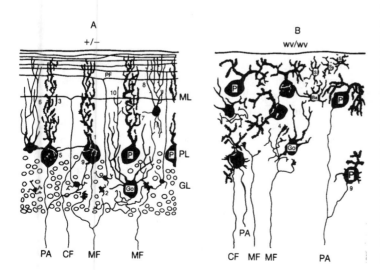

Figure 19.5 The structure of normal (+/−) and weaver (wv/wv) cerebella. (A) Normal mouse cerebellum. (B) Cerebellum of mouse homozygous for the weaver mutation. Both cerebella are taken from three-week-old mice and stained by the Golgi technique. The granule cell layer is completely missing from the weaver cerebellum and the other cells and cell processes are highly disorganised. P= Purkinje cell; Gii=Golgi cell; Ba=basket cell; PA=Purkinje axon; CF=climbing fibre; MF=mossy fibre; GL=granule cell layer; PL=Purkinje cell layer; ML= molecular layer; PF= parallel fibre; the numbers refer to specific types of synapse. From Rakic (1979), in *The Neurosciences: Fourth Study Program*, eds. F.O. Schmitt and F.G. Worden, Cambridge, MA: MIT Press, pp. 109–127; with permission.

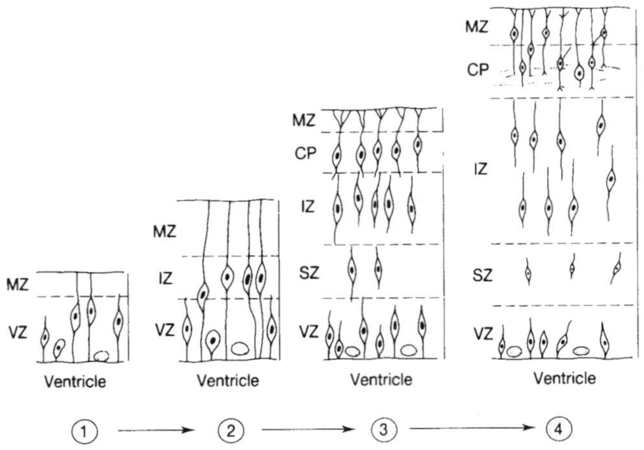

Figure 19.6 Maturation of the neural tube. (1) At an early stage the neural tube consists of two layers: the marginal zone (MZ) consisting of processes and the ventricular zone (VZ) consisting of proliferating cells. (2) Cells that have finished proliferating migrate upwards to form the intermediate zone (IZ). (3) In regions of the brain that are to develop a cortex the cells continue migrating upwards to form a cortical plate (CP). (4) Finally the cortex differentiates from the MZ and the CP; the IZ forms susbcortical white matter; glial cells continue proliferating in the SZ; the VZ becomes the ependymal lining of the ventricle. Further explanation in text. After Jacobson (1978), *Developmental Neurobiology*, New York: Plenum; with permission.

that a component essential for the proper differentiation of Bergmann glia is lacking. On the other hand, it has been shown that granule cells transplanted from weaver into normal mice cerebella appear to migrate normally. This suggests that the mutation also affects Bergmann glia. Whether the mutation affects only granule cells or both granule cells and Bergmann glia remains to be seen. But in either case granule cells lose the means of climbing to their proper position in the cerebellum. The structure and functioning of the cerebellum is consequently severely disrupted.

Next, let us consider the migration of neurons from ventricular to pial surface in a little more detail. This migration is shown in Figure 19.6. It brings about a progressive thickening of the neural tube. The original proliferative wall is shown at A. Two layers are recognisable. A **ventricular zone (VZ)** in which proliferation is proceeding and a **marginal zone (MZ)** in which only the outermost processes of the embryonic neurons are ever present. The first cells to lose their DNA-replicative ability (i.e. to stop proliferating) move up from the ventricular zone to form an **intermediate zone (IZ)**. In the forebrain we find that further migrating cells pass through the IZ to form a **cortical plate (CP)**. The CP later differentiates to give the typical six- or three-layered structure of the neo- or allocortex.

Ultimately the cerebral cortex is formed from the original marginal zone (MZ) plus the cortical plate (CP). The intermediate zone (IZ) beneath the CP becomes the subcortical white matter whilst the original VZ is transformed into the ependymal

lining of the ventricle. Above the ependyma a thin layer – the subventricular zone (SZ or SVZ) – remains as a proliferative region where glia cells and to a much smaller extent, neurons, continue to multiply.

The migration of neurons along the guidance fibres of radial glia goes far to ensure that neurons originating in the same region of the neural tube remain together in the adult cortex. It is likely that this developmental process underlies the **columnar** organisation of the cerebral cortex. It appears, in other words, that the cortical modules which we discussed in Chapter 1 (see Figure 1.23) consist of cells that have been together since their birth. Perhaps they all share similar membrane characteristics (see Section 19.7) which ensure that they stick together in a more than figurative fashion.

The **differential stickiness** of neurons has indeed been shown in numerous experiments. Disaggregated neocortical and hippocampal neurons cultured in appropriate conditions reaggregate into distinct clumps consisting of either cortical or hippocampal cells. Such clumps even show the beginnings of lamination. This indicates that some sorting of the dissociated cells into their definitive positions occurs. Similar experiments have been carried out on the retina. Disaggregated retinal cells tend to re-establish their original organisation. Furthermore it can be demonstrated that disaggregated chick retinal cells adhere preferentially to those parts of the optic tectum to which they would normally project. We shall return to the topic of 'differential stickiness' in later parts of this chapter – especially in Section 19.8, where we review some of the features of **cell adhesion molecules (CAMs)**.

Finally, another behavioural mutation in the mouse has allowed the beginnings of an insight into the molecular bases of some of these post-migratory events. This is the **reeler** mutation (*rl/rl*). In mice homozygous for this mutation the normal number of cells are formed during the proliferative stage and the initial migration along radial glia to the cortex occurs normally. But then something goes badly wrong. The cerebellar cortex, in particular, is badly disrupted. The cells do not form their correct layers or connections. In addition, the cerebral cortex and other laminated parts of the CNS also show faulty layering and other misalignments. The behavioural outcomes of these disrupted architectonics are tremors, ataxia and dystonia.

The reeler gene (*reelin*) is located on mouse chromosome 4 and its homologue has been traced to human chromosome 7. Reelin cDNA has been isolated and sequenced. It encodes a 388 kDa protein (**Reelin**) which is about 25% identical to F-spondin, a morphopoietic protein secreted by the floor plate of the spinal cord. Reelin is first expressed on day 11.5 postcoitus and is located in cells that have finished their radial migration to the cortical plate. It is not expressed in radial and Bergmann glia. It seems likely that Reelin is secreted into the extracellular space of the cortical plate where it perhaps acts as a contact substrate ensuring the correct organisation of ingrowing neurons. Concentrations of Reelin remain high throughout embryological development and then decline to a low steady level in the adult. Although, as mentioned above, its homologue is present in the human genome no neuropathologies have yet been traced to it. Perhaps any mutation is so disruptive of the more complex human cortex that it is lethal before birth.

19.2 NEURAL STEM CELLS

The conventional wisdom has long been that, after the early proliferative stage, neurons are no longer generated in the nervous system. We are born with as many neurons as we are ever to have and we can only look forward to a life-long loss of those which we do have. In recent years a reaction to this dismal

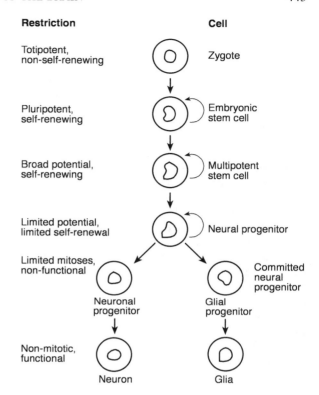

Figure 19.7 Stages in development of neural cells in mammals. Starting from the zygote, cells in the nervous system go through a series of steps in each of which their fate is more and more closely circumscribed. Modified from Gage (2000), *Science*, **287**, 1433–1438.

scenario has gathered pace. Is it possible that, in some regions of the brain, cells retain their proliferative capacity into the adult? Largely due to the hope of brain repair in neurodegenerations, injury and other conditions there has been a great increase in interest in this possibility, and evidence has accumulated that small populations of **neural stem cells (NSCs)** persist in certain regions of the adult brain. Indeed it has been suggested that the brain's core has some resemblance to the marrow of long bones and comparisons have been drawn between haemopoiesis and 'neuropoiesis'. First, however, let us be clear what we mean. Figure 19.7 shows a sequence of developmental stages ranging from the totipotent zygote to the non-mitotic differentiated adult cell.

Pluripotent cells (=embryonic stem cells) are used for creating transgenic animals (see Section

5.17). It is the prospect of using **multipotent cells** from adult tissues to repair damage in adult organisms that has created so much interest in recent years. In particular, the discovery of multipotent stem cells in the adult brain has not only opened the way to possible therapies but also overturned the received wisdom that adult neurons (in mammals) are all postmitotic.

Where are these cells to be found? There are three principal locations in the adult brain: the **olfactory bulb**, the **subventricular zone (SVZ)** and the **subgranular layer (SGL)** of the **dentate gyrus** of the hippocampus. It may also be that NSCs originate in non-neural tissue: either in the ependymal cells which line the walls of the cerebral ventricle or in the vascular system. Clusters of **bromodeoxyuridine** (**BrdU**) labelled cells (BrdU marks dividing cells) have been identified beneath the granule layer of the dentate gyrus close to dividing endothelial cells at the tips of capillaries. In addition, it has been reported that in Old World primates (macaques) new neurons are added not only to olfactory bulb and dentate gyrus, but also to three neocortical areas (but not to the striate cortex, a primary sensory area). This report has been treated with some scepticism and at the time of writing the jury is still out on the presence or absence of neurogenesis in primate association cortex.

It has proved possible to cultivate stem cells from the olfactory bulb, SVZ and SGL *in vitro* and show that they are indeed multipotent. The division of these cells may be **symmetrical** or **asymmetrical**. In the first case both daughter cells resemble their parent in remaining multipotent stem cells or, alternatively, both daughter cells are committed to differentiate into a glial or neural cell. In the second case one daughter cell remains multipotent and the other is committed to differentiate (in the same way as the *Drosophila* neuroblasts discussed in Section 18.10). *In vivo* the committed cell migrates to its proper position in the brain. Many of the NSCs in the SVZ, for example, migrate long distances to the **olfactory bulb** where they integrate into the neural network. NSCs in the SGL of the **hippocampus** migrate upwards into the granule cell layer and differentiate into granule cells.

The reproductive activity of NSCs appears to be responsive to exogenous and endogenous circumstances. It has been shown, for example, that mice raised in an enriched environment (large cage, exercise wheels, toys, social interaction, etc.) have a significantly higher ongoing neurogenesis than those confined to a solitary, impoverished environment. In one set of experiments it was shown that over a 12-day period an average of 2490 new granule cells appeared in the dentate gyrus of the enriched mice compared with an average 1330 in the impoverished group. Interestingly, also, there is evidence (still controversial) that stress diminishes and learning enhances this adult neurogenesis.

The best evidence for the persistence of NSCs into the adult brain comes from studies of the brains of **song birds**. A group of small song birds common in the forests of temperate North America, black-capped chickadees (*Parus atricapillus*), has attracted a great deal of attention. These birds, like their close relatives the willow tits (*Parus montanus*), store nuts and seeds in the autumn in various caches to provide nutriment during the winter months. This makes considerable demands on spatial memory, especially as at this time of the year, the time of leaf fall and early snow, the look of the landscape changes dramatically. Although it is known that new neurons appear in the chickadee brain and especially in the **hippocampal complex** at all times of the year, experiments show that this renewal reaches a peak during October. This peak is especially marked in juvenile birds reflecting, perhaps, the fact that young chickadees have more to learn than their more experienced elders. The NSCs responsible for this production of new hippocampal cells appear to be located in a 350 μm wide band of tissue beneath the hippocampal complex and bordering the lateral ventricle. It is suggested that these cells may be homologous with the SGZ cells of mammals. It may be that the spatial and other information memorised in the hippocampus is passed on to other regions of the brain to form a more permanent record. It is also interesting, in this connection, to note the work which shows that London cab drivers with 'the knowledge' (i.e. total recall of the London A to Z street map) have hippocampi significantly larger than average.

Whereas chickadees show increases in neuron number during the autumn foraging and storing season, many song birds show similar or yet larger neuron increases in the spring when breeding territories are being established and marked out by song. The volumes of a number of song nuclei in the brain are significantly larger in the spring than in the autumn or winter. The number of neurons in the neostriatal region of wild song sparrows (*Melospiza melodia morphna*), sometimes known as the **high vocal complex (HVc)**, increases from about 150 000 in the autumn/winter to 250 000 in the spring. It has been shown that there is a continuous turnover of neurons with new cells replacing old and dying cells throughout the year. In the spring, however, elevated levels of sex steroids slow down the death of old cells. It is clear that this work on the avian brain puts paid to the old idea that the vertebrate brain is anatomically fixed in the adult. Although mammals cannot compete with birds in these reproductive stakes (perhaps because birds are proverbially small-brained) there is accumulating evidence that replicative cells, stem cells, persist into the adult in all mammalian brains.

How, then, are stem cells recognised? Although markers are being developed and, as we noted above, BrdU is much used to detect dividing cells, the only certain way at present is retrospectively, through culture. In correct *in vitro* conditions CNS stem cells multiply producing large colonies containing glia, neurons and further stem cells. They can also be induced to form floating multicellular **neurospheres**. There is evidence that the cells responsible for generating these neurospheres are **astroglia** from the SVZ. The cells from floating neurospheres can, when transferred to culture media containing appropriate growth factors, be differentiated into specific neuronal phenotypes. It may be that when the techniques have been perfected these cells can be transplanted into damaged brains and spinal cords. Stem cells derived from neural crest form adherent colonies of peripheral neurons, glia and yet more stem cells. In the early embryo up to 50% (rat, spinal neural tube) and up to 20% (mouse, telencephalon) of the cells are stem cells. Thereafter, numbers decline rapidly as pluripotent cells differentiate into postmitotic cells so that only a very small percentage is left (as noted above) in the adult brain.

Neural stem cells are not truly multipotent. It can be shown that NSCs derived from spinal cord generate spinal cord cells; those derived from basal forebrain develop into cells expressing GABA, which is more characteristic of this region of the brain than others. NSCs, in other words, are

marked by the region from which they come. They also carry temporal information. The first divisions of cultured NSCs from the embryonic brain produce neurons and afterwards glia, thus reproducing the natural order. NSCs derived from earlier embryonic forebrain produce more generations of neurons than those derived from later forebrain, as is the case in the development of the cerebrum. We saw in Section 18.10 that *Drosophila* neuroblasts sequentially express a series of transcription factors that specify sequences of neurons. We have noted throughout this book the striking homologies at the molecular level throughout the living world. It seems likely, therefore, that what applies in *Drosophila* also applies in mammals and, in particular, in mammalian NSCs.

Finally, can adult NSCs, which, as we have seen, are multipotent rather than pluripotent (Figure 19.7), be induced to return to their embryonic pluripotency and rediscover their early ability to form all types of cell? There is some evidence that this is a possibility. When SVZ cells are cultured as neurospheres their abilities are broadened to the extent that when injected into chick neural crest they develop into cells of the peripheral nervous system. Similarly cells which would normally be fated to differentiate into oligodendrocytes in the optic nerve can be converted, *in vitro*, into neurospheres whose cells regain their multipotency. As in every other area of neurobiology, there is still far to go and much research to be done. The practical benefits of successful stem cell therapy are of course huge ranging from replenishment of lost cells in Parkinson's and Huntington's diseases, to the repair of spinal cord injuries and the repopulation of ischaemic areas caused by stroke or the myelination deficits of multiple sclerosis.

19.3 TRACING NEURONAL LINEAGES

In Section 19.1 we saw how embryonic neurons find their positions by following the guides provided by radial glia and by settling where they find compatible membrane 'stickinesses'. But, as we continually remind ourselves, the brain is an overwhelmingly complex organisation. Is there any way of following the life lines of cells within this complexity? Can we track their wanderings from place of birth (proliferative zones) to their final position in the mature brain? On the face of it, with a jungle of 10^{11} cells, this seems a very tall order indeed. If we inject stem cells in the proliferative zones with a marker such as horseradish peroxidase subsequent cell divisions will so dilute the histochemical that it will soon be impossible to locate. However, some of the newer techniques of molecular biology have come to our rescue. In this section we shall consider two of these techniques: retrovirus tagging and enhancer trapping.

19.3.1 Retrovirus Tagging

We saw in Chapter 5 that retroviruses provided effective means of introducing specific genes into eukaryotic cells. If a β-galactosidase gene is incorporated into the viral RNA and a dilute suspension of the retrovirus injected into the cerebral ventricle some of the cells in the proliferative zone will be infected. Providing the suspension is sufficiently dilute only one or at most a few cells will incorporate the β-galactosidase gene. The gene will replicate at each division of the infected cell. Ultimately the experimental animal is sacrificed and frozen sections prepared of appropriate parts of the nervous system. Any β-galactosidase activity can be detected by means of a chemical, X-gal, which is converted to a blue dye by the enzyme.

This technique confirms the observation that cells which originate together in the proliferative zone tend to stick together throughout life. It is interesting that in the retina it can be shown that clones deriving from a single marked cell differentiate into both rod and bipolar cells.

19.3.2 Enhancer Trapping

Another way of tagging neurons has been provided by the technique of enhancer trapping. We outlined this technique in Section 5.18. If we refer back to that section and to Section 3.3.1 we are reminded that enhancers are DNA sequences which control the transcription of structural genes. They are usually located many hundred or even thousand nucleotides 'upstream' of the structural gene. They rely on the flexibility and bendiness of the DNA double helix to exert their control. This allows a regulatory protein attached to the enhancer sequence to come into contact with the general transcription factors on the promoter region and

'enhance' their activity (Figure 19.8). It is also found that transposons can often have a similar effect. As we saw in Section 4.1.3, these so-called 'jumping genes' frequently carry their own regulatory regions with them. Integration into a chromosome may lead to these regulatory sequences with their attached proteins coming into contact with the promoters of foreign genes.

The use of transposons has proved a useful means of marking and tracing cells during the development of the nervous system. Although the technique is difficult and time-consuming the essence is simple to understand. A marker gene is selected, often the β-galactosidase gene mentioned above. It is incorporated into a promoterless transposon and large numbers inserted at random into the embryo. Some of these transposons will be inserted into a chromosome region influenced by a powerful enhancer. If this region also contains genes controlling a specific cell type it will be inherited in tandem. It is then easy to pick out the movements and distribution of these cells in the developing embryo by detecting the presence of the β-galactosidase gene by the histochemical technique described above.

19.4 MORPHOGENESIS OF NEURONS

It is a truism that neurons have intricate shapes. Yet we left them in Section 19.1 as simple bipolar forms clambering along the guidance fibres of the radial glia. How do they achieve their adult morphology?

Figure 19.9 shows the progressive development of neurons in the cerebral cortex of the human brain. It is clear that at birth the neurons are in a very immature state. They may be in the right place but their dendritic trees are hardly formed and in many cases their axons have yet to find their correct destination.

Each class of neurons has its own distinctive morphology. Pyramidal cells are not to be mistaken for stellate cells; Purkinje cells are completely different from granule cells, etc. These differences persist even when the cells are cultured outside the nervous system. Such major morphologies are evidently genetically determined. The genetic determination, however, probably goes further. Physiologically identified neurons can be filled with dye and their dendritic arborisations carefully studied

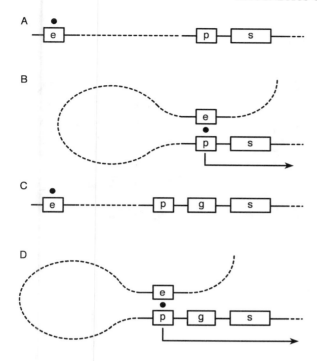

Figure 19.8 Enhancer trapping. (A) A transcription factor (represented by a dot) binds to an enhancer some distance from the promoter (p) and cell-specific structural gene (s). (B) The intervening DNA strand bends so that the enhancer and transcription factor come into the vicinity of the promoter, leading to transcription of the structural gene. (C) A marker gene (g) is inserted near the promoter and structural gene. (D) The enhancer causes transcription of both marker and structural gene. e=enhancer; g=marker gene; p=promoter; s=structural gene.

by serial sectioning. It turns out that similar neurons of genetically identical animals show remarkably similar dendritic branching patterns and the initial routes (at least) of their axons are also remarkably the same. Thus, at this level, genetic determination evidently plays a major role.

Some twenty or so genes have, to date, been shown to influence the growth and/or branching patterns of dendrites. Many of these genes have been analysed in *Drosophila*. Lateral branching is influenced by *kakapo*, which encodes a 5385 residue protein that probably serves to link together cytoskeletal proteins. The extent of dendritic branching is affected by two genes, *shrub* (*shrb*) and *tumbleweed* (*tum*). Mutation of the first, as its name suggests leads, leads to a stunted arborisation and mutation of the latter to an over-elaborate dendritic tree. Another couple of *Drosophila* genes, *flamingo* and *sequoia*, affect the size of the territory which the arborisation colonises. The trees of homologous neurons do not normally overlap, a phenomenon also well known in the vertebrate retina. Mutation of either of the latter two *Drosophila* genes leads to the dendritic tree extending across the dorsal midline of the brain into foreigner's territory. *Flamingo* has been shown to

encode Flamingo, a 7TM protein which, like other 7TM proteins, may well function as a receptor. If so, it may interact with a ligand secreted by other (homologous) midline arbors which prevents it (or, rather, the dendrite in which it is expressed) invading their territory. Although these genetic screens have been carried out in *Drosophila* it is likely that similar morphopoietic genes are at work in the mammalian CNS. Indeed, Flamingo homologues have been detected in humans (hFmi1; hFmi2) and rodents (MEGF2; Celsr1; mFmi1). The likelihood that the *Drosophila* work will find parallels in mammals increases as we recognise more and more the astonishing molecular homologies that extend across the living world.

The fine detail of synaptic connexity and dendritic and telodendritic 'twig' pattern is, however, under environmental control. Young neurons can be seen to undergo a period during which an extravagant growth of dendrites occurs. Many of these hopeful sprouts are withdrawn. This phase of exuberant arborisation is timed to coincide with the arrival of afferent fibres. It is likely that these dendritic sprouts require stabilisation by the establishment of synaptic junctions if they are not to be resorbed.

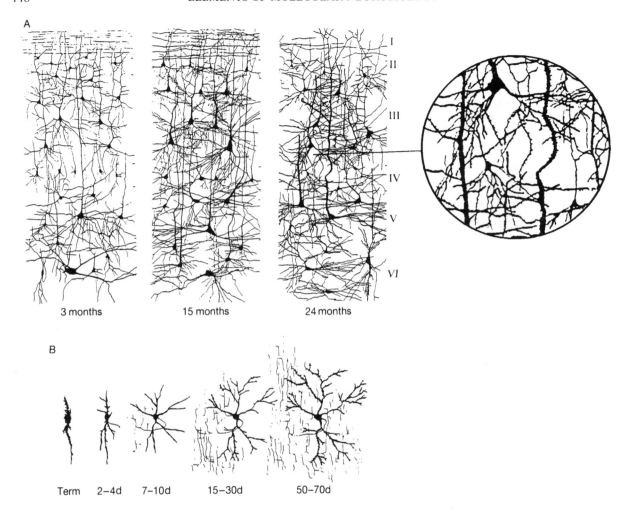

Figure 19.9 Morphogenesis of cortical neurons. (A) Tracings from Golgi-stained sections of the temporal cortex of 3-month, 15-month and 24-month human brains. The increased dendritic arborisation and in the 24-month preparation the strong development of dendritic spines is clearly visible. (B) Tracings from Golgi-stained sections of the cerebellar cortex of kittens at various ages. The dendritic processes of the granule cells show great development during the first 70 days of life and in the 50–70 day section dendritic spines are again clearly visible. Reproduced from Altman (1967), in *The Neurosciences*, eds. G.C. Quarton, T. Melnechuk and F.O. Schmitt, by copyright permission of the Rockefeller University Press.

Similarly with dendritic spines (see Box 20.1). Their formation and stabilisation has recently been investigated in rat hippocampal slices using sophisticated histochemical techniques. In essence presynaptic varicosities on incoming fibres were labelled by coupling fluorescent stains to presynapsin 1 and synaptophysin whilst postsynaptic locales were labelled by attaching another fluorescent molecule, GFP (green fluorescent protein), to PSD-95 (see Section 17.2). It could then be shown that dendrites were continuously sending out hopeful sprouts, or filopodia (Figure 19.10), but these were only stabilised into, first, protospines and, second, mature spines when they made contact with a passing axon. There seems to be a two-way influence. The touch of a searching

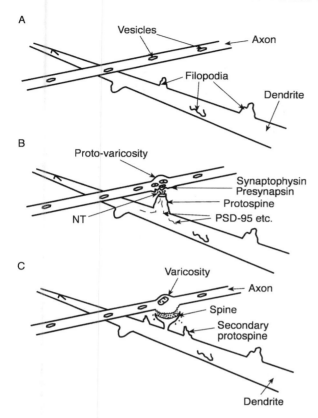

Figure 19.10 Formation of spine-varicosity synapses. (A) The dendrite is continuously forming searching filopodia. When an axon crosses its path the closest filopodium is stimulated. (B) The filopodium grows toward the axon and transforms into a protospine. At the same time its influence causes a varicosity to start forming on the passing axon. This collects synaptic elements (synaptophysin, presynapsin, etc.) and transmitter vesicles. (C) The release of neurotransmitter attracts PSD-95 and other elements (including receptors) to aggregate in the postsynaptic membrane of the spine. The spine may now be regarded as mature. If sufficient neurotransmitter is released other protospines may start growing. Further explanation in text. NT=neurotransmitter. After Matus (2001), *Nature Neuroscience*, **4**, 967–968.

filopodium leads to an accumulation of vesicles and, no doubt, other presynaptic material in the axon so that a varicosity begins to form. The forming varicosity begins to release transmitter on to the underlying filopodium and this leads to recruitment of PSD-95 and the construction of the characteristic GluR-rich postsynaptic density of a mature spine. This process is remarkably rapid. The tissue culture experiments showed that mature spine synapses formed in less than two hours after the initial contact between searching filopodium and axon. Moreover, if the overlying varicosity is particularly active its rain of glutamate may lead to the sprouting and establishment of further spines in the near vicinity of the first.

The formation of the pattern of connections in the developing brain can thus be seen as a continuation of the process of **'negative sculpting'** we noted at work in Chapter 18. The final distribution and number of dendritic spines is selected by the incoming axons from a huge population of transiently forming, withdrawing and reforming filopodia. At a larger scale the fine detail of dendritic trees is also sculpted by incoming afferents from an originally inchoate dendrite thicket. We shall consider evidence for these long-range effects on dendritic form in Section 19.11. Genetics, as we noted above, provides an outline plan; the finished article is shaped by epigenetic forces.

The concept of 'negative sculpting', of chipping away unwanted material to reveal an appropriate pattern, is fundamental. We have seen it at work in the control of transcription and we shall find it at work not only in our considerations of neural epigenesis but also in our considerations of the molecular basis of memory (Chapter 20). It has also been used to explain the way in which the cerebellum moulds the pattern of impulses by which it coordinates muscular activity. 'Negative sculpting', indeed, might almost be seen as the Darwinism of the nervous system.

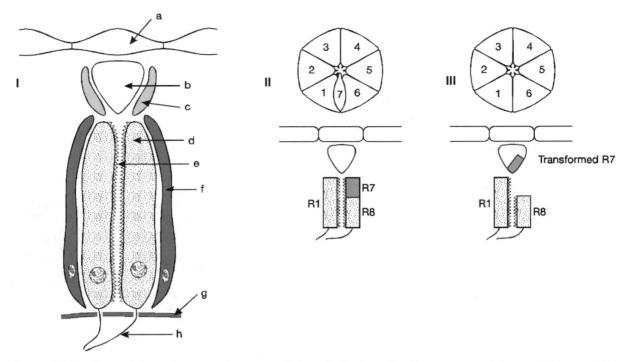

Figure 19.11 Ommatidium of a compound eye. (I) Longitudinal section through one of the c. 800 ommatidia constituting the compound eye of *Drosophila*. (a) corneal facet; (b) crystalline cone; (c) pigment cell; (d) retinula cell; (e) rhabdome (formed by the inwardly projecting rhabdomeres of all the retinula cells; (f) pigment cell; (g) fenestrated basement membrane; (h) sensory nerve fibre. (II) Normal ommatidium. Top: transverse section through top of ommatidium showing the position of the seven retinula cells. Bottom: longitudinal section to show position of R8 and R7. (III) Sevenless ommatidium. Transverse sections and longitudinal sections as in part II showing the transformation of R7 into cone. Parts II and III simplified from Palka and Schubiger (1988).

Finally, in addition to genetics and the influence of neighbouring neurons, the 'local' environment also plays a part in determining neuronal morphology. There is good evidence, for instance, that the position of a neuron in the cortex affects its pattern of branching. For example, the position of a pyramidal cell in the cerebral cortex – whether in the comparatively thin cortex of a sulcus or the thicker cortex of a gyrus – influences its morphology.

19.5 MORPHOGENESIS OF THE *DROSOPHILA* COMPOUND EYE

Drosophila's compound eye provides a valuable preparation in which to study the differentiation of neural cells. Like all compound eyes that of *Drosophila* consists of a large number (about 800)

units known as **ommatidia**. Each ommatidium consists of eight photoreceptors (**retinula cells**) grouped in a tight cylinder beneath a 'crystalline' cone which focuses light. The cylinder is surrounded by pigment cells which isolate it from neighbouring ommatidia (Figure 19.11A).

The retinula cells are numbered 1 to 8. Each has a slightly different spectral sensitivity from that of its neighbours because it expresses a different opsin. R7 is sensitive to ultraviolet light. In the mid-1970s mutant flies defective in UV sensitivity were discovered. The ommatidia of these flies lacked R7. The R7 precursor cell had instead differentiated into a cone cell. These mutants were consequently called **sevenless (sev)**. The *sev* gene was ultimately isolated and shown to encode a 280 kDa membrane-bound **tyrosine kinase (trk)**. Structural analysis of this protein suggested that it

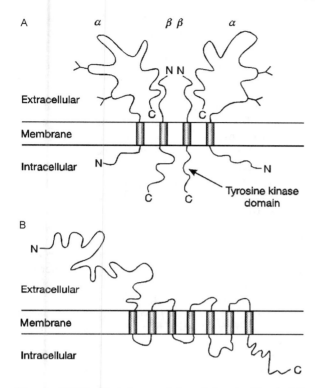

Figure 19.12 Developmental proteins in *Drosophila* retinula cells. (A) The sevenless protein. The $\alpha_2\beta_2$ tetramer is shown. Further explanation in text. After Simon *et al.* (1989), *Proceedings of the National Academy of Sciences, USA*, **86**, 8333–8337. (B) The bride of sevenless (Boss) protein.

has a large (c. 2000 amino acid) extracellular domain. The 280 kDa protein is processed to form two subunits: a 220 kDa α-subunit and a 58 kDa β-subunit. Four of these subunits aggregate to form an $\alpha_2\beta_2$ membrane-bound tetramer which looks very much like a membrane receptor (Figure 19.12A). Subsequently trk was confirmed as a membrane receptor. Many such trk receptors have subsequently been identified. We shall meet important members of the family when we discuss eph receptors and ephrins in Box 19.1. Like the trk receptor in *Drosophila* compound eye they also play crucial roles in morphopoiesis, in this case of the vertebrate nervous system.

It is clear that this system provides a powerful tool for the analysis of cell differentiation in the nervous system: the R7 precursor faces a choice, it can develop into either a retinula or a cone cell. The

full panoply of *Drosophila* genetics, biochemistry, cell and molecular biology has been deployed in its analysis. The putative ligand for the Sev receptor was eventually found. It proved to be a 90 kDa 7TM protein exclusively expressed in retinula cell 8 (R8). Hydropathic analysis suggests that in addition to its seven transmembrane segments it has a lengthy (c. 500 amino acid) extracellular N-terminal (Figure 19.12B). For obvious reasons this protein is called '**bride of sevenless**' or **Boss**. It is only when Boss interacts with Sev that the biochemical processes are switched on that determine that the R7 precursor develops into a photoreceptor. It turns out that whereas Boss only appears in the R8 membrane, Sev is expressed in several other retinula cells. It follows that the specific photosensitivity of R7 depends on delicate and precise spatial and temporal interaction at its membrane with R8 Boss. It has been reassuring to find through immunohistology that Boss is indeed only expressed at the apical tip of R8 (with which only R7 is in contact) and only during a temporal window when the fate of R7 is undecided.

The 'downstream' biochemistry in R7 after the activation of Sev has been analysed in detail by both genetic and biochemical techniques. It involves an adaptor protein, **Drk (downstream of receptor kinase)**, which transfers the signal from activated Sev to a GNRP (see Section 8.3) known as **son of sevenless, Sos**. Sos catalyses the phosphorylation of $p21^{Ras}$. Phosphorylated $p21^{Ras}$ then initiates the biochemical cascade which leads to R7 developing into a retinula cell. It is also believed, finally, that activated Sev inhibits a GAP protein, GAP1, which otherwise dephosphorylates $p21^{Ras}$ back to its resting state. This intricate biochemical cascade is schematised in Figure 19.13.

These detailed analyses (of which only a highly simplified account has been given) show how subtle the epigenetic processes leading to the differentiation of neural tissues is likely to be. On the other hand, hope of progress in the far more forbidding realm of vertebrate nervous systems is provided by the evolutionary conservation which was so prominent a part of the story told in Chapter 18. Most of the proteins mentioned above have their mammalian homologues. Sev is 47% identical to c-Ros; Sos is 45% identical to its mouse homologue; Drk is 64% identical to human GRB2. Once again it seems that analysis of evolutionarily remote forms

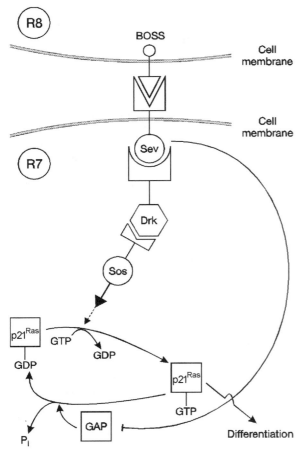

Figure 19.13 Interaction of R8 and R7 in the differentiation of R7. Cell R8 is shown in the upper part of the diagram. Boss activates Sev in the membrane of R7. Activated Sev in turn activates Drk to switch on Sos. Penultimately Sos catalyses the phosphorylation of p21^{Ras} and this initiates the cascade of biochemical events which leads to the differentiation of R7 into a retinula cell. Activated Sev also inhibits the GAP which otherwise dephosphorylates p21^{Ras} to its inactive form. See Section 8.3 for detail of the p21^{Ras} cycle. Partly after Zipursky and Rudin (1994), *Annual Review of Neuroscience*, **17**, 373–379; with permission, © 1984 by Annual Reviews; www.AnnualReviews.org.

is very far from irrelevant to mammalian and human neurobiology.

19.6 GROWTH CONES

In brains, unlike compound eyes, neurons not only have to differentiate but their processes have also

to establish appropriate connections. Neural processes (neurites) – axons and dendrites – grow by means of **growth cones**. Growth cones, as Figure 19.14 shows, are rather like outspread hands, palms down, at the ends of neurites. They only form when the neurite is in contact with an appropriate substratum. The 'palm' of the growth cone is typically about 5 mm in diameter and the numerous fingers (**microspikes**) are long (up to 50 mm) and narrow (diameter: 0.1–0.2 mm). They are in constant movement – stretching out, waving to and fro, retracting back into the 'palm', etc. The 'palm' itself frequently divides, each half seemingly searching out the best route. The least successful half is usually resorbed although, especially in the case of dendrites, branch points may be established.

If the ultrastructure of a growth cone is carefully examined it can be seen to contain innumerable small vesicles sometimes running together to give larger profiles resembling smooth endoplasmic reticulum. The cone also contains mitochondria and the characteristic filamentous elements of neurites: neurotubules, microfilaments (actin) and neurofilaments.

The growth of the neurite is probably accomplished by the synthesis of neurotubules and neurofilaments in the cell body and their continuous extrusion into the process. Labelling techniques have shown that these elements move out along an axon (and thus presumably also along dendrites) at 1 or 2 mm per day, i.e. a rate similar to slow axoplasmic transport (see Section 15.3). Whilst the neurite grows there is no disassembly of the tubules and filaments, just a gradual elongation. Superimposed upon this slow steady elongation is a more rapid transport of membranous vesicles. These presumably carry new membrane to the cone where it collects as the small vesicle population and smooth ER which, as we noted above, is characteristic of growth cone ultrastructure. The membranous material is ultimately added to the ever-growing plasmalemma of the growth cone by the processes of exocytosis (see Section 15.4). The vesicle membrane fuses with the membrane of the growth cone and its lipoprotein constituents quickly diffuse into the bilayer.

The restless probing activity of the growth cone depends on the actin microfilaments. If cytochalasin B, which inhibits actin polymerisation, is added to the culture medium the ceaseless exploratory

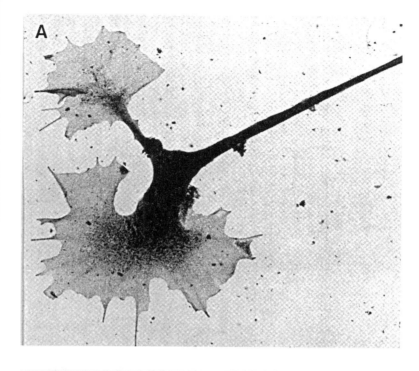

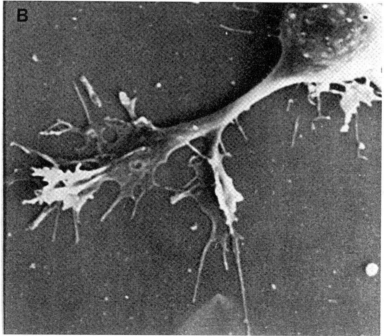

Figure 19.14 Growth cones of neurons in tissue culture. (A) Transmission electron micrograph of two growth cones at the end of the axon of a cultured rat sympathetic cell. (B) Scanning electron micrograph of cultured fetal-rat hippocampal cell showing growth cones at the tip of a dendrite. Courtesy of Dr R.S. Rothman.

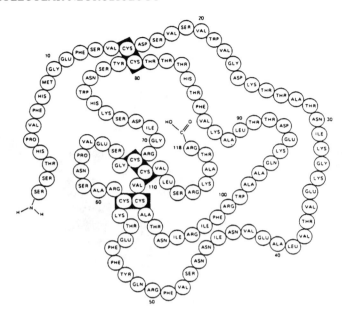

Figure 19.15 Molecular structure of a single unit of the major β-NGF dimer. There are 118 amino acids in the polypeptide chain and three disulphide linkages between cysteine residues (black rectangles) link different parts of polypeptide together. From Angelletti and Bradshaw (1971), *Proceedings of the National Academy of Sciences, USA*, **68**, 2417–2420.

movements of the growth cone cease. It no longer puts out its microspike fingers and remains quiescent, fixed to the substratum.

19.7 PATHFINDING

Elongation of a neurite is not, of course, enough. It must find its way to its destination. The way in which this vital pathfinding job is accomplished is as yet still poorly understood. Two mechanisms are believed to be at work: growth along a chemical gradient and contact guidance. One or other, and probably a combination of both, mechanisms probably underlie neurite pathfinding.

By far the best-known neurotropic substance is **nerve growth factor (NGF)**. Although NGF was first extracted from mouse sarcoma cells, its richest source is (for some unknown reason) the salivary gland. NGF exists in several different forms with various different molecular weights. Its most active form (**β-NGF**), however, has a molecular weight of 26.5 kDa and consists of a non-covalently bound dimer of two 118 amino acid units (Figure 19.15). The other forms of NGF are weakly bound associations of this fundamental structure.

The major effect of NGF is on nerve cells derived from the neural crest. These, it will be recalled, are autonomic and sensory neurons. Neurons belonging to the **sympathetic nervous system** (but not the

parasympathetic system) are particularly dependent on NGF. If anti-NGF antibodies are injected into newborn mice the sympathetic nerve fibres are selectively destroyed. Vice versa, immunological techniques show that NGF is present on the target cells of sympathetic fibres.

The **neurotropic** effects of NGF are readily demonstrated in tissue culture. Figure 19.16 shows an experiment in which sympathetic neurons are plated on the central division of a culture dish. Scratches in the substratum ensure that neurites grow only to left or right. A silicon grease barrier divides the central plateau from the plateaux on either side. Silicon grease prevents the diffusion of NGF which is introduced into the medium of one of the side plateaux but not the other. The figure shows that the sympathetic neurites only grow into the NGF-containing compartment. Other experiments have shown that cultured sympathetic neurons will 'follow' a micropipette filled with NGF through a 180° turn.

We shall return to NGF in Section 19.9 for, in addition to the neurotropic (guidance) effect which we have been considering in this section, it also has a **neurotrophic** (or nutritional) effect. This, as we shall, see is of very considerable importance in stabilising synaptic connections with target cells once the latter have been 'found'.

The second major means by which neurites find their target cells is by **contact guidance**. Growth

A

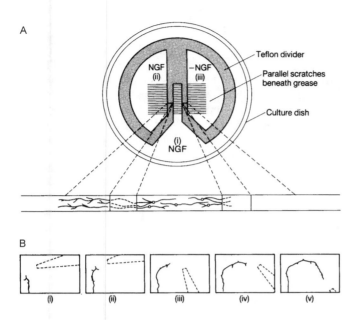

Teflon divider

Parallel scratches beneath grease

Culture dish

NGF (ii)   –NGF (iii)

(i) NGF

B

(i)   (ii)   (iii)   (iv)   (v)

Figure 19.16 Neurotropic effect of NGF. (A) The culture dish has three compartments ((i), (ii) and (iii)) separated from each other by a Teflon wall which is sealed to the dish by a layer of grease. A culture of rat sympathetic ganglion cells is placed in compartment (i). Compartment (ii) contains culture medium plus NGF whereas compartment (iii) contains medium without NGF. NGF is present in compartment (i). Parallel scratches (lined with collagen) under the grease prevent random growth of neurites. Neurites only grow into side chambers which contain NGF and if NGF is removed they regress. This is shown in the cross-sectional view of the culture dish and neurites in the bottom part of the diagram. After Campenot (1982), *Developmental Biology*, **83**, 1–21. (B) The growing neurite is on the left in this series of illustrations. It can be seen that the growth follows the nozzle of a pipette (dotted outline) containing NGF. With permission from Ribchester (1986), *Molecule, Nerve and Muscle*, Glasgow: Blackie; after Gundersen and Barrett (1979), *Science*, **206**, 1079–1080. Copyright 1979 by the AAAS.

cones show a preference for substrata to which they can adhere strongly. Indeed no growth occurs at all if a suitable surface is not supplied. One surface to which growth cones adhere, both *in vivo* and *in vitro*, is that provided by other neurons. Thus when one pathfinder neuron has found its target others can 'feel' their way along it to the same or neighbouring targets. This helps to explain the neuroanatomical fact that nerve fibres frequently run in bundles, or **fasciculi**.

In tissue culture nerve fibres can be induced to grow along tracks of poly-L-lysine and poly-L-ornithine. These polypeptides have positively charged side chains (Table 2.1) which probably interact with negative charges on the cell membranes of the growing neurites. Indeed so 'sticky' is the substratum that in these cases not only the growth cone but also the rest of the neurite remains attached and thus records the track which the neurite has taken (Figure 19.17).

Other experiments have indicated that type 1 and type 4 collagens, fibronectin and especially the protein constituent of basal laminae – **laminin** (Figure 19.18) – when bound to a polyornithine substratum are extremely potent neurotropic substances. Yet other work has shown that the plant lectin, **concanavalin A (Con A)**, also acts as a

powerful contact guide – at least for leech neurons. These tissue culture experiments hold great promise of determining the nature of the molecules which growth cones (indeed the growth cones of different types of neuron) recognise.

Having arrived at its target, a growth cone must, of course, stop. This is another puzzle. It is not known what switches off the elongation process. One can only speculate that, once again, some chemical in the target cell membrane switches on a neurotubule/neurofilament disassembly process in the growth cone. If and when this happens the neurite will assume its adult condition. Neurotubules and neurofilaments will be assembled at the perikaryal end of the process and disassembled at the synaptic end (see Figure 15.12). Axoplasmic flow will continue (as we noted in Chapter 15) but no elongation of the process will occur.

We have seen in the preceding paragraphs that the elongation and direction finding of neurites depends on contact adhesion and the following of chemical pathways and 'signposts'. We can see, in a general sort of way, that the ceaseless probing of the microspike 'fingers' could very well be a searching out of such chemical route markers. Space has only allowed mention of a few of what is turning out to be a large and diverse population of

BOX 19.1 Eph receptors and ephrins

We saw the importance of receptor-bound tyrosine kinases when discussing the development of *Drosophila* compound eye in Section 19.5. Eph receptors are another group of such kinases. They are found throughout the vertebrate body and, in particular, in the nervous system. They are deeply involved in generating and maintaining the organisation of the CNS. They show a cysteine-rich extracellular domain, a transmembrane domain and an intracellular domain with tyrosine kinase activity. They have been classified into two groups, EphA and EphB. Their ligands, the ephrins, were only discovered some time after the receptors had been isolated. They, too, are membrane-attached molecules. Like their receptors they are classified into two groups. In this case the classification is according to their mode of membrane attachment: either through a glycosylphosphatidylinositol (GPI) linkage (class A) or through a transmembrane section and short cytoplasmic domain (class B).

It has been shown that solubilised ephrins bind to Eph receptors but do not trigger activation unless aggregated. It seems then that it is only when held together in clusters on adjacent cell membranes that they interact with Eph receptors so that the latter's intracellular activity can be released. This clustering is thought to be assisted by the presence of distinct and welcoming microdomains in the phospholipid membrane (see Section 7.5). There is also a 'back reaction': binding to an Eph receptor triggers the tyrosine kinase activity of the ephrin. Eph-A receptors bind preferentially to class A ephrins and Eph-B receptors interact preferentially with class B ephrins.

The reaction between ephrins and their receptors can be either one of attraction or one of repulsion. The biochemical mechanisms responsible for this dichotomy are at present obscure. It seems, however, that as ephrin increases in concentration an initial attraction changes to repulsion. The two classes of ephrin also sometimes work antagonistically, one class bringing about adhesion the other repulsion. Their interactions clearly form an intricate 'force-field' through which the fibres of the developing nervous system have to find their way.

The understanding of how ephrins and their receptors interact has been greatly clarified by the recent X-ray crystallography at 2.7 Å resolution of an Eph receptor–ephrin complex. In essence the crystallography shows that a loop presented by ephrin inserts into a channel in the surface of the receptor. This alters the conformation of the ephrin exposing a receptor surface for another ephrin–receptor dimer. The resulting tetrameric complex activates the tyrosine kinase domains of both the ephrins and the eph receptors thus creating the forward and back reactions which, as we noted above, are such a significant feature of their biochemistry.

Ephrins are believed to play significant roles throughout the development of the nervous system. We noted in Section 18.8 that cells in the developing hindbrain were restricted to their native rhombomeres. There is evidence to show that the repulsive action of ephrin-B and Eph-B receptor is one of the factors involved in this early patterning. This antagonism also seems to be at work in the patterning of the motor neuron innervation of the somites. There are many other examples but one of particular interest in this chapter concerns the development of the visual pathway. It has been shown that early in the development of *Xenopus* visual pathway all axons pass to the contralateral optic tectum. After metamorphosis, however, due to an upregulation of ephrin-B in the chiasma, some optic fibres turn back and proceed to the ipsilateral tectum. Similarly ephrin A2 and A5 both form anterior–posterior gradients in the chick tectum and mouse superior colliculus. These gradients are sensed by ephrin receptors in the growth cones of ganglion cell axons from the retina leading to the formation of an accurate retino-tectal (retino-collicular) 'map'. Support for this conclusion comes from homozygous null mutants of ephrin A5 or ephrin A2 where the map is scrambled in the way expected.

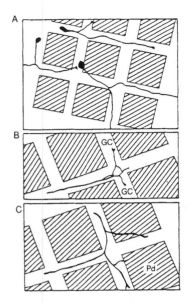

Figure 19.17 Track of cultured neurons on polyornithine. A dish has first been coated with polyornithine and then had patches of palladium deposited on top (cross-hatched). Sensory neurons from chick embryos grow along the polyornithine lanes and only very rarely cross the palladium. Pd=palladium; GC=growth cone. From Ribchester (1986), *Molecule, Nerve and Muscle*, Glasgow: Blackie; after Letourneau (1975), *Developmental Biology*, **44**, 77–91; with permission.

these markers. **Ephrins** and their receptors play important roles and are outlined in Box 19.1. Other signalling molecules include the appropriately named **semaphorins**, which have been shown in some cases to attract and in other cases repel sympathetic and other neurons. Indeed, evidence has recently been presented which shows not only

the attractive and aversive effect of semaphorins on growth cones but also that in some cases they may act to antagonise each other. Other route markers include the **neuropilins** and **plexins**. Some of these molecular 'beacons', especially some of the semaphorins, have molecular structures resembling the immunoglobulins. In this they resemble the cell adhesion molecules (CAMs). It is to these important molecules that we turn next.

19.8 CELL ADHESION MOLECULES (CAMS)

Three groups of cell adhesion molecule (CAM) are presently recognised:

1. CAMs related to immunoglobulins (Igs) with Ca^{2+}-independent binding.
2. CAMs with Ca^{2+}-dependent binding, often called cadherins.
3. Miscellaneous.

It is only the first group (some thirty members) that is of major importance in the CNS although cadherins play a minor role. Within this Ig-related family only the N-CAMS (neural-cell adhesion molecules) and the Ng-CAMs (neuron–glia cell adhesion molecules) are significant in the nervous system. P_0 and MAG are also members of this Ig-related family and we noted their important role in the structure and development of myelin in Section 7.7.

Both N-CAMs and Ng-CAMs are membrane-embedded glycoproteins. N-CAMs are found in the membranes of all neurons; Ng-CAMs appear to be 'secondary' CAMs as they are not found in the membranes of neurons during the proliferative

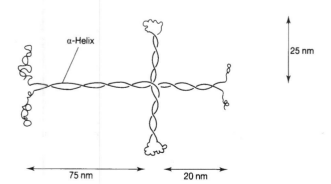

Figure 19.18 Molecular structure of laminin. Laminin is a cross-shaped molecule with a molecular weight of some 900 kDa. The arms and stem of the cross are formed of twin α-helices wound around each other to form a two-stranded rope. At each end of the cross the polypeptide chain forms a more disordered globular type region which is believed to possess cell adhesion and/or neurite growth-promoting characteristics. Partly after Davis *et al.* (1985), *Trends in Neurosciences*, **8**, 528–532.

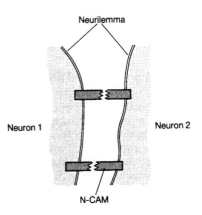

Figure 19.19 Homotypic binding between N-CAMs. The schematic shows the principle of homotypic binding between N-CAMs of two neighbouring neurons.

stage and only appear after mitosis has ceased. The great importance of neuron–glia interactions was emphasised in Section 19.1 above.

The two types of neural adhesion molecule also have somewhat different binding properties. N-CAMs show '**homotypic**' binding, i.e. N-CAMs in one neural membrane bind to N-CAMs in a neighbouring neural membrane (Figure 19.19). Ng-CAMs, in contrast, show '**heterotypic**' binding. Ng-CAMs in a neural membrane bind to a different (so far undiscovered) CAM, or some other receptor, in a glial cell's membrane.

The first detailed structure of a neural cell adhesion molecule was published in 1987 and since then a number of other CAMs have been sequenced. The first N-CAM gene to be sequenced contains 19 exons and extends over 80 kbp. It encodes a huge polypeptide of 1072 residues. The primary transcript is processed by alternative splicing to yield three major variants. Although these variants are of very different sizes they all share a common N-terminal configuration (Figure 19.20). This configuration is well known in other immunoglobulins (Igs). It consists of five homologous repeats, perhaps derived by duplications of

an ancestral gene. Each homologous repeat forms a loop whose ends are joined by a cysteine disulphide linkage. The polypeptide then assumes another configurational motif, the fibronectin motif (see Section 4.2.2), and this is repeated before inserting itself into the membrane. All three polypeptides are identical up to this point (682 amino acids). But, as Figure 19.19 shows, they differ considerably thereafter. The largest, the so-called 'large-domain' (ld) (115.5 kDa), variant has an extensive intracellular domain, the next largest (89.5 kDa) has a comparatively short intracellular domain (sd) and the smallest (ssd) (78 kDa) has no intracellular domain at all and is attached to the membrane by a phosphatidylinositol anchor.

More recently another N-CAM gene, called L1, has been located on the X chromosome. It encodes a 200 kDa protein with six Ig domains and five fibronectin motifs. The protein is present in two isoforms due to differential splicing of the gene's 28 exons. During development the isoforms are expressed on the surfaces of axons and growth cones. L1 is of particular interest because mutations cause a number of X-linked neuropathologies. Affected males show several types of mental retardation, IQs between 20 and 50, HSAS syndrome (hydrocephaly due to stenosis of the aqueduct of Sylvius), and sometimes MASA syndrome (mental retardation, aphasia, shuffling gait, adducted thumbs). These pathologies can be traced to defective pathfinding in the developing brain. The isolation of cDNA for the human L1 gene in 1991 has allowed a start to be made on analysing the part played the L1 CAM in pathfinding and hopefully to rational therapies for the pathology.

When isolated and viewed in the electron microscope N-CAMs often show the striking three-armed 'Isle of Man' motif we met in clathrin triskelions (Figure 19.21). Further analysis of these images suggests that CAMs have a 'hinge' where the fifth Ig domain attaches to the first fibronectin motif (Figure 19.22A). When isolated for EM visualisation the ld and sd N-CAMs attach at their transmembrane domain to form the

Figure 19.20 Major features of CAMs. Structures of N-CAMs of importance in the nervous system are shown. The Ig domains are represented by broken circles. The break is in fact bridged by a disulphide linkage. Fibronectin (fibronectin type 3) repeats are shown as rectangles. Polysialic acid is symbolised by 'cocktail sticks' and N-linked oligosaccharides by Ys. Further explanation in text. Modified from Edelman and Crossin (1991), *Annual Review of Biochemistry*, **60**, 155–190.

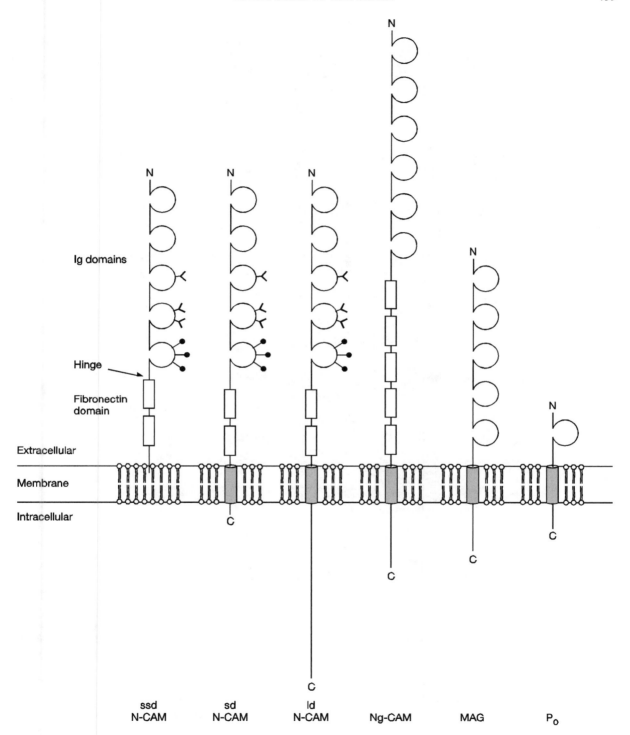

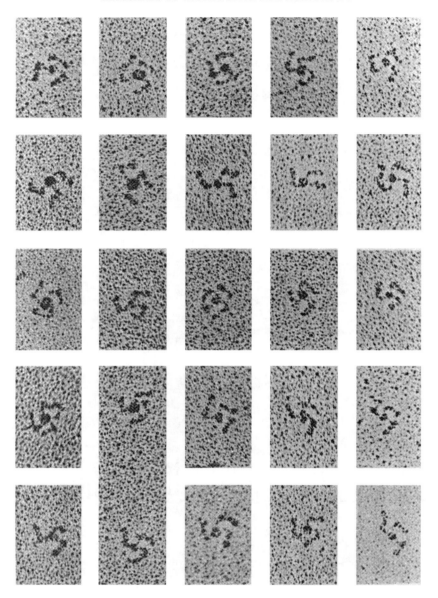

Figure 19.21 N-CAM triskelions Electron micrographs of N-CAM triskelions from embryonic chicken. The triskelions were rotary-shadowed and viewed at a magnification of ×200 000. Each arm measures about 40 nm. Reproduced with permission, from Edelman (1984), *Annual Review of Neuroscience*, **7**, 339–377, © 1984 by Annual Reviews; www.AnnualReviews.org.

three-armed structure (Figure 19.22B). This analysis also suggests the conformation of N-CAMs in neuronal membranes. This conformation is shown in Figure 19.22C. Because the structure is hinged it allows a fair amount of flexibility

between cells in relative motion or with different aspects to each other.

Figure 19.20 shows that Ng-CAM has a very similar structure to N-CAM. The figure shows that it is an even larger protein, developing six Ig-like

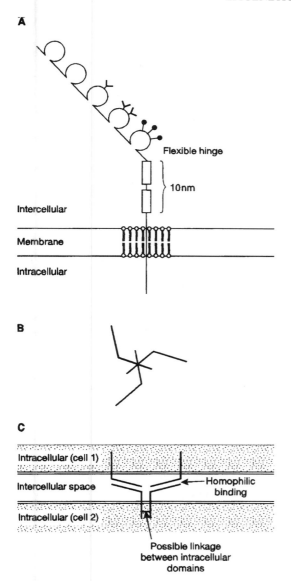

Figure 19.22 Conformation of N-CAMs. (A) Orientation of N-CAM in a neuronal membrane. Symbolism as in Figure 19.20. (B) Formation of 'Isle-of-Man' triskelions when isolated. The individual N-CAMs appear to attach together at their transmembrane domains or, perhaps, at their intracellular domains. (C) Possible conformation of N-CAMs holding two cells together. The immunoglobulin domains are responsible for homophilic binding. Also shown is a possible bond between the intracellular domains of two adjacent N-CAMs (see part B). After Becker *et al.* (1989), *Proceedings of the National Academy of Sciences, USA*, **86**, 1088–1092.

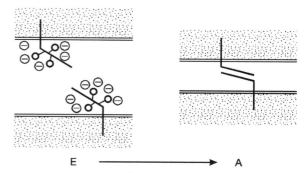

Figure 19.23 E to A conversion. In the embryonic (E) state the N-CAMs repulse each other by electrostatic forces between negative charges on polysialic acid and other oligosaccharides. The 'hinge' is also rather inflexible due to the presence of three bulky polysialic acid units. Loss of polysialic acid greatly reduces the number of negative charges and also allows greater flexibility at the hinge. Homophilic binding between the Ig domains can now take place.

homologous repeats at its N-terminal followed by five fibronectin motifs before plunging into the membrane and emerging on the cytoplasmic side with a short intracellular domain. Ng-CAM also differs from N-CAM in its carbohydrate content. Figure 19.20 shows that all the forms of N-CAM, but not Ng-CAM, bear oligosaccharides on their third, fourth and fifth Ig domains. This is not a mere detail. For N-CAM differs from Ng-CAM in showing a radical change in its carbohydrate composition during embryology. This change, which is termed the **E to A conversion** (i.e. embryonic to adult conversion), consists in a reduction by at least a third in the polysialic acid content of the molecule (Figure 19.23). Ng-CAM lacking polysialic acid cannot undergo this transition. Figure 19.20 shows that polysialic acid is attached to N-CAM at three sites on the fifth Ig domain. The loss of negatively charged polysialic acid has two effects. First, it greatly increases the mutual adhesion between adposed N-CAMs. Second, the loss of polysialic acid from the fifth Ig domain removes what is thought to be a hindrance to movement of the 'hinge'. Both effects lead to a tighter binding between adjacent cells.

Differential splicing of the primary transcript provides different N-CAMs at different developmental stages and in different brain regions. This mechanism for producing variation bypasses the perhaps more cumbersome transcriptional control which, as we saw in Chapter 18, is so important in the early stages of neuroembryology. It enables the growth cone to react rapidly to changing environments as it wends its way to its final position. Furthermore, variations in polysialic acid loss will vary the strength of the adhesion between neighbouring N-CAMs. Indeed, it has been reported that in parts of the brain where plasticity persists into the adult, for instance the hippocampus, polysialic acid is never totally lost, in other words the E to A conversion is never fully complete. Finally, it can be shown that there are great variations in the quantity of N-CAM mRNA present in the brain during development. In the mouse the quantity of N-CAM messenger peaks near birth and falls by about 80% during development.

It is believed that about 100 000 N-CAM molecules are developed in the neurilemma of an average neuron. There is thus much opportunity for homophilic binding between neurons. But, it might reasonably be asked, how does this help us with understanding the contact-adhesion theory of pathfinding? Would not all paths look the same to the probing microspikes of a growth cone? The answer to this question seems to be that different neurons have different spatial and temporal patterns for N-CAM expression. Moreover, different timings and degrees of E to A conversion in different regions of the brain ensure that different adhesions are expressed in different brain areas.

19.9 GROWTH FACTORS AND DIFFERENTIAL SURVIVAL

The phenomenon of neurogenesis is far from over when neurons and their processes reach their definitive positions. There is good evidence that there is very considerable over-production of neurons followed, once again, by a process of 'negative sculpting'.

In those systems that have been carefully studied (spinal cord of *Xenopus*, various parts of chick CNS, etc.) the over-production ranges from 40% to well over 100%. The cell loss, moreover, does not occur at random, spread over a lengthy period of time. It occurs at a definite period in development which can be predicted with some certainty. In the lateral motor column of the chick's spinal cord, for

instance, a population of over 20 000 neurons is reduced to just over 12 000 in 72 hours (Figure 19.24A); in *Xenopus* spinal cord a reduction from 4000 to 2000 cells occurs in about the same period of time.

It can be shown that the degenerating cells are not merely the last to arrive, denied their 'place in the sun'. Rather the evidence points to the causal agent being at the far end of the axons, at the growth cones. If, for instance, the target field is partially or completely extirpated the loss of cell bodies is proportionately greater, and this loss, moreover, occurs at the same time as naturally occurring cell loss. Vice versa, if the target field is increased in area (by, for instance, implanting a supernumerary limb bud into the limb field of a chick embryo) then the cell loss in the appropriate motor area of the spinal cord is much reduced.

There is evidence that this initial over-production and subsequent selective pruning is a feature of all neural systems. How are the 'successful' neurons selected? What is the 'causal agent' at the growth cone's destination? Is it some **'trophic factor'** necessary for the continuing life of the neuron? Is this 'trophic factor' rationed in such a way that only a fraction of the questing growth cones can be satisfied?

We have already met one trophic factor in the guise of nerve growth factor (NGF). As we saw in Section 19.7 this factor, in addition to being a **tropic** factor, is also a **trophic** factor. Not only does it guide sympathetic neurites to their destination, but it is also taken up by synaptic terminals and transported in a retrograde direction towards the perikaryon. If this take-up is prevented by axotomy or by injection of anti-NGF antibodies, sympathetic neurons undergo a series of metabolic and structural changes and ultimately die. These consequences can be reversed and/or prevented if NGF is presented to the cut end of the axon.

It can be shown that sympathetic terminals have specific **NGF receptors** in their membranes. When a receptor–NGF complex is formed endocytosis occurs at the terminal and the endocytotic vesicle so formed is transported in the retrograde stream along the axon to the perikaryon (Figure 19.25). Once in the perikaryon the NGF is released and can there exert its various metabolic effects. One such effect is to induce the synthesis of **tyrosine hydroxylase (TH)**. It will be recalled from Chapter

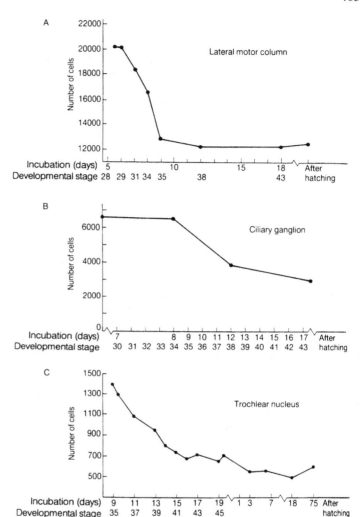

Figure 19.24 Cell death in embryonic chick central nervous system. In each region a great loss of neurons occurs at particular developmental stages and particular times of incubation and after hatching. Part A from Hamburger (1975), *Journal of Comparative Neurology*, **160**, 535–546; reproduced by permission of Wiley-Liss, Inc., a subsidiary of John Wiley & Sons, Inc., © 1975. Part B from Landmesser and Pilar (1974), *Journal of Physiology*, **241**, 737–749; reproduced by permission of the Physiology Society. Part C from Cowan and Wengler (1967), *Journal of Experimental Zoology*, **164**, 267–280; reproduced by permission of Wiley-Liss, Inc., a subsidiary of John Wiley & Sons, Inc., © 1967. After Purves and Lichtman (1985), *Principles of Neural Development*, Sunderland, MA: Sinauer.

16 that this is the rate-limiting enzyme in the synthesis of noradrenaline in adrenergic neurons.

Although, as mentioned in Section 19.7, NGF's major activity is on sympathetic neurons it has also, more recently, been shown to have an effect in the CNS. Somewhat surprisingly, it does not affect central catecholaminergic neurons. Instead it has been shown that if NGF is injected into the brains of adult rodents it is taken up and transported in a retrograde direction by **cholinergic** neurons. It causes an increased synthesis of choline acetyltransferase (CAT) and thus a general activation of cholinergic pathways. Endogenous NGF can be detected in rat hippocampus and neocortex and it can be shown to

be transported from there to cholinergic cell bodies in the basal nuclei of the forebrain.

NGF is by far the best-known trophic factor. Other trophic factors have, however, been detected in the CNS. The pig's brain has, for instance, yielded a 12.3 kDa factor which maintains the growth of sensory neurons in tissue culture. This factor is called **brain derived neurotrophic factor (BDNF)**. Analysis of BDNF shows that large parts of its primary structure are identical with that of NGF. This allowed oligonucleotide probes to be synthesised which, in turn, allowed the isolation of a small family of related neurotrophins: neurotrophins 2, 3 and 4 (NT-2; NT-3; NT-4). *In vitro*

BOX 19.2   Neurotransmitters as growth factors

In Chapter 16 we noted that the old concept of neurotransmitters as involved solely in discrete, punctate transmission was beginning to give way to a broader view. Neurotransmitters not only usually had intricate and long-lasting effects on the subsynaptic cell but also frequently diffused from varicosities through the intercellular space to influence the activity of distant cells. It is nowadays becoming apparent that in many instances, especially in the developing brain, these non-synaptic influences include acting as growth factors.

Early indications that neurotransmitters doubled as growth or trophic factors came from the observation that they were to be found in the embryonic brain before synaptic connections had been established. This observation has been supported by results from work with cultured cells derived from the embryonic nervous system. It can be shown that neurotransmitters affect cell division, neuronal sprouting, growth cone motility and the phenotypes of both neurons and glia. In some cases these effects are direct, in other cases indirect where the neurotransmitter acts on glial cells which in turn produce substances which affect neurons.

It can be shown that serotonin, acetylcholine and noradrenaline affect cell division of proliferating neurons and glia. These conclusions follow mainly from tissue culture studies where either neurotransmitter agonists or antagonists were applied. Agonists stimulated and antagonists hindered DNA synthesis and cell division. Neurotransmitters, especially those which affect Ca^{2+} channels, are also important in controlling the drop-out of at least some neurons in the developing brain (Section 19.9). It appears that internal Ca^{2+} concentrations are essential if cerebellar granule cells are to survive through a particular developmental stage. This concentration is normally provided by glutamate released from mossy fibre terminals opening NMDA and kainate gates in subsynaptic membranes belonging to the granule cells. These and other findings emphasise that afferent activity is essential if neurons are to survive into the mature brain.

This activity-dependent development of neurons also extends to the development of neurites. In some cases (granule cells) glutamate promotes extension and multiplication of neurites, in other cases (hippocampal pyramidal cells) it inhibits neurite sprouting (low doses) or kills the cells altogether (high doses). The situation *in vivo* is, however, complex. Other neurotransmitters (ACh, GABA, DA) interact with glutamate either to potentiate or to ameliorate its action. Furthermore, not only can neurotransmitters affect the development of neurons but there is beginning to be evidence that they can also control the expression of particular phenotypes. If, for instance, a noradrenaline uptake blocker (desmethylimipramine) is applied to presumptive adrenergic neurons from the neural crest a marked reduction in the proportion of adrenergic cells in the culture, and their marker enzymes (tyrosine hydroxylase, dopamine hydroxylase), is observed.

As might be expected, the neuropeptides provide many striking examples of trophic activity. Like the smaller neurotransmitters they have been shown to influence cell division during the proliferative stage, to control the population sizes of some types of neuron (e.g. retinal ganglion cells), to stimulate the development of neurites and to affect neuronal phenotype.

Both non-peptide and peptide neurotransmitters often work through glial cells to affect neurons. It has, for instance, been shown that noradrenaline stimulates astrocytes to produce NGF whilst 5-HT causes astrocytes to release S-100b, a trophic factor for serotoninergic neurons. VIP has also been shown to act through astrocytes to produce a neurotrophic factor for spinal neurons. In addition, astrocytes in the developing brain have been shown to synthesise some neuropeptides conventionally associated with neurons. Both proenkephalin and somatostatin have been detected and it may be that both are released to act as trophic factors for certain populations of neurons.

The trophic action of neurotransmitters has been most thoroughly studied in cultures derived from the developing brain. There is, however, the exciting possibility that it may also be involved in repair processes after brain injury. It is well known that excitatory amino acids can be neurotoxic. After damage glial cells release glutamate and aspartate hence potentiating the effect of EAAs already present in opening Ca^{2+} channels. The sudden 'deluge' of Ca^{2+} deranges the metabolism in the subsynaptic cell. Glia may, however, also act in the opposite sense: to help repair damage. There is a trickle of evidence to support this hypothesis. It has been shown that surgical damage to the cerebral cortex increases a growth factor mRNA in glial cells. This factor was later shown to reduce the toxic effect of glutamate on hippocampal cells.

studies subsequently showed that different types of neuron required different cocktails of these growth factors.

Receptors for neurotrophins have been detected in neuronal membranes. Most of these turn out to be members of the **tyrosine kinase (trk)** family which, as we noted in Section 19.5 and Box 19.1, are well known in other systems and, indeed, in other parts of the body. Activation of these receptors by a growth factor, as we saw in Section 19.5, initiates differentiation and survival of specific classes of neurons. In non-neural tissue the result

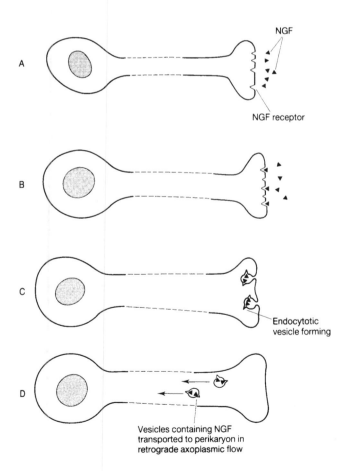

Figure 19.25 Endocytosis and axoplasmic transport of NGF in a sympathetic fibre. Schematic to show the uptake and retrograde transport of NGF in sympathetic axons. Further explanation in text.

may be cancerous proliferation. The functions of both neurotrophins and their trk receptors have been subjected to analysis by knockout genetics. In both cases (where the knockout did not prove immediately lethal) the experimental animals (mice) were defective in specific types of neurons. The situation is, however, still somewhat confused: the experimental mice were not as badly damaged as would have been expected. It has been suggested that either the brain or the genome is in some way able to compensate for what would otherwise be a dramatic loss.

In addition to centrally synthesised growth factors it has been found that the neuromuscular junctions of vertebrate skeletal muscles synthesise a factor (**cholinergic nerve growth factor (CNGF)**) which supports cholinergic neurons. It is not unreasonable to suppose that many other specific trophic factors are present in minute amounts at potential synaptic appositions in the developing brain. Indeed, some neurotransmitters, especially neuropeptide neurotransmitters, are believed to double up as neurotrophins in the developing brain (see Box 19.2).

Finally, much attention has been devoted recently to a superfamily of growth factors first isolated from human blood platelets and placenta, and from bovine kidney. This superfamily, the **TGF-β (transforming growth factor-β)** superfamily, is now known to consist of at least 25 members. Members of the superfamily are secreted into the extracellular space and detected by cell membrane receptors. They have been shown to have many functions. They play important roles in early embryology, especially in axis formation and the induction of differentiation in tissue cells. As with the molecular-genetic systems discussed in Chapter 18, the TGF-β molecules are of great antiquity and play broadly similar parts in a wide variety of organisms and tissues. So far as the nervous system is concerned, it has been demonstrated that they stimulate cultured astrocytes to synthesise NGF and increase NGF synthesis in neonatal brain. Furthermore, it is found that TGF-β mRNA levels are increased in the cerebral cortex after penetrating brain injury. Finally, there is evidence that TGF-β has neuroprotective activity in cultures of dorsal root ganglion cells. It is likely that this neuroprotection is due to the action of NGF-β in inducing the synthesis of NGF. With the great and increasing medical interest in the neurodegenerative diseases of old age it is possible that NGF-β will offer a route towards therapy or (at least) to reducing the rate at which these conditions progress.

However, as we noted at the outset of this chapter, the wiring of the vertebrate brain – especially the mammalian brain – is not completely pre-ordained. Much of the fine detail is left to environmental 'moulding'. In other words, the detail depends on the brain's activity. It depends, ultimately, on which pathways and synapses are most heavily used. This neural 'plasticity' can be studied in various of the brain's systems but it has been most extensively examined in the visual pathways. We shall examine the plasticity of these systems in the next two sections of this chapter.

19.10 MORPHOPOIETIC FIELDS

The large number of semiotic molecules that have in recent years been detected in the brain – the ephrins and their receptors, the semaphorins, the neuropilins and plexins – form a many-contoured morphopoietic landscape through which growing fibres have to wend their way. One of the best-known examples of such a landscape is provided by the **retinotectal** visual system in the lower vertebrates. In infra-mammalian vertebrates the retina does not project to a visual cortex developed from the telencephalon but to an **optic tectum** which develops in the roof of the midbrain (Figure 19.26). In fish and amphibia, moreover, regeneration of the optic nerve fibres occurs if they are sectioned. Because the optic nerve fibres have their cell bodies (the ganglion cells) in the retina this regeneration occurs **from** the retina **to** the tectum.

The original experiments on this system were performed by Roger Sperry in the 1940s and 1950s. He showed that if the optic nerve of a frog was sectioned regrowth would occur in such a way that within a few weeks normal vision (judged by appropriate visual reflexes) was restored. Electrophysiological recording also showed that photic stimulation of a patch of retina elicited activity in the correct part of the tectum. Now this accurate regrowth might well be explained by the pathfinding mechanisms we have already discussed. Indeed Sperry's interpretation – called the **chemoaffinity** hypothesis – suggested just this. Figure 19.27 schematises the experiment.

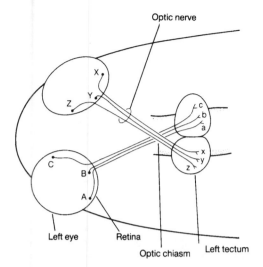

Figure 19.26 Visual pathway in fish and amphibia. In the fish and amphibia the optic nerve fibres cross completely at the chiasm to innervate the contralateral tectum. The figure shows (diagrammatically) that a 'map' of the contralateral retina is formed in each tectum. Stimulation at A gives tectal activity at a, stimulation at Y gives activity at y, etc.

Sperry followed up his original experiments by showing that if, after sectioning the optic nerve, he cut the extrinsic eye muscles and rotated the eyeball the optic fibres regrew in an anatomically 'correct'

but behaviourally 'incorrect' way. As Figure 19.28 shows, the tectal map is 180° out of phase with the retina; motor outflow from this map thus leads to behaviourally absurd responses. A frog, always striking 180° out of true, can never catch a nutritious fly and hence is apt to starve to death in the midst of plenty. Once again the chemoaffinity hypothesis seems adequate to explain this finding.

To make sure that the nerve fibres were not merely growing back along the old myelin sheaths left behind after degeneration Sperry introduced a barrier into the path of the regenerating fibres. The fibres had to find their way around this barrier before they could make their way to the tectum. In spite of this scrambling Sperry was able to show that the fibres found their way back to their anatomically correct place in the tectum. It seemed, therefore, that retinal ganglion cell and optic tectal cell were in some way chemically matched. The axon from the ganglion cell would grow back to the tectum and search around until it met its complement.

However, more recent work has shown that things are not quite as simple as this. First of all it can be shown that the apparently rigid point-for-point relation between retina and tectum is not established until a certain **critical** or **sensitive** period in the amphibian's development has been passed. If the rotation experiments are done before this

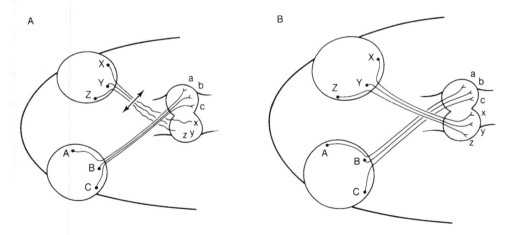

Figure 19.27 Schematic to show the principle of optic nerve section experiment. (A) The optic nerve from the right eye is sectioned (double-headed arrow). The optic nerve fibres degenerate between the section and the tectum. (B) Regrowth occurs from the cut ends of the optic nerve fibres. Electrophysiological testing shows that the fibres have reformed the tectal map as it was before the nerve was sectioned.

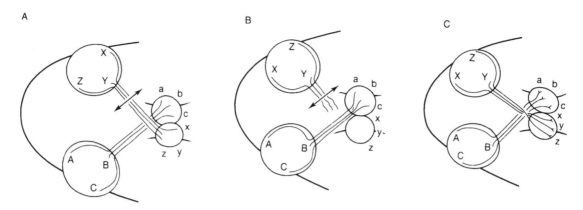

Figure 19.28 Rotation of the frog eye through 180°. (A) The optic nerve from the right eye is sectioned (double-headed arrow). (B) The eye is rotated through 180°. (C) The optic nerve is allowed to grow back to the tectum. Electrophysiological testing shows that the original map has been regenerated. Behaviour is now 180° out of line. Further explanation in text.

period has passed, the retinotectal projection regenerates in a behaviourally appropriate manner. If a chemical complementarity between individual ganglion and tectal cells exists, it must develop after this period has elapsed. Critical or sensitive periods are found in the development of all vertebrate nervous systems. We shall note the existence of similar periods in the development of the mammalian nervous system in the next section.

Secondly, Sperry's original interpretation has been subverted by experiments in which either the retina or the tectum has halved. If the optic nerve is sectioned and the tectum is halved we may legitimately wonder what the regrowing optic fibres will do. The chemoaffinity hypothesis in its simplest (perhaps 'simplistic') form would suggest that half the regrowing axons, finding their tectal destinations eliminated, would despair and die. This, however, is not what is found to happen. The regenerating axons in fact all tend to cram into the remaining half tectum (Figure 19.29B). They cram in, moreover, in a regular manner. It is as if the half tectum holds the original sensory map but at twice the density.

Vice versa, if the optic nerve is sectioned and half the retina removed, then the remaining optic nerve fibres grow back and initially innervate half the tectum. Slowly, however, the map from the half retina expands until (in a few months) the whole

tectum is innervated (Figure 19.29C). Once more the hypothesis of a precise point-for-point 'targeting' cannot be sustained. The notion of a 'morphopoietic' landscape or field begins to seem more appropriate.

Much recent work in the general field of embryology indicates that field effects are fundamental to development. The nervous system is thus in no way unusual in displaying this feature during its development. Indeed it has been possible to transplant an embryonic eye into the flank of a developing tadpole and show that the polarity characteristic of the flank is imposed on the retina. When the eye is replaced in the orbit its ganglion cells form connections with the tectum that reflect this imposed polarity rather than the polarity appropriate to its position in the orbit.

Unfortunately it is still too early to give a molecular explanation of embryonic morphopoietic fields. Box 19.1 hints that the ephrins and their receptors play important roles and we have noted the growing number of other semiotic molecules in the CNS. However, we must at present leave this topic at the phenomenological level and pass on to the penultimate section of this chapter. In doing so, however, we shall not leave behind our consideration of the visual system. For we shall find that this system also provides intriguing examples of functional sculpting.

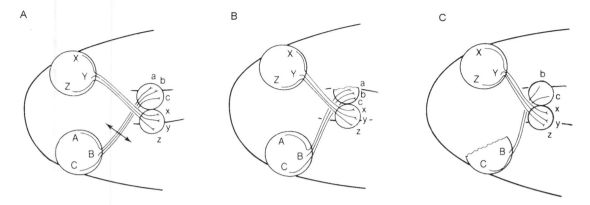

Figure 19.29 Expansion and compression of tectal maps of the retina. (A) The left optic nerve is sectioned (double-headed arrow). (B) Following section of the optic nerve half the right tectum is removed. The optic nerve fibres grow back to the remaining part of the tectum to create a miniaturised map. (C) Following section of the optic nerve half the left retina is removed. The optic nerve fibres grow back to eventually fill the whole right tectum with a map of the remaining part of the retina. Further explanation in text.

19.11 FUNCTIONAL SCULPTING

We noted in Section 19.9 that although many neurons are formed rather few are ultimately chosen. In this section we shall see that even those that are chosen depend on **use** for the persistence of their synaptic connections. Although the overall plan of neuronal connexity is determined by 'nature' the relative 'resistance' of the connections is controlled by 'nurture'. Some of the best-studied examples of this activity-dependent sculpting are found in mammalian visual systems.

As a first example let us consider the mammalian **primary visual cortex** (Brodmann area 17). A great deal is now known about the anatomy and physiology of this region of the brain. Following the pioneering studies of Hubel and Wiesel in the 1960s it has become clear that this cortex, like other parts of the neocortex, is subdivided into a large number of functional columns (see Chapter 1).

Electrophysiological recording from a column shows that the neurons all respond to visual stimuli falling on the same small patch of retina (Figure 19.30). For different neurons the patch may be of a different size – but it is always in approximately the same place in the retina. This patch of retina is referred to as the **receptive field (RF)** of the nerve cell under investigation. A column of neurons in the primary visual cortex thus 'looks' at a particular small patch of retina. The sum of all the receptive fields of the neurons in that column is called the '**aggregate field**' of the column.

As one progresses from column to column the position (and size) of the aggregate field shifts. The sensory surface – the retina – is thus mapped on to the primary visual cortex. The map is **not** isomorphous. Those regions of the retina which are of greatest biological importance – the fovea, for example – have many more columns devoted to them than do areas of lesser importance at the retina's periphery.

Now one of the first things to be found when the electrophysiology of the neurons in the visual cortical columns was investigated was that many of them responded best to **bar stimuli** flashed – about 1 second exposure – to the retina. The stimuli could be strips of light or dark bars. Circular spots of light presented to the retina provoked no response from these cortical cells (except in layer IVc). Furthermore the orientation of these bar stimuli was crucially important. All the cells in a given column respond to bar stimuli of a particular orientation. Cells in neighbouring columns will respond to bars of different orientation.

A great deal more fascinating neurophysiology has been done on the visual systems of mammals. Interested readers should once again consult references listed in the Bibliography. For our purposes it is sufficient to understand that in the normal cortex cells responding to all possible orientations of a bar

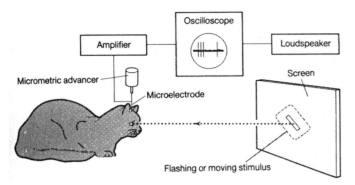

Figure 19.30 Technique for investigating the response of cells in the cat's primary visual cortex. A spectacle lens keeps the eye of the anaesthetised cat open. A flashing or moving stimulus is displayed on the oscilloscope screen. A microelectrode stereotactically inserted into the visual cortex records the response of cortical neurons. From Blakemore (1973), in *Illusion in Art and Nature*, ed. Gregory and Gombrich, London: by permission of Duckworth.

stimuli will be present – in approximately equal quantities.

Let us now return to our epigenetic concerns. How much of the organisation underlying this visual physiology is inborn, and how much is due to visual experience?

Kittens are born with their eyes closed. They do not open for about 10 days. They remain cloudy for another couple of weeks and there is little sign of accurate visuomotor coordination at this early period. Electrophysiological recording does not reveal responses to specifically orientated bar stimuli. Soon after the ocular cloudiness clears up, however, the receptive fields become tuned to bar stimuli in the way described above. This immediately suggests that visual experience plays a significant role in moulding the neural pathways underlying vision.

It is not difficult to design experiments to test the significance of visual experience. Kittens can be reared in the dark or in environments of fixed (or moving) stripes of a particular orientation. The results confirm the importance of sensory experience. Kittens reared in the dark show little sign of possessing orientation-selective cells in the primary visual cortex. Kittens reared in visual environments of, say, vertical stripes develop cortices in which the majority of cells respond to vertical bar stimuli.

It should be noted that these experiments are only effective on kittens during the **sensitive** or **critical** period (see above) of their development. They are ineffective with adult cats. During the sensitive period the neural circuitry responsible for the orientation detector cells in the primary visual cortex is in a 'plastic' state: open to environmental moulding; responsive to use and/or disuse.

The kitten's visual system can be used for another interesting investigation of the nature/nurture issue. It can be shown, firstly, that the neurons in an adult cat's primary visual cortex (except those in layer IVc) can be 'driven' from both eyes. This so-called '**binocular drive**' plays a central role in stereoscopic vision. It is thus particularly well developed in highly 'visual' animals such as felines and primates.

Now just as the primary visual cortex is built of a series of 'orientation' columns so it is also subdivided into a series of 'ocular dominance' columns. The neural pathway from the retina is arranged in such a way that fibres originating in the left eye are directed to layer IVc of one column and from the right eye to layer lVc of the adjacent column. Above and below IVc transverse connections ensure that activity from **both** eyes is delivered to a cortical cell. This arrangement is shown in Figure 19.31.

Once again this system lends itself to experimental manipulation. If one eye is patched during the kitten's sensitive period it is found that binocular drive is lost. It seems that the neural pathway from the patched eye weakens. If, however, both eyes are patched no great harm is done. After removal of the patches binocular drive is present almost as normal. It seems, therefore, that there is some form of **competition** between pathways. It seems that if one is active and the other inactive the synapses of the latter atrophy or detach (see Figure 19.32).

The experimental analysis of this system can be taken further. It can be shown that if the patched eye is opened during the third or fourth week after birth and the previously open eye patched then reversal of the pathways occurs. The previously

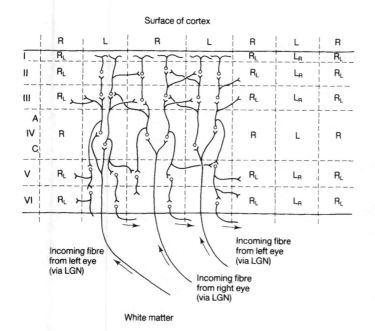

Surface of cortex

White matter

Figure 19.31 Schematic to show ocular dominance columns in cat primary visual cortex. This schematised diagram shows that fibres from the lateral geniculate nucleus (LGN) of each eye terminate in alternate columns of layer IVc. The fibres from each eye are kept separate in the LGN, which is a relay station on the route from the retina to the primary visual cortex. The diagram shows that above and below IVc transverse connections ensure that the subdominant eye has some influence on the activity of the cortical neurons.

inactive pathway is **reconstructed** and the previously active pathway fades. Clearly during this early period of the kitten's life the visual pathways are very open to environmental sculpting.

But how often can the patches be reversed? What happens if both eyes are allowed visual experience but never at the same time? Does binocular drive develop in these circumstances?

It can be shown that if the patches are reversed every 24 hours binocular drive does not develop.

Only when pathways from both eyes are operational at (approximately) the same time does binocular drive become established. Indeed it can be shown that if input from the two eyes is separated by 10 seconds or more binocular drive tends to be lost. It looks as though, in the visual system at least, two synapses have to fire at nearly the same instant if they are to stabilise each other. Any substantial length of time between firing enables one synapse to establish itself at the

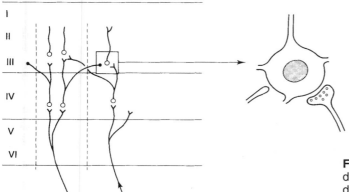

Left eye patched   Right eye unpatched

Figure 19.32 Effect of eye patches on ocular dominance in the visual cortex. The schematic diagram shows that if the left eye is patched its ability to control cells in the right eye column (i.e. to induce binocular drive of cells in layer III) is eliminated or greatly reduced.

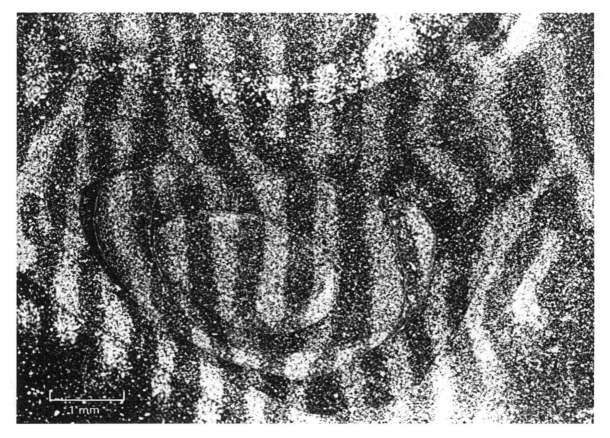

Figure 19.33 Tangential section through radiolabelled ocular dominance columns. Radiolabelled proline is injected into one of the macaque's eyes and after 10 days the animal is sacrificed and autoradiographs prepared of tangential sections of the visual cortex. The light stripes show where the radiolabelled proline has accumulated. It can be seen that the light stripes (representing input from the labelled eye) and the dark stripes (input from the unaffected eye) are approximately equal in width. From Hubel, Wiesel and LeVay (1977), *Philosophical Transactions of the Royal Society (B)*, **278**, 377–409; with permission.

expense of the other. We shall return to this observation in the next chapter for it suggests a possible mechanism for associative memory.

A further interesting experimental manipulation of the binocular drive system has used the macaque visual system rather than that of the cat. It has proved possible to visualise ocular dominance columns by injecting a radiolabelled amino acid into one eye. This is taken up by the ganglion cells of the retina and passed back to the visual cortex. After allowing an appropriate length of time for this transport to occur the animal is sacrificed and autoradiographs prepared of tangential sections of its visual cortex. As expected, the ocular dominance

columns show up as a complex pattern of stripes. The light stripes represent the radiolabelled amino acid from the injected eye. The dark stripes represent the input from the unlabelled eye (Figure 19.33).

If this experimental procedure is combined with the patch procedures described above, the electrophysiological analysis is confirmed. Figure 19.34 shows the result of rearing a macaque through its sensitive period with one eye patched and then injecting the unpatched eye with the radiolabelled amino acid. The figure shows that the columns innervated from the unpatched eye (white) have expanded greatly at the expense of the columns innervated by the patched eye (dark).

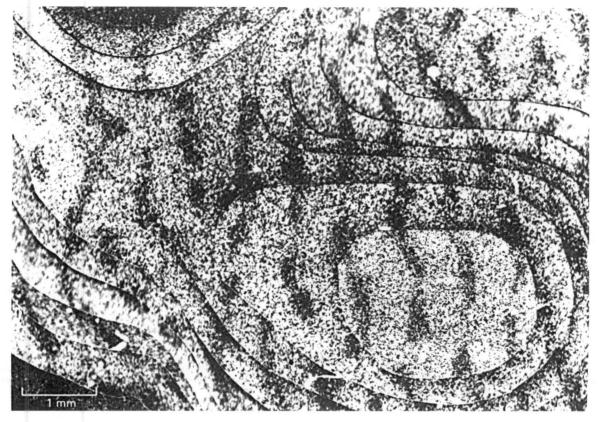

Figure 19.34 Tangential section through radiolabelled ocular dominance columns after patching one eye during the sensitive period. One of the macaque's eyes is patched during the sensitive period of the monkey's development. Radiolabelled proline is then injected into the unpatched eye and autoradiographs of the visual cortex prepared. It can be seen that the ocular dominance columns connected to the unpatched eye are much more extensive than those driven from the patched eye. From Hubel, Wiesel and LeVay (1977), *Philosophical Transactions of the Royal Society (B)*, **278**, 377–409; with permission.

Let us consider what the result of this experiment means. The size of the visual cortex has not diminished, nor has the number of cells. What has happened is that the synaptic terminals from the unpatched eye have invaded territory normally occupied by the terminals of fibres from the patched eye. Once again the conclusion seems to be clear. Synaptic **activity** is required for synaptic stability. Inactive synapses lose the struggle for survival.

Let us conclude this section by turning from mammalian visual systems back to the amphibian systems we discussed in Section 19.10. For we can find a striking analogy in an experimentally

manipulated frog to the experimental analysis of the mammalian visual cortex we have just discussed.

It has proved possible to transplant an embryonic frog eye to an anterior position on another embryonic frog. This frog thus has three eyes: two of its own and a transplant. But a frog only has two optic tecta. The optic nerve from the third eye grows back to innervate one of these tecta. In other words one of the frog's optic tecta is innervated by two eyes. Both sets of optic nerve fibres attempt to expand to fill the whole of the tectum. But they find themselves in unexpected competition with the second set of fibres. It turns out that they sort themselves out into stripe-like territories which are

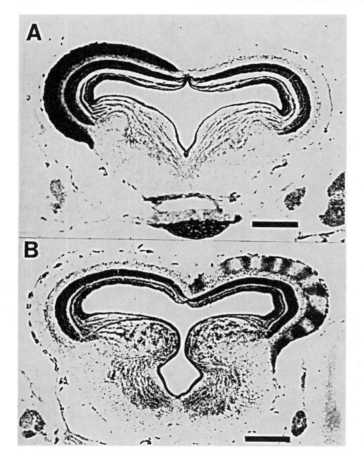

Figure 19.35 Coronal sections through optic tecta of normal and three-eyed frogs. (A) Normal otpic tecta. The frog's left eye was injected with radiolabelled proline three days before sacrifice. The autoradiograph shows that the entire superficial layer of the right tectum is filled with developed silver grains indicating the space occupied by the terminals of optic fibres from the left eye. (B) Optic tecta of three-eyed frog. The frog's left eye was once again injected with radiolabelled proline. The supernumerary eye also projects to the right tectum. It is clear that the synaptic territories in the tectum have sorted themselves into regularly spaced columns. Scale bar=400 mm. Further explanation in text. From Constantine-Paton (1981), in *The Organization of the Cerebral Cortex*, ed. F.O. Schmitt *et al.*, Cambridge, MA: MIT Press, pp. 47–67; with permission.

highly reminiscent of mammalian ocular dominance columns (Figure 19.35).

The explanation of this striking finding is identical to that which we have discussed above. Synapses with simultaneous (or near-simultaneous) activity stabilise. Hence all synapses developed on the terminals of fibres from one eye, being simultaneously activated, stabilise each other. Gradually, by competition, the territories shown in Figure 19.35 become established.

The molecular mechanism(s) underlying this selective stabilisation are as yet unknown. One attractive speculation suggests that postsynaptic cells are induced by active presynaptic terminals to secrete an antagonist which inhibits the formation of other synapses. Only active terminals are protected from the effects of this antagonist. What this antagonist might be there is as yet not

known. Very recently, however, experiments on the mouse neuromuscular junction have given substance to this speculation.

The neuromuscular junction is, in fact, far more complex than the simplified diagram of Figure 15.16 suggests. The motor nerve terminal does not make one but a large number of appositions (sometimes likened to a bunch of grapes (*en grappe*)) on the muscle fibre (Figure 19.36). But the NMJ has one great advantage for the experimentalist: it is readily accessible. It has, indeed, proved possible to accomplish the difficult task of finding and following the life of a single junction over a period of up to two months. Consider the difficulty of attempting to locate and relocate a specific synapse (*in vivo*) in the 10^{13} synapse mouse brain! The NMJ has thus allowed some very revealing studies to be made.

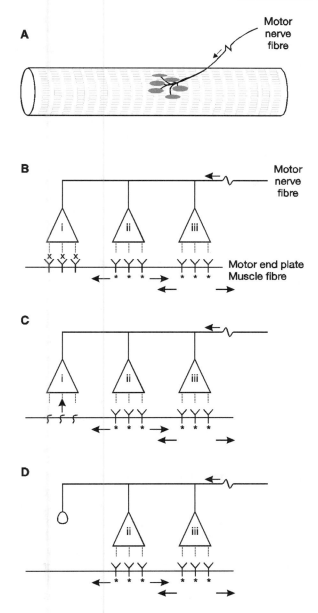

Figure 19.36 Use it or lose it at the NMJ. (A) '*En grappe*' nerve terminal on mammalian skeletal muscle fibre. (B) Schematic diagram to show the effect of blocking the nAChRs under apposition (i). $\times = \alpha$-BuTX; $\leftarrow$ and $\rightarrow$ = spread of antagonist; $*$ = defence against antagonist. (C) Later: the AChRs beneath apposition (i) have been eliminated. (D) Later still (6–8 weeks): presynaptic ending (i) has atrophied and withdrawn. Further explanation in text.

In order to examine the effects of inactivating a synaptic apposition the nAChRs in the subsynaptic membrane (motor end plate) were inhibited by the long-lasting neurotoxin a-BuTX (see Section 10.1.1). It was found that over a period of up to eight weeks all the blocked nAChRs disappeared. In addition the overlying presynaptic ending also atrophied and withdrew. Neighbouring synapses were unaffected. In a second experiment all the synapses in the *en grappe* ending were inactivated with a-BuTX. This time none of the synapses appeared to be affected. It is not difficult to see (Figure 19.36) that these experiments support the idea that an antagonist molecule is induced by activated nAChRs which diffuses (up to 50 mm) in the subsynaptic cell to attack inactive appositions. At the same time active nAChRs are defended against the antagonist and consequently survive unharmed. If all the nAChRs are inactivated then no antagonist is induced and consequently no synapses atrophy and withdraw.

This scenario, it is evident, also requires the manufacture of a 'retrograde messenger', for not only do the postsynaptic nAChRs disappear but also the presynaptic terminals. What this retrograde messenger might be and how it exerts its effects we still do not know although alterations of the basal lamina have been suspected. The NMJ preparation should, in due course, allow these signal molecules and the systems they affect to be isolated and characterised.

Although it is orders of magnitude more difficult to locate and relocate a single synapse amongst the 10^{13}–10^{14} present in the mammalian brain, some progress has been made at the biochemical and pharmacological levels. It is of interest, for example, to note that N-methyl-D-aspartate (NMDA) receptors (see Section 10.4) are involved in the competitive exclusions of the optic tecta. It has been shown that agents which block the NMDA receptor (e.g. D-AP5) also prevent the clear demarcation of optic fibre territories in the tectum of three-eyed frogs. A similar effect can be demonstrated in the monocular kitten. Could it be that the second messenger activity of the Ca^{2+} ions which flow through the NMDA receptor channel induce the synthesis of a receptor antagonist similar to that proposed for the NMJ? It will be recalled from Chapter 10 that only when a subsynaptic membrane has been depolarised by

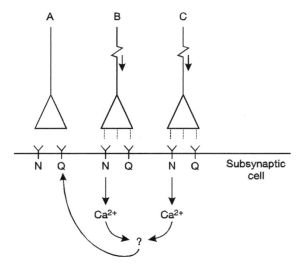

Figure 19.37 Coincidence sculpting in the visual cortex. Three afferent fibres (A, B, and C) make synaptic contact with one subsynaptic membrane. B and C fire approximately synchronously; A fires at random. Activation of the quisqualate receptor (Q) causes depolarisation of the membrane and NMDA is then able to open the NMDA receptor (N). The entry of Ca^{2+} through the latter receptor either directly or through an unknown intermediate (?) inactivates the Q receptor under terminal A. The mechanism in the postsynaptic cell is analogous to that of Figure 19.36.

some 30 mV is the Mg^{2+} blockade of the NMDA channel lifted. This might well require two or more closely set excitatory (non-NMDA) synapses to be (nearly) simultaneously active. It is clear that the NMDA channel provides a possible mechanism (Figure 19.37). We shall return to a similar speculation in Chapter 20 where we shall see that the NMDA receptor has also been implicated in the plastic change responsible for memory. We have to bear in mind, however, the hugely intricate nature of the feedforward and feedback systems in the synapse. It is likely that we have done little more than scratch the surface and that it will take a considerable time to work out the full intricacy of what is actually happening in the functional sculpting of synapses.

These speculations provide a suitable end point for this chapter. They show how molecular neurobiology is beginning to provide a theory of the brain. They show, in other words, how molecular investigations in one part of the subject (neuropharmacology; the NMJ) are beginning to impinge on and illuminate quite a different part (neuroembryology). It must be borne in mind, however, that these ideas are as yet only tentative. Like all hypotheses they have many competitors which might also account for the facts. Like the synapses themselves, only use (i.e. further research) will winnow out the false and reveal the truth.

19.12 CONCLUSION

We observed at the outset of this chapter that the brain is perhaps the ultimate challenge for the developmental biologist. It is clear from the remainder of the chapter that we still have far to go. Yet the molecular approach here, as elsewhere, is beginning to yield dividends. It begins to seem feasible that increasing knowledge of the selective stickinesses of membranes, of the subtleties of a cell's control of N-CAMs, of the distribution, uptake and effect of tropic and trophic factors and their distribution in morphopoietic fields, and of the fine tuning resulting from use and disuse, will begin to provide a framework for an understanding. Neuroembryology, furthermore, does not stand apart from the rest of neuroscience. Its growth depends on the development of our knowledge of synapses and their pharmacology, on the biophysics and molecular biology of channel proteins, as well as on behavioural biology and neuropathology. The prize is a great one. In Section 19.2 we noted the promise of neural stem cells. If we can begin to understand how the nervous system develops its harmonious and intricate structure we shall be in a better position to repair it when it is disrupted by injury or disease.

20

MEMORY

The neurobiological profundity of memory – underlying processes reach through all the layers of neurological complexity – centrality to human life – tragedy when lost in old age – significance in education of young. **Definitions of learning and memory**: habituation – sensitisation – associative; classical and operant conditioning – terminology – similarities and differences; trial and error learning – insight learning- exploratory behaviour. **Short- and long-term memory (STM and LTM)** – biological significance of distinction – retrograde amnesia – retrieval. **Location of memory trace** – Lashley and 'mass action' – specific brain regions – air-puff conditioning and the cerebellar cortex – STM, spatial memory and the hippocampus – radial and water mazes – amygdala, entorhinal and parietal cortices – pharmacological study – neural bases of memory – central role of the synapse. **Invertebrate models**: *C. elegans* – thermal conditioning – genetic analysis – Ca^{2+} sensors; *Drosophila* – chemosensory learning – genes for learning and memory – cAMP – mushroom bodies – control of dynamin via the shi gene – discrimination between learning and recall – cAMP, PKA, CREB – IEGs and K^+ channels – prospects for the dissection of memory into its molecular elements; *Aplysia* – magnitude and constancy of ganglionic cells – siphon withdrawal reflex – habituation – sensitisation – classical conditioning – neural circuitry – homosynaptic depression – inhibition of presynaptic Ca^{2+} channels – heterosynaptic facilitation – short-term sensitisation, serotonin and I_{KS} channels – long-term sensitisation, translation and transcription – biochemical cascade – associative learning – ionic and molecular mechanisms – cAMP, PKA, CREB – similarity to *Drosophila*. **Mammalia**: hippocampus – post-tetanic and long-term potentiation (PTP and LTP) – fibre pathways in hippocampus – NMDA-Rs – significance of voltage sensitivity – second messengers affect genome and/or cytosolic enzymes – induction of NOS – NO possibly the retrograde factor – the CRE site and CREB proteins – similarities between mammals, molluscs and flies. Local mechanisms – spine synapses – CaMK2 – NMDA-Rs. Power of invertebrate systems for researching problems in mammalian CNS – learning provides another case of negative sculpting

More than sixty years ago a great neuroscientist, E. R. Hilgard, wrote: 'It is a blot upon our scientific ingenuity that after so many years of search we know as little as we do about the physiological accompaniments of learning.' More than half a century has passed since those words were written. Sixty years in which a vast amount of research has been directed to discovering 'the physiological accompaniments of learning'. Yet, by and large, the blot remains. We still know surprisingly little

about what happens in the brain when learning occurs and memories are 'laid down'. In quite recent years, however, a number of openings have appeared. The horizon, to change the metaphor, begins to seem a little brighter. Perhaps in the first decade of the twenty-first century we are really beginning to see a glimmer of a solution to the perplexity. This glimmer, moreover, comes from studies at the molecular and biophysical levels. It comes, in addition, from studies on invertebrate organisms which might not have immediately commended themselves to the neuroscientists of the 1930s.

The solution to the problem of memory requires one of the great syntheses of neurobiology. That is why I have left it to the twentieth of our twenty-one chapters. It involves ion flows, second messengers, gene derepressions, channel proteins, membrane structure. It also involves levels above the molecular and ultrastructural. In mammals, and after all it is human memory which ultimately concerns us, the intricate anatomy of the brain is crucially involved. Neural networks involving thousands if not millions of neurons, millions if not billions of synapses, are at work. The study of memory concerns psychologists and ethologists, cognitive and computer scientists, as well as neurobiologists.

Yet the effort is well worthwhile. Memory is very central to our life. To an extent we are all bundles of memories. If we suffer total amnesia we lose touch with who we are. There can be few greater personal tragedies or sufferings. Consciousness itself loses its depth. All that is left is the thin patina of the existential present.

Yet to a degree we all face at least the beginnings of this tragedy. Old age customarily brings with it loss of memory. The worst scourge of old age – dementia of the Alzheimer type – is characterised by ever-deepening amnesia.

To a degree, also, we could all benefit by an improvement in the memory we have. Certainly the multibillion-pound world-wide educational profession could hardly help benefiting from a deeper knowledge of the mechanisms at work when learning is occurring and memories are being established.

Finally, the world of computer science undoubtedly has much to learn from a system which can store as much information as a Shakespeare, a Darwin or an Einstein and retrieve it so effortlessly and so rapidly. Human brains have taken a thousand million years to evolve: it is likely that they have a few lessons to teach the parvenu advocates of artificial intelligence.

20.1 SOME DEFINITIONS

The terms 'learning' and 'memory' are closely allied. Perhaps in the common usage both have a connection with consciousness. Yet this need not be

so. We are nowadays quite accustomed to the idea of 'unconscious' memories and we are quite prepared to say of someone that he learnt no end of a lesson without supposing that he was aware at the time that he was doing so. Thus even in the common usage 'learning' and 'memory' tend to become generalised. In the world of biology they are generalised yet further. Certainly the notion of conscious awareness plays no essential role in the concepts.

In its most general form we can define learning and memory in behavioural terms as a **changed response to a repeated stimulus**. This very broad definition allows us to attach the terms 'learning' and 'memory' to many non-neurological systems. For instance both bacteria and protozoa show behaviour which falls under this definition. Behavioural adaptation to repeated stimuli is a very general feature of all living forms. The immune system is another instance where, according to our definition, learning and memory occur. Everyone knows that the immunological response is very different the second time the system is challenged.

In this book, however, we are restricting ourselves to neurobiological instances. Even so our definition will include phenomena that may not immediately commend themselves as instances of learning and memory. **Habituation** and **sensitisation** are two such instances. We all experience habituation to monotonous stimuli: we may, for example, become so accustomed (habituated) to the ticking of a clock that we only become aware of it when it stops. Vice versa, the smallest intimation that a painful experience may be about to be repeated is likely to induce a vivid avoidance reaction: we are sensitised.

Both habituation and sensitisation are variations in the intensity of an already existing response to a stimulus. In this regard they show a resemblance to the immune response. In contrast to these phenomena, however, is **associative learning**. Here a response is developed to a previously 'neutral' stimulus. Some authorities insist that this type of learning is the only type worthy of the name.

Associative learning can also be classified into various subtypes: Pavlovian (=classical) conditioning, instrumental (=operant) conditioning, imprinting, insight, etc. In all cases a stimulus (or constellation of stimuli) which initially did not induce response A is brought to induce that response. Details of these different forms of associative learning and the 'training' procedures involved are to be found in texts of physiological psychology.

Here we shall very briefly describe the two fundamental subtypes – **classical** and **operant** conditioning – as they are both crucial to experimental approaches to the molecular basis of memory.

20.1.1 Classical Conditioning

The paradigmatic examples of classical conditioning are those carried out at the beginning of the twentieth century by Ivan Pavlov (Figure 20.1). Pavlov's initial interests were in digestive physiology and from this interest grew his classical research into the response of his experimental animals – dogs – to the sight or other clue of dinner. One of the inborn alimentary reflexes to the presence of food in the mouth is salivation. By collecting the saliva produced it is easy to measure the strength of this reflex response. This reflex salivation would, of course, not normally occur in response to some neutral stimulus such as a bell, a red light, or a gentle touch. However, Pavlov was able to show that if the neutral stimulus **preceded** the placing of food (or dilute acid) in the mouth sufficiently frequently then salivation would occur in response to the neutral stimulus alone.

The following terminology has been developed to describe this type of reflex. The 'neutral' stimulus is termed the **conditioned stimulus (CS)**, the food or dilute acid is called the **unconditioned stimulus (UCS)**, the inborn salivary response to the food or acid is called the **unconditioned reflex (UCR)** and the same response when elicited by the CS is called the **conditioned reflex (CR)**.

Because, as we saw above, Pavlov was able to quantify the CR he was able to work out some important properties of classical conditioning. One of the most important of these properties has to do with the time interval between the CS and the UCS. **The CS must precede the UCS**. In Pavlov's case the optimal interval between the CS and the UCS (the **interstimulus interval (ISI)**) turned out to be about 0.5 s. This time period seems to obtain in many organisms – vertebrate and invertebrate. However, there are well-known instances, for instance taste aversion, where the ISI may be a matter of hours. To establish the CR the CS–UCS pairing must be repeated a number of times. This is called **reinforcement**. If sequencing of the stimuli is reversed, if in other words the UCS precedes the CS, then the CS weakens and ultimately disappears. This is known as **extinction**. Extinction will also occur if the CS is presented a number of times on its own.

20.1.2 Operant Conditioning

We saw in the preceding section that Pavlovian CRs are in a sense 'elicited' from the animal. In

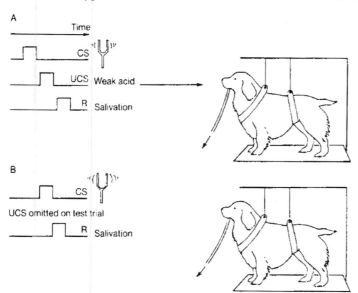

Figure 20.1 Pavlovian (classical) conditioning. (A) Conditioning. The conditioned stimulus (CS) and the unconditioned stimulus (UCS) are paired and the response (R), salivation, measured. (B) Eliciting the conditioned response. The UCS is omitted and the reflex salivation is obtained to the CS alone. After Racklin (1976), *Introduction to Modern Behaviourism*, New York: Copyright © 1976 by W.H. Freeman; with permission.

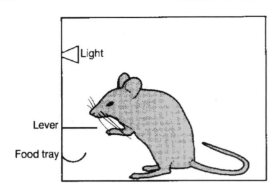

Figure 20.2   Skinner box. Further explanation in text.

Pavlov's case the dog is held in a harness, appropriate stimuli presented and the salivary response measured. In a sense the *dog is* not involved at all. This is not the case in operant conditioning. Here the whole animal responds. The initial investigations of this second type of CR were carried out by B. F. Skinner. The piece of apparatus he devised is now well known – the **Skinner box** – and operant conditioning itself is often known as Skinnerian conditioning (Figure 20.2).

The essence of Skinnerian conditioning is that the animal has to press a lever, or (if a bird) peck at a key, or perform some other '**operant**' once or a number of times to receive a reward – food or drink. The reward **reinforces** the operant behaviour. It is said to be **contingent** upon that behaviour. The strength of the operant behaviour can be measured in terms of the number of bar presses, pecks, etc. The operant, moreover, is open to different types of reinforcement. **Negative reinforcement** occurs, for instance, when an animal emits the operant in order to **avoid** a painful stimulus. **Secondary reinforcement** is induced when, for example, every time a food pellet is to be delivered into a Skinner box a light switches on. Frequency and duration of bar pressing will then be increased when the light is switched on and reduced when it is off.

Skinnerian conditioning shares some characteristics with the Pavlovian type but differs in others. In particular, operant conditioned reflexes are much more stable than Pavlovian reflexes, which require frequent reinforcement if they are not to fade to extinction.

It is not difficult to see that Skinnerian reflexes merge into trial and error learning (including that sort of trial and error learning 'done in the head' which is sometimes called 'insight learning') and into behaviours such as maze learning and exploratory behaviour.

20.2 SHORT- AND LONG-TERM MEMORY

One of the ground rules of scientific theorising was summed up by William of Occam in the thirteenth century: '*Entia non sunt mutiplicanda praeter necessitatem*' – 'entities are not to be multiplied beyond necessity' – an aphorism that has come to be known as **Occam's razor**. This principle holds as strongly in the study of memory as elsewhere in science. Neurobiologists hope to find a single mechanism at work underlying the phenomenon. Yet the facts may be otherwise. It may be that memory is a multifaceted function of the brain. At the time of writing the issue is still unsettled. We, however, shall follow Occam's recommendation and assume that memory is the outcome of one underlying process.

One very common distinction made by students of the formation of memory is to divide the process into a 'labile', **short-term memory (STM)** and a more 'stable', **long-term memory**. We shall make use of this distinction in our consideration of putative molecular bases of memory as we go through this chapter.

Short-term memory (STM) endures for seconds, minutes or hours. It is the type of memory we use when we look up a telephone number and then dial it. It is just as well that the number does not remain in the memory more than a few minutes. If we stored all the numbers we had looked up in a lifetime we might well not be able to concentrate on more important matters. Indeed this is likely to be the biological significance of this type of memory. It enables an animal to forget the trivialities of its life. Its transience and lability suggest that it has a 'physiological' basis – perhaps a transient alteration in the synaptic 'resistance' of certain pathways in the brain.

Long-term memory (LTM) endures for days, years, decades. This holds the significant data of an animal's life – where to find food or a mate, where to avoid a predator or a poison. In the lives of most

English people the date of the Battle of Hastings is of sufficient significance to be in the LTM. 1066 AD may remain firmly embedded for seventy or eighty years. This permanence argues for a permanent alteration in a cerebral pathway (or pathways). Whenever William the Conqueror, Hastings, or the Bayeux tapestry are voiced, read or otherwise understood, AD 1066 becomes available.

20.2.1 Relation Between STM and LTM

It is generally believed that information to be memorised passes first into the STM and then into the LTM. This transfer is called consolidation. Now we have already noted that STM may be distinguished from LTM by its greater 'lability'. It is not too difficult to disrupt STM. LTMs, however, are very difficult to displace. It follows that we can prevent consolidation by disrupting the STM phase. This can be done by exposing the brain to any one of a number of different 'insults': concussion, electroconvulsive shock (ECS), hypothermia, certain drugs, etc. (Figure 20.3). Many of us may have had personal experience of this. It is quite common to find that after severe concussion the period immediately before the concussion cannot be remembered. This phenomenon is known as **retrograde amnesia (RA)**.

The phenomenon of retrograde amnesia is amenable to experimental investigation in laboratory rodents. If the brain is insulted in one or other of the ways mentioned shortly after operant conditioning the chances of the operant behaviour becoming established are much reduced.

We should not finish this section without observing that not all investigators accept the STM–

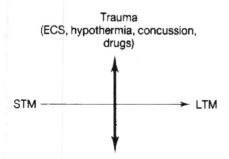

Figure 20.3 Trauma affects transfer of information from STM to LTM.

LTM scheme outlined above. Some workers suggest that it is not so much the differential laying down of memories that we are looking at but differential **retrieval** of memories. There is some evidence for this interpretation both from animal studies and from human experience. For it is well known that with time the period of retrograde amnesia shrinks. More and more of the events preceding the concussion become available though probably never all. However, it would take us too far from the subject of molecular neurobiology to pursue the pros and cons of this argument further.

We can conclude, therefore, by noting that the problem set to molecular neurobiologists by their behaviourist and psychological colleagues is to provide a molecular mechanism that can account for STM, LTM and the transfer (consolidation) between one and the other.

20.3 WHERE IS THE MEMORY TRACE LOCATED?

In Sections 20.1 and 20.2 we have been discussing memory in a very general sense. The experiments described have been behavioural and they have provided some parameters which any molecular mechanism must obey. But in order to investigate events at the molecular level we must know **where** in the brain they are occurring.

This is much easier said than done. Karl Lashley, one of the great pioneers in the neurobiology of memory, wrote at the end of his research career in the 1950s that 'I sometimes feel, reviewing the evidence on the localisation of the memory trace, that the necessary conclusion is that learning is just not possible!'

The best-known outcome of Lashley's research is the so-called law of '**mass action**' or '**equipotentiality**'. Working with rats he was able to show that their ability to memorise a maze depended not so much on which fragment of the cerebral cortex was excised but on **how much** was removed. According to this interpretation memory of a maze was 'smeared' throughout the cortex, not located in one area more than another. Later, in the 1970s, electrophysiological recording from unrestrained rats undergoing operant training in a Skinner box provided similar results. It seemed that cells throughout the brain were activated during the learning procedure.

In spite of these results it is now beginning to seem more likely that 'memory circuits' are localisable within the brain. It may well be that the 'mass action' observations merely indicate a general arousal throughout the brain when learning is occurring (memories being laid down) but that the actual pathways or 'circuits' where the changes are occurring are quite discrete.

It is nowadays believed that the areas primarily involved in the laying down of memory traces are the **cerebellum**, **hippocampus**, **amygdala** and **cerebral cortex**. Cognitive scientists studying humans make a distinction between 'explicit' or 'declarative' modes of learning (people, places, things) and 'implicit' or 'procedural' modes (motor skills, procedural strategies). Explicit learning is believed to primarily involve the cortex, especially the temporal lobe, and the hippocampus, whilst implicit learning involves specific sensory and motor systems. It is likely that both types of learning make use of the same underlying biochemical mechanisms. Let us consider, first, an example of implicit learning.

There is good evidence that a number of adaptive conditioned reflexes to aversive stimuli involve the cerebellum. One of the favourite reflexes in these studies has been the eye-blink (UCR) in response to a puff of air (UCS). This can be coupled to a tone or other conditioned stimulus (CS). Pairing of CS and UCS ultimately elicits the eye-blink in response to the CS alone. This reflex has all the characteristics of a classical Pavlovian conditioned reflex.

Careful experiments involving lesioning, microinfusion of drugs, electrophysiological recording and electrical microstimulation show that certain regions of the cerebellum are crucial to the establishment and maintenance of this conditioned reflex. These regions include the **ipsilateral interpositus nucleus** (in the base of the cerebellum) and identifiable **Purkinje cells** in the cerebellar cortex. Figure 20.4 shows that inputs from both the UCS and the CS meet on Purkinje cell dendrites and on cells of the interpositus nucleus. We shall see (especially in Section 20.5) that synaptic conjunctions of this type are believed to play an essential role in memory processes. There is overwhelming evidence that the cerebellum plays a crucial role in the performance of skilled movements. It is thus not surprising to find that it is also deeply involved in learning adaptive responses to aversive stimuli.

It is known that Purkinje cells express high levels of mGluR1 messenger RNA. We have already seen (Section 8.10) that mGluR1 is linked via G-proteins to a phosphoinsiotol (PI) second messenger system. It may be that the Ca^{2+} ions released by this second messenger system are involved in establishing the long-lasting change underlying memory (see Section 20.5). In this regard it is interesting to note that, in recent years, mouse knockout genetics have been used to eliminate the mGluR1 gene. The resulting mice show interesting behavioural defects which have been attributed to the loss of mGluR1s in the hippocampus and cerebellum. In particular, such mice show profound defects in motor coordination and a modest diminution of the eye-blink reflex. Perhaps most significant of all, however, is the substantial or total loss of long-term depression (LTD) of Purkinje cell synapses. In the normal cerebellum repetitive and simultaneous activity in both parallel and climbing fibres (Figure 20.4) leads to a persisting depression in the response of the Purkinje cell. This so-called LTD is clearly a candidate for the persistent change which we need as the physical basis of memory. It is the obverse of the long-term potentiation (LTP) which we shall discuss in Section 20.5. These knockout mouse experiments are as yet far from conclusive or, indeed, fully repeatable, but they do show the developing power of the new genetic techniques in neurobiology.

Another region of the brain that is deeply involved in memory is the **hippocampus**. It has been known for many years that injury to human hippocampi causes a profound upset of STM. If their attention is distracted, patients are unable to remember lists of words or names of objects that they have memorised just a few moments previously. They cannot remember what day it is or what they had for breakfast. Yet their long-term memories are comparatively intact. We shall see in Chapter 21 that Alzheimer's disease in which the hippocampi are particularly affected is characterised by a profound loss of short-term memory.

In rats it can be shown that the hippocampus is crucially involved in behaviours that involve spatial orientation. Means of testing such behaviour are shown in Figure 20.5. A rat may, for instance, be placed on the central platform of a radial maze. Each arm of the maze is baited with food. The optimal strategy is to visit each arm in turn thus

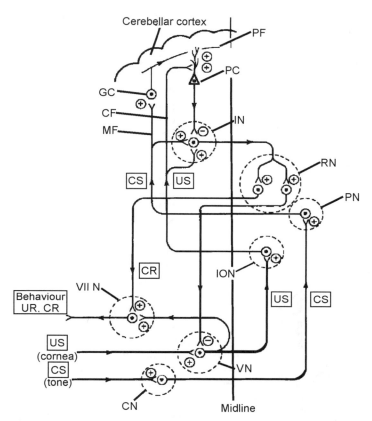

Figure 20.4 Anatomy of the air-puff conditioned reflex. This complicated diagram shows the neuronal pathways underlying the eye-blink reflex. Impulses generated by an air puff to the cornea (unconditioned stimulus, US) make their way via the fifth cranial nerve to its nucleus in the midbrain (V N) and thence via the inferior olivary nucleus (ION) to the dendritic tree of a Purkinje cell (PC) in the cerebellar cortex. A branch from this pathway leads to the sixth and seventh cranial nerve nucleus (VI and VII N) and thence to the muscles of the eyelid. Impulses generated by a tone (the conditioned stimulus, CS) make their way from the ear via the cochlear nucleus (CN) to the pontine nucleus (PN) and thence via a synapse with a granule cell (GC) to the dendritic tree of the Purkinje cell. The output from the Purkinje cell passes first to cells in the interpositus nucleus (IN), then to cells in the red nucleus (RN) and finally to the output to the eyelid muscles from the nuclei of the sixth and seventh cranial nerves. CF=climbing fibre; CN=cochlear nucleus; CR=conditioned response; CS=conditioned stimulus; GC=granule cell; IN=interpositus nucleus; ION=inferior olivary nucleus; MF=mossy fibre; PC=Purkinje cell; PF=parallel fibre; PN=pontine nucleus; RN=red nucleus; US=unconditioned response; V N=nucleus of fifth cranial nerve; VI and VII N=nuclei of sixth and seventh cranial nerves; +=excitatory synapse; − = inhibitory synapse. After Thompson (1986), *Science*, **233**, 941–947.

ensuring that time and energy is not wasted in entering an empty arm. If rats can orientate themselves by observing visual cues above and around the maze they often do not visit each arm in turn but run to each apparently at random, remembering from the visual cues which arm they have visited and which they have not. Similarly if a rat is forced to find a platform concealed beneath

cloudy water it can only do so by learning the visual cues in the room surrounding the tank. It is said that the animals carry a 'cognitive map' in their brains. Rats with lesioned hippocampi perform significantly less well in this task. It is as if their ability to form and/or store a cognitive map has been damaged. Current evidence suggests, however, that although the hippocampus is

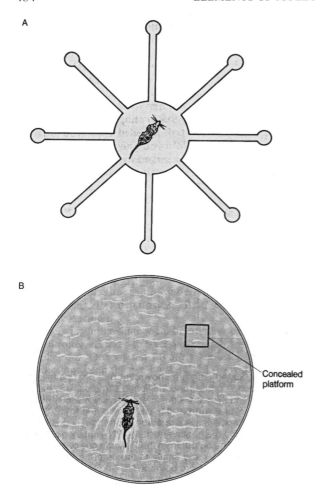

Figure 20.5 Radial and water 'mazes'. (A) Radial maze. The rat is placed on the central platform. Food or water is placed at the end of each arm. If spatial memory is unimpaired the rat should not enter the same passageway twice. (B) Water 'maze'. The rat is placed in a large circular tank filled with water. The water is commonly made cloudy by adding milk. At some point in the tank, just below the surface, is a platform. The rat is required to memorise its surroundings (the positions of lights, doors, tables, etc.) in order to navigate rapidly to the platform.

Concealed platform

manipulations on a number of regions of the rat brain. The behavioural test used in these experiments was fear conditioning. In this test an innocuous stimulus, such as a light or an audible tone, is paired with a foot shock. The association is quickly learnt and after a small number of pairings the CS alone elicits the UCR: freezing, cardio-acceleration, startle response, etc. It may be that many brain regions are involved in this conditioning but one intensively investigated route leads from the **hippocampus** to the **amygdala** to the **entorhinal** cortex and finally to the **parietal** cortex.

Infusion of the pharmacological agents AP5 (an NMDA-R antagonist (Section 10.4)) and/or muscinol (a $GABA_A R$ agonist (Section 10.2)) into either hippocampus and amygdala immediately after training caused profound retrograde amnesia. This is taken to indicate that circuits using NMDA and GABA receptors in those regions were involved in fear conditioning. Yet more interestingly, it was found that if the AMPA receptor antagonist CNQX (Section 10.4) was injected into hippocampus and/or amygdala one day after training and immediately before testing, fear conditioning was lost; if it was injected into the entorhinal cortex one day or 31 days after training and immediately pre-testing, conditioning was lost; and finally if it was injected into the parietal cortex 60 days after training and immediately pre-testing, the fear conditioning was blocked. These experiments were interpreted as showing that the memory trace decayed fairly rapidly in the hippocampus and amygdala, more slowly in the entorhinal cortex and most slowly of all (if at all) in the parietal cortex. They indicate that a number of different brain regions are differentially involved in the laying down and retrieval of fear conditioning. Support for differential involvement of brain regions, both spatial and temporal, comes from many different directions not least from brain imaging studies and the study of 'knockout' mice lacking biochemical pathways in specific brain regions.

However, to follow these fascinating leads further would take us outside the bounds of a text on molecular neurobiology. Interested readers should consult the Bibliography for further information. We shall return to the mammalian brain later. It is time now, however, to turn our attention to the

massively involved in learning and STM memory the long-term traces are held elsewhere.

The evidence for believing that hippocampus is not the only region involved in long-term memory consolidation comes from experiments with pharmacological blockers introduced into and genetic nature of the memory trace. We have seen something of what it is required to do; we have looked briefly at where it may be laid down; but what form does it take?

The physical basis of memory, we have seen, must be some alteration in nerve pathways within the brain. The connection between input and output is altered: either facilitated or inhibited. Perhaps, as in Chapters 18 and 19, we should envisage the brain as initially comparatively undifferentiated and over-provided with pathways between sensory input and motor output. Learning selects between these pathways. It ensures that some become more or less permanently differentiated from others. It ensures that impulses travel preferentially (or the opposite) through these pathways. Thus, in a sense, learning could be seen as an extension into the adult of the lability which we have seen the brain to possess during development.

In previous chapters we have noted that action potentials are all-or-nothing events. The initial segment fires or does not fire. Once the impulse is initiated it propagates without decrement. There is no calling it back, no increasing or decreasing its amplitude or velocity. This is not the case with synapses. We have seen that postsynaptic potentials are graded events. We have seen that the positions where synapses occur on the neuronal soma have a crucial bearing on their effect on the postsynaptic cell. We have seen that the size of the synapse, the amount of transmitter released when an action potential arrives, the type of transmitter, the type and number of receptors in the subsynaptic membrane, all influence the magnitude and duration and type of postsynaptic event. It is not surprising, therefore, that neurobiologists regard the synapse as the most likely locus for the changes underlying and responsible for memory.

And this is where we have, for the moment, to take leave of mammalian systems. They are too complicated. If memory is ultimately an affair of synaptic resistances, which of the mammal's 10^{14} synapses should we look at? There is much evidence to show that neuronal activity leads to biochemical changes indicative of increased protein biosynthesis. There is increased activity of DNA-dependent RNA polymerase and the appearance of increased amounts of mRNA in the perikaryal cytoplasm. Following on from this, increased incorporation of

radiolabelled amino acids into protein can be detected. It can also be shown that the density of spines on dendritic trees is significantly increased, just as, in the opposite case, inactivity leads to loss of dendritic spines. These consequences have all been well established – but do they have anything to do with the memory trace? Might they not be just consequences of increased overall cerebral activity? To be sure we have to find systems where learning is occurring in identifiable neurons; systems in which we can compare the same neuron, the same synapse, when learning is occurring and when it is not. We have thus to turn, initially, to the far simpler systems of the invertebrates.

20.4 INVERTEBRATE SYSTEMS

A number of invertebrate nervous systems are proving themselves invaluable in neurobiological research. Leaving aside the tadpole-like larva of *Ciona* with its sub-100-cell nervous system (see Section 1.1) the best-known simple nervous system is that of the small nematode worm *Caenorhabditis elegans* (Figure 20.6), which consists of just 302 cells and some 5000 synapses.

Turning from these simplest of metazoan nervous systems to some of the most highly evolved, we find that the insects provide many useful preparations for the neurobiologist. *Apis mellifera*, the honey bee, *Schistocerca gregaria*, the locust, *Periplaneta americana*, the cockroach, *Phormia regina*, the blow fly, and, finally, *Drosophila melanogaster*, the fruit fly, all have their devotees. *Drosophila*, in particular, because of its extremely well-known genetics, has received much attention.

A number of molluscan nervous systems have also proved valuable to neurobiologists interested in learning and memory. They include those of *Limax maximus*, the garden slug, *Pleurobranchaea calfornica*, a marine opisthobranch, *Lymnaea stagnalis*, the pond snail and, most important of all, *Hermissenda crassicornis*, a marine nudoibranch and *Aplysia californica*, the sea hare, a tectibranch.

Space, however, only allows discussion of three members of the Invertebrata, albeit three very different members. We shall begin by reviewing some very recent work on *Caenorhabditis elegans* and then proceed to some of the fascinating work carried out on the fruit fly *Drosophila melanogaster* and finish by discussing the equally interesting work on the mollusc *Aplysia californica*.

20.4.1 Thermal Conditioning in C. elegans

The structure of *C. elegans'* nervous systems has, so to speak, been 'run into the ground' by reconstruction from serial electron micrographs. All its neurons, their shapes, sizes and connections, have been mapped. Like many other invertebrates the neurons and their synaptic contacts are identical in isogenic animals. The genetics of the worm are also well researched and many mutants affecting the development and structure of the nervous system have been isolated.

The world of *C. elegans* is very different from ours. One of its most important parameters is temperature. *C. elegans* has long been known to move along isothermal lines with great precision, deviating from one side to the other by less than 0.1°C. This thermal sensitivity has been used in conditioning experiments. It has been shown that worms can learn to associate particular temperatures with abundance or paucity of nutriment. Because the nervous system is so well known the neural circuits underlying this thermal learning have been worked out. A class of thermosensory neurons (AFD) synapse with a network of interneurons leading to behavioural output. Because the genetics are also well understood the analysis can be taken further. It can be shown that manipulation of

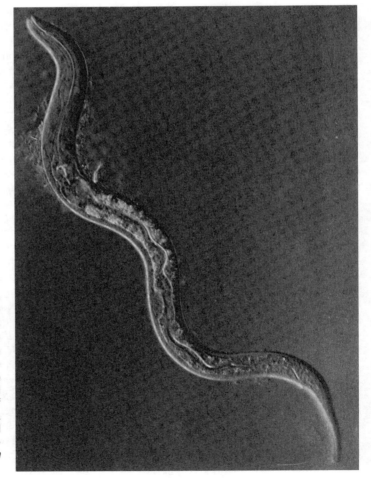

Figure 20.6 *Caenorhabditis elegans*. The worm is about 200 μm in length. It consists of 959 somatic cells of which 302 constitute the nervous system. Many of the nerve cells are large and as the body is translucent can be seen in the living animal. Although there are only 302 neurons they have been grouped into 118 classes on the basis of their shape and connectivity. In spite of this rich variety of different types there does not seem to be a clear distinction between sensory, motor and internuncial neurons: most neurons are polyfunctional. From White (1985), *Trends in Neurosciences*, **8**, 277–283; with permission.

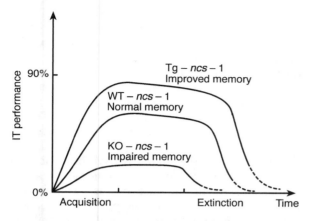

Figure 20.7 Memory and NCS-1. The figure shows the effect of over-expressing and knocking out the gene for NCS-1 on the association of isothermal tracking behaviour (IT) with food. Normal memory is shown by wild type (WT). Impaired learning and memory are shown by knockout (KO) of *ncs-1* gene. Improved acquisition and memory retention are shown by transgenic (Tg) worms in which NCS-1 is over-expressed. Further explanation in text. After Gomez *et al.* (2001), *Neuron*, **30**, 241–248.

a gene (*Ce-ncs-1*) encoding a small Ca^{2+} binding protein (neuronal calcium sensor 1 (Ce-NCS-1)) can enhance or degrade association of isothermal tracking (IT) with food. When *Ce-ncs-1* was knocked out both acquisition and extinction of the association (=memory) were diminished; vice versa, when the gene was over-expressed acquisition was improved and time to memory loss was extended (Figure 20.7). Other features of the worm's olfactory and locomotor behaviour were unchanged.

Ce-NCS-1 is expressed in both the AFD thermosensory neurons and in the interneurons (AIY interneurons) with which they synapse. It is clearly significant in the formation and retention of memory. Acting as a Ca^{2+} sensor, its presence in lesser or greater amounts controls the activity of the synapses between AFD and AIY neurons. It is suggested that the AIY interneuron acts as an integrator for inputs from AFD thermosensory neurons and olfactory neurons detecting nutriment. It is not yet known whether the Ce-NCS-1 Ca^{2+} sensor protein acts pre- or postsynaptically although the balance of evidence suggests a presynaptic location. It is interesting, also, to note that the Ce-NCS-1 Ca^{2+} sensor is a member of a large family of highly conserved NCS-1 Ca^{2+} sensors found across the biological world including yeast, *Drosophila*, rodents and humans. Clearly the existence of a learning paradigm in the deeply understood and easily manipulable *C. elegans* is likely to prove of great value in running the phenomenon of memory into the biochemical and biophysical ground.

20.4.2 *Drosophila*

The experimental approach to learning in *Drosophila* is rather similar to that used in *C. elegans* but distinctively different from that for higher organisms. Because of its small genome and the highly developed state of *Drosophila* genetics it is possible to search for single mutations that affect learning and then to search for the gene product involved. The publication of the complete genome sequence in March 2000 has greatly expedited this approach.

The associative learning test most used with *Drosophila* is to pair an olfactory or electrical stimulus with a gustatory stimulus (Figure 20.8). The gustatory stimulus (the UCS) is usually sucrose, which 'untrained' *Drosophila* will approach and ingest. The olfactory stimulus (the CS) may be one of a number of olfactants to which the insect is sensitive. The aversive electrical stimulus (also a CS) is a grid delivering about 90 V, AC. It is not difficult to condition the fly so that it will approach a source of odorant previously paired with sucrose and avoid an odorant source previously paired with an electric shock.

This test has been used to isolate mutants deficient in the learning and/or memory mechanisms required for this task. A number of such mutants have been found. Each causes a specific deficit.

Dunce (*dnc*), the first such mutant to be isolated, shows appreciable learning when tested 30 s after training, but after that the memory quickly decays. This mutation can be traced to the X chromosome and appears to affect a **cAMP phosphodiesterase (PDE 2)**. In addition to their memory defect, dnc flies also show reductions in both **habituation** and **sensitisation**. This perhaps suggests that in *Drosophila*, at least, associative and non-associative memory share a similar mechanism. We shall

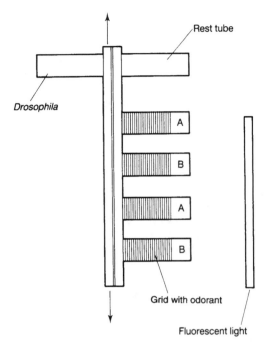

Figure 20.8 Apparatus for conditioning *Drosophila*. The fruit flies are housed in a tube which can be slid back and forth across the openings to a number of other tubes. In the 'rest' position the 'home' tube is opposite a 'rest' tube which is perforated to allow adequate air. The 'test' tubes – A, B, A, B – contain grids coated with odorant. The grids are also connected to a supply to give 90 V, AC, shock. The flies are induced to crawl into the tubes by making use of their phototaxis towards light. Conditioning may then be carried out by using either an attractive stimulus (sucrose) or an aversive stimulus (electric shock). Memory may then be tested in the second set of tubes (A, B). After Dudai *et al.* (1976), *Proceedings of the National Academy of Sciences, USA*, **73**, 1684–1688.

return to this when we consider *Aplysia* in the next section.

Dopadecarboxylase (*Ddc*) mutants in contrast to *dnc* mutants fail to learn the associations in the first place. The DOPA decarboxylase enzyme (or l-aromatic amino acid decarboxylase) is, as we saw in Chapter 16, required for the synthesis of both **dopamine** and **serotonin**. Flies deficient in these neurotransmitters thus appear incapable of learning the odour–shock and odour–sucrose pairings. It is moreover possible to breed flies with partial lesions of the gene and to show that they have a

reduced ability to learn the two associations. There seems, in other words, to be a **dosage** effect. More interesting still, it can be shown that such partial (or mild) mutants **do not forget** what they have learned. The *Ddc* mutation thus affects learning but not recall. It appears, therefore, that dopamine and/or serotonin are necessary for learning the odour–sucrose, odour–shock paradigm in *Drosophila* but not for retaining or expression/retrieval of that memory.

A number of other mutants have been isolated which affect this and other learning paradigms in *Drosophila*. As in the case of the two mutants described above it is becoming possible to connect the genetic lesion to the biochemical defect and this to very specific deficiencies in either learning or memory. The *rutabaga* mutant, for instance, which also shows learning difficulties, has been shown to carry a defect in Ca^{2+}-sensitive adenylyl cyclase. This finding provides a clue to a possible underlying mechanism. For, as with the *Dnc* mutation, cAMP metabolism is affected. Could cAMP, significant in so many areas of cell biology, also play a central role in memory?

Which part of the fly's brain is involved in this associative learning and the laying down of memory traces? Evidence is accumulating that the most important regions are two large laterally placed bodies in *Drosophila*'s brain known as the corpora pedunculata or **mushroom bodies**. Each of these bodies contains about 2500 neurons (Kenyon cells) and is divided into five lobes. Flies in which these bodies have been ablated by hydroxyurea are still able to perform visual learning but are totally unable to form the olfactory conditioned reflexes described above. It has also been shown that mutations which affect the integrity of these bodies also affect olfactory learning and memory.

A creative use of molecular biology has been used to discriminate between learning and retrieval in the mushroom bodies. This technique makes use of the *shibshire* (*shi*), a gene that encodes the insect homologue of **dynamin**. It will be recalled from Chapter 15 that dynamin is involved in the recycling of synaptic vesicles. If it is inactivated, this recycling does not happen and the synapse is closed down. Temperature-sensitive mutants of *shi* have been isolated in which the dynamin ceases to operate at 30°C and it is found that flies exposed to this temperature are paralysed. When returned to 20°C full recovery occurs in a few minutes. The primary cause of the paralysis is inactivation of neuromuscular junctions.

However, a clever piece of molecular biology allows the mutant *shibshire* gene to be targeted to specific parts of *Drosophila*'s nervous system. This molecular biology involves coupling the temperature-sensitive *shibshire* gene (shi^{ts1}) to a GAL4-dependent promoter. GAL4 is a yeast transcriptional activator which can be inserted into *Drosophila* genome (see Section 5.18). Breeding protocols were used to limit the expression of shi^{ts1} to the mushroom bodies in *Drosophila*'s brain. Locomotion remained normal when these targeted flies were exposed to 30°C and it could be shown that their sensorimotor responses to electric shock and odours remained normal. Experiments showed, however, that while these flies could learn perfectly normally an association between electric shock and odour at the restrictive temperature and recall that learning 30 min to 3 h later, at a lower temperature (say, 20°C) the opposite did not obtain. If the flies were exposed to the learning situation at 20°C and then tested for recall at 30°C, they invariably failed dismally. These experiments are interpreted as dissecting the laying down of memory from its recall. It appears that when the mutant *shibshire* gene is switched on the memory trace is confined to the mushroom bodies and is consequently unable to make its presence known. The synaptic networks necessary for its output to influence the rest of the nervous system and thus affect the fly's behaviour had been degraded.

In addition to the distinction between learning and retrieval of what has been learnt, *Drosophila* memory also follows the general rule of being divisible into two phases: short-term and long-term. Normal flies can retain an olfactory conditioned reflex for more than a week. Flies showing the mutation *amnesiac* (*amn*) showed a total absence of long-term memory. When tested for an olfactory conditioned reflex a few seconds after training they were indistinguishable from normal controls. When tested an hour later their memory had declined to zero whilst control flies still performed well. Memory acquisition had not been affected, memory retention had. It has been shown that the *amnesiac* gene encodes a preproneuropeptide which stimulates the synthesis of cAMP. Once again the experimental evidence points to the same

suspect! *Amn*, moreover, is strongly expressed in two large neurons (dorsal paired medial neurons) which project and ramify throughout the mushroom bodies. If the recycling of synaptic vesicles in these neurons is blocked, long-term memory is disrupted in the same way as found in the *amnesiac* mutant.

This indication from the *amnesiac* gene that long-term memory is in some way dependent on supplies of cAMP is consistent with biochemical investigations. Experiments have indicated that long-term memory is disrupted by cycloheximide, a protein synthesis inhibitor. Investigators were led to consider the possibility of a biochemical cascade involving PKA and CREB proteins (see Sections 3.3.1 and 17.6). To establish the validity of the hypothesis a gene coding for a defective CREB protein was inserted into *Drosophila* under the control of a heat-shock promoter. It was then possible to show that when the inserted gene was switched on by heat shock there was no effect on short-term but a strong disruptive effect on long-term memory.

The interpretation of these sophisticated experiments, which make full use of the armamentarium of techniques now available to the molecular biologist, is to suggest that the cascade through cAMP and PKA is to phosphorylate a CREB protein which then binds to the CRE site (Figure 3.9B). This leads via IEGs to the switching on of late response genes coding for proteins involved in the modulation of synaptic function (see Figure 20.9). The biochemical mechanisms responsible for the behavioural defects of *Dnc*, *rutabaga* and *amn* mutants are accordingly accounted for: all three mutations reduce cAMP synthesis, the first step in the cascade. In other words, it is argued (see Figure 20.9) that if one or both of the dorsal paired medial (DPM) neurons releases its Amn neuropeptide on to a mushroom body Kenyon cell already activated by olfactory input from the antennoglomerular tract (AGT), at least one of the conditions for memory formation is satisfied. If the conjunction of AGT and DPM activity on the Kenyon cell occurs only a small number of times, PKA phosphorylation will alter the biophysics of K^+ channels and thus the activity of the cell (STM); if, however AGT and DPM activity are conjoined a significant number of times, the CREB mechanism for switching on nuclear genes is activated. In this case a longer lasting alteration of Kenyon cell biophysics

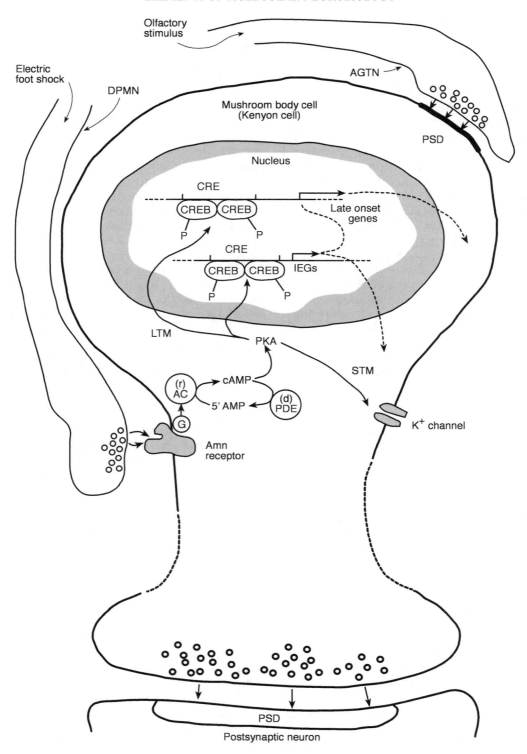

ensues (LTM). We have already noted that cAMP cascades to the CRE site have been detected at work in the long-term memory mechanisms of other organisms, in particular, in *Aplysia* and mammals. We shall return to these comparators later in this chapter.

We should not leave the topic of memory and the mushroom bodies without mentioning the intriguing hints that the activity of the 2500 cells in each mushroom are coordinated in some way to give rise to slow oscillations in Ca^{2+} concentration. These Ca^{2+} oscillations can be detected by the fluorescence of a Ca^{2+}-sensitive protein, aequorin, which can be expressed in the mushroom body neurons of transgenic flies. The Ca^{2+} levels oscillate with a mean period of 4 minutes and the amplitude is increased in *amn* mutants. It has been suggested that these oscillations are in some way related to the biochemistry of odour memory consolidation.

Summing up, we can see that approaches through molecular biology, genetics and biochemistry promise a deep understanding of the processes responsible for memory in *Drosophila*. Mutants screened out in the odour conditioning apparatus, such as *dnc*, *Ddc*, *rut*, *amn*, etc., provide insight into essential biochemical components of memory cascades and into the parts they play in consolidation and retrieval. Clever molecular biology using temperature-controlled mutations acting in specific parts of the nervous system allows the anatomical location of these cascades to be pinpointed.

Combined with more conventional biochemistry, neuroanatomy and neurohistology, these techniques have every prospect of dissecting memory into its elements. Forty years ago when the fruit fly was first proposed as a model organism for memory studies, few thought it optimal or that it could have anything to say about those concerned with the central problem: human memory. The molecular biological revolution of our times has brought home as never before the deep underlying unity between all organisms, even organisms as widely separated as *Drosophila melanogaster* and *Homo sapiens*. What we learn in the fly has every prospect of being relevant to the hugely more complex memory processes in humans.

Nevertheless, although the neural elements of *Drosophila* nervous systems are becoming better known, especially those of the mushroom bodies, they are nevertheless all very small and the neuroanatomy difficult to disentangle. In order to investigate the molecular biology of learning and memory at the cellular and subcellular level by traditional biochemical techniques it has been useful to turn to systems whose neurons are orders of magnitude larger. These systems are provided by the third of the invertebrate phyla we shall consider in this section – the **Mollusca**.

Although much interesting work has been done on *Hermissenda* we shall concentrate our attention on the best-known and most instructive of these organisms: *Aplysia californica*.

Figure 20.9 *(opposite)* Conceptual diagram to show biochemical pathways which could account for *Drosophila* behavioural genetics. A mushroom body neuron (Kenyon cell) receives input from two sources. At the top a fibre from the olfactory system, the antennoglomerular tract (AGT), releases its excitatory transmitter in response to olfactory stimulation. On the left a fibre of one of the dorsally paired medial neurons (DPMN) releases the amnesiac neuropeptide (amn) on to its receptor. This activates a G-protein which stimulates a *rutabaga*-encoded adenylyl cyclase ((r)AC) and cAMP is synthesised. The cAMP is later transformed back into 5′ AMP by the *dunce*-encoded PDE ((d)PDE). Before this happens the cAMP activates PKA. At this point two things may happen. PKA alters the conductivity of K^+ channels or other membrane biophysics. If sufficient PKA is activated, however, its catalytic subunits will diffuse through the nuclear membrane and phosphorylate CREB binding proteins and activate the CRE site which, in turn, causes the transcription of IEGs. This will be the case when both the AGT and DPM pathways are activated simultaneously. The transcripts of IEGs are often themselves transcription factors, thus leading to switching on of late onset genes which may code for cytoskeletal proteins, neurotrophins, receptors, neurotransmitter enzymes, etc. The figure shows that PKA may also affect the CRE site controlling late onset genes directly. These biochemical cascades provide in the first case for short-term memory (STM) and in the second for long-term memory (LTM). Further explanation in text. AGTN=antennoglomerular tract neuron; Amn=amnesiac; (d)PDE=*dunce*-encoded PDE; PSD=postsynaptic density; (r)AC=*rutabaga*-encoded AC. Other abbreviations as in previous figures. After Sokolowski (2001), *Nature Reviews Genetics*, **2**, 879–890; Waddell and Quinn (2001), *Annual Review of Neuroscience*, **24**, 1283–1309.

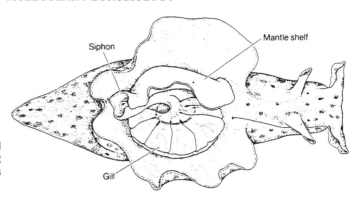

Figure 20.10 *Aplysia californica.* From 'Small systems of neurons' by E. R. Kandel. Copyright © 1979 by Scientific American Inc. All rights reserved.

20.4.3 *Aplysia* and the Molecular Biology of Memory

Figure 20.10 shows a dorsal view of *Aplysia*. It is a large marine mollusc measuring several hundred millimetres from head to tail. The figure shows the position of the **siphon** and **gill** which play so large a role in the behaviour we are about to examine.

When a weak mechanical stimulus is applied to the siphon both siphon and gill tend to withdraw beneath the mantle shelf. This is a protective reflex. Habituation, sensitisation and also classical conditioning can be demonstrated (Figure 20.11).

Habituation

If a sequence of 10–15 weak mechanical stimuli are applied to the siphon the gill withdrawal is much reduced. The response has habituated. Recovery to the original withdrawal magnitude takes about an hour. If, however, the habituating sessions are repeated four or five times then the 'memory' persists for several weeks. During this period the withdrawal is significantly less than normal. All of these characteristics show that *Aplysia* habituation is identical to habituation in vertebrate systems.

Sensitisation

If a noxious stimulus is applied to the animal's head or tail immediately before the siphon is touched then the withdrawal is much enhanced. This sensitisation again lasts about an hour. Long-term sensitisation can be induced in the same way as long-term habituation: by repeating the training sessions a number of times.

Classical Conditioning

Several conditioning paradigms have proved effective. A weak tactile stimulus to the siphon (such as would normally induce only a tiny gill retraction) can serve as the CS whilst a strong tactile stimulus to the gill itself (normally causing a major retraction) serves as the UCS. After a number of classical CS–UCS pairings the unpaired CS can be shown to elicit an enhanced gill withdrawal. This CR shows all the features of the Pavlovian type of CR. Another means of eliciting a classical CR from the *Aplysia* siphon–gill system is to pair a weak stimulus (CS) to the siphon with a strong stimulus (UCS) to the tail. Once again a number of pairings of CS and UCS leads to a strong withdrawal when the CS is applied alone.

Now all this, the reader may think, is all very well and good – but what's new? What's new is that the neural connections underlying these behavioural responses are very well understood. *Aplysia*'s nervous system is very simple compared with that of the vertebrates. Like that of *Caenorhabditis* it consists of identifiable cells and connections. Unlike *Caenorhabditis* or, for that matter, *Drosophila*, these cells are very large. The largest perikarya are up to 0.5 mm in diameter. Figure 20.12 shows a diagram of the abdominal ganglion in which some of the identifiable cells are labelled.

Some of the cells labelled in Figure 20.12 are involved in the gill withdrawal reflexes we discussed above. It can be shown that the skin of the siphon contains just 24 sensory neurons and that these make direct connections to six motor neurons which control the muscles of the gill. The cell bodies of four of the latter neurons (L7, LD$_{G1}$,

A

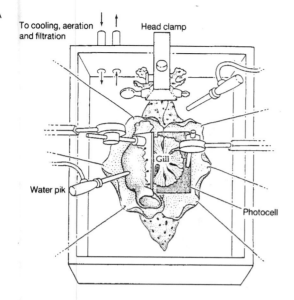

To cooling, aeration
and filtration

Head clamp

Gill

Water pik

Photocell

B   Habituation and recovery

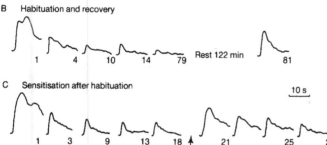

1   4   10   14   79   Rest 122 min   81

C   Sensitisation after habituation

10 s

1   3   9   13   18 ↑   21   25   27

Figure 20.11 Experimental arrangement for the study of the gill withdrawal reflex in *Aplysia*. (A) *Aplysia* is held in a small aquarium in such a way that a water jet can be directed at the siphon (Water pik) and the response of the gill detected by the amount of light falling on to a photocell beneath it (dense stipples). (B) Habituation: during some eighty water-jet stimuli to the siphon the gill response gradually habituates. The numbers indicate the number of the trial. After 122 minutes of rest the response has completely recovered. (C) Sensitisation: habituation is again induced by eighteen trials (at one-minute intervals). A strong and prolonged tactile stimulus is then delivered to the neck – time of application indicated by upwardly pointing arrow. The habituated response is immediately restored. This is sensitisation. From Purves and Lichtman (1985), *Principles of Neural Development*, Sunderland, MA: Sinauer; after Kandel (1976), *Cellular Basis of Behaviour: An Introduction to Behavioral Neurobiology*, New York: Freeman; with permission.

LD$_{G2}$, RD$_G$) can be seen in Figure 20.12. In addition to this exceedingly simple circuit, the sensory cells also synapse with a small number of interneurons interposed between them and the motor neurons. The circuit is shown in Figure 20.13.

First let us consider **habituation**. It will be recalled that this is a reduced gill withdrawal in response to repeated stimuli to the siphon. This response can be mimicked by stimulating the sensory neurons from the siphon and recording from the motor neurons to the gill. In the *Aplysia* system this can be made precise by stimulating one known sensory fibre and recording from the perikaryon of a known motor neuron.

When the experiment shown in Figure 20.14 is carried out it is clear that although the first stimulus results in a large EPSP in the motor neuron's perikaryon, subsequent stimuli elicit smaller and smaller EPSPs. It can, furthermore,

be shown that this is not due to any decrease in sensitivity of the subsynaptic membrane. The habituation is due to a **decrease in the quantity of neurotransmitter** released on arrival of an action potential at the presynaptic terminal. There is evidence that this is due to a persistent inactivation of the Ca^{2+} current into the presynaptic terminal. This, as we noted in Chapter 15, would reduce the release of transmitter from synaptic vesicles. Because the effect occurs without the intervention of any second synapse the phenomenon is termed **homosynaptic** depression (Figure 20.15).

Homosynaptic depression can also be studied in the Petri dish. The monosynaptic pathway between sensory and motor cells can be reconstituted in a dissociated culture of *Aplysia* neural tissue. This allows a variety of biochemical and biophysical investigations to be made. Some of these will be discussed below.

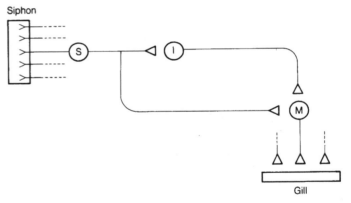

Figure 20.12 Abdominal ganglion of *Aplysia*. The abdominal ganglion consists of about 1500 cells. Identifiable cells are labelled in the figure. Cells in the left hemiganglion are prefixed by 'L' and those in the right hemiganglion by 'R'. From 'Small systems of neurons' by E. R. Kandel. Copyright © 1979 by Scientific American Inc. All rights reserved.

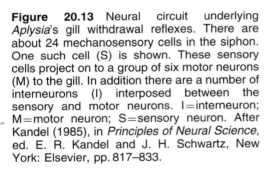

Figure 20.13 Neural circuit underlying *Aplysia*'s gill withdrawal reflexes. There are about 24 mechanosensory cells in the siphon. One such cell (S) is shown. These sensory cells project on to a group of six motor neurons (M) to the gill. In addition there are a number of interneurons (I) interposed between the sensory and motor neurons. I=interneuron; M=motor neuron; S=sensory neuron. After Kandel (1985), in *Principles of Neural Science*, ed. E. R. Kandel and J. H. Schwartz, New York: Elsevier, pp. 817–833.

Next let us turn to **sensitisation**. Here, it will be remembered, the gill retraction is markedly increased on a second stimulation. The molecular mechanisms underlying this behaviour are, as we shall see, more fully understood than are those underlying habituation. It is, moreover, a more intricate process. Instead of just two neurons – sensory and motor – being involved at least **three** neurons are implicated.

First let us consider **short-term** sensitisation. Once again the phenomenon can be investigated by stimulating a sensory neuron and recording the response of the subsynaptic motor neuron. In this case the EPSP, instead of being smaller when the motor nerve is stimulated a second time, is very

much larger. Once again, however, it can be shown that this is not due to any increased sensitivity of the subsynaptic membrane but to a **greatly increased release of neurotransmitter**.

How has this come about? The answer is shown in Figure 20.16. The figure shows that when a sensitising stimulus is delivered to the head (or tail) of *Aplysia* 'facilitatory' interneurons are activated. These synapse on the **presynaptic** terminals of the sensory neurons. It is believed that these presynaptic synapses release **serotonin** on to the presynaptic endings of the sensory neurons.

This mechanism can be confirmed in the reconstituted sensorimotor dyad mentioned above. In the presence of serotonin the quantity

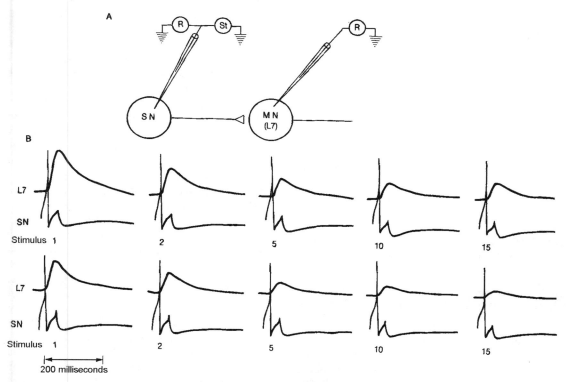

Figure 20.14 Experimental arrangement to investigate habituation in *Aplysia*. A sensory neuron (SN) which synapses on motor neuron L7 (MN L7) (see Figure 20.12) is stimulated (St) at 10-s intervals. The response in both the sensory and the motor neurons is measured (R) with an intracellular electrode. Five records from two sessions separated by 15 min are shown (the numbers indicate the number of the trial). Whilst the response in the sensory neuron remains constant that in the motor neuron diminishes dramatically. After the interval the habituation of the motor neuron is still apparent. From 'Small systems of neurons' by E. R. Kandel. Copyright © 1979 by Scientific American Inc. All rights reserved.

of neurotransmitter released by the presynaptic terminal of the sensory neuron is increased. Furthermore the effect of externally applied serotonin can be mimicked by injection of cAMP into the sensory neuron. This gives us a first hint of the underlying biochemical mechanisms. This hint is developed by the finding that inhibitors of cAMP-dependent protein kinase (PKA) block the effect of both serotonin and injected cAMP. Immediately we are reminded of the cascades that are believed to underlie *Drosophila* short- and long-term memory (Figure 20.9).

How, then, does serotonin increase the release of transmitter by the sensory neuron? If we look back to Section 11.2.3 we see that serotonin affects certain potassium channels in the membrane (K_S or S channels). In *Aplysia* it is believed that this is

achieved by serotonin activating membrane-bound adenylate cyclase. The cAMP thus produced initiates a biochemical cascade which via an increased activity of protein kinase A (PKA) leads to phosphorylation of the S channel protein (or a closely associated protein). The upshot of all this is that the S channel is blocked and the **outflow of K+ is reduced**. In Chapter 14 we noted that the outflow of potassium ions was responsible for the 'downward' or recovery phase of the action potential. If the S channel is blocked the duration of this recovery phase is increased. Finally, turning back to Chapter 15, we recall that the inward flux of Ca^{2+} ions is strongly voltage-dependent. Ca^{2+} channels only open when the membrane is depolarised. Hence if the recovery phase behind an action potential is elongated an **increased amount of**

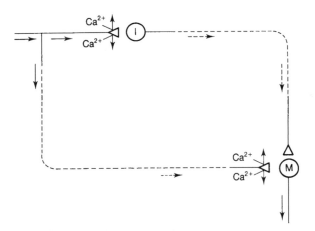

Figure 20.15 Homosynaptic depression causes habituation. A sequence of impulses in the sensory neuron leads to a partial closure of Ca^{2+} gates in the presynaptic terminal. This, in turn, diminishes the amount of transmitter released on to the motor neuron (M). Hence the decreased magnitude of the EPSP in M.

Ca^{2+} can enter. Hence, remembering that vesicle exocytosis is dependent on intraterminal Ca^{2+}, transmitter release is increased. This sequence of events is schematised in Figures 20.17 and 20.18.

Because more than one synapse is involved in sensitisation the neural process is called **hetero-**

synaptic facilitation. It depends on an activation of the presynaptic serotoninergic synapse immediately before the arrival of an action potential along the sensory fibre. It is schematised in Figure 20.18.

Next let us turn to **long-term** sensitisation (Figure 20.19). It can be shown that the same nerve pathways are involved and that a similar neurological explanation in terms of heterosynaptic facilitation holds. The biochemistry, however, is different. It is possible to show that whereas a single application of serotonin to the presynaptic terminal induces sensitisation of the reflex for a period not exceeding an hour or so, four or more repeated applications of serotonin lead to a sensitisation lasting for more than 24 hours. If the 'half-lives' of the biochemicals involved in the heterosynaptic mechanism outlined above are taken into account it is difficult to see how they could be responsible for so long lasting an effect.

It has also been found that long-term (but not short-term) sensitisation can be blocked by both **transcriptional** and **translational inhibitors**. This suggests that long-term sensitisation involves not only the activation of PKA and a consequent phosphorylation of presynaptic S channels but also some involvement of the genome. It has also been shown that two hours after repeated applications of serotonin the concentration of PKA catalytic

Figure 20.16 'Wiring diagram' for sensitisation in *Aplysia*. Superimposed on the neuronal circuit of Figure 20.13 is a system of facilitatory interneurons (FI). These neurons are activated by sensory fibres from the tail or head. They make synaptic contact with the presynaptic terminals of the sensory neurons from the siphon. These synapses are believed to be serotoninergic (represented by the cored vesicle). After Kandel (1985), in *Principles of Neural Science*, ed. E. R. Kandel and J. H. Schwartz, New York: Elsevier, pp. 817–833.

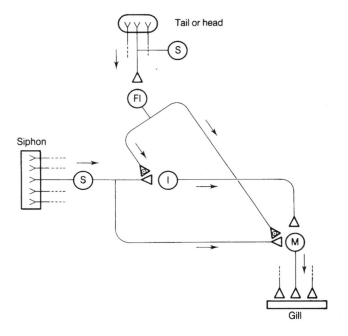

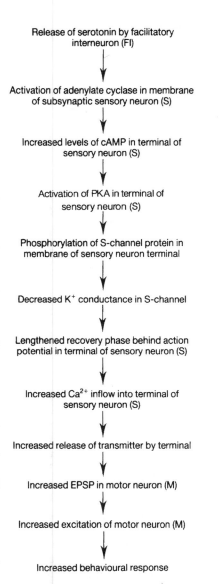

Release of serotonin by facilitatory interneuron (FI)

↓

Activation of adenylate cyclase in membrane of subsynaptic sensory neuron (S)

↓

Increased levels of cAMP in terminal of sensory neuron (S)

↓

Activation of PKA in terminal of sensory neuron (S)

↓

Phosphorylation of S-channel protein in membrane of sensory neuron terminal

↓

Decreased K^+ conductance in S-channel

↓

Lengthened recovery phase behind action potential in terminal of sensory neuron (S)

↓

Increased Ca^{2+} inflow into terminal of sensory neuron (S)

↓

Increased release of transmitter by terminal

↓

Increased EPSP in motor neuron (M)

↓

Increased excitation of motor neuron (M)

↓

Increased behavioural response

Figure 20.17 Sequence of biochemical events underlying sensitisation in *Aplysia*. Further explanation in text. After Byrne (1985), *Trends in Neurosciences*, **8**, 478–482.

subunits is increased in the nucleus. Again, we are aware of analogies with *Drosophila*. Indeed, there is evidence for the involvement of the CRE binding site and CREB proteins in *Aplysia* neurons. Other second messengers than cAMP (e.g. Ca^{2+}, IP_3, cGMP) may, of course, also be involved.

Which mRNA transcript does the cAMP induce? The short answer is that we do not at present know. It can, however, be shown that a transcription factor known as C/EBPβ is induced by either serotonin or cAMP. Disruption of this factor in *Aplysia* sensory fibres blocks serotonin-induced long-term sensitisation. It is interesting to note that, *in vivo*, C/EBPβ synthesis appears to depend on prior CREB activation. We shall find in Section 20.5 that a similar interconnection has been found in the mammalian hippocampus. It is tempting to suspect that we may have the beginnings of the biochemical cascade underlying and responsible for at least one type of memory: cAMP $\Rightarrow$ PKA $\Rightarrow$ CREB activation $\Rightarrow$ C/EBPβ $\Rightarrow \ldots$?

The question mark, however, remains. The next steps in the biochemical cascade remain obscure. Nevertheless the end result is clear. It can be shown that the number, size and vesicle complement of sensory neurons involved in long-term sensitisation are all significantly increased. Perhaps the transcript (or transcripts) programs the synthesis of appropriate enzymes and structural elements. These induced biochemicals would have to be packaged in such a way that they reach and adhere to the appropriate synaptic terminals. For it has to be remembered that neurons in general have many different synaptic terminals. But this, as we saw in Section 15.2, is no new problem in cell biology. How packages budding off the Golgi apparatus find their appropriate location in the cell remains very largely unknown. Recent work on cholinergic vesicles in the electric fish *Torpedo* suggests, however, that these, at least, contain synapse-specific glycoproteins.

The implications of this work on *Aplysia* sensitisation for the relationship between STM and LTM will not have escaped the reader.

Finally let us consider **associative learning**. A number of experiments have been carried out in an attempt to discover the molecular basis of associative memory in *Aplysia*. In general it appears that associative memory may make use of the same basic molecular mechanisms as sensitisation. This, it will be remembered, would be consistent with the findings in the *Drosophila dnc* mutants.

One of the more recent experiments used a depolarising stimulus as the CS and an application of serotonin as the UCS. After a very few pairings it was possible to show that there was a

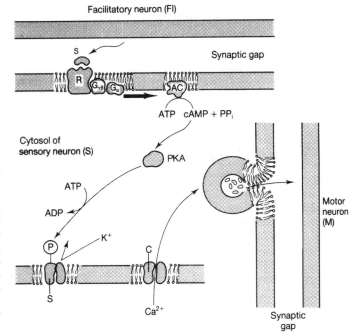

Figure 20.18 Heterosynaptic facilitation in *Aplysia*. The facilitatory interneuron (FI) releases a facilitatory transmitter (S) on to the presynaptic terminal of the sensory neuron (S). A collision-coupling mechanism generates cAMP which activates PKA to phosphorylate the I_{KS} channel (S). This reduces the efflux of K^+ ions after any action potential in the sensory neuron. The prolonged depolarisation of the membrane allows the Ca^{2+} channels (C) to stay open longer and hence increases the likelihood of synaptic vesicles moving to the active zones in the presynaptic membrane, fusing and releasing their contents into the synaptic gap above the motor neuron.

significantly increased level of cAMP in the sensory cells. Neither the CS alone nor the UCS alone nor random applications of CS and UCS led to this outcome.

The model shown in Figure 20.20 builds on the molecular biology worked out for long-term sensitisation (Figure 20.19). It suggests that the UCS (in this case an action potential) arriving at the presynaptic terminal will activate the underlying motor neuron (M) in the normal way. It suggests, next, that the CS arriving at the serotoninergic synapse will operate through its conventional G-protein to downregulate the K_S channel in the underlying UCS neuron. On its own, however, it is unable to cause release of sufficient transmitter from the UCS terminal to activate the underlying motor neuron (M).

If, however, the CS occurs always a short time before the UCS the levels of Ca^{2+} in the terminal will tend to rise. This can lead, via the calmodulin mechanism, to increased activity of adenylyl cyclase. The resulting enhanced quantity of cAMP activates sufficient PKA to act on genomic CRE sites (compare Figure 20.9). Activation of these sites leads to a switching on of genes programming factors which lead to a more perma-

nent increase in synaptic size, vesicle population, etc. It is then possible that activity in the serotoninergic CS neuron will be able, on its own, to release sufficient transmitter from the synaptic terminal to activate the underlying motor neuron (M). Thus, and in short, a new 'low-resistance' path through *Aplysia*'s nervous system is established.

It is, however, clear that this can only be a hint of what is happening at the molecular level. For it is usually the case that more than one synaptic ending will be present on a terminal. The mechanism described suggests that all such endings would be potentiated. This would vitiate the precision of the neural arcs which must underlie CRs and memory.

20.5 THE MEMORY TRACE IN MAMMALS

Let us, in conclusion, return from the simple systems of invertebrates to our 'proper study': mammals and humankind. Does the elucidation of invertebrate memory systems help us to understand what is happening in mammals? We started this chapter by alluding to Occam's razor. Let us use it again here. Let us suppose that the molecular mechanisms underlying memory are much the same

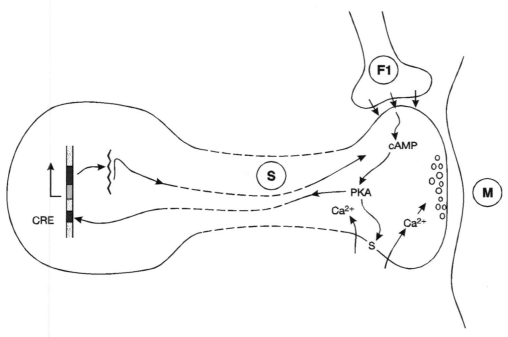

Figure 20.19 Putative mechanisms underlying long-term sensitisation. The same triadic system of facilitatory interneuron (FI), sensory neuron (S) and motor neuron (M) is shown as in previous figures. The same biochemistry obtains as in Figure 20.18. It is suggested, however, that increased release of facilitatory transmitter on to the sensory terminal generates sufficient second messenger to affect the genome. The figure shows the catalytic subunits of PKA being carried in the retrograde axoplasmic flow to the perikaryon. Here they switch on the protein synthetic machinery so that a factor (or factors) enhancing synaptic activity are manufactured and carried in the anterograde flow to the terminal. Further explanation in text.

throughout the animal kingdom. Let us suppose that the spectacular development of memory in the mammals and especially amongst the primates is due to the vastly increased size and complexity of the nervous system. This great intricacy of structure (we can suppose) allows interaction between neural circuits each of which relies, at bottom, on much the same molecular biology we have seen at work in worm, fly and mollusc. The upshot is the vastly increased range and subtlety of the memory process in the higher animals. In the case of *Homo sapiens*, of course, the situation is complicated yet further by the development of symbolic representation and communication – language.

In previous sections of this chapter we have noted the distinction between short-term and long-term memory and have also given some consideration to where in the mammalian brain the process of consolidation – whereby STM is transformed

into LTM – occurs. We noted, in particular, that there is good reason to believe that so far as 'explicit' or 'declarative' memory is concerned the **hippocampus** is deeply implicated. It is thus especially interesting to find that two of the neurophysiological processes that have been thought to underlie consolidation were first detected in the hippocampus – though they have subsequently been shown to occur elsewhere as well. These processes are called **post-tetanic potentiation (PTP)** or **short-term potentiation (STP)** and **long-term potentiation (LTP)**.

20.5.1 Post-tetanic Potentiation and Long-term Potentiation

Post-tetanic potentiation has been a candidate for the physical basis of memory for a number of years. It can be shown that after a volley of impulses has

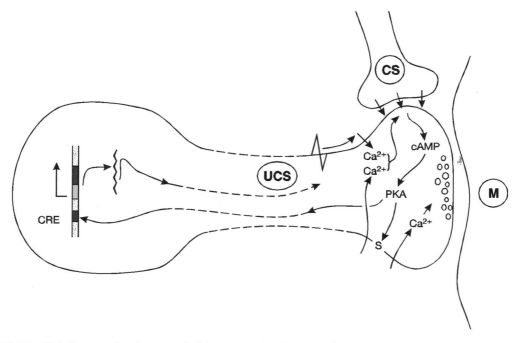

Figure 20.20 Putative mechanisms underlying consolidation of conditioned reflex. The biochemical flow is essentially similar to that in Figure 20.20. The conditioned stimulus (CS) releases serotonin. The unconditioned stimulus (UCS) is a depolarisation. After a number of pairings of CS and UCS (the CS always preceding the UCS) a significant increase in the level of cAMP in the UCS terminal is observed. It is proposed that this is due to the increased CA^{2+} in the terminal acting through a Ca^{2+}/calmodulin mechanism to potentiate adenylyl cyclase. The increased quantities of cAMP then act on PKA causing increased activation (i.e., dissociation of the catalytic subunits). The catalytic subunits are transported in the retrograde axoplasmic stream to the genome where they activate the CRE site. The protein products encoded by the structural gene find their way back to the terminal in the anterograde axoplasmic stream. Further explanation in text.

been delivered to certain synapses the excitability of the subsynaptic membrane is significantly increased. In other words subsequent impulses in the presynaptic cell induce larger EPSPs in the subsynaptic cell and thus have a greater chance of depolarising its initial segment to threshold and initiating an action potential (Figure 20.21). PTP generally persists for a few hours and then the subsynaptic membrane returns to normal.

Long-term potentiation differs from PTP in the length of time for which it persists. Instead of a few hours it can be shown to remain for days, weeks and even months. To induce LTP the stimulation must be of a greater intensity than that required for PTP. It has been found, furthermore, that volleys of impulses impinging on the subsynaptic cell from several brain regions give optimal results. It is clear that these

electrophysiological phenomena are consistent with a memory mechanism.

20.5.2 Fibre Pathways in the Hippocampus

Before going any further it is important to remember that the hippocampus is orders of magnitude more intricate than the entire nervous systems of *Drosophila* and *Aplysia*. In particular, it is important to note that there are three major elements in the excitatory pathway: perforant fibres from the entorhinal cortex; mossy fibres from the dentate gyrus to a region designated CA3; Schaffer collaterals to a region designated CA1 (Figure 20.22). It is found that LTP generated by repetitive stimulation of the perforant and Schaffer fibres depends on the presence of NMDA-Rs at their synapses; in

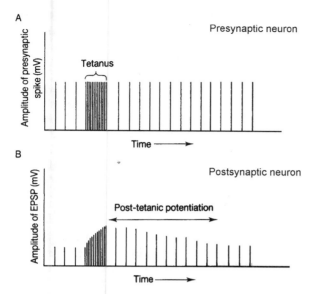

A

Presynaptic neuron

Tetanus

Time →

B

Postsynaptic neuron

Post-tetanic potentiation

Time →

Figure 20.21 Post-tetanic potentiation. (A) Impulses in presynaptic neuron. (B) Amplitude of EPSP in postsynaptic neuron. After the rapid tattoo of tetanic stimulation the magnitude of the EPSP is increased and remains increased for a considerable period of time.

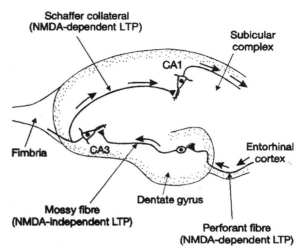

Schaffer collateral
(NMDA-dependent LTP)

Subicular
complex

CA1

Fimbria   CA3

Entorhinal
cortex

Dentate gyrus

Mossy fibre
(NMDA-independent LTP)

Perforant fibre
(NMDA-dependent LTP)

Figure 20.22 Horizontal slice through hippocampus to show major excitatory pathways. Nerve impulses enter the hippocampus via the perforant pathway from the entorhinal cortex. A synapse is made with a granule cell in the dentate gyrus. The axon (mossy fibre) of the granule cell carries the excitation to a pyramidal cell dendrite in CA3. The axon of the CA3 pyramidal cell is known as a Schaffer collateral fibre and carries excitation to a pyramidal cell in CA1. Further explanation in text. After Huang, Li and Kandell (1994), *Cell*, **79**, 69–80.

contrast, LTP generated by activity in the mossy fibres does not depend on the presence of NMDA-Rs.

20.5.3 Perforant and Schaffer Collateral Fibres

Let us consider the NMDA-R-dependent pathways first. It has been found that Ca^{2+} fluxes are once again deeply implicated. If a calcium chelator, EGTA, is injected into subsynaptic cells it can be shown that LTP is inhibited. Note, however, that it is the **subsynaptic cell** that is affected, not (as in the case of *Aplysia*) the presynaptic ending. This distinction is confirmed by the finding that blocking glutamate receptors on these subsynaptic membranes also blocks LTP. Some form of 'back-reaction' from subsynaptic to presynaptic membrane cannot, however, be ruled out. It is not impossible that the biochemical events induced in the subsynaptic cell by the rain of transmitters from a tetanised presynaptic terminal may lead to some trophic material being released into the synaptic cleft which affects the presynaptic membrane. For,

as we shall shortly see, there is also evidence that LTP is associated with an enhanced release of EAAs from presynaptic terminals. Let us, however, look first at the nature of the subsynaptic response to tetanising stimulation.

It will be recalled from Chapter 10 that there are at least three pharmacologically distinct types of i-glutamate receptor: AMPA, KA and NMDA receptors. It will also be recalled that the NMDA receptor controls a Ca^{2+} channel which can only be opened by glutamate when the membrane has been depolarised by some 30 mV. A number of antagonists are known for the NMDA receptor system. Several of the latter, for instance D-2-amino-5-phosphonovalerate (D-AP5) and phenecyclidine (PCP), have been used in an analysis of the molecular basis of LTP in the hippocampus.

It can be shown that both D-AP5 and PCP prevent the induction of LTP in the CA1 and dentate gyrus (DG) regions of the hippocampus. It can be shown, in short, that whilst NMDA antagonists block the response to high-frequency (tetanising) stimulation, the response to low-frequency, non-tetanising, stimuli (presumably mediated by KA and AMPA receptors) is unaffected. These findings have suggested an NMDA-receptor-mediated mechanism for LTP and hence short-term memory.

The suggested mechanism is based on the well-established observation that Mg^{2+} ions block the NMDA receptor at resting potential (Chapter 10). It is suggested, therefore, that low-frequency stimulation, releasing glutamate from presynaptic terminals, whilst activating the non-NMDA receptors and inducing EPSPs in the underlying cell, does not cause sufficient subsynaptic depolarisation to remove the Mg^{2+} blockade on the NMDA receptors. This blockade is, however, removed by the significantly greater subsynaptic depolarisation induced by high-frequency (tetanising) stimulation of the afferent fibres. In this case the presynaptic terminals will release large quantities of glutamate which via the KA and AMPA receptors will cause a large depolarisation of the subsynaptic membrane.

The opening of the NMDA receptor Ca^{2+} channel is thus dependent on tetanic stimulation. The consequent influx of calcium ions could have a number of consequences. It might affect the excitability of the subsynaptic membrane by acting on Ca^{2+}-dependent ion channels in the membrane, or it could activate protein kinase A, which might in turn affect membrane excitability by phosphorylating a membrane protein or, lastly, it could act as a 'second messenger' to trigger some deeper cytosolic or genomic biochemistry perhaps through inducing the synthesis of nitric oxide synthase (NOS) or activating NOS already present in the cytosol.

20.5.4 The CRE Site Again

Recently evidence has been provided which implicates the CRE site and CREB proteins. This evidence has excited speculation that a common pathway underlies the consolidation of STM into LTM in both vertebrates and invertebrates. It is suggested that phosphorylation of the CREB protein is induced by the influx of Ca^{2+} through the NMDA-R either directly or mediated by PKA. This proposal is supported by evidence that NMDA-R inhibitors (APV and MK801) block CREB-dependent transcription in hippocampal slices.

The evidence for involvement of the CRE site and the IEG system of Figure 20.9 in the hippocampus derives from experiments with mice containing defective CREB proteins. In these experiments two main isoforms were eliminated, CREBα and CREBΔ, although a third isoform, CREBβ, was unaffected. The CREB$^{-\alpha\Delta}$ mice, which developed and appeared normal in every way, were tested in three different memory tasks: contextual fear conditioning (mice associate an electrical shock to the foot with the context in which it is delivered), olfactory conditioning and the Morris water maze. In the first two cases it could be shown that long-term memory was severely affected whilst short-term memory was spared. In the Morris water maze, CREB$^{-\alpha\Delta}$ mice showed profound deficiency in spatial learning with the submerged platform but had no difficulty in learning the position of a visible platform. Lesion experiments showed that whilst the former task depends on the hippocampus the latter task does not.

These behavioural observations were supported by electrophysiological recordings from hippocampal slices. Whilst LTP remained stable for two hours in controls it decayed to baseline in ninety minutes in CREB$^{-\alpha\Delta}$ slices. Experiments using transgenic mice with a β-galactosidase reporter gene under the control of a CRE promoter also indicate the involvement of the CRE site in long-term but not short-term memory. Short-term potentiation (1 s, 100 Hz) induces no CRE activity whereas LTP, especially if three tetanising events are delivered at five-minute intervals, is strongly reported by the synthesis of β-galactosidase.

The involvement of the CRE site is also supported by experiments using C/EBPβ inhibitors. It will be recalled from Section 20.4.3 that C/EBPβ mRNA is detectable in *Aplysia* sensory neurons during long-term sensitisation. In the mammalian case rats were subjected to an inhibitory avoidance test in which they had to learn to avoid a dark compartment (which they would otherwise favour) by remembering that on entry they received a foot shock. The memory was tested 48 hours after training. It was found that the memory was disrupted if antisense C/EBPβ oligonucleotide (β-ODN) was injected into hippocampus either 5 or 24 hours after training. No effect was found if β-ODN was injected 1 hour or 46 hours after training. This is clear evidence that C/EBPβ is involved in memory consolidation. As in *Aplysia* we are beginning to glimpse the first steps of the biochemical cascades underpinning memory consolidation (learning): NMDA-R activation $\Rightarrow$ Ca^{2+} influx $\Rightarrow$ PKA activation $\Rightarrow$ CREB phosphorylation $\Rightarrow$ C/EBPβ activation $\Rightarrow$ ...?

Research in this field of neurobiology is at present very active and many seemingly contradictory and inconclusive results are reported. It would be surprising, however, if the biochemical processes so well documented in *Drosophila* and *Aplysia* do not find counterparts in mammals. It begins to look as if at least some elements, or some pathways, in what is probably a highly complex and highly heterogeneous biochemistry are evolutionarily conserved across widely separated species.

20.5.5 Mossy Fibre Pathway

We noted above that LTP in the mossy fibre pathway was not dependent on an NMDA-R mechanism. It has been shown that LTP in this pathway has two phases: early (LTP-E: decaying to baseline in about three hours) and long (LTP-L: remaining at over 200% of baseline at six hours and more). In addition to independence of NMDA-Rs the mossy fibre pathway differs from the other two pathways in that at least LTP-E is induced presynaptically. The question naturally arises: does a similar biochemistry underlie this rather different anatomy?

The answer is, yes, probably. It turns out that whereas LTP-E is unaffected by inhibitors of protein and mRNA synthesis, LTP-L is severely upset. Both phases make use of the cAMP/PKA system but only the LTP-L phase involves the synthesis of new protein. It seems likely that PKA will phosphorylate a CREB-protein leading to a protein synthetic sequence similar to that shown in Figure 20.9. If this does indeed turn out to be the case it will begin to seem as if memory throughout the animal kingdom is mediated by the same underlying biochemistry.

What is the new protein synthesised in response to activation of the CRE site? We saw above that after the activation of C/EBPβ we left a question mark. One interesting set of candidates to fill in this question mark are the **neurotrophins** (NTs) we discussed in Chapter 19. A number of different NTs have been isolated (see Section 19.9). The different NTs and their receptors are selectively expressed in different regions of the brain. A good deal of evidence has accumulated to show that mRNA levels of NTs and their trk receptors vary with neuronal activity. In the hippocampus it can be shown, furthermore, that while NT mRNA levels increase in response to stimulation of **glutaminergic** cells, the opposite occurred (decreased NT mRNA) when **GABA-ergic** cells are stimulated. At present there is controversy over whether the NTs are released pre- or postsynaptically or, perhaps, both. Whichever turns out to be the case it seems that the NTs released into the synaptic cleft increase the size of the synaptic bouton and initiate dendritic, and dendritic spine, sprouting. Further work will no doubt confirm or disconfirm this attractive mechanism for creating favoured pathways through the brain in response to neuronal activity.

20.5.6 Histology

Although neuroscientists (like all scientists) search always for unifying principles underlying diversity, the fate of predecessor theories and hypotheses must make them cautious. Although the CREB protein theory makes a very satisfying story we should always bear in mind the morphology of mammalian neurons. In Section 1.3.1 we noted how very large neurons are compared with most of the body's cells. In Box 20.1 the nature of dendritic spines, upon which most of the hippocampal synapses are made, is outlined. It is a constricted microdomain often only tenuously connected to the rest of the neuron. Is it reasonable to suppose that diffusion will lead PKA to phosphorylate a CREB protein at the nucleus far away at the bottom of the dendritic tree? Should we not look for a more local change, perhaps in the micro-compartment of the spine itself?

20.5.7 Non-genomic Mechanisms

What form could these local changes take? Numerous theories and hypotheses compete for attention. Lynch and Baudry, for example, provide evidence which suggests that the influxes of calcium ions at hippocampal synapses activate a membrane-bound enzyme – calpain. This enzyme, they further propose, breaks up the fodrin

BOX 20.1   Dendritic spines

We noted in Chapter 1 that the dendrites of many of the neurons in the cerebral cortex are covered with minute protuberances: the dendritic 'spines' or 'thorns'. These protuberances form the major synaptic surfaces of the principal input/output cells of the cerebral cortex. Although they are reasonably permanent structures they have been shown to vary in size, shape and number in response to brain activity. We saw something of their dynamism in the developing brain in Section 19.4. They have long been thought to be crucially involved in the plastic changes underlying long-term memory.

The dimensions of spines vary by an order of magnitude from one brain region to another. The smallest have a length of $0.2\,\mu m$ and volume of about $0.04\,\mu m^3$ and the largest a length of $6.5\,\mu m$ and volume of $2\,\mu m^3$. A typical spine is shown in Figure 1.21B, but there are many variations and spines frequently branch. Typically, however, the mature spine has a bulbous 'head' attached to the dendrite by narrow 'neck' (see Figure A). The apical dendrite of a large pyramidal neuron in the cerebral cortex may develop many thousands of spines.

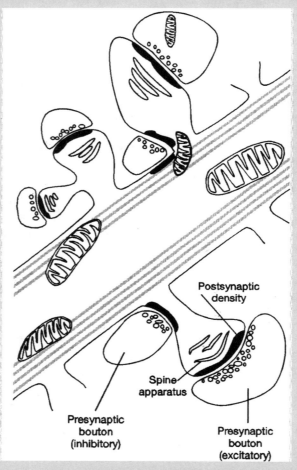

Figure A   Spine structure. High magnification of spines on apical dendrite of cortical pyramidal cell (cf. Figure 1.21).

Spines normally possess a complex ultrastructure. The membrane beneath a synaptic bouton (and spines may make up to three synaptic contacts) develops a characteristic postsynaptic density (PSD) (Section 17.2). Freeze-fracture electron microscopy shows a large number of particles (c. $2800\,\mu m^{-2}$) which are believed to be glutamate-triggered channels. In addition more than 30 different proteins have been detected in subcellular fractions enriched in PSDs. These include receptor molecules (especially glutamate receptors), protein kinases (including CaM Kinase 2), cytoskeletal proteins (e.g. actin, MAPs, neurofilaments, etc.), the important scaffolding protein PSD-95, and some metabolic enzymes. These proteins are associated with the quite complex organellar structure visible in the electron microscope.

This organellar structure constitutes the 'spine apparatus'. It is largely built of smooth endoplasmic reticulum (SER). Like other SERs it stores Ca^{2+} ions which can be released by metabotrobic second messengers such as IP_3 (see Section 8.5). Polyribosomes are usually present but mitochondria are absent, except in the very large spines of some cerebral cortical cells. In addition to SER the spine apparatus contains a well-developed filamentous network consisting largely of actin. In the neck of the spine the actin is arranged in parallel bundles, in the head it forms a loose basket-work.

The spines constitute the major excitatory input into the cells of the cerebral and cerebellar cortices and the hippocampus. Usually, there is just one presynaptic bouton at the head of the spine, occasionally other (modulating) boutons can be observed at the neck. There has been much interest in the role of spines in long-term potentiation and other changes underlying brain plasticity. It has been suggested that the constriction at the neck of the spine concentrates EPSPs and subsequent events to the head. If the head contains AMPA, KA and NMDA receptors then release of glutamate will have a better chance of reducing the potential sufficiently to release the Mg^{2+} block on the NMDA receptor. There is evidence to support this hypothesis. Developmental studies show that long-term potentiation (LTP) depends not only on the existence of NMDA channels but also on the number of spines present. Moreover, when spines mature their necks tend to constrict and LTP is easier to establish.

So, in conclusion, it is perhaps possible to conceive of the spine as a separate compartment, or 'micro-domain' (see Box 15.1), in its own right. With a narrow neck any changes on its synaptic surface leading to a changed intrabulbar state are less likely to be diluted by activity in the rest of the neuron. Plastic changes are hence more likely to be long lasting. These changes may indeed involve growth and changed morphology. We saw in Section 19.4 how rapid these changes can be. In tissue culture mature spines can form in less than 2 hours. The EM pictures of spines, showing them much as thorns on the stems of rose trees, should not delude us into thinking of them as permanent, static structures; there is much evidence to show that they are constantly changing shape and position. Such changes have been observed during learning and their inverse have been seen in the pathologies of brain dysfunction, especially the neurodegenerations of old age. Dendritic spines constitute an organisational level between macromolecular complexes such as ribosomes and membrane scaffolds and the neuron.

subsynaptic cytoskeleton (see Chapter 7), thus **unmasking** fresh glutamate receptors. The subsynaptic membrane is thus sensitised (potentiated) for further synaptic activity in response to EAAs. Unfortunately it has been impossible to demonstrate an increased binding of labelled glutamate to subsynaptic receptors during LTP as the theory predicts. As so often in this field, and elsewhere in molecular neurobiology, one is reminded of T. H. Huxley's lament over beautiful theories being skewered by ugly facts.

Another hypothesis suggests that the influx of Ca^{2+} ions affects a molecule known to be present at the synapse: **Ca^{2+}/calmodulin-dependent kinase 2 (CaMK2)**. This molecule is a major constituent of the postsynaptic density (PSD). A feature of this molecule is auto-phosphorylation. Auto-phosphorylation increases during LTP and continues for some time afterwards. Once this sustained auto-phosphorylation is set in train it will continue when the LTP stimuli are turned off. A recent hypothesis suggests that before LTP the PSD contains (among many other things) a core 'memory' element consisting of NMDA-R anchored to PSD-95. During LTP the Ca^{2+} influx through NMDA-Rs causes autophosphorylation of CaMK2, which leads it to bind securely to NMDA-Rs. This, in turn, causes recruitment of AMPA-Rs to the 'memory' complex. This com-plex (NMDA-R, CaMK2, AMPA-R) anchored through the NMDA-R to PSD-95 remains in place in the PSD whilst CaMK2 remains phosphoryl-ated without need of translational or tran-scriptional changes. Through the increased concentration of AMPA-Rs, the complex raises the excitability of the synapse and thus establishes a memory trace through the brain. This model has both biochemical and mathematical modelling support.

A final possible mechanism also makes use of the observation that NMDA receptors may be involved in the stabilisation of synchronously active synapses. In Section 19.11 evidence was provided to show that synapses which fired together stayed together; those which were not synchronised were eliminated by some (as yet unknown) antagonist. It was suggested that NMDA-controlled channels were instrumental in inhibiting the synthesis of this antagonist. In the case of LTP it is similarly suggested that NMDA receptors in tetanised synapses induce (perhaps by increasing cytosolic Ca^{2+} levels) a factor which increases the release of transmitter from the presynaptic terminal. It could be that this is the 'trophic factor' mentioned above which is required to account for the enhanced release of EAAs from tetanised terminals. It could be, in other words, that we have here a case of NMDA-induced 'temporal summation' to put beside the NMDA-induced 'spatial summation' of Chapter 19. It has been suggested (see Chapter 19) that nitric oxide is the looked-for retroactive factor. Other investiga-tors have provided evidence which implicates platelet activating factor (PAF) (1-O-alkyl-2-acetyl-sn-glycero-3-phosphocholine). This is widely distributed in the brain and elsewhere in the body and has many biochemical functions. Yet other workers favour carbon monoxide.

The student will recognise, once again, that the search for the mechanisms underlying memory at the synapse is every bit as open and controversial (and as exciting) as the search for mechanisms at the level of the genome. It is quite possible that, as in other areas of neurobiology, our quest for unity is misplaced. In spite of the seductive power of unifying hypotheses, especially those based on a growing understanding of evolutionary molecular biology, it may still turn out that was once taken for a single underlying process, labelled with a single name, memory, is in fact multifarious. It may still turn out that many different biochemical mechanisms are responsible for subtly different forms of learning and of retaining traces of past experience. As we begin the twenty-first century the problem of memory remains unsolved.

20.6 CONCLUSION

Whether the CREB cascade, the CaM kinase theory, the 'NMDA trophic factor' theory, some or all, or some totally different interpretation can be sustained only time will tell. Whether LTM involves transcription at the genome, or whether its biochemistry is confined to the synapse, or both, or varies in different instances, or in different anato-mical locations, remains open to debate. We are far further on than when Hilgard gave his despairing opinion in the middle of the last century, but we still have far to go.

The strategy of using ideas generated in the simpler and more easily manipulated nervous systems of invertebrates to set up questions to investigate in the immensely more complex nervous systems of vertebrates has in the past proved very fruitful. It is only necessary to remind ourselves of the significance of the squid giant axon in funda-mental neurophysiology to assure ourselves of this. *C. elegans*, *Drosophila* and *Aplysia* suggest what we should be looking for in the brains of mammals including ourselves. Their molecular neurobiology provides insight into our own and may allow us to see how the 'negative sculpting' which we found at work in the embryology of the brain (Chapter 19) may still be at work in the mature mammalian brain, ensuring that a continually updated 'model' of the significant features of the environment crystallises out of myriad differently weighted synaptic interactions.

21

SOME PATHOLOGIES

Personal, social, economic dimensions – fuzzy distinction between neuroses and psychoses – mind/brain dichotomy – differing therapies. **Prions**: many different neurodegenerations – presently incurable – coded by gene on chr. 20 – aetiology – prions are replicative proteins – replicative mechanism – proliferative routes – transmissibility – can prions cross species barrier? – therapies? **PKU**: many metabolic defects – PKU due to defect in phenylalanine-hydroxylase co-factors – lethal build-up of phenylpyruvate – therapy. **Fragile X (FraX)**: deficiency of FRM1 gene on X chr. – trinucleotide repeat pathology – autism and mental defect. **Neurofibromatoses (NF)**: NF1 gene on chr. 17 – encodes GAP protein – biochemistry; NF2 gene on chr. 22 – encodes protein involved in cytoskeleton. **Motor neuron disease (MND)** **(amyotrophic lateral sclerosis (ALS))** – disease of motor outflow – SOD1 – protects against superoxide radicals – probably multifactorial. **Huntington's disease (HD)**: basal ganglia – movement disorder – genetics – search for gene – located in chr. 4 – trinucleotide repeat pathology – rodent model. **Depression**: 'upward' and 'downward' causation; **endogenous**: monopolar and bipolar – genetic basis (Old Order Amish) – multigene aetiology – imbalances in 'cocktails' of transmitters; **exogenous**: stress – animal models; **neurochemistry**: many transmitters – imbalances between monoamines and ACh – 'up' and 'down' regulation of receptors – animal models – therapies. **Parkinsonism (PD)**: movement disorder – neuroanatomy – pathology of dopamine pathway – therapy – cause of nigral cell death – MPTP, MPP^+ – environment – genes – tissue grafting – ethics. **Alzheimer's disease (AD)**: human tragedy – economic cost – diagnosis – aetiology – plaques and tangles – BAPtist and Tauist controversy – chr. 21 and the APP gene – three types of APP – normal functions of APP – mutations of APP gene – chr. 1, chr. 14 and the presenilins – early-onset AD – the secretases – amyloidogenesis. Chr. 19 and the APOE gene – isoforms of APOE and their function – cholesterol – interaction with amyloid and tau. Environmental influences, Al and concussion. The BAPtist hypothesis – an amyloid cascade – Meynert's nucleus and the cholinergic system – hippocampus. Prospects for therapy – prevention and/or cure. **Conclusion**: the medical and other importance of good molecular neurobiology

'Can'st thou not minister to a mind diseased' Macbeth adjures his physician and, hearing that nothing can be done, retorts: 'Throw physic to the dogs, – I'll none of it.'

This is certainly a major reason for studying molecular neurobiology. Some 10% of the UK population will at some time in their lives be hospitalised for 'mental illness'. For most, the period in hospital is short. For some, however, it is lengthy and perhaps permanent. The magnitude of the personal tragedy, social and familial dislocation, and economic cost can hardly be over-estimated. Can an understanding of the molecular biology of the brain help us toward a treatment or, better, a cure?

21.1 NEUROSES, PSYCHOSES AND THE MIND/BRAIN DICHOTOMY

First of all let us subdivide the term 'mental illness' into two: **neurological** ('organic') and **psychological** (or 'functional'). This subdivision is by no means clear cut. Taking the neurological category first, we find it is often useful to subdivide it further into conditions that have a clear anatomical substratum (e.g. multiple sclerosis, parkinsonism) and those where that substratum is more subtle (e.g. depression, schizophrenia). Turning to the second category (the psychological) we find that it is also customarily divided into two: the **neuroses** and the **psychoses**. Again the basis of this subdivision is far from clear cut. It is generally held that in the neuroses (such as depression) the sufferer shares the same 'world' as the rest of us but sees it through whatever is the opposite of rose-tinted spectacles. In psychotic conditions (e.g. schizophrenia) the patient seems to live in a different world altogether. He or she seems to be overcome with delusions and hallucinations. On this classification Lady Macbeth would nowadays be said to be suffering from a (fairly mild) psychotic condition – a species of schizophrenia – induced by 'having known what she should not'.

To those who accept a 'dual-aspect identity' interpretation of the mind/brain relationship there are no sharp discontinuities between neurological and psychotic illnesses. Instead there is a continuum of conditions which extend from phenylketonuria (PKU), say, to paranoid schizophrenia. All have their psychological and their neurological 'aspects'. Just as (to use a classical metaphor) a curved mathematical surface has a convex face ('aspect') and a concave face ('aspect'). Remember: a 'mathematical' surface has no thickness.

This is not, of course, to say that the entire spectrum of brain/mind disease can be best tackled in the same way. It would be as absurd to try the 'talking cure' (psychoanalysis) with an individual suffering from Huntington's disease as to treat one of Freud's Viennese ladies by genetic engineering. The dual-aspect identity interpretation would have it, however, that the talking cure has an effect on the brain's physiology, albeit a subtle and at present untraceable effect, just as altering the neuronal genome would have a massive effect on the brain's functioning – and hence its 'owner's' subjectivity.

In this chapter we shall look at some of the brain/mind pathologies that have an identifiable molecular aetiology. We have already met a number of these, especially the channelopathies, as we have proceeded through the pages of this book. In this chapter we shall discuss nine more 'global' conditions. We shall start with a consideration of prions and **prion diseases**. The molecular biology of these conditions is quite unusual and is still being worked out. Nevertheless they cause terrible neurodegenerations in both humans and animals. We shall then go on to consider two of the pathologies of the developing brain, **phenylketonuria (PKU)** and **fragile X (FraX)** syndrome, then move on to some of the conditions that affect the mature brain, **neurofibromatosis, Huntington's disease (HD)** (=Huntington's chorea), **motor neuron disease (MND)** and **depression**, and end by considering two of the degenerative conditions that affect the elderly: **Parkinson's disease (PD)** and **Alzheimer's disease (AD)**. There are, of course, a multitude of other ills which can affect the brain: birth mishaps, infections by various agents, dietary deficiencies, cardiovascular accidents, etc. In particular, we have noted in several places in this book that elevation of internal Ca^{2+} levels caused, in many cases, by abnormal release of excitatory amino acids, can overwhelm the delicate Ca^{2+} homeostatic balance within a neuron, and lead to its death. But for these other ills the reader should consult one of the standard textbooks in neurology.

21.2 PRIONS AND PRION DISEASES

Prions and prion diseases have attracted much attention in recent years due to the economic significance of **bovine spongiform encephalopathy (BSE)** or 'mad cow disease' (MCD) and the fear that it might cross the species boundary and induce similar degenerations in humans. BSE, however, is just one example of a spectrum of neurodegenerative diseases ranging from **Creutzfeldt–Jakob disease (CJD)** in humans to **scrapie** in sheep (Table 21.1). The pathological signs of all these diseases include **spongiform degeneration** of the grey matter (numerous vacuolar cavities in the cells), **plaques** (insoluble protein deposits) and **reactive gliosis**. The diseases are at present incurable and lead to progressive loss of neural function and ultimately

Table 21.1 Prion diseases

Human diseases
Creutzfeldt–Jakob disease (CJD)
Variant Creutzfeldt–Jakob disease (vCJD)
Gerstmann–Sträussler–Scheinker syndrome (GSS)
Kuru
Fatal familial insomnia (FFI)
Fatal sporadic insomnia (FSI)

Animal diseases
Scrapie (sheep)
Bovine spongiform encephalopathy (BSE)
Feline spongiform encephalopathy (FSE)
Transmissible mink encephalopathy (TME)
Exotic ungulate encephalopathy (EUE) (nyala, oryx)
Chronic wasting disease (CWD) (mule, deer, elk)

Modified from Prusiner and DeArmond (1994), *Annual Review of Neuroscience*, **17**, 311–339; Prusiner (1998), *Proceedings of the National Academy of Sciences, USA*, **95**, 13 363–13 383.

death. In this respect they resemble the neuro-degeneration of Alzheimer's disease (see Section 21.10). In other respects they differ. Most importantly they differ from Alzheimer's in their transmissibility. CJD, for instance, can be transmitted by poor medical practice (growth hormone from infected pituitaries, infected surgical instruments, etc.) or by cannibalism (kuru: New Guinea). Many (not all) animal encephalopathies can (like kuru) be transmitted, often by ingesting contaminated food.

There has been (and still is) considerable controversy over the aetiology of prion diseases. For some time they were thought to be caused by a 'slow' virus but despite intensive and extended research no nucleic acid has ever been isolated. It is nowadays believed that they are the exception that proves the rule of Watson–Crick molecular biology: replicative proteins. For this reason the term 'prion' has been invented. It is furthermore believed that this protein, the **prion protein**, PrP^C, is a constituent of normal cell membranes, especially neural cell membranes. An abnormal isoform of the prion protein, **PrP^{Sc}**, is believed to be responsible for the neurodegenerations of the prion diseases. PrP^{Sc} is so called because it was first isolated from scrapie-infected sheep; some workers use the acronym in a general way to apply to any abnormal prion protein, others prefer to be more specific and refer, for example, to the CJD prion as PrP^{CJD}.

As already noted PrP^C is a glycoprotein expressed on the surface of many cell types, not neurons alone. Nevertheless it is very highly concentrated at synapses (both pre- and post-synaptically) and motor end plates in striated muscle. Its function at the synapse (or indeed in other cell membranes) remains something of a mystery. There is evidence that its lengthy extra-cellular N-terminal domain binds copper atoms. Free copper atoms are known to catalyse various biochemical reactions leading to the production of hydroxyl radicals leading to **oxidative stress**. It has been proposed, therefore, that PrP^C in mopping up free copper atoms protects cells from this stress. A number of other possible functions have been proposed for this ubiquitous membrane-embedded protein but at the time of writing its functional significance remains obscure.

The gene coding for PrP^{CJD} is located on chromosome 20 and has been named ***PRNP***. Mutation causes the inherited form of CJD. The development of transgenic (Tg) mice expressing mutant PrP genes and other molecular biological techniques have greatly aided the search for an understanding of prion diseases. These new techniques and careful investigation of the family histories of CJD and Gerstmann–Sträussler–Scheinker syndrome (GSS) patients have resulted in the identification of over 20 mutations in the *PRNP* gene which segregate with inheritance of a prion disease. But how do these mutant forms of PrP cause the prion pathology?

Radiolabelling studies show that both PrP^C and PrP^{Sc} are synthesised in the rough endoplasmic reticulum (RER). Post-translational glycosylation occurs in the normal way in the lumen of the ER and the Golgi apparatus (see Figure 15.6). But this is where differences between the isoforms begin to appear. Whereas PrP^C is carried to and inserted into the plasma membrane the PrP^{Sc} isoform remains in the cytosol. It looks, therefore, as though the infectivity of PrP^{Sc} is conferred during this post-translational processing. But what sort of molecular alteration differentiates PrP^{Sc} from PrP^C?

At this point we are very much at the frontier of research. Recent evidence suggests that PrP is very delicately balanced between α-helical and β-pleated sheet conformations. It is believed that the slight amino acid differences between PrP^C and PrP^{Sc} and

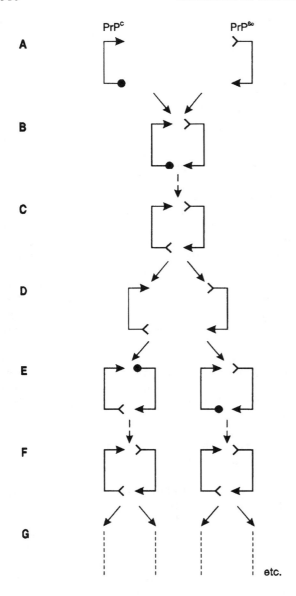

Figure 21.1 Multiplication of PrP^{Sc}. (A) Schematic of PrP^C and PrP^{Sc} in the cytosol of an infected cell. (B) PrP^C and PrP^{Sc} come together to form a dimer. (C) PrP^{Sc} transforms PrP^C into another copy of PrP^{Sc} by inducing a conformational change, perhaps a transition from α-helix to β-sheet. (D) The dimer separates into monomers and (E) each finds a further PrP^C molecules with which to dimerise so that (F) the conformational change is induced and (G) duplication can occur again, and so on ...

the consequent variation in post-translational glycosylation tips the balance one way or the other. Structural analysis of PrP^C suggests that the amino acid chain in the extracellular domain contains three **α-helical** segments; infra-red spectroscopy of these predicted segments in PrP^{Sc} showed instead a three-stranded **β-pleated sheet**. But how could this subtle alteration in three-dimensional conformation account for PrP^{Sc} infectivity?

The best answer is that PrP^{Sc} transforms existing PrP^C into replicas of itself. The schematic of Figure 21.1 shows how this may occur. Remembering how delicately balanced the conformational transition from PrP^C to PrP^{Sc} is, it is not difficult to translate the symbolism of the figure into a plausible molecular transformation. We have here, as suggested in Section 3.2, an instance of replication without information storage. So long as fresh supplies of PrP^C are available PrP^{Sc} multiplies. It is also clear from the schematic how PrP^{Sc} is infective. Providing it finds an appropriate isoform of PrP^C in the host cell it can replicate until supplies run out. In so doing it causes the typical neuropathological lesions of prion disease.

So far we have seen how PrP^{Sc} is believed to proliferate within a single cell. But how does it spread throughout the brain to engender the terrible and fatal cerebral degeneration? In order to answer this question it was first necessary to localise the PrP^{Sc} and then follow its spread. This proved difficult to do until in 1992 a new technique, **histo-blotting**, was introduced by Taraboulous and colleagues. This consisted in cutting 10 μm frozen sections of infected brain and pressing them on to nitrocellulose paper. Use is then made of the fact that proteinase K digests PrP^C but not PrP^{Sc}. Exposure of the blot to this enzyme thus leaves only PrP^{Sc} *in situ*. This can then be visualised by immunohistochemical staining with antibodies raised against prion proteins. This technique has proved very successful in locating and quantifying (by intensity of stain) PrP^{Sc} in the brain. It has been possible to show that when PrP^{Sc} is injected into experimental animals it first creates the characteristic prion pathology around the point of entry. The infection can then be shown to spread via the cerebrospinal fluid to other parts of the brain. But the widespread infection caused by this mode of transmission is not random. Different brain regions are differently affected. The thalamus, medial septal nucleus, diagonal band of Broca, all close to sources of CSF, are particularly strongly infected. Furthermore, not all transmission is via the CSF. Colonisation of the cerebral cortex is by antero-grade transport along axons from the thalamus.

But can prions jump the species barrier? The answer appears to be 'yes, but'. It depends on the height of the barrier. BSE, for instance, is believed to be caused by scrapie-infected sheep brains being used in the preparation of cattle cake. In the laboratory, injections of BSE-infected brain extracts can cause prion encephalopathies in mink, cats, antelope. It is, moreover, easy to inoculate laboratory mice with BSE and show that most (if not all) succumb to the disease. It should be noted, however, that it takes 1000 times more intracerebral inoculate of BSE to kill a mouse than a cow. Furthermore, although scrapie has been present in sheep for centuries, it has never been known to infect shepherds or humans in general. The barrier is too high. Thus the likelihood of BSE jumping the species barrier from cattle to humans is, hopefully, small.

Can we give any scientific rationale for this hope? What is the nature of the barrier? The best answer, in our present state of knowledge, has to do with PrP diversity. We have already noted that there are at least twenty different mutations in the *PRNP* gene associated with human prion diseases (CJD, GSS, FFI). This genetic diversity is greatly increased by post-translational processing, especially glycosylation. If the mechanism of prion propagation is as sketched above (Figure 21.1) then PrP^{Sc} from evolutionarily distant species may be too different in molecular structure to react with human PrP^C to initiate the pathological cascade. There is nowadays a great deal of evidence from transgenic mice expressing differently mutated prion genes to support this hypothesis. Neverthe-less, as is now well known, well over a hundred humans (all teenagers or young adults) have suffered from incurable vCJD derived from the 1990s BSE pandemic in UK cattle and that appalling number may grow in the future.

Finally, are there prospects for successful ther-apy? The effort being put into prion research holds out hope of therapeutic intervention in the years to come. One promising approach has emerged from work with genetically engineered mice. The prion gene in the mouse resides on chromosome 2 and is designated *Prn-p*. The techniques of molecular genetics allow, as mentioned above, the insertion of mutated genes coding for PrP^{Sc} and mice dutifully express the characteristic neural and behavioural symptoms of prion disease. Knockout genetics allow the removal of the *Prn-p* gene altogether. Mice lacking the gene are known as *Prn-p^{00}* mice and appear to be healthy. This suggests that PrP^C is not critical for normal life. Indeed the function of PrP^C in normal cells remains unknown. Furthermore, as would be expected from the PrP^{Sc} propagation mechanism outlined above (Figure 21.1), *Prn-p^{00}* mice are resistant to prion infections. These observations provide a hope for therapy. It is possible that individuals diagnosed in early stages of prion infection (and the disease progresses over a number of years) could be treated with antisense oligonucleotides directed against PrP^C mRNA. This is obviously at present only a speculation. It is a hope for the future: switching off the source of PrP^C would eliminate the sub-strate upon which PrP^{Sc} propagates.

21.3 PHENYLKETONURIA (PKU)

A large number of congenital defects in amino acid metabolism are now known to affect the brain. The most important of these are shown in Table 21.2. These metabolic defects are expressed in many of the body's organs but the major symptoms are neurological: mental retardation, convulsions, ataxia. The infant brain is particularly sensitive to any abnormal biochemistry. We noted the signifi-cance of sensitive periods in Chapter 19 when we discussed the brain's development. In children this period lasts for the first few years of life and different brain regions enter and leave their sensi-tive periods independently of each other. Because the defects are widespread the biochemical abnormality can normally be detected in the urine and/or the blood. An early diagnosis generally means that a satisfactory treatment is possible.

The best known of these congenital defects is that responsible for **phenylketonuria (PKU)**. The defect is in the enzyme system catalysing the transformation of **phenylalanine** to **tyrosine**. We saw in Chapter 16 (Figure 16.17) that this trans-formation requires **phenylalanine hydroxylase** and **tetrahydrobiopterin (BH4)** as a co-factor. In

Table 21.2　Amino acid metabolism and mental defects

Disease	Amino acid	Enzyme defect	Symptom
Arginosuccinic aciduria	Arginine	Arginosuccinase	Severe mental retard-ation, abnormal EEG
Cystathioninuria	Methionine	Cystathioninase	Mental retardation
Histidinaemia	Histidine	Histidase	Mental retardation
Homocystinuria	Methionine	Cystathionine synthetase	Lens dislocations, osteporosis, mental retardation
Hyperammonaemia	Ornithine	Ornithine transcarbamylase, arginosuccinase	Nausea, vomiting, mental retardation
Maple syrup urine disease (MSUD)	Leucine Isoleucine Valine	Oxidative decarboxylases	Seizures, death in infancy
Phenylketonuria	Phenylalanine	Phenylalanine hydroxylase co-factors	Severe mental defects
Tyrosinaemia	Tyrosine	Tyrosine transaminase	Mental retardation

Many of these lesions affect other bodily systems; the central nervous system (perhaps because it is the most delicately poised) is commonly the system that is most severely upset.

phenylketonuriacs phenylalanine hydroxylase is defective.

Figure 21.2 shows that the failure to transform phenylalanine to tyrosine leads to the build-up of **phenylpyruvic acid**. This can readily be detected in the urine. PKU is inherited in a Mendelian fashion.

It is carried in the heterozygous form in 1–2% of the white population of the UK. Adult hetero-zygotes show some penetrance of the defect. It can be shown that after ingesting a quantity of phenylalanine the quantity of the amino acid detectable in the blood is significantly greater

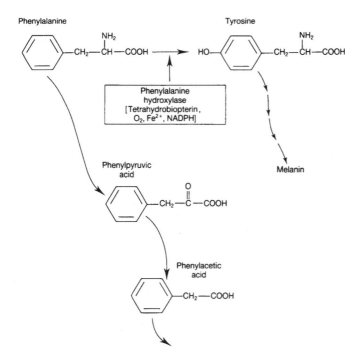

Figure 21.2　Metabolic defect in PKU. The enzyme system responsible for catalysing the step from phenylalanine to tyrosine is defective in phenylketonuriacs. In consequence the little-used pathway to phenylpyruvic acid is opened up and the latter can soon be detected in the urine. As tyrosine is a precursor for melanin it is found that phenyl-ketonuriacs are usually albino.

than in normal individuals. PKU, for reasons we shall see below, is hardly present at all in black populations.

The transformation of Phe to Tyr is, of course, particularly important in catecholaminergic neurons (see Chapter 16). The metabolic defect, however, is believed to have a general effect throughout the brain. As tyrosine is a starting point in the synthesis of the pigment **melanin** it is clear that pigmentation will be affected. Phenylketonuriacs consequently tend to be flaxen-haired and blue-eyed. This relationship between the PKU lesion and pigmentation explains its vanishingly small representation amongst black populations.

It must not be supposed, however, that the neurological consequences of PKU are due to deficiencies in catecholaminergic or indoleaminergic transmitter systems. Such deficiencies, of course, occur but the neurological defect – profound mental retardation – is due to the accumulation of phenylpyruvate in the brain. The defect in phenylalanine hydroxylase makes itself felt most acutely in the liver. Excessive phenylpyruvate carried from this source by the vascular system seems to have a number of metabolic effects which include interference with **mitochondrial oxidation**, **DOPA decarboxylase** and **5-HT decarboxylase**. At the anatomical level it can be shown that **myelination** is defective in the brains of phenylketonuriacs.

Fortunately treatment of the condition is routine – provided an early diagnosis (before six months of age) has been made. Restriction of dietary phenylalanine has proved effective. Some phenylalanine must, of course, be allowed as it is an essential amino acid. The restricted diet can be discontinued when the child reaches an age of 4–10 years. The age range for discontinuation is due to the fact that although there seem to be a number of varieties of PKU once the brain's development has passed beyond the 'sensitive period' elevated levels of phenylalanine are no longer harmful.

One final and general point should be made before we pass on to the next pathology. It is that PKU provides a good, clear and simple example of a very general issue: the interaction of heredity and environment in the development of the brain. The environment, in this instance, is represented by the quantity of phenylalanine in dietary protein, the genetics by the defective phenylalanine hydroxylase gene. Neither on its own is sufficient to cause

the pathology. The interaction between environment and heredity is quite general in the development of the brain, but the two 'forces' are seldom so easily and clearly disentangled. It is also important to note that the combination exerts its deadly effect on the infant brain, whilst it is growing, labile, through its sensitive period.

21.4 FRAGILE X SYNDROME (FraX)

Fragile X syndrome is, after Down's syndrome, the commonest inherited mental deficiency. It develops in one in every 2000 males and one in every 4000 females. In addition to low intelligence, it is often associated with autistic behaviour patterns. As its name indicates, it is due to a deficiency on the X chromosome. This is why it is twice as prevalent in males as in females. One of the intriguing features of the syndrome is that its severity tends to increase and its time of onset decrease over the generations. Although this was at first put down to increasingly effective diagnosis and monitoring it is now known to have a real genetic basis.

The fragile X gene, designated the *FMR1* gene (fragile X mental retardation 1), is located on the long arm of the X chromosome (Xq27.3) and contains a CGG repeat in the first exon. The mean repeat number in normal individuals is 29. In individuals with what is known as a 'premutation' there are 55–200 repeats whilst in fully affected individuals the number of copies of the trinucleotide is higher than 200 and may be up to 600 and even 1000 and beyond. When the number of repeats exceeds 230 the entire region is methylated, each cytosine in a CG pair coming to bear a methyl group.

The protein encoded by the *FRM1* gene is known as FRMP. It is expressed not only in the brain but also in other parts of the body and especially in the testicles. FraX males exhibit macro-orchidism: enlarged testicles. Both males and females also present with connective tissue problems including ear infections, and skeletal problems such as flat feet and double-jointed fingers. Whilst the precise role of FRMP remains unknown there is evidence that it is involved in the regulation of protein synthesis. This is likely to account for the male macro-orchidism mentioned above. High levels of FMR1 mRNA are also found in cerebellar granule cells, in the hippocampus and in the cerebral cortex.

It has been shown that individuals carrying the premutation of up to 200 extra trinucleotides suffer no mental retardation. This is related to the fact that these repeats are free of the cytosine methylation. The earlier onset and increased severity of the syndrome through subsequent generations is due to the trinucleotide repeats getting ever longer. This in some as yet unknown way attracts methylation. This methylation invades a CpG regulatory site near the *FMR1* gene and prevents transcription. This is the root cause of the syndrome.

It is interesting to note, finally, that a number of other neurological diseases states, including Friedreich's ataxia, spinobulbar muscular atrophy, myotonic dystrophy, and, as we shall see, Huntington's disease, have also been shown to involve trinucleotide repetitions (see also Section 4.1.1).

21.5 NEUROFIBROMATOSES

It is unfortunate that two clinically distinct diseases have been given the same name: neurofibromatosis. Neurofibromatosis 1 (NF1) (von Recklinghausen disease) is characterised by *café-au-lait* spots, excessive freckling of unexposed skin, nodules on the iris and peripheral neurofibromas. In contrast, patients suffering from neurofibromatosis 2 (NF2) present with bilateral acoustic neuronomas and intracranial meningiomas. Whereas NF1 is relatively common, having an incidence of 1 in 3500 worldwide, NF2 is rare with an incidence of 1 in 40 000.

Because NF2 is a relatively rare condition it is not surprising that the first neurofibromatosis gene to be isolated was that for NF1. The gene was located on the long arm of chromosome 17. It turned out to be huge stretch of DNA (350 kbp) consisting of 51 exons. The gene encodes a 2818 residue protein which was named **neurofibromin**. As mentioned in Box 8.1, an approximately 450 residue segment of this protein shows strong similarity to the catalytic domain of GTPase activating protein (GAP). This sequence is accordingly known as the GAP-related domain (GRD). Experiment shows that neurofibromin does indeed have GAP activity. Like other GAP proteins it accelerates the dephosphorylation of p21^{ras}. This has given a clue to its role in the formation of neural cancers. For it was found that not only do gene inactivations lead to neurofibromatosis but also that amino acid deletions in the GRP had the same effect. In both cases it seems that the tight binding between neurofibromin and p21^{ras} was disrupted leading to loss of p21^{ras} dephosphorylation.

The NF1 gene can thus be seen as a 'tumour suppressor' gene. Unmutated neurofibromin interacts with p21^{ras} to initiate differentiation. Mutated neurofibromin cannot so interact and p21^{ras} is then captured by growth factors which lead to undifferentiated growth (Figure 21.3). This interpretation is strengthened by the distribution of neurofibromin in the body's tissues. It is found throughout the CNS, but especially in dorsal root ganglia, oligodendroglia and Schwann cells. Within these cells neurofibromin is associated with the smooth endoplasmic reticulum and elements of the cytoskeleton, especially tubulin. These cells are known to show p21^{ras}-induced differentiation and they are also the cells most susceptible to neurofibromatosis growths.

The neurofibromatosis type 2 gene is (as mentioned above) less well known than the type 1 gene. It is located on chromosome 22 and encodes a protein of some 595 amino acids which has been called **schwannomin** (or **merlin**). There is no evidence that it has any relation to the NF1 gene, nor does it have any GAP activity. However, it is structurally related to a family of cytoskeletal proteins (moesin, ezrin, radixin: hence the name 'merlin') which, like protein 4.1 of the erythrocyte (see Section 7.8), help to link the cytoskeleton to the membrane. It is believed that, like neurofibromin, it has a tumour-suppressive role.

21.6 MOTOR NEURON DISEASE (MND): AMYOTROPHIC LATERAL SCLEROSIS (ALS)

MND is a generic term which covers a number of disorders that affect the motor outflow from the CNS. These include **amyotrophic lateral sclerosis (ALS)**, often known as **Lou Gehrig's disease** (after the US baseball player), **Kennedy's disease** (X-linked spinobulbar muscular atrophy), and four types of **spinal muscular atrophy**. In this section we shall only consider ALS. This is by far the commonest of the motor neuron diseases, having a prevalence of

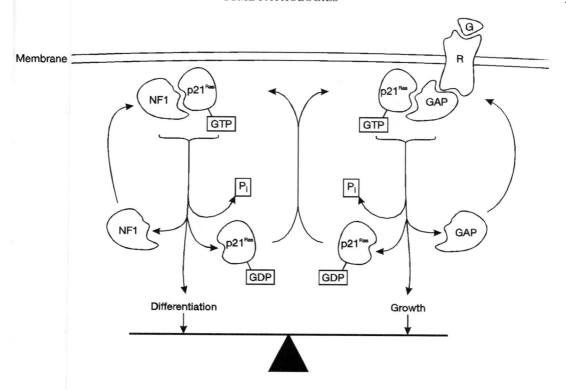

Figure 21.3 Inactivation of neurofibromin affects the balance between growth and differentiation. The figure shows how NF1 and receptor-activated GAP compete for p21^{ras}. The NF1 route leads to differentiation; the receptor-activated GAP route leads to growth. The balance is represented by a set of scales at the bottom of the figure. G=growth hormone; NF1=neurofibromin; R=receptor. Other abbreviations and acronyms as before. Further explanation in text.

about 5 per 100 000 and the risk increases by an order of magnitude after the age of 60.

There is no internationally agreed definition of ALS. In the US it tends to be applied to a number of adult-onset conditions characterised by degeneration of motor neurons. In the UK the term is restricted to conditions in which there is pathology of both upper and lower motor neurons. The loss of motor neurons leads to progressive atrophy of the skeletal musculature. Patients suffer from muscular weakness (hence 'amyotrophic'), which typically starts in hands and/or legs and may cause slurred speech. The condition is progressive but affects only the motor outflow; sensory inflow and mentality are normally unaffected and remain clear. There is at present no known cure and the condition proceeds to death in three to five years.

Five to ten per cent of ALS cases are familial, the rest are sporadic. About 20% of the familial cases are due to mutations in the gene which encodes the important antioxidant enzyme **superoxide dismutase 1 (SOD1)**. This gene is located on chromosome 21 (21q22.1). More than 90 mutations have been detected, affecting 40 of the 153 amino acids in the encoded enzyme, with the most common being A4V. The remaining 80% of familial ALS is caused by mutations of genes located in at least five other chromosomes. Sporadic ALS, which constitutes 98% of the cases, has, so far, no established cause and SOD is not affected. Nevertheless, because the symptoms of sporadic and familial ALS are so similar it has seemed sensible to use the 2% SOD cases as models to examine the aetiology of all ALS.

The normal function of SOD is to protect cells from superoxide radicals: O_2^{-} Molecular oxygen is a powerful oxidising agent and readily removes electrons from biological molecules to form the superoxide radical. SOD mops up these radicals:

$$SOD1 - Cu^{2+} + O_2^{-} \Rightarrow SOD - Cu^{+} + O_2$$

$$SOD1 - Cu^{+} + O_2^{-} + 2H^{+} \Rightarrow SOD1 - Cu^{2+} + H_2O_2$$

However, it is not the case that mutated SOD fails in its duty to protect cells from superoxide. The total elimination of SOD1 in SOD1 knockout mice does not result in ALS. Rather, it seems, the mutation results in '**gain of function**': the mutated SOD1 is even more effective in removing superoxide radicals from a cell's vicinity. The route from over-expressed SOD1 to the debilitating symptoms of ALS remains unclear at the time of writing. It may be that mutated and hence misfolded SOD1 proteins aggregate, forming toxic clumps, and/or it may be that the abnormal protein causes abnormal tyrosine nitration and peroxidation. Injection of mutant SOD1 (but not normal SOD1) into motor neurons in tissue culture leads to intracellular aggregates. Another theory makes use of the finding that mutated SOD is significantly less able to bind **zinc** than the normal enzyme. Deprived of Zn, SOD1 becomes a less efficient superoxide scavenger and the rate of tyrosine nitration increases. Zn detachment by other routes might also account for some or all of the sporadic ALS cases. The Zn detachment theory does, however, seem to run contrary to the finding, mentioned above, that mice totally lacking SOD1 show no ALS symptoms. It seems more likely, at present, that the disease is multifactorial but that misfolded mutant protein is a primary factor. This would lead to the protein clumping seen in cultured motor neurons and this in turn to neuronal death. The inexorable march of the disease would then be due to the slow spread of these clumps through the motor neuron system as more and more neurons passed a threshold allowing the aggregations to form. Alternatively, could it be that ALS, like the prion diseases discussed in Section 21.2, is propagated by a misfolding cascade?

It is obvious that much more work needs to be done before a proper understanding of the aetiology and then possibly a therapy can be achieved. How far are the different forms of the disease, the familial and the sporadic, due to the same causes? Perhaps the new DNA chip technology (Section 5.20) will allow us to compare the different genetic profiles of normal subjects and sufferers and gain some hints toward an answer. What is it, too, that confines the pathology to the million or so motor neurons out of the billions of others in the CNS? Again genetic profiling may help. As yet, however, we are stumbling in the dark. A large number of double-blind, placebo-controlled tests have been tried in the last few decades with scant if any success. Perhaps stem cell therapy provides a hope for the future, albeit a rather distant future. Advances in molecular neurobiology remain the best hope of finding a cause and a cure for ALS and the other motor neuron diseases. It is noteworthy, finally, that a gene responsible for a very rare form of **juvenile-onset ALS**, ALS2, has recently been tracked to a locus on chromosome 2q33. The molecule that it encodes appears to be a GTPase regulatory protein. It may be that this discovery will help us understand the biochemistry at work in the much more common ALS1.

21.7 HUNTINGTON'S DISEASE (=CHOREA) (HD)

Huntington's disease (=Huntington's chorea) is one of a group of disorders of the basal ganglia which lead to jerky, rapid involuntary movements. The word 'chorea' is, indeed, derived from a Greek root meaning 'dance' and an earlier name for the condition was 'St Vitus' dance'. The **striatum** is particularly affected with many prominent neurons degenerating. The striatum, along with the other basal nuclei, is central to the neurophysiology of behavioural movements.

It has been found that in many cases agents such as L-dopa aggravate the symptoms whilst drugs that block dopaminergic synapses, for instance haloperidol, or inhibit dopamine synthesis, for example reserpine, are able to reduce the symptoms. It seems, therefore, that the symptoms of HD are the consequence of an excess of dopamine. We shall see in Section 21.9 that parkinsonism is, in contrast, due to lack of dopamine. The choreas are thus often regarded as **inverse parkinsonisms**.

Huntington's disease is one of the most tragic of this group of disorders. It is tragic because it is inherited. It is doubly tragic as its clinical onset is usually during the fourth or fifth decade of life. The other choreas are usually associated with the degenerations of old age in the seventh and eighth decades. Finally it is triply tragic as it usually only shows itself when a family has been started. The sufferer may thus have unwittingly passed on this inexorable and incurable disease to his/her own sons and daughters.

Huntington's disease normally starts in a very small way as a facial twitch but this gradually spreads to all parts of the body. Speech and walking gradually become more difficult and then impossible. Ultimately loss of memory, hallucinations, delusions, disorientations, disorders of mood and emotion make the sufferer's life intolerable. Death normally intervenes about 15 years after the onset of the first symptoms.

In recent years considerable progress has been made in determining the genetic basis of the disease. It affects 5–10 people in every 100 000 and is inherited as an autosomal dominant condition. But because of its late onset it proved very

difficult to locate the gene. Moreover as the primary biochemical defect is as yet unknown it was not possible to 'fish' for the gene with synthetic oligonucleotide probes.

However, other techniques have become available. The first depends on the finding that human DNA is highly polymorphic. Indeed it has been calculated that the human haploid genome contains one variant (polymorphism) for every 100 base-pairs. Many of these variant sequences will fall at the cleavage sites of one or other of the 200+ different restriction endonucleases. In consequence these polymorphisms will reveal themselves as differences in the lengths of the resultant restriction fragments. This is shown in Figure 21.4.

The different restriction fragment lengths shown in Figure 21.4 are called **restriction fragment length polymorphisms (RFLPs)**. They are inherited in a Mendelian fashion. If an RFLP can be shown to be inherited in the same way as the gene under investigation – in this case the putative gene for HD – then the location of the latter can be

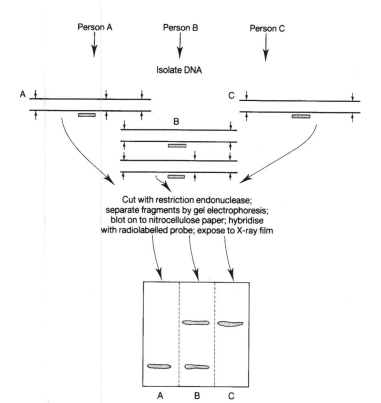

Figure 21.4 Detection of DNA polymorphisms in the human genome. In this example it is supposed that DNA is isolated from three individuals. Person C possesses a mutant DNA lacking a restriction site; person A is normal; person B is heterozygous: one chromosome possesses the restriction site, the other does not. Restriction sites are represented by arrows. The DNA is exposed to the restriction enzyme which cuts it at the arrowed sites. The different sized fragments are separated by gel electrophoresis. A radiolabelled probe is made from a clone of the critical region (shown by stippled rectangle) and hybridised with the blot on the nitrocellulose paper. The position of the gene under investigation can thus be narrowed down to a manageable fragment of the genome. Further explanation in text. After Shows, Sakaguchi and Naylor (1983), in *Advances in Human Genetics*, **12**, Ch. 5, New York: Plenum Press.

determined. This was achieved for HD by Gusella and co-workers in 1983. The HD gene was gene was mapped to the short arm of chromosome 4.

This did not mean that the gene itself had been found – only that it was linked to a particular RFLP. The situation was, moreover, complicated by the finding that the RFLP exists in a number of different varieties – 'haplotypes'. However, further analysis using some of the newer techniques of molecular biology ultimately enabled scientists to home in on the gene itself. It is located on chromosome 4 (4p16.3) and has been named *HD*. It encodes a protein designated **huntingtin**. The role of this protein is somewhat obscure. It is not confined to striatal cells nor even to the brain but found ubiquitously throughout the anatomy. Recent evidence, however, suggests that it is involved in the regulation of transcription. Furthermore it is beginning to seem that one of the transcriptional systems with which huntingtin is involved in the brain is that for neurotransmitter receptors. It will, however, require further research to determine why mutant huntingtin is particularly deadly in the striatum compared with other parts of the CNS or elsewhere in the body.

It has been fascinating to find that *HD*, like the *FRM1* gene of FraX syndrome, is subject to trinucleotide repeats. In this case the trinucleotide subject to stuttering is CAG. Normal individuals exhibit 11–34 repeats (with a median of 19); Huntington's patients show 37–86 repeats (with a median of 45).

Again, as with FraX syndrome, the age of onset diminishes as the repeats become more extensive. It has been shown that while the age of onset in the offspring of affected mothers is about the same as that of their mother, the age of onset in offspring of affected fathers is dramatically reduced: on average to about 10 years earlier than had occurred in the father. This is related to the much greater increase in length of the repeated section in the chromosome inherited from the father. In a third of the cases the length is almost double that in the paternal gene. This suggests that the defect occurs during spermatogenesis.

Now that the gene and its product have been discovered it remains to understand its function(s) in the basal nuclei, and how mutations lead to the human disaster of HD. There is accumulating evidence that huntingtin plays a central role in axonal transport and membrane trafficking and also with the transcriptional processes in the nucleus. It has been argued that the loss of these functions by mutant huntingtin is the cause of the disaster. But huntingtin is widespread in the CNS and other parts of the body. Why should the disease only affect the striatum? There is still much to be learnt before it becomes possible to develop rational therapies for treatment, amelioration of symptoms and, hopefully, in the fullness of time, cure.

It is interesting, finally, to note that laboratory experiments suggest that there is some flexibility in age of onset. Transgenic mice whose symptoms closely mimic human HD were divided into two groups. One group was exposed to an enriched, the other to an impoverished environment. It was shown that the onset of HD-like symptoms was significantly delayed in the enriched group when compared to their impoverished colleagues. Although HD is at present inexorable, these experiments show that the externalities can have some ameliorative effect. This is yet another instance of the complex interplay between genotype and environment.

21.8 DEPRESSION

The cases considered so far in this chapter have been fairly clear-cut instances of so-called **'upward' causation**. A defect in metabolism is magnified up through the intricate pathways of the brain to affect the whole. But, of course, as we noted in the introduction to this chapter this is not always the case. In normal life mood does not change because of random variations in the brain's chemical base but usually because of changes in the circumstances of our everyday life. In other words in normal life causation often starts at the 'top' and works 'downwards'. It is important to recognise that this is the case. Too exclusive a concentration on upward causation, from molecule to man, leads to the absurdity of extreme **'reductionism'**: that we are 'nothing but' molecules.

One of the most interesting cases of the complex interaction between upward and downward causation is provided by the linkage between stress, anxiety and depression (=**affective disorder**). This topic has engendered a vast literature spanning the subject areas of psychiatry, neuroanatomy and neuropharmacology. Before embarking on this subject, however, we need to make some important distinctions. First we need to separate **endogenous** from **exogenous** or **reactive** depression. Both may involve the same neurochemical mechanisms but clearly the endogenous type is less dependent upon environmental triggering. Second, within each category of the forgoing classification, it is important to distinguish between '**monopolar**' and '**bipolar**' affective disorder. The first condition is commonly known simply as major depression or major depressive disorder (MDD) and the second as 'manic-depression'.

21.8.1 Endogenous Depression

Endogenous depressions seem to afflict the sufferer without any obvious cause in his or her external circumstances. Frequently (by no means always) they are of the 'bipolar' or 'manic-depressive' type. The mood oscillates between extreme elation and extreme depression. The depression is often so deep that suicide is attempted.

There is evidence that bipolar depression has a genetic basis. One study involved the **Old Order Amish** society of Pennsylvania. This very tightly knit society has strict rules of inbreeding so that the gene pool is effectively closed. Most members of the society are equable and well balanced. But every so often manic-depressives appear – often ending in suicide. Because the society is so closely knit the family trees can be examined in detail. It can be shown that the trait is inherited as if it were determined by an autosomal dominant gene. The penetrance, however, is only 63%. In other words an individual inheriting the gene only has a 63% chance of expressing it, i.e. of suffering bipolar depression.

We noted in Section 21.7 how recombinant DNA techniques had allowed the gene coding for Huntington's disease to be traced to chromosome 4. A similar technique has been used to search for the locus of the gene responsible for endogenous depression in the Old Order Amish people. The RFLP markers indicated that the gene was located on chromosome 11. However, similar techniques applied to other populations (three Icelandic and three North American families) did not home in on the same locus. It is likely, therefore, that more than one gene is responsible. Yet another study

suggests that a third gene for the condition is located on the X chromosome. This would be consistent with the finding (noted below) that twice as many women as men suffer monopolar depression.

The implication that three (or perhaps more) genes are involved in affective disorder need not surprise us. We still have no idea at all how the defective gene or genes exert their catastrophic effects. We shall see below that the biochemical causation of depressions is thought to lie in imbalances of a 'cocktail' of neurotransmitters. It is likely that the defective genes affect one or other of the many enzymes and co-enzymes in the synthesis and degradation of these neurotransmitters. In this regard it is suggestive to find that the segment of chromosome 11 marked in the Old Order Amish study is close to the gene for **tyrosine hydroxylase (TH)**. We shall see below that the catecholamines are deeply implicated in the pharmacology of depression. TH, it will be recalled from Chapter 16, is a rate-limiting enzyme in the synthesis of these transmitters.

21.8.2 Exogenous Depression

Exogenous depression again may be either monopolar (i.e. 'major depression') or bipolar (i.e. 'manic-depression'). It should not be thought, however, that the symptoms and symptom patterns are at all clear cut. They have, in fact, been long known to constitute 'an extraordinarily heterogeneous syndrome'. It is interesting to note, however, that major studies in the US have shown that whereas male and female Americans suffer bipolar depression in about equal numbers, twice as many females as males suffer monopolar depression.

Exogenous depressions are customarily associated with major events in the patient's socio-cultural environment: bereavement, divorce, redundancy, postnatal, postmenopausal, etc. All these events are more or less 'stressful'. Obviously what is stressful to one individual may not be stressful to another. Stress, however (to use the physicist's definition), produces 'strain'. This strain may also take many forms. But if the stress is long continued, the strain very frequently takes the form of anxiety and/or depression.

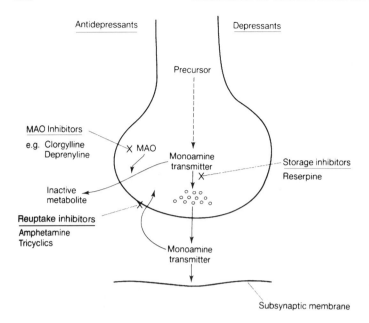

Figure 21.5 Neuropsychopharmacology of the monaminergic synapse. Reference back to Chapter 16 shows that monoaminergic synapses can be influenced at three major points. The quantity of monoamine transmitter at the synapse can be increased by inhibiting the breakdown enzyme, MAO, or by inhibiting the reuptake system. Vice versa, the concentration of the transmitter may be reduced by interfering with the mechanism leading to the monoamine becoming sequestrated within synaptic vesicles. If this step is inhibited the resulting free monoamine in the terminal is exposed to MAO.

Animal experimentation can to some extent mimic the human experience. If an animal is subjected to noxious but escapable stimuli it reacts vigorously and makes its escape. If it is subjected to the same stimuli in a situation from which there is no escape it ultimately becomes lethargic and 'despairing'. We shall return to these animal experiments later when we have looked briefly at the underlying neurochemistry.

21.8.3 Neurochemistry of Depression

There is little doubt that a neurotransmitter imbalance is associated with both types of endogenous and exogenous depression. Noradrenaline, dopamine, serotonin and acetylcholine are all probably implicated in one way or another. The evidence for this belief comes mainly from neuropharmacology (Figure 21.5).

Agents that deplete monoamines such as reserpine (see Chapter 16) tend to induce depressive illness whilst agents that have the opposite effect, most importantly the MAO inhibitors (MAOIs) such as clorgyline and deprenyline, are antidepressants. The important tricyclic antidepressants (e.g. imipramine and amitryptiline) act by blocking reuptake of monoamines from the synaptic cleft.

This evidence suggests that depression occurs when the subsynaptic membranes of monoaminergic synapses are exposed to insufficient concentrations of their accustomed neurotransmitter. Remember that many monoaminergic synapses are made '*en passant*' and thus release their transmitter to influence comparatively large volumes of grey matter (see Figure 16.19).

Acetylcholine, on the other hand, tends to act in an opposite sense to the central monoamines. Most antidepressants seem to have an anticholinergic action; vice versa, reserpine potentiates central cholinergic activity. This could further enhance the depression.

The suggestion is, therefore, that depression may result from an imbalance between central monoaminergic and cholinergic systems (Figure 21.6). It may also be supposed that the cycling of manic-depressive illness is due to first one and then the other system gaining the upper hand.

However, the situation is not quite as simple as this analysis suggests. The major difficulty is that whereas the drugs exert their effects within seconds or minutes the psychological depression may take several weeks to lift. Perhaps this is not surprising for, as we have seen, the neuropharmacology is very far from being simple and straightforward.

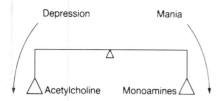

Figure 21.6 Neurotransmitter balance and depression. In manic-depressive illness first one neurochemical system then the other gains an ascendancy.

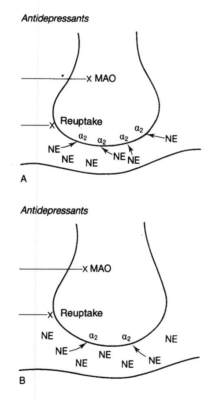

Figure 21.7 Antidepressants downregulate presynaptic monoamine receptors. This schematic diagram shows (A) that in the presence of antidepressants the concentration of noradrenaline (norepinephrine; NE) in the synaptic gap increases and tends to saturate the α_2 presynaptic receptors. It will be recalled that these have an inhibitory effect on the presynaptic terminal. It is suggested (B) that over a period of time the number and/or sensitivity of the α_2 receptors decreases, thus reducing the inhibition on the presynaptic terminal. Further explanation in text.

It has been proposed, however, that the difference in time scales between the pharmacological action and the psychological consequence can be accounted for if the notion of denervation supersensitivity is invoked. It will be recalled that we introduced this idea in Section 16.6 when we discussed the withdrawal symptoms addicts experience when an exogenous opiate is removed. The same argument applies here, only the other way around. Just as decreased impulse traffic leads to a compensatory increase in synaptic sensitivity, so an overworked pathway is likely to adapt by **decreasing** the number of receptors in its synaptic membranes. It is found that most antidepressants **'downregulate'** monoamine receptors after a few weeks of treatment (Figure 21.7). If these monoamine receptors are comparable to the α_2-receptors in the **presynaptic** membranes of adrenergic neurons then their 'downregulation' would increase the quantity of monoamine in the synaptic cleft. For it will be remembered from Section 16.4 that these presynaptic receptors respond to high synaptic concentrations of noradrenaline by inhibiting the presynaptic terminal from releasing any more. The time period for this type of effect coincides with the lifting of the depression.

Finally, it must be emphasised that transmitters other than catecholamines and acetylcholine are involved in the brain's neurochemical balance. We noted in Section 16.3 that fluoxetine, marketed as Prozac, a powerful antidepressant with few side effects, acts by reducing the reuptake of that other monoamine, serotonin. This neurotransmitter, like the catecholamines, is deployed in a global fashion throughout the brain (Figure 16.14). Like the catecholamines, therefore, it can be expected to have a widespread role in controlling the 'tone' of the cortex and hence its subjective correlative, mood. Furthermore it has recently been shown that 'knockout' mice, lacking the gene for one of the 5-HT receptors, are uncommonly aggressive.

21.8.4 Stress and Depression

We noted at the end of the section on 'exogenous depression' above that animal experimentation suggests that it is **inescapable** stress which leads to the symptoms of reactive depression. This would seem to be a good model for exogenous depression in humans. If stress is long continued and unavoidable, depression is often the outcome.

In animal experiments it appears that during the initial stress the monoaminergic synapses are overworked. The brain contains more than its usual quantity of monoamines. This readies the animal to react quickly and vigorously to its difficult situation. If the stress is continued, however, the monoaminergic synapses are depleted. Their output, so to speak, exceeds their synthetic capacity. They fall into comparative quiescence. But (it is speculated) this period of quietude has the inverse effect to that described above. If the α_2-adrenoreceptors respond to unusually large quantities of noradrenaline in the synaptic gap by a 'down-regulation' then might they not respond to unusually small quantities by an 'upregulation'? If this is the case then when monoamines are once more released into the synaptic gap they will have more than their usual inhibitory feedback on the presynaptic terminal. Hence the onset and continuance of depression.

21.9 PARKINSON'S DISEASE (PD)

The symptoms of 'the shaking palsy' were first described by James Parkinson in 1817. The disease most usually appears in the sixth decade of life though it may occur earlier. If the symptoms have not appeared by the age of 70 they are unlikely to do so. It is relatively common, affecting some 200 individuals per 100 000.

The symptoms (like those of Huntington's disease) normally start in a small way. Often the initial sign is a tremor in a hand. But they progress to give uncontrollable shaking of (sometimes) all the limbs. The tremor is due to the alternate contraction of opposing muscle groups. In addition to tremor, patients normally suffer from rigidity about a joint (sometimes referred to as 'cogwheeling' where a joint under an applied force seems to jerk from one position to another) and, most difficult of all, an inability to initiate a voluntary movement (akinesia). Furthermore, once a movement has been started, it often shows 'festination' – the walk, for instance, becomes more and more a shuffling run, the steps becoming shorter and shorter.

Clearly all these symptoms and the many others which textbooks of neurology describe indicate a defect in the neurological control of movement. It has been known since the end of the nineteenth century that the control centres affected are deep in the brain stem and basal ganglia. The principal sites involved are the **substantia nigra**, **corpus striatum**, **globus pallidus**, **subthalamic nucleus** and **thalamus**. Most important is the pathway from the substantia nigra to the corpus striatum.

As is only to be expected (think of the ice dancer or trapeze artiste) the 'wiring' responsible for so subtle a function as behavioural movement involves numerous feedback and feedforward loops (Figure 21.8). Furthermore, at least four neurochemical systems play a part: dopaminergic, cholinergic, GABA-ergic and glutamatergic. The substantia nigra, located in the midbrain, is subdivided into two parts, the **pars reticulata** and the **pars compacta**. From the pars compacta dopaminergic fibres ascend to the corpus striatum and to the globus pallidus. Both these basal ganglia are located in the thalamencephalon and both are crucial in orchestrating muscular movement. Both, moreover, receive powerful input from the cerebral cortex. This input is largely glutamatergic. Figure 21.8 shows that the output from the corpus striatum is by way of GABA-ergic fibres. These project downwards to the external and inner layers of the globus pallidus and to the pars reticulata of the substantia nigra. These GABA-ergic cells are of two types: those that project back to the substantia nigra express D1 dopamine receptors and contain dynorphin and SP as peptide co-habitants (Figure 21.8: loop 3). The others, which project to the two layers of the globus pallidus (Figure 21.8: loops 1 and 2), express D2 dopamine receptors and contain enkephalin as peptide co-habitant. From the external layer of the globus pallidus glutaminergic fibres project to the subthalamic nucleus. A chain of three neurons, two GABA-ergic and a final glutaminergic, projects back up to the cerebral cortex (Figure 21.8A).

It is unnecessary to go into the full complexity of the various loops and feedback systems. Three such loops are, however, outlined in Figure 21.8. In loop 1 glutaminergic neurons descend from the cortex to make excitatory synapses with GABA-ergic neurons running to the internal layer of the globus pallidus. These fibres make inhibitory synapses with other GABA-ergic neurons running to the thalamus. These in turn inhibit glutaminergic neurons

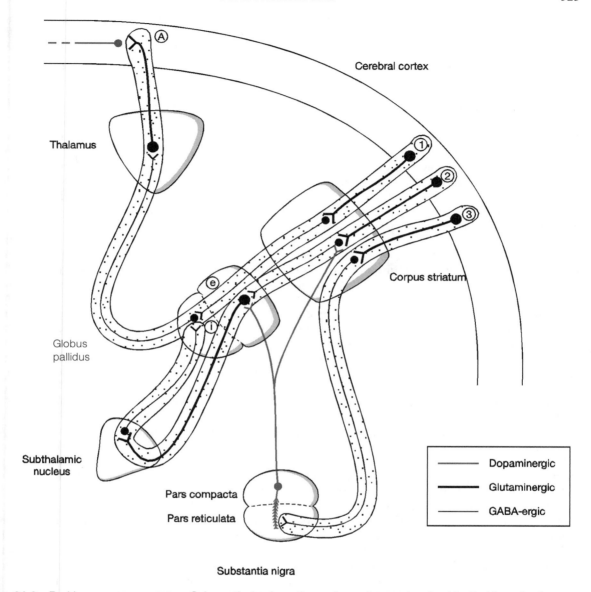

Figure 21.8 Parkinson neuroanatomy. Schematic to show the major pathways involved in Parkinson's disease. Explanation in text. e=external stratum of globus pallidus; i=internal striatum of globus pallidus. Adapted from Kopin (1993), *Annual Review of Pharmacology and Toxicology*, **33**, 467–495.

running back up to the cerebral cortex. The final outcome of activity in this loop is thus to **disinhibit** fibres running to activate the motor cortex at A. A second loop is also clear. This is loop 2 in the figure. It consists of six neurons (the final two common with the first loop). Glutaminergic neurons descend

from the cortex to synapse with GABA-ergic neurons running out to the external layer of the globus pallidus. From there glutaminergic fibres run to the subthalamic nucleus to synapse with GABA-ergic fibres running back to the internal striatum of the globus pallidus. Following through the sequence of excitatory and inhibitory synapses it can be seen that activity in the first four neurons of this loop **inhibits** the disinhibition which the first pathway achieves. The proper control of behavioural movements depends on a correct balance between activity in these two pathways.

The final loop (loop 3 in the figure) consists of glutaminergic fibres from the cortex synapsing with GABA-ergic fibres to the pars reticulata of the substantia nigra. Here they make inhibitory synapses with the dendrites of dopaminergic fibres running back up to the corpus striatum and globus pallidus. Activity in this loop is clearly going to modulate impulse traffic in loop 2 and this, in turn, will affect loop 1 and hence the functional state of neuron A.

The neural control of behavioural movement involves much more of the brain and many more circuits than shown in Figure 21.8. The circuits shown in the figure should be regarded as constituting just one module. Nevertheless, if this module is significantly deranged, muscular activity is disorganised. This is what happens in parkinsonism. It is now known that the nigro-striatal pathway is severely affected – up to 70% of nigral cells disappearing in severe cases. We noted above that the GABA-ergic neurons in all three loops express both dopamine receptors. Lack of dopamine due to loss of dopaminergic cells in the substantia nigra has a catastrophic effect.

Could the distressing symptoms be treated by supplying extra dopamine? Unfortunately it turns out that dopamine does not pass the blood–brain barrier. On the other hand, it is found that L-dopa (levodopa) (a precursor, see Figure 16.17) does. Hence, in the late 1960s, a treatment became available. L-Dopa can be given orally. The effects on the symptoms are dramatic. In some cases it is only when the drug is removed that the patient appears anything other than normal.

However, the treatment is only of the symptoms. The underlying cause – the death of the nigral cells – remains untouched. Moreover, as the disease progresses larger and larger doses of L-dopa are required to control the symptoms. Furthermore unpleasant side effects commonly manifest themselves: nausea, anorexia, odd writhing movements, tachycardia. Some of these side effects can be alleviated by treatment with dopamine agonists such as apomorphine and bromocriptine. Another

approach is to use MAO inhibitors (such as selegiline (deprenyl)) which block the breakdown of dopamine. But all these procedures are vain attempts to shore up a disintegrating system. Beneath the alleviated symptoms the disease takes its inexorable course.

What, then, causes the death of the nigral cells? Here one of the most remarkable instances of serendipity has provided a clue. The breakthrough happened in one of the most unlikely places – the Californian 'drug culture'. Addicts taking an illegally manufactured synthetic heroin developed severe neuropathological symptoms which strikingly resembled those shown in advanced states of Parkinson's disease. When samples of the synthetic heroin were analysed they were shown to contain a contaminant – **1-methyl-4-phenyl-1,2,3-tetrahydropyridine (MPTP)**. When tested on experimental animals (squirrel monkeys and mice) it was shown that MPTP induces symptoms that closely mimic the symptoms of human PD. Furthermore although there are some differences of detail (no Lewy bodies) the neuropathology of MPTP intoxication is remarkably similar to true PD – the selective elimination of nigral cells.

It was quickly shown that it is not MPTP itself which does the damage but its oxidation product MPP$^+$. In the brain MPTP is oxidised by MAO B first to MPDP$^+$ and this rearranges to form the stable metabolite MPP$^+$. This oxidation appears to take place in the supporting astrocytes and the resultant MPP$^+$ is carried into the nigral cell terminals by their own dopamine uptake system (see Box 16.3). Alternatively, and in addition, the MPP$^+$ may be bound by the pigment which has given the nigral cells their name. It is known that MPP$^+$ is strongly bound by neuromelanin which is present in these cells. Within the nigral neurons the MPP$^+$ enters mitochondria where it inhibits energy metabolism by blocking oxidative phosphorylation (Figure 21.9). Goodbye neuron.

Is all this just a lucky accident (lucky for some, unlucky for others) or does it point to the cause of parkinsonism? The jury is still out. But one very interesting observation has been made. This is the marked similarity of MPP$^+$ to the herbicide paraquat (Figure 21.10).

Could at least some parkinsonism have an environmental cause? There is some evidence that

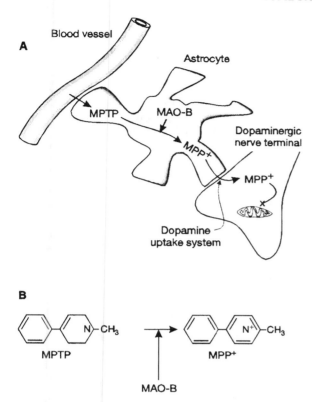

Figure 21.9 Oxidation of MPTP to MPP⁺ and pathway of MPP⁺ into dopaminergic terminals. (A) The pathway of MPTP/MPP⁺ from cerebral capillary to nerve terminal. MPP⁺ is taken up by the high-affinity dopamine uptake system and is concentrated by mitochondria, where it inhibits oxidative phosphorylation. (B) Oxidation of MPTP to MPP⁺. MAO-B catalyses this reaction. Further explanation in text.

this might indeed be the case. A survey in the province of Quebec showed a remarkably high correlation (0.967) between the incidence of the disease and the use of herbicides and pesticides. Paraquat is, of course, not the only pyridine compound in the environment – either as a pesticide or as a by-product of the chemical industry. It has thus been further suggested that PD is a disease

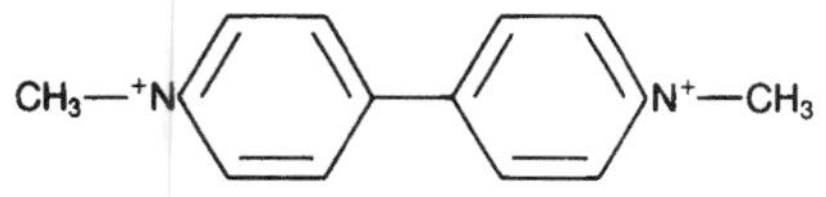

CH₃—⁺N⟨⟩—⟨⟩—N⁺—CH₃

Figure 21.10 Paraquat.

of industrialisation. Although cases of the 'shaking palsy' have been described as far back as Galen in the second century AD it has only become well known since the onset of the industrial revolution. We noted at the beginning of this section that it was first characterised in 1817. This interpretation is strengthened by the recent report that the insecticide **rotenene**, a compound that readily penetrates cell membranes and is known to inhibit mitochondrial enzymes, not only causes locomotor difficulties in experimental rats but also induces nigro-striatal degeneration including brain deposits similar to the Lewy bodies seen in Parkinsonism. The case against the environment is, however, not yet proven. There is also evidence for genetic and developmental defects which predispose towards PD.

In recent years three genes have been found to be implicated in familial Parkinsonism. These genes encode the proteins **parkin**, **ubiquitin carboxy-terminus hydrosase L1 (UCH-L1)** and **α-synuclein**. The role of these proteins in the onset of the disease remains obscure. UCH-L1 is one of the most abundant proteins in the brain. All three proteins play roles in the biochemistry of nigral cells. There is evidence that they are involved in the processing and removal of damaged and toxic proteins through a ubiquitin-dependent pathway. Damaged and toxic proteins might well be present in nigral cells subjected to the biochemical stressors described above. When the controlling genes mutate the ubiquitin-dependent pathway to removal is lost and the damaged proteins accumulate in the form of inclusion bodies. Such bodies, **Lewy bodies**, are a well-known histological feature of Parkinson brains.

Research over the next few years is likely to disentangle the environmental and genetic causes and, hopefully, point towards a means of curing sufferers and protecting those at risk. The MPTP work has, for example, suggested that antioxidants such as vitamin E might act as protectants. A very different approach also seems promising for the more distant future. This involves the replacement of degenerating nigral cells by transplantation of fresh dopaminergic cells into the brain. This has already been tried with some success in experimental animals, and there have already been some attempts with humans. In the case of experimental animals embryonic tissue has been used to transplant into brains made parkinsonian by MPTP. The

BOX 21.1   α-Synuclein

We noted in Section 21.9 that α-synuclein was one of the major constituents of Lewy bodies which are themselves characteristic inclusions in Parkinson brains. α-Synuclein is not restricted to Parkinson brains, however, but is also found in several other neurodegenerations of the elderly. These have, accordingly, been grouped together as α-synuclein diseases. In addition to parkinsonism, these include dementia with Lewy bodies, Lewy body dysphagia, pure autonomic failure, inherited Lewy body disease, striato-nigral degeneration, Shy–Drager syndrome, olivopontocerebellar atrophy.

α-Synuclein is strongly immunoreactive with Lewy bodies and there is nowadays little doubt that it is the major constituent of these bodies. It is in fact a part of a small family of synucleins consisting (at present) of three members, α-, β- and γ-, ranging from 127 to 140 amino acids in length. Although synucleins are common in the brain their normal function is unclear. Immunohistchemistry shows that α- and β-synucleins are localised at nerve terminals whilst γ-synuclein is distributed throughout a neuron. Only α-synuclein has been localised in Lewy bodies.

Although, as we noted in Section 21.9, most parkinsonism has an environmental cause a small number of families have been identified which show hereditary onset. Genetic analysis of one of these families showed that the defect was on chromosome 4q21–23 and that a transition in the gene (A53T) coded a defective α-synuclein. Although as mentioned above the function of α-synuclein in the brain remains mysterious it is known that when it binds to lipid membranes it undergoes a conformational change from amorphous to α-helical. These helices can then line up to form filaments and when this happens they undergo a further conformational change to form β-sheets. These β-sheets stack together (rather like the amyloid we shall consider in Section 21.10) to form bodies very similar to the Lewy bodies seen in pathology. The A53T transition increases the rate of this filament assembly and may thus be the cause of Lewy body spread in the cortex. It is also interesting to note that chronic intravenous injection of rotenone (Section 21.9) into the rat leads to progressive degeneration of nigro-striatal neurons and the appearance of inclusions very similar to Lewy bodies. These bodies were immunoreactive to α-synuclein and ubiquitin. It may be that the mitochondrial damage known to occur in pesticide-affected nigral cells leads to the assembly of α-synuclein into filaments, β-sheets and ultimately Lewy bodies.

grafts have 'taken' and in some cases the dopamine metabolites have recovered to within 5% of pre-MPTP levels. In the case of humans tissue has been taken from the patient's own adrenal medulla, a good source of dopaminergic cells, and surgically implanted into the brain. More recently disaggregated cells from aborted human fetuses have been introduced into the brains of parkinsonian patients. It is too early to be sure how effective these treatments will turn out to be. Results have so far proved disappointing. No long-lasting reversals of parkinsonism have been reported. However, the recent upsurge of interest in neural stem cells (Section 19.2) must surely give hope of some therapy in the not too distant future. In contrast to transplants from aborted human fetuses, the use of neural stem cells should raise no significant ethical questions. The human shipwreck of parkinsonism impels future research.

21.10 ALZHEIMER'S DISEASE (AD)

The first description of AD was given by Alois Alzheimer in 1907. His words are worth quoting:

A woman of 51 years old, showed jealousy towards her husband as the first noticeable sign of the disease. Soon a rapidly increasing loss of memory could be noticed. She could not find her way around in her own apartment. She carried objects back and forth and hid them. At times she would think that someone wanted to kill her and would begin shrieking loudly.

In the Institution her entire behaviour bore the stamp of utter perplexity. She was totally disorientated to time and place. Occasionally she stated she could not understand and did not know her way around. At times she greeted the doctor like a visitor, and excused herself for not having finished her work; at other times she shrieked loudly that he wanted to cut her, or she repulsed him with indignation, saying that she feared something against her chastity. Periodically she was totally delirious, dragged her bedding around, called her husband and her daughter, and seemed to have auditory hallucinations. Frequently, she shrieked with a dreadful voice for many hours ...

Her ability to remember was severely disturbed. If one pointed to objects, she named most of them correctly, but immediately afterwards she would forget everything. When reading she went from one line to another, reading the letters or reading with a senseless emphasis ... When talking she used perplexing phrases and some periphrastic expressions (milk-pourer instead of cup). Sometimes one noticed her getting stuck. Some questions she obviously did not understand. She seemed no longer to understand the use of some objects ...

The generalised dementia progressed. After 4½ years of the disease death occurred. At the end the patient was completely stuporous; she lay in her bed with her legs drawn up under her, and in spite of all precautions she acquired decubitus ulcers.

Alzheimer's disease is without doubt one of the most terrible afflictions of late middle age to old age. It has often (on the analogy of heart failure) been termed 'brain failure'. Others have referred to it as amentia – death of the mind. At present there is nothing that can be done. The disease must run its course. It is debatable whether the patient or his/her carers suffer most: for both it is a near intolerable condition.

In the West the demographic trends point to an increasingly elderly population. In 1950 there were about 214 million people older than 60 world-wide; in 2025 there will be 1000 million. Alzheimer's disease shows a penetrance of about 7–10% in those over 65 and 40% in those over 80. In the mid-1980s in the USA alone some $25 billion was spent per year on the institutional care of demented patients. At the end of the twentieth century more than half the budget of the US National Institute of Ageing was spent on AD. After heart disease, cancer and stroke it is the largest cause of death in the developed world. These statistics show the immensity of the problem. But there is hope. Advances at the molecular level are beginning to clarify the aetiology. If the causation can be understood relief, protection and even cure may become possible.

21.10.1 Diagnosis

Diagnosis has long presented a problem. The condition rarely presents itself in quite so clear-cut a way as Alzheimer described above. There are often confounding factors. Frequently there is some parkinsonism. Often there is (quite understandably) exogenous depression. In most cases the only certain diagnosis is post-mortem. Here the classical signs are **neurofibrillary tangles (NFTs)** in cortical pyramidal cells and elsewhere; areas of degenerating neurites known as **plaques** and **congophilic angiopathy**, i.e. changes in blood vessel walls such that they take up Congo red, a histological stain. The necrotic tissue attracts reactive gliosis. In addition, single electron-dense granules within neuronal perikarya termed **granulovacuolar degeneration (GVD)** and rod-shaped eosinophilic inclusions known as **Hirano bodies** are sometimes found. Finally, the gross appearance of the brain shows considerable atrophy and an enlargement and coarsening of the sulci (Figure 21.11).

21.10.2 Aetiology

The condition appears to be restricted to humans and perhaps the higher primates. This makes it difficult to study. There are no good animal models although, as we shall see below, genetic engineering has very recently achieved a breakthrough by inserting the human gene responsible for plaque formation into the mouse. There are, however, a number of analogies, and a number of different causes have been suggested.

Is it, like Huntington's disease, an **inherited** condition? Or, like Parkinson's disease, can we suggest an **environmental cause**? Alternatively could it, like the prion diseases of Section 21.2, be due to an **infective agent**? Or again, like PD, is it due to the degeneration of a particular population of neurons, in this case the **cholinergic neurons** in Meynert's and other brain stem nuclei? Or, finally, could there be some breakdown of the blood–brain barrier and some **autoimmune** reaction occurring? It will probably turn out that the cause is multifactorial. It is likely that more than one of the above mechanisms is at work and, quite probably, some additional ones we have not thought of. Indeed it is probable that there is more than one variety of the disease itself.

21.10.3 Molecular Pathology

Significant progress has been made at the molecular level. It was indicated above that one of the consistent histological signs of AD is an accumulation of patches of degenerating axon terminals and

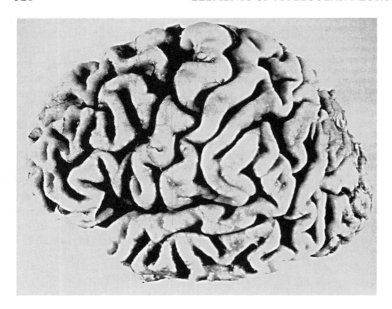

Figure 21.11 Alzheimer brain and Alzheimer histology. Lateral view of the cerebrum of a patient with Alzheimer's disease. Note the greatly enlarged sulci. From Terry (1985), in Davis and Robertson eds, *Textbook of Neuropathology*, Baltimore: Williams & Wilkins; with permission.

dendritic branches (neurites) – the so-called plaques (Figure 21.12). These occur in large numbers in the cerebral cortex, in the hippocampus and elsewhere. They also occur in smaller numbers in the brains of all aged humans. They start as diffuse densifications and end as compact cored structures. The core consists of **β-amyloid**. Amyloid is an insoluble protein (the name was coined by Virchow to mean 'starch') and there has been much controversy concerning its primary or secondary nature. Is amyloid deposition the initiating event or is it the consequence of some more fundamental pathology?

Another 'hallmark' feature of AD is, as we have already noted, the presence of neurofibrillary tangles (NFTs) in the cell bodies and processes of many neurons. Electron microscopical examination of these NFTs shows them to consist largely of pairs of 10 nm filaments twisted around each other with a periodicity of 80 nm. These helices of old age are called **paired helical filaments (PHFs)** (Figure 21.13).

Not only are NFTs found within neurons but they are also much in evidence in the plaques. It has been suggested that the amyloid protein at the centre of plaques is identical (though different in three-dimensional conformation) to the protein of NFTs. However, careful immunological studies have indicated that this is not the case and that the original identifications were due to contamina-

tion of NFTs with amyloid. The PHFs which form the major constituents of NFTs have turned out to be highly phosphorylated **tau** proteins. It will be recalled from Section 15.3 that tau proteins form important elements of the axonal cytoskeleton. Along with the MAPs they help to stabilise tubulin microtubules.

There is continuing controversy over which comes first, β-amyloid or NFTs. Is the formation of β-amyloid plaques the primary cause of the neurodegeneration and the development of NFTs secondary and perhaps caused by the amyloid deposition? Or is it the other way round, with the phosphorylation of tau and the development of NFTs first and causing a secondary formation of β-amyloid plaques? Alzheimer researchers are thus often divided into **BAPtists** and **Tauists**. The successful engineering of a mouse expressing the amyloid gene (as mentioned above) which shows large numbers of amyloid plaques in its cerebral cortex tilts the balance of the argument towards an the amyloid cascade hypothesis (see Figure 21.16). The BAPtist case is further strengthened by the recent finding that injection of β-amyloid Aβ42 (see below) fibrils into mice engineered to express mutant human tau resulted in fivefold increases in NFT formation. In consequence it is this interpretation that will be mostly followed in the following pages.

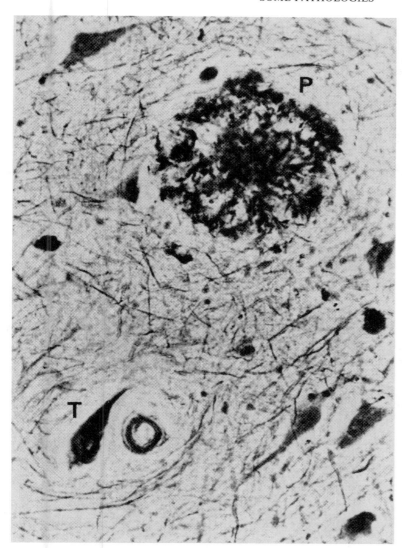

Figure 21.12 Alzheimer plaque. The large, dark, circular object at the top right-hand corner is an Alzheimer plaque. It consists of degenerating nerve processes and often has a central core of amyloid. The black triangular objects are nerve cell bodies filled with neurofibrillary tangles (NFTs). Some unaffected neurons can also be seen. The preparation is from the hippocampus of an Alzheimer patient and has been stained with a silver stain. From Wischik and Crowther (1986), *British Medical Bulletin*, **42**, 51–56; with permission.

APP and Aβ

The β-amyloid protein – called the **β-A4 protein or Aβ** – has been isolated and sequenced. It consists of 39–42 amino acid residues and is closely similar (if not identical) to the amyloid protein that causes congophilic angiopathy of cerebral blood vessels in AD and Down's syndrome patients. In amyloid deposits the Aβ strands line up in the β-pleated sheet conformation (see Chapter 2): hence the β suffix. These deposits form a core around which other proteins and ultimately astrocytes and microglial cells accumulate.

Using the recombinant DNA techniques described in Chapter 5 an oligonucleotide probe complementary to a seven-residue sequence of the Aβ protein was prepared. This was used to screen a cDNA 'library' constructed from the total mRNA from the cerebral cortex of a five-month aborted human fetus. The probe picked out a large cDNA stretch of some 3353 base pairs. This cDNA, without its leader and tail sequences, coded for a

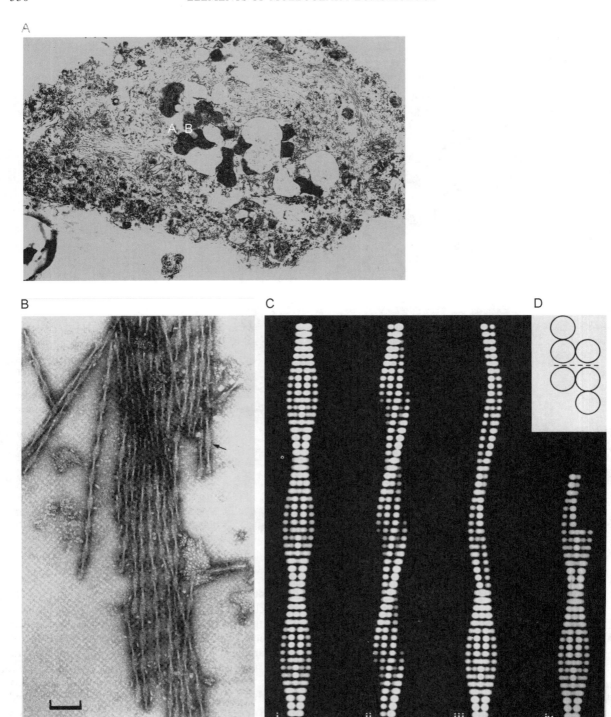

695-residue protein with a molecular weight of about 79 kDa.

Hydrophobic analysis of this 695-residue **amyloid precursor protein (APP)** shows that it contains a membrane-spanning segment from residue 625 to 648. The C-terminal is in the cytoplasmic domain, the lengthy N-terminal sequence in the extracellular domain. The 39/42 amino acid β-A4 protein (the cDNA sequence for which was recognised by the oligonucleotide probe) is found incorporated in the APP towards the C-terminal end. It begins 28 residues from the E-face of the membrane and continues 14 to 15 residues into the membrane (Figure 21.14). The Aβ fragment, on this analysis, is thus one-third in and two-thirds out of the membrane – two-thirds in and one-third out of the extracellular domain.

Computer searches of data bases in which the amino acid sequences of proteins are stored has not revealed any similar sequences. The amyloid precursor protein thus seems to be a new membrane-spanning protein. Indeed it has proved to be the first member of a small family of such proteins. The first tentative conclusion from this detailed molecular biology was that in AD the amyloid precursor protein is disrupted and the Aβ fragment is released to stack up in the β-pleated sheet conformation of amyloid.

Three Varieties of APP

Further research has shown that APP comes in at least four different varieties: the 695-residue isoform already described (APP$_{695}$), a 714-residue isoform (APP$_{714}$), a 751-residue isoform (APP$_{751}$) and a 770-residue isoform (APP$_{770}$). This is due to alternative splicing of some of the 18 exons in the APP gene. This differential splicing of the primary transcript gives three different mRNAs which vary from tissue to tissue. The most abundantly and widely expressed of these isoforms is APP$_{751}$. This isoform includes a sequence of 56 residues interpolated at residue 289. This sequence is strongly homologous to the active sites of certain well-known protease inhibitors – the protease nexins. It is known that these inhibitors promote outgrowth of neurites from neurons. It is thus speculated that the break-up of the β-amyloid precursor protein during the formation of a plaque deposits this domain in the plaque core. This would then account for the common observation that such cores are frequently invaded by neurites from the surrounding tissue.

What are the Normal Functions of APP?

APP is expressed in many cells and is abundant in neurons and glia. The part it plays in the life of the cell long remained something of a mystery. This mystery is slowly yielding to molecular biological investigation and it is now beginning to seem that APP plays many roles in the normal brain.

We noted in Section 15.3 that APP has been implicated in the kinesin-motivated axoplasmic transport of vesicles. We saw that there is evidence that it links the membrane of the transport vesicle to the kinesin motor. It has been shown that these vesicles contain presenilin 1 and β-secretase (of which more later). It will be recalled that tau is also involved in the axoplasmic ultrastructure underlying transport. Could the underlying pathology of AD be a pathology of axoplasmic transport? It is interesting, in this regard, to recall (Section 21.7) that huntingtin is also believed to be involved in axoplasmic transport.

Figure 21.13 Paired helical filaments. (A) Electron micrograph of a thin section of a nerve cell body at ×10 000. Numerous neurofibrils can be seen weaving through the cytoplasm. The black objects are lipofuchsin granules and the white spaces are vacuoles. The vacuoles often contain a small densely staining body (only the remnants of this can be seen in the figure) and this 'granulovacuolar degeneration' is also characteristic of Alzheimer's disease. (B) The neurofibrils can be extracted from Alzheimer neurons and viewed in isolation. This negative contrast electron micrograph shows that they consist of two strands twisted around each other forming the so-called 'paired helical filaments' (PHFs). Scale bar 100 nm. (C) Computer modelling of the PHFs of part B shows that the best fit is achieved if two strap-like strands are twisted around each other. (D) Each 'strap' of the PHF is believed to consist of three spherical subunits. When these are twisted around each other they yield the image shown in part C. Parts C and D reproduced from Wischik *et al.* (1985), *Journal of Cell Biology*, **100**, 1905–1912, by copyright permission of the Rockefeller University Press.

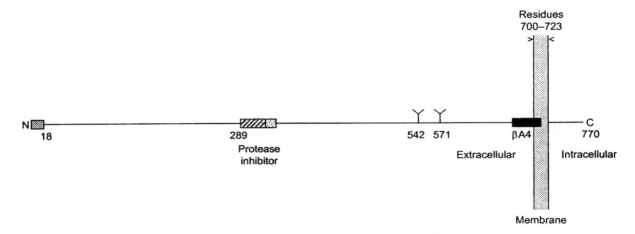

Figure 21.14   Schematic of the βA4 protein and the A4 precursor protein (APP) in a neuronal membrane. The figure is drawn to scale and shows the 770-residue isoform of APP. The βA4 extends from residue 672 to residue 714; the first 28 residues are in the extracellular space and the remaining 14 embedded in the membrane. Residues 542 and 571 provide glycosylation sites. Alternative splicing at residue 289 allows the insertion of a 56-residue protease (hatched) and/or a 19-residue sequence (stippled). The first 17 residues at the N-terminal (shaded) constitute a signal sequence which is excised in the two smaller isoforms. Adapted from Selkoe (1991), *Neuron*, **6**, 487–496.

Both APP and Aβ are found at presynaptic terminals and are released into the synaptic gap when the neuron is stimulated. Both APP and Aβ stimulate postsynaptic cells when infused on to postsynaptic membranes. Comparisons have been made with neuroactive peptides such as the preproenkephalin and POMC polyproteins of Section 4.2.3. Like POMC, APP is cut into different-sized neuroactive fragments: Aβ has been likened, for example, to Met-enk. Whether or not this comparison can be sustained, APP and aβ do appear to have at least some synaptic activity.

At a more cell biological level the protease nexin site of APP_{751} could act to inhibit extracellular serine proteases and internalise them for intracellular digestion; both APP_{695} and APP_{751} promote the proliferation of fibroblasts suggesting that more than the nexin site contributes to cell growth; there is also evidence that all the isoforms of APP, but especially APP_{751}, promote the outgrowth of neurites and neuron–neuron adhesion. In sum, therefore, it begins to seem that the various isoforms of β-APP are somehow involved in growth and adhesion in the cells of the central nervous system.

APP has also been shown to have several other functions in the normal brain including protective actions against excitotoxins and in maintaining plasticity of the hippocampus and other regions after injury or ischaemia.

The APP Gene

The APP gene is located on chromosome 21. Chromosome 21, it will be recalled, is the chromosome triplicated in **Down's syndrome** (trisomy 21). Individuals suffering from Down's syndrome thus have three copies of the APP gene. It has long been known that practically all individuals afflicted with Down's syndrome seem to show the symptoms of AD if they live beyond the age of about 40. It looks as though three copies of the APP gene predispose the brain towards the Alzheimer condition.

We noted above that the APP gene consists of at least 18 exons. Alternative forms of the APP are due to alternative splicing of exons 7 and 8. Interestingly the Aβ sequence is coded by two exons. The N-terminal 17 amino acids are coded by exon 16 and the remainder by exon 17. It was initially suggested that the pathological amyloidosis of AD might be due to aberrant splicing between these two exons. Further analysis has shown that this cannot be the case. Mis-sense mutations in the vicinity of the Aβ segment of the

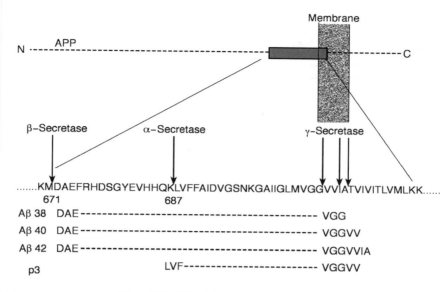

Figure 21.15 Secretase enzymes and the APP. The Aβ segment of the APP is expanded into full amino acid sequence (in single-letter code). The numbers below the sequence are counted from the APP N-terminal. The position at which the Aβ sequence is cleaved by the three secretases is shown and the resultant Aβ 38, 40 and 42 as well as p3 shown. Further explanation in text. After De Strooper and König (2001), *Nature*, **414**, 159–160.

APP do, however, show strong association with early-onset familial AD. For example, the transition $Lys_{670} \Rightarrow Asn$ (K670N) and $Met_{671} \Rightarrow Leu$ (M671L) in APP_{770}, just before the Aβ domain, have been found in a large Swedish pedigree showing early-onset familial AD.

Although the APP gene is crucial, it is not the only gene implicated in Alzheimer's disease. At least three other chromosomes are now known to carry genes involved in the disorder: the **presenilins** on chromosomes 1 and 14 and **APOE** on chromosome 19.

The Presenilins

There is good evidence that a gene on the long arm of **chromosome 1** is responsible for early-onset AD (early fifties) in a pedigree of Americans descended from a German colony living in the Volga region of Russia in the eighteenth and nineteenth centuries. Subsequently an Italian family was found to suffer from the same disruption. There is similarly good evidence that an even earlier onset (early forties) is associated with a gene on the long arm of **chromosome 14**. Both these genes encode rather similar proteins (67% identity) known as **preseni-**

lins. In both cases the protein consists of about 450 amino acid residues and each has six to eight transmembrane segments: very different from the amyloid precursor protein.

So far only two mutants of the chromosome 1 gene (***presenilin 2**, PS2*) have been found. In the German family an $Asn_{141} \Rightarrow Ile$ (N141I) substitution had occurred and in the Italian pedigree a $Met_{239} \Rightarrow Val$ (M239V) substitution was found.

In contrast, more than 30 mis-sense mutations of the gene on chromosome 14 (designated ***presenilin 1**, PS1*) have been detected. All are associated with early-onset AD. Many of these mutations occur within the transmembrane domains or close to the point of entry or exit from one or other of those domains.

Although the presenilins are widely distributed in peripheral tissues as well as the brain their normal function, at the time of writing, remains uncertain. They appear to be localised in the endoplasmic reticulum and also the Golgi body. One intriguing pointer has, however, come to light. A presenilin homologue has been found in *Caenorhabditis elegans*. It has been designated sel-12 and so similar are the two proteins that PS can be used to rescue

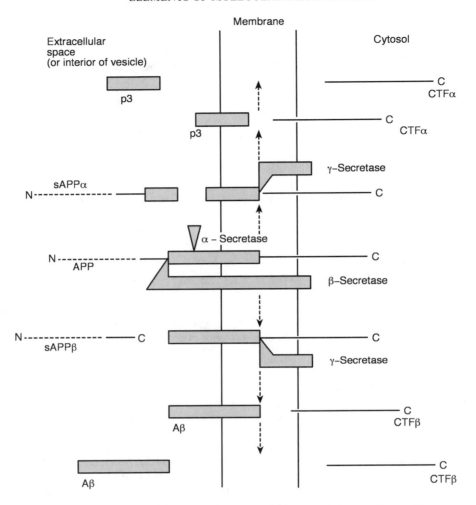

Figure 21.16 Proteolytic processing of APP. APP is shown in the centre of the figure. The sequence of proteolytic steps below the APP, starting with β-secretase (BACE) cleavage, leads to the release of Aβ into the extracellular space (or, if occurring in the membrane of a transport vesicle, into the lumen of the vesicle) and CTFβ into the cytosol. The sequence of steps above the APP, starting with α-secretase cleavage, leads to the release of the harmless p3 polypeptide into the extracellular space (lumen of vesicle) and CTFα into the cytosol. The CTFs in both cases are broken down by intracellular enzymes. CTF=C-terminal fragment; sAPP=soluble APP. Further explanation in text. After Howlett *et al.* (2000), *Trends in Neurosciences*, **23**, 565–570.

otherwise lethal mutants of the *sel-12* gene. In normal circumstances sel-12 influences membrane trafficking during the embryological development of the worm. It is suggestive, moreover, that both presenilins are similar to another *C. elegans* protein, SPE-4, a 465-residue protein expressed in the sperm. This protein is known to be a constituent of the Golgi body and may, consequently, also be involved in membrane trafficking.

The mutant forms of both *PS1* and *PS2* are associated with greatly increased amyloidosis: greater indeed than that which is associated with mutation of the chromosome 21 APP gene. Perhaps they affect the intracellular packaging and processing of APP and enhance the Aβ cascade leading to amyloid deposition. And perhaps, more intriguingly, one or both act as secretases (see below) which cut the Aβ fragment free from the APP.

Secretases, Presenilins and Aβs

The existence of three different secretases has been demonstrated: α, β and γ (Figure 21.15). **α-Secretase** cuts the APP 16 residues from the N-terminal end of Aβ sequence. **β-Secretase** cuts the APP at the N-terminal end of the Aβ sequence. **γ-Secretase** cuts the membrane-embedded C-terminal end of the APP.

Much interest has focused on β-secretase which is also known as β-site APP cleaving enzyme (**BACE**). BACE is a membrane-bound aspartyl protease located in the Golgi body and endosomes. It has been shown that mutant APP is a better substrate for this enzyme than normal or 'wild-type' APP.

γ-Secretase is another membrane-bound enzyme. The position at which it cleaves the APP varies (Figure 21.15). The length of the Aβ fragments varies accordingly: Aβ38, Aβ40 and Aβ42. It has been found that the 42-residue form, Aβ42, is significantly more aggressive in forming plaques than the others.

It has been shown that γ-secretase activity is the responsibility of a multiprotein complex containing the presenilins. It is further known that mutations in the APP and/or in the presenilins lead to increased proportions of the aggressive Aβ42 fragment.

Putting all this together we arrive at the picture of amyloidogenesis shown in Figure 21.16. In the top half of the figure a benign sequence is shown. Starting with the membrane-embedded APP, α-secretase first cuts off soluble APPα (sAPPα) which diffuses harmlessly away into the extracellular space. γ-Secretase then cuts off a similarly harmless C-terminal fragment (CTFα) leaving the innocuous p3 peptide which escapes to the exterior. The bottom half of the figure, however, shows the malign sequence leading to Aβ. Starting once more with APP, β-secretase cuts away sAPPβ which diffuses away and then γ-secretase cuts away a C-terminal fragment (CTFβ) leaving Aβ. The latter is the cause of all our woes. Mutant forms of APP and presenilin increase the proportions of Aβ42.

This biochemical processing mostly occurs in the secretory vesicle being carried in the fast axoplasmic stream to the neuronal membrane. Hence the extracellular space of the previous paragraph is, in fact, the interior of the vesicle. It will be recalled that these vesicles are now thought to be linked to kinesin by APP and are known to contain both β-secretase and

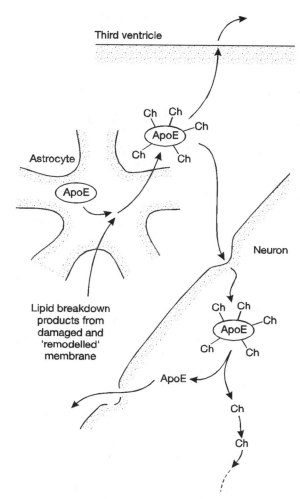

Figure 21.17 The role of ApoE in cholesterol recycling. Lipid breakdown products from damaged or 'remodelled' membrane are taken up by the astrocyte. High levels of cholesterol induce the synthesis of ApoE, which is then secreted into the intercellular space with cholesterol. A complex of ApoE and cholesterol is formed which may then pass through the ependyma into the third ventricle or may be taken up by neighbouring neurons. This has been shown to be the case in the granule cells of the dentate gyrus (hippocampus). ApoE is divested of its cholesterol, which can then be transported to regions of membrane synthesis and repair. Simplified after Poirier (1994), *Trends in Neurosciences*, **17**, 525–530.

presenilin 1. When the secretory vesicle fuses with the neuronal membrane Aβ is voided along with sAPP. Once in the extracellular space it stacks up with other Aβ42s to form the core of an amyloid plaque.

APOE and its Alleles

The *APOE* gene, located on the long arm of chromosome 19, has three alleles: **APOE-ε2, APOE-ε3** and **APOE-ε4**. The protein, **apolipoprotein E (ApoE)**, encoded by *APOE* has an M_r of 37 and consists of 299 amino acid residues. Corresponding to the three APOE alleles there are three major isoforms: ε2, ε3 and ε4. They differ only in amino acids at positions 112 and 158 (ε2: Cys_{112}, Cys_{158}; ε3: Cys_{112}, Arg_{158}; ε4: Arg_{112}, Cys_{158}). In spite of this seemingly insignificant difference the presence of the ε4 isoform represents a strong risk factor for late-onset sporadic and familial AD. Indeed the presence of two copies of APOE-ε4 increases the risk of AD from 20 to 90% and reduces the mean age of onset from 84 to 68 years.

In the CNS the gene is expressed most strongly in astrocytes. The 37 kDa ApoE protein appears to play a number of significant roles. First, it is important in the uptake and redistribution of phospholipids and especially cholesterol during repair of brain damage and possibly also in the remodelling of synaptic connections during plastic changes in the brain (see Chapters 1, 19 and 20). This ApoE-dependent recycling of lipids is schematised in Figure 21.17. There are differences in the efficiency with which the three isoforms accomplish this important process with ε4 being least effective. There are also differences in the efficiency with which the three isoforms support neurite growth in culture (with ε4 showing less support than the others). Finally, it can be shown that the quantity of CAT, the cholinergic marker, diminishes in the hippocampus and other brain regions as ε4 copy number increases. This latter finding is suggestive. The **cholinergic system** was one of the first suspected to be at fault in AD. It consists of very lengthy fibres running from the diagonal band and Meynert's nucleus to the hippocampus and elsewhere (see Section 16.1). If the ε-isoform of ApoE is less effective than the other isoforms in repairing breakdown of membrane in aged brains, this defect may be especially dangerous to these lengthy fibres. This would account for the observed loss of CAT in the hippocampus.

In addition to its role in lipid recycling ApoE has also been shown to interact with both Aβ and tau proteins. It has been shown that *in vitro* ApoE-ε4 binds avidly to soluble single copies of Aβ causing them to aggregate into β-sheet amyloid within seconds. The other two isoforms of ApoE, ε2 and ε3, are much less avid and much slower in causing this amyloidogenesis. There is similarly a difference between the ApoE isoforms in their interaction with tau. ApoE-ε3 binds to the microtubule linking domain of tau thus perhaps preventing phosphorylation. ApoE-ε4 binds much more weakly. Hence, again, the possession of the ε4 isoform may tilt the balance towards the hyperphosphorylation of tau and hence the development of those other hallmarks of the disease, PHFs and NFTs.

Finally, it is worth remembering that the APP degradation is membrane bound. Any slight variation in membrane integrity might well be sufficient to affect this biochemistry and thus initiate a cascade resulting in the deposition of amyloid.

21.10.4 Environmental Influences: Aluminium

Let us now turn from hereditary to environmental factors. Aluminium has often been implicated. Electron probe microanalysis and nuclear magnetic resonance (NMR) have shown aluminosilicate deposits in the centres of Alzheimer plaques. Natives of Guam and some of the other islands of the western Pacific who show a type of early-onset dementia resembling AD and/or PD can be shown to live in areas where aluminium concentrations in the soil are unusually high and calcium and magnesium levels unusually low. Aluminium is a

Figure 21.18 A possible Alzheimer cascade. The APP is shown inserted in a neuronal or glial membrane. The Aβ42 fragment is shown by the rectangle partly in and partly out of the membrane. At the top left-hand side of the figure chromosomes 21 and 14 are shown to have an effect on the secretase excision of APP and release of the Aβ42 fragment. At the top right-hand side, homozygous *APOE-ε4* on chromosome 19 is transcribed into ApoE-ε4, which is less effective than the products of the other two APOE alleles in maintaining the lipid integrity of the membrane. Arrows show that it also affects β-sheet formation from Aβ42 monomers and the hyperphosphorylation of tau. A dosage effect is also present in some individuals (Down's syndrome patients in particular), which leads to more APP being present in the membrane than usual. Once the Aβ42 fragment is released it aggregates to form

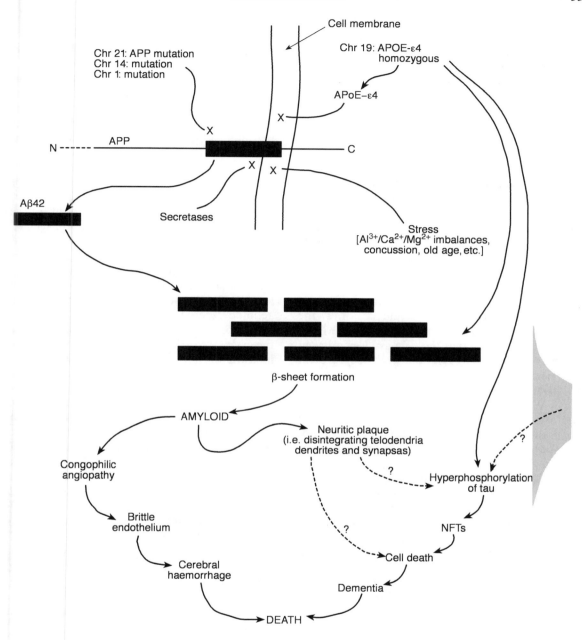

insoluble β-pleated sheet, which grows to form the microscopically visible centres of amyloid plaques. The growth of plaques destroys nerve endings and synapses. It also leads to brittleness in the endothelial linings of cerebral blood vessels and thus haemorrhages, stroke, death. The neuritic plaques lead to reactive gliosis and may result in hyperphosphorylation of the axonal tau proteins and thus NFTs. This route is not certain. So far, NFTs have not been observed in the Alzheimer mouse although characteristic amyloid plaques are abundant. Hyperphosphorylation may thus be due to other causes, especially the ApoE-ε4 route. The NFTs and/or the amyloid plaques cause neuronal death and ultimately dementia of the Alzheimer type. At present this scheme like all others is highly tentative. Broken lines and question marks indicate the more speculative connections. Further explanation in text.

well-known neurotoxin. There can be little doubt that brains already predisposed to AD could be nudged over the edge by exposure to large quantities of the element. However, whether aluminium deposits in AD plaques are primary, secondary or even artefactual remains very much an open question. Finally, there have been reports that Al^{3+} and Ca^{2+} induce neurofibrillary tangles by binding to phosphorylated C-terminal domains of neurofilaments.

21.10.5 The BAPtist Proposal: an Amyloid Cascade Hypothesis

Alzheimer's disease may have more than one cause; indeed the name may cover a variety of subtly different conditions. Nevertheless, it is often helpful to set out a hypothesis, in this case the BAPtist hypothesis, which, in the best Popperian style, is open to falsification and improvement. So, in conclusion we can perhaps begin to draw the evidence together into a tentative scheme (Figure 21.18). This suggests that the primary lesion lies in the degradation of the amyloid precursor protein and release of the Aβ42 fragment. The causes of this primary lesion are likely to be many and varied. We have seen that the BACE and the other secretases are involved, as are the presenilins, and that point mutations in the APP gene of chromosome 21 are significant. The significance of mutations on the APP gene has been emphasised by the genetically engineered Alzheimer mouse which expresses a faulty APP gene. We have seen that environmental stress (perhaps aluminium, perhaps the hard knocks which boxers endure and which induce the somewhat similar condition of **dementia pugilistica**, perhaps just the wear and tear of a long life), can also be significant. Dosage is also important. If there is more APP present in the membrane then the quantity of Aβ released will be greater. Perhaps, also, the inefficiencies of the ε4 isoform of ApoE in clearing cholesterol affects the activity of γ-secretase, thus leading to the production of more Aβ42 Whatever the cause, or combination of causes, the release of the Aβ42 fragment from the membrane initiates a cascade. When concentrations rise sufficiently first a diffuse and then a cored plaque result. Within the centre of the cored plaque the Aβ protein assumes an insoluble stack of β-pleated sheets.

The formation of a plaque (a region of necrotic neurites) has very significant consequences. The most important of these may be the induction of the neurofibrillary change which we observe as the PHFs of NFTs. We examined the filamentous and microtubular components of axons and dendrites in Chapter 15. We noted how central they were in the life of the neuron. If axoplasmic (and dendroplasmic) transport is blocked the perikaryon would be unable to obtain its trophic factors (by retrograde transport) or send out the necessary materials to axon and dendrite terminals by anterograde transport. The cell would soon die. The cause–effect relation has not, however, been established although recent work using tau transgenic mice shows that NFTs are, indeed, induced by injection of Aβ fibrils. On the other hand, the 'Alzheimer mouse' appears to have plaques but no tangles. It may be that the hyperphosphorylation of tau which leads to PHFs and then NFTs has an independent aetiology, perhaps through the action or lack of action of the different ApoE isoforms. Whatever the exact sequence, and it may of course be a mixture of both amyloid deposition and ApoE isoform, the cell dies.

The loss of the cell bodies of the cholinergic neurons in Meynert's and other brain stem nuclei is one of the recognised signs of AD. It is known that the fibre projections from cholinergic nuclei are directed to wide areas of the cortex and especially the hippocampus. The latter, we have noted, is one of the major regions of plaque development. It is also the case that choline acetyltransferase (CAT), the major marker of cholinergic terminals, is dramatically lost from the hippocampi of AD patients. At the psychological plane we recall that there is good reason to believe that the hippocampus is deeply involved in short-term memory and route finding (Chapter 20). Short-term memory and spatial orientation is dramatically destroyed in AD. It also, of course, proverbially tends to weaken in all elderly brains.

21.10.6 Therapy

As will have been gathered from the foregoing account, a huge amount of highly sophisticated molecular neurobiology has been applied in an effort to get at the roots of the scourge of Alzheimer's disease. But are we any closer to finding a cure?

The optimist will say yes. If the amyloid cascade hypothesis proves correct (and this is still an important 'if') there are two obvious ways of tackling the disease: either prevent the cascade starting or remove the end product, the amyloid plaques.

The first approach would mean avoiding the challenges to the brain represented by hard knocks and/or the ingestion of toxic materials. It could also mean developing drugs that prevent the β-secretase (BACE) and/or the γ-secretase cutting the Aβ42 fragment free from APP. Alternatively the activity of α-secretase could be increased leading to increased production of the non-toxic p3 peptide and a consequent reduction in the larger Aβ4 fragments.

The second approach relies upon removing the amyloid plaques either before or after they have begun forming in the cortex. Is it possible to develop a medicine which would wash the burdened cortex free of plaques as antibodies wash the cardiovascular system free of antigens?

It is encouraging to note that both these approaches have been tried on experimental animals and validated. A reduction in the production of Aβ42 has been demonstrated after the application of β- and γ-secretase inhibitors. Once it had been established that **β-secretase** was an aspartyl protease pharmacologists could get to work designing inhibitors. Aspartyl proteases are known to be at work in a number of other diseases (schistosomiasis, viral infections, hypertension, etc.) and biochemists have had considerable experience in synthesising inhibitors. The major problem has been targeting. The putative inhibitor has to cross the blood–brain barrier and travel deep into neurons to the Golgi bodies and endosomes. The β-secretase inhibitor approach is supported by the demonstration that BACE knockout mice (BACE$^{-/-}$) appear perfectly healthy and lack all traces of β-amyloid in the brain.

The alternative of increasing the activity of **α-secretase** has been achieved by the application of drugs such as simvastatin which work against APOE-ε4 by blocking cholesterol synthesis and thus depleting its membrane concentration.

Finally, the development of a **vaccine** against amyloid plaques has been shown to work in experimental mice. In these investigations 'Alzheimer' mice which over-express a human mutant APP were inoculated with Aβ42 either before they began to develop Alzheimer plaques or when this had already happened. It was shown that immunisation of the young animals before the development of plaques successfully prevented Alzheimer pathology developing and inoculation of the older animals markedly reduced the extent and progression of that pathology. There is a snag, however. Early trials on humans have led to symptoms consistent with brain inflammation in a small number of patients. The trials have been discontinued while further work is carried out in an attempt to avoid this side effect.

21.11 CONCLUSION

The account of progress in understanding the causes of Alzheimer's disease in the preceding section shows how unified neurobiology is becoming. In order to achieve an understanding of the neurological catastrophe which overwhelms so many of us in our later years we have to bring together data from a large percentage of all the various subdisciplines into which the subject has fragmented. Underpinning all these subdisciplines is our growing insight at the molecular level. This, therefore, forms a suitable place to finish not only this chapter but also this book.

In this and previous chapters we have touched on many of the diseases to which the nervous system is subject. We have seen how they are increasingly being traced to faults at the molecular level. The molecular approach, as we noted at the beginning of this book, provides not only hugely greater insight into the structure and functioning of the nervous system, but also hugely greater power. Used cavalierly this power could dampen some of the many quirks and idiosyncrasies which enliven human societies, and reduce some of the huge diversity of human brain/minds. It is on this diversity, like the diversity within the biosphere, that our future depends. Without it we run the danger of being reduced, as William Blake phrased it, to 'single vision and Newton's sleep'. Used wisely, however, the new-found power which molecular neurobiology provides will leave this creative diversity untouched. Used wisely it can also enable us to set right some of the undoubted ills which affect the nervous system and make far too many human lives, lives of misery. Molecular neurobiology provides the twenty-first century enormous opportunity for both good and ill.

We started this chapter with Shakespeare. Let us complete it by quoting his famous lines describing man's seventh age:

> Last scene of all,
> That ends this strange eventful history,

Is second childishness, and mere oblivion,
Sans teeth, sans eyes, sans taste, sans everything.

Our hope is that molecular neurobiology may make this seventh age (and indeed the ills of the other six) less burdensome.

APPENDIX 1

MOLECULES AND CONSCIOUSNESS

The title of Søren Kierkegaard's book *Concluding Unscientific Postscript* springs to mind. This appendix might have been similarly titled. Nonetheless, it seems inappropriate to end a text on neurobiology without a mention of what is perhaps the most pressing, as it is the most intriguing, of the brain's properties: consciousness. This topic, moreover, is receiving greatly increased attention nowadays, and some have suggested that the proper analysis lies at the molecular level.

Charles Darwin, in the notebooks he wrote immediately after his *Beagle* circumnavigation, wrestled with the problem of consciousness: what place had it in an evolutionary world? How and when did it commence? What role does it play in the struggle for existence? Ultimately he gave it up as a 'question too far' and returned to the more tractable problems of evolutionary biology. The problem has not been solved from his day to ours.

The problem of consciousness, closely related to the long-standing 'other minds' problem in philosophy, has occupied thinkers for millennia. It is important, as always, to define terms. So much of the argument in this area has been a passing rather than a meeting of minds. In this appendix we shall put aside 'intelligence' and 'intentionality' which can both be instantiated in silicon-based computers, and concentrate our attention on 'qualia' or 'raw feels': 'pain', 'redness', auditory 'tones', the 'taste' of a madeleine, sadness, grief and joy. How do these, which constitute perhaps our most pressing reality, come to be? Where and when did they 'commence' both in the evolutionary sequence and in embryological development? What is their physical correlative in the grey and/or white matter of the brain? What, indeed, to quote a seminal contribution to the debate, is it 'like' to be a bat? These, as many have remarked, are deep questions, unanswered as the twentieth century, a century otherwise so marked for its scientific advances, draws to a close.

We noted in Box 14.1 that in a famous lecture Emil Du Bois-Reymond divided unsolved questions about the world into two types: those to which one could answer ignoramus and those to which one should answer ignoramibus. The relation of mind to matter fell, according to Du Bois-Reymond, into the latter category. This has, however, not stopped research and speculation in this area. Indeed with the development of new means of neuroimaging, especially PET, fMRI and MEG, the incentive to relate our inner world to the outer world studied by science has greatly increased. At the well-attended 1996 conference, 'Towards a Science of Consciousness', in Tucson, Arizona, Du Bois-Reymond's categories were rephrased as 'easy' and 'hard' questions. In this classification problems about how the brain computes responses to a stimulus, how it responds to a coloured 'Mondrian', etc. were grouped as 'easy'. These questions are all on the physical side of the divide. We are far from an understanding as yet but we can see that, in principle, future developments in neural-network computing and neuroscience should, in the fullness of time, provide an adequate answer. But how physico-chemical activity in the cerebral cortex and our experience of, say, redness are related is something altogether more difficult: Du Bois-Reymond would answer ignoramibus; neuroscientists today, after a further century of remarkable advance, say only 'hard'.

To cut a long story very short we can say that the problem overflows the boundaries of present-day science. Although brain science is necessary it is not sufficient for a solution. We need to recognise that our languages are themselves full of presuppositions. That we may, in other words, be trying to solve a profound problem with biased instruments. In particular, the language of science has been developed to deal with the 'public world' in which observations can be checked and verified by others. This is notoriously not the case with our sensation of 'redness'. Yet we insidiously attempt to do so; our linguistic habits set up the problem as one of relationship, of relating an 'inner' to an 'outer' world.

If we say 'redness' or 'sharpness' or 'itchiness' is just how it is to be, or to 'live through', a particular cerebral state we may have got some little distance towards identifying the issue. But far from all the way. For we seem to be led back to Darwin's questions. How far down the living world can we trace this 'inwardness'? The evolutionary tree shows no discontinuities all the way down to the prokaryocytes. And beyond the prokaryocytes, the thoroughgoing evolutionist would insist, is the macromolecular chemistry of the pre-biotic ocean. Surely it is madness to conclude that molecules and atoms also have this 'inwardness'?

Some have argued that consciousness emerges 'suddenly', as a quantum leap. The best-known sponsor of this novel answer to Darwin's question is the Oxford mathematician Roger Penrose. He argues, as many have done before him, that an explanation of consciousness is not possible within orthodox Newtonian science, the paradigm within which neuroscience is presently situated. He looks to new developments in our understanding of what matter 'is' and in particular to the counter-intuitive features of quantum mechanics.

Penrose has set out his arguments in two valuable books, *The Emperor's New Mind* and *Shadows of the Mind*. In both cases he argues that the mind is more than a computer, that it must involve other and as yet unknown principles. He argues that these unknown principles derive from the post-classical physics of quantum mechanics. In *Shadows of the Mind* he takes up Stuart Hameroff's idea that microtubules are the most likely sites for large-scale quantum states to occur in the nervous system.

Why microtubules? Both Penrose and Hameroff recognise that nerve cells and synapses are too large and complicated to exhibit quantum effects. It is argued, however, that microtubules and their associated water molecules may provide conditions in which quantum coherence can develop (cf. superfluidity, superconductivity). In particular, both Penrose and Hameroff draw attention to a controlling electron situated between the α- and β-subunits of the tubulin molecule. They point out that alterations in the quantum states of this electron lead to conformational changes in the tubulin subunits and that these are propagated along the microtubule. Microtubules are, of course, not confined to nerve cells but form the cytoskeleton of most cells in the animal body and are also to be found all the way down the scala naturae to protozoa such as *Paramecium*. Does quantum coherence occur in these cells also? Do paramecia experience qualia? Do liver cells? Some will think this a *reductio ad absurdum*, others will perhaps suspend judgement.

We have met microtubules at several places in the foregoing chapters. We have seen that mutations affecting the microtubules of touch receptors in *Caenorhabditis elegans* do indeed eliminate their activity. The nematode no longer responds to touch. Microtubules also somehow 'sense' tension in the mitotic spindle when cells are about to divide. Penrose and Hamerhoff argue, furthermore, that common anaesthetics also inactivate microtubules. They suggest that the lack of consciousness and the inactivation of microtubules is more than a coincidence. The glaring weakness of this argument is, of course, that anaesthetics affect many other elements of cells, especially their membranes and receptor proteins. Indeed, we saw in Section 11.2.2 that K^+ leak channels have been suspected in this regard.

The virtue of 'microtubular consciousness' is that it helps account for the origin of consciousness in the living world. It answers, in other words, one of Darwin's questions. It provides an avenue of escape from the even more vicious *reductio* of conventional panspsychists: consciousness at the level of atoms. It should be noted, furthermore, that in Roger Penrose's treatment 'raw feels' are a necessary consequence of the functioning of vital cellular elements. Penrose believes that the inorganic computers with which we are surrounded, and any possible future development of such mechanisms, can never show any 'understanding' of the computations they undertake. In other words he goes along with the philosopher John Searle's well-known 'Chinese room' argument. It is only by introducing the ideas of quantum physics that we can begin to see how conscious understanding can come within the remit of natural science.

To the neurobiologist, however, the vice of this theory is its lack of biological insight. There is no evidence at present for quantum coherence in microtubules. There is a great deal of evidence, as we saw in Section 15.3, for their having many non-quantal functions within the neuron. Why choose microtubules? Why not some other feature of the neuron's ultrastructure. Remember, too, that the brain cells with the most highly developed system of microtubules are not neurons at all, but fibrous astrocytes. Critics of the theory conclude that we have two mysteries: quantum mechanics and consciousness. There is no good reason for believing that they are the same mystery or that by putting them together we shall have anything less than the sum of the two.

If microtubular consciousness seems a very doubtful starter in the race to account for the presence of consciousness in animal and human brains, what are the alternatives? Is it, perhaps, possible to point to the sheer complexity of matter in animal brains? An appreciation of this overwhelming intricacy should have emerged from the pages of this book. Although computer scientists have developed definitions of complexity based on program length and computer running time, these are difficult to apply to the biological world. Complexity, at present, like energy for physicists in the pre-Newtonian world, remains an intuitive concept. Furthermore neuroscientists often speak of 'levels' of complexity and this notion is also highly intuitive. What is a level for one scientist is not a level for another. The notion has heuristic value in allowing some preliminary discussion of complexity to take place, but it should not be thought anything more than a device of the human mind. Levels have little or no reality in the living brain. Nevertheless, Figure A summarises, using the device of levels, some of the complexity we have considered in this book. It

Levels of organisation in the cerebral cortex

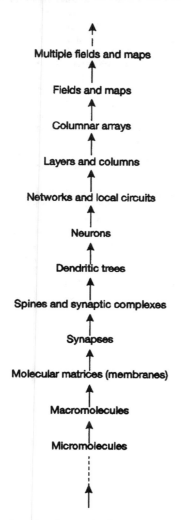

Multiple fields and maps

Fields and maps

Columnar arrays

Layers and columns

Networks and local circuits

Neurons

Dendritic trees

Spines and synaptic complexes

Synapses

Molecular matrices (membranes)

Macromolecules

Micromolecules

Figure A Levels of complexity.

emphasises that in the brain we have complexity built on complexity.

Figure A summarises some of the anatomical complexity of just one part of the brain: the cerebral cortex. At every level beyond the micromolecular there is heterogeneity. We have, for instance, devoted much space to the heterogeneity of the macromolecules constituting membrane channels and receptors. We have seen how membranes are heterogeneous, not only in their make-up of receptors, channels, G-proteins and effector enzymes, but also in their lipid constitution. We have noted the huge variability in structure and biophysical properties of

the multitude of ion channels. Synapses, again, are many and various, and Chapters 15, 16 and 17 have emphasised the intricacies of their structure and function. Beyond the synapse we meet the interacting groups and systems of synapses which we touched on briefly in Chapter 1. The dendritic spine (Box 20.1) also belongs at this level of organisation. It is often considered to be a separate functional unit in itself. The next level of organisation, the dendritic tree, so significant in the biophysics of the neuron (Chapter 12), varies greatly from one neuron to the next. No two are precisely the same.

The 'level' of the neuron is highly significant. Most biologists recognise cells as constituting a distinct organisational level. Neurocytologists recognise some 12 general types of neuron and between 50 and 500 subtypes. Each cell is a complex information processing unit in its own right and, as we have seen, each can be regarded as being an individual with a distinctive biochemical 'personality'. Yet there is evidence, as we saw in Section 7.9, that gap junctions may weld systems of cortical neurons into one continuous unit, a hyper-neuron, all of whose constituent units are functionally synchronised.

Neurons are the most heterogeneous of all the cells in the organism. They are organised into local circuits and networks. Some of the best known of these circuits are to be found in the retina, but they are also beginning to be understood in the cerebral cortex as well. The cortex is itself, as is well known, subdivided into layers (Chapter 1) and orthogonal to the layers is the columnar structure. Some indication of the neuronal 'complexity' within a column is given in given in Figure 1.23.

Finally columns, at least in the visual cortex, are themselves organised into patterned arrays although the boundary between one column and the next is seldom totally abrupt. Approaching the anatomy from the functional direction we find that the arrays of columns are organised to form what Roland calls a 'field'. These average 300 to 350 mm^2 in area (about 1 cm^3 of cortex) and consist of some 10^7 neurons and 10^{12} synapses. In brain activity, as monitored by the non-invasive techniques of rBCF, rCMR and PET, multiple fields develop in the cortex in response to sensory stimuli and learning.

In experimental animals the neuroanatomy responsible for these fields can be determined by micro-electrode and tract tracing techniques. The best-known case is the visual system of the macaque. Upwards of two dozen different visual areas, mappings and re-mappings, have been described. The interconnections run in both directions, forward and backward. Such studies emphasise, once again, if further emphasis is needed, the immense anatomical complexity of the rind of grey matter which covers the brain. It also emphasises the 'integrality' and interconnexity of the cortex.

There is accumulating evidence that the neurophysiological correlatives of 'consciousness' are widespread throughout the brain. It is a 'global' phenomenon. The exact regions involved in the mammalian brain are not important to the general problem. Is it not unacceptably 'mammalocentric' to insist that only mammals, indeed only primates, indeed only humans, are subject to 'raw feels'? In our present state of ignorance it seems far better to assume that such 'primary' consciousness is widespread in the animal kingdom. It would thus seem best to take the neurophysiological correlative to be widespread, 'global' activity. It would seem best to assume that it is the global brain state which is associated with consciousness. Friedrich Nietzsche had a telling metaphor. He likened consciousness to the Chladni figures made by sand spread over a vibrating plate. Different modes of vibration induce different figures. So with consciousness: different modes of brain activity are associated with different states of consciousness. When we experience 'redness' it is not just a specific region of the neocortex which is correlated with that raw feel, but a changed pattern of activity in the whole cortex. The 'hard problem' of course remains: why should *any* material pattern be associated with subjectivity? But that problem, as I argued above, escapes the boundaries of natural science. All we can attempt in neuroscience is to track down the physical correlatives.

Most importantly, the putative 'global' activity associated with consciousness is welded together to achieve coherent functioning. Each part of the brain affects every other part and this reaches right down to the level of molecules. In this way the 'hardware' of the brain is markedly different from a silicon-based computer. Indeed 'hardware' is an inappropriate term for we have seen how labile every part of its anatomy is. Furthermore, the brain's design is not imposed on it from the outside, but grows organically from within. It cannot be scaled up and transposed to some macroscopic contraption as Searle, in his *reductio ad absurda*, often imagines. The brain, as we have seen in the foregoing pages, is a molecular machine scaled up to macroscopic dimensions. We started in Chapter 1 with micromolecules such as amino acids, molecules which are well understood by chemists, and ascended without interruption to memory and the confusions of consciousness brought on by brain degenerations. So, rather than involve the mysteries of quantum theory it is perhaps better to investigate the dynamics of this almost unimaginable material complexity. A dynamic complexity, furthermore, which sharply distinguishes organic brains from their silicon competitors.

APPENDIX 2

UNITS

SI PREFIXES AND MULTIPLICATION FACTORS

Multiplication factor	Prefix	Symbol
$1\,000\,000\,000\,000 = 10^{12}$	tera	T
$1\,000\,000\,000 = 10^9$	giga	G
$1\,000\,000 = 10^6$	mega	M
$1000 = 10^3$	kilo	k
$100 = 10^2$	hecto	h
$10 = 10^1$	deca	da
$0.1 = 10^{-1}$	deci	d
$0.01 = 10^{-2}$	centi	c
$0.001 = 10^{-3}$	milli	m
$0.000\,001 = 10^{-6}$	micro	μ
$0.000\,000\,001 = 10^{-9}$	nano	n
$0.000\,000\,000\,001 = 10^{-12}$	pico	p
$0.000\,000\,000\,000\,001 = 10^{-15}$	femto	f
$0.000\,000\,000\,000\,000\,001 = 10^{-18}$	atto	a

SI UNITS

A (ampere) SI unit of current – the constant current which, if maintained in each of two infinitely long straight parallel wires of negligible cross-sectional area placed in a vacuum, 1 m apart, would produce between the wires a force of 2×10^{-7} newtons per metre length (definition)

Bq (becquerel) SI unit of radioactivity – one nuclear transformation per second. $1\,Bq = 2.7 \times 10^{-11}$ Ci (q.v.)

C (coulomb) SI unit of electrical charge – the quantity of electricity transported per second by a current of 1 A

Da (dalton) unit of atomic mass – defined as 1/12 the mass of ^{12}C

F (farad) the SI unit of capacitance – the capacitance of a capacitor between whose plates a potential of 1 V appears when a charge of 1 C is held

F (faraday) $9.649 \times 10^4 \, C \, mol^{-1}$

J (joule) SI unit of energy ($=4.18$ cal)

M (mole) SI unit of substance – defined as the same number of entities as there are in 0.012 kg of carbon isotope $^{12}_{6}C$ (see N_A, Avogadro's number)

m (metre) SI unit of length

N_A (Avogadro's number) the number of entities in a mole $= 6.022 \times 10^{23}$

R (gas constant) $8.314 \, J \, K^{-1} \, mol^{-1}$ $(1.987 \, cal \, K^{-1} \, mol^{-1})$

S (siemen) SI unit of conductance ($=$ a reciprocal ohm, i.e. 1/ohm). Defined as the electrical conductance between two points of a conductor when a constant potential difference of 1 V applied between these points produces a current of 1 A in the conductor

s (second) SI unit of time

V (volt) unit of potential difference between two points – the volt is defined as the potential difference between two points on a conductor when a constant current of 1 A leads to a power dissipation of 1 W between those points

Other Units

Å (Angstrom) 10^{-10} m

a (annum) year

bp (base pair) nucleotide base pair (measure of the length of a nucleic acid or polynucleotide)

cal (calorie) the quantity of heat required to raise 1 g of water from 14.5°C to 15.5°C

Ci (curie) unit of radioactivity: $1\,Ci = 3.7 \times 10^{10}$ nuclear transformations per second

M_r relative molecular mass: mass of a molecule expressed as a multiple of the mass of a hydrogen atom

S (svedberg) sedimentation coefficient in a centrifugal field of force

APPENDIX 3

DATA

BIOPHYSICS OF NEURONS

Datum	Squid	Frog (node)	Frog (myelin)	Cat (spinal motor neuron)
R_i	$30\,\Omega\,cm$	$110\,\Omega\,cm$		$70\,\Omega\,cm$
R_m	$1 \times 10^3\,\Omega\,cm^2$	$10\text{–}20\,\Omega\,cm^2$	$0.1\,M\Omega\ cm^2$	$2500\,\Omega\,cm^2$
τ	$1\,ms$			$5\,ms$
C_m	$1\,\mu F\,cm^{-2}$	$3\text{–}7\,\mu F\ cm^{-2}$	$0.004\,\mu F\,cm^{-2}$	$2\,\mu F\,cm^{-2}$

Where C_m is the capacitance of unit area of membrane, R_i is the resistance of unit volume of cytoplasm, R_m is the resistance of unit area of membrane and τ is the membrane time constant.

Data from W. Rall, 1977, in *Handbook of Physiology*, Section 1: *The Nervous System*, Vol. 1, *Cellular Biology of Neurons*, Part 1, ed. J.M. Brookhart, V.B. Mountcastle, E.R. Kandel and S.R. Geiger, Bethesda: American Physiological Society, pp. 39–97.

MOLECULAR WEIGHTS

Molecule	Molecular weight
Amino acid (average)	100
Nucleotide pair (average)	600
Rhodopsin	40 000
Haemoglobin	68 000
Amyloid precursor protein	79 000
Glycine receptor (GlyR) (subunit)	93 000
Muscarinic acetylcholine receptor (mAChR) (subunit)	51 400
Nicotinic acetylcholine receptor (nAChR)	268 000
Potassium channel (*Drosophila*, shaker) (subunit)	70 200
Sodium channel (glycosylated) (α-subunit)	260 000
Na$^+$+K$^+$ pump (α-subunit)	100 000

DENSITIES OF CHANNELS AND RECEPTORS

	per μm^2 membrane
Na$^+$+K$^+$ pump (rabbit vagus)[a]	750
Na$^+$+K$^+$ pump (inner segment of rod cell)[b]	3400
Ca^{2+} pump (rabbit SR)[c]	8700
nAChR (motor endplate)[c]	10 000
nAChR (ciliary ganglion)[d]	600
nAChR (cervical ganglion)[d]	4800
Rhodopsin (rod cells)[c]	50 000
Na$^+$ channels (squid giant axon)[c]	300
Na$^+$ channels (frog node of Ranvier)[c]	3000
Na$^+$ channels (rabbit node of Ranvier)[c]	12 000
Na$^+$ channels (rabbit unmyelinated vagus nerve)[c]	110
Na$^+$ channels (frog twitch muscle)[c]	650
K$^+$ channels (squid axon)[a]	70

Data: [a]K.H. Pfenninger, 1978, *Annual Review of Neuroscience*, **1**, 455–471; [b]W. Almers and C. Stirling, 1984, *Journal of Membrane Biology*, **77**, 169–186; [c]B. Hille, 1992, *Ionic Channels in Excitable Membranes*, Sinauer; [d]P.B. Sargent, 1993, *Annual Review of Neuroscience*, **16**, 403–443. For further discussion see D.C. Chang, *et al.*, eds, 1983, *Structure and Function of Excitable Cells*, New York: Plenum. It should be noted that as these data have been assembled from different sources which have used different counting techniques the results are not strictly comparable.

Neurons

Sizes

Smallest perikarya: $d = c. 5\,\mu m$ (cerebellar granule cells)[a]

Largest perikarya: $d = c. 100\,\mu m$ (cerebral Betz cells). Mauthner cells and some reticular formation cells are also extremely large. In the invertebrates cells in the visceral ganglia of *Aplysia* may be up to $1000\,\mu m$ in diameter[a]

A large dorsal root ganglion cell may be $120\,\mu m$ in diameter with an axon $12\,\mu m$ in diameter and up to 1 m in length. The perikaryon of such a cell may have a volume of $864\,000\,\mu m^3$ and a surface area of $43\,200\,\mu m^2$ (assuming sphericity). Its axon, however, can be calculated to have a volume of $108\,M\,\mu m^3$ and a total surface area of $36\,M\,\mu m^{2}$ [a]

Extent of dendritic arborisations[b]: pyramidal cells (mouse cortex), c. $3200\,\mu m$; stellate cells (mouse cortex), c. $5000\,\mu m$

Pyramidal dendrites develop about two spines per micrometre; stellate cells are devoid of spines[b]

Data from [a]W. Rall, 1977, *Ibid.*; [b]V. Braitenberg, 1978, 'Cortical architectonics: general and areal', in *Architectonics of the Cerebral Cortex*, ed. M.A.B. Brazier and H. Petsche, New York: Raven Press.

Numbers

Numbers of Neurons in Invertebrates

Mushroom bodies (*Drosophila*): 2500
Optic lobes of large insect: 100 000–1 000 000
Nervous system (less optic lobes): 50 000–100 000

Numbers of Neurons per Cubic Millimetre of Cerebral Cortex

	Visual cortex (mm^{-3})	Motor cortex (mm^{-3})
Mouse	100 000	60 000
Rat	47 000	25 000
Rabbit	42 000	26 000
Monkey (macaque)	115 000	18 000
Human	50 000	

Neuronal densities decrease as brain size increases; the Indian elephant, for instance, only musters about 7000 neurons per cubic millimetre of cortex.

Numbers of Synapses

	Visual cortex (mm^{-3})	Motor cortex (mm^{-3})
Mouse	6.6×10^8	8.5×10^8
Monkey (macaque)	6.2×10^8	9.6×10^8

Synapses per Neuron

	Visual cortex	Motor cortex
Mouse	7000	13 000
Monkey	5600	60 000

Well over 100 000 synaptic contacts are made on typical cerebellar Purkinje cells. Axon densities in the mouse c. 1 to 2×10^6 axons per cubic millimetre of cerebral cortex.

Data from B.G. Cragg, 1967, 'The density of synapses and neurones in the motor and visual areas of the cerebral cortex', *Journal of Anatomy*, **101**, 639–654; S.M. Blinkov and I.I. Glezer, 1968, *The Human Brain in Figures and Tables*, New York: Plenum Press; V. Braitenberg, 1978, *Ibid.*

Appendix 4

GENES

(Mentioned in text: alternative names in square brackets)

Many of these genes have close homologues in several organisms. When an organism is specified in the following list it usually only means that the gene was first detected there, or that it is best known in that organism. Hyphenated suffix 'n' indicates a number; thus *unc-n* might be *unc-86*; *brn-n* might be *brn-1, brn-2*, etc. Where appropriate the letters giving rise to the abbreviation have been highlighted.

abd-A/B: (**abd**ominal)   member of *Drosophila* HOM-C

ac: (**a**chaete)   member of *Drosophila achaete-scute* complex: four genes concerned in neural development

amn: (**amn**esiac)   *Drosophila* gene which when mutated leads to a total absence of long-term memory

Antp: (**ant**ennapedia)   member of *Drosophila HOM-C*

APOE: (**A**poli**po**protein E)   human gene for apolipoprotein E

bcd: (**b**i**c**oi**d**)   *Drosophila* A-P gene

boss: (**b**ride **o**f **s**evenle**ss** (q.v.))   gene which controls the development of ommatidium seven in *Drosophila* compound eye

brn-n: (**br**ain-n)   early development of vertebrate brain

btd: (**b**u**t**ton**d**)   *Drosophila* head development gene

bx: (**b**ithora**x**)   *Drosophila* homeotic gene

CACNA-n: gene encoding α-subunit of Ca^{2+} channel

CACNB-n: gene encoding β-subunit of Ca^{2+} channel

CACNG-n: gene encoding γ-subunit of Ca^{2+} channel

cactus: *Drosophila* dorsoventral polarity gene

cas (castor):   *Drosophila* developmental gene

ceh-6:   POU-domain gene in *Caenorhabditis elegans*

Ce-ncs-1:   encodes the **n**euronal **c**alcium **s**ensor protein in *C. elegans*

cf1a: [dp-1, D-POU]:   involved in early developmental stages (q.v.)

c-fos: (**c**ellular-**f**eline **o**steosarcoma virus)   this was first characterised as a viral oncogene (*v-fos* causing bone

tumours). Its cellular counterpart, the proto-oncogene IEG *c-fos*, encodes a transcription factor

cha: **ch**oline **a**cetyl transferase gene

c-jun: first characterised as an avian sarcoma virus (*v-jun*), its cellular counterpart the proto-oncogene IEG, *c-jun*, encodes a transcription factor

CLCN-n: genes encoding mammalian Cl$^-$ channels

Ddc: (**D**opa**de**carboxylase)   *Drosophila* behavioural gene

dfd: (**d**e**f**orme**d**)   member of *Drosophila HOM-C*

dlx: (**d**ista**l**less-related)   mouse gene family (?)

dnc: (**d**u**nc**e)   *Drosophila* behavioural gene

dorsal: *Drosophila* dorsoventral polarity gene

en: (**en**grailed)   *Drosophila* homeotic gene encoding a transcription factor (recognises 5′TCAATTAAATGA3′ and 5′TAATAATAATAA3′). Homologues found in vertebrate CNS

ems: (**em**pty **s**piracles)   *Drosophila* head development gene

emx-n: mouse homologues of *Drosophila ems* genes

ey: (**ey**eless):   *Drosophila* gene controlling early development of the compound eyes

flamingo (fmi):   gene controlling extent of dendritic branching

FMR1:   fragile X mental retardation gene 1

fos-B: member of the *c-fos* family of IEGs

fra-1:   member of the *c-fos* family of IEGs

ftz: (**f**ushi **t**arazu)   *Drosophila* homeotic gene encoding a transcription factor (recognises 5′TCAATTAAATGA3′ and 5′TAATAATAATAA3′)

hb: (**h**unch**b**ack)   *Drosophila* gap gene

HERG: gene encoding human homologue of *Drosophila* eag K$^+$ channel

hh: (**h**edge**h**og)   *Drosophila* segment polarity gene. Named for the bristly, stunted, fly embryo which develops when *hh* is 'knocked out'

int-n: gene controlling segmentation in *Drosophila*

jp: **j**im**p**y

jun-B: member of the *c-jun* family of IEGs

jun-D: member of the *c-jun* family of IEGs

kakapo: gene involved in the control of dendritic branching

KCNA-n: genes coding for mammalian *shaker* homologue K^+ channels

KCNB-n: genes coding for mammalian Shab homologue K^+ channels

KCNC-n: genes coding for mammalian Shaw homologue K^+ channels

KCND-n: genes coding for mammalian Shal homologue K^+ channels

KCNE-n: genes coding for mammalian KCNQ K^+ channels

KCNJ-n: genes coding mammalian inwardly rectifying K^+ channels

KCNK-n: genes coding for K^+ leak channels

KcsA: bacterial gene coding a K^+ channel

kni: (**kni**rps) *Drosophila* gap gene

kr: (**Kr**üpple) *Drosophila* gap gene

lab: (**lab**ial) member of *Drosophila HOM-C*

lin-n: (*lin*eage) *C. elegans* genes involved in development of lineage of cells in neurosensory systems

MASH-n: (**m**ammalian **a**chaete-**s**cute **h**omologous) *MASH-1* codes for a transcription factor in early development of vertebrate CNS and sympathetic ganglia

mass 1: **m**onogenic **a**udiogenic **s**eizure **s**usceptible

mec-n: (**mec**hanoreceptor) *C. elegans* genes involved in the development of mechanoreceptor cells

mld: **m**yelin **d**eficient

mscL: (**m**echano**s**ensitive **c**hannel: **L**arge conductance): gene coding for stretch receptor in *E. coli*

oct-n: (**oct**omer) vertebrate genes which encode transcription factors recognising the DNA 'octomer motif': 5′ATGCAAAT3′ [*oct-1*=*otf-1*; *oct-2*=*otf-2, NFlll*]

osk: (*osk*ar) *Drosophila* A-P gene

otd: (**o**rtho**d**enticle) *Drosophila* head development gene

otx-n: mouse homologues of *Drosophila otd* gene

pax-n: (**pa**ired bo**x**) family of genes involved in early eye development

pb: (**p**ro**b**oscidea) member of *Drosophila HOM-C*

pdm: (**P**OU **d**o**m**ain gene) gene involved in the differentiation of *Drosophila* CNS

pit-n: (**pit**uitary) vertebrate gene encoding a transcription factor activating prolactin and growth hormone in anterior pituitary [*Pit-1*=*gfh-1*]

POU: group of early developmental genes named from the initial letters of *pit*, *oct* and *unc* (q.v.)

PRNP: (**pr**ion **p**rotein) human prion protein gene

Prn-p: (**pr**ion **p**rotein) mouse prion protein gene

ras: (**ra**t **s**arcoma) v-ras retroviral gene inducing proliferation in rat muscle; c-ras a small family of cellular proto-oncogenes (*H-ras*, *K-ras* and *N-ras*) which, on mutation, induce proliferation in mammalian tissues

rl: (**r**ee**l**er) mouse behavioural gene

rutabaga: *Drosophila* memory gene

sc: (**sc**ute) see *ac*

SCN-nA: gene encoding α-subunit of mammalian Na^+ channel

SCN-nB: gene encoding β-subunit of mammalian Na^+ channel

scr: (**s**ex **c**ombs **r**educed) member of *Drosophila HOM-C*

sev: (**sev**enless) *Drosophila* gene controlling development of seventh ommatidium in the compound eye

sequoia: gene influencing extent of dendritic branching

sh: (**sh**aker): K^+ channel gene in *Drosophila*

shab: member of *sh* family (q.v.)

shal: member of *sh* family (q.v.)

shaw: member of *sh* family (q.v.)

shh: (**s**onic **h**edgehog) vertebrate homologue of *Drosophila hh* gene (q.v.)

shi (*shib*shire) a gene which encodes the insect homologue of dynamin

shi: (**shi**verer) *Drosophila* behavioural gene

shrub (*shrb*): gene influences extent of dendritic branching

sos: (**so**n of **s**evenless) gene involved in the development of ommatidium seven in the *Drosophila* compound eye

tap: *E. coli* gene encoding receptor-transducer protein responsive to dipeptides

tar: *E. coli* gene encoding receptor-transducer protein responsive to aspartate

toll: *Drosophila* dorsoventral polarity gene

toy (twin of eyeless (q.v.)): *Drosophila* gene controlling early development of compound eyes

trg: *E. coli* gene encoding receptor-transducer protein responsive to ribose and/or galactose

tsr: *E. coli* gene encoding receptor-transducer protein responsive to serine

tst-1: (**t**e**st**is, expressed in testis and brain); [*SCIP:* (**s**uppressed **c**AMP **i**nducible **P**OU); *Oct-6*]

tumbleweed (*tum*): gene influencing extent of dendritic branching

ubx: (**u**ltra**b**ithorax) *Drosophila* homeotic gene

unc-n: (**unc**oordinated) *C. elegans* gene involved in development of nervous system (mechanoreceptor cells); mutation leads to uncoordinated movement

vhh1: (**v**ertebrate **h**edge**h**og 1) see *shh*

wg: (**w**in**g**less) *Drosophila* segmental polarity gene

wnt-n: mammalian homologue of *Drosophila* segmentation gene, int (q.v.)

wv: **w**ea**v**er: mouse behavioural gene

Appendix 5

PHYSICAL MODELS OF ION CONDUCTION AND GATING

The schematic diagrams of ion channels in Chapter 11 are at best idealised cartoons of the molecular reality. They are inevitably drawn as if they were devices in the macroscopic everyday world of billiard balls, tubes, barrels, balls and chains, etc. Hopefully, they give a good first approximation and are certainly invaluable in fixing ideas. However, we do not have to go all the way into the counter-intuitive mysteries of quantum physics to recognise that the world of the very small (like the world of the very large) is very different from the commonplace world of human-sized objects. We must always remain on guard against being led astray by our models and diagrams. Accordingly, it is not surprising to find that the molecular models and illustrations with which we have become so familiar have their critics. New ways of explaining channel phenomena have been proposed. Let us briefly look at one of these alternative visions. It is based on some unusual properties of special types of material: superionic conductors, ferroelectrics and liquid crystals.

In Chapter 11 we described the movement of ions as if it were a flow through a water-filled pore whose walls were lined by the amino acid side chains of the H5 selectivity filter. An alternative explanation is that the channel protein itself, in its open configuration, is the ion conductor. It can do this by becoming a metalloprotein, containing a chain of permeant ions loosely bound to sites along a transmembrane path. Ions then hop from site to site, driven by the electrochemical gradient across the membrane. A number of materials which conduct in this way are known. Such so-called superionic conductors have conductivities comparable to those of ionic solutions but are highly ion selective.

How do the channels switch from their closed state to their open state? Here the explanation is based on a comparison of channel properties with those of ferroelectrics, materials that exhibit a spontaneous electric polarisation reorientable by an electric field. The similarities between channels and ferroelectrics are remarkable, including thermal, mechanical and optical properties. They include the fact that the capacitance of the sodium channel appears to obey the ferroelectric Curie–Weiss law, which describes the dependence of the dielectric permittivity on temperature. As we saw in Chapters 13 and 15, the sodium channel is closed at resting potential and opens in response to a depolarisation. The ferroelectric-superionic transition hypothesis explains channel opening simply as a shift in the transition temperature induced by the reduction in the electric field due to a depolarisation.

What kind of ferroelectric could the channel be? Certainly not a solid crystal, as many common ferroelectrics are. Since we know that the membrane can be regarded as a liquid crystal (Section 8.2), it is reasonable to 'see' the channel as if it were a liquid-crystalline molecule. The type of liquid crystal that is ferroelectric is known as **chiral smectic C**. These are elongated molecules that form layers. The layers are tilted with respect to the horizontal plane of the membrane and the molecules are not congruent to their mirror images. Structural studies suggest that ion channels fulfil these requirements.

We noted in Section 11.5 that the voltage sensor is believed to be the S4 transmembrane helix. The ferroelectric model provides an alternative explanation of the voltage-dependent opening of the channel. At resting potential the S4 segments assume an α-helical conformation. When depolarisation occurs, the ferroelectric model suggests that the hydrogen bonds holding the spirals of the helix to each other are broken. This causes a conformational change from helix to random coil and is detectable as the gating current. The sites at which the H-bonds were broken may now become sites of ion attachment, forming a salt bridge across the membrane. Further studies of these models are proceeding (see Bibliography).

ACRONYMS AND ABBREVIATIONS

A adenine

α-BuTX α-bungarotoxin

AC adenylyl cyclase

Ac acetyl

ACh acetylcholine

AChE acetylcholinesterase

AChR acetylcholine receptor

ACPD aminocyclopentane 1,3-decarboxylate

ACTH adrenocorticotrophic hormone

AD Alzheimer's disease

AD-AP Alzheimer disease amyloid polypeptide

ADP adenosine diphosphate

AHP after-hyperpolarisation

AMP adenosine monophosphate

AMPA α-amino-3-hydroxy-5-methyl-4-isoxazole propionic acid

AMPT α-methyl-*p*-tyrosine

AP-1 binding site for transcription factor AP-1 (5'TGACTCA3')

ApoE apolipoprotein E

APP amyloid precursor protein

Arc activity-related cytoskeletal protein

ATP adenosine triphosphate

α-TX α-toxin – a neurotoxin which attaches to the α-subunit of nAChR

AVEC Allen enhanced video contrast (microscopy)

BACE β-APP cleaving enzyme

β-AR β-adrenergic receptor

β-ARK β-adrenergic receptor kinase

BBB blood–brain barrier

BDNF brain-derived neurotrophic factor

BH4 tetrahydrobiopterin

BLAST basic local alignment search tool. Computer program used by bioinformaticists to determine whether a recently discovered sequence resembles those held in genome and protein databases

BP before present

bp base pair (i.e. nucleotide base pair)

BrdU bromodeoxyuridine. Label used to detect mitosing cells in a tissue. Used to detect neural stem cells

BSE bovine spongiform encephalopathy

C cytosine

CAM cell adhesion molecule

cAMP cyclic adenosine monophosphate

CAT choline acetyltransferase

CCK cholecystokinin

cDNA complementary DNA

C/EBP CAAT-enhancer binding protein: a family (α, β, δ, γ) of transcription regulators

cGMP cyclic guanosine monophosphate

CGRP calcitonin gene-related product

ChAT *see* CAT

CJD Creutzfeldt–Jakob disease

ClC chloride channel

CLIP corticotrophin-like intermediate lobe protein

CMT Charcot–Marie–Tooth disease

CNG cyclic nucleotide gated

CNGF cholinergic growth factor

CNS central nervous system

COMT catechol-*O*-methyltransferase

Con-A concanavalin A

CR conditioned response

CRE cAMP response element (5'TGACGTCA3')

CREB CRE-binding

CS conditioned stimulus

Da dalton – a unit of molecular mass defined as 1/12 the mass of ^{12}C

DA dopamine

DAG diacylglycerol

D-AP5 D-2-amino-5-phosphonovalerate

ddNTP dideoxynucleotide

DNA deoxyribonucleic acid

DOPA 3,4-dihydroxyphenylalanine

dpc days post-coitum

DSE depolarisation-induced suppression of excitation (retrograde action of cannabinoids)

DSI depolarisation-induced suppression of inhibition (retrograde action of cannabinoids)

EcoR1 much-used restriction endonuclease derived from *E. coli*

ECS electroconvulsive shock

EAA excitatory amino acid

EM electron microscope

eph from **e**rythropoietin **p**roducing **h**uman **h**epato-cellular cell line

ephrin from Eph receptor interacting protein

EPP end plate potential

EPSP excitatory postsynaptic potential

ER endoplasmic reticulum

ERG electroretinogram

FAD familial Alzheimer's disease

FAD flavine adenine dinucleotide

FAQ frequently asked question

FMN flavine mononucleotide

fMRI functional magnetic resonance imaging

FraX fragile X syndrome

G guanine

GABA γ-aminobutyric acid

GABA-T GABA aminotransferase

GAD glutamic acid decarboxylase

Gal galactose

GAP GTPase-activating protein

GDP guanosine diphosphate

GFP green fluorescent protein

Glc glucose

GluR glutamate receptor

GlyR glycine receptor

GMC ganglion mother cell (found in developing *Drosophila* CNS)

GMP guanosine monophosphate

GNRP guanine nucleotide release protein

GRP GAP-related protein (q.v. GAP)

GTP guanosine triphosphate

GVD granulovacuolar degeneration

Hb haemoglobin

HD Huntington's disease

HIAA hydroxyindoleacetic acid

HLH helix–loop–helix, a structural motif found in many transcription factors

hnRNA heteronuclear RNA

HOM-C *Drosophila*'s homeobox cluster

Hox vertebrate homeobox gene

HPRT hypoxanthine guanine phosphoribosyltransferase

hsp heat shock protein

HTH helix–turn–helix, a structural motif found in many transcription factors

HTTP hypertext transfer protocal

HUGO Human Genome Organisation

I_{Ca} calcium-dependent potassium current

I_K delayed pottassium current

I_{KA} early potassium current

I_{KM} muscarinic potassium channels

I_{KS} serotonin-dependent potassium current

ICRF Imperial Cancer Research Fund

IF intermediate filament or intitiation factor (context will decide)

Ig immunoglobulin

IEG immediate-early gene

IP$_3$ inositol triphosphate

IPSP inhibitory postsynaptic potential

IS insertion sequence

ISI interstimulus interval

KA kainate

L-AP4 L-2-amino-4-phosphonobutyrate

LGIC ligand-gated ion channel

LINE long interspersed element. A lengthy section of DNA consisting of nucleotide repeats. Particularly prevalent in the human genome (19%)

LSD lysergic acid diethylamide

LTD long-term depression

LTM long-term memory

LTP long-term potentiation

mAChR muscarinic acetylcholine receptor

MAG myelin-associated glycoprotein

MAO monoamine oxidase

MAOI monoamine oxidase inhibitors

MAP microtubule accessory protein

Mb myoglobin

MBP myelin basic protein

MEG magnetoencephalography

MPP$^+$ 1-methyl-4-phenylpyridinium ion

MPTP 1-methyl-4-biphenyl-1,2,3-tetrahydropyridine

M_r molecular weight, relative molecular mass – the ratio of the mass of a molecule to 1/12 the mass of ^{12}C

MRC Medical Research Council (UK)

mRNA messenger RNA

MS multiple sclerosis

MSH melanocyte-stimulating hormone

MT microtubule

Mt mitochondrion

nAChR nicotinic acetylcholine receptor

NAG *N*-acetylglucosamine

najaTX neurotoxin derived from the cobra *Naja naja siamensis*

NANC non-adrenergic non-cholinergic

N-CAM neural cell adhesion molecule

NCC neural crest cell

NE norepinephrine/noradrenaline

NF neurofibromatosis

NFT neurofibrillary tangle

Ng-CAM neuron–glia cell adhesion molecule

NGF nerve growth factor

NIH National Institutes of Health (US)

NMDA *N*-methyl-D-aspartate

NMR nuclear magnetic resonance

NOS nitric oxide synthase

NSC neural stem cell

NSF *N*-ethylmaleimide-sensitive fusion protein

OHDA hydroxydopamine

ORF open reading frame

P$_i$ inorganic phosphate

P$_0$ protein zero (myelin)

PAF platelet-activating factor

PAG periaqueductal grey (grey matter surrounding the central canal spinal cord and brain stem)

PAGE   polyacrylamide gel electrophoresis
pc   post coitum (see also dpc)
PCP   phencyclidine
PCR   polymerase chain reaction
PD   Parkinson's disease
PDE   phosphodiesterase
PDP   protein data bank
PDZ   from **PSD**, **D**iscs large (a *Drosophila* septate junction protein) and **ZO**-1 (the epithelial tight junction protein); a c. 90 amino acid repeat sequence found in a number of proteins and particularly in PSD-95, a protein found in the PSD
PET   positron emission tomography
PGF   polypeptide growth factor
PHF   paired helical filament
PIP$_2$ phosphatidylinositol
PKA   protein kinase A (cAMP dependent)
PKC   protein kinase C (Ca^{2+} dependent)
PKU   phenylketonuria
PLP   proteolipid protein (myelin)
PMD   Pelizaeus–Merzbacher disease
PNS   peripheral nervous system (context decides)
PNS   postive–negative selection (context decides)
POMC   pro-opiomelanocortin
POU   morphopoietic genes expressed in vertebrate forebrain (acronym from genes: Pit-1, Oct-2/Oct-3, Unc-86)
PP$_i$ inorganic pyrophosphate
PrP   prion protein
PrP^C   normal (i.e. cellular) isoform of PrP (q.v.)
PrP^{Sc}   abnormal, infective form (e.g. scrapie) of PrP (q.v.)
PSD   postsynaptic density
PTP   post-tetanic potentiation
QED   quantum emission domain
RA   retrograde amnesia
rCBF   regional cerebral blood flow
rCMR   regional cerebral metabolic rate
rd   retinal degeneration (mouse)
rds   retinal degeneration slow (mouse)
RER   rough endoplasmic reticulum
RF   receptive field
RFLP   restriction fragment length polymorphism
RNA   ribonucleic acid
rNTP   ribonucleoside triphosphate
rRNA   ribosomal RNA
RP   retinitis pigmentosa
RPE   retinal pigment epithelium
SA   sympathoadrenal

SDAT   senile dementia of the Alzheimer type
SEM   scanning electron microscope
SGL   subgranular layer
SI   Système Internationale d'Unités
SINE   short interspersed elements. Comparatively short segment of DNA consisting of nucleotide repeats. Common in human genome (about 13%)
SK   substance K
SNAP   soluble NSF attachment protein
SNARE   SNAP receptor
SNP   single nucleotide polymorphism
SP   substance P
SR   sarcoplasmic reticulum
STM   short-term memory
SV   synaptic vesicle
SVZ   subventricular zone
T   thymine or transducin (context will decide)
TEA   tetraethylammonium chloride
TEM   transmission electron microscope
Tg   transgenic animal
TGF-β   transforming growth factor β
TH   tyrosine hydroxylase
THB   tetrahydrobiopterin
7TM   seven transmembrane helices
T$_m$ the temperature at which the two strands in a double-stranded nucleic acid are 50% dissociated
TMS   transmembrane segment
Tn   transposon
TRED   triplet repeat expansion disease
trk   tyrosine kinase
tRNA   transfer RNA
TTX   tetrodotoxin
U   uracil
UCR   unconditioned reflex
UCS   unconditioned stimulus
URL   universal resource locator
VAChT   vesicular acetylcholine transporter
VAMP   vesicle-associated membrane protein
VAS   vesicle attachment site
VIP   vasoactive intestinal peptide
VMAT   vesicular monoamine transporter
V_m   resting potential across a membrane
X-gal   5-bromo-4-chloro-3-indolyl-galactoside. When hydrolysed by β-galactosidase the 'X' part of the molecule forms an insoluble green-blue precipitate. Much used as a marker when β-galactosidase is encoded by a reporter gene

GLOSSARY

aetiology   the cause of a disease

afferent   fibres conducting toward some region

after-hyperpolarisation   period during which an excitable cell's membrane is hyperpolarised in an action potential

agonist   drug which mimics a hormone or neurotransmitter, binding to the receptor and causing the normal response

allele   one of two or more alternative forms of a gene

amphipathic   applied to molecules (especially proteins) in which one part of the molecule is water-soluble and the other water-insoluble

Angelman syndrome   speech impairment, little use of words but compensated by good non-verbal communication. Balance disorders and often ataxia. Happy, laughing disposition, but short attention span

anosmia   loss of olfactory sense

antagonist   drug which binds to the receptor of a hormone or neurotransmitter and does not induce the normal response

anterograde   movement from the perikaryon towards the terminal

antibody   a protein produced by the immune system in response to a foreign substance (usually a protein). It is synthesised in such a way that it reacts with the foreign substance (antigen, q.v.) which led to its synthesis and no other

antigen   any substance (usually a protein) capable of eliciting an immune response, i.e. the synthesis of a specific antibody (q.v.)

antiport   transport across a membrane of two different small molecules and/or ions in opposite directions

anxiolytic agent   a drug such as Valium (diazepam) which reduces anxiety

aphasia   loss of power to communicate in speech or writing

apoptosis   programmed cell death. The cell nucleus shrinks, the cytoplasm shrivels, the diminished cell is scavenged cleanly by a neighbouring cell

ataxia   lack of muscular coordination

autoimmune disease   conditions (such as myasthenia) where the body's immune system mistakes its own molecules as foreign and consequently manufactures antibodies against them

autonomic nervous system   that part of the peripheral nervous system which controls the 'automatic' activity of the body's systems, e.g. heart beat, blood pressure, movement of the alimentary tract, respiratory movements, etc. It is divided into two subsystems: the sympathetic and the parasympathetic systems. The sympathetic system is generally regarded as involved in response to emergencies ('fight and flight') and operates with noradrenaline; the parasympathetic system is involved in more quiescent, vegetative functions and uses acetylcholine as a transmitter

Avogadro's number (N)   the number of particles in a mole or g ion of substance, i.e. 6.023×10^{23}

axolemma   boundary membrane of an axon

axoneme   central core of a flagellum or cilium consisting of an outer ring of nine microtubular doublets and a central pair of microtubules

baroreceptor   pressure receptor

biotinylation   attachment of biotin to a molecule

bipolar disorder   sometimes called manic-depression. Characterised by wild swings of mood from despair to elation and back again, sometimes experienced by students during the examination period

blastula early animal embryo, typically a hollow ball of cells (blastomeres) before gastrulation has commenced

bouton expansion at the termination of an axon specialised for transmitter release

capsid the proteinaceous coat of a virus

caudal towards the posterior

caveolae flask-shaped invaginations of plasma membrane

cDNA complementary DNA: a stretch of DNA which faithfully copies a particular stretch of RNA

chiasmata (sing. chiasma) a zone of contact along a chromosome pair (visible at meiotic diplotene) which represents where a cross-over is occurring

chorea involuntary jerking movements of the skeletal muscles

chromaffin cell a cell which stores adrenaline in vesicles for release in times of stress

chromatid one of the daughters of a replicated chromosome still joined to its sister chromatid at the centromere

chromosome a discrete unit of the genome carrying many genes and much nonsense DNA as well as protective protein. It is visible as an entity only during cell division. The centromere divides the chromosome into a short arm designated 'p' and long arm 'q'

clone colony of cells formed by successive division of a single parent cell

coated pit depression in the plasma membrane lined on the cytoplasmic surface with bristle-like clathrin or synapsin molecules. After endocytosis these pits form coated vesicles

consensus site a sequence of similar (or very similar) nucleotide bases which is recognised by a transcription factor or RNA polymerase

conservative substitution the substitution during evolutionary development of a protein of one amino acid by another which resembles the original in its physico-chemical characteristics

constitutive gene product produced at a constant level (contrast 'inducible', where the production varies according to external stimulus)

contig a set of overlapping sequence fragments which constitute a chromosomal region. Ultimately sufficient contigs add up to an entire chromosome

contralateral relating to the opposite side of the body

cross-over exchange of information between non-sister chromatids during meiotic division

cytoplasm material within the plasma membrane of a cell but outside (in eurkaryotic cells) the nuclear membrane

cytosol colloidal phase within the plasma membrane and outside the nuclear membrane but excluding formed elements such as mitochondria, cytoskeleton, endoplasmic reticulum, Golgi body, etc.

dementia 'an acquired global impairment of intellect, memory, and personality but without impairment of consciousness' (Lishman, W.A., 1978, *Organic Psychiatry*, Oxford: Blackwell)

depolarisation transmembrane voltage less than normal resting potential (V_m)

diencephalon (=thalamencephalon) posterior part of forebrain

dominant an allele (q.v.) which is expressed whether it is present in the homozygous or the heterozygous form (compare recessive, q.v.)

domain region of a protein which has a definite tertiary structure of its own; usually connected to other domains by flexible polypeptide sequences

downmodulation see **modulation**

efferent fibres conducting away from some region

endocytosis uptake of material from the extracellular space by membrane invagination and budding off as an intracytoplasmic vesicle

endogenous caused from within the organism

end plate area of vertebrate striped muscle lying beneath a motor nerve terminal

enhancer region of a DNA sequence to which regulatory proteins bind and thus influence transcription of a structural gene often hundreds or even thousands of nucleotides distant

euchromatin gene-rich region of a chromosome (cf. heterochromatin)

eukaryocyte cell characterised by possessing a separate membrane-bound nucleus and cytoplasmic organelles such as mitochondria

exocytosis fusion of vesicle membrane with plasma membrane so that contents of vesicle are voided to the extracellular space

exogenous caused from outside the organism

exon sequence of nucleotides in an initial mRNA transcript which is preserved in the mRNA translated at the ribosome

expression vector a means of inserting a gene into a cell so that on receipt of an appropriate signal the cell will manufacture large amounts of the protein for which the gene codes

fasciculation aggregation of a number of nerve fibres in parallel to form a bundle

festination an involuntary quickening of steps when walking

GAL4–UAS system two transgenes which can be put together to express a gene of interest in a spatially and temporally determined manner. One of the transgenes expresses the powerful transcriptional activator GAL4 in a particular tissue at a particular time. The other transgene consists of the gene of interest fused to UAS, a GAL4-sensitive promoter. The two transgenes are maintained in two different animal stocks. When interbreeding is allowed the GAL4 and UAS find themselves in the same cell and the gene of interest is expressed

gastrula the next stage in embryonic development after the blastula (q.v.) when cells have invaginated to form a primitive gut (Greek: *gaster*, stomach)

gene, structural nucleotide sequence coding for a specific polypeptide or protein

genome the total complement of genetic material in a cell or individual

glycosylation addition of monosaccharide units to a polypeptide chain – usually begins in the lumen of the endoplasmic reticulum and completed in the Golgi body

Goldman equation an equation which relates the electrical potential across a membrane to the distribution and permeability constants of the ions which that membrane separates

haplotype a particular combination of alleles (or sequence variations) that are closely linked on the same chromosome and thus likely to be inherited together

heterochromatin gene-poor region of a chromosome. Consists largely of simple base pair repeats. Originally identified by staining difference with euchromatin

heterozygous having two different alleles (q.v.) for the same trait

histone a basic protein around which DNA is coiled in eukaryotic chromosomes

homeobox a sequence of 180 base pairs in the DNA of many genes controlling early development

homeodomain the 60 amino acid sequence in the primary structure of the proteins coded for by the homeobox (q.v.)

homeosis transformation of 'something' into the likeness of 'something else'

homology; homologue similarity which is due to common evolutionary origin. Contrasted with 'analogy' where the similarity is not due to a shared ancestor

homopolymer polymer built of only one type of subunit

homopolymer tailing a procedure by which a string of identical nucleotides is added (by terminal transferase) to the end of one strand of a DNA molecule. This tail (e.g. A-A-A-A-A-) will readily stick to a complementary tail (e.g. T-T-T-T-T-) affixed to another single strand of DNA

homozygous having identical alleles (q.v.) for a given trait

housekeeping (constitutive) genes genes expressed in all an organism's cells as they provide the basics necessary for life

hydropathy disliking water – applied to those segments of a polypeptide chain that are hydrophobic and for this reason often found buried in the lipid core of biomembranes

hyperpolarisation transmembrane voltage greater than resting potential (V_m)

hypertonic a relative term to describe a solution with a higher concentration of solutes than that to which it is compared

hypotonic the converse of hypertonic

imprinted gene a gene that is marked in such a way that its maternal or paternal origin is denoted

in vitro a biological process studied outside the organism (literally 'in glass')

in vivo a biological process studied within the living organism

intron (=intervening sequence) a sequence of nucleotides in a gene which after transcription into mRNA are excised and are consequently no longer present when the mRNA is translated into polypeptide

ionotropic the action of a neurotransmitter which directly opens ion channels in a subsynaptic membrane

ipsilateral same side of the body

isoforms multiple forms of the same protein differing slightly in amino acid sequence

Klenow fragment (enzyme) fragment of *E. coli* DNA polymerase 1 which lacks the 5′–3′ exonuclease activity of the entire enzyme. Alternative to reverse transcriptase in the preparation of cDNA

Krebs cycle series of biochemical reactions whereby acetyl coenzyme A is fully oxidised to carbon dioxide and water with the release of energy. The energy is used to synthesise ATP from ADP and P_i. Also known as the tricarboxylic acid cycle and the citric acid cycle

leader sequence nucleotide sequence at the 5′ end of an mRNA primary transcript which is important in the regulation of transcription. Usually removed before translation occurs

leaflet one monolayer of a biomembrane's lipoprotein bilayer

ligand a molecule which binds to a specific site on another molecule (from Latin, *ligare*: to bind)

lumen membrane-bound cavity or space

meiosis two successive nuclear divisions but only one duplication of the chromosomes thus leading to the production of four haploid gametes (also called 'reduction division')

mesencephalon midbrain

metabotropic the action of a neurotransmitter which works through a membrane-bound collision-coupling system and usually involves second messengers such as cAMP, cGMP, PI

metencephalon anterior division of hindbrain

microsatellite markers hypervariable regions of DNA consisting of short repetitive sequence (usually CT). The length of the repetitive sequence varies between individuals and is hence a useful identifier

mitosis eukaryotic cell division whereby the chromosomes of the parent cell replicate and segregate to form two daughter nuclei

modulation alteration of a membrane channel's conductance. 'Upmodulation': increased conductance; 'downmodulation': decreased conductance

motif a group of amino acid residues which occur at binding/active sites in different proteins. The term is also applied to specific base sequences in DNA, e.g. TATA

Müller cells large neuroglial cells found in vertebrate retinas

multimeric protein a protein which consists of two or more subunits

multivesicular body vacuolated body formed by the inward budding of the membrane of an uncoated vesicle

mutation heritable change in the nucleotide sequence of a chromosome. There are many types, e.g. loss of function (usually recessive whereby the function normally controlled by the gene is lost); gain of function (increased activity, or activity in inappropriate circumstances); lethal (causes the organism to die prematurely); suppressor (suppresses the phenotypic consequence of another mutation so that both together result in a normal phenotype)

myotube a developing muscle fibre formed by the fusion of myoblasts

native normal *in vivo* conformation of a biological macromolecule

Nernst equation an equation which relates the electrical potential across a membrane to the distribution across that membrane of a single ionic species

neurilemma the plasma membrane of a neuron

neurite a nerve cell process – axon or dendrite

neuroleptic a drug which has an antipsychotic action

neuromere segment of the neural tube (see also **rhombomere**)

neuropil densely intertwined network of axons, dendrites and synapses

neurosphere a spherical mass of neural cells in different stages of maturation. Formed by a sphere-forming cell, possibly an astrocyte, in a stem cell region of the brain

nociceptive a receptor or neural pathway whose activity is sensed as pain

nucleoplasm colloidal fluid contained within the nuclear membrane

oncogene a gene which can help make a cell cancerous (usually derived from a mutated proto-oncogene, q.v.)

oocyte egg cell

open reading frame a segment of DNA which begins with a 'start' codon and finishes with a 'stop' codon, i.e. the coding sequence of a gene

orthologue, orthologous a gene that appears in two organisms by descent from a common ancestor

palindromic sequence DNA sequence which reads identically forwards and backwards (i.e. 5′⇒3′ or 3′⇒5′). An English-language analogy: 'radar'

paralogue, paralogous a gene which occurs in more than one copy in an organism due to a duplication event

paranoid showing unnecessary suspicion, fear of persecution, etc. Paranoia is often exhibited by authors when reading reviews of their books

parasympathetic nervous system see **autonomic nervous system**

phosphodiester any molecule which contains the group

$$-R-O-\underset{\underset{O}{\parallel}}{\overset{\overset{O^-}{\mid}}{P}}-O-R'-$$

where R and R' are carbon-containing groups

phosphodiesterase an enzyme which attacks the phosphate linkage joining the two R groups of a phosphodiester

phylogeny evolutionary history of an organism

pia mater the innermost of the three membranes (meninges) covering the brain and spinal cord

pial see **pia mater**

plasma membrane the boundary membrane of a cell

plasmid autonomously replicating extrachromosomal DNA

plexiform layer the vertebrate retina has two plexiform layers. The outer plexiform layer consists of the layer of synapses between the photoreceptor cells and the bipolars, the inner plexiform layer consists of the layer of synapses between the bipolars and the ganglion cells

prion coined from **pro**teinaceous+**in**fectious

prokaryocyte a cell which lacks a membrane-bound nucleus and other membrane-bound organelles, such as mitochondria

protein family proteins with primary sequences differing by less than 50%

protein kinase an enzyme which catalyses the transfer of a phosphate group from ATP to a protein to form a phosphoprotein

protein superfamily proteins whose primary sequences can be shown by statistical analysis to be significantly alike

proteome the full complement of protein coded by a genome

proto-oncogene gene concerned with the regulation of cell growth and division. Mutated proto-oncogenes (oncogenes, q.v.) often promote unregulated (cancerous) growth and division

pseudogene gene having sequence homology with an active gene but which contains mutations that prevent their expression. Derived from an ancestral active gene

recessive although the allele (q.v.) is present in the genome it is only expressed when it is in the homozygous condition. Its influence is masked by the dominant allele in a heterozygote

recombinant DNA hybrid DNA formed by fusing DNA fragments obtained from different sources

rectifier ion channels may be 'ohmic' or 'rectifier' In the first case the ion flow is directly proportional to the voltage ($I = V_m/R$); in the second case the flow is not linearly related to the voltage: the ions flow more readily in one direction than the other

refractory period the time following an action potential during which the neuronal membrane cannot carry another

resting potential electrical potential across a nerve or other cell membrane when an action potential is being conducted. Depending on cell type it may range from $-50\,mV$ to $-75\,mV$ (interior negative)

restriction map diagrammatic map showing how the DNA fragments produced by restriction endonuclease digestion could best be ordered. The map shows the relation of the sites cleaved by restriction endonucleases with respect to each other

retrograde movement from a terminal toward perikaryon

retrovirus a virus which uses RNA as its genetic material but which possesses a reverse transcriptase enabling it to synthesise a DNA copy of the whole or part of its RNA within a cell

rhombomere segment of the rhombencephalon (hindbrain)

rostral towards the anterior

sarcolemma the plasma membrane of a muscle fibre or cell

scaffold a series of contigs (q.v.) which are in the right order but are not necessarily connected in a continuous stretch or sequence

schizophrenia (=dementia praecox) a group of conditions in which the individual feels 'split off' from the rest of humanity, unable to relate to his or her fellows, sometimes apathetic, sometimes suffering from overwhelming hallucinations and delusions

soma sometimes just the perikaryon, sometimes perikaryon plus dendrites

spacer DNA lengths of DNA between genes (especially rRNA genes) which are not transcribed into mRNA

spike action potential minus the 'after-hyperpolarisation' (q.v.)

stem cell a pluripotent cell able to divide in such a way that one daughter cell remains pluripotent and the other differentiates into a post-mitotic cell

stenosis narrowing of a canal or passageway

stop codon mRNA triplets which signal end of message – UAG, UAA, UGA

structural gene see **gene**

svedberg unit for the sedimentation coefficient of macromolecules – in general the greater the particle mass the greater the sedimentation coefficient but the relationship is not linear

sympathetic nervous system see **autonomic nervous system**

symport transportation of two small molecules and/or ions in the same direction across a membrane

syncytium mass of cytoplasm surrounded by a membrane and containing many nuclei. Usually formed by successive nuclear divisions without the cell itself dividing

telencephalon anterior part of the forebrain

transcript mRNA product of DNA transcription

transcriptome the full set of activated genes (or mRNAs) in a particular tissue at a particular time

transducer a device for transforming energy from one form into another

transduction (genetics) the transfer of genes between bacteria

transfection insertion of DNA into a cell without the use of a vector (q.v.) and leading to the integration of the DNA into the host genome

transgenic an organism which has stably incorporated a gene or genes from another organism and can transmit them to succeeding generations

transposon stretch of DNA containing one or more genes which can move from one part of the genome to another rather than, like most genes, remaining fixed

tricarboxylic acid cycle see **Krebs cycle**

trophic nutritional support of cell or multicellular organism

tropic an influence that guides the growth or movement of a cell or multicellular organism

upmodulation see **modulation**

vector a vehicle used to introduce recombinant DNA into a host cell; examples of vectors are phage, plasmids, cosmids

wild-type normal, i.e. non-mutant form or, in other words, the usual type found in 'nature' or in 'the wild'

Williams–Beuren syndrome (WBS) characterised by cognitive impairment (average IQ 58, but above-average ability in facial recognition), unique personality characteristics (highly sociable, loquacious, sometimes musically gifted), distinctive facial features and cardiovascular disease

BIBLIOGRAPHY

Molecular neurobiology is a huge and rapidly advancing subject. This bibliography is intended both as a reference list for the chapters and to provide an entrée to the literature. Starred (*) publications are sources of in-text illustrations. Wherever possible review articles and papers available in most college and university libraries are cited. In this edition a number of relevant web sites have also been listed for each chapter. Students interested in following a topic in depth should be able to use these print or electronic sources as entry points.

Most scientific libraries now possess extensive bibliographic resources, especially via the internet, to help the student cope with the avalanche of scientific information. These resources range from databases such as *Medline, Excerpta Medica, Life Sciences* and *BIDS* to abstracting services such as *Index to Scientific Reviews, Biological Abstracts* and *Index Medicus* to second-order web search engines such as

Google.com and *Copernic.com*. *Current Contents (Life Sciences)* and the *Science Citation Index* make it possible to keep abreast of what is published from week to week and to follow the development of a topic once a reference has been located.

In addition to the *Annual Reviews* (*Neuroscience*; *Physiology*; *Biochemistry*; *Pharmacology and Toxicology*; *Biophysics*, etc.) students can also find valuable overviews in the *Trends* publications (*Neurosciences*; *Genetics*; *Biochemistry*, etc.), in *Current Biology*, and in *Current Opinion* (*Neurobiology, Neurology, Cell Biology*, etc.). Useful 'minireviews' are also frequently to be found in 'primary' journals such as *Cell* and *Neuron*. Finally, *Nature* (especially its offshoots, *Nature Neuroscience, Nature Cell Biology, Nature Genetics*, and *Nature Reviews Neuroscience, Nature Reviews Genetics*, etc.) and *Science* remain indispensable for those wishing to keep abreast of an explosively developing field.

CHAPTER 1:
INTRODUCTORY ORIENTATION

Web sites

A list of neuroscience images on the internet may be found at:
http://www.neuroguide.com/neuroimg_2.html
For useful pictures of development of brain from three- to five-vesicle structure see:
http://medocs.ucdavis.edu/chahph/403/syllabus/vesicle.htm
The comparative mammalian brain collections (CMBC) site contains complete brain atlases of at least 15 species (photos or movies) and photos from nearly 200 different mammals. An analysis of fossil endocasts gives access to palaeoneurology and the evolution of mammalian brains. In addition, brain sections give an insight into internal structure. The site can be viewed at:
http://brainmuseum.org/index.html

Print

1. History

Brazier, M.A.B., 1959, 'The historical development of neurophysiology' in *Handbook of Physiology*, Washington: American Physiological Society, vol. 1, 1–58

Clarke, E. and C.D. O'Malley, 1996, *The Human Brain and Spinal Cord*, Berkeley: University of California Press; 2nd edn, San Anselmo, CA: Norman Publishing

Cowan, W.M., 2000, 'The emergence of modern neuroscience: Some implications for neurology and psychiatry', *Annual Review of Neuroscience*, **23**, 343–391

Finger, S., 1994, *Origins of Neuroscience: A History of Explorations of Brain Function*, Oxford: Oxford University Press

McIlwain, H., 1958, 'Thudichum and the medical chemistry of the 1850s and 1860s', *Proceedings of the Royal Society of Medicine*, **51**, 123–127

Reichardt, L.F., 1984, 'The emergence of molecular neurobiology', *Trends in Biochemical Sciences*, **9**, 173–176

Smith, C.U.M., 1976, *The Problem of Life: an Essay in the Origins of Biological Thought*, London: Macmillan

Tower, D.B., 1958, 'Origins and development of neurochemistry', *Neurology*, **8**, Suppl. 1, 3–31

2. Molecular Neurobiology

Bradford, H.F., 1986, *Chemical Neurobiology*, New York: Freeman

Hall, Z.W., 1992, *An Introduction to Molecular Neurobiology*, Sunderland, MA: Sinauer

Kay, J., ed., 1986, *Molecular Neurobiology: Biochemical Society Symposia*, **52**, London: Biochemical Society

Siegel, G.J. *et al.*, eds, 1981, *Basic Neurochemistry*, Boston: Little Brown and Co.

Soreq, H., 1984, *Molecular Biology Approach to the Neurosciences*, Chichester: John Wiley

Watson, J.D. and R. McKay, eds, 1983, *Molecular Neurobiology*, Cold Spring Harbor, NY: Cold Spring Harbor Laboratory Press

Various, 1982, 'Trends in molecular neuroscience', *Trends in Neurosciences*, **9**, 295–322

3. Outline of Nervous Systems

(a) Invertebrate nervous systems

Bargmann, C.I., 1993, 'Genetic and cellular analysis of behaviour in *C. Elegans*', *Annual Review of Neuroscience*, **16**, 47–71

Bullock, T.H. and G.A. Horridge, 1965, *Structure and Function in the Nervous Systems of Invertebrates* (2 vols), San Francisco: Freeman

Corbo, J.C. and A. Di Gregorio, 2001, 'The Ascidian as a model organism in developmental and evolutionary biology', *Cell*, **106**, 535–538

Selverston, A.I., 1988, 'A consideration of invertebrate central pattern generators as computational data bases', *Neural Networks*, **1**, 109–117

Usherwood, P.N.R. and D.R. Newth, 1975, *Simple Nervous Systems*, London: Edward Arnold

Wiersma, C.A.G., ed., 1967, *Invertebrate Nervous Systems*, Chicago: University of Chicago Press

Wood, W., 1988, *The Nematode Caenorhabditis elegans*, Cold Spring Harbor, NY: Cold Spring Harbor Laboratory Press

(b) Vertebrate nervous systems

*Kalat, J.W., 1988, *Biological Psychology* (3rd edn), Belmont, CA: Wadsworth Publishing Co.

Kandel, E.R., J.H. Schwarz and T.M. Jessell, 1991, *Principles of Neural Science* (3rd edn), New York: Elsevier Publishing

*Nauta, W.J.H. and M. Feirtag, 1986, *Fundamentals of Neuroanatomy*, New York: Freeman

*Patten, B.M. and B.M. Carlson, 1974, *Foundations of Embryology*, New York: McGraw Hill

Shepherd, G.M., 1994, *Neurobiology* (3rd edn), Oxford: Oxford University Press

Thompson, R.F., 1993, *The Brain: A Neuroscience Primer*, New York: Freeman

*Warwick, R. and P.L. Williams, eds, 1973, *Gray's Anatomy* (35th edn), London: Longman

4. Cells of the Nervous System

Abbott, N.J., ed., 1991, 'Glial–neuronal interaction', *Annals of New York Academy of Sciences*, **633**, 1–636

Mauch, D.H. *et al.*, 2001, 'CNS synaptogenesis promoted by glia-derived cholesterol', *Science*, **294**, 1354–1357

Schmitt, F.O., F.G. Worden, G. Adelman and S.G. Dennis, eds, 1981, *The Organisation of the Cerebral Cortex*, Cambridge, MA: MIT Press

Shepherd, G.M., 1990, *The Synaptic Organisation of the Brain* (3rd edn), Oxford: Oxford University Press

Steindler, D.A., 1993, 'Glial boundaries in the developing nervous system', *Annual Review of Neurscience*, **16**, 445–470

Szentágothai, J., 1978, 'The neuron network of the cerebral cortex: a functional interpretation', *Proceedings of the Royal Society B*, **201**, 219–248

Travis, J., 1994, 'Glia: the brain's other cells', *Science*, **266**, 970–972

*Vernadakis, A., 1988, 'Neuron–glia interrelations', *Annual Review of Neurobiology*, **30**, 149–224

Watson, W.E., 1976, *The Cell Biology of Brain*, London: Chapman and Hall

5. Organisation of neurons in the brain

Braitenberg, V., 1977, *On the Texture of Brains*, Heidelberg: Springer-Verlag

Freeman, W.J., 1975, *Mass Action in the Nervous System*, New York: Academic Press

Leise, E.M., 1990, 'Modular construction of nervous systems: a basic principle of design for invertebrates and vertebrates', *Brain Research Reviews*, **15**, 1–23

Lorente de No, R., 1938, 'The cerebral cortex: architecture, intracortical connections and motor projections' in *Physiology of the Nervous System*, ed. J. Fulton, Oxford: Oxford University Press

*Pappas, G.D. and S.G. Waxman, 1972, 'Synaptic structure – morphological correlates of chemical and electrotonic transmission' in *Structure and Function of Synapses*, ed. G.D. Pappas and D.P. Purpura, Amsterdam: North Holland

*Poritsky, R., 1969, 'Two and three dimensional ultrastructure of boutons and glial cells on the motoneuronal surface of cat spinal cord', *Journal of Comparative Neurology*, **135**, 423–452

Ramon y Cajal, S., 1911, *Histologie de Système Nerveux de l'Homme et des Vertèbres*, Paris: Maloine; ed. and trs. P. Pasik and T. Pasik, 1999, 2000, '*Texture of the Nervous System in Man and Vertebrates*', 2 vols., Berlin: Springer

Smith, C.U.M., 1992, 'A century of cortical architectonics', *Journal of the History of the Neurosciences*, **1**, 201–218

Smith, C.U.M., 2000, 'Brodmann's cortical areas' in P.J. Koehler, G.W. Bruyn and J.M.S. Pearce, eds, *Neurological Eponyms*, Oxford: Oxford University Press

*Szentágothai, J., 1979, 'Local neuron circuits of the neocortex' in *The Neurosciences: Fourth Study Program*, ed. F.O. Schmitt and F.G. Worden, Cambridge, MA: MIT-Press, 399–415

6. Laboratory techniques

Miller, C., ed., 1986, *Ion Channel Reconstitution*, New York: Plenum Press

Pritchard, R.H. and I.B. Holland, 1985, *Basic Cloning Techniques*, Oxford: Blackwell

Rodriguez, R.L. and R.C. Tait, 1983, *Recombinant DNA Techniques: An Introduction*, Reading, MA: Addison-Wesley

Sakmann, B. and E. Neher, 1983, eds, *Single Channel Recording*, New York: Plenum Press

Turner, A.J. and H.S. Bachelard, 1987, *Neurochemistry: A Practical Approach*, Oxford: IRL Press

CHAPTER 2: THE CONFORMATION OF INFORMATIONAL MACROMOLECULES

Web sites

Nucleic acid sequences

www.ncbi.nlm.nih.gov/
www.ebi.ac.uk/ebi_docs/embl_db/ebi/topembl.html

Proteins and polypeptides

Primary structure (amino acid sequences):
www.expasy.ch/sprot/
www-nbrf.georgetown.edu/pir/
Secondary and higher structures:
www.rcsb.org
Hydrophobicity profiles:
http://bioinformatics.weizmann.ac.il/hydroph
Classification of protein structures:
http://scop.mrc-lmb.cam.ac.uk/scop

Print

1. General

Alberts, B., D. Bray, J. Lewis, M. Raff, K. Roberts and J.D. Watson, 1994, *MolecularBiology of the Cell* (3rd edn), New York: Garland Publishing

Avers, C.J., 1986, *Molecular Cell Biology*, Reading, MA: Addison-Wesley

Becker, W.M., 1991, *The World of the Cell* (2nd edn), Menlo Park, CA: Benjamin/Cummings Publishing

*Darnell, J., H. Lodish and D. Baltimore, 1990, *Molecular Cell Biology* (2nd edn), New York: Scientific American Books

*Rees, A.R. and M.J.E. Sternberg, 1984, *From Cells to Atoms*, Oxford: Blackwell Scientific Publications

2. Proteins

Branden, C. and J. Tooze, 1991, *Introduction to Protein Structure*, New York: Garland

Cohen, C. and D.A.D. Parry, 1986, 'Interlocking α-helices, related to the coiled-coil structure, are a common stabilising motif in proteins of all types', *Trends in Biochemical Sciences*, **11**, 245–248

*Dickerson, R.E. and I. Geis, 1969, *The Structure and Action of Proteins*, Menlo Park, CA: Benjamin/Cummings Publishing

Ellis, R.J. and S.M. van der Vies, 1991, 'Molecular chaperones', *Annual Review of Biochemistry*, **60**, 321–347

Ellis, R.J., R.A. Laskey and G.H. Lorimer, 1993, 'Molecular chaperones', *Philosophical Transactions of the Royal Society (Biological Sciences)*, **339**, 255–373

Hokfelt, T., 1991, 'Neuropeptides in perspective: the last ten years', *Neuron*, **7**, 867–879

Kay, L.E., 1997, 'NMR methods for the study of protein structure and dynamics', *Biochemistry and Cell Biology*, **75**, 1–15

Lesk, A.M., 1991, *Protein Architecture*, Oxford: IRL Press

Lesk, A.M., 2001, *Introduction to Protein Architecture: The Structural Biology of Proteins*, Oxford: Oxford University Press

Ranson, N.A., H.E. White and H.R. Saibil, 1998, 'Chaperonins', *Biochemical Journal*, **333**, 233–242

Sali, A., 1998, '100 000 protein structures for the biologist', *Nature: Structural Biology*, **5**, 1029–1032

Swindells, M.B. *et al.*, 1998. 'Contemporary approaches to protein structure classification', *BioEssays*, **20**, 884–891

Taylor, P., 1991, 'The cholinesterases', *Journal of Biological Chemistry*, **266**, 4025–4028

*Taylor, P. and Z. Radic, 1994, 'The cholinesterases: from genes to proteins', *Annual Review of Pharmacology and Toxicology*, **34**, 281–320

3. Nucleic acids

Avery, O.T, C.M. McLeod and M. McCarty, 1944, 'Studies on the chemical nature of the substance inducing transformation of pneumococcal types', *Journal of Experimental Medicine*, **79**, 137–158

Rich, A. and S.H. Kim, 1978, 'The three-dimensional structure of transfer RNA', *Scientific American*, **238**(1), 53–62

Watson, J.D. and F.H.C. Crick, 1953, 'Molecular structure of nucleic acids. A structure for deoxyribose nucleic acid', *Nature*, **171**, 737–738

Watson, J.D. and F.H.C. Crick, 1953, 'Genetical implications of the structure of deoxyribonucleic acid', *Nature*, **171**, 964–967

Watson, J.D. *et al.*, 1987, *Molecular Biology of the Gene* (3rd edn), Menlo Park, CA: Benjamin Cummings

4. Informatics

Bishop, M.J., ed., 1998, *Human Genome Computing*, New York: Academic Press

Kreil, D.P., 1999, 'DATABANKS – a catalogue data base of molecular biology data banks', *Trends in Biochemical Science*, **24**, 155–157

CHAPTER 3: INFORMATION PROCESSING IN CELLS

Web sites

Replication

A detailed account of the replication fork (including movies) can be found at:
http//chem-mgriep2.unl.edu/replic/fork.html

Transcription factors

classification: http://www.bioscience.org/tf.html

Ribosomes

http://ntri.tamuk.edu/ribosomes.html/
http://www.sciam.com/explorations/1999/092799ribo/

Roslin Institute: cloning

http://www.ri.bbsrc.ac.uk/library/research/cloning/

Print

1. General

Griffiths, A.J.F., 1998, *Modern Genetic Analysis*, New York: W.H. Freeman

Lewin, B., 2000, *Genes VII*, Oxford: Oxford University Press

Watson, J.D., N.H. Hopkins, J.W. Roberts, J.A. Steitz and A.M. Weiner, 1987, *Molecular Biology of the Gene* (4th edn), Menlo Park, CA: Benjamin/Cummings

2. The genetic code

Crick, F.H.C., 1966, 'The genetic code', *Scientific American*, **215**(4), 55–62

Dickerson, R.E., 1983, 'The DNA helix and how it is read', *Scientific American*, **249**(6), 87–102

Ycas, M., 1969, *The Biological Code*, New York: John Wiley

3. Replication

Watson, J.D. *et al.*, 1987, *Molecular Biology of the Gene* (4th edn), Menlo Park, CA: Benjamin/Cummings

4. 'DNA makes RNA and RNA makes protein'

Brown, D.D., 1981, 'Gene expression in eukaryotes', *Science*, **211**, 667–674

Darnell, J.E., 1982, 'Variety in the level of gene control in eukaryotic cells', *Nature*, **297**, 365–371

5. Transcription

Harrison, S.C., 1991, 'A structural taxonomy of DNA-binding domains', *Nature*, **353**, 715–719

Locker, J., ed., 1996, *Transcription Factors: Essential Data*, Chichester: John Wiley

*Wolberger, C., 1999, 'Multiprotein-DNA complexes in transcriptional regulation', *Annual Review of Biophysics and Biomolecular Structure*, **28**, 29–56

6. Post-transcriptional processing

Abei, M. and C. Weissmann, 1987, 'Precision and orderliness in splicing', *Trends in Genetics*, **3**, 102–107

Pabo, C.O. and R.T. Sauer, 1992, 'Transcription factors: structural families and principles of DNA recognition', *Annual Review of Biochemistry*, **61**, 1053–1095

Roeder, R.G., 1991, 'The complexities of eukaryotic transcription initiation: regulation of pre-initiation complex assembly', *Trends in Biochemical Science*, **16**, 402–408

Sommer, B. *et al.*, 1991, 'RNA-editing in the brain controls a determinant of ion flow in glutamate gated channels', *Cell*, **67**, 11–18

7. Translation

Lake, J.A., 1981, 'The ribosome', *Scientific American*, **245**(2), 56–67

Moldave, K., 1985, 'Eukaryotic protein synthesis', *Annual Review of Biochemistry*, **54**, 1109–1150

Rould, M.A. *et al.*, 1989, 'Structure of *E. coli* glutaminyl-tRNA synthetase complexed with $tRNA_{gln}$ and ATP at 2.8 Å resolution', *Science*, **246**, 1135–1142

Schimmel, P., 1987, 'Aminoacyl tRNA synthetases: general scheme of structure-function relationships in the polypeptides and recognition of transfer RNAs', *Annual Review of Biochemistry*, **56**, 125–158

Waldorf, M.M., 1989, 'The structure of the "second genetic code"', *Science*, **246**, 1122

8. Control of the expression of genetic information

Beckwith, J.J., J. Davies and J. Gallant, eds, 1983, *Gene Functions in Prokaryotes*, Cold Spring Harbor, NY: Cold Spring Harbor Laboratory Press

Brown, D.D., 1981, 'Gene expression in eukaryotes', *Science*, **211**, 667–674

Curran, T. *et al.*, 1988, 'Beyond the Second Messenger: oncogenes and transcription factors', *Cold Spring Harbor Symposia on Quantitative Biology*, **LIII**, 769–777

Hershmann, H.R., 1989, 'Extracellular signals, transcriptional responses and cellular specificity', *Trends in Biochemical Sciences*, **14**, 455–458

Hudson, L. *et al.*, 1987, 'Aberrant splicing of proteolipid protein mRNA in the dysmyelinating jimpy mutant mouse', *Proceedings of the National Academy of Sciences USA*, **84**, 1454–1458

*Karpati, G., 1984, 'Three tachykinins in the human brain', *Trends in Neurosciences*, **7**, 57–59

Lynch, D.R. and S.H. Snyder, 1986, 'Neuropeptides: multiple molecular forms, metabolic pathways and receptors', *Annual Review of Biochemistry*, **55**, 773–799

*Nawa, H., K. Kotani and S. Nakashani, 1984, 'Tissue-specific generation of two preprotachykinin mRNAs from one gene by alternative RNA splicing', *Nature*, **312**, 729–734

Ptashne, M., 1984, 'Repressors', *Trends in Biochemical Sciences*, **9**, 142–145

Stark, G.R., 1984, 'Gene amplification', *Annual Review of Biochemistry*, **53**, 447–491

Sutcliffe, J.G. *et al.*, 1984, 'Control of neuronal gene expression', *Science*, **225**, 1308–1315

Tijan, R., 1995, 'Molecular machines that control genes', *Scientific American* **272**(2), 38–45

Box 3.1: Antisense and triplex oligonucleotides

Cohen, J.S. and M.E. Hogan, 1994, 'The new genetic medicine', *Scientific American*, **271**(12), 50–55

Eguchi, Y., T. Itoh and J. Tomizawa, 1991, 'Antisense RNA', *Annual Review of Biochemistry*, **60**, 631–652

CHAPTER 4: MOLECULAR EVOLUTION

Web sites

Valuable (though advanced) tutorials on protein sequence comparison and protein evolution are at:
http://www.people.virginia.edu/~wrp/papers/ismb2000.pdf
http://www.people.virginia.edu/~wrp/prot_talk12-95.html

Print

1. General

Arbas, E.A., I.A. Meinertzhagen and S.R. Shaw, 1991, 'Evolution in nervous systems', *Annual Review of Neuroscience*, **14**, 9–38

Brockerhoff, S.E., J.E. Dowling and J.B. Hurley, 1998, 'Zebrafish retinal mutants', *Vision Research*, **38**, 1335–1339

Driscoll, M. and M. Chalfie, 1991, 'The mec-4 gene is a member of a family of *Caenorhabditis elegans* genes that can mutate to induce neuronal degeneration', *Nature*, **349**, 588–593

Jacob, F., 1977, 'Evolution and tinkering', *Science*, **196**, 1161–1166

Lesk, A.M., 2001, *Introduction to Protein Architecture*, Oxford: Oxford University Press.

Li, W-H, and D. Graur, 1991, *Fundamentals of Molecular Evolution*, San Diego: Sinauer

Patthy, L., 1999, *Protein Evolution*, Oxford: Blackwell

Watson, J.D. *et al.*, 1988, *Molecular Biology of the Gene* (4th edn), vol. 2, Menlo Park, CA: Benjamin/Cummings Publishing

2. Mutation

Jeffreys, A.J. and S. Harris, 1982, 'Processes of gene duplication', *Nature*, **296**, 9–10

Mitas, M., 1997, 'Trinucleotide repeats associated with human disease', *Nucleic Acids Research*, **24**, 2245–2253

Patthy, L., 1999, *Protein Evolution*, Oxford: Blackwell

3. Transposons

Berg, D.E. and M. Howe, 1989, *Mobile DNA*, Washington DC: American Society of Microbiology

McClintock, B., 1984, 'The significance of responses of the genome to challenge', *Science*, **226**, 792

Cold Spring Harbor Symposium, vol. XLV, 1980, *Movable Genetic Elements*, Cold Spring Harbor, NY: Cold Spring Harbor Laboratory Press.

Starlinger, P., 1984, 'Transposable elements', *Trends in Biochemical Sciences*, **9**, 125–127

4. Protein Evolution

*Avers, C.J., 1986, *Molecular Cell Biology*, Reading, MA: Addison-Wesley

Deschenes, R.J. *et al.*, 1984, 'Cloning and sequence analysis of a cDNA encoding rat preprocholecystokinin', *Proceedings of the National Academy of Sciences USA*, **81**, 726–730

Doolittle, W.F., 1985, 'The genealogy of some recently evolved vertebrate proteins', *Trends in Biochemical Sciences*, **10**, 233–237

*Doolittle, W.F., 1985, 'Protein evolution' in *The Proteins*, ed. H. Neurath and R.L. Hills, 3rd edn, vol. 4, New York: Academic Press, 1–118

*Douglass, J., O. Civelli and E. Herbert, 1984, 'Polyprotein gene expression', *Annual Review of Biochemistry*, **53**, 665–715

Hanke, W. and H. Breer, 1986, 'Channel properties of an insect neuronal acetylcholine receptor protein reconstituted in planar lipid bilayers', *Nature*, **321**, 171–174

Herbert, E. *et al.*, 1983, 'Generation of diversity of evolution of opioid peptides', *Cold Spring Harbor Symposium on Quantitative Biology*, **XLVLLL**, 375–384

Lake, J.A. and J.E. Moore, 1998, 'Phylogenetic analysis and comparative genomics', *Trends in Bioinformatics*, Cambridge: Elsevier, 22–24

Lunt, G.G., 1986, 'Is the insect neuronal nAChR the ancestral ACh receptor protein?', *Trends in Neurosciences*, **9**, 341–342

*Lynch. D.R. and S.H. Snyder, 1986, 'Neuropeptides: multiple molecular forms, metabolic pathways and receptors', *Advances in Biochemistry*, **55**, 773–799

Patthy, L., 1999, *Protein Evolution*, Oxford: Blackwell

Philippe, H. and J. Laurent, 1998, 'How good are deep phylogenetic trees?', *Current Opinion in Genetics and Development*, **8**, 616–623

CHAPTER 5: MANIPULATING BIOMOLECULES

Web sites

DNA sequence analysis

http://www.swbic.org/links/1.4.php

Gene chips

http://www.gene-chips.com/
http://cmgm.stanford.edu/pbrown/yeastchip.html

Print

Abelson, J. and E. Butz, eds, 1980, 'Recombinant DNA', *Science*, **209**, 1317–1338

Brand, A.H. and N. Perrimon, 1993, 'Targetted gene expression as a means of altering cell fates and generating dominant phenotypes', *Development*, **118**, 401–415

*Capecchi, M.R., 1989, 'Altering the genome by homologous recombination', *Science*, **244**, 1288–1292

Capecchi, M.R., 1994, 'Targeted gene replacement', *Scientific American*, **270**(3), 34–41

Emery, A.E., 1984, *An Introduction to Recombinant DNA*, Chichester: John Wiley

Griffiths, A.J.F., 1999, *Modern Genetic Analysis*, New York: W.H. Freeman

Innis, M.A. *et al.*, 1990, *PCR Protocols*, San Diego: Academic Press

Kaiser, K., 1993, 'Second generation enhancer traps', *Current Biology*, **3**, 561–562

Lewin, B., 2000, *Genes VII*, Oxford: Oxford University Press

Maniatis, T., E.F. Fritsch and J. Sambrook, 1982, *Molecular Cloning*, Cold Spring Harbor, NY: Cold Spring Harbor Laboratory Press

Marshall, A. and J. Hodgson, 1998, 'DNA-chips: An array of possibilities', *Nature Biotechnology*, **16**, 27–31

Mirnics, K. *et al.*, 2001, 'Analysis of complex brain disorders with gene expression microarrays: schizophrenia as a disease of the synapse', *Trends in Neuroscience*, **24**, 479–486

Mullis, K.B. *et al.*, 1986, 'Specific enzymatic amplification of DNA in vitro: The polymerase chain reaction' in *Cold Spring Harbor Symposia on Quantitative Biology*, **LI**, 263–273

Mullis, K.B., 1990, 'The unusual origin of the polymerase chain reaction', *Scientific American*, **262**(4), 36–43

Old, R.W. and S.B. Primrose, 1994, *Principles of Gene Manipulation* (5th edn), Oxford: Blackwell

Ptashne, M., 1986, *A Genetic Switch: Phage k and Higher Organisms* (2nd edn), Cambridge, MA and Oxford: Cell Press and Blackwell Scientific Publications

Ramsay, G. 1998, 'DNA chips: State of the art', *Nature Biotechnology*, **16**, 40–44

Smith, M., 1982, 'Site-directed mutagenesis', *Trends in Biochemical Sciences*, **7**, 440–442

Uhl, G.R., ed., 1987, *In situ Hybridisation in Brain*, New York: Plenum Press

Watson, J.D., M. Gilman, J. Witkowsi and M. Zoller, 1992, *Recombinant DNA*, New York: Scientific American Books/ W.H. Freeman

Watson, J.D., N.H. Hopkins, J.W. Roberts, J.A. Steitz and A.M. Weiner, 1987, *Molecular Biology of the Gene*, Menlo Park, CA: Academic Press

White, M.M., 1985, 'Designer channels: site-directed mutagenesis as a probe for structural features of channels and receptors', *Trends in Neurosciences*, **8**, 364–368

Williamson, R., ed., 1985, 'Techniques of genetic engineering' (Videotape set) Oxford: IRL Press

Yellen, G., 1984, 'Channels from genes: the oocyte as an expression system', *Trends in Neurosciences*, **7**, 457–458

*Zimmer, A., 1992, 'Manipulating the genome by homologous recombination in embryonic stem cells', *Annual Review of Neuroscience*, **15**, 115–137

CHAPTER 6: GENOMES

Web sites

Genomes

(a) Human:
http://www.ornl.gov/hgmis/
http://www.nature.com
(b) *Caenorhabditis elegans*:
http://www.wormbase.org/
(c) *Danio rerio* (zebra fish):
http://www.sanger.ac.uk/Projects/D_rerio/
(d) *Drosophila*:
http://fly.ebi.ac.uk
(e) *Mus musculus* (mouse):
http://www.ncbi.nlm.nih.gov/genome/guide/mouse
(f) *Xenopus*: (clawed frog):
http://vize222.zo.utexas/edu/
(g) *Saccharomyces cerevisiae* (yeast):
http://genome-www.stanford.edu/Saccharomyces/
(h) *Oryza* (rice):
http://nucleus.cshl.org/riceweb/

Other information

Human genes and disease:

http://www.ncbi.nlm.nih.gov/omim
http://www.biol.tsukba.ac.jp/~macer/index.html
Social, legal and ethical issues:
http://www.nhgri.nih.gov/ELSI;
http://www.ornl.gov/hgmis/elsi.elsi.html

Print

Adams, M.D. *et al.*, 2000, 'The genome sequence of *Drosophila melanogaster*', *Science*, **287**, 2185–2195

Baltimore, D., 2001, 'Our genome unveiled', *Nature*, **409**, 814–816

Corbo, J.C. and A. Di Gregorio, 2001, 'The ascidian as a model organism in developmental and evolutionary biology', *Cell*, **106**, 535–538

Dickson, D.W., 2001, 'α-Synuclein and the Lewy body disorders', *Current Opinion in Neurology*, **14**, 423–432

Gallagher, R. and C. Dennis, 2001, 'Unveiling your genome', *Wellcome News Supplement*, **Q1**, 13–23

International Human Genome Sequencing Consortium, 2001, 'The initial sequencing and analysis of the human genome', *Nature*, **409**, 860–921

International SNP Map Working Group, 2001, 'A map of the human genome containing 1.42 million single nucleotide polymorphisms', *Nature*, **409**, 928–933

Mirnics, K. *et al.*, 2000, 'Molecular characterisation of schizophrenia viewed by microarray analysis of gene expression in the prefrontal cortex', *Neuron*, **28**, 53–67

O'Brien, S.J., E. Ezirik and W.J. Murphy, 2001, 'On choosing mammalian genomes for sequencing', *Science*, **292**, 2264–2266

Patthy, L., 1999, *Protein Evolution*, Oxford: Blackwell

Rose, M.R., 2001, *Channelopathies of the Nervous System*, London: Butterworth-Heinemann

Sternberg, P.W., 2001, 'Working in the post-genomic *C. elegans* world', *Cell*, **105**, 173–176

Tamames, J. *et al.*, 1996, 'Genomes with distinct function composition' *FEBS Letters*, **389**, 96–101

Venter, J.C. *et al.*, 2001, 'The sequence of the human genome', *Science*, **291**, 1304–1351

CHAPTER 7: BIOMEMBRANES

Web sites

Charcot–Marie–Tooth (CMT) disease

http://www.mdausa.org/disease/cmt.html

Myelin

Myelin and Schwann cell components:
http://www.neuro.wustl.edu/neuromuscular/lab/schcell.html

Myelin disorders

http://neuro-www.mgh.harvard.edu/forum/
MyelinDisordersMenu.html
http://www.familyvillage.wisc.edu/lib_myel.htm

Tight junctions

Good pictures at:
http://www.ultranet.com/~jkimball/BiologyPages/J/
Junctions.html

Print

1. General

Alberts, B. *et al.*, 1994, *The Molecular Biology of the Cell* (3rd edn), New York: Garland Publishing

Cold Spring Harbor Symposia on Quantitative Biology, LVII, 1992, *The Cell Surface*, Cold Spring Harbor, NY: Cold Spring Harbor Laboratory Press

*Darnell, J., H. Lodish and D. Baltimore, 1986, *Molecular Cell Biology*, New York: Scientific American Books

Karp, G., 1984, *Cell Biology* (2nd edn), New York: McGraw Hill

Lee, A., 2001, 'Membrane structure', *Current Biology*, **11**, R811–R814

Robertson, R.N., 1983, *The Lively Membranes*, Cambridge: Cambridge University Press

Singer, S.J. and G.L. Nicolson, 1972, 'The fluid-mosaic model of the structure of cell membranes', *Science*, **175**, 720–731

2. Membrane fluidity

Nagle, J.F. and S. Tristam-Noble, 2000, 'Structure of lipid bilayers', *Biochimica et Biophysica Acta*, **1469**, 159–195

Quinn, A.J. and D. Chapman, 1980, 'The dynamics of membrane structure', *CRC Critical Reviews of Biochemistry*, **8**, 1–117

3. Membrane asymmetry

Rothman, J. and J. Lenard, 1977, 'Membrane asymmetry', *Science*, **195**, 743–753

4. Membrane proteins

Almers, W. and C. Stirling, 1984, 'Distribution of transport proteins over animal cell membranes', *Journal of Membrane Biology*, **77**, 169–186

Carruthers, A. and D.L. Melchior, 1986, 'How bilayer lipids affect membrane protein activity', *Trends in Biochemical Sciences*, **11**, 331–335

Jay, D. and L. Cantley, 1986, 'Structural aspects of red cell anion exchange protein', *Annual Review of Biochemistry*, **55**, 511–538

White, S.H. and W.C. Wimley, 1999, 'Membrane protein folding and stability: physical principles', *Annual Review of Biophysical and Biomolecular Structure*, **28**, 319–365

5. Mobility of membrane proteins

Brown, D. and E. London, 1998, 'Functions of lipid rafts in biological membranes', *Annual Review of Cell and Developmental Biology*, **14**, 111–136

Galbiati, F., B. Razani and M.P. Lisanti, 2001, 'Emerging themes in lipid rafts and caveolae', *Cell*, **106**, 403–411

Jacobson, K., A. Ishihara and R. Inman, 1987, 'Lateral diffusion of proteins in membranes', *Annual Review of Physiology*, **49**, 163–175

Vaz, W., F. Goodsaid-Zaldvondo and K. Jacobson, 1984, 'Lateral diffusion of lipids and proteins in bilayer membranes', *FEBS Letters*, **174**, 199–207

6. Synthesis of biomembranes

Besharse, J.C., 1986, 'Photosensitive membrane turnover: differentiated membrane domains and cell-cell interaction' in *The Retina: A Model for Cell Biology Studies*, eds R. Adler and D. Farber, Orlando, FL: Academic Press

Meer, G. van, 1993, 'Transport and sorting of membrane lipids', *Current Opinion in Cell Biology*, **5**, 661–674

7. Myelin and myelination

Lauren, R.A., M. Samiullah and M.B. Lees, 1984, 'The structure of bovine brain myelin proteolipid and its organisation in myelin', *Proceedings of the National Academy of Sciences USA*, **81**, 2912–2916

Lemke, G., 1992, 'Myelin and myelination' in Z. Hall, ed., *An Introduction to Molecular Biology*, Sunderland, MA: Sinauer

Mikoshiba, K. *et al.*, 1991, 'Structure and function of myelin protein genes', *Annual Review of Neuroscience*, **14**, 201–217

Newman, S., K. Kitamura and A.T. Campagnoni, 1987, 'Identification of cDNA coding for a fifth form of myelin basic protein in mouse', *Proceedings of the National Academy of Sciences USA*, **84**, 886–890

Takahashi, N. *et al.*, 1985, 'Cloning and characterisation of the myelin basic protein gene from mouse: one gene can encode both 14 kd and 18.5 kd MBPs by alternate use of exons', *Cell*, **42**, 138–148

8. The submembranous cytoskeleton

Bennett, V., 1985, 'The membrane skeleton of human erythrocytes and its implication for more complex cells', *Annual Review of Biochemistry*, **54**, 273–304

Bennett, V. and S. Lambert, 1991, 'The spectrin skeleton: from red cells to brain', *Journal of Clinical Investigation*, **87**, 1483–1489

Chasis, J.A. *et al.*, 1987, 'Red cell biochemical anatomy and membrane proteins', *Annual Review of Physiology*, **49**, 237–248

Fulton, A.B., 1984, *The Cytoskeleton*, London: Chapman and Hall

Mercer, R.W., 1993, 'Structure of the Na,K-ATPase', *International Review of Cytology*, **137C**, 139–168

Neher, E. and B. Sakmann, 1992, 'The patch clamp technique', *Scientific American*, **266**(3), 28–35

Pumplin, D.W. and R.J. Bloch, 1993, 'The membrane skeleton', *Trends in Cell Biology*, **3**, 113–117

9. Junctions between cells

Abbott, N.J., 1985. 'Are glial cells excitable after all?', *Trends in Neurosciences*, **8**, 141–142

Anon, 1986, 'Structure–function relationships in vertebrate retina – an introduction', *Trends in Neurosciences*, **9**, 214

Bennett, M.V.L., 1991, 'Gap junctions: new tools, new answers, new questions', *Neuron*, **6**, 305–320

Galaretta, M. and S. Hestrin, 1999, 'A network of fast-spiking cells in the neocortex connected by electrical synapses', *Nature*, **402**, 72–75

Gibson, J.R., M. Beierlein and B.W. Connors, 1999, 'Two networks of electrically coupled neurons in the cortex', *Nature*, **402**, 75–79

*Gold, G.H. and J.E. Dowling, 1979, 'Photoreceptor coupling in the retina of the Toad', *Bufo marinus*, 1, 'Anatomy', *Journal of Neurophysiology*, **42**, 292–310

Kandler, K. and L.C. Katz, 1995, 'Neuronal coupling and uncoupling in the developing brain', *Current Opinion in Neurobiology*, **5**, 98–105

Kumar, N.M. and N.B. Gilula, 1986, 'Cloning and characterisation of human and rat liver cDNAs coding for gap-junction protein', *Journal of Cell Biology*, **103**, 767–776

Marc, R.E., 1986, 'The development of retinal networks' in *The Retina: A Model for Cell Biology Studies*, eds R. Adler and D. Farber, Orlando, FL: Academic Press

Spray, D.C., C. Naus and A.C. Campos de Carvalho, eds, 2000, 'Gap junctions in the nervous system', *Brain Research Reviews*, **32**, 1–341

Stevenson, B.R. and B.H. Keon, 1998, 'The tight junctions: Morphology to molecules', *Annual Reviews of Cell Developmental Biology*, **14**, 89–109

Tsukita, S., M. Furuse and M. Itoh, 2001, 'Multifunctional strands in tight junctions', *Nature Reviews Molecular Cell Biology*, **2**, 285–293

Werner, R., ed., 1998, *Gap Junctions*, Amsterdam: IOS Press

Yang, Q and H.B. Michelson, 2001, 'Gap junctions synchronise the firing of inhibitory interneurons in guinea pig hippocampus', *Brain Research*, **907**, 139–143

10. Gap junctions and neuropathology

Fancis, P.I. *et al.*, 1999, 'Lens biology: development and human cataractogenesis', *Trends in Genetics*, **15**, 191–196

Kelsell, D.P. *et al.*, 1997, 'Connexin 26 mutations in hereditary non-syndromic deafness', *Nature*, **387**, 80–83

Kikuchi, T. *et al.*, 2000, 'Gap junction systems in mammalian cochlea', *Brain Research Reviews*, **32**, 163–166

Nance, W.E., X.-Z. Liu and A. Pandya, 2000, 'Relation between choice of partner and high frequency connexin-26 deafness', *Lancet*, **356**, 500–501

Rozental, R. and A.C. Campos de Carvalho, 2000, 'Nervous system diseases involving gap junctions', *Brain Research Reviews*, **32**, 189–191

Spray, D.C. and R. Dermietzel, 1995, 'X-linked dominant Charcot–Marie–Tooth disease and other potential gap-junction diseases of the nervous system', *Trends in Neurosciences*, **18**, 256–262

White, T.W. and D.L. Paul, 1999, 'Genetic diseases and gene knock-outs reveal diverse connexin functions', *Annual Review of Physiology*, **61**, 283–310

CHAPTER 8: G-PROTEIN-COUPLED RECEPTORS

Web sites

Valuable account of the discovery and biochemistry of G-proteins may be found at the Nobel foundation web site:
http://www.nobel.se/medicine/laureates/1994/illpres/
Well-illustrated (including movie) account of GTPases at:
http://www.mpi-dortmund.mpg.de/departments/dep1/gtpase/gtpase_detail.php3

Detailed information about the phylogenetic relationships of the G-protein coupled receptors is available at the database managed by the GPCRDB partners at:
http://www.gpcr.org/7tm/
Excellent structural and functional detail of bacteriorhodopsin can be found at:
http://an×12.bio.uci.edu/~hudel/br/

Print

1. General

Kruk, Z.L. and C.J. Pycock, 1991, *Neurotransmitters and Drugs* (3rd edn), London: Chapman and Hall

*Smith, C.U.M., 1995, 'Membrane signalling systems' in *Biomembranes*, vol. 1: *General Principles*, ed. A.G. Lee, Geenwich CT: JAI Press, 245–270

Strange, P.G., 1988, 'The structure and mechanism of neuro-transmitter receptors', *Biochemical Journal*, **249**, 309–318

Watson, S. and D. Gridlestone, 1995, 'Receptor and ion channel nomenclature supplement' (6th edn), *Trends in Pharmacological Sciences*, Cambridge: Elsevier Science

2. 7TM receptors

Dohlman, H.G. *et al.*, 1991, 'Model systems for the study of seven-transmembrane segment receptors', *Annual Review of Biochemistry*, **60**, 653–688

Fasman, G.D. and W.A. Gilbert, 1990, 'The prediction of transmembrane sequences and their conformation: an evaluation', *Trends in Biochemical Science*, **15**, 89–92

Findlay, T.B.C. *et al.*, 1993, 'Structure of the G-protein-linked receptors', *Biochemical Society Transactions*, **21**, 869–873

Iismaa, T.P. and J. Shine, 1992, 'G-protein-coupled receptors', *Current Opinion in Cell Biology*, **4**, 195–202

Strader, C.D. *et al.*, 1994, 'Structure and function of G-protein-coupled receptors', *Annual Review of Biochemistry*, **63**, 101–132

3. G-proteins and GTPases

Hall, A., 1993, 'Ras-related proteins', *Current Opinion in Cell Biology*, **5**, 265–268

Kaziro, H. *et al.*, 1991, 'Structure and function of signal transducing GTP-binding proteins', *Annual Review of Biochemistry*, **60**, 349–400

4. G-protein collision-coupling systems

Dunlap, K., G.G. Holz and S.G. Rane, 1987, 'G-proteins as regulators of ion channel function', *Trends in Neurosciences*, **10**, 241–244

Gilman, A.G., 1987, 'G-proteins: transducers of receptor-generated signals', *Annual Review of Biochemistry*, **56**, 615–649

Neer, E.J. and D.E. Clapham, 1988, 'Roles of G-protein subunits in transmembrane signalling', *Nature*, **333**, 129–134

Spiegel, A.M., 1992, 'G proteins in cellular control', *Current Opinion in Cell Biology*, **4**, 203–211

5. Effector molecules

Choi, E.J. *et al.*, 1993, 'The regulatory diversity of the mammalian adenylyl cyclases', *Current Opinion in Cell Biology*, **5**, 269–273

Tang, W.-J. *et al.*, 1992, 'Regulation of mammalian adenylyl cyclases by G-protein alpha and beta gamma subunits' in *Cold Spring Harbor Symposia on Quantitative Biology, LVII*, '*The Cell Surface*', Cold Spring Harbor, NY: Cold Spring Harbor Laboratory Press, 135–144

Worley, P.F. *et al.*, 1988, 'Lithium blocks a phosphoinositide-mediated cholinergic response in hippocampal slices', *Science*, **239**, 1428–1429

6. Networks of G-protein signalling systems

Birnbaumer, L., J. Abramowitz and A.M. Brown, 1990, 'Receptor-effector coupling by G-proteins', *Biochimica Biophysica Acta*, **1031**, 163–224

Kleuss, C. *et al.*, 1992, 'Different β-subunits determine G-protein interaction with transmembrane receptors', *Nature*, **358**, 424–426

Lamb, T.D. and E.N. Pugh, 1992, 'G-protein cascades: gain and kinetics', *Trends in Neurosciences*, **15**, 291–298

7. Adrenergic receptors

Dixon, R.A.F. *et al.*, 1986, 'Cloning of the gene and cDNA for mammalian β-adrenergic receptor and homology with rhodopsin', *Nature*, **321**, 75–79

Kobilka, B., 1992, 'Adrenergic receptors as models for G-protein-coupled receptors', *Annual Review of Neuroscience*, **15**, 87–114

8. Muscarinic acetylcholine receptors

Bonner, T.I. *et al.*, 1987, 'Identification of a family of muscarinic acetylcholine receptor genes', *Science*, **237**, 527–532

Kerlavage, A.R., C.M. Fraser and J. Venter, 1987, 'Muscarinic cholinergic receptor structure: molecular biological support for subtypes', *Trends in Pharmacological Science*, **8**, 426–431

Kubo, T., 1986, 'Cloning, sequencing and expression of complementary DNA encoding the muscarinic acetylcholine receptor', *Nature*, **323**, 411–416

Kubo, T. *et al.*, 1986, 'Primary structure of porcine cardiac muscarinic acetylcholine receptor deduced from cDNA sequence', *FEBS Letters*, **209**, 367–372

Noma, A., 1986, 'GTP-binding proteins couple cardiac muscarinic receptors to potassium channels', *Trends in Neurosciences*, **9**, 142–143

9. Neurokinin-A receptor

*Masu, Y. *et al.*, 1987, 'cDNA cloning of bovine substance-K receptor through oocyte expression system', *Nature*, **329**, 836–838

10. Metabotropic glutamate receptors

Hollmann, M. and S. Heinemann, 1994, 'Cloned glutamate receptors', *Annual Review of Neuroscience*, **17**, 31–108

11. Cannabinoid receptors

Elphick, M.R. and M. Egertova, 2001, 'The neurobiology and evolution of cannabinoid signalling', *Philosophical Transactions of the Royal Society* (Lond.) B, **356**, 381–408

12. Rhodopsin

Baehr, W. and M.L. Applebury, 1986, 'Exploring visual transduction with recombinant DNA techniques', *Trends in Neurosciences*, **9**, 198–203

Hargreave, P.A., 1986, 'Molecular dynamics of the rod cell' in *The Retina: A Model for Cell Biology Studies*, eds R. Adler and D. Farber, Orlando, FL: Academic Press, part 1, 207–237

Henderson, R. *et al.*, 1989, 'Model for the structure of bacteriorhodopsin based on high resolution electron cryo-microscopy', *Journal of Molecular Biology*, **213**, 899–929

Monon, S.T., M. Han and T.P. Sakmar, 2001, 'Rhodopsin: structural basis of molecular physiology', *Physiological Reviews*, **81**, 1659–1688

Palczewski, K. *et al.*, 2000, 'Crystal structure of rhodopsin: a G-protein-coupled receptor', *Science*, **289**, 739–745

Stryer, L., 1987, 'The molecules of visual excitation', *Scientific American*, **257**(1), 32–40

Yokoyama, S., 1995, 'Amino acid replacements and wavelength absorption of visual pigments in vertebrates', *Molecular Biology and Evolution*, **12**, 53–61

Yokoyama, S., 2000, 'Molecular evolution of vertebrate visual pigments', *Progress in Retinal and Eye Research*, **19**, 385–419

13. Cone opsins

Fryxell, K.J. and E.M. Meyerowitz, 1991, 'The evolution of rhodopsin and neurotransmitter receptors', *Journal of Molecular Evolution*, **33**, 367–378

Hunt, D.M. *et al.*, 1998, 'Molecular evolution of trichromacy in primates', *Vision Research*, **38**, 3299–3306

Kochendoerfer, G.G. *et al.*, 1999, 'How color visual pigments are tuned', *Trends in Biochemistry*, **24**, 300–305

*Nathans, J., D. Thomas and D. Hogness, 1986, 'Molecular genetics of human color vision: the genes encoding blue, green and red pigments', *Science*, **232**, 193–202

Nathans, J., 1987, 'Molecular biology of visual pigments', *Annual Review of Neuroscience*, **10**, 163–194

Box 8.1: The GTPase superfamily

Bourne, H.R., D.A. Sanders and F. McCormick, 1990, 'The GTPase superfamily: a conserved switch for diverse cell functions', *Nature*, **348**, 125–132

Bourne, H.R., D.A. Sanders and F. McCormick, 1991, 'The GTPase superfamily: conserved structure and molecular mechanism', *Nature*, **349**, 117–127

Colemam, D.E. *et al.*, 1994, 'Structure of active conformations of $G_{i\alpha1}$ and the mechanism of GTP hydrolysis', *Science*, **265**, 1405–1412

CIBA Foundation Symposium, 1993, *The GTPase Superfamily – Symposium 176*, Chichester: John Wiley

Wittinghofer, A. and E.F. Pai, 1991, 'The structure of the ras protein: a model for a universal molecular switch', *Trends in Biochemical Science*, **16**, 382–387

CHAPTER 9: PUMPS

Web sites

Na⁺+ K⁺ pump, including animation of its action

http://bio.winoma.msus.edu/berg/ANIMTNS/Na-Kpump.htm

Ca²⁺ pump

Structure:
http://www.spring8.or.jp/ENGLISH/general_info/overview/protein.html

Calmodulin

Details may be found at: **http://structbio.vanderbilt.edu/cabp_database/general/prot_pages/calmod.html**
Molecular models at:
http://www-structure.llnl.gov/Calmod/calmod.html/
http://www.scripps.edu/~wriggers/projects/calmodulin/

Print

1. General

Cantley, L., 1986, 'Ion transport systems sequenced', *Trends in Neurosciences*, **9**, 1–3
Christensen, H.N., 1975, *Biological Transport* (2nd edn), Reading, MA: Benjamin
Tanford, C., 1983, 'The mechanism of free energy coupling in active transport', *Annual Review of Biochemistry*, **52**, 399–409

2. Energetics

Christensen, H.N., 1975, *Biological Transport* (2nd edn), Reading, MA: Benjamin
Lehninger, A.L., 1982, *Principles of Biochemistry* (2nd edn), New York: Worth
Spanner, D.C., 1964, *Introduction to Thermodynamics*, New York: Academic Press

3. The Na⁺+ K⁺ pump

Bronner, F. and A. Kleinzeller, eds, 1983, 'Structure, mechanism and function of the Na/K pump', *Current Topics in Membranes and Transport*, **19**, 1–1043
Kawakami, K. et al., 1985, 'Primary structure of the α-subunit of *Torpedo californica* (Na⁺+K⁺) ATPase deduced from cDNA sequence', *Nature*, **316**, 733–736
Mercer, R.W., 1993, 'Structure of the Na,K-ATPase', *International Review of Cytology*, **137C**, 139–168
*Shull, G.E., A. Schwartz and J.B. Lingrel, 1985, 'Amino-acid sequence of the catalytic subunit of the (Na⁺+K⁺) ATPase deduced from complementary DNA', *Nature*, **316**, 691–695
Taniguchi, K. et al., 1984, 'Conformational change of sodium- and potassium-dependent adensine triphosphatase', *Journal of Biological Chemistry*, **259**, 15228–15233

4. The calcium pump

Carafoli, E., 1991, 'The calcium pumping ATPase of the plasma membrane', *Annual Review of Physiology*, **53**, 531–547

Lee. A.G. and J.M. East, 2001, 'What the structure of the calcium pump tells us about its mechanism', *Biochemical Journal*, **356**, 665–683
East, J.M., 2000, 'Sarco(endo)plasmic reticulum calcium pumps: recent advances in our understanding of structure/function and biology', *Molecular Membrane Biology*, **17**, 189–200
MacLennan, D.H. et al., 1985, 'Amino-acid sequence of Ca²⁺+Mg²⁺-dependent ATPase from rabbit muscle sarcoplasmic reticulum, deduced from its complementary DNA sequence', *Nature*, **316**, 696–700
MacLennan, D.H., 2000, 'Pumping ions', *Nature*, **405**, 633–634
Toyoshima, O. et al., 2000, 'Crystal structure of the calcium pump of the sarcoplasmic reticulum at 2.6 Å resolution', *Nature*, **405**, 647–654

5. Other pumps and transport mechanisms

Philipson, K.D. and D.A. Nicholl, 2000, 'Sodium–calcium exchange: a molecular perspective', *Annual Review of Physiology*, **62**, 111–133
Skradski, S.L. et al., 2001, 'A novel gene causing a Mendelian audiogenic mouse epilepsy', *Neuron*, **31**, 537–544
West, I.C., 1983, *The Biochemistry of Membrane Transport*, London: Chapman and Hall
Wilson, D.B., 1978, 'Cellular transport mechanisms', *Annual Review of Biochemistry*, **47**, 933–965

Box 9.1: Calmodulin

Finn, B.F. and S. Forsén, 1995, 'The evolving model of calmodulin', *Structure*, **3**, 7–11

CHAPTER 10: LIGAND-GATED ION CHANNELS

Web sites

Ligand gated ion channel databases

http://www.pasteur.fr/recherche/banques/LGIC/LGIC.html
390 entries of ligand-activated channel subunits (July 2001)
Cellular properties including those of ligand-gated receptors have been collated at:
http://senselab.med.yale.edu/senselab/CellPropDB/default.asp

nAChR

Full data on nAChRs at:
http://www.expasy.ch/cgi-bin/niceprot.pl?P02708
NMJ disorders:
http://www.neuro.wustl.edu/neuromuscular/synmg.html
Evolution of nAChRs:
http://www.pasteur.fr/recherche/unites/neubiomol/ARTICLES/LENOV1995/Lenov1995html
Useful figure (simplified) at:
http://www.emm.cbcu.cam.ac.uk/99001428h.htm

GABA_ARs

http://www.williams.edu/imput/synapse/pages/lllA9.htm

NMDARs

http://www.biochem.uni-erlangen.de/MouseDB/db/multiprot/
gluNMDA.html

Print

1. General

Neher, E. and B. Sakmann, 1992, 'The patch clamp technique', *Scientific American*, **266**(3), 28–35

Persigian, V.A., ed., 1984, 'Biophysical discussions: ionic channels and membranes', *Biophysical Journal*, **45**, 1–359

Rose, M.R., 2001, *Channelopathies of the Nervous System*, Oxford: Butterworth-Heinemann

Sakmann, B. and E. Neher, eds, 1983, *Single-Channel Recording*, New York: Plenum Press

Sakmann, B. and E. Neher, 1984, 'Patch clamp techniques for studying ionic channels in excitable membranes', *Annual Review of Physiology*, **46**, 455–472

Stevens, C.F., 1987, 'Channel families in the brain', *Nature*, **328**, 198–199

Unwin, N., 1993, 'Neurotransmitter action: opening of ligand-gated ion channels', *Cell/Neuron*, **72**(10), 31–41

2. The nicotinic acetylcholine receptor

*Barrantes, F.J., 1983, 'Recent developments in the structure and function of the acetylcholine receptor', *International Review of Neurobiology*, **24**, 259–341

*Brisson, A. and P.N.T. Unwin, 1985, 'Quarternary structure of the acetylcholine receptor', *Nature*, **315**, 474–477

Corringer, P.-J., N. Le Novère and J.-P. Changeux, 2000, 'Nicotinic receptors at the amino acid level', *Annual Review of Pharmacology and Toxicology*, **40**, 431–458

Galzi, J.-L. et al., 1991, 'Functional architecture of the nicotinic acetylcholine receptor: from electric organ to brain', *Annual Review of Pharmacology*, **31**, 37–72

Hall, Z.W. and J.R. Sanes, 1993, 'Structure and development: neuromuscular junction'. *Cell/Neuron*, **72**(10), 99–121

*Hirokawa, N., 1983, 'Membrane specialisation and cytoskeletal structures in the synapse and axon revealed by the quick-freeze, deep-etch method' in *Structure and Function of Excitable Cells*, eds D.C. Chang et al., New York: Plenum Press, 113–141

Merlie, J.P. et al., 1983, 'The regulation of acetylcholine receptor expression in mammalian muscle' in *Cold Spring Harbor Symposia on Quantitative Biology*, **XLVIII**, 135–146

Mishima, M. et al., 1985, 'Location of functional regions of acetylcholine receptor α-subunit by site-directed mutagenesis', *Nature*, **313**, 364–369

Mishima, M. et al., 1986, 'Molecular distinction between foetal and adult forms of muscle acetylcholine receptor', *Nature*, **321**, 406–411

Newsome-Davies, J., 1988, 'Autoimmunity in neuromuscular disease' in C.S. Raine, ed., Advances in Neuroimmunology, *Annals of New York Academy of Science*, **540**, 25–38

*Noda. M. et al., 1983, 'Structural homology of *Torpedo californica* acetylcholine receptor subunits', *Nature*, **302**, 528–532

*Numa, S. et al., 1983, 'Molecular structure of the nicotinic acetylcholine receptor' in *Cold Spring Harbor Symposia on Quantitative Biology*, **XLVIII**, 57–69

Phillips, H.A. et al., 1995, 'Localisation of a gene for autosomal dominant nocturnal frontal lobe epilepsy to chromosome 20q13.2', *Nature Genetics*, **10**, 117–203

Salomone, F. and M. Zhou, 1999, 'Aberrations in nicotinic acetylcholine receptor structure, function and expression: implications in disease', *McGill Journal of Medicine*, **5**, 90–97

Sargent, P.B., 1993, 'The diversity of neuronal nicotinic acetylcholine receptors', *Annual Review of Neuroscience*, **16**, 403–443

Stevens, C.F., 1985, 'AChRs: five-fold symmetry and the ε-subunit', *Trends in Neurosciences*, **8**, 335–336

*Stroud, R.M., 1981, 'Structure of an ACh receptor, a hypothesis for a dynamic mechanism of its action' in *Proceedings of Second SUNYA Conversation in the Discipline of Biomolecular Stereodynamics*, ed. R.H. Sarma, vol. 2, New York: Adenine Press

Zingsheimer, H.P. et al., 1982, 'Direct structural localisation of two toxin-recognition sites on an ACh receptor protein', *Nature*, **299**, 81–84

3. The GABA$_A$R

Barnard, E.A., M.G. Darlison and P. Seeburg, 1987, 'Molecular biology of the GABA$_A$ receptor: the receptor/channel super-family', *Trends in Neurosciences*, **10**, 502–509

Baulac, S. et al., 2001, 'First genetic evidence of GABA$_A$ receptor dysfunction in epilepsy: a mutation in the γ2-subunit gene', *Nature Genetics*, **28**, 46–48

Macdonald, R.L. and R.W. Olsen, 1994, 'GABA$_A$ receptor channels', *Annual Review of Neuroscience*, **17**, 569–602

Mehta, A.K and M.K. Ticku, 1999, 'An update on GABA$_A$ receptors', *Brain Research Reviews*, **29**, 196–217

Schofield, P.R. et al., 1987, 'Sequence and functional expression of the GABA$_A$ receptor shows a ligand-gated receptor superfamily', *Nature*, **328**, 221–227

Stevenson, F.A., 1988, 'Understanding the GABA$_A$ receptor: a chemically gated ion channel', *Biochemical Journal*, **249**, 21–32

4. The glycine receptor

Betz, H., 1992, 'Structure and function of the inhibitory glycine receptor', *Quarterly Review of Biophysics*, **25**, 381–394

Greeningloh, G. et al., 1987, 'The strychnine-binding subunit of the glycine receptor shows homology with the nicotinic acetylcholine receptor', *Nature*, **328**, 215–220

Jentsch, T.J., 1993, 'Chloride channels', *Current Opinion in Neurobiology*, **3**, 316–321

*Triller, A. et al., 1985, 'Distribution of glycine receptors at central synapses: an immunogold study', *Journal of Cell Biology*, **101**, 683–688

5. Glutamate receptors

Barnes, D.M., 1988, 'NMDA receptors trigger excitement', *Science*, **239**, 254–256

Cotman, C.W. and L.L. Iverson, 1987, 'Excitatory amino acids in the brain: focus on NMDA receptors', *Trends in Neurosciences*, **10**, 263–265

Cull-Candy, S.G. and M.M. Usowicz, 1987, 'Multiple-conductance channels activated by excitatory amino acids in cerebellar neurons', *Nature*, **325**, 525–528

Hollmann, M. and S. Heinemann, 1994, 'Cloned glutamate receptors', *Annual Review of Neuroscience*, **17**, 31–108

Jahr, C.E. and C.F. Stevens, 1987, 'Glutamate activates multiple single channel conductances in hippocampal neurons', *Nature*, **325**, 522–525

Johnson, J.W. and P. Ascher, 1987, 'Glycine potentiates the NMDA response in cultured mouse brain neurons', *Nature*, **325**, 529–531

Lerma, J. *et al.*, 2001, 'Molecular physiology of kainate receptors', *Physiological Reviews*, **81**, 971–998

MacDermott, A.B. and N. Dale, 1987, 'Receptors, ion channels and synaptic potentials underlying the integrative actions of excitatory amino acids', *Trends in Neurosciences*, **10**, 280–284

6. Purinoceptors

Barnard, E.A., G. Burnstock and T.E. Webb, 1994, 'G-protein-coupled receptors for ATP and other nucleotides: a new receptor family', *Trends in Pharmacological Science*, **15**, 67–79

Khakh, B.S., 2001, 'Molecular physiology of P2X receptors and ATP signalling at synapses', *Nature Reviews Neuroscience*, **2**, 165–174

Suprenant, A., G. Buell and R.A. North, 1995, 'P2X receptors bring new structure to ligand-gated ion channels', *Trends in Neurosciences*, **18**, 224–229

Box 10.1: Evolution of the nAChRs

Le Novère, N. and J.-P. Changeux, 1995, 'Molecular evolution of the nicotinic acetylcholine receptor: an example of multigene family in excitable cells', *Journal of Molecular Evolution*, **40**, 155–172

Ortells, M.O. and G.G. Lunt, 1995, 'Evolutionary history of the ligand-gated ion-channel superfamily of receptors', *Trends in Neurosciences*, **18**, 121–127

Box 10.2: The inositol triphosphate (IP₃ or InsP₃) receptor

Conley, E.C., 1996, *The Ion Channel Facts Book*, London: Academic Press

Furuichi, T. *et al.*, 1989, 'Primary structure and functional expression of the inositol 1,4,5-triphosphate-binding protein P400', *Nature*, **342**, 32–38

Mikoshiba, K., 1993, 'Inositol 1,4,5-triphosphate receptor', *Trends in Pharmacological Sciences*, **14**, 86–89

CHAPTER 11: VOLTAGE-GATED ION CHANNELS

Web sites

An encyclopaedic (regularly updated) review of the neuronal location and properties of VGICs can be found at:
http://senselab.medyale.edu/senselab/neuronDB/ channelGene2.htm/
Channelopathies:
http://www.neuro.wustl.edu/neuromuscular/mother/chan.html
http://www.periodicparalysis.org/

Print

1. General

Conley, E.C. and W.J. Brammar, 1999, *The Ion Channel Facts Book*, vol. IV: *Voltage-Gated Channels*, London: Academic Press

Hodgkin, A.L. and A.F. Huxley, 1952, 'Currents carried by sodium and potassium ions through the membrane of the giant axon of *Loligo*', *Journal of Physiology* (*London*), **116**, 449–472

Neher, E., 1992, 'Ion channels for communication between and within cells', *Neuron*, **8**, 605–612 (Nobel lecture)

Stevens, C.F., 1984, 'Biophysical studies of ion channels', *Science*, **225**, 1346–1350

Various, 1984, 'Biophysical discussions: ionic channels and membranes', *Biophysical Journal*, **45**, 1–337

2. Potassium channels

Bacterial

Doyle, D.A. *et al.*, 1998, 'The structure of a potassium channel: molecular basis of K⁺ conduction and selectivity', *Science*, **280**, 69–77

MacKinnon, R. *et al.*, 1998, 'Structural conservation of prokaryotic and eukaryotic potassium channels', *Science*, **280**, 106–109

Schrempf, H. *et al.*, 1995, 'A prokaryotic potassium ion channel with two predicted transmembrane segments from *Streptomyces lividans*' *EMBO Journal*, **14**, 5170–5178

Yellen, G., 1999, 'The bacterial K⁺-channel structure and its implications for neuronal channels', *Current Opinion in Neurobiology*, **9**, 267–273

Zhou, Y. *et al.*, 2001, 'Chemistry of ion coordination and hydration revealed by a K⁺channel-Fab complex at 2.0 Å resolution', *Nature*, **414**, 43–48

Neuronal 2TM

Ashcroft, F.M. and F.M. Gribble, 1998. 'Correlating structure and function in ATP-sensitive K⁺-channels', *Trends in Neurociences*, **21**, 288–294

Kubo, Y. *et al.*, 1993, 'Primary structure and functional expression of a rat G-protein-coupled muscarinic potassium channel', *Nature*, **364**, 802–806

*Nichols, C.G. and A.N. Lopatin, 1997, 'Inward rectifier potassium channels', *Annual Review of Physiology*, **59**, 171–191

Riemann, F. and F.M. Ashcroft, 1999, 'Inwardly rectifying potassium channels', *Current Opinion in Cell Biology*, **11**, 503–508

4TM(2P)

Goldstein, S.A.N. *et al.*, 2001, 'Potassium leak channels and the KCNK family of two-P-domain subunits', *Nature Reviews Neuroscience*, **2**, 175–184

6TM

Hoshi, T. and W.N. Zagotta, 1993, 'Recent advances in the understanding of potassium channel function', *Current Opinion in Neurobiology*, **3**, 283–290

Jan, L.Y. *et al.*, 1985, 'Application of *Drosophila* molecular genetics in the study of neural function – studies of the Shaker locus for a potassium channel', *Trends in Neurosciences*, **8**, 234–238

McLarnon, J.G., 1995, 'Potassium currents in motoneurones', *Progress in Neurobiology*, **47**, 513–531

Monor, D.L., 2001, 'Potassium channels: life in the post-structural world', *Current Opinion in Structural Biology*, **11**, 408–414

Phillipson, L. and R.J. Miller, 1992, 'A small K^+ channel looms large', *Trends in Pharmacological Science*, **13**, 8–11

Pongs, O., 1992, 'Molecular biology of voltage-dependent potassium channels', *Physiological Reviews*, **72**, S69–S88

Schwarz, T.L. *et al.*, 1988, 'Multiple potassium-channel components are produced by alternative splicing at the Shaker locus in *Drosophila*', *Nature*, **331**, 137–142

Tai, K.-K., K.W. Wang and S.A.N. Goldstein, 1997, 'MinK potassium channels are heteromultimeric complexes', *Journal of Biological Chemistry*, **272**, 1654–1658

Yi, B.A. and L.Y. Jan, 2000, 'Taking apart the gating of voltage-gated K^+-channels', *Neuron*, **27**, 423–425

Zhou, M. *et al.*, 2001, 'Potassium channel receptor site for the inactivation gate and quaternary amine inhibitors', *Nature*, **411**, 657–661

3. Calcium channels

Alsobrook, J.P. and C.F. Stevens, 1988, 'Cloning the calcium channel', *Trends in Neuroscience*, **11**, 1–2

Hofmann, F., M. Biel and V. Flockerzi, 1994, 'Molecular basis for Ca^{2+} channel diversity', *Annual Review of Neuroscience*, **17**, 399–418

Randall, A.D., 1998, 'The molecular basis of voltage-gated Ca^{2+} channel diversity: is it time for T', *Journal of Membrane Biology*, **161**, 207–213

*Tanabe, T. *et al.*, 1987, 'Primary structure of the receptors for the Ca^{2+} channel blockers from skeletal muscle', *Nature*, **328**, 313–318

*Varadi, G. *et al.*, 1999, 'Molecular elements of ion permeation and selectivity within calcium channels', *Critical Reviews in Biochemistry and Molecular Biology*, **34**, 181–214

4. Sodium channels

Agnew, W.S. *et al.*, 1986, 'The structure and function of the voltage sensitive Na^+ channel', *Annals of the New York Academy of Science*, **479**, 238–256

Barchi, R.L., 1987, 'Sodium channel diversity: subtle variations on a complex theme', *Trends in Neurosciences*, **10**, 221–222

*Catterall, W.A., 1992, 'Cellular and molecular biology of voltage-gated sodium channels', *Physiological Reviews*, **72**, S15–S48

Catterall, W.A., 2000, 'From ionic currents to molecular mechanisms: the structure and function of voltage-gated sodium channels', *Neuron*, **26**, 13–25

Goldin, A.L., 2001, 'Resurgence of sodium channel research', *Annual Review of Physiology*, **63**, 871–894

*Noda, M. *et al.*, 1984, 'Primary structure of *Electrophorus electricus* sodium channel deduced from cDNA sequence', *Nature*, **312**, 121–127

Numa, S. and M. Noda, 1986, 'Molecular structure of sodium channels', *Annals of the New York Academy of Science*, **479**, 338–355

*Sigworth, F.J. and E. Neher, 1980, 'Single Na^+-channel currents observed in cultured rat muscle cells', *Nature*, **287**, 447–449

Wood, J.N. and M. Baker, 2001, 'Voltage-gated sodium channels', *Current Opinion in Pharmacology*, **1**, 17–21

5. Chloride channels

Dutzler, R. *et al.*, 2002, 'X-ray structure of a ClC chloride channel at 3.0 Å reveals the molecular basis of anion selectivity', *Nature*, **415**, 287–294

Jentsch, T.J., K. Steinmeyer and G. Schwarz, 1990, 'Primary structure of *Torpedo marmorata* chloride channel isolated by expression cloning in *Xenopus* oocytes', *Nature*, **348**, 510–514

Jentsch, T.J., 1993, 'Chloride channels', *Current Opinion in Neurobiology*, **3**, 316–321

Mindell, J.A. *et al.*, 2001, 'Projection structure of a ClC-type chloride channel at 6.5 Å resolution', *Nature*, **409**, 219–223

6. Ion selectivity and gating

Catterall, W.A., 1992, 'Cellular and molecular biology of voltage-gated sodium channels', *Physiological Reviews*, **72**, S15–S48

Fedida, D and J.C. Hesketh, 2001, 'Gating of voltage dependent potassium channels', *Progress in Biophysics and Molecular Biology*, **75**, 165–199

7. Channelopathies

Ashcroft, F.M., 2000, *Ion Channels and Disease*, San Diego: Academic Press

George, A.J. *et al.*, 1993, 'Molecular basis of Thomsen's disease (autosomal dominant myotomia congenita)', *Nature Genetics*, **3**, 305–310

Hofmann, E.P., 1995, 'Voltage-gated ion channelopathies: Inherited disorders caused by abnormal sodium, chloride and calcium regulation in skeletal muscle', *Annual Review of Medicine*, **46**, 431–441

Jen, J., 1999, 'Calcium channelopathies in the central nervous system', *Current Opinion in Neurobiology*, **9**, 274–280

Wray, D., 2001, 'Inherited disorders of ion channels: An overview', *Pharmaceutical News*, **8**(2), 12–17

8. Channel evolution

Anderson, P.A.V. and R.M. Greenberg, 2001, 'Phylogeny of ion channels: clues to structure and function', *Comparative Biochemistry and Physiology, Part B*, **129**, 17–28

Hille, B., 1989, 'Ionic channels: evolutionary origins and modern roles', *Quarterly Journal of Experimental Physiology*, **74**, 785–804

Jacob, F., 1977, 'Evolution and tinkering', *Science*, **196**, 1161–1166

Ortells, M.O. and G.C. Lunt, 1995, 'Evolutionary history of the ligand-gated ion-channel superfamily of receptors', *Trends in Neurosciences*, **18**, 121–127

Rogawski, M.A., 2000, 'KCNQ2/KCNQ3 K$^+$-channels and epilepsy', *Trends in Neurosciences*, **23**, 394

Venter, J.C. *et al.*, 1988, 'Evolution of neurotransmitter receptor systems', *Progress in Neurobiology*, **30**, 105–169

9. Conclusion

Catterall, W.A., 1988, 'Structure and function of voltage-sensitive ion channels', *Science*, **242**, 50–61

Stevens, C.F., 1987, 'Channel families in the brain', *Nature*, **328**, 198–199

Box 11.1: Cyclic nucleotide-gated (CNG) channels

Zheny, J. and W.M. Zagotta, 2000, 'Gating rearrangements in cyclic nucleotide-gated channels by patch-clamp fluorimetry', *Neuron*, **28**, 369–394

CHAPTER 12: RESTING POTENTIALS AND CABLE CONDUCTION

Web sites

Two somewhat advanced sites for modelling electrotonic conductances may be found at:

http://www.neuron.yale.edu/neuron/sdsc98/zclass/zclass.htm
and
http://www.neuron.yale.edu/neuron/papers/ebench/ebench.html

Print

1. General

Aidley, D.J., 1989, *The Physiology of Excitable Cells* (3rd edn), Cambridge: Cambridge University Press

Kandel, E.R., J.H. Schwartz and T.M. Jessell, 2000, *Principles of Neural Science* (4th edn), New York: Elsevier

Katz, B., 1966, *Nerve, Muscle and Synapse*, New York: McGraw Hill

Matthews, G.G., 1986, *Cellular Physiology of Nerve and Muscle*, Palo Alto: Blackwell Scientific Publications

Murray, R.W., 1983, *Test Your Understanding of Neurophysiology*, Cambridge: Cambridge University Press

2. Measurement of the resting potential

Hodgkin, A.L., A.F. Huxley and B. Katz, 1952, 'Measurement of current voltage relations in the membrane of the giant axon of *Loligo*', *Journal of Physiology*, **116**, 424–448

3. The origin of the resting potential

Goldman, D.E., 1943, 'Potential, impedance and rectification in membranes', *Journal of General Physiology*, **27**, 37–60

Hille, B., 1977, 'Ionic basis of resting and action potentials' in *Handbook of Neurophysiology*, vol. 1, section 1, Bethesda, MD: American Physiological Society, 99–136

Hodgkin, A.L. and P. Horowicz, 1959, 'The influence of potassium and chloride ions on the membrane potential of single muscle fibres', *Journal of Physiology*, **148**, 127–160

Kandel, E.R., J.H. Schwarz and T.M. Jessell, 2000, *Principles of Neural Science* (4th edn), New York: Elsevier Science Publishing

Newman, E.A., 1985, 'Regulation of potassium levels by glial cells in the retina', *Trends in Neurosciences*, **8**, 154–159

4. Electrotonic potentials and cable conduction

Dodge, F.A., 1979, 'The nonuniform excitability of central neurons as exemplified by a model spinal neuron', in *The Neurosciences: Fourth Study Program*, eds F.O. Schmitt and F.G. Worden, New York: Rockefeller Press, 423–437

Jack, J., 1979, 'An introduction to linear cable theory' in *The Neurosciences: Fourth Study Program*, eds F.O. Schmitt and F.G. Worden, New York: Rockefeller University Press, 439–455

Meyer, E., C.O. Müller and P. Fromherz, 1997, 'Cable properties of dendrites in hippocampal neurons of rat mapped by a voltage-sensitive dye', *European Journal of Neuroscience*, **9**, 778–785

Rall, W., 1962, 'Theory of the physiological properties of dendrites', *Annals of the New York Academy of Science*, **96**, 1071–1092

Rall, W., 1977, 'Core conductor theory and cable properties of neurons' in *Handbook of Neurophysiology*, vol. 1, section 1, Bethesda, Maryland: American Physiological Society, 39–97

Shepherd, G.M., 1997, *The Synaptic Organisation of the Brain* (4th edn), Oxford: Oxford University Press

CHAPTER 13: SENSORY TRANSDUCTION

Web sites

The Chemoreception Web:
www.csa.com/crw/websites.html
Chemosensitivity in bacteria:
http://curie.che.virginia.edu/project/visualization.html
Olfaction in mammals:
www.leffingwell.com/olfaction.htm
The human olfactory repertoire (July 2001):
www.genomebiology.com/2001/2/6/research/0018/
Retinitis pigmentosa – a valuable guide to RP organisations and web sites is at:
http://www.geocities.com/retinitispigmentosa/
The British RP Society is at:
http://www.brps.demon.co.uk/

Print

1. General

Barlow, H.B. and J.D. Mollon, eds, 1982, *The Senses*, Cambridge: Cambridge University Press

Frisch, L., ed., 1965, 'Sensory receptors', *Cold Spring Harbor Symposia in Quantitative Biology*, **XXX**, Cold Spring Harbor, NY: Cold Spring Harbor Laboratory Press

Smith, C.U.M., 2000, *Biology of Sensory Systems*, Chichester: John Wiley

2. Chemoreception

(a) Prokaryotes

Adler, J., 1983, 'Bacterial chemotaxis and molecular neurobiology', *Cold Spring Harbor Symposia in Quantitative Biology*, XLVIII (vol. 2), 803–804

Bourrett, R.B., K.A. Borkovich and M.I. Simon, 1991, 'Signal transduction pathways involving protein phosphorylation in prokaryotes', *Annual Review of Biochemistry*, **60**, 401–441

Schuster, S.C. and S. Khan, 1994, 'The bacterial flagellar motor', *Annual Review of Biophysics and Biomolecular Structure*, **23**, 509–539

Simon, M.I., A. Krikos, N. Mutoh and A. Boyd, 1985, 'Sensory transduction in bacteria', *Current Topics in Membranes and Transport*, **23**, 3–15

(b) Vertebrates

Glusman, G. *et al.*, 2001, 'The complete human olfactory subgenome', *Genome Research*, **11**, 685–702

Lancet, D., 1986, 'Vertebrate olfactory reception', *Annual Review of Neuroscience*, **9**, 329–355

Lindemann, B., 1996, 'Taste reception', *Physiological Reviews*, **76**, 719–766

Malnic, B. *et al.*, 1999, 'Combinatorial codes for odors', *Cell*, **96**, 713–723

Nakamura, T. and G.H. Gold, 1987, 'A cyclic nucleotide-gated conductance in olfactory receptor cilia', *Nature*, **325**, 442–446

Schild, D. and D. Restrepo, 1998, 'Transduction mechanisms in vertebrate olfactory receptor cells', *Physiological Reviews*, **78**, 429–466

Shepherd, G.M., 1994, 'Discrimination of molecular signals by olfactory receptor neuron', *Neuron*, **13**, 771–790

Smith, C.U.M., 2000, *Biology of Sensory Systems*, Chichester: John Wiley

Teeter, J. and G.H. Gold, 1988, ' A taste of things to come', *Nature*, **331**, 298–299

Zufall, F., S. Firestein and G.M. Shepherd, 1994, 'Cyclic nucleotide-gated ion channels and sensory transduction in olfactory receptor neurons', *Annual Review of Biophysics and Biomolecular Structure*, **23**, 577–607

3. Photoreceptors

Adler, R. and D. Farber, 1986, *The Retina: a Model for Cell Biology Studies*, Orlando, FL: Academic Press

*Baylor, D.A., T.D. Lamb and K.-W. Yau, 1979, 'The membrane current of single rod outer segments', *Journal of Physiology*, **289**, 589–611

Besharse, J.C., 1986, 'Photosensitive membrane turnover: differentiated membrane domains and cell–cell interaction' in *The Retina: a Model for Cell Biology Studies*, R. Adler and D. Farber, Orlando, FL: Academic Press, vol. 1, 297–352

*Hogan, M.J., J.A. Alverado and J.E. Weddell, 1971, *Histology of the Human Eye*, Philadelphia: Saunders

Kaupp, B.U. *et al.*, 1989, 'Primary structure and functional expression from complementary DNA of rod photoreceptor cyclic GMP-gated channel', *Nature*, **342**, 762–766

Lamb, T.D., 1986, 'Transduction in vertebrate photoreceptors: the roles of cyclic GMP and calcium', *Trends in Neurosciences*, **9**, 224–228

Lamb, T.D. and E.N. Pugh, 1992, 'G-protein cascades: gain and kinetics', *Trends in Neurosciences*, **15**, 291–298

Nathans, J., 1994, 'In the eye of the beholder': Visual pigments and inherited variations in human vision', *Cell*, **78**, 357–360

4. Mechanoreceptors

(a) Bacteria

Biggin, P.C. and M.S.P. Sanson, 2001, 'Channel gating: Twist to open', *Current Biology*, **11**, R364–366

*Blount, P. *et al.*, 1996, 'Membrane topology and multimeric structure of a mechanosensitive channel protein in *Escherichia coli*', *EMBO Journal*, **15**, 4798–4805.

Chang, G. *et al.*, 1998, 'Structure of the MscL homolog from *Mycobacterium tuberculosis*: a gated mechanosensitive ion channel', *Science*, **282**, 2220–2226

Hamil, O.P. and D.W. McBride, 1994, 'The cloning of a mechano-gated membrane ion channel', *Trends in Neurosciences*, **17**, 439–443

Sukharev, S.I. *et al.*, 1994, 'A large-conductance mechanosensitive channel in *E. coli* encoded by *mscL* alone', *Nature*, **368**, 265–268

Sukharev, S.I. *et al.*, 1997, 'Mechanosensitive channels of *Escherichia coli*: the MscL gene, protein and activities', *Annual Review of Physiology*, **59**, 633–657

Sukharev, S.I. *et al.*, 2001, 'The gating mechanism of the large mechanosensitive channel MscL', *Nature*, **409**, 720–724

(b) Caenorhabditis elegans

Hong, K. and M. Driscoll, 1994, 'A transmembrane domain of the putative channel subunit MEC-4 influences mechanotransduction and neurodegeneration in *C. elegans*', *Nature*, **367**, 470–473

Huang, M. and M. Chalfie, 1994, 'Gene interactions affecting mechanosensory transduction in *Caenorhabditis elegans*', *Nature*, **367**, 467–470

*Tavernarakis, N. and M. Driscoll, 1997, 'Molecular modelling of mechanotransduction in the nematode *Caenorhabditis elegans*', *Annual Review of Physiology*, **59**, 662

(c) Vertebrates

Corwin, J.T. and M.E. Warchol, 1991, 'Auditory hair cells: structure, function, development and regeneration', *Annual Review of Neuroscience*, **14**, 301–333

Eatock, R.A., 2000, 'Adaptation in hair cells', *Annual Review of Neuroscience*, **23**, 285–314

*Flock, A., 1965, 'Transducing mechanisms in lateral line canal organ receptors' in Frisch, L., ed., *Cold Spring Harbor Symposia in Quantitative Biology*, XXX, Cold Spring Harbor, NY: Cold Spring Harbor Laboratory Press, 133–145

*Gillespie, P.G., 1995, 'Molecular machinery of auditory and vestibular transduction', *Current Opinion in Neurobiology*, **5**, 449–455

*Gillespie, P.G. and D.P. Corey, 1997, 'Myosin and adaptation by hair cells', *Neuron*, **19**, 955–958

Griffith, A.J and T.B. Friedman, 1999, 'Making sense of sound', *Nature Genetics*, **21**, 347–349

Hackney, C.M. and D.N. Furness, 1995, 'Mechanotransduction in vertebrate hair cells: structure and function of the stereociliary bundle', *American Journal of Physiology*, **268**, C215–221

Hudspeth, A.J., 1989, 'How the ear's works work', *Nature*, **341**, 398–404

Hudspeth, A.J., 1997, 'How hearing happens', *Neuron*, **19**, 947–50

Pickles, J.O. and D.P. Corey, 1992, 'Mechanoelectrical transduction by hair cells', *Trends in Neurosciences*, **15**, 254–258

Box 13.1: Retinitis pigmentosa

Armstrong, R.A., 1999, 'Ocular disease and the new genetics', *Ophthalmic and Physiological Optics*, **19**, 193–105

Berson, E.L., 1993, 'Retinitis pigmentosa: the Friedenwald lecture', *Investigative Ophthalmology and Visual Science*, **34**, 1659–1676

Cideyicin, A.V. *et al.*, 1998, 'Disease sequence from mutant rhodopsin allele to rod and cone photoreceptor degeneration in man', *Proceedings of the National Academy of Sciences USA*, **95**, 7103–7108

Humphries, P., P. Kenna and G.J. Farrar, 1992, 'On the molecular genetics of retinitis pigmentosa', *Science*, **256**, 804–808

Milam, A.H., 1993, 'Strategies for the rescue of retinal photoreceptor cells', *Current Opinion in Neurobiology*, **3**, 797–804

CHAPTER 14: THE ACTION POTENTIAL

Web sites

Useful introduction to action potentials, etc., at:
http://gwis2.circ.gwu.edu/~atkins/Neuroweb/Action.html
Electron micrograph of incisures of Schmidt–Lanterman at:
http://synapses.bu.edu/atlas/5_11.stm

Print

1. General

Aidley, D.J., 1998, *The Physiology of Excitable Cells* (4th edn), Cambridge: Cambridge University Press

*Hille, B., 2001, *Ionic Channels of Excitable Membranes* (3rd edn), Sunderland, MA: Sinauer

*Hodgkin, A.L. and A.F. Huxley, 1939, 'Action potentials recorded from inside a nerve fibre', *Nature*, **144**, 710–711

Matthews, G.G., 1986, *Cellular Physiology of Nerve and Muscle*, Palo Alto: Blackwell Scientific Publications

Ruch, T.C. and H.D. Patton, 1979, *Physiology and Biophysics: The Brain and Neural Function*, Philadelphia: Saunders

Sakmann, B. and E. Neher, eds, 1983, *Single-Channel Recording*, New York: Plenum Press

Sherrington, C.S., 1951, *Man on his Nature*, Harmondsworth: Penguin Books

2. Voltage-clamp analyses

Brismar, T. and B. Frankenhaeuser, 1981, 'Potential clamp analysis of mammalian myelinated fibres', *Trends in Neurosciences*, **4**, 68–70

*Hodgkin, A.L. and A.F. Huxley, 1952, 'A quantitative description of membrane current and its application to conduction and excitation in nerve', *Journal of Physiology*, **117**, 500–544

*Hodgkin, A.L., A.F. Huxley and B. Katz, 1952, 'Measurement of current–voltage relations in the membrane of the giant axon of *Loligo*', *Journal of Physiology*, **116**, 424–448

3. Patch-clamp analyses

Patlak, J. and R. Horn, 1982, 'Effect of N-bromoacetamide on single sodium channel currents in excised membrane patches', *Journal of General Physiology*, **79**, 333–351

Sakmann, B. and E. Nehr, 1984, 'Patch clamp techniques for studying ionic channels in excitable membranes', *Annual Review of Physiology*, **46**, 455–472

Sigworth, F.J. and E. Neher, 1980, 'Single Na^+ channel currents observed in cultured rat muscle cells', *Nature*, **287**, 447–449

4. Propagation of the action potential

Hodgkin, A.L., 1964, *The Conduction of the Nerve Impulse*, Springfield, IL: Thomas

Poo, M.M., 1985, 'Mobility and localisation of proteins in excitable membranes', *Annual Review of Neuroscience*, **8**, 369–406

5. Initiation of the action potential

Combs, J.S., D.R. Curtis and J.C. Ecles, 1957, 'The generation of impulses in motoneurones', *Journal of Physiology*, **139**, 232–249

6. Rate of impulse propagation

Hodgkin, A.L., 1964, *The Conduction of the Nerve Impulse*, Springfield, IL: Thomas

Livingstone, R.B. *et al.*, 1973, 'Specialised paranodal and interparanodal glial-axonal junctions in the peripheral and central nervous system: a freeze-etching study', *Brain Research*, **58**, 1–24

Box 14.1: Early history of the impulse

Galvani, L, 1791, *De viribus electicitatus*, Bologna; trs. M.G. Foley, 1954, Burndy Library Inc., Norwalk, Connecticut

Piccolino, M., 1997, 'Luigi Galvani and animal electricity: two centuries after the foundation of electrophysiology', *Trends in Neurosciences*, **20**, 443–448

Smith, C.U.M., 1976, *The Problem of Life*, London: Macmillan

Box 14.2: Switching off neurons by manipulating K⁺ channels

Johns, D.C. *et al.*, 1999, 'Inducible genetic suppression of neuronal excitability', *Journal of Neuroscience*, **19**, 1601–1697

Osterwalder, T. *et al.*, 2001, 'A conditional tissue-specific transgene expression system using inducible GAL4', *Proceedings of the National Academy of Sciences USA*, **98**, 12596–12601

White, B.H., 2001, Targetted attenuation of electrical activity in *Drosophila* using a genetically modified K⁺-current', *Neuron*, **31**, 699–711

CHAPTER 15: THE NEURON AS A SECRETORY CELL

Web sites

Clathrin

Electron micrographs of clathrin-coated vesicles at:
www.heuserlab.wustl.edu/Classic30Link.html
If you have QuickTime the following site provides a movie of clathrin activity:
www.hms.harvard.edu/news/clathrin

Tubulin

A 3.7 Å model can be viewed at:
www.lbl.gov/Science-Articles/Archive/3D-tubulin.html
Recent output from the Nogales tubulin laboratory including detailed molecular models may be viewed at:
http://cryoem.berkeley.edu/

Kinesin and axoplasmic transport

A very informative site which includes movies and animations is the Kinesin Home Page at:
www.proweb.org/kinesin

Synaptic exocytosis

A useful summary of the biochemical events responsible for the release of transmitter into the synaptic cleft can be found at:
www.nimh.nih.gov/events/prfusion.cfmfig6
and a video animation (for those with appropriate software) at:
http://www.drugabuse.gov/Genetics/videocast.htm

Print

1. General

*Alberts, B. *et al.*, 1994, *Molecular Biology of the Cell* (3rd edn), New York: Garland Publishing

Lodish, J.H. *et al.*, 2000, *Molecular Cell Biology* (4th edn), New York: Scientific American Books

2. Neurons and secretions

Dale, H.H., 1935, 'Pharmacology and nerve endings', *Proceedings of the Royal Society of Medicine*, **28**, 319–332

Donovan, B.T., 1970, *Mammalian Neuroendocrinology*, London: McGraw-Hill

3. Synthesis in the perikaryon

Farquhar, M.G., 1985, 'Progress in unravelling pathways of Golgi traffic', *Annual Review of Cell Biology*, **1**, 447–488

Kelly, R.B., 1985, 'Pathways of protein secretion in eukaryotes', *Science*, **230**, 25–32

Lodish, J.H. *et al.*, 2000, *Molecular Cell Biology* (4th edn), New York: Scientific American Books

Walter, P.R., R. Gilmore and G. Blobel, 1984, 'Protein translocation across the endoplasmic reticulum', *Cell*, **38**, 5–8

Wickner, W. and H.F. Lodish, 1985, 'Multiple mechanisms of insertion of proteins into and across membranes', *Science*, **230**, 400–407

4. Transport along the axon

Almenar-Queralt, A. and L.S.B. Goldstein, 2001, 'Linkers, packages and pathways: new concepts in axonal transport', *Current Opinion in Neurobiology*, **11**, 550–557

Baas, P.W. and A. Brown, 1997, 'Slow axonal transport: the polymer transport model', *Trends in Cell Biology*, **7**, 380–384

Bray, D., 1997, 'The riddle of slow transport – an introduction', *Trends in Cell Biology*, **7**, 379

Downing, K.H. and E. Nogales, 1998, 'Tubulin and microtubule structure', *Current Opinion in Cell Biology*, **10**, 16–22

Goldstein, L.S.B. and Z. Yang, 2000, 'Microtubule based transport systems in neurons: the role of kinesins and dyneins', *Annual Review of Neuroscience*, **23**, 39–71

*Hirokawa, N., 1986, 'Quick-freeze, deep-etch visualisation of the axonal cytoskeleton', *Trends in Neurosciences*, **9**, 67–71

Hirokawa, N., Y. Noda and Y. Okada, 1998, 'Kinesin and dynein superfamily proteins in organelle transport and cell division', *Current Opinion in Cell Biology*, **10**, 60–73

Kamal, A. and L.S. Goldstein, 2000, 'Connecting vesicle transport to the cytoskeleton', *Current Opinion in Cell Biology*, **12**, 503–508

Kamal, A. *et al.*, 2000, 'Axonal transport of amyloid precursor protein is mediated by direct binding to the kinesin light chain subunit of kinesin-1', *Neuron*, **28**, 449–459

*Lasek, R.J., 1986, 'Polymer sliding in axons' in *The Cytoskeleton: Cell Function and Organisation*, eds C. Lloyd, J. Hyams and R. Warn, *Journal of Cell Science*, Suppl. 5, 161–179

Miki, H. *et al.*, 2001, 'All kinesin superfamily protein, KIF, genes in mouse and human', *Proceedings of National Academy of Science*, **98**, 7004–7011

Musacchio, A. *et al.*, 1999, 'Functional organisation of clathrin in coats: combining electron cryomicroscopy and X-ray crystallography', *Molecular Cell*, **3**, 761–770

Nogales, E., S.G. Wolf and K.H. Downing, 1998, 'Structure of the tubulin dimer by electron crystallography', *Nature*, **391**, 199–203

Nogales, E. *et al.*, 1999, 'High resolution model of the microtubule', *Cell*, **96**, 79–88

Nogales, E., 2001, 'Structural insights into microtubule function', *Annual Review of Biophysics and Biomolecular Structure*, **30**, 397–420

Osborn, M. and K. Weber, 1986, 'Intermediate filament proteins: a multigene family distinguishing major cell lineages', *Trends in Biochemical Sciences*, **11**, 469–472

Sablin, E.P., 2000, 'Kinesins and microtubules: their structures and motor mechanisms', *Current Opinion in Cell Biology*, **12**, 35–41

Schnitzer, M.J., 2001, 'Molecular motors: doing a rotary two-step', *Nature*, **410**, 878–9, 881

Stelkov, S.V. *et al.*, 2001, 'Divide and conquer crystallographic approach towards an atomic structure of intermediate filaments', *Journal of Molecular Biology*, **306**, 773–781

*Weiss, D.G., 1986, 'Visualisation of the living cytoskeleton by video-enhanced microscopy and digital image processing' in *The Cytoskeleton: Cell Function and Organisation*, eds C. Lloyd, J. Hyams and R. Warn, *Journal of Cell Science*, Suppl. 5, 1–15

Zhao, C. *et al.*, 2001, 'Charcot–Marie–Tooth disease type 2A caused by mutation in a microtubule motor KIF1Bbeta', *Cell*, **105**, 587–597

5. Vesicle docking, exocytosis and endocytosis at the synaptic terminal

Augustine, G.J., 2001, 'How does Ca^{2+} trigger neurotransmitter release?', *Current Opinion in Neurobiology*, **11**, 320–326

*Akert, K. *et al.*, 1972, 'Freeze etching and cytochemistry of vesicles and membrane complexes in synapses of the central nervous system' in *Structure and Function of Synapses*, eds G.D. Pappas and G.D. Purpura, New York: Raven Press, 67–86

Baines, A.J. and V. Bennett, 1985, 'Synapsin 1 is a spectrin-binding protein immunologically related to erythrocyte protein 4.1', *Nature*, **315**, 410–413

Brunger, A.T., 2001, 'Structure of proteins involved in synaptic vesicle fusion in neurons', *Annual Review of Biophysics and Biomolecular Structure*, **30**, 157–171

Chen, Y.A. *et al.*, 1999, 'SNARE complex formation is driven by Ca^{2+} and drives membrane fusion', *Cell*, **97**, 165–174

Chi, P., P. Greengard and T.A. Ryan, 2001, 'Synapsin dispersion and reclustering during synaptic activity', *Nature Neuroscience*, **4**, 1187–1193

*Heuser, J.E. and T.S. Reese, 1979, 'Synaptic-vesicle exocytosis captured by quick freezing' in *The Neurosciences: Fourth Study Program*, eds F.O. Schmitt and F.G. Worden, Cambridge, MA: MIT Press, 573–600

Jahn, R. and T.C. Südhof, 1999, 'Membrane fusion and exocytosis', *Annual Review of Biochemistry*, **68**, 863–911

Jarouse, N. and R. Kelly, 2001, 'Endocytotic mechanisms in synapses', *Current Opinion in Cell Biology*, **13**, 461–469

Kelly, R.B., 1993, 'Storage and release of neurotransmitters', *Cell* **72**/*Neuron* **10**, 45–53

Kelly, R.B., 1995, 'Synaptotagmin is just a calcium sensor', *Current Biology*, **5**, 257–259

Kira, M.S. *et al.*, 2000, 'Three dimensional structure of the neuronal-Sec1-syntaxin-1a complex', *Nature*, **404**, 355–362

Kirchhausen, T., 2000, 'Clathrin', *Annual Review of Biochemistry*, **69**, 699–727

Littleton, J.T. and H.J. Bellen, 1995, 'Synaptotagmin controls and modulates synaptic-vesicle fusion in a Ca^{2+}-dependent manner', *Trends in Neurosciences*, **18**, 177–183

Mundigl, O. and P. De Camillo, 1994, 'Formation of synaptic vesicles', *Current Opinion in Cell Biology*, **6**, 561–567

Pevsner, J. and R.H. Scheller, 1994, 'Mechanisms of vesicle docking and fusion: insights from the nervous system', *Current Opinion in Cell Biology*, **6**, 555–560

Silver, R.B. *et al.*, 1994, 'Time resolved imaging of Ca^{2+}-dependent aequorin luminescence of microdomains and QEDs in synaptic preterminals', *Biological Bulletin*, **187**, 293–299

Sudhof, T.C., 2000, 'The synaptic vesicle revisited', *Neuron*, **28**, 317–320

Box 15.1: Subcellular geography of protein biosynthesis

Steward, O., 1995, 'Targetting of mRNAs to subsynaptic microdomains in dendrites', *Current Opinion in Neurobiology*, **5**, 55–61

Steward, O. and P.F. Worley, 2001, 'Selective targeting of newly synthesised ARc mRNA to active synapses requires NMDA receptor activation', *Neuron*, **30**, 227–240

Thomas, K.L. *et al.*, 1994, 'Spatial and temporal changes in signal transduction pathways during LTP', *Neuron*, **13**, 737–745

Box 15.2: Vesicular neurotransmitter transporters

Brenner, S., 1974, 'The genetics of *Caenorhabditis elegans*', *Genetics*, **77**, 71–94

Erikson, J.D. *et al.*, 1994, 'Functional identification of a vesicular acetylcholine transporter and its expression from a "cholinergic gene locus"', *Journal of Biological Chemistry*, **269**, 21929–21932

*Usdin, T.B. *et al.*, 1995, 'Molecular biology of the vesicular ACh transporter', *Trends in Neurosciences*, **18**, 218–224

CHAPTER 16: NEUROTRANSMITTERS AND NEUROMODULATORS

Web sites

An introductory account of neurotransmitters and neuroactive peptides accompanied by colourful animations may be found at:
http://faculty.washington.edu/chudler/chnt1.html
More detail of the biochemistry of neurotransmitters, their agonists, antagonists, etc. is available at:
http://dog.net.uk/neurotransmitters.html
Excellent rotatable 3D graphics of acetylcholine, noradrenaline, serotonin and glutamate may be viewed at:
harveyproject.science.wayne.edu/development/nervous_system/ cell_neuro/synapses/xmtrs.html
For gasnets see:
http://www.cogs.susx.ac.uk/users/toms/GasNets/

Print

1. General

Alexander, S. *et al.*, 2001, Nomenclature Supplement, *Trends in Pharmacological Sciences*, Cambridge: Elsevier

Bach-y-Rita., P., 1995, *Non-synaptic Diffusion Neurotransmission and Late Brain Reorganisation*, New York: Demos

*Bradford, H.F., 1986, *Chemical Neurobiology*, New York: Freeman

Cooper, J.R., F.E. Bloom and R.H. Roth, 1996, *The Biochemical Basis of Neuropharmacology* (7th edn), Oxford: Oxford University Press

Kruk, Z.L. and C.L. Pycock, 1991, *Neurotransmitters and Drugs* (3rd edn), London: Chapman and Hall

McIlwain, H. and H.S. Bachelard, 1985, *Biochemistry and the Central Nervous System* (5th edn), Edinburgh: Churchill Livingstone

Zoli, M. *et al.*, 1999, 'Volume transmission in the CNS and its relevance for neuropsychopharmacology', *Trends in Pharmacological Science*, **20**, 142–150

2. Acetylcholine

*Cuello, A.C. and M.V. Sofroniew, 1984, 'The anatomy of CNS cholinergic neurons', *Trends in Neurosciences*, **7**, 74–78

3. Amino acids

Fagg, G.E. and A.C. Foster, 1983, 'Amino acid neurotransmitters and their pathways in the mammalian central nervous system', *Neuroscience*, **9**, 701–719

Watkins, J.C., 1984, 'Excitatory amino acids and central synaptic transmission', *Trends in Pharmacological Sciences*, **5**, 373–376

4. Serotonin (5-HT)

Fuller, R.W., 1980, 'Pharmacology of central serotonin neurons', *Annual Review of Pharmacology and Toxicology*, **20**, 111–127

Humphrey, P.P.A., P. Hartig and D. Hoyer, 1993, 'A proposed new nomenclature for the 5-HT receptor', *Trends in Pharmacological Sciences*, **14**, 233–236

Richardson, B.P. and G. Engel, 1986, 'The pharmacology and function of 5-HT$_3$ receptors', *Trends in Neurosciences*, **9**, 424–428

5. Catecholamines

Fuller, R.W., 1982, 'Pharmacology of brain epinephrine neurons', *Annual Review of Pharmacology and Toxicology*, **22**, 31–55

Lindvall, O. and A. Björklund, 1983, 'Dopamine and nor-epinephrine containing neuron systems: Their anatomy in the rat brain', in *Chemical Neuroanatomy*, ed. P.C. Emson, New York: Raven Press, 229–255

Morrison, J.H., M.G. Molliver and R. Grzanna, 1979, 'Noradrenergic innervation of cerebral cortex: widespread effects of local cortical lesions', *Science*, **205**, 313–316

Roth, R.H., 1984, 'CNS dopamine autoreceptors: Distribution, pharmacology and function', *Annals of the New York Academy of Science*, **430**, 27–53

6. Purines

Dubyak, G.R. and J.S. Fedan, 1990, 'Biological actions of extracellular ATP', *Annals of the New York Academy of Science*, **603**, 1–542

Khakh, B.S., 2001, 'Molecular physiology of P2X receptors and ATP signalling at synapses', *Nature Reviews Neuroscience*, **2**, 165–174

7. Cannabinoids

Elphick, M.R. and M. Egertova, 2001, 'The neurobiology and evolution of cannabinoid signalling', *Philosophical Transactions of the Royal Society* (*Lond.*) B, **356**, 381–408

Montgomery, J.M. and D.V. Madison, 2001, 'The grass roots of synapse suppression', *Neuron*, **29**, 567–570

Ohno-Shosaku, T., T. Maejima and M. Kano, 2001, 'Endogenous cannabinoids mediate retrograde signals from depolarised postsynaptic neurons to presynaptic terminals', *Neuron*, **29**, 729–738

Schlicker, E. and M. Kathmann, 2001, 'Modulation of transmitter release via presynaptic cannabinoid receptors', *Trends in Pharmacological Sciences*, **22**, 565–572

Voth, E.A. and R.H. Schwartz, 1997, 'Medicinal applications of delta-9-tetrahydrocannabinol and marijuana', *Annals of Internal Medicine*, **126**, 791–798

8. Peptides

Hökfelt, T., 1991, 'Neuropeptides in perspective: the last ten years', *Neuron*, **7**, 867–879

Iverson, L.L., 1984, 'Amino acids and peptides: Fast and slow chemical signals in the nervous system?', *Proceedings of the Royal Society B* (*London*), **221**, 245–260

Lundberg, J.M. and T. Hökfelt, 1983, 'Coexistence of peptides and classical neurotransmitters', *Trends in Neurosciences*, **6**, 325–332

Mansour, A. *et al.*, 1988, 'Anatomy of CNS opioid receptors', *Trends in Neurosciences*, **11**, 308–314

Miller, R.J., 1986, 'Peptides as neurotransmitters: focus on the encephalins' in *Neuropeptides and Behaviour*, vol. 1, eds D. De Wied, W.H. Gispen and Tj. B. Van Wimersma Greidamus, Oxford: Pergamon Press

Palkovits, M., 1984, 'Distribution of neuropeptides in the central nervous system: A review of biochemical mapping studies', *Progress in Neurobiology*, **23**, 151–189

Stefano, G.B. *et al.*, 2000, 'Endogenous morphine', *Trends in Neurosciences*, **23**, 436–442

Wahlestedt, C. and D.J. Reis, 1993, 'Neuropeptide Y-related peptides and their receptors – are the receptors potential drug targets?', *Annual Review of Pharmacology and Toxicology*, **32**, 309–352

9. Nitric oxide

Garthwaite, J., S.L. Charles and R. Chess-Williams, 1988, 'Endothelium-derived relaxing factor release on activation of NMDA receptors suggests role as intracellular messenger in the brain', *Nature*, **336**, 385–388

Marletta, M.A., 1994, 'Nitric oxide synthase: aspects concerning structure and catalysis', *Cell*, **78**, 927–930

Nathan, C. and Q.-W. Xie, 1994, 'Nitric oxide synthases: rules, tolls and controls', *Cell*, **78**, 915–918

Schmidt, H.H.H.W. and U. Walter, 1994, 'NO at work', *Cell*, **78**, 919–925

Schuman, E.M. and D.V. Madison, 1994, 'Nitric oxide and synaptic function', *Annual Review of Neuroscience*, **17**, 153–183

Box 16.2: Otto Loewi and Vagusstoff

Cannon, W.B., 1934, 'The story of the development of our ideas of chemical mediation of nerve impulses', *American Journal of Medical Science*, **188**, 145–159

Loewi, O., 1935, 'The Ferrier Lecture: On problems connected with the principle of humoral transmission of nervous impulses', *Proceedings of the Royal Society, B*, **118**, 299–316

Box 16.3: Reuptake neurotransmitter transporters

*Amara, S.G. and M. Kuhar, 1993, 'Neurotransmitter transporters', *Annual Review of Neuroscience*, **16**, 73–93

Guastella, J. *et al.*, 1990, 'Cloning and expression of a rat brain GABA transporter', *Science*, **249**, 1303–1306

CHAPTER 17: THE POSTSYNAPTIC CELL

Web sites

EM of postsynaptic density at:
http://synapses.bu.edu/anatomy/chemical/psd.htm
The European Pineal and Biological Rhythms Society (EPBRS) from which other Pineal sites may be accessed is at:
http://www.eps.mai.ku.dk/

Print

1. General

Bradford, H.F., 1986, *Chemical Neurobiology*, New York: W.H. Freeman

Matthews, G.G., 1986, *Cellular Physiology of Nerve and Muscle*, Palo Alto: Blackwell

2. The postsynaptic density

*Froehner, S.C., 1986, 'The role of the postsynaptic cytoskeleton in AChR organisation', *Trends in Neurosciences*, **9**, 37–41

Kennedy, M.B., 1993, 'The post-synaptic density', *Current Opinion in Neurobiology*, **3**, 732–737

Pathy, L. and K. Nikolics, 1993, 'Functions of agrin-related proteins', *Trends in Neurosciences*, **16**, 76–81

Sheng, M. and C. Sala, 2001, 'PDZ domains and the organisation of supramolecular complexes', *Annual Review of Neuroscience*, **24**, 1–29

3. Electrophysiology of the postsynaptic membrane

Eccles, J.C., 1964, 'Ionic mechanisms and post-synaptic inhibition', *Science*, **145**, 1140–1147 (Nobel lecture)

4. Ion channels in the postsynaptic membrane

*McBurney, R.N., 1983, 'New approaches to the study of rapid events underlying neurotransmitter action', *Trends in Neurosciences*, **6**, 297–302

Sakmann, B., 1992, 'Elementary steps in synaptic transmission revealed by currents through single ion channels', *Neuron*, **8**, 613–629 (Nobel lecture)

5. Modulation of ion channels by second messenger systems

Brown, D.A., 1983, 'Slow cholinergic excitation – a mechanism for increasing neuronal excitability', *Trends in Neurosciences*, **6**, 301–306

Siegelbaum, S.A. and R.W. Tsien, 1983, 'Modulation of gated ion channels as a mode of transmitter action', *Trends in Neurosciences*, **6**, 307–313

6. Other effects of second messengers

*Dubner, R. and M.A. Rudner, 1992, 'Activity-dependent neuronal plasticity following tissue injury and inflammation', *Trends in Neurosciences*, **15**, 96–103

Miller, R.J., 1986, 'Protein kinase C: a key regulator of neuronal excitability?', *Trends in Neurosciences*, **9**, 538–541

Morgan, J.I. and T. Curran, 1989, 'Stimulus-transcription coupling in neurons: role of cellular immediate-early genes', *Trends in Neurosciences*, **12**, 459–462

7. Pineal

Carter, D.A., 1993, 'Differential intracellular mechanisms mediate the co-ordinate induction of *c-fos* and *jun-B* in the rat pineal gland', *European Journal of Pharmacology: Molecular Pharmacology Section*, **244**, 285–291

Krause, D.N. and M.L. Dubocovich, 1990, 'Regulatory sites in the melatonin system of mammals', *Trends in Neurosciences*, **13**, 464–570

Morrisey, J.J., 1981, 'The neuronal induction of pineal gland n-acetyl-transferase activity' in *Essays in Neurochemistry and Neuropharmacology*, vol. 5, eds M.B.H. Youdin, W. Lovenberg, D.F. Sharman and J.R. Lagnano, Chichester: John Wiley

Box 17.1: Cajal, Sherrington and the beginnings of synaptology

Finger, S., 2000, *Minds Behind the Brain: A History of the Pioneers and their Discoveries*, Oxford: Oxford University Press

Ramón y Cajal, S., 1892, 'La fine structure des centres nerveux', *Proceedings of the Royal Society, B*, **55**, 444–468

Shepherd, G.M., 1991, *Foundations of the Neuron Doctrine*, New York: Oxford University Press

Smith, C.U.M., 1996, 'Sherrington's legacy: Evolution of the concept of the synapse, 1894–1994', *Journal of the History of Neuroscience*, **5**, 1–14

CHAPTER 18: DEVELOPMENTAL GENETICS OF THE BRAIN

Web sites

http://anatomy.med.unsw.edu.au/CBL/embryo provides regularly updated and well-illustrated embryology site with plenty of links to other more detailed sites.

Print

1. General

Bonhoeffer, F. and J.R. Sanes, 1995, eds, 'Development', *Current Opinion in Neurobiology*, **5**, 1–112
Gilbert, S.F., 1994, *Developmental Biology* (4th edn), Sunderland, MA: Sinauer
Hall, B.K., 1997, 'Germ layers and the germ layer theory revisited: Primary and secondary germ layers, neural crest as a fourth germ layer, homology, demise of the germ-layer theory', *Evolutionary Biology*, **30**, 121–186
Laufer, E. and V. Marigo, 1994, 'Evolution in developmental biology: of morphology and molecules', *Trends in Genetics*, **10**, 261–263
Romanes, G.J., 1901, *Darwin and After Darwin*, London: Open Court
Russell, E.S., 1916, *Form and Function: A Contribution to the History of Animal Morphology*, London: John Murray
von Baer, K.E., 1828, *Entwicklungsgechichte der Thiere: Beobachtung und Reflexion*, Konigsberg: Borntrüger

2. *Drosophila*

*Finkelstein, R. and E. Boncinelli, 1994, 'From fly head to mammalian forebrain: the story of *otd* and *Otx*', *Trends in Genetics*, **10**, 310–315
Ingham, P.W., 1988, 'The molecular genetics of embryonic pattern formation in *Drosophila*', *Nature*, **335**, 25–34
Mahajan, S. and L. Cooley, 1994, 'Intercellular cytoplasm transport during *Drosophila* oogenesis', *Developmental Biology*, **165**, 336–351
Yuste, R. *et al.*, 1995, 'Neuronal domains in developing neocortex: mechanisms of co-activation', *Neuron*, **14**, 1–11

3. Early development of vertebrate brain

Lemke, G. *et al.*, 1991, 'Expression and activity of the transcription factor SCIP during glial cell differentiation and myelination', *Annals of the New York Academy of Sciences*, **633**, 189–195
Lumsden, A. and R. Keynes, 1989, 'Segmental patterns of development in chick hindbrain', *Nature*, **337**, 424–428
Puelles, L. and J.L.R. Rubenstein, 1993, 'Expression patterns of homeobox and other putative regulatory genes in the embryonic mouse forebrain suggest a neuromeric organisation', *Trends in Neurosciences*, **16**, 472–480
Riji, F.M. *et al.*, 1993, 'A homeotic transformation is generated in the rostral branchial region of the head by disruption of Hoxa-2 which acts as a selector gene', *Cell*, **75**, 1333–1349
Rubenstein, J.L.R. *et al.*, 1994, 'The embryonic vertebrate forebrain: the Prosemeric model', *Science*, **266**, 578–580

Ruiz i Altaba, A., 1994, 'Pattern formation in the vertebrate neural plate', *Trends in Neurosciences*, **17**, 233–243
Simon, H., A. Hornbruch and A. Lumsden, 1995, 'Independent assignment of antero-posterior and dorso-ventral positional values in the developing chick hindbrain', *Current Biology*, **5**, 205–214
Smith, J.C., 1994, 'Hedgehog, the floor plate and the zone of polarising activity', *Cell*, **76**, 193–196
Wilkinson, D.G. *et al.*, 1989, 'Segmental expression of Hox-2 homeobox-containing genes in the developing hindbrain', *Nature*, **341**, 405–409

4. Homeobox genes

*Botas, J., 1993, 'Control of morphogenesis and differentiation by HOM/Hox genes', *Current Opinion in Cell Biology*, **5**, 1015–1022
Gehring, W.J., M. Affolter and T. Birglin, 1994, 'Homeodomain proteins', *Annual Review of Biochemistry*, **63**, 487–526
Kenyon, C., 1994, 'If birds can fly, why can't we? Homeotic genes in evolution', *Cell*, **78**, 175–180
*Keynes, R. and R. Krumlauf, 1994, 'HOX genes and the regionalisation of the nervous system', *Annual Review of Neuroscience*, **17**, 109–132
Krumlauf, R., 1993, 'Hox genes and pattern formation in the branchial region of the vertebrate head', *Trends in Genetics*, **9**, 106–112
Krumlauf, R., 1994, 'Hox genes in vertebrate development', *Cell*, **78**, 191–201
McGinnis, W. and R. Krumlauf, 1992, 'Homeobox genes and axial patterning', *Cell*, **68**, 283–302
Salser, S.J. and C. Kenyon, 1994, 'Patterning of *C. elegans*: homeotic cluster genes, cell fates and cell migrations', *Trends in Genetics*, **10**, 159–164

5. *POU* and other genes

Rosenfeld, M.G, 1991, 'POU-domain transcription factors: pou-er-ful developmental regulators', *Genes and Development*, **5**, 897–907
Treacy, M.N., X. He and M.G. Rosenfeld, 1991, 'A POU-domain protein that inhibits neuron-specific gene activation', *Nature*, **350**, 577–584
Isshiki, T. *et al.*, 2001, '*Drosophila* neuroblasts sequentially express transcription factors which specifiy the temporal identity of their neuronal progeny', *Cell*, **106**, 511–521

6. *pax-6* and the developmental genetics of eyes

Callaerts, P., G. Halder and W.J. Gehring, 1997, '*Pax-6* in development and evolution', *Annual Review of Neuroscience*, **20**, 483–532
Heyman, I. *et al.*, 1999, 'Psychiatric disorder and cognitive function in a family with an inherited novel mutation of the developmental control gene *PAX6*', *Psychiatric Genetics*, **9**, 85–90
Kurusu, M., 2000, 'Genetic control of development of the mushroom bodies, the associative learning centers in *Drosophila* brain by *eyeless, twin of eyeless* and *Dachshund* genes', *Proceedings of the National Academy of Sciences USA* **97**(5), 2140–2144

Tomarev, S.I. *et al.* 1997, 'Squid *pax-6* and eye development', *Proceedings of the National Academy of Sciences USA*, **94**, 2421–2426.

CHAPTER 19: EPIGENETICS OF THE BRAIN

Web sites

Growth cones

QuickTime movies of growth cone movements in response to EphB are at:
http://www.ucsf.edu/neurosc/faculty/SretAvan/Eph-timelapse. html

Morphogenesis of Drosophila compound eye

Detailed information including QuickTime animation at:
http://sdb.bio.purdue.edu/fly/aimorph/eye.htm

Stem cells

Information about neural stem cells may be found at:
http://stemcells.alphamedpress.org/
and
http://www.dtemcellresearchnews.com/

Print

1. General

*Cowan, W.M., 1979, 'The development of the brain', *Scientific American*, **241**(3), 107–117

Jacobson, M., 1991, *Developmental Neurobiology* (3rd edn), New York: Plenum

Malimski, G. and S. Bryant, eds, 1984, *Pattern Formation: A Primer in Developmental Biology*, London: Macmillan

*Purves, D. and J.W. Lichtman, 1984, *Principles of Neural Development*, Sunderland, MA: Sinauer

Ribchester, R.R., 1986, *Molecule, Nerve and Embryo*, Glasgow: Blackie

Various, 1985, 'Developmental neurobiology – special issue', *Trends in Neurosciences*, **8**, 229–300

Waddington, C.H., 1957, *The Strategy of the Genes*, London: Allen and Unwin

2. The origins of neurons and glia

Briscoe, J. *et al.*, 2000, 'A homeodomain protein code specifies progenitor cell identity and neuronal fate in the ventral neural tube', *Cell*, **101**, 435–445

D'Arcangelo, G. *et al.*, 1995, 'A protein related to extracellular matrix proteins deleted in the mouse mutant reeler', *Nature*, **374**, 719–723

Hatten, M.E and C.A. Mason, 1986, 'Neuron-astroglial interactions *in vitro* and *in vivo*', *Trends in Neurosciences*, **9**, 168–174

Marquardt, T. and S.L. Pfaff, 2001, 'Cracking the trancriptional code for cell specification in the neural tube', *Cell*, **106**, 651–654

3. Morphogenesis of neurons

*Altman, J., 1967, 'Postnatal growth and differentiation of the mammalian brain, with implications for a morphological theory of memory', in *The Neurosciences*, eds G.C. Quarton, T. Melnechuk and F.O. Schmitt, New York: Rockefeller University Press

Gao, F.-B. *et al.*, 1999, 'Genes regulating dendritic outgrowth, branching and routing in *Drosophila*', *Genes and Development*, **13**, 2549–2561

Gao, F.-B. *et al.*, 2000, 'Control of dendritic field formation in *Drosophila*: the roles of Flamingo and competition between homologous neurons', *Neuron*, **28**, 91–101

Marrs, G.S., S.H. Green and M.E. Dailey, 2001, 'Rapid remodelling of postsynaptic densities in developing dendrites', *Nature Neuroscience*, **4**, 1006–1013

Matus, A., 2001, 'Moving molecules make synapses', *Nature Neuroscience*, **4**, 967–968

*Rakic, P., 1979, 'Genetic and epigenetic determinants of local neuronal circuits in the mammalian central nervous system' in *The Neurosciences, Fourth Study Program*, eds F.O. Schmitt and F.G. Worden, Cambridge, MA: MIT Press, 109–127

Scott, E.K. and L. Luo, 2001, 'How do dendrites take their shape', *Nature Neuroscience*, **4**, 359–365

4. Neural stem cells

Anderson, D.J., 2001, 'Stem cells and pattern formation in the nervous system: The possible versus the actual', *Neuron*, **30**, 19–35

Barnea, A. and F. Nottebohm, 1994, 'Seasonal recruitment of hippocampal neurons in adult free-ranging black-capped chickadees', *Proceedings of the National Academy of Sciences USA*, **91**, 11217–11221

Barnea, A. and F. Nottebohm, 1996, 'Recruitment and replacement of hippocampal neurons in young and adult chickadees: an addition to the theory of hippocampal learning', *Proceedings of the National Academy of Sciences USA*, **93**, 714–718

Gage, F.H., 2000, 'Mammalian neural stem cells', *Science*, **287**, 1433–1438

Gould, E. *et al.*, 1999, 'Neurogenesis in the neocortex of adult primates', *Science*, **286**, 548–552

Gould, E. *et al.*, 1999, 'Learning enhances adult neurogenesis in the hippocampal formation', *Nature Neuroscience*, **2**, 260–265

Kempermann, G., H.G. Kuhn and F.H. Gage, 1997, 'More hippocampal neurons in adult mice living in an enriched environment', *Nature*, **386**, 493–495

Marshak, D.R., R.L. Gardner and D. Gottlieb, 2001, *Stem Cell Biology*, Woodbury, NY: CSHL Press

Rakic, P., 2002, 'Neurogenesis in adult primate neocortex: an evaluation of the evidence', *Nature Reviews Neuroscience*, **3**, 65–71

Scheffler, R. *et al.*, 1999, 'Marrow-mindedness: a perspective on neuropoiesis', *Trends in Neurosciences*, **22**, 348–356

Temple, S., 2001, 'The development of neural stem cells', *Nature*, **414**, 112–118

Tramontin, A.D. and E.A. Brenowitz, 2000, 'Seasonal plasticity in the adult brain', *Trends in Neurosciences*, **23**, 251–258

5. Morphogenesis of *Drosphila* compound eye

Simon, M.A., D.D.L. Botwell and G.M. Rubin, 1989, 'Structure and activity of the sevenless protein required for photoreceptor development in *Drosophila*', *Proceedings of the National Academy of Sciences USA*, **86**, 8333–8337

Simon, M.A., G.S. Dobson and G.M. Rudin, 1993, 'An SH3-SH2-SH3 protein is required for p21^{Ras1} activation and binds to Sevenless and Sos *in vivo*', *Cell*, **73**, 169–177

Zipursky, S.L. and G.M. Rudin, 1994, 'Determination of the neuronal cell fate: lessons from the R7 neuron of Drosophila', *Annual Review of Neuroscience*, **17**, 373–397

6. Growth cones

Bray, D. and P.J. Hollenbeck, 1988, 'Growth cone motility and guidance', *Annual Review of Cell Biology*, **4**, 43–62

7. Pathfinding

Angelletti and Bradshaw, 1991, in *Principles of Neural Science*, eds Kandel, E.R., J.H. Schwartz and T.M. Jessell, 3rd edn, New York/London: Elsevier

Campenot, R.B., 1982, 'Development of sympathetic neurons in compartmentalised cultures', *Developmental Biology*, **93**, 1–21

Chen, H, Z. He and M. Tessier-Lavigne, 1998, 'Axon guidance mechanisms: semaphorins as simultaneous repellants and anti-repellants', *Nature Neuroscience*, **1**, 436–439

Davis, G.E., S. Varon, E. Engvall and M. Manthorpe, 1985, 'Substrate-binding neurite-promoting factors: relationships to laminin', *Trends in Neurosciences*, **8**, 528–532

Edgar, D., 1985, 'Nerve growth factors and molecules of the extracellular matrix in neuronal development', *Journal of Cell Science*, Suppl. 3, 107–113

Gundersen, R.W. and J.N. Barrett, 1979, 'Neuronal chemotaxis: chick dorsal root axons turn toward high concentrations of nerve growth factor', *Science*, **206**, 1079–1080

*Letourneau, P.C., 1975, 'Cell-to-substratum adhesion and guidance in neuronal morphogenesis', *Developmental Biology*, **44**, 77–91

Levi-Montalcini, R. and P. Calissano, 1986, 'Nerve growth factor as a paradigm for other polypeptide growth factors', *Trends in Neurosciences*, **9**, 473–477

8. Cell adhesion molecules (CAMs)

Becker, J.W. *et al.*, 1989, 'Topology of cell adhesion molecules', *Proceedings of the National Academy of Sciences USA*, **86**, 1088–1092

Cunningham, B.A., 1986, 'Cell adhesion molecules: a new perspective on molecular embryology', *Trends in Biochemical Sciences*, **11**, 423–426

Cunningham, B.A. *et al.*, 1987, 'Neural cell adhesion molecule: structure, immunoglobulin-like domains, cell surface modulation, and alternative RNA splicing', *Science*, **236**, 799–806

*Edelman, G.M., 1984, 'Modulation of cell adhesion during induction, histogenesis and perinatal development of the nervous system', *Annual Review of Neuroscience*, **7**, 339–377

Edelman, G.M., 1986, 'Cell adhesion molecules in neural histogenesis', *Annual Review of Physiology*, **48**, 417–430

Edelman, G.M. and K.L. Crossin, 1991, 'Cell adhesion molecules: implications for molecular histology', *Annual Review of Biochemistry*, **60**, 155–190

Rutishauser, U., 1986, 'Differential cell adhesion through spatial and temporal variations of NCAM', *Trends in Neurosciences*, **9**, 374–378

Rutishauser, U. and G. Goridis, 1986, 'N-CAM the molecule and its genetics', *Trends in Genetics*, **2**, 72–76

Rutishauser, U. and T.M. Jessell, 1988, 'Cell adhesion molecules in vertebrate neural development', *Physiological Review*, **68**, 819–857

Rutishauser, U. (1993), 'Cell adhesion molecules', *Current Opinion in Neurobiology*, **3**, 709

Wong, E.V. *et al.*, 1995, 'Mutations in cell adhesion molecule L1 cause mental retardation', *Trends in Neurosciences*, **18**, 169–172

9. Growth factors and differential survival

Barde, Y.A., 1989, 'Trophic factors and neuronal survival', *Neuron*, **2**, 1523–1584

Chao, M.V., 1992, 'Neurotrophic receptors: a window into neuronal differentiation', *Neuron*, **9**, 583–593

Clarke, P.G.H., 1985, 'Neuronal death in the development of the vertebrate nervous system', *Trends in Neurosciences*, **8**, 345–349

Johnson, J. and R. Oppenheim, 1994, 'Neurotrophins: keeping track of changing neurotrophic theory', *Current Biology*, **4**, 662–665

Snider, W.D., 1994, 'Function of the neurotrophins during nervous system development: what knock-outs are teaching us', *Cell*, **77**, 627–638

Sporn, M.B. and A.B. Roberts, 1992, 'Transforming growth factor-β: recent progress and new challenges', *Journal of Cell Biology*, **119**, 1017–1021

10. Morphopoietic fields

Steindler, D.A., 1993, 'Glial boundaries in the developing nervous system', *Annual Review of Neuroscience*, **16**, 445–470

Wolpert, L., 1978, 'Pattern formation in biological development', *Scientific American*, **239**(4), 153–164

11. Functional sculpting

Balice-Gordon, R.J. and J.W. Lichtman, 1994, 'Long-term synapse loss induced by focal blockade of post-synaptic receptors', *Nature*, **372**, 519–524

Barlow, H.B., 1975, 'Visual experience and cortical development', *Nature*, **258**, 199–204

Blaisdel, G.G. and J.D. Pettigrew, 1979, 'Degree of interocular synchrony required for maintenance of binocularity in the kitten's visual cortex', *Journal of Neurophysiology*, **42**, 1692–1710

*Constantine-Paton, M., 1981, 'Induced ocular-dominance zones in tectal cortex' in *The Organisation of the Cerebral Cortex*, eds F.O. Schmitt, F.G. Worden, G. Adelman and S.G. Dennis, Cambridge, MA: MIT Press: 47–67

*Hubel, D.H., T.N. Wiesel and S. Le Vay, 1977, 'Plasticity of ocular dominance columns in monkey striate cortex', *Philosophical Transactions of the Royal Society (B)*, **278**, 377–409

Hubel, D.H. and T.N. Wiesel, 1979, 'Brain mechanisms of vision', *Scientific American*, **241**(3), 130–144

Box 19.1: Eph receptors and ephrins

Davis, S. *et al.*, 1994, 'Ligands for EPH-related receptor tyrosine kinases that require membrane attachment or clustering for activity', *Science*, **266**, 816–819

Flanagan, J.G. and P. Vanderhaeghen, 1998, 'The Ephrins and Eph receptors on neural development', *Annual Review of Neuroscience*, **21**, 309–345

Himanen, J.-P. *et al.*, 2001, 'Crystal-structure of Eph receptor-ephrin complex', *Nature*, **414**, 933–938

Wilkinson, D.G., 2001, 'Multiple roles of EPH receptors and ephrins in neural development', *Nature Reviews Neuroscience*, **2**, 155–164

Box 19.2: Neurotransmitters as growth factors

Mattson, M.P. *et al.*, 1989, 'Fibroblast growth factor and glutamate – opposing roles in the generation and degeneration of hippocampal neuroarchitecture', *Journal of Neuroscience*, **9**, 3728–3740

Schwartz, J.P., 1992, 'Neurotransmitters as neurotrophic growth factors: a new set of functions', *International Review of Neurobiology*, **34**, 1–23

CHAPTER 20: MEMORY

Web sites

Drosophila: detailed anatomy of *Drosophila* brain, including mushroom bodies, antennal lobes, etc., can be found at:
http://flybrain.neurobio.arizona.edu/
Aplysia: detailed information may be accessed at the *Aplysia* database project:
http://mollusc.med.cornell.edu/
Dendritic spines: EM pictures of various types of spine may be viewed at:
http://synapses.bu.edu/atlas/1_4_1_2.stm

Print

1. General

Bower, G.H. and E.R. Hilgard, 1981, *Theories of Learning* (5th edn), Engelwood Cliffs, NJ: Prentice-Hall

Frank, D.A. and M.E. Greenberg, 1994, 'CREB: a mediator of long term memory from mollusks to mammals', *Cell*, **79**, 5–8

Hilgard, E.R. and D.G. Marquis, 1940, *Conditioning and Learning*, New York: Appleton-Century-Crofts

Martin, S.J., P.D. Grimwood and R.G.M. Morris, 2000, 'Synaptic plasticity and memory: an evaluation of the hypothesis', *Annual Review of Neuroscience*, **23**, 649–711

Shulz, D.E., 2000, 'Memories of memories: the endless alteration of the engram', *Neuron*, **28**, 25–29

*Thompson, R.F., 1986, 'The neurobiology of learning and memory', *Science*, **233**, 941–947

2. Some definitions

Bower, G.H. and E.R. Hilgard, 1981, *Theories of Learning* (5th edn), Englewood Cliffs, NJ: Prentice-Hall

3. Short- and long-term memory

Entingh, D. *et al.*, 1975, 'Biochemical approaches to the biological basis of memory' in *Handbook of Psychobiology*, eds M.S. Gazziniga and C. Blakemore, New York: Academic Press, 201–238

Goelet, P., V.F. Castellucci, S. Schacher and E.R. Kandel, 1986, 'The long and the short of long-term memory – a molecular framework', *Nature*, **322**, 419–422

4. Where is the memory trace located?

Brambilla, R. *et al.*, 1997, 'A role for the Ras signalling pathway in synaptic transmission and long-term memory', *Nature*, **390**, 281–286

Conquet, F. *et al.*, 1994, 'Motor deficit and impairment of synaptic plasticity in mice lacking mGluR1', *Nature*, **372**, 237–243

Izquierdo, I. *et al.*, 1997, 'Sequential role of hippocampus and amygdala, entorhinal cortex and parietal cortex in formation and retrieval of memory for inhibitory avoidance of pain in rats', *European Journal of Neuroscience*, **9**, 786–793

Lashley, K., 1950, 'In search of the Engram', *Symposium of Society of Experimental Biology*, **4**, 454–482

Thompson, R.F. *et al.*, 1983, 'The engram found? Initial localisation of the memory trace for a basic form of associative learning', *Progress in Psychobiological Physiological Psychology*, **10**, 167–196

5. Invertebrate systems

Alkon, D.L., 1984, 'Calcium-mediated reduction in ion channels: a biophysical memory trace (*Hemissenda*)', *Science*, **226**, 1037–1045

Carew, T.J. and C.L. Sahley, 1986, 'Invertebrate learning and memory: from behaviour to molecules', *Annual Review of Neuroscience*, **9**, 435–487

DeZazzo, J. and T. Tully, 1995, 'Dissection of memory formation: from behavioural pharmacology to molecular genetics', *Trends in Neurosciences*, **18**, 212–218

Hammer, M. and R. Menzel, 1995, 'Learning and memory in the honeybee', *Journal of Neuroscience*, **15**, 1627–1630

(a) Caenorhabditis elegans

Gomez, M. *et al.*, 2001, 'Ca^{2+} signalling via the neuronal calcium sensor-1 regulates associative learning and memory in *C. elegans*', *Neuron*, **30**, 241–248

*White, J.G., 1985, 'Neuronal connectivity in *Caenorhabditis elegans*', *Trends in Neurosciences*, **8**, 277–283

(b) Drosophila

Adams, M.D. *et al.*, 2000, 'The genome sequence of *Drosophila melanogaster*', *Science*, **287**, 2185–2195 de Belle, J.S. and M. Heisenberg, 1994, 'Associative odor learning in *Drosophila* abolished by chemical ablation of mushroom bodies', *Science*, **263**, 692–695

Dubnau, J. and T. Tully, 1998, 'Gene discovery in *Drosophila*: new insights into learning and memory', *Annual Review of Neuroscience*, **21**, 407–444

Dubnau, J. *et al.*, 2001, 'Disruption of neurotransmission in *Drosophila* mushroom body blocks retrieval but not acquisition of memory', *Nature*, **411**, 476–480

McGuire, S.E., T. Le Phong and R.L. Davis, 2001, 'The role of *Drosophila* mushroom body signalling in olfactory memory'. *Science*, **293**, 1330–1333

Quinn, W.G., P.P. Sziber and R. Booker, 1979, 'The *Drosophila* memory mutant amnesiac', *Nature*, **277**, 212–214

Rosay. P. *et al.*, 2001, 'Synchronised neural activity in *Drosophila* memory centers and its modulation by amnesiac' *Neuron*, **30**, 759–770

Sokolowski, M.B., 2001, '*Drosophila*: genetics meets behaviour', *Nature Reviews Genetics*, **2**, 879–890

Tully, T. *et al.*, 1994, 'Genetic dissection of consolidated memory in *Drosophila*', *Cell*, **79**, 35–48

Waddell, S. *et al.*, 2000, 'The *amnesia* gene product is expressed in two neurons in the *Drosophila* brain that are critical to memory', *Cell*, **103**, 805–813

Waddell, S. and W.G. Quinn, 2001, 'Flies, genes and learning', *Annual Review of Neuroscience*, **24**, 1283–1309

Yin, J.C.P. *et al.*, 1994, 'Induction of a dominant-negative CREB transgene specifically blocks memory in *Drosophila*', *Cell*, **79**, 49–58

(c) Aplysia

Alberini, C.M. *et al.*, 1994, 'C/EBP is an immediate-early gene required for the consolidation of long-term facilitation in *Aplysia*', *Cell*, **76**, 1099–1114

Bacskai, B.J. *et al.*, 1993, 'Spatially resolved dynamics of cAMP and Protein kinase A subunits in *Aplysia* sensory neurons', *Science*, **260**, 222–226

Bailey, C.H. and M. Chen, 1983, 'Morphological basis of long-term habituation and sensitisation in *Aplysia*', *Science*, **220**, 91–93

Byrne, J.H., 1985, 'Neural and molecular mechanisms underlying information storage in *Aplysia*: implications for learning and memory', *Trends in Neurosciences*, **8**, 478–482

Dale, N., S. Schacher and E.R. Kandel, 1988, 'Long-term facilitation in *Aplysia* involves increase in transmitter release', *Science*, **239**, 282–284

Frost, W.N., V.F. Castellucci, R.D. Hawkins and E.R. Kandel, 1985, 'Monosynaptic connections made by the sensory neurons of the gill- and siphon-withdrawal reflex in *Aplysia* participate in the storage of long-term memory for sensitisation', *Proceedings of the National Academy of Sciences USA*, **82**, 8266–8269

*Kandel, E.R., 1976, *Cellular Basis of Behaviour: An Introduction to Behavioural Neurobiology*, New York: Freeman

*Kandel, E.R., 1979, 'Small systems of neurons', *Scientific American*, **241**(3), 60–70

Montarolo, P.G. *et al.*, 1986, 'A critical period for macromolecular synthesis in long-term heterosynaptic facilitation in *Aplysia*', *Science*, **234**, 1249–1254

Ocorr, K.A., E.T. Walters and J.H. Byrne, 1985, 'Associative conditioning analog selectivity increases cAMP levels of tail sensory neurons in *Aplysia*', *Proceedings of the National Academy of Sciences USA*, **82**, 2538–2552

6. The memory trace in mammals

Barnes, D.M., 1988, 'NMDA receptors trigger excitement', *Science*, **239**, 254–256

Bourtchuladze, R. *et al.*, 1994, 'Deficient long-term memory in mice with a targetted mutation of the cAMP-responsive element binding protein', *Cell*, **79**, 59–68

Collingridge, G.L. and T.V.P. Bliss, 1987, 'NMDA receptors – their role in long-term potentiation', *Trends in Neurosciences*, **10**, 288–293

Deisseroth, K., H. Bito and R.W. Tsien, 1996, 'Signalling from synapse to nucleus: postsynaptic CREB phosphorylation during multiple forms of hippocampal synaptic plasticity', *Neuron*, **16**, 89–101

Huang, Y.-Y., X.-C. Li and E.R. Kandel, 1994, 'cAMP contributes to mossy fiber LTP by initiating both a covalently mediated early phase and a macromolecular synthesis-dependent late phase', *Cell*, **79**, 69–80

Lisman, J., 1994, 'The CaM Kinase 2 hypothesis for storage of synaptic memory', *Trends in Neurosciences*, **17**, 406–412

Lisman, J.E. and A.M. Zhabotinsky, 2001, 'A model of synaptic memory: a CaMKII/PP1 switch that potentiates transmission by organising an AMPA receptor anchoring assembly', *Neuron*, **31**, 191–201

Lynch, G. and M. Baudry, 1984, 'The biochemistry of memory: a new and specific hypothesis', *Science*, **224**, 1057–1063

Martin, S.J., P.D. Grimwood and R.G.M. Morris, 2000, 'Synaptic plasticity and memory: an evaluation of the hypothesis', *Annual Review of Neuroscience*, **23**, 649–711

Schiner, A.E. and M.-M. Poo, 2000, 'The neurotrophin hypothesis for synaptic plasticity', *Trends in Neurosciences*, **23**, 639–645

Silva, A.J. *et al.*, 1998, 'CREB and memory', *Annual Review of Neuroscience*, **21**, 127–148

Taubenfeld, S.M. *et al.*, 2001, 'The consolidation of new but not reactivated memory requires hippocampal C/EBPβ', *Nature Neuroscience*, **4**, 813–818

Waterman, M., G.H. Murdoch, R.M. Evans, M.G. Rosenfeld, 1985, 'Cyclic AMP regulation of eukaryotic gene transcription by two discrete mechanisms', *Science*, **229**, 267–269

Box 20.1: Dendritic spines

Harris, K.M. and S.B. Kater, 1994, 'Dendritic spines: cellular specialisations imparting both stability and flexibility to synaptic function', *Annual Review of Neuroscience*, **17**, 341–371

Hering, H. and M. Sheng, 2001, 'Dendritic spines: structure, dynamics and regulation', *Nature Reviews Neuroscience*, **2**, 880–888

Lisman, J.E. and K.M. Harris, 1993, 'Quantal analysis and synaptic anatomy – integrating two views of hippocampal plasticity', *Trends in Neurosciences*, **16**, 141–147

Llinas, R., 1995, 'Cerebellar purkinje cells, calcium channels', *Nature*, **373**, 107–108

CHAPTER 21: SOME PATHOLOGIES

Web sites

General

The whole brain atlas gives many excellent images of disease-challenged brains. It may be found at:
http://www.med.harvard.edu/AA_/home.html

Prions

Useful introduction with links to other prion sites at:
http://www.microbe.org/news/prions.asp
Good detail and illustrations at:
http://www-micro.msb.le.ac.uk/335/Prions.html
The Official Mad Cow Disease Home Page is at:
http://www.mad-cow.org/

Fragile X syndrome

Home page of the National Fragile X foundation at:
http://www.fragilex.org/home.htm
Fragile X facts page which gives scientific detail at:
http://www.ich.ucl.ac.uk/cmgs/frax98.htm

Neurofibromatoses

The National Neurofibromatosis Foundation provides extensive information for sufferers and their carers at:
http://www.nf.org/
Information about the UK association, support groups, etc. at:
http://www.nfa.zetnet.co.uk/

Motor neuron disease

The international society of ALS/MND may be found at:
http://www.alsmndalliance.org/

Huntington's disease

Home page with links to other resources at:
http://www.interlog.co/~rlaycock/2nd.html
Comprehensive guide to Huntington's disease at:
http://www.lib.uchicago.edu/~rd13/hd
The National Institute of Neurological Disorders and Stroke provide an information sheet with links to other sites at:
http://www.ninds.nih.gov/health_and_medical/pubs/huntington_disease-htr.htm

Parkinson's disease

The Parkinson's Disease Foundation home page is at:
http://www.pdf.org
The World Parkinson Disease Association (WPDA) web site provides links to detailed information on all aspects of Parkinsonism at:
http://www.wpda.org/

Alzheimer's disease

The Alzheimer's Association provides information for caregivers and others and is at:
http://www.alz.org/

A valuable gateway to up-to-the-minute research is provided by the Alzheimer web at:
http://home.mira.net/~dhs/ad.html

Print

1. General

Barondes, S.H., 1993, *Molecules and Mental Illness*, New York: Freeman/Scientific American

Davis, R.L. and D.M. Robertson, 1985, *Textbook of Neuropathology*, Baltimore: Williams & Wilkins

Dubinsky, J.M., 1993, 'Examination of the role of calcium in neuronal death' in 'Markers of Neuronal Injury and Degeneration', *Annals of the New York Academy of Science*, **679**, 34–42

Kramer, P.D., 1994, *Listening to Prozac*, London: Fourth Estate

Mandel, J.-L., 1994, 'Trinucleotide diseases on the rise', *Nature Genetics*, **7**, 453–455

Martin, J.B., 1987, 'Molecular genetics: Applications to clinical neurosciences', *Science*, **238**, 765–772

Sacks, O.W., 1973, *Awakenings*, London: Duckworth

Sarason, I.G. and B.R. Sarason, 1984, *Abnormal Psychology*, Englewood Cliffs, NJ: Prentice-Hall

Smith, C.U.M., 1988, 'Biology and psychiatry', *Journal of the Royal Society of Medicine*, **81**, 439–440

Strange, P.G., 1992, *Brain Biochemistry and Brain Disorders*, Oxford: Oxford University Press

Walton, J., 1993, *Brain's Diseases of the Nervous System* (10th edn), Oxford: Oxford University Press

2. Prions and prion diseases

Brown, D.R., 2001, 'Prion and prejudice: normal protein and the synapse', *Trends in Neurosciences*, **24**, 85–90

Collinge, J., 2001, 'Prion diseases of humans and animals: their causes and molecular basis', *Annual Review of Neuroscience*, **24**, 519–550

Prusiner, S.B. and S.J. DeArmond, 1994, 'Prion diseases and neurodegeneration', *Annual Review of Neuroscience*, **17**, 311–339

Prusiner, S.B., 1995, 'The prion diseases', *Scientific American*, **272**, 30–37

Prusiner, S.B., 1998, 'Prions' (Nobel lecture)', *Proceedings of the National Academy of Sciences USA*, **95**, 13363–13383

Taraboulos, A. *et al.*, 1992, 'Regional mapping of prion proteins in brains', *Proceedings of the National Academy of Sciences USA*, **89**, 7620–7624

3. Phenylketonuria (PKU)

McIlwain, H. and H.S. Bachelard, 1985, *Biochemistry and the Central Nervous System*, Edinburgh: Churchill Livingstone

Walton, J., 1993, *Brain's Diseases of the Nervous System* (10th edn), Oxford: Oxford University Press

4. Fragile X syndrome

Gaskey, C.T. *et al.*, 1992, 'Triplet repeat mutations in human disease', *Science*, **256**, 784–788

Jin, P. and S.T. Warren, 2000, 'Understanding the molecular basis of fragile X syndrome', *Human Molecular Genetics*, **9**, 901–908

Ross, C.A. *et al.*, 1993, 'Genes with triplet repeats: candidate mediators of neuropsychiatric disorders', *Trends in Neurosciences*, **16**, 254–260

Warren, T. and D.L. Nelson, 1993, 'Trinucleotide repeat expansions in neurological diseases', *Current Opinion in Neurobiology*, **3**, 752–759

5. Neurofibromatoses

Gutmann, D.H., 1994, 'New insights into neurofibromatoses', *Current Opinion in Neurology*, **7**, 166–171

Viskochil, D., R. White and R. Cawthon, 1993, 'The neurofibromatosis type 1 gene', *Annual Review of Neuroscience*, **16**, 183–205

6. Motor neuron disease (MND)/amyotrophic lateral sclerosis (ALS)

Beckman, J.S. *et al.*, 2001, 'Superoxide dismutase and the death of motoneurons in ALS', *Trends in Neurosciences*, **11** (Suppl.), S15–S20

Cleveland, D.W. and J.D. Rothstein, 2001, 'From Charcot to Lou Gehrig: Deciphering selective motor neuron death in ALS', *Nature Reviews Neuroscience*, **2**, 806–819

Hadano, S. *et al.*, 2001, 'A gene encoding a putative GTPase regulator is mutated in familial amyotrophic lateral sclerosis', *Nature Genetics*, **29**, 166–174

Julien, J.-P., 2001, 'Amyotrophic lateral sclerosis: unfolding the toxicity of the misfolded', *Cell*, **104**, 581–591

Rosen, D.R. *et al.*, 1993, 'Mutations in the Cu/Zn superoxidase dismutase gene are associated with familial amyotrophic lateral sclerosis', *Nature*, **362**, 59–62

Rowland, L.P. and N.A. Shneider, 2001, 'Amyotrophic lateral sclerosis', *New England Journal of Medicine*, **344**, 1688–1700

7. Huntington's disease (=chorea) (HD)

Cattaneo, E. *et al.*, 2001, 'Loss of normal huntingtin function: new developments in Huntington's disease research', *Trends in Neurosciences*, **24**, 182–188

Cha, J.-H.J., 2000, 'Transcriptional dysregulation in Huntington's disease', *Trends in Neurosciences*, **23**, 387–392

Gusella, J.F. *et al.*, 1983 'A polymorphic DNA marker genetically linked with Huntington's disease', *Nature*, **306**, 234–238

Gusella, J.F., 1986, 'DNA polymorphism and human disease', *Annual Review of Biochemistry*, **55**, 831–854

Harper, P.S., 1984, 'Localisation of the gene for Huntington's chorea', *Trends in Neurosciences*, **7**, 1–2

Huntington's Disease Collaborative Research Group, 1993, 'A novel gene containing a trinucleotide repeat that is expanded and unstable on Huntington's disease chromosomes', *Cell*, **72**, 971–983

Van Dellen, A. *et al.*, 2000, 'Delaying the onset of Huntington's in mice', *Nature*, **404**, 721–722

Wexler, N.S., E.A. Rose and D.E. Houseman, 1991, 'Molecular approaches to hereditary diseases of the nervous system: Huntington's disease as a paradigm', *Annual Review of Neuroscience*, **14**, 503–529

8. Depression

Egeland, J.A. *et al.*, 1987, 'Bipolar affective disorders linked to DNA marker on chromosome 11', *Nature*, **325**, 783–787

McGuffin, P. and R. Katz, 1989, 'The genetics of depression and manic-depressive disorder', *British Journal of Psychiatry*, **155**, 294–304

Paykel, E.S., ed., 1992, *Handbook of Affective Disorders* (2nd edn), Edinburgh: Churchill Livingstone

Shows, T.B., A.Y. Sakaguchi and S.L. Naylor, 1983, 'Mapping the human genome, cloned genes, DNA polymorphisms and inherited disease' in *Advances in Human Genetics*, 12, New York: Plenum Press: Chapter 5

Van Praag, H.M., 1982, 'Neurotransmitters in CNS disease: Depression', *Lancet*, **ii**, 1259–1264

9. Parkinson's disease (PD)

Betarbet, R. *et al.*, 2000, 'Chronic systemic pesticide exposure reproduces features of Parkinson's disease', *Nature Neuroscience*, **3**, 1301–1306

Kenny, K.K. *et al.*, 2001, 'The role of the ubiquitin proteasomal pathway in Parkinson's disease and other neurodegenerative disorders', *Trends in Neurosciences* (Suppl.), S7

Kitada, T. *et al.*, 1998, 'Mutations in the parkin gene cause autosomal recessive juvenile parkinsonism', *Nature*, **392**, 605–608

Kopin, I.J., 1993, 'The pharmacology of Parkinson's disease therapy: an update', *Annual Review of Pharmacology and Toxicology*, **33**, 467–495

Leroy, E. *et al.*, 1998, 'The ubiquitin pathway in Parkinson's disease', *Nature*, **395**, 451–452

Lewin, R., 1985, 'Parkinson's disease: an environmental cause', *Science*, **229**, 257–258

Marsden, C.D., 1990, 'Parkinson's disease', *Lancet*, **335**, 948–952

10. Alzheimer's disease (AD)

Bayreuther, K. and C.L. Masters, 1991, 'Amyloid genes and neuronal dysfunction', *Brain Research Reviews*, **16**, 86–88

Carrell, R.W., 1988, 'Alzheimer's disease: enter a protease inhibitor', *Nature*, **331**, 478–479

Corder, E.H. *et al.*, 1993, 'Gene dose of apolipoprotein E type 4 allele and the risk of Alzheimer's Disease in late onset families', *Science*, **261**, 921–923

De Strooper, B. and G. König, 2001, 'An inflammatory drug prospect', *Nature*, **414**, 159–160

Delabar, J.-M., 1987, 'β-Amyloid gene duplication in Alzheimer's disease and karyotypically normal Down syndrome', *Science*, **235**, 1390–1392

Fassbender, K. *et al.*, 2001, 'Simvastatin strongly reduces levels of Alzheimer's disease-amyloid peptides A42 and A40 *in vitro* and *in vivo*', *Proceedings of the National Academy of Sciences USA*, **98**, 5856–5861

Games, D. *et al.*, 1995, 'Alzheimer type neuropathology in transgenic mice over-expressing V717F β-amyloid precursor protein', *Nature*, **373**, 523–527

Goldgaber, D. *et al.*, 1987, 'Characterisation and chromosomal localisation of a cDNA encoding brain amyloid of Alzheimer's disease', *Science*, **235**, 877–880

Götz, J. *et al.*, 2001, 'Formation of neurofibrillary tangles in P301L tau transgenic mice induced by A42 fibrils', *Science*, **293**, 1491–1495

Hayflick, L., 2000, 'New approaches to old age', *Nature*, **403**, 365

Howlett, D.R. *et al.*, 2000, 'In search of an enzyme: the β-secretase of Alzheimer's disease is an aspartic proteinase', *Trends in Neurosciences*, **23**, 565–570

Kamal, A. *et al.*, 2001, 'Kinesin-mediated axonal transport of a membrane compartment containing secretase and presenilin 1 requires APP', *Nature*, **414**, 643–648

Katzman, R., 1994, 'Apolipoprotein E and Alzheimer disease', *Current Opinion in Neurobiology*, **4**, 703–707

Kosic, K.S., 1992, 'Alzheimer's disease: a cell-biological perspective', *Science*, **256**, 780–783

Lee, V. M.-Y., 1991, 'A68: a major subunit of paired helical filaments and derivatised forms of normal tau', *Science*, **251**, 675–678

Luo, Y. *et al.*, 2001, 'Mice deficient in BACE1, the Alzheimer's β-secretase have normal phenotype and abolished β-amyloid generation', *Nature Neuroscience*, **4**, 231–232

Panegyres, P.K., 2001, 'The functions of the amyloid precursor protein gene', *Reviews in the Neurosciences*, **12**, 1–39

Poirier, J., 1994, 'Apolipoprotein E in animal models of CNS injury and in Alzheimer's disease', *Trends in Neuroscience*, **17**, 525–530

Roth, M. and L.L. Iverson, eds, 1986, 'Alzheimer's disease and related disorders', *British Medical Bulletin*, **42**(1), 1–115

Schenk, D. *et al.*, 1999, 'Immunisation with amyloid-β-attenuates Alzheimer-disease-like pathology in the PDAPP mouse', *Nature*, **400**, 173–177

Scheuner, D. *et al.*, 1996, 'Secreted amyloid β-protein similar to that in senile plaques of Alzheimer's disease is increased *in vivo* by presenilin 1 and 2 and APP mutations linked to familial Alzheimer's disease', *Nature Medicine*, **2**, 864–870

Selkoe, D.J., 1991, 'The molecular pathology of Alzheimer's disease', *Neuron*, **6**, 487–498

Selkoe, D.J., 1994, 'Normal and abnormal biology of the β-amyloid precursor protein', *Annual Review of Neuroscience*, **17**, 489–517

St George-Hyslop, P.H., 1987, 'The genetic defect causing familial Alzheimer's disease maps on chromosome 21', *Science*, **235**, 885–889

*Wischik, C.M. *et al.*, 1985, 'Subunit structure of paired helical filaments in Alzheimer's Disease', *Journal of Cell Biology*, **100**, 1905–1912

*Wischik, C.M. and R.A. Crowther, 1986, 'Subunit structure of the Alzheimer tangle', *British Medical Bulletin*, **42**, 51–56

Yan, R. *et al.*, 1999, 'Membrane-anchored aspartyl protease with Alzheimer's disease β-secretase activity', *Nature*, **402**, 533–537

Box 21.1: α-Synuclein

Betarbet, R. *et al.*, 2000, 'Chronic systemic pesticide exposure reproduces features of Parkinson's disease', *Nature Neuroscience*, **3**, 1301–1306.

Goedert, M., 2001, 'Alpha-synuclein and neurodegenerative diseases', *Nature Reviews Neuroscience*, **2**, 492–501

Appendix 1: Molecules and Consciousness

Chalmers, D.J., 1995, *The Conscious Mind*, Oxford: Oxford University Press

Darwin, C., c. 1837, 'Old and Useless Notes', in *Metaphysics, Materialism and the Evolution of Mind: Early writings of Charles Darwin*, transcribed by P.H. Barrett, Chicago: Chicago University Press, 35–36,

Felleman, D.J. and D. van Essen, 1991, 'Distributed hierarchical processing in the Primate cerebral cortex', *Cerebral Cortex*, **1**, 1–47

Globus, C.G., 1973, 'Unexpected symmetries in the "world-knot"', *Science*, **180**, 1129–1136

Hameroff, S., 1994, 'Quantum coherence in microtubules: A neural basis for emergent consciousness?', *Journal of Consciousness Studies*, **1**, 91–118

Nagel, T., 1974 'What is it like to be a bat?', *Philosophical Review*, **LXXXIII**, 435–450

Nietzsche, F., 1872, 'The Philosopher: Reflections on the struggle between art and knowledge', para. 64, in D. Breazeale, 1979, *Philosophy and Truth: Selections from Neitzsche's notebooks of the early 1870s*, Sussex: Harvester Press

Penrose, R., 1989, *The Emperor's New Mind*, Oxford: Oxford University Press

Penrose, R., 1994, *Shadows of the Mind: A Search for the Missing Science of Consciousness*, Oxford: Oxford University Press

Roland, P.E., 1993, *Brain Activation*, New York: Wiley-Liss

Searle, J.R., 1980, 'Minds, brains and programs', *Behavioural and Brain Sciences*, **3**, 417–424

Smith, C.U.M., 1983, 'Anatomical concepts and the problem of mind', *Journal of Social and Biological Structures*, **6**, 381–392

Smith, C.U.M., 1994, 'The complexity of brains: A biologist's view' in *Complex Systems*, eds R.J. Stonier and X.H. Yu, Oxford: IOS Press, 93–100

Smith, C.U.M., 2000, 'Evolutionary biology and the "hard problem"', *Evolution and Cognition*, **6**, 162–175

Wilczek, F., 1994, 'A call for a new physics (review of Penrose: Shadows of the Mind)', *Science*, **266**, 1737–1738

Wooldridge, M.J. and N.R. Jennings, 1994, 'Agent theories, architectures and languages: a survey' paper presented at Complex Conference, Rockhampton, 1994

Appendix 5: Physical models of ion conduction and gating

Leuchtag, H.R., 1994, 'Long-range interactions, voltage sensitivity, and ion conduction in S4 segments of excitable channels', *Biophysical Journal*, **66**, 217–224

Leuchtag, H.R. and V.S. Bystrov, 1999, 'Theoretical models of conformational transitions and ion conduction in voltage-dependent ion channels: bioferroelectricity and superionic conduction', *Ferroelectrics*, **220**, 157–204

INDEX OF NEUROLOGICAL DISEASES

INDEX

F=figure; T=table